# Ravine Lands: Greening for Livelihood and Environmental Security

Jagdish Chander Dagar • Anil Kumar Singh
Editors

# Ravine Lands: Greening for Livelihood and Environmental Security

Springer

*Editors*
Jagdish Chander Dagar
Indian Council of Agricultural Sciences
New Delhi, India

Anil Kumar Singh
RVS Krishi Vishwa Vidayalaya
Gwalior, Madhya Pradesh, India

ISBN 978-981-10-8042-5 ISBN 978-981-10-8043-2 (eBook)
https://doi.org/10.1007/978-981-10-8043-2

Library of Congress Control Number: 2018935227



Printed on acid-free paper

This Springer imprint is published by the registered company Springer Nature Singapore Pte Ltd.
The registered company address is: 152 Beach Road, #21-01/04 Gateway East, Singapore 189721, Singapore

# Foreword

Soils are one of the world's biogeophysical treasures. They provide structure and function to our landscapes. They sustain habitats and biodiversity, as well as provision more than 95% of the human diet. It is alarming to consider that it takes 100 years to form 2cm of topsoil, but this can be lost in a single rainfall event. Although soil erosion and subsequent sedimentation are natural processes caused by water and wind, it is their heightened frequency that raises concern. Much of this is attributed to agricultural activities – numerous ones. Thriving civilizations collapsed because of the severe degradation of their soil resource base on which they had flourished. Although soil erosion prediction technology began over 70 years ago when Austin Zingg published a relationship between soil erosion by water and land slope and length, it took another 20 years to evolve the Universal Soil Loss Equation (USLE). The USLE is widely heralded as the foremost achievement in soil erosion prediction in the last century. Yet despite these predictions, they are meaningless if we do nothing to address the drivers of soil loss. Globally, the total land area affected by water erosion is estimated to be about one billion hectares, of which 750 million hectares is severely affected. There are also regional hot spots of erosion including the Himalayan–Tibetan ecosystem in South Asia, the Loess Plateau in China, sub-humid and semiarid regions of sub-Saharan Africa, highlands of Central America, the Andean region, Haiti, and the Caribbean. Soil erosion by water usually begins when raindrops first strike the land's surface and detach and transport soil particles which are further transported downslope by raindrop splash. Once rill erosion begins, it typically progresses upslope by a series of intensely erodible head cuts, at places where concentrated runoff from a slope is sufficient in volume and velocity to cut deep trenches or where concentrated water continues cutting in the same groove (such as rill) resulting in a deep soil incision, commonly referred to as a gully. The advance stage of gully erosion leads to ravine erosion. There may be both natural (neotectonic) and anthropogenic factors responsible for ravine formation. Recent studies indicate that gully erosion is one of the most dangerous threats to agricultural lands and represents an important sediment source in a range of environments. Well-constructed gullies can also be effective means for transferring runoff and sediment from uplands to valley bottoms.

During the last two decades, a significant progress in the understanding of gully erosion and its controlling factors has been made. A plethora of research papers have been brought out on various aspects of gully erosion including modelling for predicting the impact of gully erosion. Yet, there remain several research gaps and unanswered questions related to gully erosion, which include historical studies on gully erosion for better understanding of different stages of their formation, appropriate measuring techniques for monitoring or better understanding of the geological and hydrological parameters and processes of gully development, interaction of hydrological and other soil degradation processes, impact of climate and environmental changes on soil erosion processes, development of effective and reliable gully erosion models, and effective gully prevention and control measures and watershed-based management options.

Ravines occupy almost 4 million hectares area in India and have a huge potential for carbon sequestration, agroforestry, and improving the socio-economic status of the inhabitants. This book entitled *Ravine Lands: Greening for Livelihood and Environmental Security* is a timely publication, in which various aspects of ravine ecology and rehabilitation of these degraded lands, particularly through biological approaches, have been dealt with. By linking up this sound theory and evidence with increasing remote sensing and GIS capabilities, the prospects for tackling the insidious erosion of India's land are huge. The contributors are renowned experts in their relevant fields, and their experiences have added enormous value to this publication. We should rightly applaud the high-quality contributions of all authors and editors who have ensured this superb publication. I trust that this publication will be very useful for the scientists working in the field of soil conservation/rehabilitation, as well as students, environmentalists, educationists, and policy makers. I congratulate the editors for their commendable efforts for bringing out an excellent publication containing valuable information related to the reclamation of the most degraded land resources. When we consider that land health is national wealth, it should be on every policy maker's bookshelf.

World Agroforestry Centre (ICRAF) Member
Global Restoration Council,
Nairobi, Kenya

Anthony Simons

# Preface

The role that soils play in the day-to-day existence of human beings can never be overemphasized as they provide for almost all our basic needs, be it food, fuel, fodder, fibre and even housing materials (furniture, bricks, etc.). Yet this resource has always been mismanaged, which has led to erosion on a vast scale. The total land area affected by water erosion is estimated to be about 1 billion hectares, of which 750 million hectares are severely affected and terrain deformation (rills and gullies) occurs on 175 million hectares. Most available literature shows that the annual sediment transport into the ocean by the world's rivers is 15–20 billion Mg per year. There are also regional hot spots of erosion, including the Himalayan-Tibetan ecosystem in South Asia, the Loess Plateau in China, the subhumid and semiarid regions of sub-Saharan Africa, the highlands of Central America, the Andean region, Haiti and the Caribbean. Ravines, which are caused by water erosion, are perhaps the most degraded and vulnerable ecosystems with severe on-site and off-site implications and have caused major disasters worldwide. Formation of ravines has resulted in loss of precious, virtually nonrenewable resource and adversely impacted rural economy leading to several socio-economic problems. In addition to the above, they impact the total landscape, adversely affecting human and animal population, physical infrastructure and agricultural lands. In several developing countries, many villages and communities have been displaced and disappeared as a result of scourges of gully erosion. Climate change is likely to further accelerate the whole process of erosion. Ravines are rich in biodiversity with species which have adapted to this ecosystem over centuries. They are a treasure house of genes which can help scientists in developing multiple stress-tolerant products.

Considering the importance and need for rehabilitation of ravines, it was decided to organize an International Conference on "Managing Ravines for Food and Livelihood Security" at Gwalior, India, from March 7–10, 2016. Probably this was the first time an International Conference was organized in India focusing specifically on ravines and their management. It was decided to hold it in Gwalior (Madhya Pradesh) because of its proximity to the Chambal ravines which are (in)famous internationally. Madhya Pradesh state has the second largest area under ravines in India next to Uttar Pradesh. In India, there is about 4 million ha area under ravines.

Considering the quality of invited and voluntary papers, it was decided to bring out a publication in the form of a book entitled *Ravine Lands: Greening for Livelihood and Environmental Security*. It includes 28 chapters addressing core issues related to *general features of ravines – extent, geology, soil, hydrology, formation, classification and biodiversity*; *rehabilitation and management technologies*; *exploration of ravine lands for livelihood and environmental securities*; and *socio-economic and policy issues.* Care has been taken to select relevant quality papers presented in the conference and invite experts to contribute on unaddressed issues so that the publication is complete on all aspects. More importantly, it deals with rehabilitation of one of the most degraded ecosystems ensuring the livelihood of resource-poor farming and landless families living in harsh ecologies which are more vulnerable to climate change.

We would like to express our sincere gratitude to all the authors who have painstakingly worked on their manuscripts on a short notice and contributed in making this a very informative and useful publication. We are also thankful to Dr Anthony Simons, Director General, World Agroforestry Centre (ICRAF), Nairobi, Kenya for writing the foreword to this unique publication.

We trust that this will be a very useful publication for the scientists working in the field of soil conservation/rehabilitation, students, environmentalists, educationists and policymakers.

New Delhi, India — Jagdish Chander Dagar
Gwalior, Madhya Pradesh, India — Anil Kumar Singh

# Contents

1 **Introduction** ... 1
Jagdish Chander Dagar and Anil Kumar Singh

2 **Ravines: Formation, Extent, Classification, Evolution and Measures of Prevention and Control** ... 19
Jagdish Chander Dagar

3 **Perspectives of Vegetation Ecology and Biodiversity for Management of Ravine Lands** ... 69
Jagdish Chander Dagar

4 **Plant Biodiversity of Ravine Ecosystem: Opening New Vistas for Enhancing Productivity** ... 119
A. R. Uthappa, S. B. Chavan, M. N. Ramesha, S. Kala, Veeresh Kumar, A. K. Handa, and O. P. Chaturvedi

5 **Nutrient Management Strategies for Improving Soil Health and Productivity of Ravine Lands** ... 143
S. K. Verma and Anil Kumar Singh

6 **Soil, Landscape and Nutrient Management of Ravine Areas for Enhancing Crop Productivity and Livelihood Security** ... 161
P. Dey, S. K. Verma, K. Ramesh, J. Somasundaram, and M. Vassanda Coumar

7 **Some Practices of Gully Rehabilitation in Central Brazil** ... 183
Silvio Carlos Rodrigues

8 **Gully Erosion and Land Degradation in Brazil: A Case Study from São Luís Municipality, Maranhão State** ... 195
Antonio Jose Teixeira Guerra, Michael Augustine Fullen, José Fernando Rodrigues Bezerra, and Maria do Carmo Oliveira Jorge

9 **Technological Interventions for Managing Ravine Lands for Livelihood and Environmental Security** ... 217
A. K. Sikka, P. K. Mishra, R. K. Singh, B. Krishna Rao, and Adlul Islam

10 **Formation, Control, and Management of Deep (Chambal) Ravines: An Energy Concept** ........ 237
V. S. Tomar and S. K. Verma

11 **Management of Chambal Ravines for Income Enhancement** ........ 253
G. P. Verma, Y. P. Singh, Anil Kumar Singh, and S. K. Verma

12 **Reclamation of Ravine Lands for Higher Production** ........ 279
M. L. Soni, V. Subbulakshmi, P. S. Renjith, J. C. Dagar, and N. D. Yadava

13 **Soil and Water Conservation in Ravinous Watersheds: Case Studies from Uttar Pradesh in India** ........ 309
Suraj Bhan and Sanjay Arora

14 **Agroforestry Potential for Higher Productivity from Degraded Ravine Watersheds** ........ 335
O. P. Chaturvedi, J. C. Dagar, A. K. Handa, Rajesh Kaushal, and V. C. Pandey

15 **Forest and Fruit Trees-based Agroforestry Systems for Productive Utilization of Ravine Lands** ........ 361
A. K. Parandiyal, R. Kaushal, and O. P. Chaturvedi

16 **Silvopastoral System: A Key Component for Rehabilitation of Ravine Lands** ........ 385
P. K. Ghosh and S. K. Mahanta

17 **Bamboo-based Bioengineering Interventions for Rehabilitation of Ravines** ........ 397
B. Krishna Rao, V. C. Pande, R. S. Kurothe, A. K. Singh, and A. K. Parandiyal

18 ***Leucaena leucocephala*****: A Leguminous Tree Suitable for Eroded Habitats of Hawaiian Islands** ........ 413
Kazue L. Ishihara, Michael D. H. Honda, Ahmed Bageel, and Dulal Borthakur

19 **Ravines: Prospective Zone for Carbon Sequestration** ........ 433
J. Somasundaram, A. K. Parandiyal, Pramod Jha, Brij Lal Lakaria, R. K. Singh, B. L. Mina, S. Kala, and Shakir Ali

20 **Sequestering Carbon in Ravine-prone and Eroded Landscapes** ........ 445
Rattan Lal

21 **Greening of Ravines: A Strategy for Viable Production and Mitigation to Climate Change** ........ 455
Anil Kumar Singh, Y. P. Singh, and S. K. Verma

22 **Management of Ravines Through Anicuts and Afforestation** ........ 477
Akhilesh Singh and S. K. Verma

**23 Rehabilitation of Degraded Lands in Semiarid and Subhumid Ecologies in India** ... 505
V. P. Singh, L. Verchot, C. Martius, D. Nayak, and J. Rizvi

**24 Impact Assessment of Management Techniques Implemented in Chambal Ravines, Madhya Pradesh, India** ... 523
V. P. Singh, R. B. Sinha, Y. P. Singh, D. Nayak, A. K. Singh, and J. Rizvi

**25 Modeling Soil Erosion by Water** ... 549
M. K. Hardaha

**26 Gender Challenges and Socio-economic Strategies for Livelihood Security in Ravine Areas** ... 573
Reeta Mishra, Yagya Dev Mishra, P. P. Singh, and B. P. S. Raghubanshi

**27 Socio-economic and Conservation Measures in Ravine-affected Areas of Gujarat: Policy Interventions** ... 591
V. C. Pande, R. S. Kurothe, H. B. Singh, B. Krishna Rao, Gopal Kumar, and P. R. Bhatnagar

**28 Greening Ravine Lands: Policy Issues and the Way Forward** ... 601
Jagdish Chander Dagar and Anil Kumar Singh

**Index** ... 617

## About the Editors

**Jagdish Chander Dagar** has been well recognized both nationally and internationally and has grown professionally through strong pursuit of agricultural, ecological and agroforestry research which is evident from his more than 280 research papers published in peer-reviewed journals, book chapters and papers in conferences/symposia proceedings, 12 books authored/edited, 10 research bulletins and more than 50 popular articles. He was the Assistant Director General at the headquarters of ICAR and Emeritus Scientist. His research interest has been in the areas of biosaline agriculture, agroforestry, management of natural resources, rehabilitation of degraded lands, biodrainage, ethnobotany, plant ecology, climate change, sustainable agriculture and policy. Recognizing the research contributions, Dr. Dagar has been conferred with several awards and honours: Sajjad Memorial Gold Medal, Hari Om Ashram Trust Award of ICAR, Swami Pranavananda Saraswati National Award of UGC, CSSRI Excellence Award on Soil Salinity and Water Management and Dr. K.G. Tejwani Award for Excellence in Agroforestry Research and Development. Further, he is Fellow of the National Academy of Agricultural Sciences and Fellow and Life Member of several other professional societies: International Society of Tropical Ecology, Indian Botanical Society, National Institute of Ecology, Range Management Society of India, National Environmental Science Academy, Andaman Science Association, and Indian Society of Soil Salinity and Water Quality. He has also been consulting several national and international agencies, such as FAO of UN, CIMMYT, and Haryana Forest Department. Dr. Dagar has travelled in several countries in different capacities and had been member of several Research Advisory Committees and Management Committees of ICAR Institutes and Board of Studies in Universities. He is the Chief Editor of *Journal of Soil Salinity and Water Quality* and Member Board of Editors of many Journals and is also active in social services.

**Anil Kumar Singh** has been the Vice Chancellor of the Rajmata Vijayaraje Scindia Krishi Vishwavidyalaya, Gwalior, Madhya Pradesh, India. He was the Deputy Director General (Natural Resource Management) of ICAR, 2007–2012; acting Director of IARI, 2005–2006; and Project Director at Water Technology Centre, 1999–2007. His areas of research interest were soil-water-nutrient-plant interactions, crop growth simulation, irrigation water management and integrated watershed management. Dr. Singh had been a DAAD Fellow at the University of Hohenheim, Germany, and served in IRRI between 1993 and 1995. He is Fellow of several professional societies, including the National Institute of Agricultural Sciences. Dr. Singh is a recipient of several awards/hours: National Science Talent Scholarship, 1967–1976; Gold Medal in MSc, 1972; IARI Best Teacher Award, 2005; Soil Conservation Society of India Gold Medal, 2006; Dr. C Subramaniam Outstanding Teacher Award, 2007–2008; 10th Shri Hari Krishna Shastri Memorial Award, 2009; and several Memorial Lecture Awards.

# Introduction

# 1

Jagdish Chander Dagar and Anil Kumar Singh

**Abstract**

The problem of accelerated erosion has been in existence since the advent of agriculture and thought to be responsible for the collapse of numerous, once-thriving civilizations that vanished because of the degradation of the very resource base on which they flourished. Soil erosion prediction technology began over 70 years ago when Austin Zingg published a relationship between soil erosion by water and land slope and length, followed shortly by a relationship by Dwight Smith that expanded this equation to include conservation practices. But, it was nearly 20 years before this work led to the Universal Soil Loss Equation (USLE), perhaps the foremost achievement in soil erosion prediction in the last century. The total land area affected by water erosion is estimated to be about 1 billion ha, of which 750 million ha are severely affected. There are also regional hot spots of erosion including the Himalayan–Tibetan ecosystem in South Asia, the Loess Plateau in China, the subhumid and semiarid regions of sub-Saharan Africa, highlands of Central America, the Andean region, Haiti and the Caribbean. Soil erosion by water usually begins when raindrops first strike the land's surface and detach and transport soil particles which cause sealing and compacting of the soil surface. Eroded soil is transported downslope by raindrop splash. When runoff water concentrates, and flows through fingerlike channels (rills) from upland areas carrying soil particles, the erosion is called rill erosion, and the soil eroded from between rills is called inter-rill erosion. Once rill erosion begins, it typically progresses up slope by a series of intensely erodible head cuts. Places, where concentrated runoff from a slope is sufficient in volume and velocity to cut deep trenches or where concentrated water continues cutting in the same groove

J. C. Dagar (✉)
Indian Council of Agricultural Sciences, Krishi Anusandhan Bhavan-II, Pusa, New Delhi, India

A. K. Singh
RVS Krishi Vishwa Vidyalaya, R. V. S., Gwalior, India



J. C. Dagar, A. K. Singh (eds.), *Ravine Lands: Greening for Livelihood and Environmental Security*, https://doi.org/10.1007/978-981-10-8043-2_1

(such as rill) making a deep soil incision, are referred to as gully. The advance stage of gully erosion leads to ravine erosion. There may be both natural (neo-tectonic) and anthropogenic factors responsible for ravine formation. Recent studies indicate that gully erosion is one of the most dangerous threats to agricultural lands and represents an important sediment source in a range of environments; and gullies are effective links for transferring runoff and sediment from uplands to valley bottoms and permanent channels where they aggravate off-site effects of water erosion. During the last two decades, a significant progress in the understanding of gully erosion and its controlling factors has been made, and bulk of research papers have been brought out on various aspects of gully erosion including on modelling predicting the processes of gully erosion. Yet, there remain several research gaps and unanswered questions related to gully erosion, which include historical studies on gully erosion for better understanding of different stages of their formation; appropriate measuring techniques for monitoring or better understanding of the geological and hydrological parameters and processes of gully development; interaction of hydrological and other soil degradation processes; impact of climate and environmental changes on soil erosion processes; development of effective and reliable gully erosion models; and effective gully prevention and control measures and watershed-based management options. In the present book, an attempt has been made to address some of these aspects along with impacts of ravine formation on socio-economic and policy issues.

The problem of soil erosion, land wasting and soil depletion is age old, in fact as old as the first man who cut the first tree to practise arable farming. As per the Random House Dictionary (© Random House, Inc. 2017), the word "erosion" was first recorded in 1535–1545 and is from the Latin word *ērōsiōn-* (stem of *ērōsiō*) or *erodere*, which means to eat away or to excavate, and may be defined as the process by which the surface of the earth is washed away by the action of water, glaciers, winds, waves, etc. In science, the word was first used in geology by Albrecht Penck in his famous book *Morphologie der Erdoberflache*, i.e. "Morphology of Earth Surface", published in two volumes (Penck 1894) for describing the hollows created by water and the wearing away of solid material by action of river water, while surface wash and precipitation erosion were called "ablation" (*abatio*—to carry away). In addition to *erosion* and *ablation*, a number of other terms such as "corrosion" (*crradere*—to chew together), "corrosion" (*coroder*—to chew to pieces), "abrasion" (*abrodere*—to scrape off) and "denudation" (*denundere*—to strip) were also used (Hardaha 2015). The first attempt in the direction of morphological analysis was the theory of the *Cycle of Erosion* developed by WM Davis (c.f. Walther Penck 1953), which had a purely geographical aim, viz. the systematic description of land forms on a genetic basis. Soil erosion, displacement of soil from the place of its formation by causative agents (e.g. raindrop, runoff, wind, gravity, etc.) and its deposition at a depressional and/or protected site, is a natural geologic phenomenon (Lal 2003).

The slow geologic erosion is a constructive process, which has created vast tracts of fertile soils of alluvial floodplains and loess plateaus around the world. These soils, with built-in soil fertility renewal mechanisms, have supported ancient civilizations (e.g. in the valleys of the Nile, Euphrates, Indus, Yangtze) and thriving cultures for millennia (Lal 2003). In contrast, the accelerated soil erosion, exacerbated by anthropogenic perturbations, is a destructive process. It depletes soil fertility, degrades soil structure, reduces the effective rooting depth and destroys the most basic of all-natural resources. The problem of accelerated erosion has been prevalent since agriculture began (~10,000 years ago), and it is opined that the collapse of numerous, once-thriving civilizations occurred because of the degradation of the resource base on which they arose (Lowdermilk 1953; Eckholm 1976; Olson 1981; Brown 1981, 1984; Dregne 1982; Rozanov et al. 1993).

All over the world, wherever man indulged in agricultural activities, the problems of soil erosion and land degradation were initiated. For example, Mesopotamia civilization, which stretched across modern-day Syria and Iraq, maintained a progressive civilization from ~4000 BC to the mid of 1200s AD; and its decline is believed to be related to erosion and land degradation due to deforestation and overgrazing in the Armenia highlands in eastern Turkey. Polynesian people most likely settled on Easter Island (local name *Rapa Nui*) in a Chilean territory sometime between 700 and 1100 CE and created a thriving and industrious culture as evidenced by the island's numerous enormous stone *moai* and other artefacts. However, the anthropogenic activities led to gradual deforestation and extinction of natural resources which severely weakened the *Rapa Nui* civilization (Hunt 2006). By the time of European arrival in 1722, the island's population had dropped to 2000–3000 from an estimated high of approximately 15,000 just a century earlier. European diseases and Peruvian slave trading in the 1860s further reduced the *Rapa Nui* population to a low of only 111 inhabitants in 1877 (Peiser 2005).

In India, Mohenjodaro and Harappa civilizations came to their end because of problems of soil erosion and land degradation. In addition, the watershed areas of several rivers like Saraswati have disappeared because of their exploitation and mismanagement (Suresh 1997). After conquering the Achaemenid Empire of Persia, the Macedonian King (then the great king of the Persian Empire), Alexander the Great, invaded India in 326 BC. In those days, there were *sal* (*Shorea robusta*) forests sufficient in extent and so thick in density that he could conceal his entire army into these forests. Only some traces of these forests can be found today. After the eleventh century, Mohammedan invaders had no respect for forests, and under the Mughals, forests were destroyed indiscriminately. Many areas today bear evidence of their, having been productive and fertile as attested by the presence of centuries-old buildings and centres of culture (Tejwani et al. 1975b). The destruction of land resource had more than local effects. The land erosion has caused rivers to silt up and affected water supplies and transportation. Many cities and villages have relocated due to encroachment by ravines as are the examples along the banks of river Chambal. At present, there is very little area in India free from the hazard of soil erosion or danger of flooding and waterlogging. With the destruction of vegetation in the highly erosive soils in Siwaliks, these hills spew their *chos* with much sand and damage agricultural lands. The outer Himalayas send down huge amount of debris and sediment choking up the rivers, causing floods, depositing sediment on fertile agricultural lands, silting up hydroelectric reservoirs and causing landslides and landslips and thus, destroying life, biodiversity and property.

In earlier literature, the pioneer work on soil erosion by water is well documented (Bennet 1939; Zingg 1940; Ellison 1944a, b, 1947; Musgrave 1947; Erasmus et al. 1970; Beasley 1972; Kinnell 1974). Soil erosion by water usually begins when raindrops first strike the land's surface and detach soil particles (*splash erosion*). Falling drops break down soil aggregates, detach and transport soil particles which cause sealing and compacting of the soil surface. This reduces the ability of the soil to absorb water and increase surface runoff. Eroded soil is transported downslope by raindrop splash. Net downslope movement by splash increases as slope steepens. Most eroded soil is transported downslope by concentrated runoff which begins when the soil surface storage is filled and rainfall intensity exceeds the infiltration rate of the soil.

Soil erosion by water usually on landscape of gentle slopes is known as *sheet erosion*, where the top fertile layer of soil is ripped off in thin films every year by the running water from the entire surface of such lands. When runoff water concentrates, and flows through fingerlike channels (rills) from upland areas carrying soil particles, the erosion is called *rill erosion*, and the soil eroded from between rills is called "inter-rill erosion". Rill erosion is caused primarily by runoff, whereas inter-rill erosion results primarily from raindrop impact. Rills usually develop because of tillage marks, surface roughness or topographic variations and later by flow patterns of runoff. Once rill erosion begins, it typically progresses up slope by a series of intensely erodible head cuts. Place, where concentrated runoff from a slope is sufficient in volume and velocity to cut deep trenche or where concentrated water continues cutting in the same groove (such as rill) making a deep soil incision, is called gully.

Gullies often have their beginning in slight depressions in fields or just below fields where runoff water so often concentrates. Gullies are usually formed when many rills join and further extended with more than 30 cm depth. The earliest known usage of the term "gully" is from 1657, and it originates from the French word *goulet*, a diminutive form of *goule* which means *throat* (https://en.wikipedia.org/wiki/Gully). It is possible that the term was derived from a type of knife at the time, a *gully knife*, because hills that have gullies look as if they are cut open with a sharp knife. A gully is known as a "cárcava" in Spain, "ravine" in France, "lavaka" in Madagascar, "wadi" in Arabic, "donga" in South Africa, "voçoroca" in Brazil and "barranco" in Argentina (Castillo and Gómes 2016). *Gully erosion* is defined as the erosion process by which runoff water cuts new unstable channels into erodible soil and weathered rock. It causes severe land degradation and is a major source of sediment and associated nutrients to river systems and ultimately to inland basins or coastal waters.

Bennet (1939), one of the most pioneer workers, studied gully erosion in USA and categorized them into small, medium and large. Without mentioning the specifications, he explained that the small gullies could be checked by ploughing and smoothening the surface and adopting contour strip cropping or putting under pasture; medium gullies could be checked by suitable mechanical means and putting them under pasture and forest; large gullies were defined as those which had gone beyond the stage where reclamation for cropping or meadow was practicable. Ayres (1939) classified gullies as small (cross section up to 90 × 90 cm), medium (3 × 3 m) and large (greater than 3 × 3 m), while Ireland et al. (1939) classified gullies based on their form, gully head characteristics, length, breadth and catchment. All the American workers did not have to tackle the problem of reclamation of large gullies for cultivation. In fact, the best land use of gullies is to put them under permanent vegetation.

Despite historical evidence (as pointed out earlier) that human-induced erosion has occurred worldwide, whenever land has been denuded from vegetation or cultivated, awareness of erosion-induced problem developed slowly. The study of soil erosion as a science started only approximately 140 years ago, between 1877 and 1895, by German soil scientist Ewald Wollny (Dotterweich 2013, Laflen and Flanagan 2013 ; Flanagan et al. 2013). Although his early work was often overlooked until the 1930s (Nelson 1958), Ewald Wollny is generally credited as being a pioneer in soil and water conservation research (Baver 1938). His early work included the effect of soils and topographic properties on runoff and erosion, including factors such as slope steepness, plant cover and soil type (Wollny 1888). The earliest quantitative erosion measurements in USA began in 1912 on overgrazed rangeland and measured by the US Forest Service in Central Utah (Sampson and Weyl 1918; Chapline 1929; Forsling 1931; Stewart and Forsling 1931). Duley and Miller (1923) were the first in USA to report scientific measurements of soil erosion. Miller and colleagues at the Missouri Agricultural Experiment Station Columbia were generally credited with the concept of erosion plot research (Miller 1926; Miller and Krusekopf 1932; Duley and Ackerman 1934; Miller 1946a, b). In 1928, the US Dept. of Agriculture published a circular on "Soil Erosion—A National Menace" (Bennett and Chapline 1928). Bennett (1939) indicated that the publication of this bulletin plus the educational campaign by the USDA were critical elements in securing public and political attention to soil erosion. Hugh Bennett undoubtedly had more influence on soil conservation efforts in USA than any other single individual, and his early work led to his recognition rightly as the "Father of Soil Conservation". His crusades concerning soil erosion as a natural menace along with his evangelistic zeal to start needed research led to Congressional action in 1929 which established ten experimental stations dealing with soil and water conservation research in USA (Renard 1985).

Several workers (mentioned in above para) had made measurements of soil erosion and of the factors that affected it; no one had published any mathematical relationship between these factors. The first to do so was Austin W. Zingg in 1940, who evaluated data from field experiments under natural rainfall and from a rainfall simulation experiment on a Shelby loam soil in Missouri (Laflen and Flanagan 2013). The following year, Smith (1941) expanded Zingg's work, and Browning et al. (1947) presented a full soil erosion prediction technology based on Smith's work that included a soil erodibility factor. They developed soil erodibility factors and permissible soil loss limits for a suite of Iowa soils and used Smith's equation to compute slope length limits for management of these soils; and soil erosion process models were developed. From 1954, onwards, the focus was on analysing the existing data sets and developing an overall scheme to support a broader prediction technology built on the previous work. Major works published by Wischmeier and his colleagues were related to the accomplishment of this goal as stated below:

- Factors affecting sheet and rill erosion (Smith and Wischmeier 1957)
- Rainfall energy and its relationship to soil loss (Wischmeier and Smith 1958)
- A rainfall erosion index for a Universal Soil Loss Equation (Wischmeier 1959)
- Cropping management factor evaluation for a Universal Soil Loss Equation (Wischmeier 1960)
- First publication of the USLE in an ARS Special Report. A universal equation for predicting rainfall erosion losses—an aid to conservation farming in humid regions (Wischmeier and Smith 1961)
- Soil erodibility evaluations for soils on the runoff and erosion stations (Olson and Wischmeier 1963)
- First publication of the USLE in a USDA Agriculture Handbook—Predicting rainfall erosion losses from cropland east of the Rocky Mountains: Guide for selection of practices for soil and water conservation (Wischmeier and Smith 1965)
- Second publication of USLE in a USDA Agriculture Handbook—Predicting rainfall erosion losses: A guide to conservation farming (Wischmeier and Smith 1978)
- Publication of RUSLE in a USDA Agriculture Handbook—Predicting soil erosion by water: A guide to conservation planning with the Revised Universal Soil Loss Equation (RUSLE) (Renard et al. 1997)

The most effective model on the soil loss is considered to be modified Soil Loss Equation developed by Wischmeir and Smith (1978), popularly known as Universal Soil Loss Equation (USLE). At the same time, Elwell (1978) developed Soil Loss Equation model for South Africa (SLEMSA) to predict mean annual soil loss, rising from sheet erosion on area of arable land. Bhargav (1999) modified it for Indian conditions for conservation practices in use by incorporating conservation practice factor. Morgan et al. (1984) developed a model for estimating annual soil loss from field size area on hill slopes, while Nearing et al. (1989) developed Water Erosion Prediction Project (WEPP) model having capability of predicting spatial and temporal distribution of net soil loss for entire hill slope for any period of time. A soil erosion model (Soil Erosion Model for Mediterranean-SEMMED regions) was developed for the test site Ardeche, France, by De Jong (1997), comprising several modules, each of which described a part of the erosion process such as soil particle detachment, moisture storage in top soil and transport of soil particles by overland flow. Victor Demidov (2001) used quasi three-dimensional runoff model for soil erosion modelling, which allows to simulate the temporal and spatial variations in erosion by raindrop impact and overland flow, sediment transport and deposition. For a more complete picture, the readers may also wish to read Meyer (1984), Meyer and Moldenhauer (1985), Renard (1985), Reich et al. (2001), Laflen and Moldenhauer (2003), Flanagan et al. (2007), Hughes and Prosser (2012), Shellberg et al. (2012) and Laflen and Flanagan (2013).

The LISEM model (De Roo et al. 2001) is one of the first examples of a physically based model that is completely incorporated in a raster geographical information system. A process-based model for channel degradation applicable to ephemeral gully erosion was developed by Casali et al. (2003); and Daba et al. (2003) assessed gully erosion in eastern Ethiopia using photogrammetric techniques. Thus, over the last decades, most research dealing with soil erosion by water has concentrated on sheet and rill erosion processes operating at the plot scale. Relatively few studies have been conducted on gully erosion operating at larger spatial scales. Poesen et al. (2003) indicated that gully erosion represents an important sediment source in a range of environments and gullies are effective links for transferring runoff and sediment from uplands to valley bottoms and permanent channels where they aggravate off-site effects of water erosion. Consequently, there is need for monitoring, experimental and modelling studies of gully erosion as a basis for predicting the effects of environmental change (climate and land use changes) on gully erosion rates. Further, only a few studies have quantified the hydrogeomorphic mechanisms responsible for the propagation or initiation of alluvial gully erosion.

Recently, Shellberg et al. (2012) investigated the hydrogeomorphic influences on alluvial gully erosion along the Mitchell River fluvial megafan, and longitudinal changes in floodplain inundation were quantified using river gauge data, local stage recorders and HEC-RAS modelling based on LiDAR topographic data. Intra- and interannual gully scarp retreat rates were measured using daily time-lapse photographs and annual GPS surveys. Erosion was analysed in response to different water sources and associated erosion processes across the floodplain perirheic zone, including direct rainfall, infiltration, excess runoff, soil-water seepage, river back water and overbank flood inundation. Hughes and Prosser (2012) predicted the patterns of gully density across large area ($1 \times 10^6$ km$^2$) of Murray–Darling Basin in Australia, using aerial photograph mapping of gullies across part of the basin and a multivariate statistical model of a range of environmental factors. The model is reasonably successful at predicting the variations in mapped gully density compared with similar attempts to predict erosion processes at this scale. Hillslope gradient and mean annual rainfall are the most important single factors across the region. The predicted mean gully density across the basin is 0.08 km$^{-2}$, and gullies contribute up to $27 \times 10^6$ Mg year$^{-1}$ of sediment to the river network. This is more than the amount that has been estimated from the combined contribution of hillslope ($14 \times 10^6$ Mg year$^{-1}$) and riverbank ($8.6 \times 10^6$ Mg year$^{-1}$) erosion by other studies within the Basin. Such methods also improve our understanding of the environmental controls on gully erosion.

The importance of protecting and restoring the soil resource from degradation is increasingly been recognized by the world community (Lal 2003; Poesen et al. 2003; Valentin et al. 2005). UNEP (1986) estimated that 2 billion ha of land that was once biologically productive has been irreversibly degraded since 1000 AD. Further, Rozanov et al. (1993 reported that more productive soil may have been irreversibly lost in the past 10,000 years than is currently under agricultural production (estimated at about 1500 million ha). With minor modifications of the GLASOD Project data, Middleton and Thomas (1997) estimated land area susceptible to dryland water erosion to be 467.4 million ha. Reich et al. (2001) reported the global assessment of water and wind erosion in a simplified model considering only soil and climate variables.

It has been emphasized that the actual intensity of erosion is a function of many factors, the most important of which is the level of management as determined partly by the socio-economic status of the land users. The relationship between erosion vulnerability and the inherent land quality is employed to compute magnitude of global soil loss through water erosion. They estimated the annual potential yield of sediment through water erosion from 72.5 million $km^2$ of global land area to be 130 billion Mg. In the arable lands of the world, water erosion may contribute about 67 billion Mg of sediment; and in susceptible drylands, water erosion could yield about 92 billion Mg, which are about 71% of the total global soil loss. Yet most available statistics on the extent and severity of soil erosion is subjective, qualitative, obsolete, crude and unreliable. There are several uncertainties in these available statistics on the extent and severity of soil erosion (Lal 2003). The most widely used statistics on soil erosion is that by Oldeman (1994) who has reported that total land area affected by water erosion is 1094 million hectares (Mha), of which 751 Mha is severely affected, and that by wind erosion is 549 Mha, of which 296 Mha is severely affected.

There are also regional hot spots of erosion including the Himalayan–Tibetan ecosystem in South Asia, the Loess Plateau in China, the subhumid and semiarid regions of sub-Saharan Africa, highlands of Central America, the Andean region, Haiti and the Caribbean (Scherr and Yadav 1996). Like the land area affected, there are also uncertainties about the extent of sediment transported into the oceans. Estimates of global rates of soil erosion have been made at 75 billion Mg per year by Pimentel et al. (1995), assuming an average erosion rate of 100 Mg $ha^{-1}$ on 751 Mha of area affected by severe erosion. The first attempt at estimating the sediment transport by rivers was that by Fournier (1960) followed by Strakhov (1967) who differed from that by Fournier both in trends and sediment yield by several orders of magnitude. Since then, several attempts at estimating sediment transport have been made including those by Jansson (1988), Walling (1996, 1997), Walling and Webb (1996), Xiang-zhou et al. (2004) and Hughes and Prosser (2012).

The annual sediment transport to the ocean is about 20 billion Mg, but the mass of sediment in motion is about 30 billion Mg per year (Milliman and Syvitski 1992). Most available literature shows that the annual sediment transport into the ocean by the world's rivers is 15–20 billion Mg per year (Walling and Webb 1996; Lal 2003). Erosion from stream and gully banks can generate up to 90% of the total sediment yield in Australian catchments (Olley et al. 1993; Prosser and Winchester 1996; Wasson et al. 1998) and in parts of other continents (Simon 1989; Poesen et al. 1996). The estimates of soil degradation in India varied from a minimum of 53.28 to 187.70 million ha depending upon the agency involved. An attempt was made by the Indian Council of Agricultural Research (ICAR) in 2010 to harmonize the area statistics, and it was agreed upon by all the relevant agencies that the degraded lands/wastelands in India occupy 120.72 million ha out of which water erosion affected area (>10 Mg $ha^{-1}$ $year^{-1}$) is about 73.27 million ha. It is estimated that the annual soil loss is more than 5 billion Mg yearly causing a nutrient loss of 6 million

Mg annually. Twenty-nine per cent of the soil is permanently lost to sea which is of a major concern.

Brice (1966) distinguished rills from gullies by a criterion of minimum width of 0.3 m and a minimum depth of about 0.6 m. As to the upper limit of gullies, no clear-cut definition existed. In the 1980s, the term *ephemeral gully erosion* was introduced to include concentrated flow erosion larger than rill erosion but less than classical gully erosion and now being recognized a major part of the erosional systems on croplands. These are small channels eroded by concentrated overland flow that can be easily filled by normal tillage, only to reform again in the same location by additional runoff events. Soil Science Society of America in 2001 defined *permanent gullies* for agricultural land in terms of channels too deep to easily ameliorate with ordinary farm tillage equipment, typically ranging from 0.5 m to as much as 25–30 m depth (Poesen et al. 2003). It must be acknowledged that the transition from rill erosion to ephemeral gully erosion to classical gully erosion and to river channel erosion represents a continuum, and any classification of hydraulically related erosion that forms into separate classes, such as micro-rills, rills, mega-rills, ephemeral gullies and gullies, is, to some extent, subjective (Poesen et al. 2003).

In Australia, two types of gullies, i.e. colluvial or hillslope gullies and alluvial gullies, have been described (Shellberg and Brooks 2012). Most of the research about gully erosion processes and rehabilitation management in Australia is based on *hillslope* or *colluvial gullies* typically found on the tablelands and mid-slopes of coastal south-eastern Australia (Crouch 1990) and northern Australia (Hancock and Evans 2006). Hillslope gullies typically erode into colluvium (i.e. accumulated rock and soil at the base of hillslopes from long-term gravitational processes and overland flow), but this colluvium can also be mixed with some minor alluvium (i.e. sediment transported, abraded and sorted by flowing water in channels).

The advance stage of gully erosion is formation of ravines. The word "ravine" means a deep gorge and represents the last stage of water erosion. Ravine-cut landscapes occur in many areas and known by different names (Haigh 1984). The French call them *lavaka* and the Brazilians *vossoroca* or *bocorroca*, and these landscapes are reported from Zimbabwe, Madagascar, central Russia, Australia and many parts of the American Southwest, where they are associated with the name *arroyo*. In most of the literature across the world, gully and ravine erosion has not been distinguished, particularly for management point of view; but in India the reclamation approach is different for two types of structures. These are reclaimed as per the reclaimability classes of ravine/gully erosion.

In countries like India, where there is great pressure on agricultural lands for food production, these lands cannot be ignored as such; hence, the pioneering work on soil and water conservation in eroded habitats was initiated in the mid-1950s, and all research centres became a part of the Indian Council of Agricultural Research in 1967, when the Council was reorganized. By this time, these centres had been engaged in soil and water conservation research for periods varying from 10 to 12 years. To give impetus to coordinated research on soil and water conservation in India and to compile the work done so far and make it available to research, educational and development agencies, a comprehensive monograph on "Soil and Water

Conservation Research 1956–71" was brought out by Tejwani et al. (1975a, b), which also included classification, formation, control and reclamation of ravine lands in India. Later, Bali and Karale (1977), Chinnamani et al. (1980), Sharma et al. (1980), Dhuruva Narayana and Ram Babu (1990), Verma and Chinnamani (1990), Dhuruva Narayana (1993), Dagar (1999), Sethy et al. (2011), Rao et al. (2012, 2015), Chaturvedi et al. (2014), Verma et al. (2015, 2016) and many others brought out very useful publications on formation and reclamation of ravine lands in India.

Gullies and ravines are common features of mountainous or hilly regions with steep slopes as is evident from several examples reported from the Peninsular India (Rama Mohan Rao 1990); Himalayan region of India (Tejwani et al. 1975a; Khybri et al. 1982; Gupta et al. 1990; Chaturvedi et al. 2014); the French Alps (Esteves et al. 2005; Mathys et al. 2005); Slovakia (Stankoviansky 2003); Morocco (Naimi et al. 2003); Ethiopia (Daba 2003; Nyssen et al. 2002; Bewke and Sterk 2003; Nyssen et al. 2004a, b, c); Kenya (Jungerius et al. 2002); Laos (Chaplot et al. 2004, 2005a, b); Peninsular Malaysia (Sidle et al. 2004); New Zealand (Gomez et al. 2003a, b); Idaho (Istanbulluoglu et al. 2003); and Ecuador (Podwojewski et al. 2002; Vanacker et al. 2003). Steep slopes favour high runoff velocity and thus rill and gully initiation, but, given climatic conditions, they can produce lower runoff volumes than gentle slopes, as recently shown in northern Thailand (Janeau et al. 2003). Hillslope or colluvial gullies tend to be linear erosional features, where their lateral and vertical erosion is confined by bedrock and their upslope migration tends to be self-limiting as a function of the catchment area, slope and the availability of colluvium to erode. Integrated approach involving engineering work and bringing the land under perennial vegetation, particularly grasses and trees, can help in reducing soil erosion and sedimentation.

*Alluvial gullies* and *ravines* on the other hand originate at steep stream banks and erode into adjacent, relatively flat, alluvial floodplains and terraces. They have been inconsistently described in the international literature as bank gullies, ravines, valley-bottom gullies and alluvial breakaways from locations around the world including Europe, India, Africa, USA and Australia (Brooks et al. 2009; Shellberg 2011; Shellberg et al. 2013). Due to a lack of bedrock confinement, alluvial gullies are often as wide as they are long and expand longitudinally and laterally until they develop new equilibrium channel slopes and consume massive volumes of floodplain alluvium. This alluvial material tends to be much finer sediment than most colluvial deposits, thereby contributing a higher proportion of fine sediment to river suspended sediment loads. They are also highly connected to the stream network, delivering their sediment load directly to the main channel. Alluvial gullies are widely distributed in northern Australia across floodplain environments covering Normanby, Mitchell, Gilbert, Leichhardt, Gregory, Victoria and Fitzroy catchments (Shellberg and Brooks 2012). They likely exist along additional floodplain rivers in the Australian tropics. However, they are not restricted to the tropics and exist in other locations across the continent and the world. Alluvial gullies have been inconsistently described in the literature as valley-bottom gullies, bank gullies, ravines and alluvial breakaways from locations around the world (e.g. USA, Thomas et al. 2004; Australia, Pickup 1991; Pringle et al. 2006; Brooks et al. 2008, 2009; McCloskey 2010; Europe, Poesen 1993; Vandekerckhove et al. 2000, 2001a, b, 2003; Africa, Oostwoud Wijdenes and Bryan 2001; China, Hui and Mingan 2000; India, Tejwani et al. 1975a, b; Dagar 1999; Singh and Dubey 2000; Yadav and Bhushan 2002; Joshi 2006; Verma et al. 2015).

Gully/ravine erosion due to river channel trenching is a problem that threatens vast tracts of the world's agricultural land. The damage is greatest in the alluvial plains of the semiarid and arid zones. Gullies not only occur in merely badlands and mountainous or hilly regions but also more globally in soils subjected to soil crusting such as loess (European belt, Chinese Loess Plateau, North America) and sandy soils (Sahelian zone, north-east Thailand) or in soils prone to piping and tunnelling such as dispersive soils. They are known to occur over alluvium or colluvium and in a variety of environments including on the planet Mars. Recent findings indicate that gullies are widespread at mid to high latitudes on the surface of Mars and are some of the youngest features observed on that planet, probably forming within the last few 100,000 years (https://en.wikipedia.org/wiki/Gully). There, they are one of the best lines of evidence for the presence of liquid water on Mars in the recent geological past, probably resulting from the slight melting of snow packs on the surface (Christensen 2003) or ice in the shallow subsurface (Costard et al. 2007) on the warmest days of the Martian year. Flow as springs from deeper-seated liquid water aquifers in the deeper subsurface is also a possible explanation for the formation of some Martian gullies (Mallin and Edgett 2000).

During the last two decades, a significant progress in the understanding of gully erosion and its controlling factors has been made, and over hundred research papers have been brought out in international journals and books and from the international conferences on gully erosion which include Poesen et al. (2003), proc. Leuven symposium in 2000; Li et al. (2004), proc. Chengdu symposium in 2002; Valentin et al. (2005), proc. Chengdu symposium in 2002; Römkens and Bennett (2005), proc. Oxford, MS, USA symposium in 2004; Casali et al. (2009), proc. Pamplona symposium in 2007; and Verma et al. (2016), Gwalior, India International Workshop in 2016. In addition, between 2000 and 2016, over 650 papers discussing various aspects of gully erosion have been reported in the Web of Science. Yet, there remain several research gaps and unanswered questions related to gully erosion, which include historical studies on gully erosion for better understanding of different stages of their formation; appropriate measuring techniques for monitoring or better understanding of the geological and hydrological parameters and processes of gully development; interaction of hydrological and other soil degradation processes; impact of climate and environmental changes on soil erosion processes; development of effective and reliable gully erosion models; and effective gully prevention and control measures and watershed-based management options. In the present book, an attempt has been made to address some of these aspects along with impacts of ravine formation on socio-economic and policy issues.

## References

Ayres QC (1939) Soil erosion and its control. McGraw Hill Book Co, New York

Bali YP, Karale RL (1977) Reclamability classification of ravines for agriculture. Soil Conserv Digest 5(2):40–47

Baver LD (1938) Ewald Wollny-a pioneer in soil and water conservation research. Soil Sci Soc Am Proc 3:330–333

Beasley RP (1972) Erosion and sediment pollution control. The Iowa State University Press, Ames

Bennet HH (1939) Soil conservation. McGraw Hill Book Co, New York

Bennett HH, Chapline WR (1928) Soil erosion-a national menace. US Dept of Agric. Circular 33, Washington, DC

Bewke W, Sterk G (2003) Assessment of soil erosion in cultivated fields using a survey methodology for rills in the Chemoga watershed, Ethiopia. Agric Ecosys Environ 97(1–3):81–93

Bhargav KS (1999) A modified SLEMSA model for Naurar subcatchment of Ramganga river. Master's Thesis submitted to GB Pant Agric Univ Pantnagar

Brice JC (1966) Erosion and deposition in the Loess-Mantled Great Plains, Medicine Creek Drainage Basin, Nebraska. US Geological Survey, Professional Paper 352-H, Reston, Virginia, pp 255–339

Brooks AP, Spencer J, Shellberg JG, Knight J, Lymburner L (2008) Using remote sensing to quantify sediment budget components in a large tropical river – Mitchell River, Gulf of Carpentaria. In: Sediment dynamics in changing environments. Proceedings of a symposium held in Christchurch, New Zealand; December 2008, IAHS Publ No 325 pp 225–236

Brooks AP, Shellberg JG, Spencer J, Knight J (2009) Alluvial gully erosion: an example from the Mitchell fluvial megafan, Queensland, Australia. Earth Surf Process Landf 34:1951–1969

Brown LR (1981) Eroding the base of civilization. J Soil Water Conserv 36:255–260

Brown LR (1984) The global loss of topsoil. J Soil Water Conserv 39:162–165

Browning GM, Parish CL, Glass J (1947) A method for determining the use and limitations of rotation and conservation practices in the control of soil erosion in Iowa. J Am Soc Agron 39(1):65–73

Casali J, Lopez JJ, Giraldez JV (2003) A process-based model for channel degradation: application to ephemeral gully erosion. Catena 50:435–447

Casali J, Giménez R, Bennett S (2009) Gully erosion processes and modelling. Proc. fourth international Symposium on Gully Erosion, Pamplona, Spain, September 2007. Earth Surf Process Landf 34:1839–1984

Castillo C, Gómes JA (2016) A century of gully erosion research: urgency, complexity and study approaches. Earth Sci Rev 160:300–319

Chapline WR (1929) Erosion on rangeland. J Am Soc Agron 21:423–429

Chaplot V, Giboire G, Marchand P, Valentin C (2004) Dynamic modeling for gully initiation and development under climate and land-use changes in northern Laos. In: Li Y, Poesen J, Valentin C (eds) Gully erosion under global change. Sichuan Science and Technology Press, Chengdu, pp 233–245

Chaplot V, Coadou le Brozec E, Silvera N, Valentin C (2005a) Spatial and temporal assessment of linear erosion in catchments under sloping lands of northern Laos. Catena 63:167–184

Chaplot V, Giboire G, Marchand P, Valentin C (2005b) Dynamic modelling for linear erosion initiation and development under climate and land-use changes in northern Laos. Catena 63:318–328

Chaturvedi OP, Kaushal R, Tomar JMS, Prandiyal AK, Panwar P (2014) Agroforestry for wasteland rehabilitation: mined, ravine, and degraded wasteland areas. In: Dagar JC, Singh AK, Arunachalam A (eds) Agroforestry systems in India: livelihood security and ecosystem services, advances in agroforestry, vol 10. Springer, New Delhi, pp 233–272

Chinnamani S, Venkataraman C, Tejwani KG (1980) Landslides and land slips in peninsular India, Bull No. 4. Central Soil & Water Conservation Research & Training Institute, Dehradun

Christensen PR (2003) Formation of recent Martian gullies through melting of extensive water-rich snow deposits. Nature 422(6927):45–48. https://doi.org/10.1038/nature01436.PMID12594459
Costard F, Forget F, Mangold N, Peulvast JP (2007) Formation of recent Martian debris flows by melting of near-surface ground ice at high obliquity. Science 295:110–113
Crouch RJ (1990) Rates and mechanisms of discontinuous gully erosion in a red brown earth catchment, New South Wales, Australia. Earth Surf Process Landf 15(3):277–282
Daba S (2003) An investigation of the physical and socioeconomic determinants of soil erosion in the Hararghe highlands, eastern Ethiopia. Land Degrad Dev 14(1):69–81
Daba S, Rieger W, Strauss P (2003) Assessment of gully erosion in eastern Ethiopia using photogrammetric techniques. Catena 50:273–291
Dagar JC (1999) Soil erosion parameters with special reference to Kshipra ravines in Madhya Pradesh. Indian J For 22(3):210–221
De Jong SM (1997) MeMon-Satellite-based desertification monitoring in the Mediterranean Basin. www.geog.uu.nl/fg/demon.html
De Roo APJ, Wesseling CG, Cremers NHDT, Offermans RJE, Ritsema CJ, van Ostindie K (2001) Lisem: a physically-based hydrological and soil erosion model incorporated in a GIS. www.odyssey.maine.edu/gisweb/spatd6/egis/eg94023.html
Demidov V (2001) Modeling soil erosion and sediment transport on watersheds with the help of quasi three-dimensional runoff model. www.epa.gov/OWOW/watershed/proced/demidov.html
DhuruvaNarayana VV (1993) Soil and water conservation research in India. Indian Council of Agricultural Research, New Delhi, p 454
DhuruvaNarayana VV, Babu R (1990) Social and economic considerations in the management of soils of hilly regions. In: Abrol IP, DhuruvaNarayana VV (eds) Technologies for wasteland development. ICAR, New Delhi, pp 273–284
Dotterweich M (2013) The history of human-induced soil erosion: geomorphic legacies, early descriptions and research, and the development of soil conservation a global synopsis. Geomorphology 201:1–34 Available online: http://www.Sciencedirect. com/science/article/pii/s0169555x1300370x
Dregne HE (1982) Historical perspective of accelerated erosion and effect on world civilization. Determinants of soil loss tolerance. ASA Spec Publ, vol. 45. Madison, WI pp 1–14
Duley FL, Ackerman FG (1934) Runoff and erosion from plots of different lengths. J Agric Res 48(6):505–510
Duley FL, Miller MF (1923) Erosion and surface runoff under different soil conditions. Res Bull 63:50. Univ Mo Agric Exp Sta Columbia
Eckholm EP (1976) Losing ground. Norton, New York
Ellison WD (1944a) A rain drop, surface flow and erosion. Soil Conserv 10:128–129
Ellison WD (1944b) Studies on rain drop erosion. Agric Eng 25(131–136):181–182
Ellison WD (1947) Soil erosion studies. Part I–VII. Agric Eng 28:145–146. 197–201, 245–248, 297–300, 353, 402–405, 408, 442–444, 450
Elwell HA (1978) Modelling soil loss in southern Africa. J Agric Eng Res 23:117–127
Erasmus LI, Madan-Lal, Raghunath B, Mathur PS (1970) Evaluation and erosion potential from rainfall data. Indian For 96:817–825
Esteves M, Descroix L, Mathys N, Lapetite JM (2005) Soil hydraulic properties in a marly gully catchment (Draix, France). Catena 63:282–298
Flanagan DC, Gilley JE, Franti TG (2007) Water Erosion Prediction Project (WEPP): development history, model capabilities and future enhancements. Trans Am Soc Agric Bio Eng 50(5):1603–1612
Flanagan DC, Frankenberger JR, Cochrane TA, Renschler CS, Elliot WJ (2013) Geospatial application of the Water Erosion Prediction Project (WEPP) model. Trans Am Soc Agric Bio Eng 56(2):591–601
Forsling CL (1931) A study of the influence of herbaceous plant cover on surface runoff and soil erosion in relation to grazing on the Wasatch Plateau in Utah. U S Dept Agr Tech Bull 220:72
Fournier F (1960) Climate erosion. Presses Universitaires de France, Paris

Gomez B, Banbury K, Marden M, Trustrum NA, Peacock DH, Hoskin PJ (2003a) Gully erosion and sediment production, TeWeraroa stream, New Zealand. Water Resour Res 39(7):1187
Gomez B, Trustrum NA, Hicks DM, Rogers KM, Page MJ, Tate KR (2003b) Production, storage, and output of particulate organic carbon, Waipaoa River basin, New Zealand. Water Resour Res 39(6):ESG 2-1–ESG 2-8
Gupta RK, Bhardwaj SP, Mittal SP (1990) Management of degraded lands in the Himalayas. In: Abrol IP, DhuruvaNarayana VV (eds) Technologies for wasteland development. ICAR, New Delhi, pp 196–213
Haigh MJ (1984) Ravine erosion and reclamation in India. Geoforum 15(4):543–561
Hardaha MK (2015) Modeling soil erosion from watershed. In: Verma SK, Trivedi SK, Singh A, Tomar AS (eds) Management of degraded lands and soil health. Biotech Books, New Delhi, pp 67–83
Hughes AO, Prosser IP (2012) Gully erosion prediction across a large region: Murray–Darling Basin, Australia. Soil Res 50:267–277
Hui S, Mingan S (2000) Soil and water loess from the Loess Plateau in China. J Arid Environ 45:9–20
Hunt T (2006) Rethinking the fall of Easter Island. Am Sci 94(5):412. https://doi.org/10.1511/2006.61.1002
Ireland HA, Sharpe CFS, Eargle DG (1939) Principles of gully erosion in the piedmont of South Carolina. US Dept Agric Tech Bull 633
Istanbulluoglu E, Tarboton DG, Pack RT, Luce C (2003) A sediment transport model for incision of gullies on steep topography. Water Resour Res 39(4):ESG 6-1–ESG 6-15
Janeau JL, Bricquet JP, Planchon O, Valentin C (2003) Soil crusting and infiltration on steep slopes in northern Thailand. Eur J Soil Sci 54(3):543–554
Jungerius PD, Matundura J, van de Ancker JAM (2002) Road construction and gully erosion in West Pokot, Kenya. Earth Surface Processes and Landforms 27(11):1237–1247
Jansson MB (1988) A global survey of sediment yield. Geogr Ann Ser A 70:81–98
Joshi VU (2006) Morphological adjustments of gullies on the anthropogenic interference in the landscapes. In: Chandrashekhar N (ed) Quaternary climatic changes and landforms. Tirunelveli Publication, pp 327–356
Khybri ML, Prasad SN, Ram S (1982) Effect of loss of top soil on the yield and economics of maize-wheat rotations. Indian J Soil Cons 10(1):58–62
Kinnell PIA (1974) Splash erosion-some observations on the splash cup-technique. Soil Sci Soc Am Proc 38:657–660
Laflen JM, Flangen DC (2013) The development of US soil erosion prediction and modelling. Int Soil Cons Res 1(2):1–11
Laflen JM, Moldenhauer WC (2003) Pioneering soil erosion prediction-the USLE story. World Association of Soil and Water Conservation, Special Publication #1
Lal R (2003) Soil erosion and global carbon budget. Environ Int 29:437–450
Li Y, Poesen J, Valentin C (eds) (2004) Gully erosion under global change, Proc. Second International Symposium on Gully Erosion, Chengdu, China, May 2002. Sichuan Science and Technology Press, Chengdu, p 354
Lowdermilk WC (1953) Conquest of the land through seven thousand years. SCS Agric Inform Bull, vol 99, Washington, DC, 30 pp
Malin MC, Edgett KS (2000) Evidence for recent groundwater seepage and surface runoff on Mars. Science 288(5475):2330–2335
Mathys N, Klotz S, Esteves M, Lapetite JM (2005) Runoff and erosion in the Black Marls of the French Alps, observations and measurements at the plot scale. Catena 63:261–281
McCloskey GL (2010) Riparian erosion morphology, processes and causes along the Victoria river, northern territory, Australia. Charles Darwin University, PhD Thesis, p 200
Meyer LD (1984) Evolution of the universal soil loss equation. J Soil Water Conserv 39(2):99–104
Meyer LD, Moldenhauer WC (1985) Soil erosion by water: the research experience. Agric Hist 59(2):192–204
Middleton N, Thomas D (1997) World atlas of desertification, 2nd edn. Arnold Pub, London, p 182

Miller MF (1926) Waste through soil erosion. J Am Soc Agric 18:153–160
Miller MF (1946a) Early measurements of runoff and erosion. U S Dept Agric Soil Conserv 11(11):255–257
Miller MF (1946b) Early investigations dealing with water runoff and soil erosion. Agronomy J8:657–660
Miller MF, Krusekopf HH (1932) The influence of systems of cropping and methods of culture on surface runoff and soil erosion. Res Bull 177. Mo Agr Exp Sta Columbia:32
Milliman JD, Syvitski JPM (1992) Geomorphic/tectonic control of sediment discharge to the ocean: the importance of small mountainous rivers. J Geol 100:325–344
Morgan RPC, Morgan DDV, Finney HJ (1984) A predictive model for the assessment of erosion risk. J Agric Eng Res 30:245–253
Musgrave GW (1947) The quantitative evaluation of factors in water erosion a first approximation. J Soil Water Conserv 2(3):133–138. 170
Naimi M, Tayaa M, Ouzizi S, Ilha CR, Kerby M (2003) Dynamique de l'e'rosion par ravinementdans un bassin versant du Rif occidental au Maroc. Se'cheresse 14(2):95–100
Nearing MA, Foster GR, Lane LJ, Finkner SC (1989) A process based soil erosion model for USDA Water Erosion Prediction Project. Trans ASAE 32(5):1587–1593
Nelson LB (1958) Building sounder conservation and water management research programs for the future. Soil Sci Soc Am Proc 22:355–358
Nyssen J, Poesen J, Moeyersons J, Luyten E, Veyret Picot M, Deckers J, Mitiku H, Govers G (2002) Impact of road building on gully erosion risk, a case study from the northern Ethiopian highlands. Earth Surf Process Landf 27(12):1267–1283
Nyssen J, Poesen J, Moeyersons J, Deckers J, Mitiku H, Lang A (2004a) Human impact on the environment in the Ethiopian and Eritrean highlands—a state of the art. Earth-Sci Rev 64(3–4):273–320
Nyssen J, Veyret Picot M, Poesen J, Moeyersons J, Mitiku H, Deckers J, Govers G (2004b) The effectiveness of loose rock check dams for gully control in Tigray, northern Ethiopia. Soil Use Manag 20(1):55–64
Nyssen J, Poesen J, Moeyersons J, Deckers J, Mitiku H (2004c) Erosion et conservation des sols en montagnesahe'lienne, le cas de l'Ethiopie du Nord. Se'cheresse 15(1):33–39
Oldeman LR (1994) The global extent of soil degradation. In: Greenland DJ, Szabolcs I (eds) Soil resilience and sustainable land use. CAB International, Wallingford, pp 99–118
Olley JM, Murray AS, Mackenzie DM, Edwards K (1993) Identifying sediment sources in a gullied catchment using natural and anthropogenic radioactivity. Water Resour Res 29:1037–1043
Olson GW (1981) Archaeology: lessons on future soil use. J Soil Water Conserv 36:261–264
Olson TC, Wischmeier WH (1963) Soil erodibility evaluations for soils on the runoff and erosion stations. Soil Sci Soc Am Proc 27:590–592
OostwoudWijdenes DJ, Bryan RB (2001) Gully-head erosion processes on a semi-arid valley floor in Kenya: a case study into temporal variation and sediment budgeting. Earth Surf Process Landf 26:911–933
Peiser B (2005) From genocide to ecocide: the rape of Rapa Nui (PDF). Energy Environ 16(3&4). https://doi.org/10.1260/0958305054672385
Penck A (1894) Morphologie der erdoberflache, veriag von J. EngethornNachf, Stuffgart
Penck W (1953) Morphological analysis of land forms-a contribution to physical geology. McMillan & Co. Ltd, London, p 47
Pimentel D, Harvey C, Resosudarmo P, Sinclair K, Kurz D, McNair M et al (1995) Environmental and economic costs of soil erosion and conservation benefits. Science 267:1117–1123
Podwojewski P, Poulenard J, Zambrana T, Hofstede R (2002) Overgrazing effects on vegetation cover and properties of volcanic ash soil in the paramo of Llangahua and La Esperanza (Tungurahua, Ecuador). Soil Use Manag 18(1):45–55
Poesen J (1993) Gully typology and gully control measures in the European loess belt. In: Wicherek S (ed) Farm land erosion in temperate plains environment and hills. Elsevier, Amsterdam, pp 221–239

Poesen J, Vandaele K, van Wesemael B (1996) Contribution of gully erosion to sediment production in cultivated lands and rangelands. IAHS Publications 236:251–266

Poesen J, Nachtergaele J, Verstraeten G, Valentin C (2003) Gully erosion and environmental change: importance and research needs. Catena 50(2–4):91–133

Pringle HJR, Watson IW, Tinley KL (2006) Landscape improvement, or ongoing degradation – reconciling apparent contradictions from the rangelands of Western Australia. Landsc Ecol 21:1267–1279

Prosser IP, Winchester SJ (1996) History and processes of gully initiation and development in eastern Australia. Zeitschriftfu¨ r Geomorphol N F Suppl.bd 105:91–109

Rama Mohan Rao MS (1990) Approaches to the management of denuded lands in the Peninsular India. In: Abrol IP, DhuruvaNarayana VV (eds) Technologies for wasteland development. ICAR, New Delhi, pp 232–244

Rao BK, Kuruthe RS, Singh AK, Parandiyal AK, Pande VC, Kumar G (2012) Bamboo plantation based technological interventions for reclamation and productive utilization of ravine lands, Technical Bull T-62/V-4. 2012 Central Soil & Water Conservation Research & Training Institute, Dehradun, p 30

Rao BK, Mishra PK, Kuruthe RS, Pande VC, Kumar-Gopal (2015) Effectiveness of *Dichanthiumannulatum* in water courses for reducing sediment delivery from agricultural watersheds. Clean Soil Air Water 43(5):710–716

Reich P, Eswaran H, Beinroth F (2001) Global dimensions of vulnerability to wind and water erosion. In: Stot D, Mohtar RH, Steinhardt GC (eds) Sustaining the global farm. Selected papers from the 10th International Soil Cons Organisation Meeting at Purdue Univ & USDA ARS Erosion Res Lab, pp 838–846

Renard KG (1985) Rainfall simulates and USDA erosion research: history perspective and future. In: Lane LJ (ed) Erosion on rangelands: emerging technology and data base. Society of Range Management, Denver

Renard KG, Foster GR, Weesies GA, McCool DK, Yoder DC (1997) Predicting soil erosion by water: a guide to conservation planning with the Revised Universal Soil Loss Equation (RUSLE). Agricultural Handbook No. 703, US Dept. of Agr. Washington DC, p 384

Römkens MJM, Bennett SJ (2005) Gully erosion. Proc. third international symposium on Gully Erosion, Oxford, MS, USA, April 2004. Int J Sediment Res 20(3):157–280

Rozanov BG, Targulian V, Orlov DS (1993) Soils. In: Tumer BL, Clark WC, Kates RW, Richards JF, Mathews JT, Meyers WB (eds) The earth as transformed by human action: global and regional changes in the biospere over the past 300 years. Cambridge University, Cambridge, pp 203–214

Sampson AW, Weyl LH (1918) Range preservation and its relation to erosion control on western grazing lands, Bull 675. USDA, Washington, DC, p 35

Scherr SJ, Yadav S (1996) Land degradation in the developing world: implications for food, agriculture and the environment to 2020. IFPRI, Food, Agric. and the Environment Discussion Paper 14, Washington, DC, 36 pp

Sethy BK, Parandiyal AK, Ali S, Kumar A, Singh RK (2011) Cost effective conservation measures for management of medium and deep ravenous lands. Annual report. CSWCRTI, Dehradun, pp 73–74

Sharma AK, Pradhan IP, Neema JP, Research Centre Vasad, Gujarat, Tejwani KG (1980) 25 years of research on soil and water conservation in ravine lands of Gujarat. Central Soil & Water Conservation Research & Training Institute, Dehradun

Shellberg J (2011) alluvial gully erosion rates and processes across the Mitchell River Fluvial Megafan in Northern Queensland, Australia. Griffith University, PhD Thesis, Brisbane, Australia, November 2011, 283 pp

Shellberg J, Brooks A (2012) Alluvial gully erosion: a dominant erosion process across tropical Northern Australia. Australian Rivers Institute, Griffith University, Tropical Rivers Coastal Knowledge (TRaCK), Darwin NT 0909, Australia p 18

Shellberg JG, Brooks AP, Spencer J, Ward D (2012) The hydrogeomorphic influences on alluvial gully erosion along the Mitchell River fluvial megafan, northern Australia. Hydrol Process. https://doi.org/10.1002/hyp.9240

Shellberg JG, Brooks AP, Rose CW (2013) Sediment production and yield from an alluvial gully in northern Queensland. Aust Earth Surf Proc Landf. https://doi.org/10.1002/esp.3414

Sidle RC, Sasaki S, Otsuki M, Noguchi S, Rahim NA, Onda Y (2004) Sediment pathways in a tropical forest, effects of logging roads and skid trails. Hydrol Process 18(4):703–720

Simon A (1989) The discharge of sediment in channelized alluvial streams. Water Resour Bull 25:1177–1188. https://doi.org/10.1111/j.1752-1688.1989.tb01330.x

Singh S, Dubey A (2000) Temporal variations in the network of man-impacted gully basins in the sub-humid tropical alluvial riverine environment – a case study of Deoghat gullies of Allahabad District, India. Z Geomorphol 44:175–194

Smith DD (1941) Interpretation of soil conservation data for field use. Agric Eng 22(5):173–175

Smith DD, Wischmeier WH (1957) Factors affecting sheet and rill erosion. Trans Am Geophys Union 38:889–896

Stankoviansky M (2003) Historical evolution of permanent gullies in the Myjava Hill land, Slovakia: geomorphic responses to land use changes. Catena 51(3–4):223–239

Stewart G, Forsling CL (1931) Surface runoff and erosion in relation to soil and plant cover on high grazing lands of Central Utah. J Am Soc Agron 23:815–832

Strakhov NM (1967) Principles of lithogenesis, vol 1. Oliver and Boyd, Edinburgh

Suresh R (1997) Soil and water conservation engineering. Standard Publishers, New Delhi

Tejwani KG, Dayal R, Verma B, Gidwani HM, Gupte SC (1975a) Classification, control and reclamation of ravine lands. In: Tejwani KG, Gupta SK, Mathur HN (eds) Soil and water conservation research (1956–1971). ICAR Pub, New Delhi, pp 254–310

Tejwani KG, Gupta SK, Mathur HN (eds) (1975b) Soil and water conservation research (1956–1971). ICAR Pub, New Delhi, p 358

Thomas JT, Iverson NR, Burkart MR, Kramer LA (2004) Long-term growth of a valley-bottom gully, western Iowa. Earth Surf Process Landf 29:995–1009

UNEP (1986) Sands of change: why land becomes desert and what can be done about it. UNEP Brief #2, United Nations Environment Programme, Nairobi, Kenya. 8 pp

Valentin C, Poesen J, Li Y (2005) Gully erosion: a global issue. Proc. second international symposium on Gully Erosion, Chengdu, China, May 2002. Catena 63(2–3):129–330

Vanacker V, Govers G, Poesen J, Deckers J, Dercon G, Loaiza G (2003) The impact of environmental change on the intensity and spatial pattern of water erosion in a semiarid mountainous Andean environment. Catena 513(4):329–347

Vandekerckhove L, Poesen J, Oostwoud Wijdenes D, Nachtergaele J, Kosmas C, Roxo MJ, De Figueiredo T (2000) Thresholds for gully initiation and sedimentation in Mediterranean Europe. Earth Surf Process Landf 25:1201–1220

Vandekerckhove L, Muys B, Poesen J, De Weerdt B, Coppe N (2001a) A method for dendrochronological assessment of medium-term gully erosion rates. Catena 45:123–161

Vandekerckhove L, Poesen J, OostwoudWijdenes D, Gyssels G (2001b) Short-term bank gully retreat rates in Mediterranean environments. Catena 44:133–161

Vandekerckhove L, Poesen J, Govers G (2003) Medium-term gully headcut retreat rates in Southeast Spain determined from aerial photographs and ground measurements. Catena 50:329–352

Verma B, Chinnamani S (1990) Rehabilitation of ravine lands. In: Abrol IP, DhuruvaNarayana VV (eds) Technologies for wasteland development. ICAR, New Delhi, pp 263–272

Verma SK, Trivedi SK, Singh A, Tomar AS (eds) (2015) Management of degraded lands and soil health. Biotech Books, New Delhi, p 249

Verma SK, Singh AK, Kandalkar VS, Ranade DH, Bhadauria UPS, Singh A, Srivastava SC, Tomar AS (eds) (2016) Global ravine conference on managing ravines for food and livelihood security-souvenir. Rajmata VijayarajeScindiaKrishiVishwaVidyalaya Gwalior, Madhya Pradesh, p 147

Walling DE (1996) Erosion and sediment yield in a changing environment. In: Branson J, Brown AG, Gregory KJ (eds) Global continental changes: the context of paleohydrology, vol 115. Geol Soc Spec Publ, London, pp 43–56

Walling DE (1997) The response of sediment yield to environmental change. In: Walling DE, Probst J-L (eds) Human impact on erosion and sedimentation, vol 245. IAHS Publ, Wallingford, pp 77–89

Walling DE, Webb BW (1996) Erosion and sediment yield: a global overview. Erosion and sediment yield: global and regional perspectives. Proc. Exeter Symp, July 1996. IAHS Publ, vol. 236. pp 3–19

Wasson RJ, Mazari RK, Starr B, Clifton G (1998) The recent history of erosion and sedimentation on the southern Tablelands of south-eastern Australia: sediment flux dominated by channel incision. Geomorphology 24:291–308

Wischmeier WH (1959) A rainfall erosion index for a universal soil loss equation. Soil Sci Soc Am Proc 23:247–249

Wischmeier WH (1960) Cropping management factor evaluations for a universal soil loss equation. Soil Sci Soc Am Proc 23:322–326

Wischmeier WH, Smith DD (1958) Rainfall energy and its relationship to soil loss. Trans Am Geophys Union 39(2):285–291

Wischmeier WH, Smith DD (1961) A universal equation for predicting rainfall erosion losses – an aid to conservation farming in humid regions. ARS Special Report 22–66 Agricultural Research Service, US Dept. of Agric. Washington DC. 11 pp

Wischmeier WH, Smith DD (1965) Predicting rainfall erosion losses from cropland east of the rocky mountains. USDA Handbook No. 282, Washington, DC

Wischmeier WH, Smith DD (1978) A guide to conservation planning. USDA Handbook 537, Washington, DC

Wollny E (1888) Untersuchungenuber die sickerwassermengen in verschicdenen. Bordenartenforsch. a.d. Gebiete d. Agr physic Band II, Scitepp 1–68

Xiang-Zhou X, Hong-wu Z, Ouyang Z (2004) Development of check-dam systems in gullies on the Loess Plateau, Chona. Environ Sci Policy 7:79–86

Yadav RC, Bhushan LS (2002) Conservation of gullies in susceptible riparian areas of alluvial soil regions. Land Degrad Dev 13:201–219

Zingg AW (1940) Degree and length of land slope as it affects soil loss in runoff. Agric Eng 21(2):59–64

# Ravines: Formation, Extent, Classification, Evolution and Measures of Prevention and Control

2

Jagdish Chander Dagar

**Abstract**

Ravine and gully erosion threatens large tracts of the world's agricultural land and contributes to huge amount of soil loss and sediment production under different climatic and land use conditions. The damage is greatest in the alluvial plains of the semiarid and arid zones. In recent times, global attention has been paid to solve the problem of land restoration, and several techniques and models have been used for measuring the extent of gully erosion and their management for livelihood and environmental security. The objective of this chapter is to review historical evidences of gully formation and recent studies on extent, impacts, factors, processes, control methods and management of gully/ravine erosion in different ecologies across the globe. Many techniques have proved to be effective for gully prevention and control including vegetation cover, minimum tillage, terracing and check dams. However, these techniques are rarely adopted by farmers and other stakeholders in the long run because of several bottlenecks in extension services and policy implementation. Rehabilitation of these lands requires an integrated approach of using appropriate site-specific soil and water conservation measures and putting land under vegetation cover involving afforestation, horticulture, pasture and energy plantations depending upon the land suitability class following watershed approach. There is an urgent need of policy interventions at different levels to handle this menace.

J. C. Dagar (✉)
Indian Council of Agricultural Sciences, Krishi Anusandhan Bhavan-II, Pusa, New Delhi, India



J. C. Dagar, A. K. Singh (eds.), *Ravine Lands: Greening for Livelihood and Environmental Security*, https://doi.org/10.1007/978-981-10-8043-2_2

## 2.1 Introduction

Soil erosion caused by water is considered one of the most serious problems in the world. It has caused major agricultural, economic, social and environmental disasters. Vast tracts of agricultural lands have been engulfed by gullies and ravines severely affecting and threatening the agricultural production and livelihood security, particularly in developing countries. The soil erosion is considered to be a severe geomorphic hazard influencing long-term effects on soil productivity and agricultural sustainability (Morgan 2005). Further, it leads to global environmental problems through sedimentation, pollution, flooding and increasing 75–80% of carbon content into the atmosphere (Lal 2003; Blanco-Canqui and Lal 2012).

The problem of soil erosion has been addressed since long, but only in recent years, more attention has been paid to this menace, and many workers across the globe have generated information related to the process of soil erosion and gully formation (Haigh 1998; De Baets et al. 2009; Mathys et al. 2003; Prasad and Römkens 2004; Martinez-Murillo et al. 2013; Shellberg et al. 2013; Zhao et al. 2013; Gabarron-Galeote et al. 2013; Mandal and Sharda 2013; Lieskovsky and Knderessy 2014; Mekonnen et al. 2015; Dai et al. 2015; Erkossa et al. 2015; Prosdocimi et al. 2016; Novara et al. 2016). It is estimated that out of about 2 billion ha land affected by various forms of human-induced land degradation, out of which ~1.1 billion ha land is eroded by water (Oldeman 1994). Middleton and Thomas (1997) reported that about 467.4 Mha dryland areas susceptible to water erosion occur in all the continents of which 141.0, 213.2 and 113.2 Mha fall in dry subhumid (precipitation/potential evapotranspiration P/PET ratio 0.5–0.65), semiarid (P/PET ratio <0.5) and arid (P/PET ratio <0.2) regions of the world, respectively. Almost 80% of the terrain affected by water erosion has a light to moderate degree of degradation, and around 225 Mha of it are degraded to the extent that these are no more suitable for agricultural land use. Many workers have estimated the extent and severity of soil erosion, but Lal (2003) has stated that most of the available statistics are subjective, qualitative, obsolete, crude and unreliable. The most widely used data is that given by Oldeman (1994) as placed in Table 2.1.

**Table 2.1** Global extent of water erosion (Mha)

| Region | Light | Moderate | Strong + extreme | Total | Dry zone[a] | Humid zone[a] |
|---|---|---|---|---|---|---|
| Africa | 58 | 67 | 102 | 227 | 122 | 105 |
| Asia | 124 | 242 | 73 | 441 | 165 | 276 |
| South America | 46 | 65 | 12 | 123 | 35 | 88 |
| Central and North America | 15 | 68 | 23 | 106 | 38 | 68 |
| Europe | 21 | 81 | 12 | 114 | 48 | 66 |
| Oceania | 79 | 3 | 1 | 83 | 70 | 13 |
| Total | 343 | 526 | 223 | 1094 | 478 | 615 |

Source: Oldeman (1994)

[a]Dry zone is defined as the climatic region with an annual precipitation/evapotranspiration ratio of 0.65 or less; the humid zone >0.65

Though the process of soil erosion by water is ancient, soil and water conservation as a science is just 140 years old. Only during the early twentieth century, some pioneer workers worked on the process of soil erosion at plot scale (Sampson and Weyl 1918; Duley and Miller 1923; Miller 1926; Chapline 1929; Stewart and Forsling 1931); but the unique and most important contribution based on field experiences was published by Bennet (1939) in the form of a book *Soil Conservation*, and he was rightly recognized as "father of soil conservation". This work was most conceptual as well as field-oriented and could help in understanding the process of soil erosion and formulating further research programs for handling the menace. This is also relevant and referred even today. Soil erosion by water usually begins when raindrops first strike the land's surface and detach soil particles. Eroded soil is transported downslope by raindrop splash; and the water erosion usually on landscape of gentle slopes is known as *sheet erosion*, where the top fertile layer of soil is ripped off in thin films every year by the running water from the entire surface of such lands. When runoff water concentrates, and flows through finger-like channels (rills) from upland areas carrying soil particles, the erosion is called *rill erosion*. These rills usually develop because of tillage marks, surface roughness or topographic variations and later by flow patterns of runoff (Figs. 2.1 and 2.3). Places, where concentrated runoff from a slope is sufficient in volume and velocity to cut deep trenches or where concentrated water continues cutting in the same groove making deep soil incision, are called gully (Fig. 2.2). Gully erosion is an obvious form of soil erosion consisting of an open, incised and unstable channel generally from a few to many meters deep. Once a gully is formed, it acts like a wild fire and goes on engulfing more and more area of the table land till the entire area is destroyed and becomes a network of gullies of various sizes and shapes in various stages of development. The advance stage of gully erosion is formation of ravines, which are a function of the depth of river and are invariably confined to the vicinity of rivers

**Fig. 2.1** Formation of rills and gullies in cold desert in Sangla valley in Himachal Pradesh, India (Photo courtesy Pulkit Dagar)

**Fig. 2.2** Gully heads extending in table lands (Photo courtesy AK Sikka)

**Fig. 2.3** A network of typical alluvial ravines in Chambal catchment area in India

and tributaries and proceed from the river or tributaries to the agricultural lands in the form of a network (Fig. 2.3).

Many of the earlier workers (mentioned above) had made measurements of soil erosion and the factors that affected it, but no efforts were made to publish any mathematical relationship between these factors. The first to do so was Austin W. Zingg in 1940, who evaluated data from field experiments under natural rainfall and from a rainfall simulation experiment derived from the relationship between different parameters (c.f, Laflen and Flangen 2013). The following year, Smith (1941) expanded Zingg's work, and from the year 1954 onward, his focus was on analysing the existing data sets and developing an overall scheme to support a broader prediction technology built on the earlier work. Later, Wischmeier and Smith (1965) developed the most significant model in the form of Soil Loss Equation, which was further modified by them (Wischmeier and Smith 1978), and the model became famous by the name of Universal Soil Loss Equation, which is relevant even today. The same was adopted widely, particularly for eroded lands. This was followed by several models explaining various aspects of gully erosion (e.g. Nearing et al. 1989; Sidorchuk 1999; Casali et al. 2003; Prasad and Römkens 2003; Souchére et al. 2003; Torri and Borselli 2003; Sidorchuk 2005; Marzolff and Poesen (2009); Poesen et al. 2011; Ghosh and Guchhait 2016 are among a few).

During the last two decades, various aspects of gully erosion have been explored in detail and attracted increasing attention from scientists across the globe (Poesen and Valentin 2003; Poesen et al. 2003; Li et al. 2004; Valentin et al. 2005; Romkens and Bennett 2005; Casali et al. 2009; Verma et al. 2016). During this period, more than 700 papers on a wide range of topics have appeared in the Web of Science. Many workers have elaborated the impact of environmental conditions leading to gully erosion including historical data (Gábris et al. 2003; Strunk 2003; Boardman et al. 2003; Lang and Bork 2006; Poesen 2011) and the processes and factors of gully formation (e.g. Valcarcel et al. 2003; Oygarden 2003; Faulkner et al. 2003; Verma et al. 2016). Authors such as Ries and Marzolff (2003) and Vandekerckhove et al. (2003) demonstrated the use of large-scale aerial photos for monitoring gully development. Billi and Dramis (2003) and Ionita (2003) discussed morphological and hydraulic characteristics of gullies; and Moeyersons (2003), Morgan and Mngomezulu (2003) and Kirkby et al. (2003) described the topographical threshold conditions while describing gully head initiation. Casali et al. (2003), Torri and Borselli (2003), Souchére et al. (2003) and Sidorchuk et al. (2003) dealt with various process-based models to predict spatial and temporal changes in gully dynamics, whereas Mathys et al. (2003) explained a model capable of predicting sediment yield from intensively gullied uplands. Rey (2003) and Bochet et al. (2009) explored the influence of erosion processes on vegetation establishment, describing erosion as an ecological driver that affects vegetation composition, structure and spatial pattern. Lal (2003) reported the impact of water erosion on global carbon pool. Corona et al. (2011), Saez et al. (2011), Bollati et al. (2012), Jain et al. (2012) and Gallart et al. (2012) reported exclusive work on gullies based on long-term studies on Badlands describing physical to vegetational approaches. Based on the sedimentological and morphological analysis and incorporation of a few radiometric dates, Joshi et al. (2013) reported that gully formations in Deccan Province of India and rangeland of Colorado, USA, owe their origin to the Holocene climate change events, but their responses to the climate change were different in two ecologies. Recently, Ghosh and Guchhait (2016) applied the modern concept of geomorphic threshold on the issue of gully erosion hazard to identify the critical slope of gully head and upstream drainage area with a core relationship in them.

Thus, no doubt that during the last two decades, much attention has been paid to the problems arising from gully/ravine erosion across the globe, but several research gaps and unanswered questions related to gully erosion remain there. There is need of historical studies on gully erosion for better understanding of different stages of their formation. Some of the concerns have been expressed by Poesen et al. (2003) which include appropriate and affordable modern measuring techniques for monitoring or better understanding of the geological and hydrological parameters and processes of gully development and interaction of hydrological and other soil degradation processes especially in ravenous lands. We also need to study the impact of climate and environmental changes on soil erosion processes, carbon sequestration in eroded habitats and draining out the carbon with sedimentation. Soil erosion models are not sufficient, and we need more effective and reliable gully erosion models. Further, effective gully prevention and control measures need to be developed based on watershed-based management options. In this chapter, an attempt has been made to address some of these concerns along with impacts of ravine formation on socio-economic and policy issues.

## 2.2 Historical Studies of Gully Erosion

For better assessment and analyses of the erosion processes involved in ravine formation through water erosion and the interactions between environmental change and land degradation processes, the knowledge of the history of the event is very important. But very little information is available or generated on this aspect, particularly related to ravine formation in different ecologies of the world. The quotes by Lang and Bork (2006) cited by Poesen (2011) will not be out of place to mention here for understanding the importance of history, which state, "Those who cannot remember the past are condemned to repeat it" and "The past is the key to the present and the future". In this reference, it is clear that gullies and ravines are mentioned among the few morphological evidences of past history related to soil erosion periods reflecting impacts of environmental changes in the landscape. The ravines are the unique geomorphic features to trace human-environment interactions and their influence on socio-economic conditions in historical times. Applying the modern tools, we can easily trace the links of past and present climate on geological formations in the areas of gully and ravine formation. Therefore, detailed studies on historical aspects of gullies and ravines and their formation are crucial, not only to reconstruct the past history but also to learn from it for their better management in future.

Sometimes, the archaeological evidences, particularly art and cultural aspects of a particular period, may shed light on significant development in a relevant field. For example, Poesen (2011) cited unique examples of historical paintings revealing the gully erosion on northern wall of the choir in the Holy Trinity Chapel of Lublin Castle of Poland, painted by Andrei in 1416, which have depicted landforms resembling the gully channels clearly suggesting that gully erosion also occurred in the past and the concept had significant impact on socio-economic condition of the people. Moreover, there are ample other physical evidences of many other historical gully erosion phases in various parts of the world. It has been further revealed that the erosion and gully formation events had occurred sometimes in combination with extreme rain events induced by socio-economic changes, as is evident from the conversion of forest to cropland as already available in prehistoric and Roman periods in Belgium, Germany and Poland. Historic examples are also available related to overgrazing in eroded mountain areas in France and Spain, expansion of vineyards in Spain at the cost of natural vegetation, conversion of Mediterranean shrub land to cropland for cultivation of wheat and almond to generate improved pastures and expansion of maize cultivation in cropland in the European loess belt (Poesen 2011).

In some parts of the world, there are historic records available for channel trenching as is evident from the reports of Cooke and Reeves (1976) and Sheridan (1981) who mentioned the exact dates of the channel trenching in southern Arizona. The reports are available from the late nineteenth and early twentieth centuries, and in one report it has been depicted that there was at least one *arroyo* (synonym to gullies/ravines and known by name of Greens Canal), which was completely artificial and was created around 1910 as an irrigation canal. It evolved into an *arroyo* following floods in 1914 which carved a deep incision into the ditch's floor. This indicates the process involved in gully formation. Further, Soviet workers, who face similar problems in the central Russian uplands and Dnieper Highlands, have independently reached the same conclusion which is mentioned in the reports prepared by Bogolyubova and Karaushev (1979). Based on extensive study, Schumm et al. (1984) stated that a gully is an incised, steep-sided channel with an eroded head-cut and slumping sidewall and formed due to anthropogenic causes. These statements were strengthened by the work of Gabris et al. (2003) who studied the impact of land use change in gully formation over the last 200 years in hill catchment of northeastern Hungary. They found that the increase of gully length per unit area calculated for different time periods showed the trends: (i) until 1860, when more than 50% of the catchment was forested, it was 5 m $year^{-1}$ $km^{-2}$; (ii) between 1860 and 1920, when forest area dropped to almost 25% and agricultural land use was extended to slopes steeper than 25%, it still remained at roughly same rate; (iii) after 1920, with 24–25% forest cover and with the extension of farming activity to the steepest slopes, it reached 10 m $year^{-1}$ $km^{-2}$. It could be shown that gully erosion on cultivated slopes leads to the development of gully systems in 50–60 years even if slope gradient is below 12%. After analysis of these studies, an inference was drawn by them that to prevent further gully development, at least 30% of the area should have been forested and slopes steeper than 17% must not be cultivated at all.

Erosion-related historical accounts are also available from Chinese literature. Chen et al. (2001) mentioned in their studies that the Loess Plateau in ancient China was an area with luscious grasslands and dense forests, which later was degraded into non-productive watershed. According to a report (CAS 1992a, b), the accelerated erosion started 250,000–300,000 years ago and was further accelerated in the last 50,000 years, which was before mankind started to impact on the environment. This suggests that the ultimate determinant in causing severe soil and water erosion on the Loess Plateau is not human activities, but the arid climate and the unfavourable geology and topography. This is in contrast with Fu and Gulinck (1994) who stated that these forests were destroyed by severe soil and water erosion due to undue deforestation by humans. Meng (1996) also reported that in the past forest cover had been over 50% greater on the Loess Plateau; hence soil and water conservation could be achieved by planting trees and grasses on the slopes. In the middle reaches of the Yellow River, especially in the areas that presently suffering with soil and water losses most severely and with the same severe environment lasting for a long time, accelerated erosion started 250,000–300,000 years ago, and was further accelerated in the last 50,000 years, which was before mankind started to impact on the environment (CAS 1992a, b). This suggests that the ultimate determinant in causing severe soil and water erosion on the Loess Plateau is not human activities but the arid climate and the unfavourable geology and topography. It is believed that the present situation arrived long back. Anyway, this is one of the past historic evidences on water erosion and gully formation; the interpretation for reasons may vary.

Some of the historical facts related to gully erosion have been studied in detail from Europe by Dotterweich et al. (2003) and Gábris et al. (2003), who have revealed that there was enhancement in the area of gully erosion as consequence of particular land use changes. The formation of ravines was traced into the silty-sandy material of the Triassic bench lands from the Wolfsgraben site in northern Bavaria in Germany. In one observation, nine main phases of landscape evolution, caused by extreme rainfall, runoff and soil erosion in agriculturally used areas, were identified since the Medieval period, and the highest amount of soil loss occurred during fourteenth, eighteenth and nineteenth centuries as a result of intensive land use. Climate change affected the intensity of rainfall and the magnitude of soil loss during these periods. It was concluded that the gullies were initiated or enlarged by floods in virtually all the morphogenetic regions.

Shellberg et al. (2010) and Shellberg and Brooks (2012) observed that in the Mitchell catchment in the Northern Australia, historic gully areas increased between 1.25 and 10 times at different sites from their initial area of the year 1949, mostly due to anthropogenic causes. The extrapolation of gully area growth in this catchment suggested that the current phase of extensive gullying was initiated between the years 1880 and 1950 and the alluvial gullies are expected to continue to be chronic erosion features on this landscape for several hundred to several thousand years, growing 10–50 times their initial size.

Recently, Colgan (2016) has summarized the geological history of ravine formation in Grand Valley State University over the last 15,000 years. According to his theory, the land on which the university is located is broad and gently sloping, and the upland was formed during the Ice Age by a glacier lobe depositing sediment in a large glacial lake. Following this glaciation, the Grand River began eroding its valley through this landscape. According to him, the ravines probably began forming at about the same time, and by about 6000 years ago, they had grown to roughly their present dimensions. He has cited several lines of evidences for this perception which include a log radiocarbon dated to 5720–5920 calendar years old making him to conclude that since the valley had to be there before the sediment and wood could fill the valley.

From these observations and studies conducted elsewhere, it can be concluded safely that there is a need to combine detailed field observations and process-based knowledge of contemporary gullying and dating techniques to reconstruct conditions leading to gully channel incision, development and infilling. Yet this process is equally important to better understand gully dynamics. More such information need to be collected from different ecological zones across the globe, and with the help of modern tools, the linkages between the patterns of gully/ravine formation in different ecologies may be established for better understanding of different processes and management aspects.

## 2.3 Processes of Gully/Ravine Development

### 2.3.1 Initial Stages of Soil Erosion

The pioneer work on soil erosion caused by water is well documented in earlier literature (Bennet 1939, 1955; Ellison 1944a, b, 1947; Wischmeier 1959; Erasmus et al. 1970; Beasley 1972; Kinnell 1974; Tejwani et al. 1975a, b). It is well understood that soil erosion by water usually begins when raindrops first strike the land's surface and detach soil particles. In most humid areas of the world, several quadrillion raindrops fall at speeds up to 32 km $h^{-1}$, and it has been reported that more than 247 Mg of soil per ha may be detached by the raindrops in a single rain and the splashed particles may reach a height of 60–90 cm covering a radius of as much as 150 cm (Ellison 1944a, b, 1947).On sloping land, the force of raindrops on the surface is such that more than half of the splashed soil is moved downslope as it falls back to the surface. It has become widely accepted that drop impact is a major erosive agent in causing sheet erosion and that the erosive power of rainfall is function of the kinetic energy of the drops at impact (Young 1989). Earlier, rainfall erosion index (EI) values were used for estimating soil loss due to rain, having accurate classification of rainstorm according to their soil eroding potential, for accurate evaluation of erosivity of different soils, and estimating cover and management effect. After recording rainfall (in short duration for measuring intensity) contentiously, the rainfall energy could be computed using the equation derived by Wishmeir (1959):

$$Y = 210.3 + 89\log_{10} X$$

(where $Y$ being the kinetic energy in meter tons per hectare centimetre and $X$ the rainfall intensity in cm per hour)

In India, during the 1960s, EI values of each storm having more than 12.5 mm rainfall were compiled for monthly and annual values for many research stations of Central Soil and Water Conservation Research Institute, Dehradun, by Gupta and Ram Babu (1967). EI values of 2, 5 and 10 years' frequencies were also recorded, and the rainfall energy was found to be linearly related to rainfall amount by Das et al. (1967) using the following equation:

$$E = 63.2 + 69.87P$$

(where $E$ = rainfall energy in meter tons and $P$ = rainfall in cm)

Further, Dagar (1999) observed that rain intensity of 6.6 mm $h^{-1}$ eroded more soil due to splash than 3 mm $h^{-1}$, while the total rainfall was almost the same. Loss in soil due to splash depends on rain intensity and total rainfall, and based on studies conducted using splash cups, the following equation could be drawn:

$$Y = -0.9665 + 0.0736X_1 + 0.7910X_2 \left(R^2 = 0.9525, p \le 0.05\right)$$

(where $Y$ being the soil splashed (g), $X_1$ the total rainfall (mm) and $X_2$ the rain intensity in mm $h^{-1}$).

The developments in the field of hydrology resulted in the evolution of soil loss equation ($A$ = RKLSCP) by Wischmeier and Smith (1978), where $A$ is the computed soil loss and the six terms on the right are factors for rainfall, soil erodibility, slope length, slope steepness, cover factor and erosion control practice, respectively. In this equation, rainfall factor "R" is considered an important component.

Erasmus et al. (1970) and Dagar (1999) while working on rainfall pattern found highly significant relationship between product of rainfall amount and intensities for various time intervals with erosion index. The later could be calculated by the following equation:

$$\text{Erosion index}(\text{EI}) = \text{KE} \times I / 100,$$

(where $I$ being the average rainfall intensity in cm $h^{-1}$, i.e. total rainfall for the day/24); and KE is the kinetic energy in Mg $m^{-1}$ $ha^{-1}$)

Further, it was also found that the precipitation index (PI), i.e. daily rainfall in cm × rainfall intensity (I), and EI values showed a significant linear relationship, which could be explained by the linear equation (Dagar 1999):

$$Y = 0.02 + 2.47X$$

(where $Y$ being the daily EI and $X$ the daily PI)

Falling raindrops break down soil aggregates and detach and transport soil particles which cause sealing and compacting of the soil surface. This reduces the ability of the soil to absorb water and increase surface runoff. Eroded soil is transported downslope by raindrop splash. Net downslope movement by splash increases as slope steepens. Most eroded soil is transported downslope by concentrated runoff which begins when the soil surface storage is filled and rainfall intensity exceeds the infiltration rate of the soil. The quantity and size of material that can be transported by runoff is a function of runoff velocity and turbulence, which increases as the slope steepness and flow increases.

In one study, Dagar (1999) observed that when the slope length was increased from 3 to 6 m, the soil splashed also increased from 2.45 to 6.51 g when rainfall and rain intensity were 54.8 and 6.8 mm $h^{-1}$, respectively, and the soil loss increased almost four times at rainfall 140 mm with rain intensity of 14 mm $h^{-1}$ (Table 2.2). The regression equations at two different lengths showed $R^2$ values 0.9957 ($p \leq 0.05$) and 0.9845 ($p \leq 0.05$), respectively. The vegetation cover, particularly grasses, controlled the splash up to a large extent. Time of exposure to rain also affects the soil loss which was highest in first 5 min and was reduced abruptly when exposed for longer period.

Soil erosion, usually on landscape of gentle slopes, is known as *sheet erosion*, where the top fertile layer of soil is ripped off in thin films every year by the running water from the entire surface of such lands, and it occurs where the velocity of surface runoff is about 0.3–0.6 m per second. More commonly, however, the direct

**Table 2.2** Soil splashed without and with vegetation cover along a slope (Dagar 1999)

| Rainfall of a single day (mm) | Rainfall intensity (mm $h^{-1}$) | Soil splashed (g) | | Soil cover rating for Sn-Da community |
|---|---|---|---|---|
| | | Without cover | Sn-Da community | |
| 32.4 | 4.0 | 6.34 | 0.53 | 91.64 |
| 12.5 | 2.1 | 0.73 | Traces only | 100.00 |
| 70.8 | 11.8 | 19.56 | 3.62 | 81.49 |
| 40.7 | 10.2 | 12.42 | 1.93 | 84.46 |
| 38.6 | 6.4 | 8.21 | 0.87 | 89.40 |
| 50.3 | 5.1 | 9.56 | 1.08 | 88.7 |

Soil cover rating = Soil splashed under cover/soil splashed without cover × 100

Sn-Da community stands for *Sehima nervosum-Dichanthium annulatum* grass community

impact of raindrops on soil particles causes their detachment and gradual downhill movement – splash erosion. Sheet erosion is barely detectable in the short term because it is a gradual process. Every time when it rains, the running water becomes muddy and flows down to the lower reaches, rivulets and river. The erosion capacity of flowing water depends upon its velocity, turbulence and the amount and type of abrasive material it transports. Soil particles are detached by a combination of rolling, lifting and abrasive actions, while soil particles are transported by a combination of surface creep, siltation and suspension (Beasley 1972). The amount of material transported depends upon the transporting capacity of the runoff and the transportability of soil. The latter is influenced by the size, density and retaining effect of vegetation, soil binding capacity of roots (Dagar 1986, 1987d, 1989, 1995) or any other obstacle. This also depends on soil texture. The root development is very important parameters in soil conservation studies.

When runoff water concentrates, and flows through finger-like channels (rills) from upland areas carrying soil particles, the erosion is called rill erosion and the soil eroded from between rills is called inter-rill erosion. Rill erosion is caused primarily by runoff, whereas inter-rill erosion results primarily from raindrop impact. Rills usually develop because of tillage marks, surface roughness, or topographic variations and later by flow patterns of runoff. Once rill erosion begins, riling typically progresses upslope by a series of intensely erodible head cuts, and this is very much obvious in Chambal command areas in India. Places, where concentrated runoff from a slope is sufficient in volume and velocity to cut deep trenches or where concentrated water continues cutting in the same groove (such as rill) making deep soil incision, are called gully. Gullies often have their beginning in slight depressions in fields or just below the fields where runoff water so often concentrates. Gullies are usually formed when many rills join and further extended with more than 30 cm depth.

To describe the development of gullies, Geyik (1986) explained that a gully is developed in three distinct stages of erosion, namely, waterfall erosion, channel erosion along the gully bed and landslide erosion on gully banks. In waterfall erosion, first of all sheet erosion occurs on tableland followed by development of finger-like rills, which start gaining depth and gradually reach to lower B-horizon of soil. Further in the second stage, the gully reaches further deeper in C-horizon of soil, and the weak parent material such as sand is removed. In the third stage, the water from the gully head falls and also starts carving a hollow at the bottom of the gully by direct action as well as by splashing. This further deepens with more rains. When the excavation has become too deep, the steep gully head wall collapses. This process is repeated time and again, so that the gully head progresses backwards to the upper end of the watershed. This process is called gully head advancement. As a result of this, new gully branches develop in backward and lateral directions. This branching of the gullies may continue until a gully network or multiple-gully system is developed covering the entire tableland. Gradually this network extends in the watershed area with a greater dimension beyond control. During the rainy season, when the soil becomes saturated, big soil blocks start sliding down the banks and are washed away through the gully channel. Some gullies in Chambal catchment in India are over 20 m deep and still growing deeper and extensive. These may be designated as ravines. The extreme and most dangerous form of soil erosion is ravine erosion which is considered a major sediment-producing process, generating between 10% and 95% of total sediment mass at catchment scale, whereas gully channels often occupy less than 5% of total catchment area of the river (Poesen 2011). Therefore, the best way of erosion management must be focussed on the management of watershed.

In cold regions, landslide erosion on gully banks also occurs when temperature alternates between freezing and thawing. Luna (2006) has explained the process of gully formation in temperate climate. When the temperature drops below zero Celsius, wet gully banks freeze. After the temperature rises above zero, the banks thaw, the soil loosens, and the loose gully banks easily slide during the first rainy season. After landslides have occurred on all gully banks, a considerable number of new branch gullies may begin along the disturbed banks. During the third stage of gully development, gullies become deeper and longer as well as wider. The three stages of gully development, i.e. waterfall erosion, channel erosion along the gully bed and landslides on gully banks, will continue unless the gully is stabilized by structural control measures and revegetation. If left undisturbed, the gullies may be stabilized through natural vegetation. Sometimes gullies erode entire soil profiles exposing the rocks below and often cut into unconsolidated material beneath the soil. Gullies between 1 and 5 m deep are considered medium sized. Those outside this range may be labeled large and may be called ravines.

### 2.3.2 Factors Affecting Gully Erosion

The factors affecting gully formation can broadly be categorized into two groups, namely, physical and anthropogenic factors. The main physical factors governing the soil erosion due to water and also effect runoff may include precipitation, topography, soil properties and nature of vegetative cover. The intensity of rainfall plays more important role than the total but smooth and uniform rainfall. Less rainfall with higher intensity can create more runoff than more rainfall with low intensity (Tejwani et al. 1975a, b; Dagar 1999).

In continental and temperate-climate countries, prolonged precipitation of moderate intensity for duration of several days causes less effect on soil erosion as compared to short and intensive rainstorms lasting shorter duration of a few minutes. The later may cause landslides, formation of gullies and floods because of the increased runoff in the watersheds and lesser infiltration in soil. Here of course nature of vegetation will also play an important role. Torrential floods, which generally occur after the short, intensive rainstorms, destroy agricultural lands, residential areas, roads, irrigation ditches and canals at the base of the valley below a deteriorated watershed. In temperate regions, snowmelts turn into high runoff. This increased surface runoff acts as a cutting agent and produces gullies (see Fig. 2.1). The prolonged rains of moderate intensity as well as short intensive rainstorms and rapid snowmelts cause destructive floods if the natural vegetation is in depleting stage. The dense forests and green cover of grasses help in conserving both soil and water.

The physiographic or topographic features such as the size and shape of a drainage area, the length and gradient of slopes and the nature of soil structure also influence the runoff rate and amount of surface water, which in turn affect the soil erosion intensity. The steeper the slope, the higher the velocity and erosive power of the runoff. Among soil types (sand, sandy loam, loam, silt loam, clay loam, clay), the sandy soils are more prone to soil erosion as compared to clay soil. The infiltration rate increases from clay to sand, but resistance against erosion decreases. Further, the role of vegetative cover is to intercept rainfall, keep the soil covered with litter, maintain soil structure and pore space and create openings and cavities by root penetration. This is best achieved by an undisturbed multistory dense forest cover. Under special conditions, however, a well-protected, dense grass cover may also provide the necessary protection. In general, it is the management and protection rather than the type of the vegetative cover which determines its effectiveness in gully control. Any vegetation, which is well-adapted to local and stress conditions, develops deep root system and shows vigorous growth in eroded habitats which may be used for soil conservation (Dagar and Mall 1980; Dagar and Singh 1980; Geyik 1986; Dagar 1995). The local flora has better adaptation to biotic and erosion forces and found to regenerate frequently in eroded habitats (Dagar et al. 1978; Dagar and Mall 1980; Dagar 1987c, 1999, 2001). The fibrous roots of grasses bind soil more efficiently than many woody species (Dagar 1995).

Among anthropogenic factors, denudation of vegetation for meeting the needs of ever-increasing population, overgrazing, forest and grass fires, mining and developmental activities are the major factors influencing soil erosion and ravine formation. Gullies may also be formed on livestock and vehicle trails that run along hillsides. This is because the traffic on them compacts the soil and reduces the water holding capacity, hence increasing runoff and accelerating soil erosion. In forest regions, logging with tractors downslope may also lead to gully erosion, because the runoff becomes concentrated along the skid trails. Dagar (1999, 2001), Mathys et al. (2003), Rey (2003) and Chaturvedi et al. (2014) underlined the importance of vegetation cover pattern in reducing sediment yield at the gully outlet by interrupting the connectivity in the catchment. In this book, Dagar (2018) and Uthappa et al. (2018), while dealing with plant ecology and biodiversity of ravine lands, have highlighted the potential of biodiversity in controlling soil erosion and increasing the productivity of these degraded habitats.

### 2.3.3 Evolution and Formation of Gully Channels and Ravines

It has been a well-recognized fact that gully channels are the product of superficial and subterranean flow concentrations and are of two types, i.e. continuous and discontinuous gullies (Heede 1974; Haigh 1984). The continuous gully channels evolve at the ground surface because of the concentration of overland runoff in upper catchment area. This concentration of flow may be governed by many biotic and topographic factors. These channels begin due to merging of rills and gradually become deeper down the slope. At times, conical holes are formed and may be due to the burrowing activities of wild animals such as hare, mouse and reptiles or cracking in the upland soil. Discontinuous gullies in turn begin as an abrupt head cut and usually become shallower down the slope often ending as a mid-slope alluvial fan. These are most haphazard and associated with the subterranean concentration of flow of water in soil pipes. This fact is the major cause of local lowering of the water table which in turn increases the hydraulic gradient. Once established, a soil pipe extends headwards up the hydraulic gradient and expands its sides, and roof collapses exposing the pipe at the soil surface initially as a steep-sided oval depression and then later as a trench formation (Haigh 1984). Then the sides and depths of gullies extend forming dense net of gullies in the long run. Formation of gullies and ravines is explained as mismanagement of surface runoff mainly due to overgrazing, deforestation and ill-managed tillage in non-congenial environment which is particularly susceptible to erosion. This susceptibility is in part due to the intensity of rainfall and concentration of runoff during the monsoon and in part due to the erodibility of the deep, alluvial soil where most of the ravines develop. Alluvial soils are more prone to erosion. There are varying opinions regarding the process of evolution of gullies/ravines. In the past, workers such as Sobolev (1948), Ollier and Brown (1971), Tejwani et al. (1975a), Sharma (1980), Schwab et al. (1981) and Haigh (1984) preferred to explain ravine evolution in terms of surface runoff and expressed that each gully during development process passes through different stages before stabilization (Table 2.3).

Unlike many other countries, there is no concrete historical record available in Indian subcontinent to explain when the ravine erosion started. However, there is common opinion among the workers that indiscriminate land use practices and anthropogenic factors leading to a disturbance of the hydrological balance are the main reasons of ravine formation. Western theories, which usually relate ravine and gully erosion to climate, do not find any support within the Indian scientific community. Ahmad (1973), however, is exception who did not give anthropogenic reasons of ravine formation, but rather argued for the possibility of peripheral uplift of the Peninsular Shield by pressure against the Himalayas, and the discontinuous pattern of incision has been explained due to differential rates of disturbance. To some extent, this concept may explain the formation of gullies and ravines in Himalayan regions (Fig. 2.1). Recently, Goswami and Mishra (2014) stated that the formation of Himalayan foreland basin has led to one of the most extensive alluvial plains in the world, i.e. Indo-Gangetic plains, which represent the foredeep part of the system and might be the result of extensive erosion in uplands. According to these authors, the southern margin of foreland basin is marked by a regional gentle forebulge in the form of Bundelkhand-Vindhyan Plateau, while the northern margin is marked by the Siwalik Hills. From north to south, the Gangetic Plain can be subdivided into three geomorphic units, namely, Piedmont Zone, Central Alluvial Plain and Marginal Alluvial Plain (Agarwal et al. 2002; Goswami and Mishra 2014). The Marginal Alluvial Plains are incised with the most severe intricate network of gullies (see Fig. 2.3), thus forming badlands, also known as ravines along the Chambal, the Betwa and the Yamuna rivers and their tributaries.

**Table 2.3** Evolution of gully channels and ravine systems as expressed by different workers in different countries

| | Different stages of development | | | |
|---|---|---|---|---|
| Reference | Stage 1 | Stage 2 | Stage 3 | Stage 4 |
| Sobolev (1948); (undivided USSR) | Rills form, often on scarp face due to human activities, land sliding or different slope retreat | Small gully formation on valley side slope, due to "accidental" concentration of flow due to natural or artificial causes. Gully head cuts upslope | Gully deepens towards the development of an "equilibrian" longitudinal profile involving no erosion or deposition beyond head-cut advance and sidewall retreat | Either the gully achieves a stable long profile and heals as its banks retreat to a gentle angle or it continues to grow to become the size of a ravine |
| Ollier and Brown (1971); New Guinea (study based on a young volcanic cone of pumice and ash) | Rills begin 10–30 m from slope head along the entire length of a water course | Major rills grow into gully channels; relict rills are left as small hanging tributaries | Gully heads extend by "particle extrusion"; pore-water causes particles to be extruded from a free-face in saturated porous material; this process is enhanced by gully wall undermining and collapse | Colonization by dense tropical vegetation fossilizes the gully system; erosional features persist, relatively unaltered |
| Tejwani et al. (1975a); India | Rills form in a sheet-washed surface and cut through topsoil to subsoil beneath | Small gully cut down into sandy layers below subsoil; these are readily washed out, and the gully widens by undercuttings and collapse, forming vertical head cuts and sidewalls | Incision stopped by encounter with stable horizon or water table; head cuts and sidewalls retreat laterally and become less steep | Either vegetation or shrubs colonize the sidewalls, then stabilize the gully channel; cycle is repeated on a bigger scale |
| Sharma (1980); India | Subterranean solution of calcium carbonate nodules creates a swallow hole | Swallow holes become linked to walls on incising master channel by soil pipes and a steepening of the local hydraulic gradient | Subterranean pipe system is opened by collapse | Sidewall retreat causes the channel to get wider and more shallow as, and if, sediments accumulate on channel bed |

(continued)

**Table 2.3** (continued)

| Reference | Different stages of development | | | |
|---|---|---|---|---|
| | Stage 1 | Stage 2 | Stage 3 | Stage 4 |
| Schwab et al. (1981); USA | Channel created by incision into topsoil (the process may be slow where the topsoil resists erosion) | Gully channel enlarges in width and depth; head cut advances upstream; channel cut into C-horizon and the weak weathered parent material is removed; a waterfall develops where flow plunges from upstream segment to the eroded channel below | Healing begins with vegetation beginning to grow in the channel and on its margins | Gully is established; the channel reaches a stable gradient; the gully walls have a stable slope; vegetation colonization proceeds sufficiently to anchor the soil and permit the development of a new topsoil |

Source: Modified from Haigh (1984)

In the lateritic region of West Bengal, which is about 7700 km$^2$, land degradation is considered quite acute because of aberrant weather, frequent drought, ferruginous crusting, rill and gully erosion, nutrient deficiency in soil, low water holding capacity of soil and wide range of land use conversion without any scientific planning (Jha and Kapat 2009, 2011; Shit and Maiti 2012, 2014; Shit et al. 2014). The major districts of lateritic *Rarh* plain include Birbhum, Barddhaman, Bankura, Purulia and West Medinipur, which have 731,000 ha of degraded land intensively affected by gully/ravine erosion. In light of appropriate measures for gully prevention and restoration in the lateritic region, it is important to know the threshold conditions in which gullies develop (Ghosh and Guchhait 2016).

While studying morphotectonics along the lower reaches of the Chambal River, Mishra and Vishwakarma (1999) observed a 74-km-long palaeochannel on the left flank of the river and stated that this was a tributary to the Chambal River which dried up before the incision of the latter. During the satellite interpretations, many similar toned dark linear/curvilinear features were observed running parallel to the lower reaches of the Chambal River, on its right flank. Interestingly, these features seem broken links and appeared to come from and diminish into the badlands.

The neotectonic theory also has gained momentum. Among various evidences of these activities, the incision of rivers and the presence of badlands among them are most striking evidences (Mishra and Vishwakarma 1999; Agarwal et al. 2002). It has been argued by Tandon et al. (2006) and Joshi (2014) that along with the upwarping of the area, intensification of south-west monsoon in the late Pleistocene-Holocene period is also considered a possible reason for badland formation. However, Gibling et al. (2005), after dating a gully fill in the Kalpi region, found that the youngest gully fill sediments are ca. 35 ± 4 ka (ka is kiloannus; and 1 ka = 1000 years) old. Since these sediments are gully fills, hence, gullies and badlands must have formed prior to that period (Tandon et al. 2006).

Jain et al. (2007) have explained the phenomenon of badland development in detail. We know the Chambal River originating from the Vindhya Range is overlaid by Deccan Traps and flows through Malwa Plateau, running a total length of 960 km with a total basin area of about 143,219 $km^2$ and a discharge of 387 $m^3\ s^{-1}$ and 22 million Mg per year of sediment load, measured at Dholpur station (Bawa et al. 2014). Due to neotectonic activities, the Chambal River has undergone many changes before reaching to its present planform. This study reported 113 palaeochannels on the right flank of the river along its lower reaches. At palaeochannels, soil has rather small particles (silt and clay), while in its surrounding areas, soil has larger amount of sand particles. These channels have significantly influenced the development of badlands along the lower Chambal River and have given distinct and conspicuous spatial patterns. In the light of evidences, a modified schematic geomorphic evolution of badland development is also proposed starting from a pre-incision scenario till the present-day situation. The study adds a concept of multichannel planform of an un-incised Chambal River. While studying geomorphological evolution of badlands, Ranga et al. (2015a, b) reported the dynamics of palaeochannels along the lower Chambal valley which represents the worst case of water erosion in India.

The phenomenon of tectonic movements has also been reported by Xiang-Zhou et al. (2004) in the Loess Plateau in China, capped with the deepest loess in the world and is one of the most active areas of tectonic movements, accelerating erosion. In the Quaternary loess, the crust has been uplifted by 150–200 m. Soil erosion in the valleys is aggravated by the increased energy of the water. Further, earthquakes also bring about gravitational erosion and destroy landform structure and lead to the collapse and sliding of areas on the Loess Plateau (Hui and Mingan 2000). Similarly, cold desert of Ladakh (Fig. 2.4) has been formed by the combination of several factors such as tectonic plate subduction, volcanism, magmatism, igneous proliferation, fossiliferous and non-fossiliferous sedimentation, accretion and uplifting. In this region two different stratifications of rock formation fused or "sewn together" at fault lines are found. Other activities such as folding; severe erosion; cracking and splitting by freezing water, invasive igneous rock and earthquakes (still very active); and polishing, grading and carving by a glaciation are common (http://www.quora.com). Instead, its available water is from melting snowpacks and glaciers.

Contrary to the above discussion, in western countries, there is difference of opinion regarding the origin of ravines or *arroyos*, trenched channels and badlands. Some workers believe that change in climate influences many other factors such as ground vegetation cover, concentration of runoff and changes to the level of groundwater tables, which in turn effect gully formation. Others believe that erosion is caused by human activities which have induced the runoff and erosion (Cooke and Reeves 1976; Graf 1979; Valentin et al. 2005). However, Brooks et al. (2009) and Shellberg and Brooks (2012) explained both natural and anthropogenic factors influencing alluvial gully erosion in Australia. Natural phenomenon such as the elevational relief or vertical height between the river beds and high floodplain or terrace provides the potential energy for erosion. The kinetic energy for alluvial gully erosion may come from multiple water sources such as rainfall, runoff from

**Fig. 2.4** General features of soil erosion in cold desert of Leh (above) and Stock Valley, Ladakh (below) (Photo courtesy Pulkit Dagar)

watershed, groundwater, river backwater, floodwater, etc. This explanation differentiates the hillslope or colluvial gully formations that have standard rainfall-runoff responses and easily definable catchment areas. As most of the gullies in alluvial soils are often situated on floodplains or terraces near rivers and water bodies, therefore, river backwater or overbank flood water can also contribute significantly to gully erosion. Vegetation on banks and adjacent floodplains play the key mitigating role in resisting alluvial gully erosion. The native deep-rooted perennial grass performs best at protecting and binding surface soils to prevent erosion initiation, compared to tree, shrub or exotic weed species. Both natural factors such as drought and fire and anthropogenic factors such as grazing, fire, weed invasion, fencing, deforestation and agriculture can influence vegetative cover which in turn influences erosion.

It is a well-known fact that besides above-mentioned factors, gully formation depends on the runoff-producing characteristics of the watershed, the drainage area, soil characteristics and the alignment and the gradient of the channel. Only a few studies have quantified the hydrogeomorphic mechanisms responsible for the propagation or initiation of alluvial gully erosion. Thomas et al. (2004) reported "valley-bottom" gullies in Iowa, which are eroded by both direct rainfall-runoff and groundwater seepage enhancing mass failure. In Kenya, Oostwoud-Wijdenes and Bryan (2001) observed that both rainfall and runoff processes dominate erosion in a watershed-controlled alluvial gully system, but somehow, they did not quantify hydrological processes of gullies along the floodplain rivers they studied. Some others (Oostwoud-Wijdenes et al. 2000; Vandekerckhove et al. 2000, 2001a, b, 2003) working with gully erosion in Spain also found that alluvial bank gullies were affected by rainfall and runoff processes enhancing agriculture activities. Brooks et al. (2009) suggested that hydrological mechanisms influencing gully initiation or propagation could span the full continuum of surface and subsurface erosion models in northern Australia. From these studies it could be concluded that if we develop predictive models of gully initiation and evolution and incorporate these into sediment budgets, it is imperative that we understand the relative importance of the various hydrological drivers and associated erosion processes of alluvial gully erosion.

Shellberg et al. (2012) undertook a detailed study in Australia. They reported that the frequency of inundation of alluvial gullies due to flooding influenced gully erosion longitudinally as per river incision and confinement. It was observed that near the top of the mega-fan, flood water was contained within the macro-channel up to the 100-year recurrence interval, but river backwater still partially inundated adjacent gullies eroding into Pleistocene alluvium. The remaining variability of the facts could be explained by development of cracks in topsoil, pore-water pressure in soil, wetting and drying of soil and heterogeneity in soil properties. Another influencing factor was seasonal vegetation cover. It is pertinent to observe here that management of grazing and vegetation cover has direct impact on runoff, rainfall erosion, water infiltration, runoff volume, water concentration along tracks and the resistance of highly dispersible soils to gully initiation and their further expansion under intense rainfall.

Therefore, an inference can be drawn that though the processes of gully erosion and their extension in agricultural fields are well worked out and understood, the gully development in other types of environment is not well studied. For example, piping, impact of gullies in urban areas, and road or railway track gullies, and process of gully infilling need more attention. For instance, in pasture land of northern Europe or abandoned cropland on Neogene marls in the Mediterranean region, discontinuous gullies have been reported to occur regularly. These gullies along with bank gullies are often preceded by intense soil piping in these areas. The areal extent of the land is seriously at risk from piping erosion in Western Europe, and it has been estimated to exceed 260,000 $km^2$ (Faulkner 2006). The limited amount of data also shows that soil losses by piping may often exceed soil loss tolerance values (Poesen 2011). So far, factors controlling soil piping and incipient gullying are poorly understood, and these processes deserve more attention through controlled laboratory experiments as well through detailed field monitoring studies.

Many inhabited areas across the world including developing countries are in danger of gullies formed along roads or railway tracks, at times even inside villages. Many of the urban areas in the world are at risk because of extended gullies causing damage to infrastructure and private property. Likewise, improper drainage of rural roads in tropical environment often results in the rapid development of large gully channels turning the often-unpaved roads into road gullies. Very few studies have focussed on environmental conditions leading to gully infilling. Yet we know from many case studies (e.g. in Belgium, Romania) that several cycles of gully cutting and infilling have taken place in a time span of decades to centuries (Poesen 2011). If we want to better predict gully evolution and to develop proper gully management strategies, we need to better understand factors controlling gully infilling. It is pertinent to mention here that in most of the ravine habitats the biodiversity is at stake both due to natural and biotic factors. Special programs are needed towards this aspect.

### 2.3.4 Classification of Gullies and Ravines

The international literature contains a substantial number of different approaches for classification of gullies and ravines. In pioneer work, Bennet (1939, 1955) defined a field gully as a newly eroded channel at least about 45 cm wide and 25 cm deep and a woodland gully as a newly eroded channel deep enough to expose the main lateral roots of large trees. He further described two principal kinds of gullies, viz. those which have sloping sides with narrow, V-shaped bottom and those which have straight, vertical sides with broad U-shaped bottoms. The V-shaped gully may be considered the normal type of gully trenching, except the soils with soft, easily cut substrate, where U-shaped type is common. The reason why broad-bottomed gullies usually are more serious is that the soft, unstable materials commonly found in their depths, such as sand, loose gravel or soft rotten rock, are easily incised and cut out by flood waters. This undercutting near the bottom causes the banks to cave off from above, often splitting off in great vertical blocks thus favouring rapid expansion, with consequent outright destruction of land. When caved-in material is washed out of the gully, the trench is left box-shaped with straight walls and a bottom width equal to that at the top. The gully remains at this size until the next flooding takes off additional chunks from the sides.

Frevert et al. (1955) classified gullies based on their size into small (<1 m deep, drainage area <2 ha), medium (1–5 m deep, drainage area 2–20 ha) and large (>5 m deep with drainage area >20 ha). Based on the shape, Weidelt (1976) classified gullies into U-shaped, V-shaped and trapezoidal gullies. The last type of gullies is stated to be formed where the gully bottom is made of more resistant material than the topsoil. Based on continuation, Geyik (1986) in his FAO manual classified gullies into two types, namely, continuous (having many branch gullies, mainly alluvial) and discontinuous (mainly on hillsides). Pani and Mohapatra (2001) and Pani (2016) classified the ravines of the Chambal valley according to the multiple criteria of shape, depth, side slope and head slope into three different types of ravines: *young* (narrow base width with steep side slopes), *mature* (wide base width with moderate side slopes) and *old* (very wide base width with gentle side slopes). The ravine system in the region is dynamic and changing in terms of size, pattern and aerial extent.

From field research and remote sensing mapping across northern Australia, alluvial gullies eroding into floodplain systems come in a variety of forms depending on local processes. Shellberg and Brooks (2012) explained that most alluvial gullies which drain directly into main channels may be called *proximal* gullies, while some gullies draining away from the main channel towards distally draining creeks or lagoons are known as *distal* gullies. Often both *distal* and *proximal* gullies may be found at a single location, consuming the floodplain from both directions. They further mentioned that in planform *linear* alluvial gullies are often young incipient gullies, commonly associated with land use disturbances such as stock tracks, roads and fences that tend to concentrate overland flow. Another type of gullies includes d*endritic* gullies, which are associated with well-defined drainage networks, separated by distinct interfluves with often less distinct or continuous head scarps; a*mphitheatre* gullies, which are often as wide as or wider than they are long due to the lack of structural control and have well-developed head scarps that drain into relatively narrow outlet channels; and c*ontinuous scarp front* gullies, which are mature in phase, located parallel with main channels, and develop from the coalescence of numerous laterally expanding amphitheatre gullies and/or from river bank erosion on meander bends.

In India, despite the large amount of work, no consolidate system of classification could be established. It was, however, agreed that gullies vary in size along a regular scale that could be expressed in terms of depth, bed width and side slope steepness (Tejwani et al. 1975a). The classification differed from agriculture department to forest department from state to state and has been linked with treatment of gullies for different land uses like agricultural and non-agricultural ones. It considers the influence of soil characteristics and gully dimensions on suitability of reclaiming gullies for cropping, horticulture, forestry, growing grasses, etc. Various gully classifications are in use in the ravine-infested states of northern India for evolving suitable methods of ravine erosion control and management. Tejwani et al. (1975a) classified ravines based on depth, bed width and side slopes of the gullies (Table 2.4), which may be applied for all Indian ravines.

Bali and Karale (1977), however, proposed more consolidated ravine land classification which included gully depth; width; slope with drainage density; soil texture; depth to caliche; hardpan or rock, especially on the inter-gully tableland; and depth to ground water. Verma (1981) and Verma and Bhushan (1986) proposed a

**Table 2.4** Classification of ravines popularly followed in India

| | Very small ($G_1$) | Small ($G_2$) | Medium ($G_3$) | Deep ($G_4$-a) | Narrow ($G_4$-b) |
|---|---|---|---|---|---|
| Bed width (m) | <18 | >18 | >18 | <18 | Vary |
| Depth (m) | <3 | <3 | 3–9 | 3–9 | >9 |
| Side slope (%) | Vary | Vary | 8–15 (uniform) | Vary | Vertical (most steep) |

Source: Tejwani et al. (1975a)

detailed classification of ravines for optimum utilization in which gullies have been classified in the six reclaim ability classes (for details see Chaturvedi et al. 2014). In the gullies having mild erosion (Classes 1–3), all local crops such as sorghum (*Sorghum durra*), maize (*Zea mays*), pearl millet (*Pennisetum typhoides*), wheat (*Triticum aestivum*), green gram (*Vigna radiata*), black gram (*V. mungo*), pigeon pea (*Cajanus cajan*), gram (*Cicer arietinum*), mustard (*Brassica juncea*) and tobacco (*Nicotiana tabacum*) can be grown in gully beds and on bench terraces made on gully sides. Moderately eroded gullies (Class 4) may be put under arable crops along with fruit trees such as lemon (*Citrus limon*), mango (*Mangifera indica*), ber (*Ziziphus mauritiana*) and gooseberry (*Emblica officinalis*) at required spacing in agriculture crops in gully beds. Highly eroded gullies (Classes 5 and 6) may be planted under perennial vegetation consisting of forest trees and fodder-producing grasses.

### 2.3.5 Extent and Rate of Gully Expansion

Water erosion occurs on all continents and under all climatic conditions, although somewhat more in humid climates. Middleton and Thomas (1997) reported 467.4 Mha dryland area susceptible to water erosion is found occurring in all the continents of which 141.0, 213.2 and 113.2 Mha fall in dry subhumid, semiarid and arid regions of the world, respectively. Though enormous data have been generated on soil erosion rates across the world, yet no consolidated figures are available regarding country-wise total area covered by ravines. The only data available is as mentioned by Oldeman (1994) who stated that most widespread form of water erosion is the loss of the topsoil on 920 million hectares (Mha), while terrain deformation in the form of rills and gullies occurs on 175 Mha.

Iacob and Laura (2011) reported that in Europe up to the year 2000, an area of about 115 Mha (~12% of the Europe's surface) was estimated to be affected by water erosion. In Romania, the weighted average erosion intensity was reported to be 16.28 Mg $ha^{-1}$ $year^{-1}$ (Motoc and Vatau 1992); however, in the bend area of the Carpathian Mountains (Buzau and Vrancea), it was 40 and 35 Mg $ha^{-1}$ $year^{-1}$, respectively. Gully erosion in different parts of the world has varied data of soil erosion (Gardner 1996; Meng 1996; Mircea 1999; Iacob and Laura 2011). The Loess Mesa Ravine Region and the Loess Hill Ravine Region of Loess Plateau of

China which together cover 30% of the total area (524,000 km$^2$) are considered to be the most severe area of soil and water erosion in the world, where average erosion reaches 5000–10,000 Mg km$^{-2}$ year$^{-1}$, sometimes even up to 20,000–30,000 Mg km$^{-2}$ year$^{-1}$. In Romania, a network totalling over 25,000 km of gully erosion in formations has been reported. In Ethiopia, about half of the country's highland area is significantly eroded, and over 1900 million Mg of soil are lost from the highlands annually. If the trend continues, shortly about 60,000 km$^2$ area will have a soil depth of 10 cm, below which the soil will be too shallow to support cropping (http:www.fao.org/docrep/x5318e/x5318e02.htm).

Based on recent GPS surveys at 18 sites in the Mitchell catchment in Northern Australia, Brooks et al. (2009) reported average linear rates of scarp retreat between <0.1 and 0.8 m year$^{-1}$, with a median rate of 0.23 m year$^{-1}$. However, maximum local rates at active lobes ranged between 2 and 15 m year$^{-1}$. Using historical air photos between the 1940s and 2000s, similar average linear rates of scarp retreat have been documented in the Mitchell (median 0.37 m year$^{-1}$; range <0.1–1.2 m year$^{-1}$) by Shellberg et al. (2012), in Norman (median 0.46 m year$^{-1}$; range 0.1–0.7) by Brooks et al. (2013) and in Victoria (average 0.86 m year$^{-1}$; range 0.3–1.6 m year$^{-1}$) catchments by McCloskey (2010). Further, Hughes and Prosser (2012), using aerial photograph mapping of gullies, predicted patterns of gully density across the $1 \times 10^6$ km$^2$ Murray-Darling Basin in Australia. At a 10-km grid resolution, gully density has been predicted to vary from 0 to 1.2 km km$^{-2}$, with 22% of the basin having a gully density >0.1 km km$^{-2}$ and 3% a density >0.5 km km$^{-2}$. The predicted mean gully density across the basin is 0.08 km km$^{-2}$, and gullies contribute up to $27 \times 10^6$ Mg year$^{-1}$ of sediment to the river network. This is more than the amount that has been estimated from the combined contribution of hill slope ($14 \times 10^6$ Mg year$^{-1}$) and river bank ($8.6 \times 10^6$ Mg year$^{-1}$) erosion by other studies within the basin.

Dhuru Narayana and Babu (1983) estimated that in India soil erosion is taking place at an average rate of 16.35 Mg ha$^{-1}$ year$^{-1}$, and about 29% of total eroded soil is usually lost permanently to the sea, and 10% of it is deposited in reservoirs. Singh et al. (1992), however, reported that soil erosion is found at a rate of exceeding 40 Mg ha$^{-1}$ year$^{-1}$ in the ravines and badlands of India. It is estimated that severe soil loss of Damodar River catchment (eastern part of India) increases sediment yield from 1729 to 2387 m$^3$ km$^{-2}$ year$^{-1}$, and it escalates the siltation of Panchet reservoir at a rate of 0.033–0.047 cm per year (Majumder et al. 2012). In the humid subtropical region of India, soil erosion (about 15 million Mg per year) leads to low crop productivity and an annual loss of 13.4 million Mg in the production of crops due to water erosion equivalent to about $2.51 billion (Bhattacharyya et al. 2007; Sharda et al. 2010; Sharda and Dogra 2013).

While working with Chambal ravines, Verma and Singh (2015) reported that the area under ravines of Chambal increased from 228,000 ha in 1943 to 234,000 ha in 1950 and to 311,000 ha in 1975 (but no recent estimates are given) showing that area under ravines is increasing with fast rate. They further reported that area under gullies and ravines in 66 villages of Morena district of Madhya Pradesh in India increased to 218.7 km$^2$ in 1994 from 130.15 km$^2$ in 1969, i.e. an increase of 88.55 km$^2$ in 25 years only.

### 2.3.6 Sedimentation

The last stage of soil erosion process is the movement of finer soil particles along with runoff water and finally deposition along banks of the river or tributary. The soil particles come to rest when resistant forces are greater than erosive forces. The first to be deposited are the heavier particles of lowest transportability, whereas the lighter particles of highest transportability are deposited farther downstream, while colloidal materials may remain in suspension indefinitely. Deposition of erosion debris in places where it is not wanted is one of the major damages resulting from erosion. The clogged channels and reservoirs and pollution by sediment affecting the biodiversity adversely are the major concerns to be addressed.

Like the data on land area affected by water erosion, there are also uncertainties about the extent of sediment transported by the rivers into the oceans. This is obvious by various reports mentioned here. For example, Milliman and Syvitski (1992) estimated that the annual sediment transport to the ocean is about 20 billion Mg, but the mass of sediment in motion is about 30 billion Mg year$^{-1}$. Contrary to this, Pimentel et al. (1995) estimated the global rates of soil erosion to be 75 billion Mg year$^{-1}$ by assuming an average erosion rate of 100 Mg ha$^{-1}$ on 751 Mha of area affected by severe erosion. Reich et al. (2001) reported 130 billion Mg as annual potential yield of sediment from 72.5 million km$^2$ global land area subjective to severe erosion. Global maps of suspended sediment yield on continental basis have been compiled by Walling and Webb (1983) and Lvovich et al. (1991), while Walling and Webb (1996) further stated that the most available literature reflects that the annual sediment transport into the ocean by the world's rivers is 15–20 billion Mg. For estimation of the actual soil erosion using the sediment yield data, it requires information on the delivery ratio, which is defined by Walling (1983) as the fraction of the sediment originating over the land delivered to the major drainage or river system. Assuming a delivery ratio of 10% (Walling 1983) and 13–20% (Lvovich et al. 1991), estimates of gross erosion on continental scale are summarized by Lal (2003) and presented in Table 2.5, which appears to be realistic. We may note the discrepancy in data that the global erosion rate of 201–394 billion Mg (Table 2.5) is about 2.7–5.2 times that reported by Pimentel et al. (1995).

When all other factors such as soil type, climate, vegetation, etc. were common, the rate of sedimentation load of Kshipra river (a tributary of Chambal river in Madhya Pradesh) catchment (1140 km$^2$) was observed to be 2384, 1281, 61 and 34 m$^3$ km$^2$, when rainfall during observation period was 1827, 1127, 930 and 650 mm, respectively (Dagar 1999). In India, rivers carry huge sediment load, particularly during rainy season. Dhuru Narayana (1990) reported sediment deposition in some Indian reservoirs (Table 2.6), because of that major river basins regularly experience severe floods with very high sediment loads.

In a well-documented review, Poesen et al. (2003) compiled sediment production rates due to gully erosion at 56 locations of different countries (Table 2.7). Soil loss rate varied from 0.1 to 64.9 Mg ha$^{-1}$ year$^{-1}$which was 10–94% of the total sediment at a particular site depending upon catchment area and environmental factors. These studies indicated that gully erosion is an important sediment source in a range of

**Table 2.5** The global pattern of suspended sediment yield and the estimations of gross erosion rate

| Region | Land area ($10^9$ ha) | Suspended load[a] ($10^6$ Mg) | | Gross erosion[a] ($10^9$ Mg year$^{-1}$) assuming delivery ratio of | | Mean erosion rate[a] (Mg ha$^{-1}$ year$^{-1}$) | |
|---|---|---|---|---|---|---|---|
| | | | | 10% | 13–20% | | |
| | | 1 | 2 | 1 | 2 | 1 | 2 |
| Africa | 3.01 | 3891.3 | 2342.1 | 38.9 | 23.4 | 12.9 | 7.8 |
| Asia | 4.46 | 7401.3 | 22088.3 | 74.0 | 220.9 | 16.6 | 49.5 |
| South America | 1.78 | 3942.6 | 5599.7 | 39.4 | 56.0 | 22.1 | 31.5 |
| North America | 2.28 | 2808.6 | 4534.2 | 28.1 | 45.3 | 12.3 | 19.9 |
| Europe | 0.98 | 1312.7 | 3288.2 | 13.1 | 32.9 | 13.4 | 33.6 |
| Oceania | 0.80 | 763.8 | 1525.5 | 7.6 | 15.3 | 9.5 | 19.1 |
| Total/mean | 13.26 | 20120.3 | 39378.0 | 201.1 | 393.8 | 15.2 | 26.7 |

[a]Calculated by Lal (2003) from the estimates of 1 (Walling and Webb 1983) and 2 (Lvovich et al. 1991)

**Table 2.6** Sediment deposition in some Indian reservoirs

| Reservoir | Catchment area (000 ha) | Annual rate of silting (Mg ha$^{-1}$) |
|---|---|---|
| Bhakra Nangal | 5688 | 9.21 |
| Hirakud | 8265 | 5.37 |
| Maithon | 521 | 19.65 |
| Panchet | 982 | 15.00 |
| Machkund | 196 | 3.85 |
| Tungabhadra | 2583 | 9.81 |
| Mayurakshi | 179 | 24.64 |
| Gandhinagar | 2187 | 5.56 |
| Ramganga | 300 | 27.28 |
| Ghod | 363 | 22.86 |
| Dhantiwad | 286 | 4.78 |

Source: Dhuru Narayana (1990)

environments; and these are effective links for transferring runoff and sediment load from upland watershed to valley bottom as well as off-site effects which include environmental effects.

## 2.4 Appropriate Techniques for Measuring Gully Erosion Processes

Since the inception of soil conservation as science, several methods have been used to measure soil erosion right from the impact of raindrops to the quantification of sedimentation load. Contrary to sheet and rill erosion, where standardized

**Table 2.7** Contribution of gully erosion to overall soil loss rates and sediment production rates by water erosion in different locations across the world

| Country | Location | Soil loss due to gully erosion (Mg $ha^{-1}$ $year^{-1}$) | Percent (by gully) of total erosion (%) |
|---|---|---|---|
| United States of America | New York | 11.3 | 18 |
| | Wisconsin[a] | 0.5–2.0 | 18–36 |
| | Wisconsin | 9.4 | 35 |
| | Iowa[a] | 2–18.2 | 19–35 |
| | Iowa | 8.7 | 19 |
| | Iowa | 6.7 | 24 |
| | Iowa | 11.9 | 45 |
| | Michigan | 2.7 | 21 |
| | Louisiana | 13.5 | 25 |
| | Kansas | 17.9 | 27 |
| | Georgia | 12 | 28 |
| | Rhode Island | 8.3 | 29 |
| | Mississippi[a] | 7.7 | 60 |
| | Mississippi | 16.8 | 30 |
| | Maine | 11.5 | 31 |
| | North Dakota | 8.0 | 32 |
| | Alabama | 20.8 | 37 |
| | Alabama[a] | 19.7–35.9 | 50–60 |
| | Pennsylvania | 4.0 | 41 |
| | Illinois | 11.6 | 42 |
| | Maryland | 9.0 | 43 |
| | New Jersey | 11.6 | 44 |
| | Virginia | 28.7 | 50 |
| | Vermont | 13.7 | 58 |
| | Arizona | 1.3–3.9 | 60–81 |
| | South California | 36.8 | 71 |
| | Delaware | 5.6 | 71 |
| | Washington | 4.2 | 73 |
| Spain | North-West | 1.5 | 26 |
| | Guadalentin | 37.6 | 51 |
| | Catalonia | n.a. | 58 |
| | – | 1.2 | 59 |
| | – | 4.2 | 73 |
| | Almeria | 9.7 | 83 |
| Belgium | Central[a] | 22.3 | 10 |
| | Central | 3.6 | 44 |
| | Central | n.a. | 60 |
| France | North | n.a. | 10–45 |
| | Normandy | n.a. | 21–56 |
| | North | n.a | 46–55 |

(continued)

**Table 2.7** (continued)

| Country | Location | Soil loss due to gully erosion (Mg $ha^{-1}$ $year^{-1}$) | Percent (by gully) of total erosion (%) |
|---|---|---|---|
| Australia | – | n.a. | 50 |
| | New South Wales | 0.1 | 60 |
| | North-West | n.a. | 80 |
| Germany | South | n.a. | 12–29 |
| | South-West | n.a. | 36 |
| China | North | 25–43 | 40–70 |
| | Guangdong | n.a. | 87 |
| Portugal | Braganca | 16.1 | 47 |
| | Alentejo | 3.2 | 80 |
| Ethiopia | Tigray | 4.7–12.1 | 33–55 |
| Kenya | Baringo | 3.4 | 53 |
| Niger | Ader Dutchi Massif | 32 | 75 |
| Romania | – | n.a. | 37 |
| Norway | Leira basin | 12.7 | 55 |
| Argentina | Northeast Patagonia | n.a. | 58 |
| Lesotho | – | 15 | 94 |

Compiled by Poesen et al. (2003) from different sources
*n.a.* denotes "not available"
[a]Variation in data at same location is due to different sources at different periods

procedures for assessment of erosion rates exist, no standardized procedures are available for measuring gully erosion rates and the controlling factors, particularly for large areas. Poesen et al. (2003) reviewed techniques available and commonly followed for measuring gully erosion processes and rates in different regions of the world. Some workers (Auzet et al. 1995; Casali et al. 1999; Oostwoud-Wijdenes and Bryan 2001; Nachtergaele et al. 2001a, b, 2002; Valcárcel et al. 2003; Øygarden 2003) have used both ground-based and airborne techniques to assess eroded volumes of gully erosion based on the short timescale of a few months to few years up to 10 years. Hancock and Evans (2006) applied geomorphic threshold technique for positioning of the gully heads. In some short-time studies, photogrammetric techniques for sequential measures and large-scale aerial photographs to determine the volume of soil lost by concentrated flow erosion have been applied with success (Thomas et al. 1986; Ries and Marzolff 2003). Most of these studies have been tested in a wide range of environments.

At the medium timescale (10–70 years), many workers (Burkard and Kostaschuk 1995; Derose et al. 1998; Nachtergaele and Poesen 1999; Daba 2003; Daba et al. 2003; Gábris et al. 2003; Martinez-Casasnovas 2003) have analysed aerial photographs to measure temporal changes in morphological features such as length, area or volume of various types of gullies. Vandekerckhove et al. (2001a) developed a framework using dendrochronological methods for estimating medium-term gully erosion rates based on the analysis of roots exposed by gully erosion, remaining parts/scars due to browsing by animals and exposed roots and regenerating parts from root suckers, fallen branches of trees and patterns of tree found growing in gullies. These days, tracers such as carbon, nitrogen, the nuclear bomb-derived radionuclide 137 Cs, magnetics and the strontium isotopic ratio are increasingly used to fingerprint sediment (Valentin et al. 2005). Recent studies conducted by many workers (Gábris et al. 2003; Poesen et al. 2003; Xiang-Zhou et al. 2004; Valentin et al. 2005; Brooks et al. 2008; Shellberg et al. 2012) in Australia, China, Hungary, Ethiopia and the USA showed that the major part of the sediment in reservoirs might have come from gully erosion.

In the field, it is very difficult to measure the relative importance of various subprocesses in gully development; therefore, laboratory experiments were conducted by some workers (Govers et al. 1990; Robinson and Hanson 1996; Bennett et al. 2000; Bennett and Casali 2001), who have generated some useful information in understanding the basics of processes involved in gully erosion. Recently, Ghosh and Guchhait (2016) using geomorphic threshold estimation for gully erosion in the laterite soils have examined a quantitative approach to predict critical conditions and locations where gully heads might develop. Although a significant number of studies dealing with the measurement of gully erosion rates exist, the possibilities and limitations of the various monitoring and experimental approaches used in these studies are not always clear, and due to lack of standardization, the data obtained in various experiments are not always comparable.

As is evident from literature, we do not have any consolidated global data regarding the area under gully and ravine erosion. Though satellite and air photo remote sensing technique has been used in many places, it has its limitations. For example, Shellberg and Brooks (2012) used the technique in Queensland, Northern Australia, and mapped extensive areas of floodplains degraded by active alluvial gully erosion in many catchments. These estimates are based on the mapping of bare, denuded surfaces eroded by alluvial gullies. The quantification is towards lower side because of masking of gullies under tree canopy. James et al. (2007), however, could study gully channels under forest cover using LiDAR data. Further, Marzolff and Poesen (2009) explored the potential of 3D gully monitoring with GIS using high-resolution (low-altitude) photography and a digital photogrammetric system to document gully head evolution over a 6-year period in Germany, Spain and the USA. Saey et al. (2008) have made significant advances in the detection of in-filled gullies on top of tertiary landscapes in Belgium by using electromagnetic induction sensors. There is a need to further develop and to apply these new techniques to a wider range of gully environments to define their possibilities and limitations. We also need to develop techniques to correlate the geological advances with other climatic factors.

There are only a few studies which have led to quantify the hydrogeomorphic mechanisms responsible for understanding the processes involved in alluvial gully erosion (Oostwoud-Wijdenes et al. 2000; Vandekerckhove et al. 2003; Thomas et al. 2004). Some years back, Shellberg et al. (2012), while studying the hydrogeomorphic processes influencing alluvial gully erosion, evaluated these processes at multiple spatial and temporal scales across the Mitchell River fluvial mega-fan in tropical

Queensland, Australia. They quantified longitudinal changes in floodplain inundation using river gauge data, local stage recorders and HEC-RAS modelling based on LiDAR topographic data. Several erosion processes were analysed which included direct rainfall, infiltration-excess runoff, soil water seepage, river backwater and overbank flood inundation. To develop predictive models of gully initiation and evolution, it is imperative that we understand the relative importance of the various hydrological drivers and associated erosion processes of alluvial gully erosion. As is evident from recent examples from all over the world, land use change is expected to have a greater impact on gully erosion than climate change. Modern historical data are also of great significance in making us to understand the sequence of procedures. Reconstructions of historical causes of gully erosion, using high-resolution stratigraphy, archaeological dating of pottery and $^{14}C$ dating of wood and charcoal show that the main gully erosion periods identified in Europe correspond to a combination not only of deforestation and overuse of the land but also to periods with high frequency of extreme rainfall events (Valentin et al. 2005; Colgan 2016).

## 2.5 Effective Gully Erosion Models

Though there had been many approaches to measure the soil erosion rate, only a few studies could be utilized for predicting long-term strategies for controlling soil erosion. There are only a few models claiming to be capable of predicting ephemeral gully erosion rates (Table 2.8). In these models, concentrated flow detachment rate is proportional to the difference between (i) flow shear stress exerted on the bed material and the critical shear stress and (ii) the transport capacity of the flow and the sediment load. Although these models are claimed to have great potential in predicting soil losses by gully erosion, they have never been thoroughly tested for this erosion process (Poesen et al. 2003).

Some of the workers (Howard 1997; Sidorchuk et al. 2003; Sidorchuk 1999, 2005; Casali et al. 2003; Ghosh and Guchhait 2016) have developed models that predict rapid changes in gully morphology, particularly during the early stages of gully development. These models are also static for calculation of final metamorphic parameters of permanent and stable gullies/ravines in different environments. Gutiérrez et al. (2009) developed a model for large area (>40 × $10^3$ $km^2$) in Spain

**Table 2.8** Some models for predicting gully erosion rates

| Model | Explanation | Reference |
|---|---|---|
| CREAMS | Chemicals, runoff and erosion from agricultural management systems | Knisel (1980) |
| GLEAMS | Groundwater loading effects of agricultural management systems | Knisel (1993) |
| EGEM | Ephemeral gully erosion model | Woodward (1999) and Nachtergaele et al. (2001a, b) |
| WEPP | Water erosion prediction project | Nearing et al. (1989) and Flanagan and Nearing (1995) |

using multivariate regression splines to determine that lithology and soil type best explained the location of gullies. Oostwound-Wijdenes et al. (2000) and Vandekerckhove et al. (2003) attempted to quantify and predict process-based gully head-cut retreat models, whereas Prasad and Römkens (2003) presented a holistic and energy-based conceptual framework for modelling head-cut dynamics. Further, Poesen et al. (2011) reviewed all details dealing with the prediction of gully erosion processes and concluded that, although there are several studies on models, still no reliable and validated models are available allowing one to predict the effects of environmental change on gully erosion or gully infilling rates at various temporal and spatial scales and the impacts of gully erosion on sediment yield, hydrological processes and landscape evolution. This is a major research area requiring more efforts. Recently, however, Shellberg et al. (2012) attempted to correlate hydrogeological impacts and other parameters influencing gully erosion giving somewhat clear understanding of erosion processes. These studies need to be extrapolated in different ecological regions.

## 2.6 Controlling and Management of Ravine/Gully Lands

It is a well-known fact that anthropogenic factors and mismanagement of rainwater are responsible for the formation of gullies and ravines. Besides controlling biotic factors such as overgrazing, deforestation and over-exploitation, we need to minimize surface runoff in gully control. The watershed development technology is locality specific and thus must be used involving the participation of local people and other stakeholders. In gully control, the following steps can be applied according to the given order:

- Improvement of watershed area/gully catchments to reduce and regulate the run-off rates and peak flows
- Raising bunds along gully heads and stabilizing them by planting suitable shrubs and perennial grasses
- Diversion of surface water above the gully area
- Stabilization of gullies by structural measures and accompanying revegetation
- Smoothing the land on both sides of the gully (between two bunds or gully plugs) to reduce the slope of land
- Treatment of gullies (reclamation measures)

### 2.6.1 Factors Controlling Gully Erosion

For controlling gully erosion, we must have deep knowledge of factors controlling gully erosion so that the appropriate planning may be developed for efficient handling of rehabilitation projects. As discussed earlier, gully erosion is controlled by a wide range of factors such as topography (slope, critical drainage area, shape of watershed area), soil and lithological factors (soil types, soil crusting, lithology, geomorphology, piping, etc.), natural vegetation, land use changes, environmental factors, biotic factors and climate change. Many workers (Poesen et al. 2003; Xiang-Zhou et al. 2004; Valentin et al. 2005; Shellberg et al. 2012; Chaturvedi et al. 2014; Verma et al. 2015) have reviewed the work done related to most of these parameters. Some aspects have already been discussed earlier in this chapter; hence only brief account will be given under this head. We know that gullies and deep ravines are common features of mountainous or hilly regions with steep slopes. Steep slopes favour high runoff velocity, and thus rill and gully initiation takes place, but, given the climatic and vegetation conditions, they can produce lower runoff volumes than gentle slopes, as has been shown in northern Thailand (Janeau et al. 2003). In some areas, soil crusting has an ambivalent effect on gully formation, and it can delay the initiation of gullies (Zhu and Cai 2004). Because soil crusts favour runoff generation and concentration downslope, soils prone to crusting are generally eroded not only by sheet but also by gully erosion. Even for very gentle slope gradients, rills and gullies develop in the regions where crusting is a common problem, such as in the loess belts of China (Wu and Cheng 2005), Europe (Vanwalleghem et al. 2005a, b) and North America (Thomas et al. 2004). Because of the scarcity of vegetation, soils of the arid and semiarid regions are subjected to crusting and thus runoff production and formation of gullies and ravines.

Soil properties such as texture, moisture, bulk density, water holding, infiltration capacity, etc. also play an important role in development of gullies and ravines and also their distribution pattern. Field observations have shown that landform development could often be related to tectonically induced compressional or tensional forces. These can form fractures or cracks in rocks without causing actual tectonic movement or displacement. These depressions, which can also be old landslide scars, become sites for surface flow concentration and thus gully erosion. It has been found that during dry period, clay soils (usually saline vertosols in India) develop deep cracks. These cracks act as structurally weakened starting points for weathering processes. During rains, water percolates through these cracks and birth of rill erosion takes place. As explained by Ranga et al. (2015a, b), swallow holes are developed in soils of badlands of Chambal River frequently, and these are interconnected to each other. On the course of time, these holes will connect to the nearby gully forming a tunnel gully. Although we do not have data to support this interconnection, but internationally, piping has been accepted as a critically important soil erosion process which often results in gullies and rills (Faulkner 2006) and has been reported in a number of different climatic and soil conditions (Faulkner 2008; Verachtert et al. 2009).

There is little information on how gully systems may respond to climatic change (Li et al. 2004). Where annual rainfall is known to have decreased significantly over the last decades, no concurrent decrease has been observed for high-intensity rain events (Yu and Neil 1993). During the protracted droughts or long periods of dry season, the associated vegetation decays leaving large areas unprotected from splash and subjected to soil crusting. In such conditions, when rainfall occurs, the runoff will increase promoting gully erosion. Under cold conditions, global warming is expected to increase the frequency of freeze–thaw cycles exacerbating the risk of gullying as shown in southern Saskatchewan, Canada, (Archibold et al. 2003) and in southern Norway (Øygarden 2003). More generally, land use change is expected to have a greater impact on gully erosion than climate change. However, more data need to be generated on impact of climate change on soil erosion and gully formation.

The international literature available on gully erosion indicates that different direct intervention (engineering methods) and passive rehabilitation actions using biological, chemical and physical methods are available to rehabilitate gully erosion (Shellberg 2011; Chaturvedi et al. 2014). The experiences with trying to engineering stability in gullies have proven to have mixed results, and failure of many projects is common due to naïve attempts to fully engineer nature, lack of initial rigorous experimentation, poor design and implementation and lack of adaptive management or maintenance. Usually direct methods in combination with vegetation options lead to more success. Xiang-Zhou et al. (2004) advocated plantations and check-dam systems in gullies on the Loess Plateau of China to control the sediment pouring into the Yellow River. The former is not suitable due to the arid climate and the barren soil, while the latter system in gullies is one of the most effective ways to conserve soil and water in the Loess Plateau. In northern Australia and elsewhere, trees and native and exotic grasses can readily and progressively recolonize alluvial gully floors once eroded. Proactive vegetation seeding and planting of appropriate species in gullied catchments has proven to be an effective management approach to stabilizing large areas of alluvial and colluvial gullies around the world including in India, China, Africa, New Zealand, USA and to a less extent Australia (Dagar 2001; Shellberg and Brooks 2012; Chaturvedi et al. 2014), and some of these options are dealt in portion to follow.

### 2.6.2 Measures for Checking Extension of Gullies

It has been commonly observed that the uninterrupted high-velocity flow of runoff water from the arable lands of watershed area in to the adjoining channels or ephemeral gullies leads to the extension of earlier initiated or formed gullies in to the arable lands. It is essential that the runoff may be diverted to some other storage option or allowed to reach the gullies at controlled velocity preferably passing through vegetative path at designated points only where safety measures for controlling the damages are installed. Peripheral bunds may be erected around the arable lands at a distance from the gully heads that is equal to double the depth of gullies. These bunds must be stabilized by planting of perennial grasses such as *Dichanthium annulatum, Chrysopogon fulvus, Heteropogon contortus, Panicum antidotale, Cynodon dactylon, Saccharum munja* and even oil-yielding aromatic *Cymbopogon* grass, which has been found quite effective in stabilizing the bunds in Chambal ravines (Fig. 2.5).

The construction of these bunds also leads to moisture in the adjoining lands. At the places where the excess runoff enters the gullies from the peripheral bunds, spillways like drop spillways and chute spillways may be constructed for safe disposal of excess runoff from the adjoining arable lands through the peripheral

**Fig. 2.5** *Cymbopogon* grass found effective in stabilization of bunds in Chambal ravines (Photo courtesy SK Verma)

bunds to the gully beds (Chaturvedi et al. 2014). Construction of the brushwood check dams also helps in reducing the velocity of runoff and soil loss. The silt deposited behind these structures leads to stabilization of gully beds. Xiang-Zhou et al. (2004) have reported that laying down of wattles of plant species easily sprouting at a vertical interval of 3 m up to 66% slope and 5–7 m up to 33% slope helped in holding the soil and reducing the raindrop impact.

For the control of deep gullies, permanent gully control structures may be needed. These are designed depending on the catchment's area, peak flow and grade of gullies. These structures prevent the extension of gully and help in retaining large quantity of runoff and silt behind them and allow more water to infiltrate. In the small gullies where runoff from the catchment is not high, live vegetative checks of locally available shrubs and perennial vegetation are effective for retention of soil and reducing the grade of gully bed. Once established, they add to the productive value of ravines by providing fuel, fodder and/or another usable biomass. Depending on the moisture availability, shrubs like *Agave americana, Lawsonia alba, Balanites roxburghii* and *Capparis decidua* and tall grasses like *Arundo donax* and *Erianthus (Saccharum) munja* having protective as well as economic value are planted. Seeding of fodder grasses behind these check dams may improve fodder availability.

Several studies (Dagar and Mall 1980; Dagar and Singh 1980; Geyik 1986; Dagar 1987a, c, 2001; Sidorchuk and Grigorev 1998; Chow et al. 1999; Nogueras et al. 2000; Rey 2003; Dagar and Pathak 2005) have demonstrated worldwide the crucial role of vegetation in stabilizing beds of gullies and ravine and in valley bottoms. Overgrazing of ravine lands has drastic impact on root development, particularly of grasses, which play a key role in soil binding. Protection from grazing and selective harvesting helped a lot in improving root development and biomass production in these habitats (Dagar 1987a, b, 1995).When the area is closed from grazing, the natural succession takes place, and more palatable species appear and establish replacing the grazing-resistant species (Dagar 2001). Where possible, natural vegetation with well-developed root mats should be established in disturbed concentrated flow zones affected by gully erosion (Sidorchuk and Grigorév 1998; Dagar 2001; Morgan and Mngomezulu 2003). In doing so, soil loss and sediment production will be cut down, and the connectivity in the landscape will be interrupted resulting in a smaller sediment delivery to valley bottoms or river channels.

### 2.6.3 Management of Arable Land (Mostly in Farmers' Fields) Near the Gully Heads

The area near gully head in catchment area falls in farmers' fields and requires special attention because this area is highly vulnerable to erosion. The shallow ravines can be reclaimed for cultivation of field crops by levelling to some extent and also putting earthen embankments. The provision must be made for runoff water so that it goes in gullies through grass beds or is diverted for storage for other purposes. Earthen marginal bunds must be stabilized by growing perennial grasses which develop fine roots and protect the bunds. Desirable mechanical measures like contour or graded bunds need to be employed, depending on the soil type and rainfall. The area between bunds should be levelled and proper disposal of excess runoff be ensured. For reducing erosion from cultivated areas, cover crops and intercropping of cereals with legumes are widely recommended. Runoff water can also be harvested into dug out ponds for crop use during soil moisture stress period. Wherever possible agroforestry systems involving multipurpose trees and arable crops or fodder crops must be developed. Depending on the tree architecture and species, spacing of planting may vary from 2 × 2 to 8 × 8 m. With a view to utilize the interspaces in widely planted trees, planting of palatable grasses is a better option. If possibilities of developing irrigation facilities exist, the installation of drip irrigation facilities may be beneficial. Life-saving irrigation may be essential for the fruit cultivation in ravines at least during summers.

Besides arable crops, the catchment area may be put under agri-horticultural system involving fruit trees such as lemon (*Citrus limon*), mango (*Mangifera indica*), ber (*Ziziphus mauritiana*) and gooseberry (*Emblica officinalis*). These can successfully be grown with agricultural crops in humps and gully beds (Verma et al. 1986). These fruit trees can also be grown successfully with fodder grasses in ravine areas (Fig. 2.6).

Contrary to this approach, Xiang-Zhou et al. (2004) reported that the check-dam project has been given great emphasis ever since the founding of People's Republic of China as one of the primary measures to conserve water and soil, and by 2002, about 113,500 check dams had been built, creating 3200 $km^2$ of farmlands with high productivity and intercepting a total of 700 million $m^3$ of sediments that pour into the Yellow River. In the lower reaches of this river, the amount of retained sediment is between 16,680 and 72,840 Mg $ha^{-1}$. The sediments retained by check dams in watershed-7127 $mm^3$ are held up by check-dam system, which is 67% of the total

**Fig. 2.6** Goose berry (*Emblica officinalis*)-based horti-pastoral system in ravine areas (Photo courtesy AK Sikka)

amount; and 3812 km$^2$ of farmlands were formed because of measures taken. Thus, the check-dam system is the most important in China to conserve soil and water particularly in dry regions where survival chances of vegetation are remote.

### 2.6.4 Management of Non-arable Land Covered Under Gullies and Ravines

One of the prerequisites for reclamation of ravine lands is to effectively protect these areas by fencing from biotic agencies such as overgrazing and harvesting of fuelwood and forage leaves and grasses. This will result in substantial increase in vegetation cover and yield of grasses. Most of the native trees will regenerate and form an effective cover. Dagar (1987a, c), Dagar and Pathak (2005) and Pathak and Dagar (2015) reported the ecology and management of grazing lands in India. They argued that even protection from grazing helps to a greater extent in biomass production of grasses (used as fodder) and conserving soil and moisture. Closure should be accompanied by soil conservation measures such as construction of check dams and disposal of runoff. More details on this aspect are available in next chapter. The gullies of Chambal River can successfully be brought into good grass cover within a period of 5 years (Fig. 2.7). The author has the experience of witnessing a good tropical thorn forest developed after about 40 years of protection having trees, shrubs, grasses and leaf litter with almost no runoff and soil loss, high infiltration and enriched wildlife.

The selection of tree species is of important factor in ravine areas. It has been found that only those native trees will sustain which have potential of tolerating stress and can adapt to soil erosion stress through regeneration from root stock or exposed roots. Trees such as *Acacia nilotica*, *A. eburnea*, *A. catechu*, *A. tortilis*, *Carissa carandas*, *Emblica officinalis*, *Dalbergia sissoo*, *Dendrocalamus strictus*, *Dichrostachys cinerea*, *Azadirachta indica*, *Albizia amara*, *Leucaena leucocephala* and *Prosopis juliflora* are among the promising species. Shrubs such as *Capparis*

**Fig. 2.7** Vegetation cover after protection from grazing in ravines (Photo courtesy Dr. SK Verma)

*decidua*, *Carissa carandas* and *Balanites roxburghii* are found regenerating from the root stocks and hence are most successful in eroded habitats. The highly eroded soils coupled with extremes of temperature and limitations of moisture hamper the survival and growth of many trees. Fruit trees such as *E. officinalis*, *C. carandas*, *Ziziphus mauritiana*, *Punica granatum*, *Aegle marmelos*, *Annona squamosa* and *Morus alba* along with many medicinal and aromatic species may find place which will also enhance the economy of the farmer. Among grasses *Dichanthium annulatum*, *D. caricosum*, *Bothriochloa pertusa*, *Cenchrus ciliaris*, *Cenchrus setigerus*, *Panicum antidotale*, *Panicum maximum*, *Pennisetum purpureum* and *Brachiaria mutica* are important species suitable for improving the fodder availability in ravine regions. Grass species such as *Heteropogon contortus*, *Themeda triandra*, *Iseilema laxum* and *Apluda mutica* along with legumes like *Alysicarpus*, *Tephrosia* and *Stylosanthes* can successfully be grown in eroded habitats as is the case in Kshipra ravines in Madhya Pradesh, India, but these do not stand grazing pressure and will perform in protected environment. Silvopastoral system is most ideal for gully or ravine rehabilitation.

In a success story (http://www.icar.org.in/node/5101), an indigenous bamboo and Anjan grass (*Cenchrus ciliaris*)-based silvopasture system for enhancing productivity of ravines was established at Vasad in ravines of Gujarat (Fig. 2.8). It was proved that the ravine lands, under unproductive use, could be successfully reclaimed by planting bamboo, a fast-growing plant species, on the gully beds and Anjan grass on the side slopes and the interspaces of gully bed for economic utilization of gullied land. Planting of grass also protected the ravine slopes and reduced soil loss, and green fodder could be obtained from the stabilized slopes. The grass

**Fig. 2.8** Bamboo-*Cenchrus ciliaris* silvopastoral system in ravine lands (Source: http://www.icar.org.in/node/5101)

yield from interspaces of bamboo planted on bed portion was about 10 Mg $ha^{-1}$ $year^{-1}$ during the initial 5 years. The grasses fetched an income of ₹3000–6000 per hectare over the period of 5 years at Vasad Regional Research Centre. About 300 clumps of bamboo with 3000–4000 old and 1000–1500 new culms were available in 1 ha of bamboo plantation of over 7 years old.

About 30% of old culms could be harvested easily per year. At 2008 prices, the bamboo planted on the ravine bed could fetch up to ₹27,000 $ha^{-1}$ $year^{-1}$. The watershed under this system absorbed more than 80% of rainfall that was either utilized by the plant or percolated deep to recharge the ground water. Due to less runoff, soil loss was reduced to less than 1 Mg $ha^{-1}$ $year^{-1}$ only from about 20 Mg $ha^{-1}$ $year^{-1}$ from degraded ravines prior to plantation. Bamboo is a fast-growing plant and establishes well along the slopes and beds of the gullies (Fig. 2.9). Many agencies put together utilized this technology for reclamation of nearly 1000 ha of community and wastelands in Mahi river stretch in Gujarat. Sethy et al. (2011) utilized the ravine land by digging trenches in micro-watersheds at Kota in India and planted gooseberry fruit trees on ravine humps, bamboo on gully beds and interspaces with *Cenchrus ciliaris* grass. After 6 years of planting, 3.9–8.5 Mg $ha^{-1}$ $year^{-1}$ fruits could be harvested in different micro-watersheds and the production of grass ranged from 7.71 to 9.91 Mg $ha^{-1}$. The trenches helped in controlling runoff and soil erosion and conservation of moisture to greater extent.

In temperate regions, for controlling gully erosion, watershed-based integrated approach has been found most suitable. Trees such as *Robinia pseudoacacia*, *Grevillea robusta* and *Fraxinus ornus*; species of *Salix*, *Juglans*, *Celtis*, *Quercus* and *Prunus*; *Elaeagnus angustifolia*, *Pinus nigra* and *P. brutia* and *Eucalyptus robusta*,

**Fig. 2.9** Establishing bamboo along the slopes (Left) and bottom (right) of ravines (Photo courtesy SK Verma)

*E. saligna* and *E. camaldulensis*; and grasses such as species of *Agropyron*, *Dactylis*, *Arrhenatherum* and *Cynodon* with legumes such as *Onobrychis*, *Trifolium* and *Alfalfa* are suitable for stabilizing the gully beds and providing wood and forages.

In hilly regions, landslides adversely affect utility services, human settlements, agricultural production and several developmental and cultural activities. The causes of landslides are weak geology (sedimentary rocks), seismic activity and improper land use including overgrazing, fire and deforestation. Protective, vegetative and economic measures are needed to stabilize and control landslides. Protective measures may be keeping the human activities minimum. Structural measures are mainly gabion drop spillways, spurs and retaining walls. These measures control the flow in the torrent and prevent further slope failure. Contour formation and terracing along with perennial vegetation are quite effective measure in stabilizing the slopes (Fig. 2.10).

Vegetative measures to stabilize the slopes include planting trees, shrubs and grasses and mulching and contour wattling. Dealing with hillside erosion of Himalayas, Gupta et al. (1990) advocated multipurpose tree species such as *Albizia lebbeck*, *A. chinensis*, *Bauhinia purpurea*, *Grewia optiva*, *Leucaena leucocephala*, *Eucalyptus globulus*, *E. grandis*, *E. tereticornis*, *Alnus nepalensis* (for eastern Himalayas), *Salix alba*, *Acacia mearnsii*, etc., and grasses like *Andropogon pumilus*, *Arundinella nepalensis*, *Bothriochloa intermedia*, *Chrysopogon fulvus*, *Heteropogon contortus*, *Pennisetum orientale*, *P. clandestinum*, *Cynodon dactylon*, *C. plectostachyus*, *Themeda anathera*, *Setaria sphacelata*, *Brachiaria decumbens*, *B. mutica*, *Cenchrus ciliaris*, *Chloris gayana*, *Dichanthium annulatum*, *Melinis*

**Fig. 2.10** Terracing to check soil erosion and productivity enhancement (Source Chaturvedi et al. 2014)

*minutiflora* and *Panicum coloratum* for cultivation in degraded habitats. These form the most suitable silvopastoral systems. Bare slopes may be protected through contour wattling at suitable intervals (5–7 m) and mulching to stop moisture evaporation from the soil surface and to check direct impact of raindrops. The mulched area may be planted with fast-growing bush and grass species such as *Ipomoea carnea*, *Vitex negundo* (medicinal), *Salix tetrasperma*, *Arundo donax* and *Pennisetum purpureum*. Other species such as *Woodfordia fruticosa*, *Lannea grandis*, *Erythrina suberosa*, *Alnus nepalensis*, *Wendlandia exserta*, *Boehmeria rugulosa*, *Moringa oleifera*, *Bauhinia retusa*, *Azadirachta indica* and *Populus ciliata* are recommended depending upon the situation. Several grasses like *Eulaliopsis binata*, *Chrysopogon fulvus*, *Eragrostis curvula*, *Cymbopogon* sp., *Dactylis glomerata*, *Lolium perenne* and *Cenchrus* spp. are found to be very promising. These areas must be protected from grazing, and only cut-and-carry system must be adopted (Dagar 2001). For more details also see Gupta et al. (1990), Verma and Chinnamani (1990), Chaturvedi et al. (2014) and Kashyap et al. (2014).

## 2.7 Conclusions

Ravine rehabilitation program requires an integrated approach of treatment of table, marginal lands and catchment contributing runoff to the gullies on watershed basis. Both engineering and conventional vegetation approaches need to be followed. Using gullies according to land capability classes, adopting soil and water conservation measures and bringing land under permanent vegetation cover involving afforestation, fodder and fruit tree-based pasture systems and energy plantations are some of the simple and cost-effective ways of rehabilitation of ravine area. Though several viable technologies have been developed by the scientists related to ravine land rehabilitation, their adoption by the farmers have not been very encouraging. Thus, there is immediate need for capacity building, technology transfer and policy initiatives. In addition, intangible benefits due to ravine rehabilitation also need to be demonstrated.

Reclamation of ravine lands is a complex task which requires adequate technological, financial and policy support. There is a need to develop senses among beneficiary groups, developmental agencies and policy makers for the need and urgency of initiatives and developmental potential of ravine lands. However, the success of ravine reclamation program is linked with careful micro-level planning, adaption of economically viable production systems, effective community mobilization and efficient monitoring and policy support. Ravine area development will not only contribute to bridging the growing demand and supply gap for food, fibre and woody produces; but also, it will help in restoring livelihood security for resource constraint communities and mitigate climate change impacts.

## References

Agarwal KK, Singh IB, Sharma M, Sharma S, Rajagopalan G (2002) Extensional tectonic activity in the craton -ward parts (peripheral bulge) of the Ganga Plain foreland basin, India. Int J Earth Sci 91(5):897–905

Ahmad E (1973) Soil erosion in India. Asia Publishing House, Bombay

Archibold OW, Levesque LMJ, de Boer DH, Aitken AE, Delanoy L (2003) Gully retreat in a semi urban catchment in Saskatoon, Saskatchewan. Appl Geogr 23(4):261–279

Auzet AV, Boiffin J, Ludwig B (1995) Concentrated flow erosion in cultivated catchments: influence of soil surface state. Earth Surf Process Landf 20:759–767

Bali YP, Karale RL (1977) Reclamability classification of ravines for agriculture. Soil Conserv Dig 5(2):40–47

Bawa N, Jain V, Shekhar S, Kumar N, Jyani V (2014) Controls on morphological variability and role of stream power distribution pattern, Yamuna River, western India. Geomorphology 227:60–67

Beasley RP (1972) Erosion and sediment pollution control. The Iowa State University Press, Ames

Bennet HH (1939) Soil conservation. McGraw Hill Book Co, New York

Bennet HH (1955) Elements of soil conservation. McGraw Hill Book Co, New York

Bennett SJ, Casali J (2001) Effect of initial step height on headcut development in upland concentrated flows. Water Resour Res 37:1475–1484

Bennett SJ, Casali J, Robinson KM, Kadavy KC (2000) Characteristics of actively eroding ephemeral gullies in an experimental channel. Trans ASAE 43(3):641–649

Bhattacharyya T, Babu R, Sarkar D, Mandal C, Dhyani BL, Nagar AP (2007) Soil and crop productivity model in humid sub-tropical India. Curr Sci 93(10):1397–1403

Billi P, Dramis F (2003) Geomorphological investigation on gully erosion in the Rift valley and the northern Highlands of Ethiopia. Catena 50:353–368

Blanco-Canqui H, Lal R (2012) Principles of soil conservation and management. Springer, Netherland, pp 1–4

Boardman J, Parsons A, Holland R, Holmes PJ, Washington R (2003) Development of badlands and gullies in the Sneeuberg, Great Karoo, South Africa. Catena 50:165–184

Bochet E, Garcia-Fayos P, Poesen J (2009) Topographic thresholds for plant colonization on semi-arid eroded slopes. Earth Surf Process Landf 34:1758–1771. https://doi.org/10.1002/esp.1860

Bogolyubova IV, Karaushev AV (1979) Problems of gully formation and development. Sov Hydrol Sel Pap 18:255–264

Bollati M, Seta M, Pelfini M, Del-Monte MP, Fredi E, Palmieri L (2012) Dendrochronological and geomorphological investigations to assess water erosion and mass wasting processes in the Apennines of Southern Tuscany (Italy). Catena 90:1–17

Brooks AP, Spencer J, Shellberg JG, Knight J, Lymburner L (2008) Using remote sensing to quantify sediment budget components in a large tropical river – Mitchell River, Gulf of Carpentaria. In: Sediment dynamics in changing environments. Proceedings of a symposium held in Christchurch, New Zealand; December 2008, IAHS Publ No 325. pp 225–236

Brooks AP, Shellberg JG, Spencer J, Knight J (2009) Alluvial gully erosion: an example from the Mitchell fluvial megafan, Queensland, Australia. Earth Surf Process Landf 34:1951–1969

Brooks AP, Spencer J, Olley J, Pietsch T, Borombovits D, Curwen G, Shellberg J, Howley C, Gleeson A, Simon A, Bankhead N, Klimetz D, Eslami-Endargoli L, Bourgeault A (2013) Sediment sources, sinks and drivers in the Normanby catchment, Cape York. Griffith University, Australian Rivers Institute, Draft Final Report for Australian Government Reef Rescue Program, October 2012

Burkard MB, Kostaschuk RA (1995) Initiation and evolution of gullies along the shoreline of Lake Huron. Geomorphology 14:211–219

CAS (1992a) Resource, environment and social economy data. Report of comprehensive survey on Loess Plateau of Chinese Academy of Sciences (CAS) of Loess Plateau zone. Economics Press of China, Loess Plateau, p 14. (in Chinese)

CAS (1992b) Brief report on comprehensive control and development on the Loess Plateau, by of Chinese Academy of Sciences (CAS). Economics Press of China, Loess Plateau, pp 445–464. (in Chinese)

Casali J, López JJ, Giraldez JV (1999) Ephemeral gully erosion in southern Navarra (Spain). Catena 36(1–2):65–84

Casali J, López JJ, Giraldez JV (2003) A process-based model for channel degradation: application to ephemeral gully erosion. Catena 50:435–447

Casali J, Giménez R, Bennett S (2009) Gully erosion processes and modelling. Proc. fourth international symposium on gully erosion, Pamplona, Spain, September 2007. Earth Surf Process Landf 34:1839–1984

Chapline WR (1929) Erosion on rangeland. J Am Soc Agron 21:422–429

Chaturvedi OP, Kaushal R, Tomar JMS, Prandiyal AK, Panwar P (2014) Agroforestry for wasteland rehabilitation: mined, ravine, and degraded wasteland areas. In: Dagar JC, Singh AK, Arunachalam A (eds) Agroforestry systems in India: livelihood security and ecosystem services, Advances in agroforestry, vol 10. Springer, New Delhi, pp 233–272

Chen L, Wang J, Fu B, Qiu Y (2001) Land-use change in a small catchment of Northern Loess Plateau, China. Agric Ecosyst Environ 86:163–172

Chow TL, Rees HW, Daigle JL (1999) Effectiveness of terraces/grassed waterway systems for soil and water conservation: a field evaluation. J Soil Water Conserv 3:577–583

Colgan PM (2016) A brief history of the ravines. http://www.gvsu.edu/cms3/assets/56DCA.9CB-EC2F-5B8F...colgan_article.pdf. Accessed on 8 Apr 2017

Cooke RU, Reeves RW (1976) Arroyos and environmental change in the American South West. Oxford Research Studies in Geography, Calerandon Press, Oxford

Corona C, Saez JL, Rovera G, Stoffel M, Astrade L, Berger F (2011) High resolution quantitative reconstruction of erosion rates based on anatomical changes in exposed roots at Draix, Alpes de Haute-Provence—critical review of existing approaches and independent quality control of results. Geomorphology 125(3):433–444

Daba S (2003) An investigation of the physical and socioeconomic determinants of soil erosion in the Hararghe highlands, eastern Ethiopia. Land Degrad Dev 14(1):69–81

Daba S, Rieger W, Strauss P (2003) Assessment of gully erosion in eastern Ethiopia using photogrammetric techniques. Catena 50:273–291

Dagar JC (1986) Studies on root systems of some grasses growing in Kshipra ravines. J Indian Bot Soc 65(4):397–403

Dagar JC (1987a) Species composition and plant biomass of an ungrazed and a grazed grassland at Ujjain, India. Trop Ecol 28:208–215

Dagar JC (1987b) Responses of clipping treatments on nine palatable perennial grasses. Trop Ecol 28:216–221
Dagar JC (1987c) Studies on reclamation of Kshipra ravines. Indian J For 10(2):83–89
Dagar JC (1987d) Studies on root systems of some woody species growing in Kshipra ravines. Indian J Appl Pure Biol 2(1):17–21
Dagar JC (1989) Studies on root systems of some forbs growing in Kshipra ravines. Indian J Appl Pure Biol 4(1):35–47
Dagar JC (1995) Soil conservation values of some grasses and forbs. Proc Indian Natl Sci Acad B 61:163–170
Dagar JC (1999) Soil erosion parameters with special reference to Kshipra ravines in Madhya Pradesh. Indian J For 22(3):210–221
Dagar JC (2001) Biological spectrum and successional trends in Kshipra watershed areas. Indian J For 24(3):351–356
Dagar JC (2018) Perspectives of vegetation ecology and biodiversity for management of ravine lands. In: Dagar JC, Singh AK (eds) Ravine lands: greening for livelihood and environmental security. Springer, New Delhi
Dagar JC, Mall LP (1980) Studies on the vegetation of ravines and banks of the river Kshipra and its tributaries. J Indian Bot Soc 59:234–245
Dagar JC, Pathak PS (2005) Grassland dynamics and their management. Range Manag Agrofor 26(1):7–31
Dagar JC, Singh VP (1980) The ecological studies of the vegetation of river Kshipra and its tributaries. Biologia 25:43–62
Dagar JC, Singh VP, Mall LP (1978) Regeneration in plants growing on eroded soil. Ann Arid Zone 17:68–74
Dai Q, Liu Z, Shao H, Yang Z (2015) Karst bare slope soil erosion and soil quality: a simulation case study. Solid Earth 6(3):985–995
Das DC, Raghunath B, Sreenathan A (1967) Bench terracing and soil and water management in Nilgiris. J Indian Soc Agric Eng 4(2):49–59
De Baets S, Poesen J, Reubens B, Muys B, De Baerdemaeker J, Meersmans J (2009) Methodological frame work to select plant species for controlling rill and gully erosion: application to a Mediterranean ecosystem. Earth Surf Process Landf 34:1374–1392. https://doi.org/10.1002/esp.1826
Derose RC, Gomez B, Marden M, Trustrum N (1998) Gully erosion in Mangatu Forest, New Zealand, estimated from digital elevation models. Earth Surf Process Landf 23:1045–1053
Dhuru Narayana VV (1990) Land degradation due to water erosion. In: Abrol IP, Dhuruva Narayana VV (eds) Technologies for wasteland development. ICAR, New Delhi, pp 189–195
Dhuru Narayana VV, Babu R (1983) Estimation of soil erosion in India. J Irrig Drain Eng 109(4):419–434
Dotterweich M, Schmitt A, Schmidtchen A, Bork HR (2003) Quantifying historical gully erosion in northern Bavaria. Catena 50:135–150
Duley FL, Miller MF (1923) Erosion and surface runoff under different soil conditions. Res Bull 63:50. Univ. Mo. Agric. Exp. Sta. Columbia
Ellison WD (1944a) A rain drop, surface flow and erosion. Soil Conserv 10:128–129
Ellison WD (1944b) Studies on rain drop erosion. Agric Eng 25(131–136):181–182
Ellison WD (1947) Soil erosion studies. Part I–VII. Agric Eng 28:145–146. 197–201, 245–248, 297–300, 353, 402–405, 408, 442–444, 450
Erasmus LI, Madan-Lal, Raghunath B, Mathur PS (1970) Evaluation and erosion potential from rainfall data. Indian For 96:817–825
Erkossa T, Wudneh A, Desalegn B, Taye G (2015) Linking soil erosion to on-site financial cost: lessons from watersheds in the Blue Nile basin. Solid Earth 6(2):765–774
Faulkner H (2006) Piping hazard on collapsible and dispersive soils in Europe. In: Boardman J, Poesen J (eds) Soil erosion in Europe. Wiley, Chichester, pp 537–562
Faulkner H (2008) Connectivity as a crucial determinant of badland morphology and evolution. Geomorphology 100(1–2):91–103

Faulkner H, Alexander R, Wilson BR (2003) Changes to the dispersive characteristics of soils along an evolutionary slope sequence in the Vera badlands, southeast Spain: implications for site stabilisation. Catena 50:243–254
Flanagan DC, Nearing M (1995) USDA—water erosion prediction project hillslope profile and watershed model documentation, Indiana Report No. 10. National Soil Erosion Research Laboratory, West Lafayette
Frevert RK, Schwab GO, Edminister TW, Barnes KK (1955) Soil and water conservation engineering. Wiley, New York
Fu B, Gulinck H (1994) Land evaluation in an area of severe erosion: the Loess Plateau of China. Land Degrad Rehabil 5:33–40
Gabarron-Galeote MA, Martinez-Murillo JF, Quesasa MA, Ruiz-Sinoga JD (2013) Seasonal changes in the soil hydrological and erosive response depending on aspect, vegetation type and soil water repellency in different Mediterranean microenvironments. Solid Earth 4(2):497–509
Gábris GY, Kertész A, Zámbo L (2003) Land use change and gully formation over the last 200 years in a hilly catchment. Catena 50:151–164
Gallart F, Marignani M, Perez-Gallego N, Santi E, Maccherini S (2012) Thirty years of studies on badlands, from physical to vegetational approaches-a succinct review. Catena 106:4–11
Gardner G (1996) Shrinking fields: cropland loss in a world of eight billion. Wordwatch paper 131, Washington, DC, p 56. ISBN: 1-878071-33-5
Geyik MP (1986) FAO watershed management manual-gully control, FAO conservation guide 13/2. Food and Agriculture Organization of United Nations, Rome
Ghosh S, Guchhait SK (2016) Geomorphic threshold estimation for gully erosion in the lateritic soil of Birbhum, West Bengal, India. SOIL Discuss. https://doi.org/10.5194/soil-2016-2048
Gibling MR, Tandon SK, Sinha R, Jain M (2005) Discontinuity-bounded alluvial sequences of the southern Gangetic Plains, India: aggradation and degradation in response to monsoonal strength. J Sediment Res 75(3):369–385
Goswami PK, Mishra JK (2014) Morphotectonic evolution of the Piedmont Zone of the west Ganga Plain, India. Z Geomorphol 58(1):117–131. https://doi.org/10.1127/0372-8854/2013/0114
Govers G, Everaert W, Poesen J, Rauws G, De Ploey J, Lautridou JP (1990) A long flume study of the dynamic factors affecting the resistance of a loamy soil to concentrated flow erosion. Earth Surf Process Landf 15:313–328
Graf WL (1979) The development of montane arroyos and gullies. Earth Surf Process Landf 4:1–14
Gupta SK, Babu R (1967) Rainfall index of Dehradun. Indian For 93(10):734–738
Gupta RK, Bhardwaj SP, Mittal SP (1990) Management of degraded lands in the Himalayas. Abrol IP, Dhuruva Narayana VV Technologies for wasteland development, ICAR New Delhi, 196–213
Gutiérrez AG, Schnabel S, Felicísimo AM (2009) Modelling the occurrence of gullies in rangelands of southwest Spain. Earth Surf Process Landf 34:1894–1902
Haigh MJ (1984) Ravine erosion and reclamation in India. Geoforum 15(4):543–561
Haigh MJ (1998) Ravine erosion and reclamation in India. In: Pandey DC (ed) Managing agriculture for a better tomorrow: the Indian experience. MD Publications PVT LTD, New Delhi, pp 161–193
Hancock GR, Evans KG (2006) Gully position, characteristics and geomorphic thresholds in an undisturbed catchment in Northern Australia. Hydrol Process 20:2935–2951
Heede BH (1974) Stages of development of gullies in the western United States. Z Geomorphol 18:260–271
Howard AD (1997) Simulation of gully erosion and bistable landforms. In: Wang SSY, Langendoen EJ, Shields FD (eds) Proceeding of the conf. on management of landscapes disturbed by channel incision Center for Computational Hydroscience and Engineering. The University of Mississippi, Oxford, pp 516–521
Hughes AO, Prosser IP (2012) Gully erosion prediction across a large region: Murray–Darling Basin, Australia. Soil Res 50:267–277
Hui S, Mingan S (2000) Soil and water loess from the Loess Plateauin China. J Arid Environ 45:9–20

Iacob N, Laura C (2011) The gully erosion effect on the environment. In: Salampasis M, Matopoulos A (eds) Proceedings of the International Conference on information and communication technologies for sustainable agri-production and environment (HAICTA2011), Skiathos, 8–11 September, 2011. pp 621–629

Ionita I (2003) Hydraulic efficiency of the discontinuous gullies. Catena 50:369–379

Jain SK, Agarwal PK, Singh VP (2007) Hydrology and water resources of India. Springer, Dordrecht

Jain V, Tandon SK, Sinha R (2012) Application of modern geomorphic concepts for understanding the spatio-temporal complexity of the large Ganga River dispersal system. Curr Sci 103(11):1300–1319

James LA, Watson DG, Hansen WF (2007) Using LiDAR data to map gullies and headwater streams under forest canopy: South Carolina, USA. Catena 71(1):132–144

Janeau JL, Bricquet JP, Planchon O, Valentin C (2003) Soil crusting and infiltration on steep slopes in northern Thailand. Eur J Soil Sci 54(3):543–554

Jha VC, Kapat S (2009) Rill and gully erosion risk of lateritic terrain in south-western Birbhum district, West Bengal, India. Rev Soc Natureza 21(2):141–158

Jha VC, Kapat S (2011) Degraded lateritic soil scape and land uses in Birbhum district, West Bengal, India. Rev Soc Natureza 23(3):545–556

Joshi VU (2014) The Chambal Badlands. In: Kale VS (ed) Landscapes and landforms of India, 1st edn. Springer, Dordrecht, pp 143–149

Joshi VU, Daniels MJ, Kale VS (2013) Morphology and origin of valley-side gullies formed along the watersheds of Deccan Province, India and the rangeland of Colorado, USA. Trans Inst Indian Geogr 35(1):103–122

Kashyap SD, Dagar JC, Pant KS, Yewale (2014) Soil conservation and ecosystem stability: natural resource management through agroforestry in northwestern Himalayan region. In: Dagar JC, Singh AK, Arunachalam A (eds) Agroforestry systems in India: livelihood security & ecosystem services. Adv Agrofor 10:21–56

Kinnell PIA (1974) Splash erosion-some observations on the splash cup-technique. Soil Sci Soc Am Proc 38:657–660

Kirkby MJ, Bull LJ, Poesen J, Nachtergaele J, Vandekerckhove L (2003) Observed and modelled distributions of channel and gully heads—with examples from S.E. Spain and Belgium. Catena 50:415–434

Knisel WG (1980) CREAMS: a Field-Scale Model for chemicals, runoff and erosion from agricultural management systems. U.S. Department of Agriculture, Conservation Report No. 26, 640 pp

Knisel WG (1993) GLEAMS: groundwater loading effects of agricultural management systems. University of Georgia, Coastal Plains Experiment Station, Biological and Agricultural Engineering Department Publication No. 5. 260 pp

Laflen JM, Flangen DC (2013) The development of US soil erosion prediction and modelling. Int Soil Cons Res 1(2):1–11

Lal R (2003) Soil erosion and global carbon budget. Environ Int 29:437–450

Lang A, Bork HR (2006) Past soil erosion in Europe. In: Boardman J, Poesen J (eds) Soil erosion in Europe. Wiley, Chichester, pp 465–476

Li Y, Poesen J, Valentin C (eds) (2004) Gully erosion under global change, Proceeding of the second international symposium on Gully Erosion, Chengdu, China, May 2002. Sichuan Science and Technology Press, Chengdu, p 354

Lieskovsky J, Knderessy P (2014) Modelling the effect of vegetation cover and different tillage practices on soil erosion in vineyards: a case study in Vrable (Slovakia) using WATEM/SEDM. Land Degrad Dev 25(3):288–296

Luna RK (2006) Plantation forestry in India. Offset Printers and Publishers, Dehradun, pp 361–386

Lvovich MI, Karasik GY, Bratseva NL, Medvedeva GP, Maleshko AV (1991) Contemporary intensity of the world land intercontinental erosion. USSR Academy of Sciences, Moscow

Majumder A, Ghosh S, Dasgupta A, Seth D (2012) Analyzing reservoir sedimentation of Panchet dam, India using Remote Sensing and GIS. Panchakot Essays 2(3):82–95

Mandal D, Sharda VN (2013) Appraisal of soil erosion risk in the eastern Himalayan region of India for soil conservation planning. Land Degrad Dev 24(5):430–437
Martinez-Casasnovas JA (2003) A spatial information technology approach for the mapping and quantification of gully erosion. Catena 50:293–308
Martinez-Murillo JF, Nadal-Romero E, Regues D, Cerda A, Poesen J (2013) Soil erosion and hydrology of the western Mediterranean badlands throughout rainfall simulation experiments: a review. Catena 106:101–112
Marzolff I, Poesen J (2009) The potential of 3D gully monitoring with GIS using high-resolution aerial photography and a digital photogrammetry system. Geomorphology 111:48–60
Mathys N, Brochot S, Meunier M, Richard D (2003) Erosion quantification in the small marly experimental catchments of Draix (Alpes de Haute Provence, France). Calibration of the ETC rainfall-erosion model. Catena 50:527–548
Mathys N, Klotz S, Esteves M, Lapetite JM (2005) Runoff and erosion in the Black Marls of the French Alps, observations and measurements at the plot scale. Catena 63:261–281
McCloskey GL (2010) Riparian erosion morphology, processes and causes along the Victoria River, northern territory, Australia. Charles Darwin University, PhD Thesis. p 200
Mekonnen M, Keesstra SD, Baartman JE, Ritsema CJ, Melesse AM (2015) Evaluating sediment storage dams: structural off-site sediment trapping measures in northwest Ethiopia. Cuad Investig Geogr 41:7–22
Meng Q (ed) (1996) Soil and water conservation in the Loess Plateau, vol 75. Water Resource Press of Yellow River, Beijing, pp 316–317. (in Chinese)
Middleton N, Thomas D (1997) World atlas of desertification, 2nd edn. Arnold Pub, London, p 182
Miller MF (1926) Waste through soil erosion. J Am Soc Agric 18:153–160
Milliman JD, Syvitski JPM (1992) Geomorphic/tectonic control of sediment discharge to the ocean: the importance of small mountainous rivers. J Geol 100:325–344
Mircea S (1999) Evolution study of gully erosion formations in terms of arrangement and not arrangement in the Mishra MN, Vishwakarma LL (1999) Morphotectonics of the Chambal and the Yamuna valleys in the western marginal gangetic alluvial plains. Geological Survey of India, New Delhi. http://www.portal.gsi.gov.in/pls/gsipub/PKGPTLPORTALLINKS.pGetCaseStudyRegion?inpRegionId=35
Mishra MN, Vishwakarma LL (1999) Morphotectonics of the Chambal and the Yamuna valleys in the western marginal gangetic alluvial plains. Geological Survey of India, New Delhi. http://www.portal.gsi.gov.in/pls/gsipub/PKG PTLPORTAL LINKS.pGetCaseStudyRegion?inpRegionId=35
Moeyersons J (2003) The topographic thresholds of hillslope incisions in southwestern Rwanda. Catena 50:381–400
Morgan RPC (2005) Soil erosion and conservation. Wiley-Blackwell, New York
Morgan RPC, Mngomezulu D (2003) Threshold conditions for initiation of valley-side gullies in the Middle Veld of Swaziland. Catena 50:401–414
Motoc M, Vatau A (1992) Indicators of soil erosion. In: Experimental review, vol 3, no. 3, Bucharest, Romania
Nachtergaele J, Poesen J (1999) Assessment of soil losses by ephemeral gully erosion using high-altitude (stereo) aerial photographs. Earth Surf Process Landf 24:693–706
Nachtergaele J, Poesen J, Steegen A, Takken I, Beuselinck L, Vandekerckhove L, Govers G (2001a) The value of a physically based model versus an empirical approach in the prediction of ephemeral gully erosion for loess-derived soils. Geomorphology 40:237–252
Nachtergaele J, Poesen J, Vandekerckhove L, Oostwoud Wijdenes D, Roxo M (2001b) Testing the ephemeral gully erosion model (EGEM) for two Mediterranean environments. Earth Surf Process Landf 26:17–30
Nachtergaele J, Poesen J, Oostwoud Wijdenes D, Vandekerckhove L (2002) Medium-term evolution of a gully developed in a loess-derived soil. Geomorphology 46(3–4):223–239
Nearing MA, Foster GR, Lane LJ, Finkner SC (1989) A process based soil erosion model for USDA Water Erosion Prediction Project. Trans ASAE 32(5):1587–1593

Nogueras P, Burjachs F, Gallart F, Puigdefabregas J (2000) Recent gully erosion in El Cautivo badlands (Tabernas, SE Spain). Catena 40(2):203–215

Novara A, Keesstra S, Cerdà A, Pereira P, Gristina L (2016) Understanding the role of soil erosion on CO2-C loss using $^{13}C$ isotopic signatures in abandoned Mediterranean agricultural land. Sci Total Environ 550:330–336

Oldeman LR (1994) The global extent of soil degradation. In: Greenland DJ, Szabolcs I (eds) Soil resilience and sustainable land use. CAB International, Wallingford, pp 99–118

Ollier CD, Brown MJF (1971) Erosion of a young volcano in New Guinea. Z Geomorphol 15:12–28

Oostwoud-Wijdenes DJ, Bryan RB (2001) Gully-head erosion processes on a semi-arid valley floor in Kenya: a case study into temporal variation and sediment budgeting. Earth Surf Process Landf 26:911–933

Oostwoud-Wijdenes DJ, Poesen J, Vandekerckhove L, Ghesquiere M (2000) Spatial distribution of gully head activity and sediment supply along an ephemeral channel in a Mediterranean environment. Catena 39:147–167

Øygarden L (2003) Rill and gully development during an extreme winter runoff event in Norway. Catena 50:217–242

Pani O (2016) Controlling gully erosion: an analysis of land reclamation processes in Chambal valley. Dev Pract 26(8):1047–1059

Pani P, Mohapatra SN (2001) Delineation and monitoring of gullied and ravenous lands in a part of Lower Chambal Valley, India using remote sensing and GIS, vol 1. ACRS. 2001, Singapore, pp 671–675

Pathak PS, Dagar JC (2015) Indian grasslands and their management. In: Ghosh PK, Mahanta SK, Singh JB, Pathak PS (eds) Grassland: a global resource perspective. Range Management Society of India, Jhansi, pp 3–36

Pimentel D, Harvey C, Resosudarmo P, Sinclair K, Kurz D, McNair M, Crist S, Shpritz L, Fitton L, Saffouri R, Blair R (1995) Environmental and economic costs of soil erosion and conservation benefits. Science 267:1117–1123

Poesen J (2011) Changes in gully erosion research. Landf Anal 17:5–9

Poesen J, Valentin C (2003) Gully erosion and global change. Proc. first international symposium on Gully Erosion, Leuven, Belgium, April 2000. Catena 50(2–4):87–562

Poesen J, Nachtergaele J, Verstraeten G, Valentin C (2003) Gully erosion and environmental change: importance and research needs. Catena 50(2–4):91–133

Poesen J, Torri D, Vanwalleghem T (2011) Gully erosion: procedures to adopt when modelling soil erosion in landscapes affected by gullying. In: Morgan RPC, Nearing MA (eds) Handbook of erosion modelling. Blackwell Publishing Ltd, Chichester. Chapter 19: 360–386 +Plates 16–19

Prasad SN, Römkens MJM (2003) Energy formulations of head cut dynamics. Catena 50:469–488

Prasad SN, Römkens MJM (2004) Mechanic energy and subsurface hydrologic effect in head-cut processes. In: Li Y, Poesen J, Valentin C (eds) Gully erosion under global change. Sichuan Science and Technology Press, Chengdu, pp 109–120

Prosdocimi M, Cerdà A, Tarolli P (2016) Soil water erosion on Mediterranean vineyards: a review. Catena 141:1–21

Ranga V, Mohapatra SN, Pani P (2015a) Geomorphological evolution of badlands based on the dynamics of palaeo-channels and their implications. J Earth Syst Sci 124(5):909–920

Ranga V, Van Rompaey A, Poesen J, Mohapatra SN, Pani P (2015b) Semi-automatic delineation of Badlands using contrast in vegetation activity: a case study in the lower Chambal valley, India. Geocarto Int. https://doi.org/10.1080/10106049.2015.1004130

Reich P, Eswaran H, Beinroth F (2001) Global dimensions of vulnerability to wind and water erosion. In: Stot D, Mohtar RH, Steinhardt GC (eds) Sustaining the global farm. Selected papers from the 10th International Soil Cons Organisation Meeting at Purdue Univ & USDA ARS Erosion Res Lab. pp 838–846

Rey F (2003) Influence of vegetation distribution on sediment yield in forested marly gullies. Catena 50:549–562

Ries JB, Marzolff I (2003) Monitoring of gully erosion in the central Ebro Basin by large scale aerial photography taken from a remotely controlled blimp. Catena 50:309–328

Robinson KM, Hanson GJ (1996) Gully headcut advance. Trans ASAE 39(1):33–38

Römkens MJM, Bennett SJ (2005) Gully erosion. Proc. third international symposium on Gully Erosion, Oxford, MS, USA, April 2004. Int J Sediment Res 20(3):157–280

Saey T, Simpson D, Vitharana UWA, Vermeersch H, Vermang J, Van Meirvenne M (2008) Reconstructing the paleo-topography beneath the loess cover with the aid of an electromagnetic induction sensor. Catena 74:58–64. https://doi.org/10.1016/j.catena.2008.03.007

Saez JL, Corona C, Stoffel M, Rovera G, Astrade L, Berger F (2011) Mapping of erosion rates in marly badlands based on a coupling of anatomical changes in exposed roots with slope maps derived from LiDAR data. Earth Surf Process Landf 36(9):1162–1171

Sampson AW, Weyl LH (1918) Range preservation and its relation to erosion control on western grazing lands, Bull 675. USDA, Washington, DC, p 35

Schumm SA, Harvey MD, Watson C (1984) Incised channels: morphology, dynamics and control. 200 Water resource Publication, High-land Ranch, Colorado

Schwab GO, Frevert RK, Edminister TW, Barnes KK (1981) Soil and water conservation engineering, 3rd edn. Wiley, New York

Sethy BK, Parandiyal AK, Ali S, Kumar A, Singh RK (2011) Cost effective conservation measures for management of medium and deep ravenous lands, Annual Report. CSWCRTI, Dehradun, pp 73–74

Sharda VN, Dogra P (2013) Assessment of productivity and monetary losses due to water erosion in rainfed crops across different states of India for prioritization and conservation planning. Agric Res 2(4):382–392

Sharda VN, Dogra P, Prakash C (2010) Assessment of production losses due to water erosion in rainfed areas of India. J Soil Water Conserv 65(2):79–91

Sharma HS (1980) Ravine erosion in India. Concept Publishing Co, New Delhi, pp 1–96

Shellberg J (2011) Alluvial gully erosion rates and processes across the Mitchell River fluvial megafan in Northern Queensland, Australia. Griffith University, PhD thesis, Brisbane, Australia, November 2011, 283pp

Shellberg J, Brooks A (2012) Alluvial gully erosion: a dominant erosion process across tropical Northern Australia. Australian Rivers Institute, Griffith University, Tropical Rivers Coastal Knowledge (TRaCK), Darwin, p 18

Shellberg J, Brooks A, Spencer J (2010) Land-use change from indigenous management to cattle grazing initiates the gullying of alluvial soils in northern Australia. Soil Solutions for a Changing World, 19th World Congress of Soil Science, 1–6 August 2010. Brisbane, Australia, pp 9–62. http://www.mitchell-river.com.au/news-and-issues/MitchellCattleAnd%20ErosionShellberg.pdf

Shellberg JG, Brooks AP, Spencer J, Ward D (2012) The hydrogeomorphic influences on alluvial gully erosion along the Mitchell River fluvial megafan, northern Australia. Hydrol Process. https://doi.org/10.1002/hyp.9240

Shellberg JG, Brooks AP, Rose CW (2013) Sediment production and yield from an alluvial gully in northern Queensland, Australia. Earth Surf Process Landf. https://doi.org/10.1002/esp.3414

Sheridan D (1981) Desertification in the United States. Council on Environmental Quality, Washington, DC

Shit PK, Maiti R (2012) Mechanism of gully-head retreat – a study at Ganganir Danga, Paschim Medinipur, West Bengal. Ethiop J Environ Stud Manag 5(4):332–342

Shit PK, Maiti R (2014) Gully erosion control – lateritic soil 1 region of West Bengal. Indian Sci Cruiser 28(3):54–61

Shit PK, Bhunia GS, Maiti R (2014) Morphology and development of selected badlands in South Bengal (India). Indian J Geogr Environ 13:161–171

Sidorchuk A (1999) Dynamic and static models of gully erosion. Catena 37:401–414

Sidorchuk A (2005) Stochastic components in the gully erosion modelling. Catena 63:299–317

Sidorchuk A, Grigorév V (1998) Soil erosion on the Yamal Peninsula (Russian Arctic) due to gas field exploitation. Adv GeoEcol 31:805–811

Sidorchuk A, Märker M, Moretti S, Rodolfi G (2003) Gully erosion modelling and landscape response in the Mbuluzi River catchment of Swaziland. Catena 50:507–525

Singh G, Babu R, Narain P, Bhushan LS, Abrol IP (1992) Soil erosion rates in India. J Soil Water Conserv 47(1):97–99
Smith DD (1941) Interpretation of soil conservation data for field use. Agric Eng 22(5):173–175
Sobolev SS (1948) Development of erosion processes in the European USSR and their control (in Russian). Izvestiya an SSSR, Moscow
Souchére V, Cerdan O, Ludwig B, Le Bissonnais Y, Couturier A, Papy F (2003) Modelling ephemeral gully erosion in small cultivated catchments. Catena 50:489–505
Stewart G, Forsling CL (1931) Surface runoff and erosion in relation to soil and plant cover on high grazing lands of Central Utah. J Am Soc Agron 23:815–832
Strunk H (2003) Soil degradation and overland flow as causes of gully erosion on mountain pastures and in forests. Catena 50:185–198
Tandon SK, Gibling M, Sinha R, Singh V, Ghazanfari P, Dasgupta A (2006) Alluvial valleys of the Gangetic Plains, India: causes and timing of incision. In: Dalrymple RW, Leckie DA, Tillman RW (eds) Incised Valley Tulsa, vol 85. pp 15–35
Tejwani KG, Dayal R, Verma B, Gidwani HM, Gupte SC (1975a) Classification, control and reclamation of ravine lands. In: Tejwani KG, Gupta SK, Mathur HN (eds) Soil and water conservation research (1956–1971). ICAR Pub, New Delhi, pp 254–310
Tejwani KG, Gupta SK, Mathur HN (eds) (1975b) Soil and water conservation research (1956–1971). ICAR Pub, New Delhi, p 358
Thomas AW, Welch R, Jordan TR (1986) Quantifying concentrated-flow erosion on cropland with aerial photogrammetry. J Soil Water Conserv 4:249–252
Thomas JT, Iverson NR, Burkart MR, Kramer LA (2004) Long-term growth of a valley-bottom gully, western Iowa. Earth Surf Process Landf 29:995–1009
Torri D, Borselli L (2003) Equation for high-rate gully erosion. Catena 50:449–467
Uthappa AR, Chavan SB, Ramesha MN, Kala S, Kumar V, Handa AK, Chaturvedi OP (2018) Plant biodiversity of ravine ecosystem: may open new vistas for enhancing productivity. In: Dagar JC, Singh AK (eds) Ravine lands: greening for livelihood and environmental security. Springer, New Delhi
Valcárcel M, Taboada MT, Paz A, Dafonte J (2003) Ephemeral gully erosion in Northwestern Spain. Catena 50:199–216
Valentin C, Poesen J, Li Y (2005) Gully erosion: a global issue. Proc. second international symposium on Gully Erosion, Chengdu, China, May 2002. Catena 63(2–3):129–330
Vandekerckhove L, Poesen J, Oostwoud Wijdenes D, Nachtergaele J, Kosmas C, Roxo MJ, De Figueiredo T (2000) Thresholds for gully initiation and sedimentation in Mediterranean Europe. Earth Surf Process Landf 25:1201–1220
Vandekerckhove L, Muys B, Poesen J, De Weerdt B, Coppé N (2001a) A method for dendrochronological assessment of medium-term gully erosion rates. Catena 45:123–161
Vandekerckhove L, Poesen J, Oostwoud Wijdenes D, Gyssels G (2001b) Short-term bank gully retreat rates in Mediterranean environments. Catena 44:133–161
Vandekerckhove L, Poesen J, Govers G (2003) Medium-term gully headcut retreat rates in Southeast Spain determined from aerial photographs and ground measurements. Catena 50:329–352
Vanwalleghem T, Bork HR, Poesen J, Schmidtchen G, Dotterweich M, Bork H, Deckers J, Brusch B, Bungeneers J, De Bie M (2005a) Rapid development and infilling of a historical gully under cropland, central Belgium. Catena 63:221–243
Vanwalleghem T, Poesen J, Van Den Eeckhaut M, Nachtergaele J, Deckers J (2005b) Reconstructing rainfall and land use conditions leading to the development of old gullies. The Holocene 15(3):378–386
Verachtert E, Van Den Eeckhaut M, Poesen J, Deckers S (2009) Characteristics and distribution of soil piping erosion in loess-derived soils of Belgium. In: Congreso Internacional sobre Desertificacion
Verma B (1981) Twenty classification of ravines for optimum utilization. Pub. Central Soil & Water Conservation Research & Training Institute, Research Centre, Kota

Verma B, Bhushan LS (1986) Management of gullied watersheds. Indian J Soil Conserv 14(3):105–110
Verma B, Chinnamani S (1990) Rehabilitation of ravine lands. In: Abrol IP, Dhuruva Narayana VV (eds) Technologies for wasteland development. ICAR, New Delhi, pp 263–272
Verma SK, Singh A (2015) Chambal ravines, perspectives of formations and management options. In: Verma SK, Trivedi SK, Singh A, Tomar AS (eds) Management of degraded lands and soil health. Biotech Books, New Delhi, pp 31–46
Verma B, Chinamani S, Bhola SN, Rao DH, Prasad SN, Prakash C (1986) Twenty five years of research on soil and water conservation in ravine lands of Rajasthan. Central Soil & Water Conservation Research & Training Institute Pub. Research Centre, Kota, p 215
Verma SK, Trivedi SK, Singh A, Tomar AS (eds) (2015) Management of degraded lands and soil health. Biotech Books, New Delhi, p 249
Verma SK, Singh AK, Kandalkar VS, Ranade DH, Bhadauria UPS, Singh A, Srivastava SC, Tomar AS (eds) (2016) Global ravine conference on managing ravines for food and livelihood security-souvenir. Rajmata Vijayaraje Scindia Krishi Vishwa Vidyalaya Gwalior, Madhya Pradesh, p 147
Walling DE (1983) The sediment delivery problem. J Hydrol 65:209–237
Walling DE, Webb BW (1983) Patterns of sediment yield. In: Gregory KJ (ed) Background to paleohydrology. Wiley, Chichester, pp 69–100
Walling DE, Webb BW (1996) Erosion and sediment yield: a global overview. Erosion and sediment yield: global and regional perspectives. Proc. exeter symp, July 1996. IAHS Publ 236:3–19
Weidelt HJ (1976) Manual of reforestation and erosion control. German Agency for Technical Cooperation, Ltd, Eshborn, p 569
Wischmeier WH (1959) A rainfall erosion index for a universal soil loss equation. Soil Sci Soc Am Proc 23:247–249
Wischmeier WH, Smith DD (1965) Predicting rainfall erosion losses from cropland east of the Rocky Mountains. USDA Handbook No. 282. Washington, DC
Wischmeier WH, Smith DD (1978) A guide to conservation planning. USDA Handbook 537. Washington, DC
Woodward DE (1999) Method to predict cropland ephemeral gully erosion. Catena 37:393–399
Xiang-Zhou X, Hong-wu Z, Ouyang Z (2004) Development of check-dam systems in gullies on the Loess Plateau, Chona. Environ Sci Pol 7:79–86
Yongqiu W, Hong C (2005) Monitoring of gully erosion on the Loess Plateau of China using a global positioning system. Catena 63:154–166
Young A (1989) Agroforestry for soil conservation. CAB International, Wallingford
Yu B, Neil DT (1993) Long-term variations in regional rainfall in the south-west of Western Australia and the difference between average and high intensity rainfalls. Int J Climatol 13:77–88
Zhao G, Mu X, Wen Z, Wang F, Gao P (2013) Soil erosion, conservation and eco-environment changes in the loess plateau of China. Land Degrad Dev 24(5):499–510
Zhu Y, Cai Q (2004) Rill erosion processes and its factors in different soils. In: Li Y, Poesen J, Valentin C (eds) Gully erosion under global change. Sichuan Science and Technology Press, Chengdu, pp 96–108

# Perspectives of Vegetation Ecology and Biodiversity for Management of Ravine Lands

3

Jagdish Chander Dagar

**Abstract**

Importance of vegetation ecology in the abatement of soil erosion and thereby in stabilizing the gullies and ravines has been well known and widely advocated. The indigenous woody vegetation and perennial grasses are considered generally more effective in halting runoff and erosion than cultivated crops and exotics. The native species have adapted themselves against all edaphic and biotic pressure and have capacity to regenerate along eroded slopes. The phytoclimate of the watersheds of rivers having ravines indicates to be therophytic showing biotic pressure of over-exploitation and overgrazing. But at the same time, there are quite a good number of phanerophytes showing that the watershed areas favour the luxuriant growth of both phanerophytes and therophytes provided the over-exploitation of woody vegetation and overgrazing are stopped. Successional trend also indicates that the ravine lands are facing tremendous biotic pressure and the vegetation is a biotic climax. After closing to grazing, it will lead to dry deciduous or savannah forest which consequently will offer more effective check against erosion. Along with herbage cover, the underground parts mainly thin roots also play an important role in retaining the soil and moisture, infiltration of nutrient cycling and ameliorating the soil. This all depends upon the configuration, density and the strength of the root system. The extent of root systems and their soil-binding capacity vary from plant to plant. Therefore, the ecological studies of vegetation including root systems help in understanding the role of different species and plant communities in conserving the soil in watershed areas including ravines and consequently silting of reservoirs, canals, streams and rivers. In the

J. C. Dagar (✉)
Indian Council of Agricultural Sciences, Krishi Anusandhan Bhavan-II, Pusa, New Delhi, India





J. C. Dagar, A. K. Singh (eds.), *Ravine Lands: Greening for Livelihood and Environmental Security*, https://doi.org/10.1007/978-981-10-8043-2_3

present chapter, an attempt has been made to explore the ecological perspectives of vegetation and terrestrial biodiversity in controlling the ravine formation by stabilizing the soil and protecting the biodiversity for its potential and sustainable use for livelihood and environmental security.

## 3.1 Introduction

Importance of vegetation in the abatement of soil erosion and thereby in stabilizing the gullies and ravines has been widely advocated (Dagar and Mall 1980; Stocking 1984; Young 1989; Ambasht and Shankar 1993; Dagar 2001; Gallart et al. 2012). Natural shrubs and perennial grasses have been considered more effective in checking runoff and erosion than the arable crops (Anwar et al. 1989). Along with aboveground herbage cover, the underground parts also play an important role in retaining the soil intact, infiltration of nutrients and anchorage of plants. Beside these values, the vital contribution of plant roots includes absorption of soil water, initiation of nutrient cycling, supporting structure for photosynthetic material essential for energy capture and transfer, storage of food material, perennation, propagation, flood reduction, usable water yield, and soil development and stability. Different plant species have different soil and moisture conservation values depending upon their structure, anatomy, adaptation capacity and above all root ramification (Ambasht et al. 1988; Dagar 1986, 1987a, b, c, d, 1989, 1995; Kumar et al. 1990). In the past, some workers attempted the vegetational analysis of the riverbanks (Ambasht 1968; Dagar and Singh 1980), terraces (Gupta 1961; Rao 1965, 1969; Murthy 1971; Pandeya et al. 1972) and ravines (Misra 1944; Shah 1969; Dagar 1978; Dagar and Mall 1980). Most of these studies were of floristic and ecological in nature. Mueller-Dombois and Ellenberg (1974) coined the term "vegetation ecology" and defined it as the study of both the structure of vegetation and vegetation systematics, which includes the investigation of species composition and the sociological interaction of species in communities. Vegetation ecology is very much concerned with correlations between the environment and vegetation and with the causes of community formation. To select suitable species for plantation in ravine or other degraded lands, it is imperative to evaluate the exact distribution of various species, their relationship with soil, root systems and capacity of adaptation to the ravine-like hostile environment. Terrestrial animal biodiversity also plays important role in the life of people leaving in the vicinity of ravine lands. In the present chapter, an attempt has been made to explore the ecological perspectives of vegetation in controlling the ravine formation by stabilizing the soil and protecting the biodiversity for its potential and sustainable use for livelihood and environmental security.

## 3.2 General Features of Areas of Ravine Formation

### 3.2.1 Geology and Soil

As discussed earlier in detail (Chap. 2), about 1 billion hectares of land is eroded by water and the most widespread form of water erosion is the loss of the topsoil on 920 million hectares (Mha), while terrain deformation (rills and gullies) occurs on 175 Mha (Oldeman 1994). Ravine and gully erosion threatens vast tracts of the world's agricultural land and contributes to overall soil loss and sediment production at various temporal and spatial scales under different climatic and land use conditions. The damage is greatest in the alluvial plains of the semiarid and arid zones and most serious where it threatens precarious subsistence-oriented agricultural systems in the developing world. The ecological conditions such as climate, soil and geology vary from region to region. Some of the examples have already been discussed in different regional chapters in this book.

In India, the Chambal ravines are one of the worst kinds of land degradation. The region experiences sub-tropical climate where hot winds during summer flow for a greater part of the day and night temperature remains high. The hottest months are May and June (mercury touches 48 °C), and temperature drops considerably in the last week of June. The high wind velocity and temperature during summer desiccate all existing vegetation. December and January are the coldest months, and minimum temperature reaches freezing point during night. Winter rains are erratic and irregular. The average rainfall of Chambal Division is 891 mm, and majority of it is received during July to September. High-intensity rain storms (which are more frequent during last three decades) enhance land degradation due to erosion particularly where the surface is devoid of vegetation.

The region is of great geological significance and lies near the junction of two different geological systems, namely, Vindhyan system (geologically very old system belonging to the Paleozoic era) and the Aravalis system. Geologically, the Chambal Basin is a junction between the north-western lobe of the Vindhyan Basin and the south-eastern fringe of the Aravalli Range in the sense that here the younger formations of the Vindhyan period have truncated and faulted against the Gwalior formation. The north-western part comprises the intensely folded Archean formations, while the south-eastern part is composed of horizontally reposing Vindhyan system. The Vindhyans of the southern part are covered by the Deccan Trap. The lower Chambal Valley forms the extreme north-eastern portion of the Chambal Basin. So, the valley falls to the southeast of the great boundary fault (www.rajasthaninfoline.comlrinfolminerals.htm).

The Gwalior group of the valley is characterized by genesis, schists, shales, haematite slates, quartzites, jasper, beds and dolerites. It occupies most of the area in the north-west along the main fault. The Vindhyans form the border hills of Bundi district north of the Chambal River, the surface of the Dangland north-east of Banas water gap, the Kota plateau near the Chambal canyon in the south-west and the circular dome of Ramgarh between the Parbati and Kul rivers in the south-eastern portion of the Valley. The lower Vindhyan formations in the lower Chambal Valley are found between the two main faults in the north-east. The lowest beds are exposed in the double syncline. In the lower Chambal Valley, the Kaimur conglomerate is the interrupting rock formation, which separates the upper Vindhyans from the lower by well-marked unconformity. The upper Vindhyans are predominantly composed of stratified shales, sandstones and limestones. They are found in disconnected areas, which are Kaimur group, Rewa group, Bhander group, etc.

Granite, basalt and rhyolite are the types of igneous rocks which are found in upstream region of the Chambal Basin. Sandstone, shales and limestone are mainly found near Gandhi Sagar, Rawatbhata Atomic Power Plant, Kota and southeast region of the Chambal Basin. The main tributaries which confluence Chambal in MP are Kshipra and Choti Kali Sindh and in Rajasthan are Kalisindh, Parbati and Banas (www.rajirrigation.gov.in). The rock types throughout the Kshipra Valley are the basalts or Deccan Trap with no fossiliferous intertrappean bed in this series. Lavas and ash beds of the area form thicker part (1219 m) of the Deccan Trap series, and it cannot be older than Danian stage of Upper Cretaceous (Wadia 1961). Types of sedimentary rocks which dominate the region of three main tributaries (Kalisindh, Parbati and Banas) of the Chambal River are alluvium, blown sand and laterite. Amphibolite, epidiorite and unclassified crystal lines are localized in the western parts of Rajasthan. Quartzite and schists which are the metamorphic forms of sandstone lie comfortably on sedimentary rocks near Gandhi Sagar. Iron ore deposits of economic interest are in Zawar, Pali, Bhilwara, Gwalior, Gandhi Sagar, etc. Famous zinc mines are located at Dariba, Zawarmala (Zawar mines) and Rampura aghucha (opencast) near Bhilwara. Lead and silver are also found in some places of Bhilwara, Udaipur, Rajasmand, Gwalior and Jhansi. Rajasthan produces 89% of the country's asbestos, and the region lies near Pali, Ajmer, Bhilwara, Sirohi, Dungarpur, etc. Region where granite and emerald are found is mainly located in Rajasthan (Bhilwara and Tonk). Rajasthan is well known for exporting jewels like diamond, emerald, granite, etc. The largest copper smelters in the country are based in Rajasthan. The Khetri area is especially rich in copper ores. These were extracted, melted and then made into different kinds of implements.

Extensive deposits of limestone are known in the geological formations of the Aravalli Range in Rajasthan. Madhya Pradesh and Rajasthan are the largest producers of limestone in the country. Rich limestone deposits at Gotan in Rajasthan are best available in the country for white cement. High-grade limestone and chemical-grade limestone deposits are also found in abundance. Several bands of crystalline limestone varying in composition from high calcium to dolomite and siliceous type are found near Sawai Madhopur and Kota. The total reserves of gypsum in the state are estimated at 1070 million tonnes. The important localities are Bhadwasi (Nagaur), Kota and Tonk. The manganiferous formations are associated with orthoquartzite limestone and quartzite. The important manganese deposits are located between Ratlam, Banswara, Chittorgarh and Beawar (www.rajirrigation.gov.in).

Soils are sandy in texture, low in organic carbon and low to very low in nitrogen and phosphorus contents (Table 3.1). No consistency is found in sulphur and potassium content in different layers. The water-dispersive aggregates show that the soils of ravines have very weak-binding properties.

**Table 3.1** Physicochemical properties of soil of Chambal ravines (mean of three profiles)

| | Mechanical analysis | | | | | Available nutrients (kg $ha^{-1}$) | | | |
|---|---|---|---|---|---|---|---|---|---|
| Soil depth (m) | Sand (%) | Silt (%) | Clay (%) | Dispersion ratio | Organic carbon (%) | N | P | K | S |
| 0.0–0.5 | 52.9 | 18.0 | 29.0 | 0.88 | 0.20 | 95.0 | 9.8 | 317 | 12.0 |
| 0.5–1.0 | 60.0 | 20.7 | 19.0 | 0.86 | 0.16 | 77.6 | 2.8 | 298 | 9.7 |
| 1.0–1.5 | 55.7 | 11.7 | 32.6 | 0.78 | 0.21 | 99.8 | 6.7 | 285 | 9.6 |
| 1.5–2.0 | 48.3 | 22.0 | 29.7 | 0.77 | 0.19 | 83.9 | 3.8 | 320 | 11.7 |
| 2.0–2.5 | 47.6 | 24.7 | 27.7 | 0.79 | 0.17 | 81.0 | 7.3 | 350 | 13.8 |
| 2.5–3.0 | 54.3 | 22.7 | 22.7 | 0.88 | 0.11 | 52.3 | 5.5 | 297 | 14.6 |
| 3.0–3.5 | 48.1 | 23.4 | 29.4 | 0.89 | 0.16 | 76.3 | 3.5 | 362 | 11.9 |
| 3.5–4.0 | 60.8 | 14.1 | 25.0 | 0.92 | 0.18 | 83.9 | 6.2 | 390 | 14.0 |
| 4.0–4.5 | 55.9 | 23.0 | 20.1 | 0.90 | 0.19 | 99.0 | 7.1 | 428 | 7.0 |
| 4.5–5.0 | 49.5 | 17.7 | 32.7 | 0.88 | 0.20 | 95.0 | 6.9 | 314 | 10.9 |

Source: Modified from Verma and Singh (2015)

The percentage of silt and clay fractions of soil is more in disperse condition (dispersion ratio between 0.77 and 0.92) in all soil profiles and depths. More soil aggregates (50–70%) are in finer size, and only 8–14% are in coarse size. The recorded value of zeta potential (mv) showed a value of only −40 to −45 indicating very poor presence of colloidal fraction (Verma and Singh 2015).

### 3.2.2 Terrestrial Animal Biodiversity

Biodiversity inventories serve as repositories of baseline information on species occurrences, biogeography and their conservation status (Chandra and Gajbe 2005). In turn, they are essential tools for developing our knowledge and understanding of biodiversity and often the first step to undertake effective conservation action. This information is also fundamental to assess changes in species composition and distribution (Abraham 2011) in a timescale. As such very limited information is available regarding animal biodiversity of ravines; however, based on their field survey and consultation of literature, Nair and Chitanya-Krishna (2013) updated checklist of vertebrate fauna of the Chambal River Basin in north-central India with an emphasis on the National Chambal Sanctuary (lying between 24°55′–26°50′N and 75°34′–79°18′E) covering nearly 1800 km$^2$ across Rajasthan, Madhya Pradesh and Uttar Pradesh states. A total of 147 fish (32 families), 56 reptile (19 families), 308 bird (64 families) and 60 mammal (27 families) species are reported, which include 6 critically endangered, 12 endangered and 18 vulnerable species, as categorized by the IUCN Red List of Threatened Species (IUCN 1998, 2011). Much of the basin has been influenced by a long history of anthropogenic influences, chiefly in the form of sand mining; bankside cultivation; domestic activities like bathing, washing and water collection; hydrological modifications; fishing; poaching; livestock herding; grass soaking; river crossing; and temple fairs. As such, from this list only a limited number of species visit ravine areas, but still quite sizeable number of birds, reptiles and mammals can be seen at least the ravines near the river. Thomas et al. (2011) mentioned the faunal biodiversity of Piprai village ravines in Morena district consisting of 26 animal species including 5 domestic and 21 wild species. The domestic species included four mammals, viz. buffalo (*Bison bison*), cow (*Bos taurus*), goat (*Capra aegagrus*) and sheep (*Ovis aries*), and one bird, hen (*Gallus gallus domesticus*). Among wild species seven are listed as mammals, namely, deer (*Axis axis*), fox (*Valpes benghalensis*), hare (*Lepus ruficaudatus*), langur (*Presbytis entellus*), nilgai (*Boselaphus tragocamelus*) and sambar (*Rusa unicolor*). They listed 14 species of avian fauna which included common myna (*Acridotheres tristis*), peacock (*Pavo cristatus*), house sparrow (*Passer domesticus*), bater or common quail (*Coturnix coturnix*), grey partridge (*Francolinus pondicerianus*), rock pigeon (*Columba livia*), red junglefowl (*Gallus gallus*), brown fish owl (*Bubo zeylonensis*), king fisher or Indian roller (*Coracias benghalensis*), Asian koel (*Eudynamys scolopacea*), house crow (*Corvus splendens*), papiya or Jacobin cuckoo (*Clamator jacobinus*), golden sparrow (*Passer* sp.) and rose-ringed parakeet (*Psittacula krameri*). Most of the reptiles and mammals except those which are restricted to water have also relevance with ravine fauna; therefore, the checklist reported by Nair and Chitanya-Krishna (2013) has been listed here (Table 3.2) for those who are interested in biodiversity of ravines. This checklist of reptile and mammal fauna is of great ecological significance for utilization of ravine lands for livelihood security. One rare species Hardwicke's spiny-tailed lizard (*Saara hardwickii*) found in ravines of Gujarat and sometimes in Rajasthan is overexploited for its aphrodisiac properties.

**Table 3.2** Checklist of reptiles and mammals of the Chambal River Basin

| S. no. | Species | Common name | IUCN status (2009–2012) |
|---|---|---|---|
| Reptiles | | | |
| 1 | *Ahaetulla nasuta* | Common vine snake | |
| 2 | *Amphiesma stolatum* | Buff striped keelback | – |
| 3 | *Argyrogena fasciolata* | Banded racer | – |
| 4 | *Batagur dhongoka* | Three-striped roofed turtle | EN |
| 5 | *Batagur kachuga* | Red-crowned roofed turtle | CR |
| 6 | *Boiga trigonata* | Common cat snake | LC |
| 7 | *Bungarus caeruleus* | Common krait | – |
| 8 | *Calotes versicolor* | Oriental garden lizard | – |
| 9 | *Chamaeleo zeylanicus* | Indian chameleon | – |
| 10 | *Chitra indica* | Narrow-headed softshell turtle | EN |
| 11 | *Coelognathus helena* | Common trinket snake | – |
| 12 | *Crocodylus palustris* | Mugger or marsh crocodile | VU |
| 13 | *Daboia russelii* | Russell's viper | |
| 14 | *Dendrelaphis tristis* | Common bronzeback tree snake | – |
| 15 | *Echis carinatus* | Saw-scaled viper | |
| 16 | *Enhydris sieboldii* | Siebold's smooth-water snake | LC |
| 17 | *Eryx johnii* | Red sand boa | – |
| 18 | *Eutropis carinata* | Keeled grass skink | – |
| 19 | *Eutropis innotata* | Blanford's grass skink | – |
| 20 | *Eutropis macularia* | Bronze grass skink | – |
| 21 | *Gavialis gangeticus* | Gharial | CR |
| 22 | *Geochelone elegans* | Indian star tortoise | LC |
| 23 | *Gongylophis conicus* | Common sand boa | – |
| 24 | *Hardella thurjii* | Crowned river turtle | VU |
| 25 | *Hemidactylus brookii* | Brook's house gecko | – |
| 26 | *Hemidactylus flaviviridis* | Yellow-bellied house gecko | – |
| 27 | *Hemidactylus frenatus* | Common house gecko | LC |
| 28 | *Hemidactylus leschenaultii* | Bark gecko | – |
| 29 | *Hemidactylus maculatus* | Northern spotted gecko | – |
| 30 | *Hemidactylus triedrus* | Termite hill gecko | – |
| 31 | *Lissemys punctata* | Indian flapshell turtle | LC |
| 32 | *Lycodon striatus* | Barred wolf snake | |
| 33 | *Lycodon aulicus* | Common wolf snake | |
| 34 | *Lygosoma punctata* | Spotted supple skink | – |

(continued)

**Table 3.2** (continued)

| S. no. | Species | Common name | IUCN status (2009–2012) |
|---|---|---|---|
| 35 | *Macropisthodon plumbicolor* | Green keelback | – |
| 36 | *Naja naja* | Spectacled cobra | – |
| 37 | *Nilssonia gangetica* | Indian softshell turtle | VU |
| 38 | *Nilssonia hurum* | Peacock softshell turtle | VU |
| 39 | *Oligodon arnensis* | Common kukri snake | – |
| 40 | *Oligodon taeniolatus* | Russell's kukri snake | LC |
| 41 | *Ophisops jerdonii* | Jerdon's snake-eye | LC |
| 42 | *Ophisops microlepis* | Small-scaled lacerta | LC |
| 43 | *Ophisops minor nictans* | Indian dwarf lacerta | – |
| 44 | *Pangshura tecta* | Indian roofed turtle | LC |
| 45 | *Pangshura tentoria* | Indian tent turtle | LC |
| 46 | *Platyceps ventromaculatus* | Glossy-bellied racer | |
| 47 | *Psammophis leithii* | Leith's sand snake | – |
| 48 | *Ptyas mucosa* | Indian rat snake | – |
| 49 | *Python molurus* | Indian rock python | NT |
| 50 | *Ramphotyphlops braminus* | Brahminy blind snake | – |
| 51 | *Sibynophis subpunctatus* | Dumeril's black-headed snake | – |
| 52 | *Sitana ponticeriana* | Fan-throated lizard | |
| 53 | *Spalerosophis atriceps* | Black-headed royal snake | – |
| 54 | *Varanus bengalensis* | Common Indian monitor | LC |
| 55 | *Varanus griseus* | | |
| 56 | *Xenochrophis piscator* | Chequered keelback | – |
| Mammals | | | |
| 1 | *Antilope cervicapra* | Blackbuck | NT |
| 2 | *Axis axis* | Chital | LC |
| 3 | *Bandicota bengalensis* | Lesser bandicoot rat | LC |
| 4 | *Boselaphus tragocamelus* | Nilgai | LC |
| 5 | *Canis aureus* | Common jackal | LC |
| 6 | *Canis lupus pallipes* | Indian wolf | LC |
| 7 | *Caracal caracal* | Caracal | LC |
| 8 | *Cuon alpinus* | Dhole | EN |
| 9 | *Cynopterus sphinx* | Short-nosed fruit bat | LC |
| 10 | *Felis chaus* | Jungle cat | LC |
| 11 | *Felis silvestris* | Wild cat/desert cat | LC |
| 12 | *Funambulus pennantii* | Five-striped palm squirrel | LC |
| 13 | *Gazella gazella* | Chinkara | VU |
| 14 | *Hemiechinus collaris* | Collared hedgehog | LC |
| 15 | *Herpestes edwardsii* | Indian grey mongoose | LC |
| 16 | *Herpestes javanicus* | Small Asian mongoose | LC |
| 17 | *Herpestes smithii* | Ruddy mongoose | LC |
| 18 | *Hipposideros fulvus* | Fulvous leaf-nosed bat | LC |
| 19 | *Hyaena hyaena* | Striped hyaena | NT |
| 20 | *Hystrix indica* | Indian crested porcupine | LC |

(continued)

**Table 3.2** (continued)

| S. no. | Species | Common name | IUCN status (2009–2012) |
|---|---|---|---|
| 21 | *Lepus nigricollis* | Indian hare | LC |
| 22 | *Lutrogale perspicillata* | Smooth-coated otter | VU |
| 23 | *Macaca mulatta* | Rhesus macaque | LC |
| 24 | *Manis crassicaudata* | Indian pangolin | NT |
| 25 | *Megaderma lyra* | Greater false vampire bat | LC |
| 26 | *Mellivora capensis* | Honey badger | LC |
| 27 | *Melursus ursinus* | Sloth bear | VU |
| 28 | *Millardia meltada* | Soft-furred field rat | LC |
| 29 | *Mus musculus* | House mouse | LC |
| 30 | *Panthera pardus* | Leopard | NT |
| 31 | *Panthera tigris* | Tiger | EN |
| 32 | *Paradoxurus hermaphroditus* | Asian palm civet | LC |
| 33 | *Paraechinus micropus* | Indian hedgehog | LC |
| 34 | *Pipistrellus coromandra* | Indian pipistrelle | LC |
| 35 | *Pipistrellus tenuis* | Indian pygmy bat | LC |
| 36 | *Platanista gangetica* | Gangetic river dolphin | |
| 37 | *Prionailurus bengalensis* | Leopard cat | LC |
| 38 | *Prionailurus viverrinus* | Fishing cat | EN |
| 39 | *Pteropus giganteus* | Indian flying fox | LC |
| 40 | *Rattus rattus* | House rat | LC |
| 41 | *Rhinolophus lepidus* | Blyth's horseshoe bat | LC |
| 42 | *Rhinopoma hardwickii* | Lesser mouse-tailed bat | LC |
| 43 | *Rhinopoma microphyllum* | Greater mouse-tailed bat | LC |
| 44 | *Rousettus leschenaulti* | Leschenault's rousette | LC |
| 45 | *Rusa unicolor* | Sambar | VU |
| 46 | *Scotophilus heathii* | Greater yellow house bat | LC |
| 47 | *Scotophilus kuhlii* | Lesser yellow house bat | LC |
| 48 | *Scotozous dormeri* | Dormer's bat | LC |
| 49 | *Semnopithecus entellus* | Bengal hanuman langur | LC |
| 50 | *Suncus murinus* | Asian house shrew | LC |
| 51 | *Sus scrofa cristatus* | Indian wild boar | LC |
| 52 | *Tadarida aegyptiaca* | Egyptian free-tailed bat | LC |
| 53 | *Taphozous longimanus* | Long-winged tomb bat | LC |
| 54 | *Taphozous melanopogon* | Black-bearded tomb bat | LC |
| 55 | *Taphozous perforatus* | Egyptian tomb bat | LC |
| 56 | *Taphozous nudiventris* | Naked-rumped tomb bat | LC |
| 57 | *Tatera indica* | Indian gerbil | LC |
| 58 | *Tetracerus quadricornis* | Four-horned antelope | VU |
| 59 | *Viverricula indica* | Small Indian civet | LC |
| 60 | *Vulpes bengalensis* | Indian fox | LC |
| 61 | *Vulpes vulpes pusilla* | Desert fox | LC |

IUCN status: *LC* least concern, *NT* near threatened, *VU* vulnerable, *EN* endangered, *CR* critically endangered. Source: Nair and Chaitanya-Krishna (2013)

## 3.3 Vegetation Ecology

Vegetation structure for ecological purposes is considered at a number of levels that are usually integrated as physiognomy, biomass, life form, floristics, stratification, phenology and distribution (Puri et al. 1983). The distribution of species and/or communities follows many patterns such as:

(i) Under habitat conditions species or communities tend to form similar groupings in diverse geographical areas.
(ii) No two habitats can be considered exactly alike.
(iii) Diversity in richness or number of plants in a community is a feature that gives it its even or uneven (unique) character; and the stability is the persistence of a given condition in the community.
(iv) Maturity applies to the successional development of the vegetation.
(v) Heterogeneity in environmental conditions in spatial dimensions and time also influences diversity and stability of a community.

Singh et al. (1979) conducting a detailed study of nine grassland communities with special reference to importance value index of the species, phenology, biomass structure, diversity, dominance, stability and production clearly established the fact that diversity was found inversely related to stability and dominance, while the dominance made the system stable. Species number was found correlated with the dominance. Production was found inversely proportional to stability and dominance though less efficient.

### 3.3.1 Distribution of Species

There are many systems and methods in use for the classification of vegetation in different parts of the world. In India, Champion and Seth (1968) gave a revised classification, which is widely adapted in India. According to this classification, the vegetation of the country has been classified into 16 groups, each further divided into sub-groups and sub-groups into further forest types. The sub-group (Northern tropical thorn forests) of main group 6 (tropical thorn forests) consists "ravine thorn forest" (Fig. 3.1). Though all ravines do not fall in tropical region, some are part of sub-tropical regions.

Very limited efforts have been made to study the detailed vegetational aspects of ravine lands in India. While dealing with the reclamation and management issues, many species found in ravine lands have been mentioned for plantation purposes (Tejwani et al. 1975; Dagar et al. 1978; Verma and Chinnamani 1990; Chinnamani 1992; Prajapati et al. 1993; Dagar and Singh 2003; Dagar 2012, 2017; Dagar and Arunachalam 2013; Chaturvedi et al. 2014; Verma and Singh 2015; Dagar and Gupta 2016). In most of these publications, suitable tree and grass species have been mentioned based on limited experimentations without going into details of the ecological adaptations of these species. Tejwani et al. (1975) dealing with

**Fig. 3.1** Natural vegetation of ravines (Photos Yaduvendra Singh)

reclamation of ravine lands mentioned natural tree species found in Gujarat ravines, which include *Acacia nilotica*, *A. eburnea*, *A. leucophloea*, *A. senegal*, *Azadirachta indica*, *Albizia lebbeck*, *Feronia limonia*, *Holoptelea integrifolia*, *Cassia auriculata*, *Gymnosporia montana*, *Ziziphus mauritiana* and *Prosopis cineraria*. These trees are found consistently in all the ravine areas in India. Many tree species were tried in ravine lands of different regions. Out of them *Acacia nilotica*, *A. benthamii*, *A. tortilis*, *A. leucophloea*, *Ailanthus excelsa*, *Albizia lebbeck*, *Azadirachta indica*, *Dalbergia sissoo*, *Dendrocalamus strictus* (natural in most of the ravines), *Eucalyptus camaldulensis*, *E. citriodora*, *Pongamia pinnata*, *Emblica officinalis*, *Salmalia malabarica* and *Tectona grandis* were found successful in Gujarat. Most of these species were also suitable for Chambal and Yamuna ravines at Agra.

Seeds of *Prosopis juliflora* were spread aerially in ravines; therefore, in most of the ravines, this species is found distributed. In ravines, distributed along the rivers of Rajasthan, *Prosopis cineraria*, *Capparis decidua*, *Salvadora oleoides*, *Tecomella undulata*, *Balanites aegyptiaca*, *Acacia senegal*, *A. leucophloea*, *Lycium europaeum*, *Maytenus emarginata*, *Z. nummularia*, *A. jacquemontii* (mainly on beds), *Clerodendrum phlomidis*, *Leptadenia pyrotechnica*, *Sericostoma pauciflorum*, *Crotalaria burhia*, *Tephrosia purpurea*, *T. falciformis*, *Aerva persica*, *Fagonia cretica* and *Calotropis procera* are prominent trees and shrubs found almost throughout the ravines. Among bushes, *Ephedra foliata*, *Mukia maderaspatana*, *Cocculus hirsuta*, *Asparagus racemosus* and *Momordica dioica* are prominent climbers. *Citrullus colocynthis* found as frequent trailer. On gravelly soils, *Leptadenia pyrotechnica*, *Bonamia latifolia*, *Tribulus terrestris*, *T. alatus*, *Barleria acanthoides*, *Boerhaavia diffusa*, *Corbichonia decumbens*, *Fagonia cretica*, *Indigofera linifolia*, *I. enneaphylla* and *Alysicarpus minima* are frequent species. Among grasses, *Aristida adscensionis*, *A. paniculata*, *Cenchrus ciliaris*, *C. setigerus*, *C. biflorus*, *Eragrostis ciliaris*, *E. tremula*, *Eleusine compressa*, *Eremopogon foveolatus*, *Melanocenchris jacquemontii*, *Sporobolus diander*, *Oropetium thomaeum*, *Dactyloctenium sindicum*, *D. aegyptium*, *Cymbopogon jwarancusa*, *Saccharum benghalensis* (in beds), *Vetiveria zizanioides*, *Desmostachya bipinnata*, *Digitaria adscendens*, *Dichanthium annulatum*, *Bothriochloa pertusa* and *Cynodon dactylon* (along beds) are prominent.

During rehabilitation studies, a very pertinent observation was made that the first step of bio-reclamation is closure to grazing and other biotic interference. Most of the ravine lands in India are poor in vegetation, and most of the annuals and unpalatable grasses are found growing in places of desirable climax of associated species, which should normally exist in the prevailing soil climatic environments. For example, in Gujarat ravines *Aristida funiculata* and *Themeda triandra* are found growing where *Dichanthium annulatum*-*Cenchrus* spp. association (more desirable) will replace after protection. After closing the site from grazing, a tall annual grass *Apluda mutica* and perennials like *Eremopogon foveolatus*, *Heteropogon contortus*, *Dichanthium annulatum*, *Cenchrus ciliaris* and *C. setigerus* will replace *Aristida*-*Themeda* association. In this area loss of soil and water will progressively decrease, and there will be qualitative and quantitative increase in forage production. Continued closure to grazing resulted in high increase in the proportion of *A. mutica* at all sites, and in the first 2–3 years, there was considerable increase in *D. annulatum* and *Aristida royleana*, but soon this was replaced by *A. mutica*. After 10–15 years of protection from biotic disturbances, the area ultimately got converted into forest. Similar was the trend in Kota ravines. Under the upper Damodar catchment conditions, closure to grazing reduced the soil loss from 3.3 Mg $ha^{-1}$ (under overgrazed condition) to only 0.6 Mg $ha^{-1}$, and length and weight of grass roots were 44% and 184%, respectively, more than those when the plots were overgrazed.

In the Yamuna ravines at Agra, the ravine tops are occupied by *Apluda mutica*, *Cenchrus ciliaris*, *Cymbopogon jwarancusa*, *Aristida cyanantha*, *Boerhaavia stricta* and *Vicoa indica*. The slopes are found occupied by *C. jwarancusa*, *Eremopogon foveolatus*, *Apluda mutica*, *Vicoa indica*, *Crotalaria phillipsiae* and *Enicostema verticillatum* and on the ravine bottom, *Desmostachya bipinnata*, *Cenchrus ciliaris*, *Heteropogon contortus*, *Pluchea lanceolata*, *Ichnocarpus frutescens*, *Dalbergia sissoo* and *Balanites roxburghii are predominant. Grass Cenchrus ciliaris* is well adapted to slope as well as top of ravines.

Dagar and Mall (1980) studied the vegetation of ravines and banks of the river Kshipra, which is one of the most important tributaries of Chambal River. The study area was under eroded alluvial bed lands and disturbed soils. The alluvial soil near the river is ranging in thickness 2–15 m, yellowish brown, highly calcareous and with a net of gullies and ravines of various sizes and shape. The sand, silt and clay contents varied from 30 to 43%, 21 to 32% and 31 to 49%, showing more clay and silt contents and lesser sand contents as compared to Chambal ravines. The vegetation was studied with reference to topography, viz. on top of the ravines (plateau), along slopes and bed (bottom) of the ravines. The grazing pressure is very high which deteriorates the vegetal cover encouraging erosion forces.

On plateau (top) of ravines, *Balanites roxburghii* and *Acacia eburnea* are the dominant woody species followed by *Acacia nilotica*, *A. leucophloea* and *Dichrostachys cinerea* (Table 3.3). *Capparis sepiaria*, *Mimosa hamata*, *Phoenix sylvestris*, *Butea monosperma* and *Feronia limoni*a are occasionally found. Along the slopes of the ravines, *A. eburnea* is most dominating tree followed by *B.*

**Table 3.3** Importance value index of woody vegetation distributed in Kshipra ravines near Hariakheri village (Ujjain) (quadrate size: 2 × 2 m)

| Species | Plateau (top) | Slope |
|---|---|---|
| *Balanites roxburghii* | 137.39 | 56.02 |
| *Acacia eburnea* | 73.70 | 113.02 |
| *Acacia nilotica* | 27.36 | 55.08 |
| *Acacia leucophloea* | 22.99 | 25.90 |
| *Dichrostachys cinerea* | 19.71 | 23.07 |
| *Capparis sepiaria* | 9.33 | 13.19 |
| *Mimosa hamata* | 6.37 | 3.78 |
| *Phoenix sylvestris* | 2.03 | 5.80 |
| *Butea monosperma* | 0.59 | 3.57 |
| *Feronia limonia* | 0.86 | – |

*roxburghii*, *A. nilotica* and *D. cinerea.* Along the bed of the ravines, *Acacia nilotica* is the most common species. *Pongamia pinnata* and *Ipomoea fistulosa* are other frequent species found in pure patches. On some other sites, trees such as *Tamarindus indica*, *A. nilotica*, *Azadirachta indica* and *Ziziphus mauritiana* were found growing at top of ravines may be as introduced at some stage. Along slopes, besides earlier species, *Lantana camara*, *Capparis horrida* (climber in bushes), *Clerodendrum phlomidis*, *Barleria prionitis* and *Ziziphus nummularia* were found. Occasionally, *Salvadora persica*, *Salmalia malabarica*, *Jatropha gossypiifolia*, *Capparis grandis* and *Cordia rothii* were also found. *Pupalia lappacea*, *Ipomoea hederifolia*, *I. nil*, *I. obscura* and *Asparagus racemosus* were found as climbers among *Capparis* bushes.

Other woody species in ravines and along the banks of the river and its tributaries (other than the above sites) noticed include *Annona squamosa* (introduced), *Capparis decidua*, *Aegle marmelos*, *Citrus limon* (introduced), *Melia azedarach*, *Maytenus senegalensis*, *Ziziphus* sp., *Sapindus mukorossi*, *Mangifera indica*, *Moringa oleifera*, *Sesbania sesban*, *Bauhinia racemosa*, *B. variegata*, *Caesalpinia cristata*, *C. sepiaria*, *Cassia fistula*, *C. javanica*, *C. sophera*, *Delonix regia*, *Acacia catechu*, *A. farnesiana*, *A. jacquemontii*, *Albizia lebbeck*, *Mimosa rubicaulis*, *Pithecellobium dulce*, *Prosopis cineraria*, *P. juliflora*, *Anogeissus acuminata*, *Terminalia arjuna*, *T. tomentosa*, *Eucalyptus* sp., *Psidium guajava*, *Syzygium cumini*, *S. heyneanum*, *Mimusops elengi*, *Diospyros montana*, *Jasminum grandiflorum*, *Carissa carandas*, *Nerium indicum*, *Tabernaemontana divaricata*, *Thevetia peruviana*, *Ehretia acuminata*, *Millingtonia hortensis*, *Nyctanthes arbor-tristis*, *Emblica officinalis*, *Ficus benghalensis*, *F. carica*, *F. glomerata*, *F. religiosa*, *Morus indica* and *Bambusa arundinacea*. Many of these may be introduced due to *Singhast/Kumbh* religious festival being celebrated on Kshipra River at Ujjain.

Among herbaceous vegetation, in protected field on plateau of ravines, *Apluda aristata*, *Dichanthium annulatum*, *Heteropogon contortus*, *Bothriochloa pertusa*, *Dichanthium caricosum*, *Iseilema laxum*, *Cymbopogon martinii*, *Tragus biflorus* and *Themeda triandra* among grasses; and *Indigofera linifolia*, *Alysicarpus longifolius*, *Achyranthes aspera*, *Euphorbia geniculata*, *Rhynchosia capitata* and many others among legumes; and other herbaceous species are common. On grazed plateau, the most dominant species is *Bothriochloa pertusa* followed by *Dichanthium annulatum*, *Chrysopogon fulvus*, *Iseilema laxum*, *Heylandia latebrosa*, *Indigofera linifolia*, *Oldenlandia corymbosa*, *Alysicarpus monilifer*, *Sehima nervosum*, *Rhynchosia minima*, *Euphorbia hypericifolia* and *Cassia tora* showing their adaptation to biotic pressure. Along slopes, many species on plateau are common, but *Themeda triandra*, *Cassia pumila*, *Rungia repens*, *Tragus biflorus*, *Vernonia cinerea*, *Launaea asplenifolia*, *Blepharis maderaspatensis*, *Polygala chinensis*, *Setaria glauca*, *Rhynchosia capitata*, *Volutarella ramosa*, *Indigofera linnaei*, *Glossocardia bosvallia*, *Echinops echinatus*, *Boerhaavia diffusa*, *Phaseolus trilobus*, *Lagascea mollis* and *Eclipta prostrata* are more frequent on slopes (Table 3.4), where the grazing pressure is slightly low due to steep slopes.

Some species are eliminated from grazed fields and present only in protected fields. *Heteropogon contortus*, *Themeda triandra*, *Cymbopogon martinii*, *Ischaemum rugosum*, *Tragus biflorus*, *Thelepogon elegans*, *Alysicarpus rugosus*, *A. longifolius*, *Achyranthes aspera*, *Polygala chinensis*, *Setaria glauca*, *Euphorbia geniculata*, *Rhynchosia capitata*, *Rungia repens*, *Peristrophe bicalyculata*, *Paspalidium flavidum*, *Justicia diffusa*, *Striga euphrasioides*, *Phaseolus trilobus*, *Lagascea mollis*, *Scleria stocksiona* and *Solanum nigrum* were restricted to protected fields or if found in grazed fields were only occasional. Species such as *Oldenlandia corymbosa*, *Sporobolus coromandelianus*, *Tridax procumbens*, *Eragrostis pilosa*, *E. tenella*, *Cynodon dactylon*, *Enicostema verticillatum*, *Evolvulus alsinoides*, *Convolvulus pluricaulis*, *Cassia tora*, *Ocimum americanum*, *Iphigenia indica*, etc. occurred in grazed fields only or very occasional in protected fields. A relatively high plus value for grazing susceptible number (GSN) indicates that the species has some disadvantage under grazing, while conversely a relatively high minus value indicates that the species is benefited from grazing. *Bothriochloa pertusa*, *Chrysopogon fulvus*, *Iseilema laxum*, *Heylandia latebrosa*, *Rhynchosia minima* and *Euphorbia hypericifolia* had more IVI (Table 3.4) and high negative GSN (Table 3.5), showing grazing resistance. Species such as *Apluda aristata* and *Sehima nervosum* benefited maximally from protection, while *Dichanthium annulatum*, *Alysicarpus monilifer* and *Cyperus rotundus* were less affected by grazing.

At some places, *Andropogon pumilus*, *Arthraxon ciliaris*, *Melanocenchris jacquemontii*, *Sporobolus diander*, *S. marginatus*, *Iseilema anthephoroides*, *Setaria verticillata*, *Acalypha indica*, *Abutilon indicum*, *Polygala chinensis*, *Glossocardia bosvallia*, *Indigofera linnaei*, *Xanthium strumarium*, *Lagascea mollis*, *Kickxia racemosa*, *Ageratum conyzoides*, *Borreria stricta*, *Echinochloa crusgalli*, *E. colonum*, *Chloris barbata*, *Trichoderma amplexicaul*, *Blumea membrana*, *Commelina benghalensis*, *C. forskalii*, *C. kurzii*, *Justicia diffusa*, *Peristrophe bicalyculata*, *Acalypha malabaricum*, *Dipteracanthus prostratus*, *Tephrosia purpurea*, *Paspalidium flavidum*, *Cyanotis axilaris*, *Leucas urticifolia*, *Desmodium triflorum*, *Hemiadelphis polyspermus*, *Cassia occidentalis*, *Indigofera glandulosa* (in protected stand only), *I. trita*, *Mavastrum coromandelianum*, *Ziziphus nummularia*, *Tragia cannabina*, *Anisomelos indica*, *Verbina bipinnata*, *Rungia pectinata*, *Hemigraphis hirta*, *Sonchus arvansis*, *Blumea lacera*, *Acanthospermum hispidum*, *Corchorus olitorius*, *C. trilocularis*, *Launaea nudicaulis*, *Indigofera trifoliata*, *Psoralea corylifolia*, *Cocculus hirsutus*, *Lepidagathis cristata* and *Tragia*

**Table 3.4** Importance value index (IVI) of herbaceous vegetation in Kshipra ravines (quadrate size 50 × 50 cm)

| Species | Plateau | | Slope (unprotected) |
|---|---|---|---|
| | Protected | Grazed | |
| *Dichanthium annulatum* | 40.96 | 47.77 | 51.36 |
| *Apluda aristata* | 36.81 | 2.71 | 9.66 |
| *Bothriochloa pertusa* | 25.78 | 51.88 | 61.26 |
| *Heteropogon contortus* | 27.56 | – | – |
| *Dichanthium caricosum* | 22.52 | – | – |
| *Themeda triandra* | 17.35 | – | 1.05 |
| *Chrysopogon fulvus* | 12.13 | 39.83 | 5.24 |
| *Alysicarpus monilifer* | 9.36 | 11.52 | 12.56 |
| *Cymbopogon martinii* | 9.19 | – | – |
| *Iseilema laxum* | 9.02 | 22.99 | 1.19 |
| *Sehima nervosum* | 8.61 | 11.06 | 1.84 |
| *Indigofera linifolia* | 7.03 | 16.54 | 37.67 |
| *Ischaemum rugosum* | 6.05 | – | – |
| *Tragus biflorus* | 5.68 | – | 4.94 |
| *Alysicarpus longifolius* | 5.44 | – | – |
| *Achyranthes aspera* | 4.65 | – | – |
| *Heylandia latebrosa* | 4.32 | 17.36 | 9.46 |
| *Euphorbia thymifolia* | 3.84 | 2.00 | 0.72 |
| *Scleria stocksiona* | 3.26 | – | – |
| *Alysicarpus rugosus* | 3.19 | – | – |
| *Crotalaria medicaginea* | 2.99 | 1.33 | 1.65 |
| *Setaria glauca* | 2.77 | – | 3.01 |
| *Rhynchosia minima* | 2.58 | 8.97 | 11.41 |
| *Euphorbia hypericifolia* | 2.35 | 9.72 | 7.34 |
| *Euphorbia hirta* | 2.03 | 3.38 | 2.18 |
| *Euphorbia geniculata* | 1.97 | – | – |
| *Cyperus rotundus* | 1.95 | 1.91 | – |
| *Biophytum sensitivum* | 1.93 | 1.60 | 0.45 |
| *Rungia repens* | 1.93 | – | 6.35 |
| *Rhynchosia capitata* | 1.90 | – | 0.49 |
| *Indigofera linnaei* | 1.68 | – | 0.91 |
| *Polygala chinensis* | 1.59 | – | 0.09 |
| *Thelepogon elegans* | 1.56 | – | – |
| *Peristrophe bicalyculata* | 1.54 | – | – |
| *Paspalidium gracile* | 1.37 | – | – |
| *Amaranthus gracilis* | 1.21 | – | – |
| *Justicia diffusa* | 1.19 | – | – |
| *Boerhaavia diffusa* | 1.11 | – | 0.92 |
| *Striga euphrasioides* | 1.09 | – | – |
| *Phaseolus trilobus* | 1.07 | – | 0.57 |
| *Lagascea mollis* | 0.88 | – | 0.84 |
| *Solanum nigrum* | 0.19 | – | – |

(continued)

**Table 3.4** (continued)

| Species | Plateau | | Slope (unprotected) |
|---|---|---|---|
| | Protected | Grazed | |
| *Oldenlandia corymbosa* | – | 12.80 | 8.15 |
| *Sporobolus coromandelianum* | – | 8.20 | – |
| *Eragrostis pillosa* | – | 6.25 | – |
| *Cassia tora* | – | 4.68 | 1.96 |
| *Enicostemma hyssopifolium* | – | 4.28 | 4.45 |
| *Eragrostis tenella* | – | 2.42 | – |
| *Ocimum americanum* | – | 1.75 | 1.66 |
| *Iphigenia indica* | – | 0.86 | – |
| *Convolvulus microphyllus* | – | 0.69 | 1.43 |
| *Tridax procumbens* | – | 0.46 | 7.21 |
| *Evolvulus alsinoides* | – | 0.46 | 0.89 |
| *Polygala erioptera* | – | 0.45 | – |
| *Arthraxon ciliaris* | – | – | 12.63 |
| *Vernonia cinerea* | – | – | 5.45 |
| *Launaea asplenifolia* | – | – | 4.74 |
| *Blepharis maderaspatensis* | – | – | 3.96 |
| *Volutarella ramosa* | – | – | 1.89 |
| *Cocculus hirsutus* | – | – | 1.63 |
| *Glossocardia bosvallia* | – | – | 1.39 |
| *Echinops echinatus* | – | – | 1.17 |
| *Eclipta prostrata* | – | – | 0.86 |
| *Abutilon indicum* | – | – | 0.81 |
| *Trichodesma amplexicaule* | – | – | 0.78 |
| *Leucas aspera* | – | – | 0.76 |
| *Polygala telephioides* | – | – | 0.58 |
| *Triumfetta rhomboidea* | – | – | 0.46 |
| *Cayratia carnosa* | – | – | 0.44 |
| *Lepidagathis cristata* | – | – | 0.43 |
| *Tribulus terrestris* | – | – | 0.24 |
| *Ruellia tuberosa* | – | – | 0.23 |
| *Phyllanthus maderaspatensis* | – | – | 0.11 |

*cannabina* are found along slopes or among bushes. *Cynodon dactylon* is the only dominant grass at bed with some less common species such as *Cyperus rotundus*, *Phyla nodiflora*, *Cassia tora* and *Dichanthium annulatum*.

Most of the species found along Kshipra ravines came across when visited Chambal ravines except a few which were unique along Chambal ravines. For example, Guggal (*Commiphora wightii*), *Salvadora oleoides*, *Acacia catechu*, *Bombax ceiba*, *Capparis decidua*, *Lawsonia alba*, *Anogeissus latifolia*, *A. pendula*, *Jatropha curcas*, *Carissa carandas*, *Cordia rothii*, *Calotropis gigantea* and *Ziziphus mauritiana* among trees and shrubs and safed musli (*Chlorophytum arundinaceum*), *Citrulus colocynthis*, *Ephidra foliata*, *Tinospora cordifolia*, *Fagonia cristata*,

**Table 3.5** Grazing susceptibility number (GSN) of some species found in ravines

| Species | GSN[a] | Species | GSN |
|---|---|---|---|
| *Bothriochloa pertusa* | −6 | *Dichanthium annulatum* | +1 |
| *Chrysopogon fulvus* | −6 | *Alysicarpus monilifer* | +1 |
| *Iseilema laxum* | −5 | *Cyperus rotundus* | +2 |
| *Euphorbia hypericifolia* | −7 | *Apluda aristata* | +3 |
| *Heylandia latebrosa* | −6 | *Crotalaria medicaginea* | +5 |
| *Rhynchosia minima* | −6 | *Biophytum sensitivum* | +6 |
| *Euphorbia hirta* | −5 | *Euphorbia thymifolia* | +7 |
| *Indigofera linifolia* | −4 | *Sehima nervosum* | +7 |

[a]
$$\text{GSN} = \frac{\text{Density of species in ungrazed pasture} - \text{Density of species in grazed pasture}}{\text{Density in ungrazed when numerator is positive } \textbf{or} \text{ Density in ungrazed when numerator is negative}}$$

*Cymbopogon jwarancusa*, *Saccharum benghalensis* (in beds), *Vetiveria zizanioides* and *Desmostachya bipinnata* are not found in Kshipra ravines but are frequent in Chambal ravines.

Many species, which could not be covered in quantitative data and are annuals during winter season and found distributed in ravines (visual observation of the author) or along the banks of the river in the vicinity of ravines, include *Cleome viscosa*, *Polygala telephioides*, *Portulaca quadrifida*, *Abutilon asiaticum*, *Sida cordifolia*, *S. glutinosa*, *S. rhombifolia*, *Clitoria ternatea*, *Desmodium gangeticum*, *Cardiospermum halicacabum*, *Glycine max* (on dried mud), *Cajanus cajan* (escape), *Indigofera cordifolia*, *Lathyrus aphaca*, *L. sativus*, *Medicago polymorpha*, *Melilotus indica*, *M. alba*, *Crotalaria sericea*, *Psoralea corylifolia*, *Rhynchosia capitata*, *Cassia auriculata*, *C. sophera*, *Passiflora foetida* (climber in bushes), *Opuntia elatior*, *Glinus lotoides* (on dried mud), *Trianthema portulacastrum*, *Centella asiatica*, *Oenanthe javanica*, *Trichosanthes cucumerina* (among bushes), *Knoxia corymbosa*, *Oldenlandia aspera*, *Blainvillea latifolia*, *Eupatorium coelestinum*, *Cyathocline purpurea*, *Emilia sonchifolia*, *Erigeron canadensis*, *Grangea maderaspatana*, *Vicoa indica*, *Anagallis arvensis*, *Calotropis procera*, *Dregea volubilis*, *Heliotropium supinum*, *Convolvulus arvensis*, *Ipomoea pes-tigridis*, *Cuscuta reflexa* (parasite on trees and bushes), *Datura metel*, *Physalis minima*, *Solanum nigrum*, *S. surattense*, *Martynia annua*, *Andrographis paniculata*, *Blepharis asperrima*, *Hemigraphis hirta*, *Ruellia tuberosa*, *Ocimum gratissimum*, *Alternanthera pungens*, *Amaranthus gracilis*, *A. spinosus*, *Celosia argentea*, *Digera alternifolia*, *Gomphrena celosioides*, *Pupalia lappacea* (climber among bushes), *Asparagus racemosus*, *Chenopodium murale*, *Baliospermum montana*, *Phyllanthus fraternus*, *Ricinus communis*, *Aloe barbedensis*, *Alloteropsis cimicina*, *Aristida adscensionis*, *Bothriochloa intermedia*, *Brachiaria reptans*, *Cenchrus ciliaris*, *C. setigerus*, *Chloris gayana*, *Dactyloctenium aegyptium*, *D. sindicum*, *Desmostachya bipinnata*, *Digitaria setigera*, *Dinebra arabica*, *Eragrostiella bifaria*, *Eragrostis poaeoides*, *Eremopogon foveolatus*, *Eriochloa procera*, *Oropetium thomaeum*, *Panicum antidotale*, *P. trypheron* and *P. maximum*.

Dagar and Singh (1980) reported the phenology, seasonal distribution and ecological classification of 216 species belonging to 60 families of flowering plants found in water and moist banks of river Kshipra, as hydrophytes, submerged along banks in marshy places and growing as amphibians. Based on recent confirmations and observations, the important herbaceous vegetations found in wet places along the banks of river Kshipra and Chambal include *Vetiveria zizanioides*, *Saccharum spontaneum*, *Sorghum halepense*, *Pennisetum purpureum*, *Paspalum distichum*, *Sorghum halepanse*, *Paspalidium flavidum*, *Panicum antidotale*, *Ischaemum rugosum*, *Hygroryza aristata*, *Hemarthria compressa*, *Eragrostis diarrhena*, *Eleusine indica*, *Eleocharis atropurpurea*, *Echinochloa colonum*, *E. crusgalli*, *Digitaria adscendens*, *D. setigera*, *Desmostachya bipinnata*, *Dactyloctenium aegyptium*, *Hachelochloa granularis*, *Cynodon dactylon*, *Coix lacryma-jobi*, *Bambusa arundinacea*, *Arundo donax*, *Phragmites australis*, *P. karka*, *Scirpus littoralis*, *S. roylei*, *S. articulatus*, *Fimbristylis dichotoma*, *F. miliacea*, *Eleocharis atropurpurea*, *E. plantaginea*, *Cyperus compressus*, *C. exallatus*, *C. iria*, *C. rotundus*, *C. bulbosus*, *C. kyllinga*, *C. triceps*, *Carex fedia*, *Typha angustata*, *Commelina benghalensis*, *C. forskalii*, *C. kurzii*, *C. haskarlii*, *Crinum defixum*, *Euphorbia geniculata*, *Chrozophora prostrata*, *C. rottleria*, *Acalypha indica*, *Rumex dentatus*, *Polygonum glabrum*, *P. plebeium*, *Alternanthera sessilis*, *Aerva lanata*, *Achyranthes aspera*, *Anisomelas indica*, *Phyla nodiflora*, *Lindenbergia indica*, *Rungia repens*, *R. pectinata*, *Peristrophe bicalyculata*, *Justia* spp., *Asteracantha longifolia*, *Verbascum chinensis*, *Ageratum conyzoides*, *Veronica anagallis-aquatica*, *Stemodia viscosa*, *Striga euphrasioides*, *Sutera glandulosa*, *Bacopa monnieri*, *Nicotiana plumbaginifolia*, *Merremia gangetica*, *Ipomoea aquatica*, *I. fistulosa (carnea)*, *Parthenium hysterophorus*, *Xanthium strumarium*, *Laggera aurita*, *Lagascea mollis*, *Gnaphalium indicum*, *G. luteo-album*, *Eclipta prostrata*, *Caesulia axillaris*, *Blumea* spp., *Ammania baccifera*, *A. multiflora*, *Potentilla supina*, *Sesbania bispinosa*, *Indigofera glandulosa*, *I. trifoliata*, *Aeschynomene indica*, *Alhagi pseudalhagi*, *Biophytum sensitivum*, *Impatiens balsamina*, *Oxalis corniculata*, *O. martiana*, *Corchorus aestuans*, *C. bilocularis*, *Sida glutinosa*, *Argemone mexicana*, *Fumaria indica*, *Capsella bursa-pastoris*, *Sisymbrium irio* and *Portulaca oleracea.*

In another study Thomas et al. (2011) mentioned 4 edible plants, 21 medicinal plants, 13 of fodder value, 19 used as fuelwood and 5 used in furniture and handicrafts, found naturally in ravines of Pipri village in Morena district. These are also found in other localities and mentioned in Table 3.4.

Verma and Singh (2015) mentioned that during their experimentation in deep ravines of Morena, they came across 74 plant species found distributed naturally, out of which 56 were listed having medicinal properties. Recently, a final technical report of a project "Plant Diversity of Chambal Region with Special Reference to the Conservation Strategy of Threatened Taxa" has been submitted by MP Biodiversity Board (www.mpsbb.nic.in/completedproject/pdcr.pdf), in which ethnobotanical uses of plant species of northern Madhya Pradesh have been given. This report also contains three sites [Piparai, Jabuol (Sabalgarh), Bamsoli] of Morena district and three sites (Barhi, Spad-beet, Ater) of Bhind districts, where ravines are formed. From this report, 232 plant species distributed in 6 sites having ravines along with their uses have been segregated and tabulated (Table 3.6).

**Table 3.6** Distribution of plant species in ravines of Morena and Bhind districts

| Species | Local name | Localities | | | | | | Uses |
|---|---|---|---|---|---|---|---|---|
| | | Morena | | | Bhind | | | |
| | | 1 | 2 | 3 | 4 | 5 | 6 | |
| Trees (50) | | | | | | | | |
| *Acacia catechu* | Khair | | + | | | + | | T |
| *Acacia leucophloea* | Ronjh | + | + | | + | + | + | Fu, T |
| *Acacia nilotica* | Desi kikar, babool | | + | + | + | + | + | T, Fo, M, Fu, G |
| *Aegle marmelos* | Bael | | | + | | | | Fr |
| *Ailanthus excelsa* | Maharukh | | | | + | | | Fu, pulpwood |
| *Alangium salvifolium* | Ankol | | | | + | | | M |
| *Albizia lebbeck* | Siris | | + | + | + | + | | Or, Fo |
| *Albizia odoratissima* | Kala siris | | | | + | + | | Or |
| *Anogeissus latifolia* | Dhawada | + | | | | | | G, M, Fu |
| *Anogeissus pendula* | Kardhai | + | | | | | | Fu, T |
| *Artocarpus heterophyllus* | Kathal | | | + | | | | Fr |
| *Azadirachta indica* | Neem | | | + | + | + | + | M |
| *Balanites aegyptiaca* | Hingot | + | + | + | + | + | + | M, Fo |
| *Bauhinia racemosa* | Kachnar | | | | + | + | | M, V(flowers) |
| *Bombax ceiba* | Semal | + | + | | | | | Fi, M |
| *Butea monosperma* | Palash | | + | | | | | M |
| *Cassia fistula* | Amaltas | | | | + | | | Or, M |
| *Cassia siamea* | Kasud | | | + | + | | | Or (introduced) |
| *Citrus aurantifolia* | Lemon | + | | | | | | Fr |
| *Cordia dichotoma* | Lasura | | | | | + | + | Fr (pickle), Fo |
| *Dalbergia sissoo* | Sisham | | + | + | + | + | + | T |
| *Emblica officinalis* | Aonla | | | + | | + | | Fr, M |
| *Eucalyptus umbellata* | Nilgiri | | | | + | | | Oil, introduced |
| *Feronia limonia* | Kaith | | | + | | + | | Fr, Fo |
| *Ficus benghalensis* | Bargad | | | + | + | + | | R |
| *Ficus hispida* | Bhui-gular | | + | | | | | |
| *Ficus palmata* | Anjir, karkara | + | + | + | | | | Fr |
| *Ficus racemosa* | Gular | | + | + | + | + | | R, Fo |
| *Ficus religiosa* | Pipal | | | + | + | + | + | R |
| *Flacourtia indica* | Kakher | + | | | | | | Fu |
| *Grewia tenax* | Chabni | + | | | | | | M |
| *Hardwickia binata* | Anjan | | | | + | | | M, T |
| *Holoptelea integrifolia* | Chirol | | | + | + | + | | T, M |
| *Leucaena leucocephala* | Subabool | | | + | + | + | | T, Fo, Fu |
| *Madhuca longifolia var. latifolia* | Mahua | | | | | + | | M |
| *Mangifera indica* | Mango, Am | | | + | | | | Fr |
| *Melia azadirachta* | Bakan | | | | + | | | M, T |

(continued)

**Table 3.6** (continued)

| Species | Local name | Localities | | | | | | Uses |
|---|---|---|---|---|---|---|---|---|
| | | Morena | | | Bhind | | | |
| | | 1 | 2 | 3 | 4 | 5 | 6 | |
| *Miliusa tomentosa* | Kari | | | | | | + | T |
| *Moringa oleifera* | Sonjana | | + | + | + | | | V, Fo |
| *Nyctanthes arbor-tristis* | Siyari | | | | | | | Basketary |
| *Pongamia pinnata* | Karanj | + | | | + | + | | Fu, Fo, O, M |
| *Prosopis cineraria* | Chenkur, shami | + | + | + | + | + | + | Fo, Fu, V, R |
| *Prosopis juliflora* | Vilayti babool | + | + | + | + | + | + | Fu, T |
| *Psidium guajava* | Guava, Amrood | | | + | | | | Fr |
| *Punica granatum* | Pomegranate, Anar | | | + | | | | Fr (introduced) |
| *Sesbania sesban* | Jainti | + | | | | | | V, Fu |
| *Tamarindus indica* | Tamarind, Imli | | | + | | + | | Fr |
| *Tectona grandis* | Sagon | | + | | | | | T |
| *Terminalia arjuna* | Arjun | | | + | | | | M |
| *Ziziphus mauritiana* | Ber | | | + | | + | + | Fr, Fo |
| Shrubs/bushes (41) | | | | | | | | |
| *Acacia jacquemontii* | | + | | | | | | Fu |
| *Adhatoda zeylanica* | Bamsori | | + | + | + | + | + | M |
| *Agave mexicana* | Hathi-chinghad | | + | + | | | | Or, Fi, fence |
| *Boswellia serrata* | Salai | + | | | | | | Fu, fence |
| *Calotropis gigantea* | Madar | | + | + | | | + | M |
| *Calotropis procera* | | | | + | + | + | | M |
| *Cannabis sativa* | Bhang | | | | | + | | M |
| *Capparis decidua* | Kareel, Tind | + | + | + | + | + | + | Fr (pickle) |
| *Capparis sepiaria* | Hins | + | + | + | + | + | + | Fence, M |
| *Carissa carandas* | Karonda | | | + | | + | | Fr (pickle) |
| *Clerodendrum multiflorum* | Inni | + | | | + | + | + | Fo |
| [a]*Commiphora wightii* | Guggal | + | + | | + | + | + | M |
| *Datura fistuosa* | Kala dhatura | + | | | | | | M |
| *Datura innoxia* | Dhatura | + | | | | | | M |
| *Datura strumarium* | Dhatura | | | | + | + | + | M |
| *Dendrocalamus strictus* | Bans/bamboo | + | | + | | + | | Basketary |
| *Dichrostachys cinerea* | Vilayati fuel | + | + | + | | + | | Fu |
| *Grewia flavescens* | Jher kher | | + | | + | + | + | Basketary |
| *Ipomoea carnea* | Besharam, bansa | | + | + | | | | M |
| *Jasminum grandiflorum* | Chameli | | | + | | | | Or (introduced) |
| *Jasminum sambac* | Mogra | | | | | | + | Or (introduced) |
| *Jatropha curcas* | Ratanjot | | | | + | + | + | Oil, M |
| *Jatropha podagrica* | Jungli arindi | + | | | | | | M |
| *Lantana camara* | Baramasi | + | + | | | + | + | Or |

(continued)

**Table 3.6** (continued)

| Species | Local name | Localities | | | | | | Uses |
|---|---|---|---|---|---|---|---|---|
| | | Morena | | | Bhind | | | |
| | | 1 | 2 | 3 | 4 | 5 | 6 | |
| *Lantana indica* | Gundhamena | | | | | | + | M, Or |
| *Lawsonia inermis* | Mehndi | + | | | | | | M, dye |
| *Macfadyena unguis-cati* | – | + | | | | | | M |
| *Maytenus senegalensis* | | + | | | | | | |
| *Mimosa hamata* | | | + | | | | | |
| *Mimosa himalayana* | Arali | | + | | | | + | |
| *Morus alba* | Shatoot | + | | | | | | Fr, basketary |
| *Nerium indicum* | Kaner | | | + | | | | Or |
| *Opuntia dilleni* | Nagphani | + | | | | | | M |
| *Opuntia vulgaris* | Nagphani | | | | | | + | M |
| *Plumbago zeylanica* | Chitwa | + | | | | | | M |
| *Ricinus communis* | Castor, Arind | | | + | + | | + | Oil |
| *Salvadora oleoides* | Pilu | + | + | | + | | + | Fo, Fr, O, Fu, M |
| *Securinega leucopyrus* | Hartho | | | | | | + | M |
| *Sesbania bispinosa* | Dhadon | + | | | | | | |
| *Ziziphus nummularia* | Jharber | + | + | | | | | Fr, Fo |
| *Ziziphus oenoplia* | Zariya, Jhar-beri | | | | + | + | + | Fr, Fo |
| Herbs (93) | | | | | | | | |
| *Achyranthes aspera* | Chir chita | | + | | + | + | + | M |
| *Abutilon pannosum* | Kasili | + | | | | | | M |
| *Ageratum conyzoides* | | | | | | + | + | |
| *Alhagi pseudalhagi* | Javasa | + | + | | | + | | Fo, M |
| *Alysicarpus vaginalis* | | | | | | | + | |
| *Amaranthus spinosus* | Kanteli-chaulai | | | | + | + | + | V, M |
| *Amaranthus viridis* | Chaulai | | | + | | | | V, M |
| *Ammannia baccifera* | Dadmari | + | + | | | | | M |
| *Anisomeles indica* | | | + | | | | | |
| *Aponogeton natans* | | | + | | | | | |
| *Argemone mexicana* | Pili kateli | + | + | + | + | + | + | M |
| *Asphodelus tenuifolius* | Piaji | + | + | + | | | | – |
| *Atylosia platycarpa* | | | | | + | | | |
| [a]*Bacopa monnieri* | Brahmi | | + | | | | | M |
| *Barleria prionitis* | Piyabans | | | | | + | + | M |
| *Biophytum sensitivum* | | | | | | + | + | |
| *Blepharis asperrima* | | + | + | | | | | |
| *Blumea eriantha* | | | | + | | | | |
| *Blumea lacera* | Gandhili | + | | | | | | M |
| *Boerhaavia diffusa* | Punarnava | | + | | | + | + | M |
| *Buddleja asiatica* | Neemda | | + | | | | | V |
| *Canna coccinea* | | | | + | | | | Or |

(continued)

**Table 3.6** (continued)

| Species | Local name | Localities | | | | | | Uses |
|---|---|---|---|---|---|---|---|---|
| | | Morena | | | Bhind | | | |
| | | 1 | 2 | 3 | 4 | 5 | 6 | |
| *Cassia kleinii* | | | | | + | | | |
| *Cassia occidentalis* | Ban-chironta | | | | + | + | + | M |
| *Cassia tora* | Puar | | + | | + | + | + | V |
| *Celosia argentea* | Silmili | | | | + | + | + | M, V |
| [a]*Ceropegia bulbosa* | Chakhalgata | | | | + | | | M |
| *Chenopodium album* | Bathua | | | + | | | | V |
| *Chenopodium murale* | Jungli-bathua | | | + | | | | |
| [a]*Chlorophytum tuberosum* | Safed musali | + | | | | + | | |
| *Chrozophora rottleri* | | + | | | | | | |
| *Commelina benghalensis* | | + | | | | | | M |
| *Convolvulus microphyllus* | Shankahauli | + | | | | | | M |
| *Convolvulus prostratus* | Shankh booti | + | + | + | | | | M |
| *Conyza leucantha* | | | | + | | | | |
| *Corchorus aestuans* | | | | | | + | + | M |
| *Corchorus depressus* | Aundhphalli | | | | | + | + | M |
| *Crinum latifolium* | Sukhdarshan | + | | | | | | M |
| *Crotalaria medicaginea* | Gulali | | | | | | + | M, Fi |
| *Curculigo orchioides* | Kali musli | + | | | | | | M |
| *Digera muricata* | Kundhra, lesua | | | | + | + | + | Fo, V |
| *Echinops echinatus* | Unt katera | + | + | + | | | | M |
| *Eclipta prostrata* | Bhringraj | + | | | | + | + | M |
| *Elephantopus scaber* | Jungli-gobhi | + | | | | | | Fo |
| *Elytraria acaulis* | Patsak | | | | + | + | + | M |
| *Emilia sonchifolia* | | + | | | | | | |
| *Enicostema axillare* | Nai | | | | | + | + | |
| *Euphorbia dracunculoides* | | + | | | | | | |
| *Euphorbia hirta* | Dudhi | | | + | + | + | + | M |
| *Euphorbia thymifolia* | Dudhi | | | + | + | + | + | M |
| *Evolvulus alsinoides* | Shankavali | | | | | + | + | M |
| *Helicteres isora* | Marorphali | + | | | | | | M |
| *Heliotropium ellipticum* | | + | | | | | | |
| *Hibiscus caesius* | | + | | | | | | |
| *Indigofera linnaei* | Torki | | | | + | + | + | M |
| *Justicia diffusa* | | | | | + | + | + | |
| *Kickxia ramosissima* | | | + | | | | | |
| *Lathyrus aphaca* | Jungli mattar | | | + | | | | |
| *Launaea acaulis* | | | + | + | | | | Fo |
| *Lepidagathis cristata* | Serbetna | + | + | | + | + | + | M |
| *Lepidagathis hamiltoniana* | Chekna | + | | | | | | M |
| *Martynia annua* | Bicchu buti | + | | | + | + | + | M |

(continued)

**Table 3.6** (continued)

| Species | Local name | Localities | | | | | | Uses |
|---|---|---|---|---|---|---|---|---|
| | | Morena | | | Bhind | | | |
| | | 1 | 2 | 3 | 4 | 5 | 6 | |
| *Melhania hamiltoniana* | Kakai | | | | + | + | + | M |
| *Merremia gangetica* | Anchatani | + | | | | | | M |
| *Mollugo hirta* | | | | | | | + | |
| *Nepeta hindostana* | | + | | | | | | |
| *Ocimum canum* | Tulsi | | | | + | | + | M |
| *Ocimum gratissimum* | Nagad basuri | + | + | | | | | |
| *Ocimum tenuiflorum* | Tulsi | | + | + | | | | R |
| *Orthosiphone pallidus* | Diamzari | + | | | | | | M |
| *Parthenium hysterophorus* | Congress grass | | | | + | + | + | |
| *Pavonia zeylanica* | Sat | | | | | | + | M |
| *Pedalium murex* | Dakhini gokhru | + | | | | | | M |
| *Pedilanthus tithymaloides* | | | | | | | + | M |
| *Peristrophe paniculata* | Atrilal | + | | | + | + | + | M |
| *Phyllanthus niruri* | Bhui-anwala | | | | + | | | M |
| *Phyllanthus viridis* | | | | | | + | | M |
| *Portulaca quadrifida* | Loniya | | | | + | | | V |
| *Potamogeton crispus* | | | + | | | | | |
| *Pupalia lappacea* | Chit-chita | + | | + | + | + | + | M |
| *Rhynchosia minima* | | + | | | | | | M |
| *Sida cordifolia* | Madanmast | | | | | + | + | M |
| *Solanum nigrum* | Makoi | + | | | | | | M |
| *Solanum virginianum* | Bhatkataiyya | + | + | + | + | + | + | M |
| *Sphaeranthus indicus* | Gorakhmundi | + | | | | | | M |
| *Tephrosia purpurea* | Surpunkha | | | | + | + | + | M, Fo |
| *Tephrosia tinctoria* | Sarpunkha, Dhamasa | + | + | | | | | M, Fo |
| *Tribulus terrestris* | Gokhru | | + | | + | + | + | M |
| *Tricholepis radicans* | | + | + | | | | | M |
| *Urginea indica* | Ghamna | + | | | | | | M |
| *Vernonia cinerea* | Sahdevi | | | | + | + | + | M |
| *Withania somnifera* | Ashava-gandha | + | | | | | | M |
| *Xanthium indicum* | Bara gokhru | + | | + | + | + | + | M |
| *Xanthium strumarium* | | + | | | | | | M |
| Climbers among bushes (24) | | | | | | | | |
| [a]*Abrus precatorius* | Ratti | + | | | + | | | M |
| *Ampelocissus latifolia* | Panibel | + | | | + | | | M |
| [a]*Asparagus racemosus* | Shatawar | + | + | + | + | + | + | M, V |
| *Cardiospermum halicacabum* | Kanphuta | + | | | | + | | M |
| *Cuscuta reflexa* | Amarbail | | | | | + | + | |
| *Cayratia trifolia* | Tipani | | + | | + | + | + | M |

(continued)

**Table 3.6** (continued)

| Species | Local name | Localities | | | | | | Uses |
|---|---|---|---|---|---|---|---|---|
| | | Morena | | | Bhind | | | |
| | | 1 | 2 | 3 | 4 | 5 | 6 | |
| *Cissampelos pareira* | Pahadbel | + | | | | | | M |
| *Citrullus colocynthis* | Gursambha | + | | | | | | M |
| *Clitorea ternatea* | Aparjit, Kajroti | | | | | + | + | M |
| *Coccinia grandis* | Kundru | + | | | + | | | V |
| *Cocculus hirsutus* | Chirenta, Jamti ki bel | + | + | + | + | + | + | M |
| *Cryptolepis buchanani* | Naag-bel | | + | | | | | M |
| [a]*Gloriosa superba* | Kalihari | | | | | | + | M |
| *Hemidesmus indicus* | Anantmul | | + | | + | + | + | M |
| *Ipomoea hederifolia* | | | | | + | + | + | |
| *Ipomoea pes-tigridis* | Panch-patri | | | | + | + | + | M |
| *Momordica charantia var. muricata* | Ban karela | + | + | + | + | + | + | M, V |
| *Momordica dioica* | Murela | + | | | | | | M, V |
| [a]*Mucuna pruriens* | Kewanch | + | | | + | | + | M |
| *Pergularia daemia* | Sadowani | + | | | + | + | + | M |
| *Rhynchosia capitata* | Makra | | | | + | | | Fo |
| *Trichosanthes anguina* | Chichinda | | | | + | | | M |
| [a]*Tinospora cordifolia* | Giloy | + | | | | + | + | M |
| *Wattakaka volubilis* | Nakchikni | + | | | + | | | M |
| Grasses and sedges (24) | | | | | | | | |
| *Apluda mutica* | Dinanath | + | + | + | | | | Fo |
| *Aristida adscensionis* | | | | | + | | | Fo |
| *Aristida depressa* | | | | + | | | | Fo |
| *Aristida funiculata* | Lapusari | + | | + | | + | + | Fo |
| *Brachiaria ramosa* | | | | + | | | | Fo |
| *Bulbostylis barbata* | | | + | | | | | |
| *Chloris barbata* | Jarghas | + | | | + | + | + | Fo |
| *Cymbopogon citratus* | Nimbu-ghas | + | | | | | | Oil |
| *Cymbopogon martinii* | Palmarosa | + | | | | | | Oil |
| *Cynodon dactylon* | Dub | + | + | + | | + | + | Fo, M, R |
| *Cyperus rotundus* | Nagarmotha | | + | | | | | M, Fo |
| *Dactyloctenium aegyptium* | Makra-ghas | | + | | + | + | | Fo |
| *Desmostachya bipinnata* | Dabh | + | | | | | | |
| *Heteropogon contortus* | Kusal, lampa | + | | + | + | + | + | Fo |
| *Oplismenus burmannii* | | + | | + | | | | Fo |
| *Oplismenus compositus* | | | + | | + | | | Fo |
| *Paspalidium punctatum* | | | + | | | | | Fo |
| *Pennisetum purpureum* | | + | | | | | | Fo |
| *Saccharum munja* | Jhund | + | | | | | | Basketary |

(continued)

**Table 3.6** (continued)

| Species | Local name | Localities | | | | | | Uses |
|---|---|---|---|---|---|---|---|---|
| | | Morena | | | Bhind | | | |
| | | 1 | 2 | 3 | 4 | 5 | 6 | |
| *Saccharum spontaneum* | Kans | | + | | + | | | Fo, M |
| *Setaria comosa var. pillosa* | | + | | + | | | | Fo |
| *Sporobolus diander* | | + | | | + | + | + | Fo |
| *Sporobolus marginatus* | | + | | | | | | Fo |
| *Thysanolaena maxima* | | + | | | | | | |
| Total (232) | | 105 | 67 | 67 | 87 | 92 | 86 | |

Source: Modified from the reports of MP State Biodiversity Board (www.mpsbb.nic.in/completed project/pdcr.pdf)
Localities depict as 1, Piparai; 2, Jabrol (Sabalgarh); 3, Bamsoli; 4, Barhi; 5, Spad-beet; 6, Ater (deep ravines)
Letters under uses indicate *Fi* fibre, *Fr* fruit, *Fo* fodder, *Fu* fuel, *G* gum/resin, *M* medicine, *O* oil, *Or* ornamental, *R* religion, *T* timber, V vegetable
[a]Very rare in distribution

## 3.3.2 Seasonal Variations

Ravine lands have scrub vegetation and have a monotonous appearance during hot season (summer) due to high temperature and low humidity. The vast stretches of land are barren with a poor cover of dry foliage of grasses and forbs. At places *Prosopis juliflora* communities flourish and seen lush green, and *Capparis decidua* bushes are covered with mix of pink flowers and green fruits frequently harvested by local women, who sell these for pickle-making. *Salvadora persica* and *S. oleoides* also bear fruits, which are sweet but now can be explored commercially for seed oil. As discussed earlier, *Acacia eburnea*, *Balanites roxburghii*, *Acacia nilotica*, *Capparis sepiaria* and *Clerodendrum phlomidis* bushes are prominently visible. At places small trees like *Anogeissus latifolia*, *A. pendula*, *Boswellia serrata*, medicinal *Commiphora wightii*, *Acacia senegal*, *A. tortalis*, *Ziziphus nummularia*, *Dichrostachys cinerea*, *Maytenus emarginata*, *Dichrostachys cinerea* and *Prosopis cineraria* (in Rajasthan areas) are prominent. *Butea monosperma* is common in Madhya Pradesh areas while not common in ravines of Rajasthan area. *Calotropis procera* forms dense patches at some isolated places. Among herbaceous species *Cynodon dactylon*, *Sehima nervosum*, *Chrysopogon fulvus*, *Sporobolus* spp., *Dichanthium annulatum* and *Bothriochloa pertusa* are among the surviving with depleted cover. Clumps of *Saccharum munja* (*benghalensis*), *Vetiveria zizanioides*, *Heteropogon contortus* and *Cymbopogon martinii* are also found. Slopes of the ravines which are not facing the sun directly possess slightly better cover adjacent to bushes. Plants such as *Tridax procumbens*, *Vernonia cinerea*, *Echinops echinatus*, *Euphorbia thymifolia*, *Launaea asplenifolia*, *L. nudicaulis*, *Heylandia latebrosa*, *Solanum nigrum*, *S. surattense*, *Chrozophora prostrata*, *Dipteracanthus prostratus*, *Argemone mexicana* (winter annual), *Kickxia ramosissima*, *Rhynchosia minima*,

*Parthenium hysterophorus*, *Cocculus hirsutus*, *Alhagi pseudalhagi*, *Pluchea lanceolata*, *Cyperus rotundus*, *Ageratum conyzoides*, etc. are found in localised places. On bed of the ravines, *Cynodon dactylon* and *Dichanthium annulatum* are the only grasses possessing better cover. In isolated places *Saccharum spontaneum* and *S. munja* form their communities.

Low-lying areas, after drying of accumulated water, (mostly rich in clay content) are occupied by special types of plant species such as *Argemone mexicana* (forms gregarious pure patch), *Polygonum plebeium*, *Chrozophora prostrata*, *C. rottleri*, *Alhagi pseudalhagi*, *Solanum surattense*, *Cyperus rotundus*, *Polygonum glabrum* and *Rumex dentatus* (in swampy places), *Glinus lotoides*, *Euphorbia hirta*, *E. thymifolia*, *Alternanthera sessilis*, *Verbascum chinensis*, *Sutera glandulosa*, *Datura metel*, *Gnaphalium luteo-album*, *G. indicum*, *Vernonia cinerea*, *Eclipta prostrata*, *Ziziphus nummularia*, *Phyla nodiflora*, *Portulaca oleracea*, *Launaea nudicaulis* and *Parthenium hysterophorus*. On salty patches, halophytes such as *Suaeda fruticosa*, *Kochia indica*, *Heliotropium curassavicum*, *Aeluropus lagopoides*, *Leptochloa fusca*, *Trianthema portulacastrum* and *Sporobolus marginatus* are found as indicator species. Sometimes interesting plants, e.g. *Orobanche aegyptiaca*, are found as parasites on roots of *Nicotiana plumbaginifolia* and also was recorded on roots of *Parthenium hysterophorus*.

At the commencement of the rains, species such as *Argemone mexicana*, *Chrozophora prostrata*, *C. rottleri*, *Polygonum plebeium*, *Echinops echinatus* and *Alhagi pseudalhagi* (only root parts survive) disappear. After the commencement of the first showers of the rain, the seeds of annuals of rainy season germinate and remains of perennial grasses start sprouting and ravines appear green. The species which appear first include *Indigofera linifolia*, *Apluda aristata*, *Cassia tora*, *Setaria glauca*, *Oldenlandia corymbosa*, *Polygala chinensis*, *P. erioptera*, *P. telephioides*, *Achyranthes aspera*, *Xanthium strumarium*, *Parthenium hysterophorus*, *Borreria stricta*, *Enicostemma hyssopifolium*, etc. followed by other annuals of rainy season such as *Tribulus terrestris*, *Biophytum sensitivum*, *Alysicarpus* spp., *Blepharis maderaspatensis*, *Hemiadelphis polyspermus*, *Hemigraphis hirta*, *Lepidagathis cristata*, *Rungia repens*, *Cassia pumila*, *Euphorbia* spp., *Phyllanthus maderaspatensis*, *Iphigenia indica*, *Commelina* spp., *Andropogon pumilus*, *Melanocenchris jacquemontii*, *Cenchrus ciliaris*, *C. setigerus*, *Chloris gayana*, *Sporobolus* spp., etc.

Buds of many species from the exposed roots sprout and get established, particularly along slopes where many plants cannot succeed. Many woody species such as *Balanites aegyptiaca*, *Capparis decidua*, *C. sepiaria*, *C. horrida*, *Diospyros cordifolia*, *Holoptelea integrifolia*, *Pongamia pinnata*, *Clerodendrum inerme*, *Acacia eburnea*, *Alhagi pseudalhagi*, *Enicostemma hyssopifolium*, *Clerodendrum phlomidis* and even *Dalbergia sissoo* have developed the capacity to produce vegetative and reproductive buds on exposed roots and grow to produce new plants (Dagar et al. 1978). This is the basic region of their adaptability and dominance on such lands. Interestingly, some plants such as *Rhynchosia minima*, *Convolvulus microphyllus*, *Evolvulus alsinoides*, *Heylandia latebrosa*, *Tephrosia purpurea*, *Ziziphus nummularia*, *Triumfetta rhomboidea*, *Abutilon indicum* and *Alhagi pseudalhagi* develop very deep root systems in alluvial soil, and when their aerial parts are removed or grazed, these again reproduce from their rootstocks, and if only a part of root system remains in contact with soil, most of these can survive. Some species such as *Ipomoea carnea*, *Polygonum glabrum*, *Rumex dentatus* and *Pongamia pinnata* adapt themselves for marshy situations and form thick associations regenerating from stem nodes and rootstocks in muddy substratum along banks of the river. *Acacia jacquemontii* also regenerates from rootstocks and is found frequently along the banks of rivers or in beds of the ravines facing the river. Perennial grasses such as *Dichanthium annulatum*, *Chrysopogon fulvus*, *Sehima nervosum*, *Cynodon dactylon* and *Bothriochloa pertusa* sustain the grazing pressure throughout the year and can grow from their rhizomes or stem nodes. Species susceptible to heavy grazing are restricted to protected or moderately grazed fields only. These include *Tragus biflorus*, *Cymbopogon martinii*, *Heteropogon contortus* and some annuals such as *Apluda aristata*, *Themeda triandra*, *Setaria glauca*, *Indigofera glandulosa* and many others. Many climbers such as species of *Ephedra*, *Asparagus*, *Momordica*, *Pupalia*, *Cocculus* and *Cardiosperma* love shade and escape in bushes.

Some of the annuals complete their life cycle within a short period and disappear by October. These include *Setaria glauca*, *Oldenlandia corymbosa*, *Polygala* spp., *Biophytum sensitivum* and *Iphigenia indica* followed by *Echinochloa colonum*, *E. crusgalli*, *Melanocenchris jacquemontii*, *Acanthospermum hispidum*, *Cassia pumila*, *C. tora*, etc., and the rest of the species exist till the end of the rainy season. In winter a number of annuals of rainy season disappear and are replaced by annuals of winter season. These include *Vicoa indica*, *Fumaria indica*, *Spergula arvensis*, *Asphodelus tenuifolius*, *Anagallis arvensis*, *Carthamus oxyacantha*, *Cleome viscosa*, *C. brachycarpa*, *Gynandropsis pentaphylla*, *Setaria media*, *Spergula pentandra* and *Argemone mexicana*. Plants which are still found in flowering stage till late March include *Launaea nudicaulis*, *Tridax procumbens*, *Hemiadelphis polyspermus*, *Ocimum americanum*, *Sonchus asper*, *Heliotropium indicum*, *Senebiera didyma*, *Sisymbrium irio*, *Rungia repens*, *Hemigraphis hirta*, *Melilotus parviflora*, *Euphorbia hirta*, *Corchorus* spp. (also found in rainy season), *Abutilon indicum*, *Euphorbia* spp., *Heylandia latebrosa*, *Rhynchosia minima* (persist in rainy season also), *Lagascea mollis*, *Enicostemma hyssopifolium*, *Echinops echinatus* and *Eclipta prostrata*. On exposure of the flood plains, many of the winter annuals appear on low-lying land or newly deposited soil. Some of these complete their life cycle in winter itself, while some of them remain till summer and as discussed in the text above survive in summer also at localized places.

Thus, under different biotic, climatic, topographic and edaphic factors, plants adapt themselves to various adverse of the nature. The ravine lands show extreme xeric characteristics, as the river passes in the vicinity and the level of ground water of ravine lands becomes low. Therefore, besides biotic and other factors, sloping substratum, different solar insulations, direction of slope, different moisture conditions (due to change in the level of water in river), removal of soil due to erosion and deposition of silt are some of the adverse factors which change the nature of the vegetation in comparison to other places.

**Table 3.7** Biological spectrum of Kshipra watershed as compared to normal spectrum

| Life form | Percent distribution of life forms of | |
|---|---|---|
| | Raunkiaer's normal spectrum | Kshipra watershed |
| 1. Phanerophytes | 46 | 20.8 |
| Megaphanerophytes | 8 | 3.9 |
| Mesophanerophytes | – | 10.3 |
| Microphanerophytes | 18 | 3.1 |
| Nanophanerophytes | 15 | 2.7 |
| Epiphytes/succulents/parasites/lianas | 5 | 0.8 |
| 2. Chamaephytes | 9 | 8.4 |
| 3. Hemicryptophytes | 26 | 8.2 |
| 4. Geophytes (including hydrophytes) | 6 | 9.9 |
| 5. Therophytes | 13 | 52.7 |

### 3.3.3 Biological Spectrum

The phytoclimate of a region is characterized by the dominant life forms in biological spectrum given by Raunkiaer (1934) which is relevant even today. The information on biological spectrum is also valuable for expressing differences and similarities among communities (Yadava and Singh 1977; Dagar and Balakrishnan 1984). Meher-Homji (1964), Puri et al. (1983) and Dagar and Singh (1999), while comparing the biological spectrum of various regions of the country, have shown that these spectra for different regions are related to their bioclimates. Very little attempts have been made in the country to study the biological spectrum of eroded habitats. Dagar (2001) carefully listed 416 phanerogamic species distributed in 167 genera belonging to 85 families in Kshipra watershed along with their life forms and compared with Raunkiaer's normal spectrum. It was found that phanerophytes (including mega-, meso-, micro- and nano-phanerophytes and lianas and succulents), chamaephytes, hemicryptophytes, cryptophytes (including hydrophytes in river water) and therophytes represented 20.8%, 8.4%, 8.2%, 9.9% and 52.7%, respectively (Table 3.7). The trend across Chambal River was also found similar though the results were not published.

It is evident from the data that the phytoclimate of Kshipra catchment is of therophytic nature as anthropogenic factors such as grazing and human interferences greatly reduce phanerophytic flora and replaced by therophytic type. The presence of phanerophytes is lower than half of the normal and cryptophytes and chaemophytes more and equal, respectively, while therophytes is almost four times higher than the normal. The preponderance of therophytes and cryptophytes in the eroded habitats reflects periodicity in the climatic conditions and the intensity of biotic pressure such as grazing and removal of fuelwood from woody vegetation. Thus, the therophytes can survive the hot and xeric conditions of summer as seeds or clumps of vegetation; while the cryptophytes are able to withstand the grazing pressure because of the hidden, sub-surface position of their perennating buds.

It is important to mention here that both therophytes and phanerophytes share equal importance in depicting a climate of dual character having warm-dry climate and warm-moist climate. No doubt, phanerophytic climate is found in tropics with humid atmosphere, while therophytic climate is found in sub-tropical regions where most of these survive in unfavourable season in the form of seed and complete their life cycle within a single favourable season. It is interesting to note here that no doubt the watershed areas of ravine lands have therophytic vegetation, but at the same time, these are occupied by considerable number of phanerophytes also particularly in protected areas. This shows that more therophytes are present because of deforestation and overgrazing. Further, it is clear from the discussion in next section that if we protect the vegetation from grazing and other biotic interferences, the climax vegetation will be tree-dominating. Therefore, it is not safe to designate the phytoclimate in such cases as therophytic, and the biological spectrum in such cases is an indication of biotic exploitation. Therefore, the inference can be drawn that the watershed areas of the rivers having ravines favour the luxuriant growth of phanerophytes along with therophytes provided the over-exploitation, especially overgrazing, is stopped.

### 3.3.4 Successional Trends

Ecological succession is the process by which the composition of species in a community changes over time. Following a disturbance, an ecosystem generally progresses from a simple level of organization (i.e. few dominant pioneer species) to a more complex community over time. Depending on the severity of the disturbance, restoration often consists of initiating, assisting or accelerating ecological successional processes (Luken 1990). For a system, that has experienced a more severe disturbance (as is the case of ravine lands), it may require intensive efforts to restore the environmental conditions for favouring the natural successional processes. During their migration disseminules of plant species from the neighbouring areas colonize the fresh areas provided that the habitat conditions are suitable for their growth. In case the fresh area is completely devoid of vegetation, the colonizers happen to be the pioneer species, and these react upon the habitat in many ways and change it. In the process, the changed habitat becomes unsuitable for the growth and the perpetuation of the pioneer species and many invader species find a place for their establishment. The process is repeated in time and space, and at any time the old lingering species can be seen growing side by side (in limited numbers) with the more aggressive invaders. In nature, the interaction between the species develops in the final stage of vegetation, and this phenomenon of community development is called plant succession (Misra 1962).

In most of the ravine lands, there are patches of freshly exposed soil due to erosion. It has been found that many annual species such as *Cassia pumila*, *Glossocardia bosvallia*, *Verbena bipinnatifida*, *Tridax procumbens*, *Setaria glauca*, *Arthraxon ciliaris*, *Alysicarpus monilifer* and *Rhynchosia minima* appear on the base of the slopes, while *Melanocenchris jacquemontii*, *Indigofera linifolia*, *Justicia diffusa*, *Oldenlandia corymbosa*, *Oropetium thomaeum* and *Apluda mutica*, etc. appear on bare plateau (top) of ravines. These species are followed by other species such as

*Triumfetta rhomboidea*, *Echinops echinatus*, *Andropogon pumilus*, *Aristida adscensionis*, *Indigofera linifolia*, *Eragrostis* spp., *Sporobolus coromandelianus*, *S. diander*, *Chloris barbata*, *Imperata cylindrica*, *Polygala chinensis*, etc. With further protection, species like *Iseilema laxum*, *I. anthephoroides*, *Alysicarpus longifolius*, *Bothriochloa pertusa* and *Dichanthium caricosum* appear. These are followed by *Dichanthium annulatum*, *Sehima nervosum*, *Heteropogon contortus*, *Themeda triandra*, *Chrysopogon fulvus*, *Cymbopogon martinii* and *Apluda aristata*. These may form isolated or mixed communities. Among these grasses there lies high proportion of legume component consisting of species such as *Alysicarpus* spp., *Desmodium triflorum*, *Rhynchosia minima*, *Crotalaria medicaginea*, *Indigofera glandulosa*, *I. trifoliata*, etc. Most probable pattern of development of plant communities in Kshipra watershed (in Chambal also the pattern is almost similar except change in a few components in Rajasthan region) is shown in Fig. 3.2, where the climax is savannah forest, provided the area is protected from biotic interference. Soil erosion and the low capacity of soil in retaining the moisture are mainly responsible for the existence of thorn-scrub vegetation in these areas.

Due to heavy grazing, the most preferred species are always hit hard. That is the reason most palatable grasses such as *Bothriochloa pertusa* and *Dichanthium annulatum* are grazed the most and will be eliminated as compared to less palatable like *Sehima nervosum*. Similarly, in Rajasthan area, in a grassland with *Cenchrus ciliaris* and *Lasiurus sindicus*, the former will be first to go out. Closure to grazing brings the useful species such as *Apluda aristata*, *D. annulatum*, *Heteropogon contortus* and *Themeda triandra* with sufficient number of legumes.

In Rajasthan area, the sandy or sandy loam soils in ravines of Chambal and other rivers get the invasion of species such as *Aristida funiculata*, *Melanocenchris jacquemontii*, *Tephrosia purpurea*, *Indigofera cordifolia*, *Eragrostis tenella*, etc. Under protection, perennial grasses and undershrubs gain the upper hand in 3–4 years, and species such as *Ziziphus nummularia*, *Capparis* spp. and *Prosopis cineraria* establish well. At many places, *Balanites aegyptiaca*, *Calotropis procera*, *Cassia auriculata*, *Crotalaria burhia* and *Maytenus emarginata* are the prominent undershrubs and shrubs before turning into *Prosopis*-*Salvadora* type of vegetation formation. Other common species at various stages of vegetation community formation include *Cenchrus ciliaris*, *C. setigerus*, *C. biflorus*, *Aristida adscensionis*, *Dactyloctenium sindicum*, *Eleusine compressa*, *Panicum turgidum*, *Lasiurus sindicus*, *Cymbopogon jwarancusa* and *Desmostachya bipinnata* among grasses; *Calligonum polygonoides*, *Commiphora wightii*, *Lycium barbarum*, *Acacia jacquemontii*, *Mimosa hamata*, *Tecomella undulata*, *Carissa carandas*, *Tamarix* sp. and *Euphorbia* spp. among shrubs; and *Acacia senegal*, *A. nilotica*, *A. tortilis*, *Cordia rothii*, *Anogeissus pendula*, *Azadirachta indica* and *Salvadora oleoides* among trees. All these are found frequently forming stable communities.

Impact of browsing by animals is also prominent on trees and shrubs. Species such as *Acacia eburnea*, *A. nilotica*, *A. senegal*, *Balanites aegyptiaca*, *Dichrostachys cinerea* and *Capparis* species assume cushion or bush form due to browsing of tops, and *Ziziphus nummularia* is reduced to ground surface due to excess browsing. These along with many others (e.g. *Butea monosperma*, *Dalbergia sissoo*,

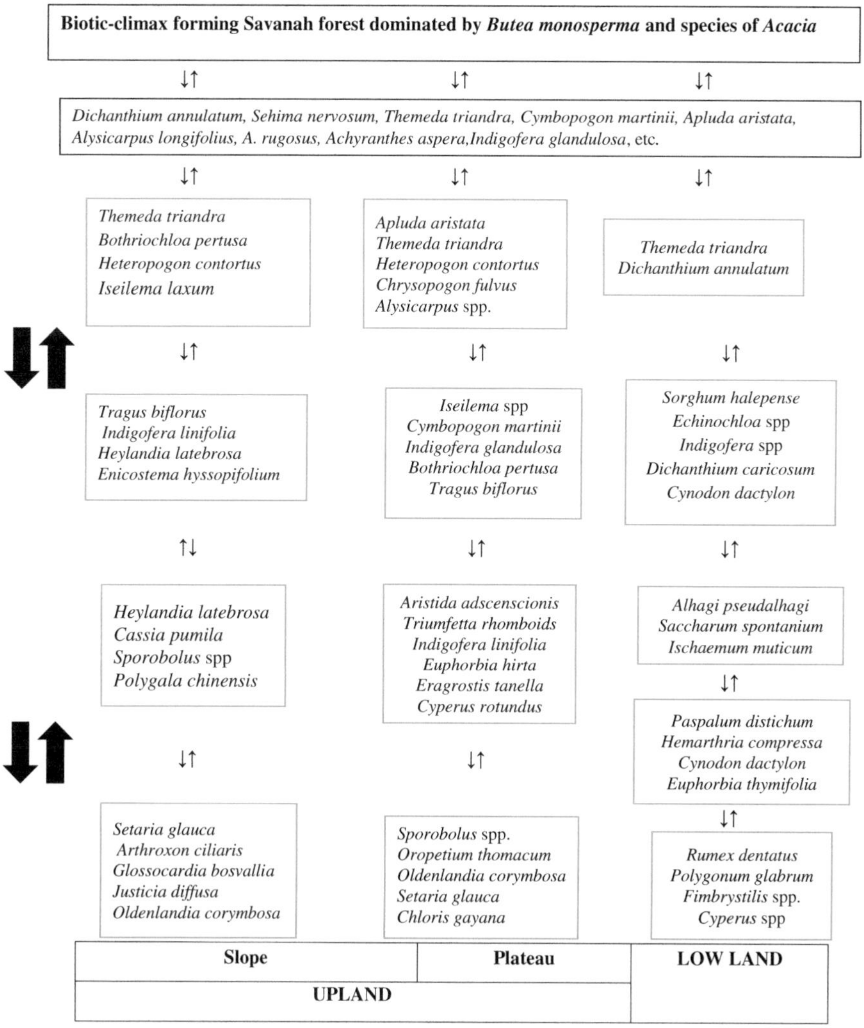


**Fig. 3.2** Probable outline of the trend of succession of vegetation in Kshipra watershed. Upward arrows show progressive (after protection), and downward arrows show retrogressive (due to biotic pressure) successional trend

*Azadirachta indica*, *Cordia dichotoma*, *Aegle marmelos* and *Feronia limonia*) are found in better forms in protected areas developing enormous coverage.

Bor (1960) considered that the climate in India is either a forest or a desert climate; therefore, the grasslands in India are considered as interim stage in the succession (seral stage) and often remain somewhat stable (preclimax) under the influence of repeated fire and grazing. Though the effect of climate on the distribution of various types of vegetation cannot be underrated, the occurrence of different communities may also be largely determined by soil moisture, soil type (texture, depth, etc.) and topography.

Many workers in India (Saxena 1977; Dagar 1978; Singh et al. 1979; Dagar 2001; Dagar and Pathak 2005; Pathak and Dagar 2015) feel that the monoclimate theory of succession cannot fulfil the criteria for the preclimax to change towards progression, unless and until the whole climate itself changes. Therefore, there is much inclination towards the polyclimax view. Once the existence of grazing and burning is admitted, the term climatic climax cannot be applied to the community under any system of terminology, and these grasslands (particularly in ravines) can, therefore, be considered only as biotic climax.

Further, with an increase in the organic matter in the soil, *Heteropogon contortus* becomes dominant and later replaced by *Bothriochloa pertusa-Dichanthium annulatum* community on premature soil which can be maintained by moderate grazing. However, when the soil is established into mature soil with high organic matter, the grassland is dominated by *Themeda-Iseilema* community. On calcium-rich soil, however, *Sehima-Chrysopogon* community may develop (Dagar 2001). In secondary succession, *Bothriochloa-Dichanthium* becomes the dominant on stabilized soil. Shankar et al. (1971) observed that the stable grassland community of *Sehima nervosum* if protected for a long time develops into a forest supporting the view that the grasslands in India represent bio-edaphic serial stages which perpetuate through human manipulation in the form of grazing, harvesting and burning.

Some species invade in the area and spread very quickly eliminating other species. For example, *Prosopis juliflora* starts as a pioneer or colonizer on denuded abandoned ravines where the natural forest is tropical thorn forest. The speed in which it proceeds to colonize is amazingly fast. In 5–10 years, the natural regeneration is complete, and one can encounter frondose habit of bushy saplings to small trees. Where it was introduced, its cover expanded in the second, third, fourth, fifth and tenth year to 1–5%, 5–15%, 10–20%, 20–50% and 35–90% of the total area, respectively. It is cut for firewood from the second or third year, and this is repeated once every 6–18 months thereafter. *P. juliflora* tolerates this heavy abuse and even appears to thrive on it, and having no competitors this results in gregarious, thick and pure patches of this species which dominates sites under severe biotic and edaphic conditions.

In the northern tropical thorn forest, *P. juliflora* occurs in private and government revenue land. In the 50–60th year after initial establishment in such lands, *P. juliflora* forms only 15–20% of the total tree composition (Chinnamani 1998; Table 3.8), there being a drastic reduction in the compositional gradient from the fifth to tenth year when *P. juliflora* accounts for 35–90% of the total. This shows that this species is a pioneer colonizer in ravines and forms better conditions for other species to colonize during passage of time. Existing as one of the associates of the northern tropical thorn forest, *P. juliflora* is eliminated by secondary successors with further time-course changes (Table 3.9), leading to dry deciduous forest in 100–200 years as species of a higher order replace it in the successional gradient (Chinnamani 1992, 1998).

From the above discussion, the inference can be drawn that the area under ravines is facing extreme biotic stress and the vegetation is also a biotic climax. After closing to grazing and other anthropogenic interferences, the entire area may be converted into a dry deciduous or savannah forest depending upon ecoclimate (rainfall), which consequently offers more effective check against erosion.

**Table 3.8** Tree species composition of northern tropical thorn forest in the ravines of North India

| Species | % Composition | Species | % Composition |
|---|---|---|---|
| *Prosopis juliflora* | 15.0 | *Tamarindus indica* | 3.0 |
| *Prosopis cineraria* | 14.0 | *Acacia senegal* | 2.5 |
| *Azadirachta indica* | 14.0 | *Cordia myxa* | 2.5 |
| *Acacia nilotica* | 10.0 | *Albizia lebbeck* | 2.0 |
| *Acacia leucophloea* | 8.0 | *Dalbergia sissoo* | 2.0 |
| *Holoptelea integrifolia* | 7.0 | *Dichrostachys cinerea* | 2.0 |
| *Ziziphus nummularia* | 5.0 | *Feronia elephantum* | 2.0 |
| *Balanites aegyptiaca* | 3.0 | Other tree species | 8.0 |

Source: Chinnamani (1998)

**Table 3.9** Successional trends in northern tropical thorn forest under two different managements (Chinnamani 1998)

| Time scale | Protected conditions | Grazing and fire |
|---|---|---|
| Dry deciduous forest | | |
| 100–200 years | *Salmalia – Terminalia* | *Holoptelea – Cordia* |
| Northern tropical thorn forest | | |
| 50–70 years | *Acacia – Albizia* | *Prosopis – Balanites* |
| 5–10 years | *Cassia – Capparis – Grewia* | *Lantana – Prosopis* |
| 4–6 years | *Dichanthium* | *Eremopogon – Prosopis* |
| 2–4 years | *Heteropogon* | *Eremopogon – Prosopis* |
| Under 2 years | *Aristida – Sporobolus* | *Eragrostis – Desmostachya* |
| | Bare ground | Bare ground |

## 3.4 Root Ecology

Revegetation strategies for erosion control rely in most cases on the effects of the aboveground biomass on reducing water erosion rates, whereas the role of the belowground biomass is often neglected. In the context of ravines in India, the aboveground biomass temporarily disappear because of overgrazing or fire, and when concentrated flow erosion occurs, roots can play an important role in controlling soil erosion rates. Plant roots play an important role in stabilizing slopes and stream banks. These tend to respond to unfavourable stress conditions in a self-correcting manner through a bio-adaptive process (Gray and Barker 2004).

In Indian ravines, as discussed earlier, the roots of many woody species such as *Balanites aegyptiaca*, *Pongamia pinnata*, *Capparis sepiaria*, *C. horrida*, *C. decidua*, *Acacia eburnea*, *Dichrostachys cinerea*, *Butea monosperma*, *Ziziphus mauritiana*, *Z. nummularia*, *Holoptelea integrifolia*, *Clerodendron phlomidis* and *Feronia limonia* growing in eroded ravines get exposed due to erosion, and numerous adventitious buds are developed on them which sprout and when attached to the soil substratum establish new shoots. This is an adaptation to eroded habitats. After travelling horizontally, most of these roots bend vertically and divert towards the river flow, due to positive response to water. Sometimes a curvature forming a knee-like structure in the main root or the secondary sinker roots of the tree is developed.

Vanicek (1973) referred this bio-adaptive feature as "edaphoecotropism" which refers to the ability of plant roots to adapt to or avoid stress, vis-à-vis mechanical obstacles, lack of moisture, light and deep burial. The mechanical reinforcing or restraining influence of roots on a slope is usually limited to a zone about 1.5 m from the surface. Shields and Gray (1993) measured root-area-ratios (RAR) (the fraction of the total cross-sectional area of a soil that is occupied by the roots) for a variety of woody plant species and found that RAR tended to decrease exponentially with depth and only negligible roots were encountered below a depth of 1.2 m. Root spreads reported in the technical literature typically range from one to three times (at least 1.5 times) the radius of the crown (Gray and Barker 2004).

In a Mediterranean context where, unfortunately, information on root characteristics of plants growing on seminatural lands and their effects on the topsoil resistance to concentrated flow erosion is lacking, de Baets et al. (2007) examined typical Mediterranean grass, herb, shrub and tree root systems of plants growing in habitats that are prone to concentrated flow erosion (i.e. in ephemeral channels, abandoned fields and steep badland slopes); and their erosion- reducing potential was evaluated. They measured root density (RD), root-length density (RLD) and root diameters of 26 typical Mediterranean species within 10–90 cm of the soil profile and then transformed into relative soil detachment rates using an empirical relationship to predict the erosion-reducing effect of root systems during concentrated runoff. RD in the 10 cm thick topsoil ranged between 0.13 kg $m^{-3}$ for *Bromus rubens* and 19.77 kg $m^{-3}$ for *Lygeum spartum*, whereas RLD ranged between 0.01 km $m^{-3}$ for *Nerium oleander* and 120.43 km $m^{-3}$ for *Avenula bromoides*. Grasses such as *Helictotrichon filifolium*, *Piptatherum miliaceum*, *Juncus acutus*, *Avenula bromoides*, *Lygeum spartum* and *Brachypodium retusum* showed the highest potential to reduce soil erosion rates by concentrated flow in the 10 cm topsoil.

Shrubs such as *Anthyllis cytisoides* and *Tamarix canariensis* having high root densities in topsoil had the potential of reducing soil erosion drastically. Among the species growing in channels, *Juncus acutus* is reported to have erosion-reducing potential, whereas *Phragmites australis* the least effective. On steep badland slopes, *Helictotrichon filifolium* and *Anthyllis cytisoides* performed the best in the analysis of erosion-reducing potential. These findings have implications for ecological restoration and management of erosion-prone gullies. Thus, root development and structure are affected initially by genetic disposition but ultimately are governed more by environment and edaphic conditions. During evolution, a great variety of root systems have adapted to overcome the many physical, biochemical and biological problems brought about by soil and anthropogenic interferences (Ranade et al. 2002; de Kroon and Visser 2003).

While investigating root ecology of ravine lands, Dagar (1978, 1986, 1987d, 1989) observed that most of the herbaceous species go for different adaptation strategies. Development of root structure of 73 forbs was compared and found that early in the rainy season, the species have tendency to develop long shoots and dwarf root systems with high moisture content in the substratum. Annuals with short life span bear shallow root systems, while the perennials produce deep root systems because at later stage they must face the moisture stress. The xeric phase of soil seems to encourage deep vertical extension. If first order of laterals increases, then root depth decreases. Root elongation rate is also inversely related to soil strength. Taylor (1976) emphasized that species are distinguished by a considerable degree of plasticity in shaping their roots which in turn depends upon ecological conditions such as type of soil, relief of the locality, soil nutrients, water and atmospheric regions of the site, aboveground conditions, surrounding community of other plants and biotic factors. In soils displaying satisfactory structures, favourable water, atmospheric balances and moderate soil nutrients, the roots are relatively longer and more strongly remitted than those developing in fertile and less oxygen-supplied soils (Vanicek 1973; Gray and Barker 2004; Reubens et al. 2007).

Biotic factors such as grazing have immense impact on structure and composition of soil, which consequently affects root stratification. In ravine habitats, due to grazing pressure and xeric conditions, comparatively roots travel deep in soil. Aboveground biomass of plants also has ramifications, and due to removal of herbal cover (due to grazing), root development is reduced. Soil-binding factor is directly proportional to the dry weight of the roots and inversely proportional to the mean radius of the roots (Bhimaya et al. 1956; Tejwani et al. 1975; Dagar 1986). Thus, length, weight and thickness of the roots play an important role in binding the soil; hence plant species with maximum spread of roots with numerous thin laterals occupying more volume and bearing more root hairs are better soil binders. Tejwani et al. (1975) and Dagar (1986) evaluated suitability of some grasses in eroded habitats to conserve soil and found that *Dichanthium caricosum*, *D. annulatum*, *Cynodon dactylon*, *Sehima nervosum*, *Apluda aristata*, *Bothriochloa pertusa*, *Iseilema laxum*, *I. rugosum* and *Chrysopogon fulvus* were found having higher soil-binding factors in ravine lands as compared to other grasses (Table 3.10).

For stabilizing earthen bunds in the ravines of Gujarat, *Dichanthium annulatum*, *Amphilophis glabra* and *Andropogon ischaemum* were found superior soil binders because of their elaborate root systems and excellent ground cover and fodder yield (Table 3.11). *Cenchrus ciliaris* (rhizomatous strain) was also found to have good soil-binding capacity with well-developed roots and ground cover. Dagar (1995) in a well-conducted experiment found that grasses such as *Dichanthium annulatum*, *D. caricosum*, *Cynodon dactylon*, *Sehima nervosum*, *Bothriochloa pertusa*, *Cymbopogon martinii*, *Chrysopogon fulvus*, *Iseilema laxum*, *Heteropogon contortus* and *Themeda triandra* are quite effective soil binders both in vegetative phase and when clipped keeping roots intact.

Ranade et al. (2002) and Reubens et al. (2007) reviewed the role of fine and coarse roots in shallow-slope stability and soil erosion control with a focus on root system architecture. The effect of soil properties and the relative importance of different root sizes and of woody versus nonwoody species are briefly discussed. Empirically and intuitively, architectural features seem to determine the effect of root systems on erosion phenomena, but, still, the research to underpin this relationship is poorly developed. A variety of methods are available for detailed root system architectural measurement and analysis. Although, generally time-consuming, a full

**Table 3.10** Characteristics of root systems of grasses growing in ravine lands of Kshipra River

| Grass species | Average root diameter (mm) | Average root depth (cm) | Soil-binding factor[a] |
|---|---|---|---|
| *Apluda aristata* | 0.42 | 24.0 | 429 |
| *Bothriochloa pertusa* | 0.43 | 65.0 | 422 |
| *Cenchrus ciliaris* | 0.62 | 43.6 | 293 |
| *Chloris barbata* | 0.41 | 16.2 | 183 |
| *Chrysopogon fulvus* | 0.62 | 96.8 | 400 |
| *Cymbopogon martinii* | 0.58 | 100.0 | 392 |
| *Cynodon dactylon* | 0.46 | 35.5 | 471 |
| *Dichanthium annulatum* | 0.46 | 94.3 | 499 |
| *Dichanthium caricosum* | 0.53 | 87.4 | 521 |
| *Digitaria adscendens* | 0.40 | 55.0 | 183 |
| *Echinochloa colonum* | 0.38 | 26.1 | 349 |
| *Echinochloa crusgalli* | 0.44 | 29.5 | 364 |
| *Eleusine indica* | o.50 | 39.4 | 198 |
| *Heteropogon contortus* | 0.48 | 87.6 | 427 |
| *Ischaemum rugosum* | 0.55 | 78.7 | 390 |
| *Iseilema laxum* | 0.44 | 23.0 | 405 |
| *Melanocenchris jacquemontii* | 0.18 | 12.8 | 266 |
| *Panicum antidotale* | 0.60 | 56.1 | 349 |
| *Sehima nervosum* | 0.62 | 63.7 | 450 |
| *Setaria glauca* | 0.44 | 18.0 | 151 |
| *Sorghum halepense* | 1.24 | 29.8 | 187 |
| *Sporobolus diander* | 0.34 | 18.4 | 327 |
| *Themeda triandra* | 0.46 | 43.6 | 261 |
| *Tragus biflorus* | 0.66 | 40.1 | 378 |
| *Vetiveria zizanioides* | 0.54 | 49.8 | 159 |

[a]Soil-binding factor = Dry root biomass (in 25 × 25 × 25 cm soil monolith)/(radius)$^2$

3D architectural description followed by analysis in software such as AMAPmod offers the possibility to extract relevant information on almost any root system architectural characteristics.

## 3.5 Ravine Restoration Ecology

The ecological restoration is an intentional activity that initiates or accelerates the recovery of an ecosystem with respect to its health, integrity and sustainability (SER 2004). The restoration practice includes a wide scope such as erosion control, reforestation, usages of genetically local native species, removal of exotics and weeds, revegetation of disturbed areas (here ravines), reintroduction of native species and improvement of habitat and range of targeted species. The natural ecosystems provide human society with food, fuel and timber and above all the ecosystem services, which have been estimated to be worth trillions of dollars

**Table 3.11** Soil-binding capacity of some grasses in ravines of Vasad under protection (core size 10 cm diam × 22.5 cm depth)

| Grass species | Average root radius (mm) | Root volume (ml) | Soil-binding factor[a] |
|---|---|---|---|
| *Amphilophis glabra* | 0.58 | 27 | 80 |
| *Amphilophis odorata* | 0.47 | 12 | 54 |
| *Andropogon ischaemum* | 0.31 | 10 | 104 |
| *Cenchrus ciliaris* (local) | 0.45 | 19 | 94 |
| *C. ciliaris* (Australia) | 0.65 | 23 | 54 |
| *Chloris gayana* | 0.43 | 9 | 49 |
| *Cynodon dactylon* | 0.31 | 5 | 52 |
| *Dichanthium annulatum* | 0.33 | 20 | 184 |
| *Eragrostis superba* | 0.45 | 16 | 79 |
| *Eulaliopsis binata* | 0.39 | 14 | 92 |
| *Panicum antidotale* | 0.43 | 19 | 103 |
| *Panicum repens* | 0.12 | 18 | 1250 |
| S.E. (%) | | | 100 |
| LSD ($p \leq 0.05$) | | | 67 |

[a]Soil-binding factor = volume/radius$^2$

annually (Daily et al. 1997). Habitat loss is the leading cause of both species extinction (Wilson 1988) and ecosystem service decline (Daily et al. 1997). The conservation of currently viable habitat and restoration of degraded ravine habitats are the main ways to reverse this trend of habitat loss.

Each year, about 75 billion Mg of soil are eroded from the world's terrestrial ecosystems (Pimentel and Kounang 1998). Most of the agricultural land in the world is losing soil at rates ranging from 13 to 40 Mg ha$^{-1}$ year$^{-1}$. Rain and wind energies are the two prime causes of erosion from tilled to bare land. Erosion occurs when the soil lacks protective vegetative cover. Revegetation strategies for erosion control and restoration of habitat mainly depends on choice of most suitable species which can easily adapt to the environment and protect soil erosion. The following strategies may be handy in restoration of ravine areas for sustainable biological production and habitat conservation:

### 3.5.1 Protection from Grazing

Most of the ravine areas are under tremendous anthropogenic influence including overgrazing; therefore, it is necessary that the ravine areas must be protected from grazing. This will increase density of existing species and better forage and soil-conserving species which are susceptible to grazing will reappear which might had been disappeared due to overgrazing (Fig. 3.3). On an average 938 and 159 g m$^{-2}$ aboveground biomass and 587 and 237 g m$^{-2}$ belowground biomass were observed in protected and grazed ravine areas, respectively, in upper 30 cm soil depth.

**Fig. 3.3** Stabilization of ravines by natural vegetation after long protection from grazing (Photo by Yaduvendra Singh)

**Table 3.12** Belowground biomass (g $m^{-2}$) at different depths of soil as measured in July (values as mean ± SE); values in parenthesis denote the percentage of total root biomass

| Field condition | 0–10 cm | 10–20 cm | 20–30 cm | >30 cm |
|---|---|---|---|---|
| Protected | 772.0 ± 83.2 (48.4) | 451.7 ± 60.8 (28.3) | 246.7 ± 33.6 (15.5) | 124.8 ± 25.6 (7.8) |
| Grazed | 232.6 ± 67.7 (54.8) | 115.2 ± 24.3 (27.1) | 56.2 ± 13.2 (13.2) | 21.0 ± 6.1 (4.9) |

Dagar (1987a) observed that in July about 78.7% and 81.8% of total belowground biomass were restricted to 20 cm of soil in protected and grazed plateau, respectively, and the belowground biomass reduced abruptly in grazed field as compared to protected (Table 3.12).

The above- and belowground biomass in protected and grazed fields was maximum in rainy season and minimum in summer, hence mainly depending upon rainfall and temperature and phenology of the species. The relationship between biomass and rainfall and temperature in protected and grazed fields was found as follows:

$$Y_1 = 2252.5844 + 4.1091\ X_1 – 69.0075\ X_2\ (R^2 = 0.41; p \le 0.05)$$
$$Y_2 = 428.7562 + 0.8341\ X_1 – 14.1307\ X_2\ (R^2 = 0.69; p \le 0.05)$$
$$Y_3 = 967.6142 + 2.1771\ X_1 – 23.7078\ X_2\ (R^2 = 0.65; p \le 0.05)$$
$$Y_4 = 483.\ 4168 + 1.0306\ X_{1\_}13.9081\ X_2\ (R^2 = 0.64; p \le 0.05)$$

where $Y_1$ and $Y_2$ are aboveground biomass ($gm^{-2}$) in protected and grazed fields, respectively, $Y_3$ and $Y_4$ are the corresponding belowground biomass values, $X_1$ is rainfall (mm per month), and $X_2$ is mean temperature (°C mean daily).

Although clipping does not simulate grazing precisely, an appreciation of grazing responses can be obtained by imposing clipping treatments. Therefore, different frequencies at different heights of clipping reveal the change in the herbage and subterranean biomass of plant species particularly in grasses (Caughenour et al. 1984; Pande and Singh 1985), and this may help in managing grazing in rangelands. Dagar (1987b) experimented during rainy season taking nine grasses found growing in ravines of different strata and clipped at different heights (depending upon the stratum of a species) and intervals of time (15, 30 and 45 days initiating after 2 months of establishment, thus clipped six, three and two times, respectively, during 5 months of growth). Of these species, *Dichanthium caricosum*, *D. annulatum*, *Chrysopogon fulvus* and *Sehima nervosum* produced more aboveground dry matter under selected conditions of clipping as compared to unclipped, while in other species, the aboveground biomass decreased, and it varied with species and frequency of clipping. Belowground biomass decreased in each case except one treatment in *S. nervosum* (Table 3.13).

The reduction in total biomass was more than 50% when clipped at the interval of 15 days (representing frequent grazing) in *D. annulatum*, *D. caricosum*, *T. triandra* and *I. laxum*, while in others it was less than 50%. The frequent clipping (grazing) and consequent recovery allow little time for the manufacture of photosynthate; hence downward translocation is limited resulting the reduction in biomass production. Moreover, during this time the earlier accumulated total nonstructural carbohydrates in these organs are utilized internally to support their maintenance respiration, which may be enhanced due to wounding effect (Evans 1972) and the increased exudation losses (Bokhari and Singh 1974). In one study Edroma (1985) observing the effects of clipping on *Themeda triandra* and *Brachiaria platynotan* found that the clipping stimulated tillering and yield but later depressed them, severely when cut fortnightly at lower heights. Dry matter production in both shoots and roots increased with rising clipping heights and interval and in all the observed species root/shoot ratio of unclipped plants was higher than the clipped ones.

Observations in the ravines of Yamuna river near Agra indicated that on protection from grazing, the quantity and quality of perennial grasses such as *Dichanthium annulatum*, *Cenchrus ciliaris* and *Eremopogon foveolatus* increased and of annuals like species of *Aristida* and *Eragrostis* decreased. On an average, the length and biomass of roots were found to be more by 45% and 20%, respectively, in protected plots as compared to grazed plots. Similarly, ecological studies at Bellary (Ootacamund) showed that after protection, species of *Eragrostis*, *Arundinella*, *Tripogon*, etc. are replaced by *Themeda*, *Ischaemum* and *Chrysopogon* as subtypes and finally *Sehima-Dichanthium*-type vegetation is developed. In these grazing lands, there lies very high proportion of fodder legumes like *Alysicarpus longifolius*, *A. rugosus*, *Desmodium diffusum*, etc. along with other grass species such as *Iseilema laxum*, *Bothriochloa pertusa* and *Eremopogon foveolatus*.

After protection, there is tremendous improvement in soil health with increment in soil organic matter, water-holding capacity, water-stable soil aggregates and infiltration rate. Somasundaram et al. (2009) observed different soil properties

**Table 3.13** Shoot and root biomass ($gm^{-2}$) of different grass species when clipped at different heights and interval of time

| Species and clipping height | Shoot biomass | | | | Root biomass | | | |
|---|---|---|---|---|---|---|---|---|
| | 15 days | 30 days | 45 days | Unclipped | 15 days | 30 days | 45 days | Unclipped |
| *Dichanthium caricosum* | | | | 183 | | | | 75 |
| Ground level* | 95 | 123 | 140 | | 21 | 35 | 41 | |
| 8 cm | 101 | 135 | 187 | | 28 | 39 | 49 | |
| 16 cm | 104 | 132 | 184 | | 28 | 36 | 47 | |
| *Dichanthium annulatum* | | | | 144 | | | | 62 |
| Ground level* | 73 | 98 | 127 | | 12 | 21 | 29 | |
| 8 cm | 80 | 112 | 149 | | 15 | 27 | 34 | |
| 16 cm** | 88 | 124 | 147 | | 18 | 31 | 37 | |
| *Heteropogon contortus* | | | | 185 | | | | 69 |
| Ground level | 112 | 131 | 274 | | 27 | 34 | 43 | |
| 8 cm | 131 | 187 | 229 | | 31 | 37 | 50 | |
| 16 cm** | 136 | 191 | 243 | | 29 | 39 | 57 | |
| *Themeda triandra* | | | | 117 | | | | 33 |
| Ground level | 66 | 79 | 87 | | 9 | 11 | 16 | |
| 8 cm | 80 | 88 | 100 | | 12 | 14 | 19 | |
| 16 cm | 72 | 76 | 106 | | 15 | 17 | 23 | |
| *Chrysopogon fulvus* | | | | 173 | | | | 98 |
| Ground level* | 152 | 159 | 163 | | 20 | 37 | 42 | |
| 10 cm** | 170 | 176 | 183 | | 23 | 51 | 53 | |
| 20 cm** | 174 | 179 | 182 | | 39 | 49 | 64 | |

| | | | | | | | | |
|---|---|---|---|---|---|---|---|---|
| *Sehima nervosum* | | | | 159 | | | | 89 |
| Ground level | 129 | 136 | 214 | | 34 | 58 | 23 | |
| 10 cm | 143 | 107 | 203 | | 39 | 60 | 97 | |
| *Bothriochloa pertusa* | | | | 133 | | | | 45 |
| Ground level*, ** | 66 | 111 | 125 | | 14 | 24 | 32 | |
| *Iseilema laxum* | | | | 122 | | | | 37 |
| Ground level*, ** | 58 | 97 | 116 | | 15 | 20 | 26 | |
| *Tragus biflorus* | | | | 142 | | | | 281 |
| Ground level* | 110 | 133 | 128 | | 154 | 216 | 195 | |

*Correlation coefficient between clipping frequency and shoot biomass significant at $p \leq 0.05$
** Correlation coefficient between clipping frequency and root biomass significant at $p \leq 0.05$

under different land use systems. They found that top of ravines under mixed forest (developed due to long protection) consisting of species such as *Acacia senegal*, *Soymida febrifuga*, and *Leucaena leucocephala* and protected grazing land with dominance of *Dichanthium annulatum* and *Heteropogon contortus* species had higher organic carbon than cropping systems and horticultural systems (Table 3.14). Micronutrient contents (especially Fe and Zn) were low in agricultural fields and ravine lands as compared to other land uses.

### 3.5.2 Afforestation and Silvopastoral Systems

As discussed at many places in this publication, several programs of afforestation were undertaken by different agencies from time to time. Even aerial seeding was done in ravine areas which included species of *Prosopis juliflora*, *Acacia nilotica*, *A. catechu*, *A. leucophloea* and *Dichrostachys cinerea*. Many afforestation studies have been conducted involving fast-growing and of economic value tree species (Fig. 3.4) showing success of afforestation programs.

Forest trees such as *Acacia tortilis*, *A. nilotica* and *Azadirachta indica* and fruit trees like *Emblica officinalis*, *Ziziphus mauritiana* and bamboo (*Dendrocalamus* sp.) along with palatable grasses like *Cenchrus ciliaris* and *Dichanthium annulatum* have been cultivated with success even in deep ravines (Fig. 3.5; Verma et al. 2015). Regional Centers of Central Soil and Water Conservation Research and Training Institute conducted several experiments for evaluation and identification of suitable tree and grass species in eroded habitats.

There is always a better scope of success with species already found growing in these habitats. Many of these have high potential of providing products of economic importance. Many fruit trees are found growing in ravine fields which include *Emblica officinalis*, *Carissa carandas*, *Aegle marmelos*, *Mangifera indica*, *Feronia limonia*, *Annona squamosa*, *Psidium guajava*, *Syzygium cumini*, *Capparis decidua*, *Punica granatum* and *Tamarindus indica*. Many trees are of timber importance or may be lopped for fodder and other useful products. These include *Acacia nilotica*, *Dalbergia sissoo* and *Eucalyptus* (introduced and found to be successful) for timber, *Acacia senegal* and *A. nilotica* for gum, *Bombax ceiba* for fibre, *Butea monosperma* for making leaf plates and as lac-host, *Moringa oleifera* for its nutritious fruits and leafy vegetables, *Pongamia pinnata* and *Salvadora oleoides* for seed oil and many others as multipurpose trees.

### 3.5.3 Selection of Herbaceous Species

Among herbaceous species leguminous and grass species are preferred over several other forbs because grasses besides having fodder value have fibrous root systems which help in binding the soil and leguminous species help in management of soil fertility. Many studies have shown that grasses are more suitable for conserving soil as compared to other crops. Tejwani et al. (1975), Dagar (1987a, b, c, d) and

**Table 3.14** Physicochemical properties of topsoils (0–15 cm) in different land uses of Chambal ravines and adjacent fields near Kota, Rajasthan

| | Org C | Clay | CEC | Available nutrients | | | | | | |
|---|---|---|---|---|---|---|---|---|---|---|
| | | | [cmol (p+) | (kg ha$^{-1}$) | | | DTPA-extractable micronutrients (mg kg$^{-1}$) | | | |
| Land use | (g kg$^{-1}$) | (%) | kg$^{-1}$] | N | P | K | Mn | Zn | Fe | Cu |
| Irrigated cropping | 5.7 | 42.0 | 28.77 | 275 | 15.5 | 435 | 15.8 | 0.77 | 2.14 | 2.49 |
| Rainfed cropping | 3.6 | 40.8 | 31.70 | 280 | 14.6 | 405 | 12.0 | 0.74 | 5.23 | 1.11 |
| Lemon-based agri-horti system | 4.2 | 38.4 | 21.15 | 350 | 18.1 | 395 | 18.8 | 0.93 | 5.77 | 0.23 |
| Grassland (10 years of protection) | 6.3 | 38.2 | 18.55 | 392 | 24.4 | 484 | 23.9 | 1.06 | 9.73 | 0.93 |
| 10-year-old *Leucaena leucocephala* plantation | 7.2 | 37.8 | 24.40 | 392 | 24.3 | 425 | 23.1 | 1.17 | 11.59 | 1.70 |
| 10-year-old *Soymida febrifuga* plantation | 8.5 | 40.0 | 21.10 | 393 | 23.9 | 474 | 20.8 | 1.41 | 9.44 | 0.80 |
| Ravines under/over 20-year-old mixed forest | | | | | | | | | | |
| Top | 7.7 | 20.2 | 16.90 | 258 | 20.5 | 430 | 8.1 | 0.44 | 2.83 | 1.93 |
| Slope | 2.8 | 18.5 | 15.51 | 252 | 17.2 | 335 | 10.6 | 0.53 | 2.54 | 1.77 |
| Bottom | 5.4 | 20.5 | 23.20 | 260 | 23.1 | 504 | 10.2 | 1.28 | 7.03 | 2.88 |
| Ravines under *Acacia senegal* | | | | | | | | | | |
| Top | 7.0 | 22.8 | 13.00 | 294 | 20.3 | 557 | 7.0 | 1.15 | 2.88 | 2.61 |
| Slope | 4.2 | 23.5 | 11.55 | 350 | 16.8 | 445 | 13.2 | 0.72 | 2.77 | 2.21 |
| Bottom | 5.6 | 22.6 | 13.30 | 350 | 24.4 | 612 | 8.7 | 1.50 | 6.43 | 1.87 |

Source: Somasundaram et al. (2009)

**Fig. 3.4** Afforestation of multipurpose vegetation in ravines (Photo by SK Verma)

**Fig. 3.5** Bamboo and other vegetation grown successfully in ravines (Photo courtesy of Yaduvendra Singh)

**Table 3.15** Aggregate analysis under some grasses found grown in ravine lands

| | Percent of different-sized aggregates | | | | |
|---|---|---|---|---|---|
| Species | >2 mm | 2–1 mm | 1–0.2 mm | 0.2–0.1 mm | <0.1 mm |
| *Dichanthium caricosum* | 19.32 | 14.21 | 26.31 | 0.40 | 40.00 |
| *Bothriochloa pertusa* | 19.08 | 9.51 | 28.11 | 0.38 | 42.30 |
| *Cynodon dactylon* | 18.53 | 17.38 | 29.67 | 1.32 | 33.10 |
| *Iseilema laxum* | 17.54 | 6.77 | 21.34 | 1.22 | 53.13 |
| *Dichanthium annulatum* | 14.51 | 12.57 | 35.34 | 0.40 | 40.00 |
| *Sehima nervosum* | 9.84 | 8.90 | 28.65 | 0.79 | 51.82 |
| *Saccharum spontaneum* | 7.85 | 11.32 | 24.13 | 1.20 | 55.50 |
| *Chrysopogon fulvus* | 6.99 | 10.13 | 27.42 | 0.97 | 54.49 |
| *Sorghum halepanse* | 5.87 | 5.55 | 25.78 | 1.60 | 61.20 |
| *Tragus biflorus* | 5.59 | 9.33 | 31.88 | 0.87 | 52.33 |
| *Heteropogon contortus* | 5.45 | 9.28 | 39.81 | 1.50 | 43.95 |
| Control (without any grass) | 2.83 | 5.00 | 20.71 | 3.10 | 69.26 |

Young (1989), working on different types of soil and climatic conditions, clearly indicated that natural fallow lands covered with grasses and shrubs or grasses alone allow least and negligible runoff and soil loss compared to cultivated fallow of crops. Grasses such as *Bothriochloa pertusa*, *Dichanthium caricosum*, *D. annulatum* and *Cynodon dactylon* were found to have better soil aggregative properties as compared to other studied grasses (Dagar 1987c, d), but the results showed that other grasses also showed capability of aggregating soil particles particularly of size up to 2 mm as compared to soil without any grass (Table 3.15).

After studying all ecological characteristics of root systems, it was concluded that perennial grasses could be preferred over annuals. *Bothriochloa pertusa*, *Dichanthium caricosum*, *D. annulatum*, *Cynodon dactylon*, *Iseilema laxum*, *Heteropogon contortus*, *Cenchrus ciliaris*, *Chrysopogon fulvus* and *Cymbopogon* spp. and *Vetiveria zizanioides* (for aromatic oil) and *Saccharum munja*, which are also resistant to grazing, can be cultivated with success. Many species such as *Achyranthes aspera*, *Aloe vera*, *Asparagus racemosus*, *Barleria prionites*, *Bacopa monnieri* (in low-lying areas), *Cardiospermum halicacabum*, *Chlorophytum arundinaceum*, *Citrullus colocynthis*, *Coccinia grandis*, *Momordica dioica*, *Pedalium murex*, *Portulaca oleracea*, *Solanum indicum*, *Tinospora cordifolia*, *Tribulus terrestris*, *Urginea indica*, *Withania somnifera*, *Xanthium strumarium* and many others have medicinal and commercial potential and can successfully be cultivated in ravine lands.

### 3.5.4 Filling of Cracks and Potholes, Checking Growth of Gullies in Adjacent Fields and Preventing Deepening of Gully Beds

Cracks and potholes formed at the gully heads should be filled with soil before commencement of rains involving community participation, and grass cover should be established over that soil. Species of *Dichanthium*, *Cenchrus*, *Chrysopogon*,

**Fig. 3.6** Temporary brushwood check dam to control runoff and soil erosion (Source: Chaturvedi et al. 2014)

*Cynodon*, *Dactyloctenium* and *Chrysopogon* were found suitable for this purpose. To check the growth of the gullies, the bunds around the upland field must be constructed and stabilized with grass cover, and water-flow channels must be planted with grass vegetation. In deep gullies where reclamation is difficult or very costly, grasses and bamboo plantation effectively check in widening and deepening of gullies. The gully plugs stabilized with grasses reduce the speed of runoff water redistributing it, increase infiltration, encourage silting and improve the soil moisture regime for establishing tree cover. *Cynodon dactylon* grass that predominantly grows at beds of the ravines can stabilize the plugs and conserve the flow of silt into river. Temporary brushwood check dams are also found effective in checking runoff and soil erosion along slopes of ravines (Fig. 3.6) Further, application of mulch on bare slopes before commencement of rains will help not only to protect soil erosion due to rain showers when the vegetation cover is minimum but will also increase organic matter in the soil which in turn will help in more infiltration and less runoff and, hence, consequently checking soil erosion.

### 3.5.5 Prevention of Bank Erosion

Bank erosion of the rivers can be prevented by various treatment methods, viz. growing live hedges of species which grow naturally in such habitats. *Ipomoea carnea*; *Arundo donax*; *Sorghum halepense*; *Phragmites australis*; *P. karka*; *Pennisetum purpureum*; *Paspalum* spp.; *Brachiaria mutica*; species of *Cyperus*, *Fimbristylis* and *Scirpus*; *Polygonum glabrum*; *Rumex dentatus*; and *Saccharum spontaneum* are found growing frequently along the banks of the rivers. *Terminalia arjuna*,

*Woodfordia fruticosa*, *Ficus glomerata*, *Pongamia pinnata*, *Sesbania sesban* and *Bambusa arundinacea* are the woody species which may be planted along banks to minimize the water currents.

Other vegetative and engineering measures and their interactions such as retards, jacks, jetted post, bank sloping, brush-mat riprap, spurs or jetties or groynes, log spurs or post-jetties, boulder spurs, reinforced with vegetation, gabion spurs, tree jetties, gully plugs, etc. in controlling the bank erosion have been discussed by many (Tejwani et al. 1975; Dagar 1987c; Chaturvedi et al. 2014; Verma et al. 2015) and elsewhere in this publication.

Along with the above-mentioned measures which are the basic natural and permanent tools in our hands, engineers also must not lose sight of the fact that temporary check dams should also be used to facilitate the establishment of vegetation. Measures, such as contour bunding, terracing, strip cropping, growing leguminous and cover crops, constructing drainage structures, planned and managed agricultural devices, etc., should also be followed with the help of engineers and soil scientists with ecological understanding of these habitats.

## References


Abraham R (2011) Puntius fascicatus. In: IUCN Red list of threatened species, version 2011.2. www.iucnredlist.org

Ambasht RS (1968) Ecology of river bank. Proc Symp Rec Adv Trop Ecol Varanasi 1:466–470

Ambasht RS, Shankar M (1993) Rehabilitation of degraded river corridors. In: Singh JS (ed) Restoration of degraded lands: concepts and strategies. Rastogi Publications, Meerut, pp 191–209

Ambasht RS, Shardendu, Srivastava AK, Shankar M, Sinha M (1988) Plant population, community structure and function of riparian and aquatic macrophytes in relation to pollution of river Ganga. Technical Report, DoEn- Ganga Project P-739 (A), pp 139

Anwar C, Baheramsyah K, Hamzah Z (1989) Effectiveness of shrubs and agroforestr in Kadipaten village (Indonesia, upper Citanduy sub-watershed) in trimming runoff and erosion. Bull Penelitian Hutan 511:1–8

Baets D, Poesen J, Knapen A, Barbera GG, Navarro JA (2007) Root characteristics of representative Mediterranean plant species and their erosion-reducing potential during concentrated runoff. Plant Soil 294(1):169–183

Bhimaya CP, Rege ND, Srinivasan S (1956) Preliminary studies on the role of grasses in soil conservation in the Nilgiris. J Soil and Water Cons India 4:113–117

Bokhari UG, Singh JS (1974) Effects of temperature and clipping on growth, carbohydrate reserves and root exudation of western wheat grass in hydrophonic culture. Crp Sci 14:790–794

Bor NL (1960) Grasses of Burma, Ceylon, India and Pakistan. Pergamon Press, London

Caughenour MB, McNaughton SJ, Wallace LL (1984) Simulation study of East African perennial graminoid responses in defoliation. Ecol Model 26:177–202

Champion HG, Seth SK (1968) A revised survey of the forest types of India. Manager of Publications, Delhi

Chandra K, Gajbe PU (2005) An inventory of Herpetofauna of Madhya Pradesh and Chhttisgarh. Zoo's Print J 20(3):1812–1819

Chaturvedi OP, Kaushal R, Tomar JMS, Prandiyal AK, Panwar P (2014) Agroforestry for wasteland rehabilitation: mined, ravine, and degraded wasteland areas. In: Dagar JC, Singh AK, Arunachalam A (eds) Agroforestry systems in India: livelihood security and ecosystem services, advances in agroforestry, vol 10. Springer, New Delhi, pp 233–272

Chinnamani S (1992) Silvipasture and natural grasslands of Mahi ravines and ravines of Gujarat. J Range Manag Agroforest 13:175–181
Chinnamani S (1998) Ecology of succession of Prosopis Juliflora in the ravines of India. In: Tewari JC, Pasiecznik NM, Harsh LN, PJC H (eds) Prosopis species in the arid and semiarid zones of India. The Prosopis Society of India & the Henry Doubleday research association. CAZRI, Jodhpur
Dagar JC (1978) Studies on soil erosion of Kshipra ravines. Ph D Thesis, Vikram University, Ujjain, India
Dagar JC (1986) Studies on root systems of some grasses growing in Kshipra ravines. J Indian Bot Soc 65(4):397–403
Dagar JC (1987a) Species composition and plant biomass of an ungrazed and a grazed grassland at Ujjain, India. Trop Ecol 28:208–215
Dagar JC (1987b) Responses of clipping treatments on nine palatable perennial grasses. Trop Ecol 28:216–221
Dagar JC (1987c) Studies on reclamation of Kshipra ravines. Indian J For 10(2):83–89
Dagar JC (1987d) Studies on root systems of some woody species growing in Kshipra ravines. Indian J Appl Pure Biol 2(1):17–21
Dagar JC (1989) Studies on root systems of some forbs growing in Kshipra ravines. Indian J Appl Pure Biol 4(1):35–47
Dagar JC (1995) Soil conservation values of some grasses and forbs. Proc Indian Nat Sci Acad B 61:163–170
Dagar JC (2001) Biological spectrum and successional trends in Kshipra watershed areas. Indian J For 24(3):351–356
Dagar JC (2012) Utilization of degraded lands / habitats and poor-quality water for livelihood security and mitigating climate change. Indian J Agroforetsry 14(1):1–16
Dagar JC (2017) Potentials for fodder production in degraded lands. In: Ghosh PK, Mohanta SK, Singh JB, Vijay D, Kumar RV, Yadav VK, Sunil K (eds) Approaches towards fodder security in India. Studera Press, New Delhi, pp 333–364
Dagar JC, Arunachalam A (2013) Management of degraded ecologies for mitigating climate change. In: Arunachalam A, Dagar JC, Singh AK (eds) Agroforestry and climate change management. Pointer Publications, Jaipur, pp 1–15
Dagar JC, Balakrishnan NP (1984) Life forms and biological spectrum of Andaman and Nicobar Islands. Bull Bot Surv India 26(3&4):147–159
Dagar JC, Gupta SR (2016) Agroforestry: potentials for rehabilitation of degraded lands, constraints and the way forward. In: Dagar JC, Tewari JC (eds) Agroforestry research developments. Nova Science Publishers, New York, pp 46–97. (ISBN: 978-1-63485-046-9)
Dagar JC, Mall LP (1980) Studies on the vegetation of ravines and banks of the river Kshipra and its tributaries. J Indian Bot Soc 59:234–245
Dagar JC, Pathak PS (2005) Grassland dynamics and their management. Range Manag Agroforestry 26(1):7–31
Dagar JC, Singh VP (1980) The ecological studies of the vegetation of river Kshipra and its tributaries. Biologia 25:43–62
Dagar JC, Singh NT (1999) Plant resources of Andaman & Nicobar Islands. Bishen Singh Mahendra Pal Singh Publishers, Dehra Dun, India. p. 985
Dagar JC, Singh G (2003) Pasture production in degraded and problematic soils- status and prospects. In: Jakhmola RC, Jain RK (eds) Sustainable animal production. Pointer Publishers, Jaipur, pp 94–119
Dagar JC, Singh VP, Mall LP (1978) Regeneration in plants growing on eroded soil. Ann Arid Zone 17:68–74
Daily GC, Alexander S, Ehrlich PR, Goulder L, Lubchenco J, Matson PA, Mooney HA, Postel S, Schneider SH, Tilman D, Woodwell GM (1997) Ecosystem services: benefits supplied to human societies by natural ecosystems. Issues Ecol 1(2):1–18
De Kroon H, Visser EJW (eds) (2003) Root ecology. Springer, Berlin
Edroma EL (1985) Effect of clipping on Themeda Triandra and Brachiaria platynotan in Queen Elizabeth National Park, Uganda. Afr J Ecol 23:45–52

Evans GC (1972) The quantitative analysis of plant growth. Univ Calif Press, Berkeley
Gallart F, Marignani M, Perez-Gallego N, Santi E, Maccherini S (2012) Thirty years of studies on badlands, from physical to vegetational approaches-a succinct review. Catena 106:4–11
Gray DH, Barker D (2004) Root-soil mechanics and interaction. In: Bennett SJ, Simon A (eds) Riparian vegetation and fluvial geomorphology-water science and application series 8. American Geophysical union, Washington, DC, pp 113–123
Gupta RK (1961) Vegetation in catchment areas of river ganga and Yamuna with reference to their soil and water conservation problems. J Indian Bot Soc 40:325–334
IUCN (1998) 1997 IUCN red list of threatened species. International Union for Conservation of Nature and Natural Resources, Gland
IUCN (2011) IUCN Red list of threatened species, version 2011.2. www.iucnredlist.org
Kumar R, Shankar M, Ambasht RS (1990) Water, soil and nutrient movement from riparian and Obra-Renukoot-Singrauti area and its impact on natural and derived ecosystems. Tech report. Banaras Hindu University, Varanasi, pp 285–304
Luken JO (1990) Directing ecological succession. Chapman and Hall, New York. ISBN 0-412-34450-5
Meher-Homji VM (1964) Life form and biological spectra as ephamonic criteria of aridity and humidity in the tropics. J Indian Bot Soc 43:424–430
Misra R (1944) The vegetation of Raighat ravines. J Indian Bot Soc 23:113–121
Misra R (1962) The science of vegetation with reference to India. Bull Bot Survey India 4:113–118
Mueller-Dombois D, Ellenberg H (1974) Aims and methods of vegetation ecology. Wiley, New York
Murthy MS (1971) Ecology of upper catchment area of river Narmada-Climo-vegetational relationships. Japanese J Ecol 21:188–192
Nair T, Chitanya-Krishna Y (2013) Vertebrate fauna of the Chambal river basin, with emphasis on the National Chambal Sanctuary, India. J Threat Tax 5(2):3620–3641. https://doi.org/10.11609/JoTT.03238.3620-41
Oldeman LR (1994) The global extent of soil degradation. In: Greenland DJ, Szabolcs I (eds) Soil resilience and sustainable land use. CAB International, Wallingford, pp 99–118
Pande H, Singh JS (1985) Influence of clipping and water stress on growth performance and nutrient value of four range grasses. Proc Indian Acad Sci (Plant Sc) 95:389–403
Pandeya SC, Pandya SM, Murthy MS (1972) Ecology of river Narmada upper catchment in Central India. I. Introduction, situation, terrain, lithology, climate, general forests and methods of vegetational analysis. J Indian Bot Soc 51:356–373
Pathak PS, Dagar JC (2015) Indian grasslands and their management. In: Ghosh PK, Mahanta SK, Singh JB, Pathak PS (eds) Grassland: a global resource perspective. Range Management Society of India, Jhansi, pp 3–36
Pimentel D, Kounang N (1998) Ecology of soil erosion in ecosystems. Ecosystems 1(5):416–426. https://doi.org/10.1007/s100219900035
Prajapati MC, Nambiar KTN, Puri DN, Singh JP, Malhotra BM (1993) Fuel and fodder production in Yamuna ravines at Agra. Indian J Soil Cons 21(3):8–13
Puri GS, Meher-Homji VM, Gupta RK, Puri S (1983) Forest ecology, 2nd ed Vol I: phytogeography and forest conservation. Oxford & IBH Publishing Co, New Delhi, p 549
Ranade DH, Mishra VK, Gupta RK, Verma SK (2002) Root distribution pattern of some grass species in black clay soil in relation to their capability for controlling erosion. J Indian Soc Soil Sci 50(3):280–282
Rao CC (1965) Ecological studies on a piece of watershed land. Trp Ecol 6:124–132
Rao CC (1969) Phytosociological studies on a piece of watershed land in the vicinity of rivers Karamnasa and Chandraprabha (Varanasi district). Ph D Thesis Banaras Hindu University, Varanasi, India
Raunkiaer C (1934) The life forms of plants and statistical plant geography-being the collected papers of C Raunkiaer. (Translated into English by Carter HG, Tansley AG, Fausboll) Oxford Clarendon Press, pp 632
Reubens B, Poesen J, Danjon F, Geudens G, Muys B (2007) The role of fine and coarse roots in shallow slope stability and soil erosion control with a focus on root system architecture: a review. Trees 21(4):385–402

Saxena SK (1977) Vegetation and its succession in the Indian desert. In: Desertification and its control. Indian Council of Agricultural Research, New Delhi, pp 176–192
SER (2004) The SER Primer on Ecological Restoration, Version 2. Society for Ecological Restoration (SER), Science and Policy Working Group. http://www.ser.org/reading_resources.asp
Shah GL (1969) A contribution to the flora of Mahi ravines in Central Gujarat. Indian For 95(4):270–277
Shankar V, Kanodia KC, Dabadghao PM (1971) Studies on plant succession in *Sehima-Dichanthium* grass cover. Ann Arid Zone 10:158–164
Shields FD, Gray DH (1993) Effects of woody vegetation on the structural integrity of sandy levees. Water Res Bull 28:917–931
Singh VP, Dagar JC, Uppadhyaya SD (1979) Analysis of structure, production dynamics and successional trends of tropical grassland communities of Ujjain (India). Sylvatrop-Philip For Res Journ 4:231–254
Somasundaram J, Singh RK, Parandiyal AK, Prasad SN (2009) Micronutrient status of soils under different land use systems in Chambal ravines. J Indian Soc Soil Sci 57(3):307–312
Stocking MA (1984) Erosion and soil productivity: a review. FAO Consultant working paper No. 1. Soil Conservation Program, Land and Water Development Division, FAO, rome, Italy
Taylor HM (1976) The effect of soil compaction on rooting pattern and water uptake of cotton. Paper in 7th Conference. Soil Intnl Soil Till. Res Org Swedon
Tejwani KG, Gupta SK, Mathur HN (eds) (1975) Soil and water conservation research (1956–1971). ICAR Pub, New Delhi, p 358
Thomas M, Sahu P, Shrivastava A, Hussain Z (2011) Biodiversity and livelihood options of people in Chambal ravine of Morena district, Madhya Pradesh, India. J Trop For 27(July–Sept, 2011):40–56
Vanicek V (1973) The soil protective role of specially shaped plant roots. Biol Conserv 5:175–180
Verma B, Chinnamani S (1990) Rehabilitation of ravine lands. In: Abrol IP, Dhuruva Narayana VV (eds) Technologies for wasteland development. ICAR, New Delhi, pp 263–272
Verma SK, Singh A (2015) Chambal ravines, perspectives of formations and management options. In: Verma SK, Trivedi SK, Singh A, Tomar AS (eds) Management of degraded lands and soil health. Biotech Books, New Delhi, pp 31–46
Verma SK, Trivedi SK, Singh A, Tomar AS (eds) (2015) Management of degraded lands and soil health. Biotech Books, New Delhi, p 249
Wadia DM (1961) Geology of India. The English Language Book Soc, and Mc Millan & Co Ltd, London
Wilson EO (1988) Biodiversity. National Academy, Washington, DC. ISBN 0-309-03739-5
Yadava PS, Singh JS (1977) Progress in ecology vol 2: grassland vegetation. Today and Tomorrow's Print and Publ, New Delhi
Young A (1989) Agroforestry for soil conservation. CAB International, Wallingford

# Plant Biodiversity of Ravine Ecosystem: Opening New Vistas for Enhancing Productivity

4

A. R. Uthappa, S. B. Chavan, M. N. Ramesha, S. Kala, Veeresh Kumar, A. K. Handa, and O. P. Chaturvedi

**Abstract**

Ravines are the worst form of land degradation and characterized by edaphic, topographic, and climatic adversities. Due to unregulated and overexploitation, the forest areas under ravines are facing severe threats of losing biodiversity. Biodiversity of an area determines the health of the ecosystem. The degraded ravine ecosystem also harbors rich diversity and provides fuelwood, fodder, medicines, and small timber to the local dwellers. The biodiversity also supports various kinds of ecosystem services. The unscientific management of ravines has escalated the problems. This chapter is an attempt to document the plant diversity of the ravines and to rehabilitate these ecosystems using agroforestry and soil and water conservation measures. An integrated approach is required for scientific management of ravines which will pave the way toward sustaining ravine ecosystem and livelihood of the people.

## 4.1 Introduction

Biological Diversity Act, 2002, defines biological diversity as the variability among living organisms from all sources and the ecological complexes of which they are part and includes diversity within species or between species and of ecosystems

A. R. Uthappa (✉) · S. B. Chavan · V. Kumar · A. K. Handa · O. P. Chaturvedi
ICAR-Central Agroforestry Research Institute, Jhansi, Uttar Pradesh, India

M. N. Ramesha
ICAR-Indian Institute of Soil and Water Conservation Research Centre, Ballari, Karnataka, India

S. Kala
ICAR-Indian Institute of Soil and Water Conservation Research Centre, Kota, Rajasthan, India



J. C. Dagar, A. K. Singh (eds.), *Ravine Lands: Greening for Livelihood and Environmental Security*, https://doi.org/10.1007/978-981-10-8043-2_4

(http://nbaindia.org). Biodiversity is essential for human survival and economic well-being and for the proper functioning of ecosystem and its stability. Its detailed knowledge has allowed farming systems to evolve since agriculture inception. Biodiversity is considered as fundamental to agricultural production and food security and is an important factor in designing climate-resilient agriculture practices. Practices that conserve, sustainably use, and enhance biodiversity are necessary at all levels in farming systems and are of critical importance for food production, livelihood security, health, and the maintenance of ecosystem (Thrupp 2000). Biodiversity directly or indirectly contributes to improved productivity, resilience in farming systems, income generation, nutritional values, and food and livelihood security for numerous societies. Biodiversity also provides ecosystem services such as soil and water conservation, pollination, fertility and nutrient cycling, and insect and disease management.

Agricultural expansion has also reduced the overall biodiversity of natural habitats, including tropical forests, grasslands, and wetland areas, through the process of land conversion. Numerous experiments have found that biodiversity influences the primary productivity of ecosystems and other aspects of ecosystem functioning (Hector et al. 1999; Tilman et al. 2001; Reich et al. 2001; Hooper et al. 2005; Balvanera et al. 2006; Cardinale et al. 2011). A change in plant diversity from 4 to 16 species caused as large an increase in productivity as addition of 54 kg $ha^{-1}year^{-1}$ of fertilizer N and was as influential as removing a dominant herbivore, a major natural drought, water addition, and fire suppression; and further, change in diversity from 1 to 16 species caused a greater biomass to increase than 95 kg $ha^{-1}$ $year^{-1}$ of N or any other treatment (Tilman et al. 2012). A new global assessment of forests suggests that, on average, a 10% loss in biodiversity leads to a 2–3% loss in the productivity, including biomass that forests can offer. The value of biodiversity in maintaining commercial productivity has been estimated to be 166–490 billion (USD), which would be considerably greater than the total cost of effective global conservation (Liang et al. 2016). Livelihood pattern of the people is directly influenced by the local diversity. The loss of biodiversity and depletion of natural resources will decline the productivity of the system and jeopardize the food security. The ravine areas are the most fragile ecosystem and subjected to various degrees of degradation enforced upon by climatic, edaphic, and anthropogenic forces causing loss of natural resources and threat to biodiversity. Due to unregulated and overexploitation, the forest area under ravines is facing severe threats of losing biodiversity (Fig. 4.1). These ravine areas are facing heavy biotic pressure mainly for agriculture expansion, fuelwood, and fodder. Therefore, in this chapter the floristic diversity of the ravine ecosystem has been dealt along with its potential utility including rehabilitation of ravine ecosystem.

**Fig. 4.1** Ravines of Morena District, Madhya Pradesh, India, showing overexploitation of biodiversity

## 4.2 Floral Diversity of Ravines

Champion and Seth (1968) classified the natural vegetation of ravines as "northern tropical ravine thorn forests." The ravines are characterized by dominance of thorny vegetation. The floristic diversity of the ravines has been documented by various workers (Sajwan 1975, 1976; Dagar and Mall 1980; Prajapati 1995; Parandiyal et al. 2000). The typical vegetation consists of low growing type of coarse grasses like *Aristida adscensionis*, *Eremopogon foveolatus*, *Heteropogon contortus*, *Themeda triandra*, *Sehima nervosum*, *Eragrostis tenella*, *E. pilosa*, *Chrysopogon fulvus*, *Eleusine aegyptiacum*, etc.; undershrubs such as *Calotropis procera*, *Clerodendrum phlomidis*, *Capparis zeylanica*, *Capparis decidua*, *Crotalaria burhia*, *Cassia tora*, *Grewia pilosa*, *Grewia tenax*, *Lantana camara*, *Tephrosia purpurea*, *Xanthium strumarium*, *Ziziphus nummularia*, etc.; and xerophytic trees like *Acacia senegal*, *Acacia leucophloea*, *Acacia nilotica*, *A. ebúrnea*, *Balanites aegyptiaca*, *Dichrostachys cinerea*, *Prosopis juliflora*, etc.

Kotia et al. (2008) described the vegetation found in ravines of Madhya Pradesh and Rajasthan. Thorny bushes or small trees commonly found in this area include *Capparis decidua*, *Capparis sepiaria*, *Balanites aegyptiaca*, *Acacia senegal*, *A. nilotica*, *A. leucophloea*, *Prosopis juliflora*, *Butea monosperma*, *Maytenus emarginata*, *Tamarix* sp., *Salvadora persica*, *S. oleoides*, *Crotalaria medicaginea*, *C. burhia*, *Clerodendrum phlomidis*, *Calotropis procera*, *Xanthium indicum*, and *Leptadenia pyrotechnica* associated with climbers such as *Maerua oblongifolia*,

*Pergularia daemia*, and *Ceropegia bulbosa* and herbs like *Argemone mexicana*, *Farsetia hamiltonii*, *Tephrosia purpurea*, *Cleome viscosa*, *Tribulus terrestris*, *Glinus lotoides*, *Sericostoma pauciflorum*, *Rivea* sp., *Ipomoea* sp., *Pedalium murex*, *Sesamum mulayanum*, *Lepidagathis* sp., *Boerhavia diffusa*, and *Chrozophora* sp. and grasses and sedges like species of *Cyprus*, *Fimbristylis*, *Brachiaria*, *Chloris*, *Echinochloa*, *Cenchrus*, *Dichanthium*, and *Sporobolus*. Dagar and Mall (1980), Dagar and Singh (1980), and Dagar (2001) gave extensive account of vegetation ecology of Kshipra ravines of Madhya Pradesh and its tributaries including the successional trends of the vegetation under biotic pressure. Ecological studies have revealed that in most ravenous tracts, the secondary succession takes 10–15 years to reach a stable tree stage, while in ravine bottom, it takes 6–7 years only. Hardy grass species, viz., *Sehima nervosum*, *Heteropogon contortus*, *Dichanthium annulatum*, *Chrysopogon fulvus*, *Cynodon dactylon*, and legumes *Alysicarpus rugosus* and *Rhynchosia minima* can withstand biotic pressure, while species such as *Themeda triandra*, *Iseilema laxum*, *I. prostratum*, and *Apluda mutica* and legumes like *Indigofera glandulosa*, *Rhynchosia capitata*, and *Alysicarpus longifolius* appear only when the area is protected from grazing.

### 4.2.1 Tree Diversity in Ravines

Diversity analysis is an important tool for biologists, ecologists, and agroforesters to understand the community structure. Various diversity indices are mathematical measures of species diversity in a community and provide important information about rarity and commonness of species in a community. Thorny and stunted vegetation (Champion and Seth 1968) of ravine areas are facing heavy biotic pressure mainly due to expansion of agricultural land and unregulated and overexploitation of flora for fuelwood and fodder. Thus, the floral diversity existing in ravines ecosystem is under severe threats. Uthappa et al. (2016) studied the species diversity, richness, and evenness using different indices like Shannon, Simpson, Margalef, and Berger-Parker (Table 4.1) in the ravines of Orai, Banda, and Datia district of Bundelkhand, India.

The diversity indices varied according to the location of ravines. The overall diversity of all the three sites was found low. The survey also reported high biotic interferences, which escalates ravine degradation. However, crop cultivation in these lands is not desirable as it leads to accelerated erosion, thus adding to their fast degradation. It could be concluded that the best scientific use of these lands is to place them under permanent vegetation through agroforestry interventions involving forest and fruit trees along with the grasses. The planting and management of indigenous tree species like, *Acacia leucophloea*, *A. nilotica*, *Ziziphus mauritiana*, *Carissa carandas*, and *Capparis decidua* will pave the way toward sustaining ravine ecosystem and livelihood of the people. Floristic diversity in the ravines of Ottangan river, Agra, India, was studied for assessing the impact of biotic disturbance on the phytosociological characters of the flora. Species richness of trees varied from 3 to 13 in unprotected to protected ravine. *Prosopis juliflora*, *Acacia nilotica*, *Dalbergia*

**Table 4.1** Diversity indices for different ravine areas of Bundelkhand region of India

| Index | Orai, Uttar Pradesh | Banda, Uttar Pradesh | Datia, Madhya Pradesh |
|---|---|---|---|
| Shannon diversity index | 1.259 | 1.871 | 1.887 |
| Shannon equitability | 0.782 | 0.899 | 0.859 |
| Simpsons diversity (D) | 0.333 | 0.183 | 0.194 |
| Dominance index D-1 | 0.667 | 0.817 | 0.806 |
| Reciprocal Simpsons diversity (1/D) | 1.499 | 1.224 | 1.241 |
| Margalef index | 0.705 | 1.351 | 1.646 |
| Berger-Parker dominance index (d) | 0.400 | 0.303 | 0.357 |
| Inverted Berger-Parker dominance (1/d) | 2.500 | 3.296 | 2.804 |

*sissoo*, and *Balanites aegyptiaca* were the dominant species in the ravines (Chauhan et al. 2015).

### 4.2.2 Fuelwood-yielding Tree Diversity in Ravines

Wood is by far the largest used fuel in India and accounts for about 60% of the total fuel in the rural areas. Most of the fuelwood has been reported to be derived from forests with some from trees growing on farmlands and common lands outside forests. With the rapid increase in population, the area under agriculture has expanded and forest area has reduced. The estimated annual consumption of fuelwood far exceeds estimates of the sustainable annual output of wood from the country's forests. The rural communities collect the fuelwood from the nearby community lands or wastelands. The ravines are also one of the sources of fuelwood for rural communities. In the Chambal ravines of Morena district, M. P., India, Thomas et al. (2011) recorded 19 tree species suitable for fuelwood. These included neem (*Azadirachta indica*), Kaith (*Feronia limonia*), semal (*Bombax ceiba*), Kanki-phul (*Pavetta indica*) (its roots also used as fuel), peepal (*Ficus religiosa*), guggul (*Commiphora wightii*), Palash (*Butea monosperma*) and Ronjh (*Acacia leucophloea*), Kardhai (*Anogeissus pendula*), dhawa (*Anogeissus latifolia*), khair (*Acacia catechu*), vilayati babul (*Prosopis juliflora*), gular (*Ficus glomerata*), karel/kair (*Capparis decidua*), Asaria (*Capparis zeylanica*), Chameli (*Jasminum arborescens*), Kalijori (*Clerodendrum multiflorum*), Jaal/Mithajal (*Salvadora persica*), and Galphusia (*Lantana camara*).

#### 4.2.2.1 Case Study: Fuelwood Consumption Survey in Ravines of Bundelkhand (Chouhan et al. 2015; Chavan et al. 2016)

In Bundelkhand region of Uttar Pradesh and Madhya Pradesh, people are more dependent on forest for livelihood. Erratic monsoon, less developed agriculture coupled with poverty, rapid population growth, economic stagnation and unemployment are the major problems of this region. In one special project study of Bundelkhand conducted in 2012, it was reported that the consumption of fuelwood is around 1.22 kg per person per day. The population of Bundelkhand is 8.232 million, and there is demand of 3.66 million Mg per annum of fuelwood, and people are dependent on wood collection from ravine tracts and nearby forests.

Using interviews and questionnaires for key respondents in the local community, the study identified the villager's dependence on their forest for fuelwood and supported with socio-economic stability of villagers in Jalaun district of Bundelkhand mainly ravenous type of forest. The survey revealed that 70% of local people were heavily dependent on forests mainly for fuelwood due to easy availability and cost-free. Out of 120 households surveyed, 93.5% respondents use fuelwood for cooking and 4% use cow dung as primary and subsidiary source for cooking. Only 2.5% households had LPG stove. Among different forest species *Acacia senegal*, *A. leucophloea*, *A. nilotica*, *A. catechu*, *Azadirachta indica*, *Eucalyptus tereticornis*, *Dalbergia sissoo*, *Anogeissus pendula*, *Leucaena leucocephala*, and *Prosopis juliflora* were selected in survey. Based on matrix and preferential ranking, ten fuelwood species selected by key informants used for scientific assessment on fuelwood properties, viz., basic density, moisture content, ash content, volatile matter, fixed carbon, lignin content, calorific value, heating value, and fuelwood value index, were carried out to select ideal fuelwood species in studied villages of Bundelkhand. Out of various laboratory tests, the fuelwood value index is an emerged tool for identifying new and reinforcing the traditionally used fuelwood species as quality fuelwood (Table 4.2).

Based on data, *Acacia catechu*, *Acacia senegal*, *Anogeissus pendula*, and *Prosopis juliflora* could be listed as desirable fuelwood species. In a holistic perspective, the comparative analysis of survey and scientific assessment indicated that *Prosopis juliflora* was most commonly preferred due to its easy availability in forest and community lands. This extensive study concluded that plantation of locally available multipurpose tree species can be promoted in the region for growing in the wastelands including ravine lands, community lands, and private lands.

### 4.2.3 Fodder Species Diversity in Ravines

Cattle rearing is a main component of people living in and around the ravine lands. These lands are also rich source of fodder trees and grass species (Table 4.3).

The open grazing system is most prevalent in this region. A diversity analysis of Chambal ravines documented about 13 species suitable for fodder (Thomas et al. 2011). Further, Uthappa et al. (2016) also identified *Acacia leucophloea*, *Anogeissus pendula*, *Balanites aegyptiaca*, *Dalbergia sissoo*, *Ziziphus* sp., and *Prosopis juliflora* as major fodder tree species in Bundelkhand region (Fig. 4.2). The floristic diversity study of Ottangan river ravines near Agra, India, documented eight species suitable for fodder extraction, viz., *P. cineraria*, *P. juliflora*, *A. nilotica*, *Cordia rothii*, *Moringa oleifera*, *Bauhinia* sp., *D. sissoo*, and *Balanites aegyptiaca* (Chauhan et al. 2015).

**Table 4.2** Physical and chemical properties of selected fuelwood species in ravine tract of Bundelkhand

| Tree species | Moisture (%) | Ash (%) | Volatile matter (%) | Fixed carbon (%) | Basic density (g $cm^{-3}$) | Heating value (MJ $kg^{-1}$) | Calorific value (kcal $kg^{-1}$) | Lignin content (%) | Fuel value index |
|---|---|---|---|---|---|---|---|---|---|
| *Acacia catechu* | 35.44 | 2.36 | 23.38 | 74.26 | 0.815 | 33.32 | 4946.06 | 34.13 | 3885.93 |
| *Acacia leucophloea* | 46.96 | 2.45 | 12.24 | 85.30 | 0.744 | 33.39 | 4968.59 | 22.13 | 1759.45 |
| *Acacia nilotica* | 42.62 | 1.58 | 12.68 | 85.73 | 0.630 | 33.68 | 4957.26 | 33.27 | 2439.04 |
| *Acacia senegal* | 22.12 | 3.51 | 19.65 | 76.83 | 0.729 | 32.96 | 4954.66 | 22.19 | 3625.41 |
| *Anogeissus pendula* | 27.25 | 3.68 | 14.80 | 81.51 | 0.701 | 32.94 | 4938.77 | 23.35 | 2969.40 |
| *Azadirachta indica* | 31.25 | 3.49 | 15.31 | 81.19 | 0.646 | 33.01 | 4536.83 | 27.50 | 2579.82 |
| *Dalbergia sissoo* | 36.93 | 1.69 | 11.74 | 86.55 | 0.646 | 33.65 | 4926.30 | 23.37 | 2014.93 |
| *Eucalyptus tereticornis* | 42.43 | 1.61 | 22.80 | 75.58 | 0.642 | 33.59 | 4835.6 | 21.78 | 1594.27 |
| *Leucaena leucocephala* | 35.29 | 1.70 | 14.51 | 83.79 | 0.647 | 33.63 | 4433.89 | 19.58 | 1593.07 |
| *Prosopis juliflora* | 36.26 | 2.25 | 13.35 | 84.39 | 0.736 | 33.45 | 4973.20 | 26.38 | 2661.66 |

**Table 4.3** Important fodder species of ravines in India

| Common name | Scientific name | Family | Plant part(s) used |
|---|---|---|---|
| Revnuja | *Acacia leucophloea* | Fabaceae | Leaves and pods |
| Babool/Kikar | *A. nilotica* | Fabaceae | Leaves |
| Khejri | *Prosopis cineraria* | Fabaceae | Leaves and pods |
| Vilayati babul | *P. juliflora* | Fabaceae | Leaves |
| Shisham | *Dalbergia sissoo* | Fabaceae | Leaves |
| Kachnar | *Bauhinia racemosa* | Fabaceae | Leaves |
| Hingota | *Balanites aegyptiaca* | Balanitaceae | Leaves |
| Kardhai | *Anogeissus pendula* | Combretaceae | Leaves |
| Neem | *Azadirachta indica* | Meliaceae | Leaves |
| Kalijori | *Clerodendrum multiflorum* | Verbenaceae | Leaves |
| Kutch | *Acacia catechu* | Mimosaceace | Leaves |
| Ber | *Ziziphus mauritiana* | Rhamnaceae | Leaves |
| Chota ber | *Ziziphus xylocarpa/Z. nummularia* | Rhamnaceae | Leaves |
| Adusa | *Adhatoda zeylanica* | Acanthaceae | Leaves |
| Birilhara | *Convolvulus arvensis* | Convulvulaceae | Leaves |
| Kans | *Saccharum spontaneum* | Poaceae | Plant twig |
| Doob grass | *Cynodon dactylon* | Poaceae | Leaves |
| Gobhi | *Elephantopus scaber* | Asteraceae | Whole plant |
| Nagarmotha | *Cyperus rotundus* | Cyperaceae | Leaves |
| Patharchatta | *Boerhaavia diffusa* | Nyctaginaceae | Leaves |
| Bathua | *Chenopodium album* | Chenopodiaceae | Leaves |
| Makraghas | *Dactyloctenium aegyptium* | Poaceae | Whole plant |

**Fig. 4.2** Lopping of *Acacia leucophloea* for fodder in ravines of Madhya Pradesh

In Kshipra ravines, Dagar and Mall (1980) and Dagar (1999) identified several suitable woody and herbaceous species for growing in ravine lands. The forage trees, which are suitable for these lands, include *Acacia eburnea*, *A. nilotica*, *A. leucophloea*, *Balanites roxburghii*, *Cordia rothii*, *Azadirachta indica*, *Pongamia pinnata* (in river beds or bottom of ravines), *Dichrostachys cinerea*, *Leucaena leucocephala*, and *Prosopis juliflora*. On plateau, *Butea monosperma* also thrives well. It was found that grasses like *Dichanthium annulatum*, *D. caricosum*, *Bothriochloa pertusa*, and *Cynodon dactylon* had very high soil conservation and aggregation values, and these are quite palatable and sustain drought situations and can form part of silvopastoral system with abovementioned trees (Dagar 1987a, b, c).

### 4.2.4 Medicinal Plant Diversity in Ravines

India has been traditionally known as a treasure house of high value medicinal plants used in both the local folk health traditions and in the Indian Systems of Medicine. The demand for medicinal plant-based raw materials is growing at the rate of 15–25% annually, and according to an estimate of WHO, the demand for medicinal plants is likely to increase more than US $5 trillion in 2050. But due to biotic pressure, the medicinal plant wealth of Indian forests is declining constantly over the years. An estimated 4000–10,000 species of medicinal plants face potential local, national, regional, or global extinction, with subsequent serious consequences for livelihoods, economies, and health-care systems (Dhyani and Kala 2005).

Kala et al. (2017) studied the diversity and distribution of medicinal plant flora in the Yamuna ravines of Agra (U.P) and Chambal ravines of Kota district (Rajasthan) (Fig. 4.3). In Agra, 63 species belonging to 37 families were reported in the rehabilitated ravine areas. The study revealed that the species distributions were mainly confined to the families of Euphorbiaceae, Apocynaceae, Solanaceae, and Caesalpiniaceae. The documented plants were categorized according to their life form, viz., trees (34.92%), shrubs (26.98%), herbs (23.8%), climbers (11.15%), and grasses (3.18%). At Chambal ravines, 106 species representing 54 families were observed with dominant families of Euphorbiaceae, Fabaceae, Asteraceae, Solanaceae, and Amaranthaceae and were categorized according to their life form, viz., herbs (42.5%), trees (23.6%), shrubs (21.6%), climbers (10.4%), and grasses (2%). The important medicinal plants documented in rehabilitated ravines are given in Table 4.4.

In other study, Singh et al. (2013) in Chambal ravines also documented a rich medicinal plant diversity. A total of 57 species belonging to 37 families were recorded. Among all the species 40% were shrubs, 21% trees, 23% climbers, and 2% grasses. The study also concluded that the rehabilitated ravine lands had rich diversity of medicinal plants as compared to outside unprotected ravines. Therefore, conservation of ravines by protecting existing flora and fauna opens new vista for in situ conservation of medicinal plants and conservation of soil and water in the ravine regions. In the context of changing socio-economic, climatic, and policy regimes, there is immediate need to evolve management system involving farmers,

**Fig. 4.3** Medicinal plants diversity in ravine ecosystem. Outer circle clockwise: (**a**) *Acalypha indica*; (**b**) *Bombax ceiba*; (**c**) *Carissa carandas*; (**d**) *Solanum nigrum*; (**e**) *Citrullus colocynthis*; (**f**) *Calotropis procera*; (**g**) *Argemone mexicana*; and ((**h**) *Adhatoda vasica.* Inner circle clockwise (**i**) *Aegle marmelos*; (**j**) *Balanites aegyptiaca*; (**k**) *Capparis decidua*; and (**l**) *Carissa carandas*

industries, financing institutions, research institutes, concerned line departments, and NGOs to deal with entire supply chain of products derived from these lands. Ecosystem service offered by ravines must be estimated and accounted for determining conservation value of these lands. The existing ravine reclamation program is to be revisited and to be redesigned with new approach to ensure perennial green cover of these lands by making payment for ecosystem services to different stakeholders similar to conservation program in the United States (https://www.fsa.usda.gov/programs-and-services/index). Conservation of ravines not only protects river banks but also ensures the quality of water entering into the river system by filtering through the root systems.

**Table 4.4** List of medicinal plants distributed in the rehabilitated Yamuna Ravines at Agra, Uttar Pradesh

| Botanical name | Local name | Family | Habit | Part used and formulation | Medicinal use |
|---|---|---|---|---|---|
| *Abutilon indicum* | Kanghil | Malvaceae | Herb | Leaf extract | Itching, wound healing |
| *Acacia catechu* | Khair | Fabaceae | Tree | Bark and root decoction | Arthritis |
| *Acacia nilotica* | Babool | Mimosaceae | Tree | Bark, gum | Burning sensation |
| *Acalypha indica* | Chauriya | Euphorbiaceae | Herb | Leaf paste | Wound healing |
| *Achyranthes aspera* | Chirchita | Amaranthaceae | Shrub | Whole plant | Tuberculosis |
| *Adhatoda vasica* | Adathoda | Acanthaceae | Shrub | Leaf paste | Piles |
| *Aegle marmelos* | Beal | Rutaceae | Tree | Fruit and leaves | Sunstroke, dysentery |
| *Aerva lanata* | Geduakichal | Amaranthaceae | Herb | Root paste | Headache |
| *Ailanthus excelsa* | Anjan | Simaroubaceae | Tree | Leaf and bark paste | Joint pain and leprosy |
| *Albizia lebbeck* | Siris | Fabaceae | Tree | Seed paste | Snake bite |
| *Aloe vera* | Gwarpatha | Liliaceae | Herb | Dried leaf decoction | Easy and normal delivery |
| *Andrographis paniculata* | Kalmegh | Acanthaceae | Herb | Whole plant | Snake bite |
| *Argemone mexicana* | Prickly poppy | Papaveraceae | Herb | Root paste | Skin disease |
| *Asparagus racemosus* | Shatavari | Liliaceae | Climber | Dried roots | Increase lactation in cattle |
| *Azadirachta indica* | Neem | Meliaceae | Tree | All parts paste | Skin disease and blood purifier |
| *Balanites aegyptiaca* | Hingot | Balanitaceae | Tree | Fruit extract | Fever, deworming |
| *Bauhinia variegata* | Kachnar | Caesalpiniaceae | Tree | Root decoction | Reducing corpulence |
| *Butea monosperma* | Palash | Papilionaceae | Tree | Root and bark decoction | Dysentery control |
| *Calotropis procera* | Madar | Asclepiadaceae | Shrub | Root paste, latex | Snake bite, toothache |
| *Capparis decidua* | Karil | Cappariaceae | Shrub | Leaves, fruits | Dysentery |

(continued)

**Table 4.4** (continued)

| Botanical name | Local name | Family | Habit | Part used and formulation | Medicinal use |
|---|---|---|---|---|---|
| *Carissa carandas* | Karonda | Cappariaceae | Shrub | Fruit | Dysentery and diarrhea |
| *Cassia fistula* | Amaltas | Caesalpiniaceae | Tree | Fruits and seeds | Dysentery (cattle) |
| *Cassia tora* | Chakora | Fabaceae | Herb | Root paste | Deworming in children |
| *Catharanthus roseus* | Sadabahar | Apocynaceae | Shrub | Extract | Honey bee bite |
| *Clitoria ternatea* | Aparajita | Papilionaceae | Climber | Root powder | Chest pain and insect bite |
| *Commiphora mukul* | Guggul | Burseraceae | Herb | Gum decoction | Scorpion bite |
| *Cordia myxa* | Lasoda | Boraginaceae | Tree | Bark powder | Cure piles |
| *Cynodon dactylon* | Doob grass | Graminae | Grass | Whole plant | Menstrual disorders and fatness reduction |
| *Datura metal* | Kali Datura | Solanaceae | Herb | Leaf paste | Cure pimples |
| *Dendrocalamus strictus* | Bans | Poaceae | Tree | Leaves, seeds, bark juice | Wound healing, stop bleeding |
| *Emblica officinalis* | Aonla | Euphorbiaceae | Tree | Fruits | Vomiting |
| *Euphorbia prostrata* | Dudhi | Euphorbiaceae | Climber | Root powder | Milk secretion |
| *Grewia asiatica* | Phalsa | Tiliaceae | Shrub | Root paste | Back pain |
| *Hemidesmus indicus* | Gurmar | Asclepediaceae | Climber | Root paste | Blood purifier |
| *Ipomoea carnea* | Bansa | Convolvulaceae | Climber | Root | Increase lactation |
| *Jatropha curcas* | Ratanjot | Euphorbiaceae | Shrub | Seeds, oil and latex | Mouth ulcer |
| *Lawsonia inermis* | Mehndi | Lythraceae | Tree | Leaves, fruit paste | Body cooling, hair dyeing |
| *Madhuca longifolia* | Mahua | Sapotaceae | Tree | Leaves, flower, and seeds | Stomach pain |
| *Melia azedarach* | Bakayan | Meliaceae | Tree | Leaf and flower paste | Pregnancy weakness |
| *Mimosa pudica* | Laajivanthi | Mimosaceae | Herb | Leaf paste | Ear pain and headache |
| *Moringa oleifera* | Sahjan | Moringaceae | Tree | Leaf, fruit, gum | Vomiting and constipation |

(continued)

**Table 4.4** (continued)

| Botanical name | Local name | Family | Habit | Part used and formulation | Medicinal use |
|---|---|---|---|---|---|
| *Morus alba* | Shahtut | Moraceae | Tree | Root decoction | Dysentery |
| *Nerium indicum* | Kaner | Apocynaceae | Shrub | Root paste | Skin disease |
| *Ocimum gratissimum* | Van tulsi | Lamiaceae | Herb | Seed paste | Fever, dysentery |
| *Ocimum sanctum* | Tulsi | Lamiaceae | Herb | Seed paste | Eye burning, dysentery |
| *Phyllanthus niruri* | Bhuianola | Euphorbiaceae | Shrub | Whole plant extract | Jaundice |
| *Pongamia pinnata* | Karanj | Leguminosae | Tree | Seed oil and root bark paste | Mouth ulcer |
| *Punica granatum* | Anar | Punicaceae | Shrub | Flower and fruit paste | Toothache, dysentery |
| *Putranjiva roxburghii* | Putranjiva | Euphorbiaceae | Tree | Leaf extract | Fever, burning sensation |
| *Ricinus communis* | Caster | Euphorbiaceae | Shrub | Root powder | Pregnancy weakness, ulcers |
| *Bombax malabarica* | Semal | Bombacaceae | Tree | Flower paste | For easy delivery |
| *Sida cordifolia* | Kharenti | Malvaceae | Shrub | Whole herb | Rheumatism |
| *Solanum nigrum* | Makoy | Solanaceae | Herb | Root decoction | Body pain externally |
| *Syzygium cumini* | Jamun | Myrtaceae | Tree | Fruits and leaves | Diabetics |
| *Tamarindus indica* | Imlli | Caesalpiiaceae | Tree | Leaf extract | Constipation blood purifier |
| *Tephrosia purpurea* | Sharpunkha | Papilionaceae | Herb | Whole plant powder | Stomach problem |
| *Terminalia arjuna* | Arjun | Combretaceae | Tree | Leaf paste | Urinary infection |
| *Tinospora cordifolia* | Giloy | Menispermaceae | Climber | Fruit paste | Blood purifier |
| *Tridax procumbens* | Nahtoota | Asteraceae | Herb | Leaf extract | Rheumatism, ear pain |
| *Vitex negundo* | Nirgundi | Verbanaceae | Shrub | Twig and Leaf paste | Toothache, cold and cough |
| *Withania somnifera* | Ashwagandha | Solanaceae | Shrub | Leaf powder | Reduce fatness |
| *Wrightia tinctoria* | Kudruku | Apocynaceae | Tree | Bark extract | Piles problem |
| *Ziziphus mauritiana* | Ber | Rhamnaceae | Shrub | Fruit pulp | Health tonic, constipation |

Source: Kala et al. (2017)

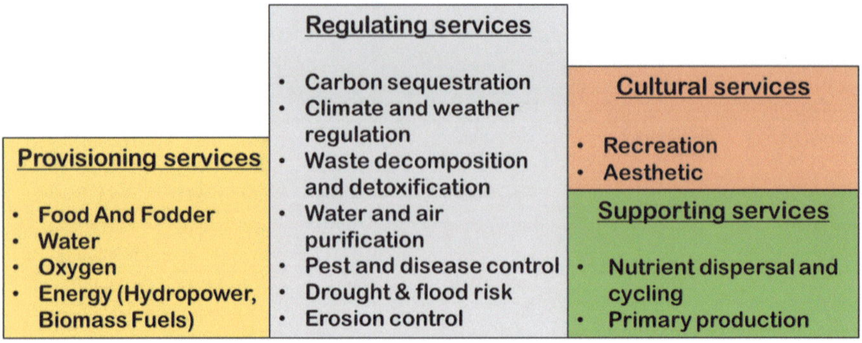


**Fig. 4.4** Ecosystem services provided by floral diversity in ravines

## 4.3 Importance of Ravine Biodiversity

Convention on Biological Diversity (CBD) was the results of the Earth Summit held in Rio de Janeiro in 1992. The Biodiversity Convention called upon parties to conserve and sustainably use biological diversity while equitably sharing the benefits of the genetic resources. These goals are the key elements of sustainable development. Biodiversity has been considered as the best bate for life insurance. In the degraded and fragile ravine ecosystem, there exists wide array of diversity which acts as supermarket for the local people. It seems equally ironic that while poverty reduction is rightly on everybody's mind, few recognize that the livelihood of the poorest (1.2 billion people) on Earth depends either entirely or at least partly on wild resources (Martin 2002). According to a World Bank study, more than two billion people rely on traditional medicine. Thus, biodiversity conservation must be an essential part of the maintenance of a healthy human environment and as crucial to the economic well-being of all societies.

Biodiversity in the ravine ecosystems provides tangible goods such as fuel, timber, fodder, food, and pharmaceutical products. It also provides the ecosystem services (Fig. 4.4), viz., purification of air and water, regulation of climate, nutrient and water cycling, pollination, pest and disease control, maintenance and regeneration of soil fertility, and the lessening of the impact of droughts and floods. Increased biodiversity is correlated with stability of ecosystems and lower rates of ecosystem collapse and extinction of species. Ravine ecosystem is a mosaic of agricultural, forest, and grassland biodiversity. The World Resource Institute (WRI) has classified the goods and services provided by different ecosystems (Table 4.5).

**Table 4.5** Goods and services provided by different biodiversity/ecosystems

| Ecosystem/ biodiversity | Goods | Services |
|---|---|---|
| Agricultural ecosystems | Food crops | Maintain limited watershed functions |
| | Fiber crops | Provide habitat for humans, birds, pollinators, soil organisms important to agriculture, maintain biodiversity and cycle nutrients |
| | Crop genetic resources | |
| | Other crops (energy, fodder, etc.) | |
| | Cultural resources | Sequester atmospheric carbon |
| | | Provide employment |
| | | Contribute to aesthetic beauty and provide recreation |
| Forest ecosystem | Timber | Remove air pollutants, emit oxygen, cycle nutrients |
| | Fuelwood | Maintain array of watershed functions (infiltration, purification, flow control, soil stabilization |
| | Drinking and irrigation water | |
| | Fodder | Sequester atmospheric carbon |
| | Non-timber forest products food (honey, mushrooms, fruit, and other edible plants) | Moderate weather extremes and impacts |
| | | Generate soil |
| | | Provide employment |
| | Aesthetic resources | Provide human, wildlife, and beneficial insect habitat |
| | Cultural resources | Contribute to aesthetic beauty and provide recreation |
| Grassland ecosystem | Livestock (food, game, hides, fiber) | Maintain array of watershed functions (infiltration, purification, flow control, soil stabilization) |
| | Drinking and irrigation, water | Cycle nutrients |
| | | Remove air pollutants, emit oxygen |
| | Genetic resources | Maintain biodiversity |
| | Cultural resources | Generate soil |
| | | Sequester atmospheric carbon |
| | | Provide employment |
| | | Provide human and wildlife habitat |
| | | Contribute to aesthetic beauty and provide recreation |

Source: Adapted from World Resources Institute (2000)

## 4.4 Ravine Rehabilitation for Increasing Biodiversity and Livelihood

The climate, topography, and soils of the ravine region are typically harsh and, thus, adversely affect the composition and growth of the vegetation. The soils of these lands are poor in fertility and physical conditions (Dagar and Mall 1980) as almost all the soil profiles are eroded leaving behind rocky, sandy, or stiff clay with poor water retention or water absorbing capacity. The land feature of ravines is rugged and characterized by three distinct parts, namely, ravine top or hump, slope, and bottom/bed. In general, the depth of ravines ranges from 1 to 20 m but may reach up to 55 m. Most of the ravenous lands are not fit for cultivation of agricultural crops and are classified as non-arable lands. Crop cultivation in these lands is not desirable as it leads to accelerated erosion thus, adding to their fast degradation. The best scientific land use for these lands is to place them under permanent vegetation involving forest and fruit trees along with forage grasses and energy plantations. Ravine rehabilitation requires an integrated approach of using soil and water conservation measures along with selection of suitable tree species.

Proper soil and water conservation structures need to be erected to check the uninterrupted flow of runoff into the ravines. Peripheral bunds, permanent drop structures like spillway, brush wood check dams, gabions, live vegetative check dams, and trenching were found suitable to rehabilitate the ravines (Chaturvedi et al. 2014). After successful checking of the ravine extension, the most important measure for reducing the risk of degradation of catchments of gullies and marginal lands along the ravines is introduction of trees and grasses in the ravine lands. Depending on the problems and needs of the area, trees may be introduced as alley, boundary plantation, or scattered tree plantation in the field. The highly eroded soils coupled with extremes of temperature and limitations of moisture hamper the survival and growth of trees. The species selected for growing along slopes must have regenerating capacity, particularly from exposed roots as is the case in *Balanites aegyptiaca* (Dagar et al. 1978). The ravines areas in Morena district was rehabilitated by Rajmata Vijayaraje Scindia Krishi Vishwavidyalaya (RVSKVV), Gwalior, using trees, grasses, and medicinal plants (Fig. 4.5). The tree and grass species (Table 4.6) selected have ability to withstand stress, are fast-growing, are easy to establish, have multipurpose uses, have fibrous roots, and have potential to ameliorate the soil. Also consult Chap. 3 of this publication.

### 4.4.1 Agroforestry Interventions

Different agroforestry practices have been applied successfully in reclamation of ravines. Under agri-horticulture systems, fruit trees such as lemon (*Citrus limon*), mango (*Mangifera indica*), *ber* (*Ziziphus mauritiana*), and goose berry (*Emblica officinalis*) can be grown with agricultural crops in humps and gully beds (Verma et al. 1986). In Yamuna ravines at Agra, Prajapati et al. (1993) observed that bamboo (*Dendrocalamus strictus*) produced 30–33 harvestable culms every 3 years after proper establishment at spacing of 3 × 3 m to 8 × 8 m producing 4000 poles $ha^{-1}$. Cultivation of grasses with the multipurpose trees is also a viable option for economic utilization of ravines. Medium and shallow gullies can be utilized under forest and fruit tree-based pastoral systems. Suitable grass species for protection and improving the fodder availability in ravine regions are *Dichanthium annulatum*, *Cenchrus ciliaris*, *Cenchrus setigerus*, *Panicum antidotale*, *Panicum maximum*, *Pennisetum purpureum*, and *Brachiaria mutica*.

(a) *Dalbergia sissoo* in ravine slopes

(b) *Albizia procera* in the ravine top

(c) *Aloe vera* in the ravine slopes

**Fig. 4.5** Rehabilitation of ravines using trees and medicinal plants at RVSKVV, Gwalior

Ravinous catchments of Chambal at Kota when planted with *Acacia* + *D. annulatum* and *D. annulatum* alone generated 5.8 and 2.6% of runoff and 1.26 and 0.62 Mg $ha^{-1}$ of soil loss, respectively, compared to 14.7% of runoff and 3 Mg $ha^{-1}$ of soil loss from agricultural catchments. Production of 4.5 Mg $ha^{-1}$ of air-dry grass + firewood from such degraded lands proved the effectiveness of grasses and trees as an alternative land use for protection and productive utilization of degraded ravine lands (Sharda and Venkateswarlu 2007). Further, Sethy et al. (2011) planted *Emblica officinalis* on gully humps and bamboo on gully bed, and interspace was planted with *Cenchrus ciliaris* grass and different trenching densities as treatment. The production of goose berry fruit ranged from 1.31 Mg $ha^{-1}$ in control (no trenches) to 6.61 Mg $ha^{-1}$ in treatment involving trenching for trapping 75% runoff, and the production of grass ranged from 7.71 to 9.91 Mg $ha^{-1}$. Ravine area development will not only contribute to bridging the growing demand and supply gap of food, fodder, and fuel but also helps in restoring livelihood security for resource constraint communities, besides providing improved ecosystem services (Sikka et al. 2016).

**Table 4.6** Choice of species for ravine rehabilitation

| Species | Planting location | Uses |
|---|---|---|
| *Acacia catechu* | Gully slopes and humps | Fuelwood, fodder, and minor forest products (MFP) |
| *Acacia leucophloea* | Gully slopes | Fodder and fuelwood |
| *Acacia nilotica* | Hump top, slope, and ravine beds | Fodder, fuelwood, small timber |
| *Acacia tortilis* | Hump top, slope, and ravine beds | Fuelwood and fodder |
| *Ailanthus excelsa* | Gully slopes and humps | Fodder, fuelwood, and timber |
| *Albizia lebbeck* | Hump top | Fodder and fuelwood |
| *Azadirachta indica* | Hump top and beds | Fodder, fuelwood, pesticides, and timber |
| *Balanites aegyptiaca* | Hump top and slope | Fuelwood and MFP |
| *Brachiaria mutica* | Hump top, slope, and ravine beds | Fodder, soil binder, and silage |
| *Cenchrus ciliaris* | Hump top, slope, and ravine beds | Fodder, hay, silage, and soil conservation |
| *Dalbergia sissoo* | Hump top and slope | Fuelwood and small timber |
| *Dendrocalamus strictus* | Gully bottom | Poles and fuelwood |
| *Dichanthium annulatum* | Hump top and slope | Fodder and mat |
| *Eucalyptus tereticornis* | Gully bottom | Poles and fuelwood |
| *Gmelina arborea* | Ravine beds | Small timber |
| *Leucaena leucocephala* | Gully slopes and humps | Fodder and fuelwood |
| *Panicum antidotale* | Ravine top and slope | Fodder and soil binder |
| *Prosopis juliflora* | Hump top, slope, and ravine beds | Charcoal, fuelwood, and fencing |
| *Soymida febrifuga* | Hump top | Fodder, fuel, and small timber |
| *Tamarix dioica* | Swampy areas in gully bottom | Reclamation of saline soils and fuelwood |
| *Ziziphus mauritiana* | Hump top and slope | Fruit and fuelwood |

Source: Modified from Sikka et al. (2016)

### 4.4.2 Case Study 2: Bamboo – Poor Man's Timber for Ravine Management and Livelihood Security

Bamboo was found to have incredible potential to provide the management of ravines and to support livelihood of ravine-dependent communities. It is a versatile non-wood forest product and a material which accompanies man from cradle to grave. The bamboo is grown on approximately 37 million ha area worldwide comprising more than 1200 species. In India, 136 species of bamboo under 36 genera are found to exist, which is second largest in world after China. Bamboo is known to be one of the fastest growing plants in the world, with a growth rate ranging from 30 to 100 cm per day in growing season. This nature makes bamboo to establish

**Fig. 4.6** *Dendrocalamus strictus* in bund stabilization (left) and bamboo stick seller (right) in Bundelkhand, India

very easily in ravine ecosystem, as availability of moisture is very limited up to 60–85 days, so fast-growing nature provides sufficient advantages over other species.

Because of extensive rhizome-root system and accumulation of leaf mulch, bamboo serves as an efficient agent in preventing soil erosion, conserving moisture, reinforcement of embankments and drainage channels, etc. With this idea, National Bamboo Mission project on "Hydrologic and Economic Evaluation of Bamboo Plantations in Gullied Lands under Major Ravine Systems of India" at Indian Institute of Soil and Water Conservation (IISWC), Dehradun, experienced the role of bamboo in environmental and livelihood security in ravines. Two hardy bamboo species *Dendrocalamus strictus* (manvel or solid bamboo) and *Bambusa bambos* (Katas or Thorny bamboo) were grown successfully in ravines under successive period of severe drought. The staggered contour trenches of 0.5 m (width) × 0.5 m (depth) × 2 m (length) were dug out at 4 × 4 m spacing in ravine beds and used for planting of bamboo. In average, the bamboo planted check dam recorded very low soil loss (0.4–1 Mg $ha^{-1}$) as compared to control, without bamboo (3–6 Mg $ha^{-1}$). The economic evaluation of these activities provided the actual livelihood improvement of local communities. The bamboo project in different ravines of Yamuna and Mahi rivers revealed that for a production span of 20 years, the net value ranged from ₹222,588 to ₹319,731 per ha $yr^{-1}$.

Further, under Bundelkhand Special Package, UP Forest Department had undertaken different practices of degraded land management activities for reducing the impact of drought through afforestation, live bunding, and check dam. These activities were evaluated by ICAR-CAFRI, Jhansi, and reported that the role of different trees and grasses in prevention of soil loss was greater than other activities. In ravine areas of Jalaun, Hamirpur, and Banda, afforestation through *Dendrocalamus strictus* was remarkable and helped to restore natural resources (Fig. 4.6). During survey, the local communities were collecting bamboo culm for preparation of baskets, mats, poles, etc. The bamboo stick business is mainly prevalent in Jhansi, Hamirpur, Mahoba, and Banda districts of Bundelkhand and chiefly sold in village fair and festivals earning ₹15,000–20,000 per year by each person selling the bamboo sticks.

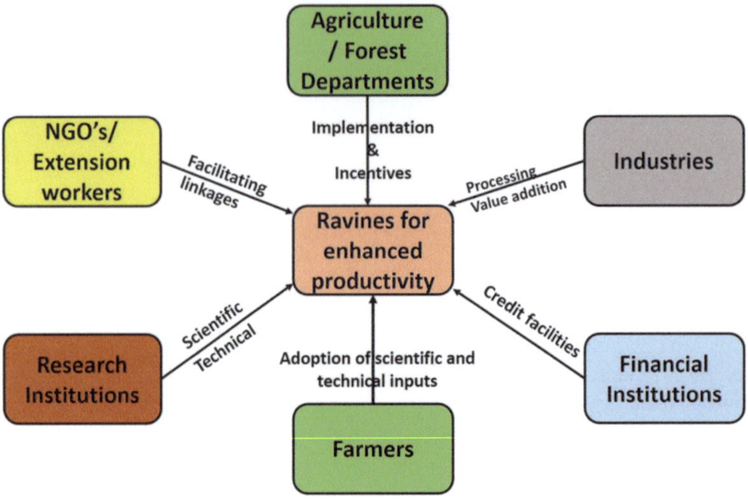


**Fig. 4.7** Role of different stakeholders in enhancing productivity in ravines

## 4.5 Way Forward

The harsh environmental condition and degraded natural resources pose a tough task to rehabilitate the ravine ecosystem. The major stakeholders in ravines rehabilitation are farmers, agricultural department, forest department, financial institutions, nongovernmental organizations, and the research institutes (Fig. 4.7). Research institutions may play important role in preparation of status report of the soil and water conservation measures, biodiversity, and biotic pressure on the ravines.

Based on the basic data, the research institutes may provide technical information regarding know-how to check soil and water erosions, suitable species (tree/shrubs/herbs), quality planting materials, planting layout, and aftercare of the plantations. Agriculture and Forest Departments must facilitate the implementation of the scientific and technical inputs given by the research institutions. The government must provide subsidies and other monetary benefits to the farmers who are keen in rehabilitating the ravines. NGOs and extension workers may play an important role in disseminating, motivating, and facilitating linkages among farmers and other stakeholders. Financial institutions like NABARD and other nationalized banks must provide credit facilities to farmers for conserving and improving the ravine ecosystem. Loans for construction of soil and water conservation structures, afforestation activities, and establishment of small-scale industries for processing and value addition of agricultural and forest products including drugs may be provided on low interest. Small-scale and cottage industries should be set up for processing and value addition of goods produced in the ravines. Creating linkages between different stakeholders will create a sustainable model, and it will be a win-win situation for all the stakeholders.

## 4.6 Conclusion

Biodiversity in ravines is declining drastically due to deforestation, encroachment, and urbanization processes in India, and it is great concern for restoration of both resources and biodiversity. The diversity of ravines provides livelihood security to different sections of society by providing food, fodder, fuelwood, medicine, fruits, etc. The different programs like National Bamboo Mission, Watershed Management, Agroforestry Mission, and drought prone programs may play critical role to rehabilitate and maintain degraded ecosystems. Ravine rehabilitation requires an integrated approach of treatment of table and marginal lands contributing runoff to the gullies on watershed basis. The different techniques like gully plugging, bunding, check dam, live fencing, vegetation barriers against soil and water erosion, and agroforestry-based farming may be implemented. Species like *Butea monosperma* may be used for lac cultivation, thus improving the livelihood security of the local people. Planting and management of indigenous plant species like *Anogeissus pendula*, *Acacia leucophloea*, *A. nilotica*, *Ziziphus mauritiana*, *Carissa carandas*, and *Capparis decidua* along with palatable fodder grasses in silvopastoral mode will pave way toward sustaining ravine ecosystem and livelihood of the people.

## References

Balvanera P, Pfisterer AB, Buchmann N, He JS, Nakashizuka T, Raffaelli D, Schmid B (2006) Quantifying the evidence for biodiversity effects on ecosystem functioning and services. Ecol Lett 9:1146–1156

Cardinale BJ, Matulich KL, Hooper DU, Byrnes JE, Duffy E, Gamfeldt L, Balvanera P, O'Connor MI, Gonzalez A (2011) The functional role of producer diversity in ecosystems. Am J Bot 98:572–592

Champion HG, Seth SK (1968) A revised survey of forest types of India. Controller of Publication, New Delhi

Chaturvedi OP, Kaushal R, Tomar JMS, Prandiyal AK, Panwar P (2014) Agroforestry for waste land rehabilitation: mined, ravine, and degraded watershed areas. In: Dagar JC, Singh AK, Arunachalam A (eds) Agroforestry systems in India: livelihood security & ecosystem services, advances in agroforestry, vol 10. Springer, New Delhi, pp 233–271

Chauhan BPS, Shrivastava A, Kumar A, Bhadauria S (2015) Floristic diversity in the ravines of Ottangan River under protection and biotic disturbance. Int J Pharma Res Bio-Sci 4(3):100–107

Chavan SB, Chauhan DS, Keerthika A, Uthappa AR, Jha A, Newaj R (2016) Fuelwood characteristics of selected tree species from Bundelkhand region of Central India. Eco Environ Cons 22(April Suppl):S95–S103

Chouhan DS, Chavan SB, Keerthika A, Jha A, Dwivedi RP, Kumar A (2015) Socio-economic fuelwood survey in Bundelkhand region of central India. In: Mishra GC, Yadav SK (eds) Energy research and environmental management: an innovative approach. Krishi Sanskriti Publications, New Delhi

Dagar JC (1987a) Species composition and plant biomass of an ungrazed and a grazed grassland at Ujjain, India. Trop Ecol 28:208–215

Dagar JC (1987b) Responses of clipping treatments on nine palatable perennial grasses. Trop Ecol 28:26–221

Dagar JC (1987c) Studies on reclamation of Kshipra ravines. Indian J For 10(2):83–89

Dagar JC (1999) Soil erosion parameters with special reference to Kshipra ravines in Madhya Pradesh. Indian J For 22(3):210–221
Dagar JC (2001) Biological spectrum and successional trends in Kshipra watershed areas. Indian J For 24(3):351–356
Dagar JC, Mall LP (1980) Studies on the vegetation of ravines and banks of the river Kshipra and its tributaries. J Indian Bot Soc 59:234–245
Dagar JC, Singh VP (1980) The ecological studies of the vegetation of the river Kshipra and its tributaries. Biologia 26(1):43–62
Dagar JC, Singh VP, Mall LP (1978) Regeneration in plants growing on eroded soil. Ann Arid Zone 17:36–74
Dhyani PP, Kala CP (2005) Current research on medicinal plants: five lesser known but valuable aspects. Curr Sci 88:335–340
Hector A, Schmid B, Beierkuhnlein C, Caldeira MC, Diemer M et al (1999) Plant diversity and productivity experiments in European grasslands. Science 286:1123–1127
Hooper DU, Chapin FS, Ewel JJ, Hector A et al (2005) Effects of biodiversity on ecosystem functioning: a consensus of current knowledge. Ecol Monogr 75:3–35
Kala S, Meena HR, Rashmi I, Prabavathi M, Singh AK, Singh RK (2017) Status of medicinal plants diversity and distribution at rehabilitated Yamuna and Chambal ravine land ecosystems in India. Int J Curr Microbiol App Sci 6(3):618–630
Kotia A, Umeshkumar LT, Rawat GS (2008) Semiarid region of India: vegetation characteristics and threatened plants. In: Rawat GS (ed) Special habitats and threatened plants of India. ENVIS bulletin: wildlife and protected areas, vol 11(1). Wildlife Institute of India, Dehradun, pp 109–116
Liang J, Crowther TW, Picard N, Wiser S, Zhou M, Alberti G, Schulze ED et al (2016) Positive biodiversity-productivity relationship predominant in global forests. Science 354(6309):aaf8957. https://doi.org/10.1126/science.aaf8957
Martin C (2002) Meeting sustainable development goals without endangering species diversity. CBD News, p 7
Parandiyal AK, Samara JS, Singh KD, Ratan S, Rathore BL (2000) Floristic diversity of Chambal Ravines under varying levels of protection. Indian J Soil Cons 28(2):160–166
Prajapati MC (1995) Floristic wealth of Yamuna ravines. Indian J Soil Cons 23(1):4–61
Prajapati MC, Nambiar KTN, Puri DN, Singh JP, Malhotra BM (1993) Fuel and fodder production in Yamuna ravines at Agra. Indian J Soil Cons 21(3):8–13
Reich PB, Knops J, Tilman D, Craine J, Ellsworth D et al (2001) Plant diversity enhances ecosystem responses to elevated CO2 and nitrogen deposition. Nature 410:809–812
Sajwan SS (1975) Ground flora and their distribution in forest plantation in Jamuna ravines. Indian For 101(5):269–275
Sajwan SS (1976) Dominant grasses and their associates in ravinous tracts of Jamuna. Soil Conserv Digest 4(1):26–35
Sethy BK, Parandiyal AK, Shakir A, Kumar A, Singh RK (2011) Cost effective conservation measures for management of medium and deep ravenous lands, Annual report. CSWCRTI, Dehradun, pp 73–74
Sharda VN, Venkateswarlu B (2007) Crop diversification and alternate land use systems in watershed management. In: Best-bet options for integrated watershed management. Proceedings of the comprehensive assessment of watershed programs in India, 25–27 Jul 2007. ICRISAT, Patancheru, pp 111–128
Sikka AK, Mishra PK, Singh RK, Krishna BK Islam A (2016) Management of ravines for food, livelihood and environment security. Paper presented at the Global ravine conference on managing ravines for food and livelihood security, RVSKVV, Gwalior, India, March 7–10, 2016. pp 1–14
Singh A, Verma SK, Singh YP (2013) Ethno-medicinally important plants from Chambal ravine region of Madhya Pradesh. Bioinfolet 10(4A):1186–1192
Thomas M, Sahu P, Shrivastava A, Hussain Z (2011) Biodiversity and livelihood options of people in Chambal ravines of Morena district, Madhya Pradesh, India. J Trop For 27:40–56

Thrupp LA (2000) Linking agricultural biodiversity and food security: the valuable role of sustainable agriculture. Int Aff 76(2):265–281

Tilman D, Reich PB, Knops J, Wedin D, Mielke T, Lehman C (2001) Diversity and productivity in a long-term grassland experiment. Science 294:843–845

Tilman D, Reich PB, Isbell F (2012) Biodiversity impacts ecosystem productivity as much as resources, disturbance, or herbivory. Proc Natl Acad Sci U S A 109(26):10394–10397

Uthappa AR, Chavan SB, Singh M, Sridhar KB, Dev I, Ram A, Sathish BN, Kumar M, Dwivedi RP, Singh R, Singh RK, Singh R, Tewari RK, Handa AK, Chaturvedi OP (2016) Tree diversity in ravines and their rehabilitation through agroforestry interventions in Bundelkhand Region of India. Indian J Agrofor 18(1):77–83

Verma B, Chinamani S, Bhola SN, Rao DH, Prasad SN, Prakash C (1986) Twenty five years of research on soil and water conservation in ravine lands of Rajasthan. Central soil and water conservation research and training institute publication. Research Centre, Kota, pp 1–215

World Resources Institute (2000) World resources 2000–2001, people and ecosystems: fraying the web of life. World Resources Institute, Washington, DC

# Nutrient Management Strategies for Improving Soil Health and Productivity of Ravine Lands

5

S. K. Verma and Anil Kumar Singh

**Abstract**

The deep ravine of Chambal is the most fragile stressed ecosystem and suffers heavily because of erosion. The soils of this region are sandy loam with very little organic matter and very poor binding capacity. There are tremendous soil (>30 Mg $ha^{-1}$ $year^{-1}$), water (>70% of rainfall), and nutrient (>350 kg of NPK $year^{-1}$) losses during rainy season, and this problem is still aggravating to the farmers by spreading to fertile lands. The loosely held soil particles favor high infiltration rate (>7 cm $h^{-1}$), and the high elevation differences between land and riverbed are catalyzing reason to create a high gradient for gushing underground water current toward the river which carries soil and water with it. The formation of ravine is a hydraulic energy-generated phenomenon, and it can be cured only after a clear understanding of how to minimize hydraulic gradient and hydrodynamic pressures. Further, advancement of ravine can be checked very easily by adopting indigenous and innovative technologies based on energy concept. Multistep leveling is one of the very recent and innovative technologies and is economically viable with almost checking all kinds of losses, whereas improvised gabion structure is a very economical and suitable conservation method to check/minimize soil and water losses under ravine conditions. Adoption of various kinds of modules also helped in minimizing natural losses among them. Silvopastoral and silvi-medicinal systems are very effective from point of conservations, while agro-horti, agroforestry, silvopastoral, and silvi-medicinal modules are most feasible under ravine conditions and are efficient in the order mentioned.

S. K. Verma (✉)
Department of Soil Science & Agricultural Chemistry, RVSKVV, Vishwa Vidyalaya, Gwalior, Madhya Pradesh, India

A. K. Singh
RVS Krishi Vishwa Vidyalaya, R. V. S., Gwalior, India





J. C. Dagar, A. K. Singh (eds.), *Ravine Lands: Greening for Livelihood and Environmental Security*, https://doi.org/10.1007/978-981-10-8043-2_5

## 5.1 Introduction

Soil-quality (health) is defined as the capacity of a soil to function, within ecosystem and land-use boundaries, to sustain biological productivity, maintain environmental quality, and promote plant and animal health (Doran and Parkin 1994). Among many environmental hazards, land degradation is of utmost importance as it results in the decline in productivity on arable and nonarable land. It is estimated that about 80% of the current degradation on agricultural land in the world is caused by soil erosion due to water (Angima et al. 2003). About 2 billion ha in the world are affected by various forms of human-induced land degradation with erosion by water being the chief contributor (1.1 billion ha; Oldeman 1994). Hydrological disaster coupled with high erosion rate has serious social, economic, and environmental implications (Pimentel 2000).

One of the major negative on-site effects of soil erosion is the loss of soil fertility leading to decline in productivity. In India, out of 120.8 million ha (Mha) degraded land, 82.6 Mha is estimated to suffer from water erosion, 24.7 Mha from chemical degradation, 12.4 Mha due to wind erosion, and 1.1 Mha from physical (mainly due to stagnation of water) degradation; and about 6.73 Mha is estimated to be salt-affected (Maji et al. 2010). It is estimated that India suffers an annual loss of 13.4 million Mg in the production of major cereals, oil seeds, and pulses due to water erosion equivalent to about 2.51 billion Indian rupees (Sharda et al. 2010). A ravine is a small valley usually carved by running water, especially the narrow excavated valley of a mountain stream {*etymology* – French, "mountain torrent" (Bates and Jackson 1984)}. The greatest challenge in the twenty-first century is to feed the ever-increasing population along with the improvement and maintenance of soil health and environmental quality. In this chapter, an attempt has been made to manage soil and water resources for increasing productivity of ravine lands.

## 5.2 Soil and Water Conservation Approaches for Management of Ravines

Ravine soils are very poor, or we can say that a kind of totally washed soils and their physicochemical and biological conditions are very poor (Table 5.1) and need a special kind of management to restore them and make them productive. These soils are deprived of all kinds of available nutrients. Ravine soils generally suffer from very huge natural resource losses, and the soil loss may go more than 25 (Mg ha$^{-1}$ year$^{-1}$) along with >80% of rainfall as runoff (Table 5.2) and loss of huge quantity of nutrients (Table 5.3) every year with a tune of 356 kg ha$^{-1}$ (N, P and K). These losses can be minimized even to zero by adopting various kinds of management systems.

Several studies were conducted by applying nutrients both macro and micro in isolation or in combination with organic and biological amendments/fertilizers. The results were very encouraging as explained below.

**Table 5.1** General soil properties of a deep ravine

| S. no. | Soil characteristics | Values |
|---|---|---|
| 1 | pH (1:2.5) | 8.08 |
| 2 | Electrical conductivity (dS $m^{-1}$) | 0.15 |
| 3 | Organic carbon (%) | 0.14 |
| 3 | Available nitrogen (kg $ha^{-1}$) | 132.72 |
| 4 | Available phosphorus (kg $ha^{-1}$) | 12.94 |
| 5 | Available potassium (kg $ha^{-1}$) | 196.92 |
| 6 | Total nitrogen (%) | 0.10 |
| 7 | Total phosphorus (%) | 0.14 |
| 8 | Total potassium (%) | 0.38 |
| 9 | Ammoniacal-N (Mg $kg^{-1}$) | 31.48 |
| 10 | Nitrate-N (Mg $kg^{-1}$) | 5.85 |
| 11 | Available zinc (Mg $kg^{-1}$) | 0.53 |
| 12 | Available Cu (Mg $kg^{-1}$) | 0.48 |
| 13 | Available Mn (Mg $kg^{-1}$) | 7.76 |
| 14 | Available Fe (Mg $kg^{-1}$) | 2.12 |
| 15 | Available sulfur (Mg $kg^{-1}$) | 15.26 |

Source: Argal (2017)

**Table 5.2** Erosion losses computed and recorded under different systems (Verma et al. 2016)

| | Computed | | Recorded | |
|---|---|---|---|---|
| Systems | Silt loss (Mg $ha^{-1}$ $year^{-1}$) | Runoff (cm) | Silt loss (Mg $ha^{-1}$ $year^{-1}$) | Runoff (cm) |
| Diversified cropping system | 18.24 | 32.42 | 28.7 | 30.8 |
| Agri-horticultural system | 07.54 | 26.94 | 16.2 | 25.6 |
| Horti-medicinal/pastoral system | 02.87 | 25.60 | 24.0 | 25.8 |
| Silvi-medicinal system-I | 14.37 | 25.21 | 18.6 | 25.9 |
| Silvi-medicinal system-II | 11.50 | 25.90 | 14.8 | 24.6 |
| Silvopastoral system | 11.50 | 26.37 | 08.4 | 24.2 |
| Multistep leveling | Nil | Nil | Nil | Nil |

### 5.2.1 Management of Soil Physical Properties

**Soil bulk density (Mg $m^{-3}$):** The bulk density values indicate compactness and porosity of the soil. The continuous application of organic material in the form of farmyard manure (FYM) and vermi-compost may lower the bulk density of soil. The results showed that (Table 5.4) continuous application of FYM at 10 Mg $ha^{-1}$ + PSB + Azotobacter ($T_7$) and 50% RDF + 5 Mg FYM $ha^{-1}$ + PSB+ $ZnSO_4$ 25 kg $ha^{-1}$ ($T_5$) maintained the soil bulk density of soil between 1.39 and 1.42 Mg $m^{-3}$ in the final stage of study period, respectively. With increased doses of RDF (50–75, 100%), the bulk density of treatments ($T_5 > T_6 > T_2 > T_4$) increased over the organic manure treatments ($T_1$ and $T_7$). The highest bulk density 1.52 Mg $m^{-3}$ was recorded under the treatment 150% RDF ($T_3$), which remained unchanged at the end of the study period.

**Table 5.3** Estimated losses (kg ha$^{-1}$) of major nutrients (N, P, and K) through runoff under different systems adopted for ravine lands (Verma et al. 2016)

| Systems | N | $P_2O_5$ | K | Total |
|---|---|---|---|---|
| Diversified cropping system | 88.97 | 8.61 | 258.30 | 355.88 |
| Agri-horticulture system | 49.36 | 7. 56 | 145.80 | 202.72 |
| Horti-medicinal/pastoral system | 74.40 | 7.20 | 216.00 | 297.60 |
| Silvi-medicinal system-I | 67.66 | 5.58 | 167.40 | 240.64 |
| Silvi-medicinal system-II | 45.88 | 4.44 | 133.20 | 183.52 |
| Silvopastoral system | 26.04 | 2.52 | 75.60 | 104.16 |
| Multistep leveling | Negligible | Negligible | Negligible | Negligible |

**Table 5.4** Bulk density (Mg m$^{-3}$), mean weight diameter (MWD in mm), and moisture content (%) as affected by nutrient management after a 2-year study (2014–2015)

| Tr. no. | Treatments | Bulk density | MWD | Moisture |
|---|---|---|---|---|
| T 1 | Farmer practices | 1.34 | 0.31 | 12.60 |
| T 2 | 100% RDF | 1.47 | 0.33 | 14.74 |
| T 3 | 150% RDF | 1.52 | 0.48 | 15.76 |
| T 4 | NPK application based on STCR equation | 1.47 | 0.43 | 14.72 |
| T 5 | 50% RDF+5 Mg FYM ha$^{-1}$ + PSB+ $ZnSO_4$ at 25 kg ha$^{-1}$ | 1.42 | 0.42 | 14.16 |
| T 6 | 75% RDF + 2.5 Mg FYM ha$^{-1}$ + PSB + $ZnSO_4$ at 25 kg ha$^{-1}$ | 1.42 | 0.50 | 18.24 |
| T 7 | FYM at 10 Mg ha$^{-1}$ + PSB + Azotobactor | 1.39 | 0.48 | 15.69 |
| | SEm± | 0.04 | 0.02 | 0.64 |
| | LSD ($p \leq 0.05$) | NS | 0.07 | 1.96 |

Source: Argal (2017)

Depictions: *RDF* recommended dose of fertilizers, *PSB* phosphorus-solubilizing bacteria, *STCR* soil test crop response, *FYM* farmyard manure

**Mean weight diameter (mm):** Mean weight diameter (MWD) of the surface layer were highest (0.50 mm) in the plots where 75% RDF + 2.5 Mg FYM ha$^{-1}$+ PSB + $ZnSO_4$ at 25 kg ha$^{-1}$ ($T_6$) were applied and were significantly higher than those of the 150% RDF ($T_3$) and farmer practices ($T_1$) treatment (Table 5.4).

**Soil moisture content (%) on dry basis:** Soil moisture content (SMC) of the surface layer was highest in the plots treated with 75% RDF + 2.5 Mg FYM ha$^{-1}$ + PSB + $ZnSO_4$ at 25 kg ha$^{-1}$ ($T_6$) and was significantly higher than those of the 150% RDF ($T_3$) and farmer practices ($T_1$) treatment (Table 5.4).

**Soil moisture content retention at different suctions:** The study of the data on soil moisture content at different suctions (bars) for 0–15 cm and 15–30 cm soil layer (Table 5.5) indicates that the top layer was affected significantly compared to the lower layer because of applied different fertility levels after 2 years. The soil moisture content at different fertility levels at suction 0.33 (field capacity) bars in the layer (0–15 cm and subsurface layer 15–30 cm) ranged from 12.89% to 21.49% and 12.42–17.17% in the 0–15 cm and 15–30 cm layer, respectively. Surface soil contained higher amount of moisture content at 0.33 bars which decreased gradually with 1 bar, 5 bars, and 10 bars, and least status was recorded in 15 bars (wilting

**Table 5.5** Soil moisture content ($cm^3\ cm^{-3}$) at different applied suctions (bars) for 0–15 and 15–30 cm depth soil as affected by nutrient management after 2 years (2014–2015)

| Tr. no. | Treatments | 0–15 cm | | | | | | 15–30 cm | | | | | |
|---|---|---|---|---|---|---|---|---|---|---|---|---|---|
| | | 0.33 | 1.00 | 5.00 | 10.00 | 15.00 | Available | 0.33 | 1.00 | 5.00 | 10.00 | 15.00 | Available |
| T 1 | Farmer practices | 12.89 | 10.61 | 5.60 | 5.54 | 5.30 | 7.59 | 12.42 | 10.64 | 6.10 | 5.94 | 5.77 | 6.65 |
| T 2 | 100% RDF | 21.10 | 13.09 | 6.74 | 6.54 | 6.48 | 14.62 | 15.54 | 10.67 | 7.65 | 7.24 | 7.06 | 8.48 |
| T 3 | 150% RDF | 20.59 | 11.82 | 7.71 | 6.83 | 5.33 | 15.26 | 16.56 | 10.74 | 7.75 | 7.33 | 7.14 | 9.42 |
| T 4 | NPK application based on STCR equation | 20.35 | 12.09 | 7.24 | 6.15 | 5.95 | 14.40 | 14.07 | 11.34 | 7.28 | 7.08 | 6.69 | 7.38 |
| T 5 | 50% RDF + 5 Mg FYM $ha^{-1}$+ PSB + $ZnSO_4$ at 25 kg $ha^{-1}$ | 21.19 | 12.77 | 7.37 | 6.84 | 5.60 | 15.59 | 15.47 | 12.84 | 6.43 | 6.14 | 5.87 | 9.60 |
| T 6 | 75% RDF+2.5 Mg FYM $ha^{-1}$ + PSB+ $ZnSO_4$ at 25 kg $ha^{-1}$ | 21.49 | 13.49 | 8.85 | 7.45 | 7.10 | 14.39 | 17.17 | 16.04 | 8.96 | 8.66 | 7.40 | 9.77 |
| T 7 | FYM at 10 Mg $ha^{-1}$ + PSB + Azotobactor | 16.16 | 11.43 | 7.02 | 7.15 | 5.81 | 10.35 | 12.79 | 12.62 | 7.35 | 7.05 | 5.83 | 6.96 |
| SEm± | | 0.15 | 0.12 | 0.25 | 0.29 | 0.31 | | 0.18 | 0.10 | 0.29 | 0.14 | 0.15 | |
| LSD ($p \leq 0.05$) | | 0.45 | 0.36 | 0.77 | 0.89 | 0.97 | | 0.54 | 0.30 | 0.90 | 0.45 | 0.48 | |

point) in all the fertility treatments in the layer (0–15 cm and subsurface layer 15–30 cm). Inorganic and organic treatment of ($T_6$) 75% RDF + 2.5 Mg FYM $ha^{-1}$ + PSB + $ZnSO_4$ at 25 kg $ha^{-1}$ shows maximum moisture content at 0.33 bars in the surface (0–15 cm) and subsurface (15–30 cm) layer and was higher over the rest of the other treatment ($T_1$) farmer practices, in both the layers. The study on moisture retention to the different suction (bars) applied to 30–60 cm and 60–90 cm soil layer (Table 5.6) was also affected significantly in similar pattern to the soil treated with different fertility levels for 2 years.

The moisture content and total available water content at suction 0.33 bars in the layer (30–60 cm and subsurface layer 60–90 cm) were ranged from 15.19% to 19.67% and 15.36–18.27% at different fertility levels, respectively. Surface soil showed higher moisture retention at field capacity (0.33 bars suction) which was decreased gradually with increasing suction at 1 bar, 5 bars, and 10 bars suctions, and least was recorded at 15 bars irrespective of fertility treatments, and this trend was observed similar in the lower depth (30–60 cm and subsurface layer 60–90 cm). Inorganic and organic fertility treatments were found favorable in increasing moisture retention capacity of soil. However, NPK application based on STCR equation shows maximum moisture content at field capacity (0.33 bars) at lower depth layers (30–60 cm and subsurface layer 60–90 cm) and which was higher over other treatments. Total available moisture content was computed with the difference in moisture at field capacity (0.33 bars) and moisture content at wilting point (15.00 bars). There was significant variation in available moisture content in plough layer soil (0–15 cm), and only minor changes were recorded in other lower depths of soil due to application of different treatments.

### 5.2.2 Management of Chemical Properties of Soil

The effect of organic and inorganic fertilization application on physicochemical properties of ravine land, viz., pH, electrical conductivity (dS $m^{-1}$), and organic carbon, was studied during 2013–2014 and 2014–2015, and results are presented in Table 5.7. The treatments showed higher variation in soil pH and ranged between 8.05–8.44 and 8.31–8.44 (1:2.5), 2013–2014 and 2014–2015, respectively. The EC ranges between 0.13–0.29 (dS $m^{-1}$) and 0.43–0.52 (dS $m^{-1}$), both the study period 2013–2014 and 2014–2015, respectively. The organic carbon values increased from 0.11 to 0.20 (%) in 2013–2014 and 0.12–0.18% in 2014–2015 study period, respectively. The regular increase in organic carbon during the 2-year study period was observed in fertility treatments 75% RDF+2.5 Mg FYM $ha^{-1}$ + PSB + $ZnSO_4$ at 25 kg $ha^{-1}$ ($T_6$) and/or treatment ($T_3$), ($T_4$), ($T_5$), ($T_6$), ($T_7$) and super-optimal dose fertility treatment ($T_2$) as compared to farmer practices. The pooled data on soil reaction (pH) (1:2.5), electrical conductivity (dS $m^{-1}$), and organic carbon (%) show that the highest pH value (8.48) was recorded under ($T_2$) 100% RDF, electrical conductivity 0.39 (dS$m^{-1}$), in application FYM at 10 Mg $ha^{-1}$ + PSB + *Azotobacter* ($T_7$); and organic carbon varied in range from 0.13% to 0.19% in pooled basis, respectively. Lowest value was recorded with treatment $T_1$ (farmers' practices).

**Table 5.6** Soil moisture content at different suctions (bars) for 30–60 and 60–90 cm soil depths as affected by nutrient management after 2-year studies (2014–2015)

| | | 30–60 cm soil layer | | | | | | 60–90 cm soil layer | | | | | |
|---|---|---|---|---|---|---|---|---|---|---|---|---|---|
| Tr. no. | Treatments | 0.33 | 1.0 | 5.0 | 10.0 | 15.0 | Available | 0.33 | 1.0 | 5.0 | 10.0 | 15.0 | Available |
| T 1 | Farmer practices | 15.19 | 10.54 | 7.55 | 7.33 | 6.27 | 8.91 | 15.36 | 11.24 | 8.06 | 7.21 | 6.15 | 9.21 |
| T 2 | 100% RDF | 16.36 | 12.55 | 9.08 | 8.05 | 6.38 | 9.98 | 17.47 | 12.97 | 9.33 | 8.36 | 6.91 | 10.56 |
| T 3 | 150% RDF | 17.26 | 13.42 | 9.20 | 8.91 | 7.27 | 9.99 | 18.27 | 13.57 | 10.00 | 9.14 | 7.78 | 10.49 |
| T 4 | NPK application based on STCR equation | 17.19 | 19.01 | 11.41 | 8.10 | 7.24 | 9.95 | 16.41 | 18.21 | 12.10 | 11.00 | 7.00 | 9.41 |
| T 5 | 50% RDF+5 Mg FYM $ha^{-1}$+PSB+ $ZnSO_4$ at 25 kg $ha^{-1}$ | 16.14 | 11.68 | 10.08 | 7.91 | 7.34 | 9.80 | 17.58 | 12.68 | 9.01 | 7.83 | 6.80 | 10.78 |
| T 6 | 75% RDF+2.5 Mg FYM $ha^{-1}$+PSB+ $ZnSO_4$ at 25 kg $ha^{-1}$ | 16.14 | 12.42 | 7.88 | 7.75 | 7.07 | 9.07 | 17.14 | 12.81 | 9.00 | 7.40 | 6.18 | 9.96 |
| T 7 | FYM at 10 Mg $ha^{-1}$+PSB + Azotobacter | 19.67 | 11.40 | 10.47 | 9.74 | 9.27 | 10.40 | 15.36 | 11.68 | 8.27 | 8.07 | 6.75 | 8.61 |
| | SEm± | 0.06 | 0.11 | 0.06 | 0.04 | 0.08 | | 0.14 | 0.09 | 0.09 | 0.10 | 0.08 | |
| | LSD ($p \leq 0.05$) | 0.17 | 0.33 | 0.20 | 0.12 | 0.24 | | 0.43 | 0.27 | 0.29 | 0.32 | 0.24 | |

**Table 5.7** Effect of nutrient management on changes of physicochemical properties in ravine soil during 2013–2014 and 2014–2015

| Tr. no. | Treatments | Physicochemical properties of soil | | | | | | | | |
|---|---|---|---|---|---|---|---|---|---|---|
| | | 2013–2014 | | | 2014–2015 | | | Pooled | | |
| | | pH (1:2.5) | EC (dS $m^{-1}$) | O.C. (%) | pH (1:2.5) | EC (dS $m^{-1}$) | O.C. (%) | pH (1:2.5) | EC (dS $m^{-1}$) | OC (%) |
| T 1 | Farmer practices | 8.05 | 0.13 | 0.11 | 8.42 | 0.52 | 0.15 | 8.40 | 0.33 | 0.13 |
| T 2 | 100% RDF | 8.22 | 0.25 | 0.13 | 8.42 | 0.49 | 0.17 | 8.48 | 0.37 | 0.15 |
| T 3 | 150% RDF | 8.07 | 0.22 | 0.16 | 8.35 | 0.43 | 0.12 | 8.10 | 0.32 | 0.14 |
| T 4 | NPK application based on STCR equation | 8.12 | 0.27 | 0.18 | 8.31 | 0.47 | 0.18 | 8.16 | 0.37 | 0.18 |
| T 5 | 50% RDF+5 Mg FYM $ha^{-1}$+ PSB+ $ZnSO_4$ at 25 kg $ha^{-1}$ | 8.18 | 0.19 | 0.19 | 8.48 | 0.49 | 0.17 | 8.14 | 0.34 | 0.18 |
| T 6 | 75% RDF+2.5 Mg FYM $ha^{-1}$ + PSB+ $ZnSO_4$ at 25 kg $ha^{-1}$ | 8.26 | 0.29 | 0.20 | 8.44 | 0.38 | 0.18 | 8.09 | 0.33 | 0.19 |
| T 7 | FYM at 10 Mg $ha^{-1}$ + PSB + Azotobactor | 8.44 | 0.28 | 0.17 | 8.29 | 0.51 | 0.16 | 8.24 | 0.39 | 0.16 |
| | SEm± | 0.20 | 0.08 | 0.02 | 0.08 | 0.09 | 0.02 | 0.21 | 0.05 | 0.02 |
| | LSD ($p \leq 0.05$) | NS | NS | NS | NS | NS | NS | NS | NS | NS |

Source: Argal et al. (2017b)

#### 5.2.2.1 Nutrient Management

The total nitrogen, phosphorous, and potassium uptake (kg ha–[1]) by wheat as influenced by different treatments of fertility levels is presented in Table 5.8. Based on 2-year pooled data, total N, P, and K uptake by wheat was observed in the range of 46.9–129.4, 4.4–21.7, and 29.3–81.4 kg $ha^{-1}$, respectively, under different INM practices. Application of 75% RDF + 2.5 Mg FYM $ha^{-1}$ + PSB + $ZnSO_4$ at 25 kg $ha^{-1}$ ($T_6$) showed maximum total N, P, and K uptake by wheat, which was significantly higher over remaining fertility treatments except $T_3$ and $T_5$ during both the years. Significantly lower Total N, P, and K uptake by wheat was recorded with treatment $T_1$ (farmer practices) over all other fertility treatments during both years. Similar trends were found on the basis of 2-year pooled data.

#### 5.2.2.2 Nutrient Use Efficiency

Nutrient use efficiency (NUE) may be defined as yield per unit input. In agriculture, this is usually related to the input of fertilizer, whereas in scientific literature, the NUE is often expressed as fresh weight or product yield per content of nutrient. Improvement of NUE is an essential prerequisite for expansion of crop production into marginal lands with low nutrient availability. The nutrients most commonly limiting plant growth are N, P, K, and S. The data computed for nutrient use efficiency (%) of N, P, and K are presented in Table 5.9 after considering farmers' practice as base. In case of N, the maximum N use efficiency (73.6%, 61.59%, and

**Table 5.8** Effect of nutrient management practices on total nitrogen, phosphorous, and potassium uptake (kg ha$^{-1}$)

| Tr. no. | Treatments | N 2013–2014 | N 2014–2015 | N Pooled | P 2013–2014 | P 2014–2015 | P Pooled | K 2013–2014 | K 2014–2015 | K Pooled |
|---|---|---|---|---|---|---|---|---|---|---|
| T 1 | Farmer practices | 51.6 | 42.2 | 46.9 | 5.6 | 3.2 | 4.4 | 28.2 | 30.4 | 29.3 |
| T 2 | 100% RDF | 109.0 | 85.0 | 97.0 | 15.6 | 13.2 | 14.4 | 52.8 | 58.8 | 55.8 |
| T 3 | 150% RDF | 117.1 | 92.0 | 104.5 | 16.1 | 15.2 | 15.6 | 62.4 | 71.0 | 66.7 |
| T 4 | NPK application based on STCR equation | 84.2 | 71.7 | 77.9 | 9.8 | 10.6 | 10.2 | 44.3 | 52.0 | 48.1 |
| T 5 | 50% RDF+5 Mg FYM ha$^{-1}$+ PSB+ $ZnSO_4$ at 25 kg ha$^{-1}$ | 95.8 | 73.8 | 84.8 | 12.1 | 10.7 | 11.4 | 48.3 | 52.8 | 50.5 |
| T 6 | 75% RDF+2.5 Mg FYM ha$^{-1}$ +PSB+ $ZnSO_4$ at 25 kg ha$^{-1}$ | 147.2 | 111.6 | 129.4 | 25.5 | 17.9 | 21.7 | 82.2 | 80.6 | 81.4 |
| T 7 | FYM at 10 Mg ha$^{-1}$ + PSB + Azotobactor | 70.0 | 57.6 | 63.8 | 11.0 | 9.7 | 10.3 | 33.2 | 37.4 | 35.3 |
| | SEm± | 4.6 | 3.3 | 3.1 | 1.0 | 1.2 | 0.7 | 1.0 | 4.4 | 2.4 |
| | LSD ($p \leq 0.05$) | 14.2 | 10.0 | 9.4 | 2.9 | 3.6 | 2.3 | 3.2 | 13.6 | 7.3 |

Source: Argal et al. (2017b)

67.68%) was recorded under the application of FYM at 10 Mg ha$^{-1}$ +PSB + Azotobactor in 2013–2014, 2014–2015, and pooled basis, respectively. Lowest N recovery was recorded with treatment $T_6$ (FYM 10 Mg ha$^{-1}$) during both the years and pooled basis.

Data regarding nutrient use efficiency of P presented in Tables 5.4, 5.5, 5.6, 5.7, 5.8, and 5.9 revealed that nutrient use efficiency of P varied in range from 8.32% to 21.49% and 8.35–26.02% in 2013–2014 and 2014–2015, respectively. The highest apparent recovery of P was registered with treatment $T_7$ followed by $T_5$, while lowest was noted from treatment $T_6$ during both the years. Maximum nutrient use efficiency of K was observed under 50% RDF+5 Mg FYM ha$^{-1}$ + PSB+ $ZnSO_4$ 25 kg ha$^{-1}$ during both the years, and it was closely followed by treatment $T_3$. The minimum value of nutrient use efficiency of K was recorded with treatments $T_6$ during both the years.

#### 5.2.2.3 Maintenance of Soil of Nitrogen, Phosphorous, and Potassium

The data regarding the balance of nitrogen, phosphorous, and potassium in soil during study period was computed based on depletion/buildup and presented in Table 5.10.

**Table 5.9** Effect of nutrient management practices on nutrient use efficiency of N, P, and K

| | | N | | | P | | | K | | |
|---|---|---|---|---|---|---|---|---|---|---|
| Tr. no. | Treatments | 2013–2014 | 2014–2015 | Pooled | 2013–2014 | 2014–2015 | Pooled | 2013–2014 | 2014–2015 | Pooled |
| T 1 | Farmer practices | – | – | – | – | – | – | – | – | – |
| T 2 | 100% RDF | 47.80 | 35.68 | 41.74 | 8.32 | 8.35 | 8.33 | 20.51 | 23.65 | 22.08 |
| T 3 | 150% RDF | 54.58 | 41.51 | 48.05 | 8.74 | 10.03 | 9.39 | 28.47 | 33.87 | 31.17 |
| T 4 | NPK application based on STCR equation | 36.18 | 32.78 | 34.48 | 4.74 | 8.22 | 6.48 | 17.86 | 23.98 | 20.92 |
| T 5 | 50% RDF+5 Mg FYM $ha^{-1}$+ PSB+ $ZnSO_4$ at 25 kg $ha^{-1}$ | 73.61 | 52.80 | 63.21 | 10.81 | 12.50 | 11.65 | 33.43 | 37.39 | 35.41 |
| T 6 | 75% RDF+2.5 Mg FYM $ha^{-1}$+ PSB+ $ZnSO_4$ at 25 kg $ha^{-1}$ | 53.11 | 38.57 | 45.84 | 11.05 | 8.18 | 9.61 | 30.00 | 27.90 | 28.95 |
| T 7 | FYM at 10 Mg $ha^{-1}$+ PSB + Azotobacter | 73.76 | 61.59 | 67.68 | 21.49 | 26.02 | 23.75 | 19.96 | 27.83 | 23.90 |
| | SEm± | 7.45 | 4.94 | 5.55 | 2.22 | 3.09 | 1.18 | 1.15 | 4.32 | 2.44 |
| | LSD ($p \leq 0.05$) | 22.97 | 15.22 | 17.10 | 6.85 | 9.52 | 3.63 | 3.53 | 13.30 | 7.50 |

Source: Argal et al. (2017b)

**Table 5.10** Effect of nutrient management practices on balance sheet of nitrogen, phosphorous, and potassium

| Tr. no. | Treatments | N (kg $ha^{-1}$) | | | P (kg $ha^{-1}$) | | | K (kg $ha^{-1}$) | | |
|---|---|---|---|---|---|---|---|---|---|---|
| | | 2013–2014 | 2014–2015 | Pooled | 2013–2014 | 2014–2015 | Pooled | 2013–2014 | 2014–2015 | Pooled |
| T 1 | Farmer practices | 25.2 | 3.8 | 14.5 | −47.0 | −45.9 | −46.5 | −47.9 | −113.6 | −80.7 |
| T 2 | 100% RDF | −40.3 | −82.8 | −61.6 | −46.3 | −47.5 | −46.9 | −36.1 | −120.0 | −78.1 |
| T 3 | 150% RDF | −11.3 | −75.6 | −43.4 | −46.1 | −43.6 | −44.9 | −3.4 | −209.7 | −106.5 |
| T 4 | NPK application based on STCR equation | −10.0 | −21.0 | −15.5 | −37.8 | −33.3 | −35.6 | −1.4 | −104.8 | −53.1 |
| T 5 | 50% RDF+5 Mg FYM $ha^{-1}$+ PSB+ $ZnSO_4$ at 25 kg $ha^{-1}$ | −1.9 | 3.4 | 0.8 | −20.2 | −21.3 | −20.7 | −9.8 | −121.6 | −65.7 |
| T 6 | 75% RDF+2.5 Mg FYM $ha^{-1}$+ PSB+ $ZnSO_4$ at 25 kg $ha^{-1}$ | −37.0 | −74.7 | −55.8 | −69.2 | −73.5 | −71.4 | −29.1 | −230.5 | −129.8 |
| T 7 | FYM at 10 Mg $ha^{-1}$ + PSB + Azotobacter | 49.2 | 33.6 | 41.4 | −39.3 | −37.7 | −38.5 | −56.8 | −117.8 | −87.3 |
| | SEm± | 16.6 | 12.30 | 13.84 | 1.62 | 2.97 | 1.66 | 25.35 | 24.03 | 14.61 |
| | LSD ($p \leq 0.05$) | 51.15 | 37.90 | 42.65 | 4.99 | 9.15 | 5.13 | NS | 74.04 | 45.02 |

Source: Argal (2017)

Management of Nitrogen (N)

The computed data clearly indicated that maximum quantitative depletion in nutrient content was recorded for potassium in both the years as well as based on pooled value, followed by phosphorus and minimum in nitrogen. There was only very marginal reduction in the first year (2013–2014) and which jumped to very high tune in the second year (2014–2015). Among different treatments there was highest in nitrogen in the treatment $T_2$ followed by $T_6$ and $T_3$, while there was a gain in the case of $T_7$ and $T_1$. Maximum depletion in phosphorus was recorded in $T_6$ and followed by $T_1$, $T_2$, and $T_3$ during both the years. In case of potassium maximum reduction was recorded in the treatment $T_6$ which was very closely followed by $T_3$ and $T_7$, and minimum was in the treatment $T_4$.

### Management of Phosphorus (P)

The data on available-P (kg $ha^{-1}$) are presented in Table 5.11. In case of available-P (kg $ha^{-1}$), the highest available-P (kg $ha^{-1}$) values 16.57, 16.00, and 16.29 were recorded under application of 75% RDF + 2.5 Mg FYM $ha^{-1}$+ PSB + $ZnSO_4$ at 25 kg $ha^{-1}$ ($T_6$) in 2013–2014, 2014–2015, and pooled basis, respectively. Lowest available-N (kg $ha^{-1}$) was recorded with treatment $T_1$ (farmer practices) during both the years and pooled basis.

### Management of Available Potassium (K)

The data on available-K (kg $ha^{-1}$) of Chambal ravine are presented in Table 5.12. Data regarding available-K (kg $ha^{-1}$) varied in range from 152.4 to 263.0 (kg $ha^{-1}$), 143.7–217.3 (kg $ha^{-1}$), and 148.1–240.2 (kg $ha^{-1}$) in 2013–2014, 2014–2015, and pooled basis, respectively. The highest available-K (kg $ha^{-1}$) was registered with treatment 75% RDF+2.5 Mg FYM $ha^{-1}$ + PSB + $ZnSO_4$ at 25 kg $ha^{-1}$ ($T_6$), while lowest was noted from treatment $T_1$, farmers' practices.

**Table 5.11** Available soil phosphorus content (kg ha–$^1$) as affected by nutrient management after harvesting the wheat (2013–2014 and 2014–2015)

| Tr. | Treatments | Available soil phosphorous (kg $ha^{-1}$) | | |
|---|---|---|---|---|
| | | 2013–2014 | 2014–2015 | Pooled |
| T 1 | Farmer practices | 5.16 | 4.82 | 4.98 |
| T 2 | 100% RDF | 15.33 | 11.71 | 13.52 |
| T 3 | 150% RDF | 13.59 | 13.54 | 13.57 |
| T 4 | NPK application based on STCR equation | 10.25 | 9.34 | 9.80 |
| T 5 | 50% RDF+5 Mg FYM $ha^{-1}$+ PSB+ $ZnSO_4$ at 25 kg $ha^{-1}$ | 16.36 | 15.85 | 16.11 |
| T 6 | 75% RDF+2.5 Mg FYM $ha^{-1}$+ PSB+ $ZnSO_4$ at 25 kg $ha^{-1}$ | 16.57 | 16.00 | 16.29 |
| T 7 | FYM at 10 Mg $ha^{-1}$ + PSB + Azotobacter | 13.25 | 10.87 | 12.05 |
| | SEm± | 0.53 | 2.33 | 1.27 |
| | LSD ($p \leq 0.05$) | 1.62 | NS | 3.92 |

Source: Argal (2017)

**Table 5.12** Available soil potassium content (kg $ha^{-1}$) as influenced by nutrient management after harvesting the wheat (2013–2014 and 2014–2015)

| Tr. | Treatments | Available soil potassium (kg $ha^{-1}$) | | |
|---|---|---|---|---|
| | | 2013–2014 | 2014–2015 | Pooled |
| T 1 | Farmer practices | 152.4 | 143.7 | 148.1 |
| T 2 | 100% RDF | 211.7 | 206.1 | 208.9 |
| T 3 | 150% RDF | 232.2 | 168.4 | 200.3 |
| T 4 | NPK application based on STCR equation | 200.1 | 185.9 | 193.0 |
| T 5 | 50% RDF+5 Mg FYM $ha^{-1}$+ PSB+ $ZnSO_4$ at 25 kg $ha^{-1}$ | 228.9 | 208.3 | 218.6 |
| T 6 | 75% RDF+2.5 Mg FYM $ha^{-1}$+ PSB+ $ZnSO_4$ at 25 kg $ha^{-1}$ | 263.0 | 217.3 | 240.2 |
| T 7 | FYM at 10 Mg $ha^{-1}$ + PSB + Azotobacter | 166.5 | 159.8 | 163.2 |
| | SEm± | 10.7 | 13.3 | 6.8 |
| | LSD ($p \leq 0.05$) | 33.0 | 41.1 | 20.9 |

Source: Argal (2017)

#### 5.2.2.4 Available Zinc (Zn) (mg $kg^{-1}$)

The data on micronutrient content in Chambal ravine in presented Table 5.13. The data revealed that Zn content in ravine land was influenced significantly due to different fertility levels during both the years as well as pooled basis. Data regarding Zn content (mg $kg^{-1}$) revealed that Zn content (mg $kg^{-1}$) varied in range from 0.061 to 0.146 (mg $kg^{-1}$), 0.093–0.316 (mg $kg^{-1}$), and 0.076–0.233 (mg $kg^{-1}$) in 2013–2014, 2014–2015, and pooled basis, respectively. The highest content (mg $kg^{-1}$) was registered with treatment 75% RDF+2.5 Mg FYM $ha^{-1}$+ PSB + $ZnSO_4$ at 25 kg $ha^{-1}$ ($T_6$), while lowest was noted from treatment $T_1$ during both the years.

#### 5.2.2.5 Available Copper (Cu) (mg $kg^{-1}$)

The data on copper are presented in Table 5.14, and highest Cu content (mg $kg^{-1}$) value (0.576, 0.406, and 0.486) was recorded under application of 75% RDF + 2.5 Mg FYM $ha^{-1}$ + PSB + $ZnSO_4$ at 25 kg $ha^{-1}$ ($T_6$), in 2013–2014, 2014–2015, and pooled basis, respectively. Lowest Cu content (mg $kg^{-1}$) was recorded with treatment $T_1$ (farmers' practices).

#### 5.2.2.6 Available Iron (Fe) (mg $kg^{-1}$)

Available iron content estimated is presented in Table 5.15. The highest Fe (mg $kg^{-1}$) content value (3.626, 5.144, and 4.386) were recorded under the application of 75% RDF + 2.5 Mg FYM $ha^{-1}$ + PSB + $ZnSO_4$ at 25 kg $ha^{-1}$ ($T_6$), in 2013–2014, 2014–2015, and pooled basis, respectively. Lowest Fe content (mg $kg^{-1}$) was recorded with treatment $T_1$ (farmers' practices).

#### 5.2.2.7 Available Mangnese (Mn) (mg $kg^{-1}$)

The data on micronutrient Mn content in Chambal ravines are presented in Table 5.16. The data revealed that Mn content in ravine land was influenced

**Table 5.13** Available-Zn content (mg $kg^{-1}$) as affected by nutrient management after harvesting the wheat (2013–2014 and 2014–2015)

| Tr. no. | Treatments | Available-Zn (mg $kg^{-1}$) | | |
|---|---|---|---|---|
| | | 2013–2014 | 2014–2015 | Pooled |
| T 1 | Farmer practices | 0.061 | 0.093 | 0.076 |
| T 2 | 100% RDF | 0.088 | 0.146 | 0.113 |
| T 3 | 150% RDF | 0.113 | 0.166 | 0.138 |
| T 4 | NPK application based on STCR equation | 0.093 | 0.133 | 0.113 |
| T 5 | 50% RDF+5 Mg FYM $ha^{-1}$+ PSB+ $ZnSO_4$ at 25 kg $ha^{-1}$ | 0.146 | 0.220 | 0.186 |
| T 6 | 75% RDF+2.5 Mg FYM $ha^{-1}$ +PSB+ $ZnSO_4$ at 25 kg $ha^{-1}$ | 0.146 | 0.316 | 0.233 |
| T 7 | FYM at 10 Mg $ha^{-1}$ + PSB + Azotobacter | 0.077 | 0.123 | 0.106 |
| | SEm± | 0.006 | 0.008 | 0.005 |
| | LSD ($p \leq 0.05$) | 0.014 | 0.025 | 0.011 |

Source: Argal et al. (2017a)

**Table 5.14** Available-Cu content (mg $kg^{-1}$) in soil as affected by nutrient management after harvesting the wheat (2013–2014 and 2014–2015)

| Tr. no. | Treatments | Available-Cu content (mg $kg^{-1}$) | | |
|---|---|---|---|---|
| | | 2013–2014 | 2014–2015 | Pooled |
| T 1 | Farmer practices | 0.133 | 0.283 | 0.213 |
| T 2 | 100% RDF | 0.458 | 0.313 | 0.386 |
| T 3 | 150% RDF | 0.453 | 0.306 | 0.381 |
| T 4 | NPK application based on STCR equation | 0.358 | 0.346 | 0.353 |
| T 5 | 50% RDF+5 Mg FYM $ha^{-1}$+ PSB+ $ZnSO_4$ at 25 kg $ha^{-1}$ | 0.462 | 0.367 | 0.413 |
| T 6 | 75% RDF+2.5 Mg FYM $ha^{-1}$ +PSB+ $ZnSO_4$ at 25 kg $ha^{-1}$ | 0.576 | 0.406 | 0.486 |
| T 7 | FYM at 10 Mg $ha^{-1}$ + PSB + Azotobacter | 0.216 | 0.288 | 0.253 |
| | SEm± | 0.008 | 0.055 | 0.028 |
| | LSD ($p \leq 0.05$) | 0.025 | NS | 0.088 |

Source: Argal et al. (2017a)

**Table 5.15** Available-Fe content (mg $kg^{-1}$) in soil as affected by nutrient management after harvesting the wheat (2013–2014 and 2014–2015)

| Tr. no. | Treatments | Available-Fe content (mg $kg^{-1}$) | | |
|---|---|---|---|---|
| | | 2013–2014 | 2014–2015 | Pooled |
| T 1 | Farmer practices | 1.913 | 3.696 | 2.808 |
| T 2 | 100% RDF | 2.666 | 4.686 | 3.673 |
| T 3 | 150% RDF | 2.963 | 4.716 | 3.841 |
| T 4 | NPK application based on STCR equation | 2.403 | 4.726 | 3.567 |
| T 5 | 50% RDF+5 Mg FYM $ha^{-1}$+ PSB+ $ZnSO_4$ at 25 kg $ha^{-1}$ | 3.066 | 4.866 | 3.966 |
| T 6 | 75% RDF+2.5 Mg FYM $ha^{-1}$+ PSB+ $ZnSO_4$ at 25 kg $ha^{-1}$ | 3.626 | 5.144 | 4.386 |
| T 7 | FYM at 10 Mg $ha^{-1}$ + PSB + Azotobacter | 2.222 | 4.733 | 3.486 |
| | SEm± | 0.097 | 0.207 | 0.107 |
| | LSD ($p \leq 0.05$) | 0.296 | 0.639 | 0.327 |

Source: Argal et al. (2017a)

**Table 5.16** Available-Mn content (mg $kg^{-1}$) in soil) as affected by nutrient management after harvesting the wheat (2013–2014 and 2014–2015)

| Tr. no. | Treatments | Available- Mn (mg $kg^{-1}$) | | |
|---|---|---|---|---|
| | | 2013–2014 | 2014–2015 | Pooled |
| T 1 | Farmer practices | 2.366 | 4.803 | 3.587 |
| T 2 | 100% RDF | 4.676 | 7.182 | 5.929 |
| T 3 | 150% RDF | 5.030 | 7.660 | 6.345 |
| T 4 | NPK application based on STCR equation | 3.282 | 4.813 | 4.046 |
| T 5 | 50% RDF+5 Mg FYM $ha^{-1}$+ PSB+ $ZnSO_4$ at 25 kg $ha^{-1}$ | 8.806 | 8.407 | 8.606 |
| T 6 | 75% RDF+2.5 Mg FYM $ha^{-1}$+ PSB+ $ZnSO_4$ at 25 kg $ha^{-1}$ | 9.360 | 8.816 | 9.083 |
| T 7 | FYM at 10 Mg $ha^{-1}$ + PSB + Azotobactor | 3.453 | 6.074 | 4.766 |
| | SEm± | 0.373 | 0.386 | 0.323 |
| | LSD ($p \leq 0.05$) | 1.144 | 1.184 | 0.995 |

Source: Argal (2017)

**Table 5.17** Available-S content (mg $kg^{-1}$) as affected by nutrient management after harvest the wheat (2013–2014 and 2014–2015)

| Tr. no. | Treatments | Available-S content (mg $kg^{-1}$) | | |
|---|---|---|---|---|
| | | 2013–2014 | 2014–2015 | Pooled |
| T 1 | Farmer practices | 1.85 | 13.32 | 7.59 |
| T 2 | 100% RDF | 2.53 | 20.49 | 11.51 |
| T 3 | 150% RDF | 5.82 | 38.01 | 21.91 |
| T 4 | NPK application based on STCR equation | 6.84 | 23.00 | 14.92 |
| T 5 | 50% RDF + 5 Mg FYM $ha^{-1}$ + PSB + $ZnSO_4$ at 25 kg $ha^{-1}$ | 9.76 | 41.67 | 25.72 |
| T 6 | 75% RDF+2.5 Mg FYM $ha^{-1}$ + PSB + $ZnSO_4$ at 25 kg $ha^{-1}$ | 15.74 | 45.87 | 30.80 |
| T 7 | FYM at 10 Mg $ha^{-1}$ + PSB + Azotobactor | 3.07 | 22.45 | 12.76 |
| | SEm± | 0.65 | 2.70 | 1.28 |
| | LSD ($p \leq 0.05$) | 2.01 | 8.33 | 3.95 |

Source: Argal (2017)

significantly due to different fertility levels during both the years as well as pooled basis. The different fertility levels were associated with increase the available-Mn in soil during both the years of study. Available-Mn ranged from 2.366 to 9.360 (mg $kg^{-1}$), 4.803–8.816 (mg $kg^{-1}$), and 3.587–9.083 (mg $kg^{-1}$) during in 2013–2014, 2014–2015, and pooled basis, respectively.

#### 5.2.2.8 Available Sulphur (S) (mg $kg^{-1}$)

The data on micronutrient content in Chambal ravine in presented in Table 5.17. In case of S content (mg $kg^{-1}$), the highest S content (mg $kg^{-1}$) value (15.74, 45.87, and 30.80) was recorded under application of 75% RDF + 2.5 Mg FYM $ha^{-1}$ + PSB + $ZnSO_4$ at 25 kg $ha^{-1}$ ($T_6$) in 2013–2014, 2014–2015, and pooled basis, respectively. Lowest S content (mg $ha^{-1}$) was recorded with treatment $T_1$ (farmers' practices) during both the years and pooled basis.

**Table 5.18** Soil microbial biomass carbon (SMBC), dehydrogenase activities (DHA), and fluorescein diacetate (FDA) after harvesting wheat crop during (2014–2015)

| Tr. no. | Treatments | SMBC (μgC $g^{-1}$) | FDA (μg $g^{-1}$ $h^{-1}$) | DHA (μg $g^{-1}$ TPF $g^{-1}$ $h^{-1}$) |
|---|---|---|---|---|
| T 1 | Farmer practices | 51.71 | 5.83 | 12.43 |
| T 2 | 100% RDF | 54.85 | 6.09 | 26.83 |
| T 3 | 150% RDF | 63.79 | 8.59 | 36.29 |
| T 4 | NPK application based on STCR equation | 56.27 | 7.22 | 28.73 |
| T 5 | 50% RDF + 5 Mg FYM $ha^{-1}$ + PSB + $ZnSO_4$ at 25 kg $ha^{-1}$ | 62.92 | 8.93 | 45.49 |
| T 6 | 75% RDF+2.5 Mg FYM $ha^{-1}$ + PSB + $ZnSO_4$ at 25 kg $ha^{-1}$ | 66.05 | 11.24 | 59.46 |
| T 7 | FYM at 10 Mg $ha^{-1}$ + PSB + Azotobacter | 65.46 | 7.26 | 47.03 |
| | SEm± | 1.84 | 0.31 | 1.03 |
| | LSD ($p \leq 0.05$) | 5.67 | 0.94 | 3.19 |

Source: Argal et al. (2017a)

### 5.2.3 Management of Biological Properties of Soil

#### 5.2.3.1 Soil Microbial Biomass Carbon (μgC/g)

Soil microbial biomass carbon increased in doses of organic sources as compared to inorganic fertilizers. It may be due firstly to an increase in microbial population (Hasebe et al. 1985) and secondary to the formation of root exudates, mucigel sloughed off cell, and underground roots of previous cut crops, which also play an important role in increasing biomass carbon (Goyal et al. 1992). The study of the data soil microbial biomass carbon (Table 5.18) revealed that the soil microbial biomass carbon (SMBC) was affected significantly due to different fertility levels after a 2-year study in ravine soils. Soil microbial biomass carbon (SMBC) was recorded highest (66.05 μg $g^{-1}$) in treatment FYM at 10 Mg $ha^{-1}$ + PSB + Azotobactor ($T_7$), and it was lowest (51.71 μg $g^{-1}$) in treatments having farmers' practices ($T_1$).

#### 5.2.3.2 Fluorescein Diacetate (μg $g^{-1}$ $h^{-1}$)

It is obvious from Table 5.18 that fluorescein diacetate (FDA) was significantly highest in treatment $T_7$ (FYM at 10 Mg $ha^{-1}$ + PSB + Azotobactor) and followed by $T_5$ (50% RDF + 5 Mg FYM $ha^{-1}$ + PSB+ $ZnSO_4$ 25 kg $ha^{-1}$) over farmers' practices, while other treatments were statistically at par to $T_3$ (150% RDF), and the treatment $T_7$ (FYM at 10 Mg $ha^{-1}$ + PSB + Azotobactor) maintained its superiority among all the treatments.

#### 5.2.3.3 Dehydrogenase Activity (μg $g^{-1}$ TPF $g^{-1}$ $h^{-1}$)

It is obvious from Table 5.18 that dehydrogenase activity was significantly higher with treatment$T_7$ where FYM at 10 Mg $ha^{-1}$ + PSB + Azotobactor was applied, and it was closely followed by ($T_5$) 50% RDF + 5 Mg FYM $ha^{-1}$ + PSB+ $ZnSO_4$ 25 kg

**Table 5.19** Different cropping systems and their productivity (in INR) in a ravine land

| System | Kharif season | | Rabi season | | Total |
|---|---|---|---|---|---|
| | Grain | Straw | Grain | Straw | |
| 1. Pearl millet- mustard | 41,612 | 16,638 | 33,916 | 7231 | 99,397 |
| 2. Pearl millet-wheat | 41,612 | 16,638 | 28,110 | 20,784 | 107,144 |
| 3. Pearl millet-taramira | 41,612 | 16,638 | 8476 | 1681 | 68,407 |
| 4. Cluster bean-gram | 65,415 | 4827 | 29,223 | 2496 | 101,961 |
| 5. Cluster bean-taramira | 65,415 | 4827 | 8476 | 1681 | 80,399 |
| 6. Black gram-wheat | 18,490 | 1667 | 28,110 | 20,784 | 69,051 |
| 7. Black gram-mustard | 18,490 | 1667 | 33,916 | 7231 | 61,304 |
| 8. Soybean-wheat | 10,931 | 1139 | 28,110 | 20,784 | 60,964 |
| 9. Soybean-mustard | 10,931 | 1139 | 33,916 | 7231 | 53,217 |
| 10. *Sesamum*-wheat | 9225 | 891 | 28,110 | 20,784 | 59,010 |
| 11. *Sesamum*-mustard | 9225 | 891 | 33,916 | 7231 | 51,263 |

Source: Verma et al. (2016)
Based on market prices (2014–2015) (equivalent to pearl millet-mustard)

$ha^{-1}$ over farmers' practices, while other treatments were statistically at par to the treatment 150% RDF ($T_3$), while treatment $T_7$ (FYM at 10 Mg $ha^{-1}$ + PSB + Azotobactor) maintained its superiority among all the treatments.

## 5.3 Cropping Systems and System Productivity

Different cropping systems were adopted to assess the best suitable cropping system that can give higher return to the farmers (Table 5.19) after reclamation of ravine land. Among various cropping systems, pearl millet-wheat was the best combination and gave return of INR 107 thousand $year^{-1}$ $ha^{-1}$ which was closely followed by cluster bean-gram (INR 102 thousand $year^{-1}$ $ha^{-1}$), pearl millet-mustard, and cluster bean-taramira (*Eruca sativa*).

## References

Angima SD, Angima DE, Stott MK, Neill O, Ong CK, Weesies GA (2003) Soil erosion prediction using RUSLE for central Kenyan highland conditions: agriculture. Ecosyst Environ 97:295–308

Argal MS (2017) Effect of organic and inorganic sources of plant nutrients on wheat production and soil health in Chambal Ravine, Ph.D. thesis submitted to Rajmata Vijayaraje Scindia Krishi Vishwa Vidyalaya, Gwalior (M.P)

Argal MS, Verma SK, Sunil-Rajput (2017a) Effect of nutrient management on soil health and wheat (*Triticum aestivum* L.) production in degraded land of Chambal ravine. Indian Agric (in press)

Argal MS, Verma SK, Tomar PS (2017b) Nutrient management options in ravine of Chambal and its effects on growth and yield of wheat (*Triticum aestivum* L.) Int J Agric Environ Biotechnol 10(2):1–10

Bates RL Julia JA (1984) Dictionary of geological terms prepared by the American Geological Institute. ISBN 0-385-18101-9. New York
Doran JW, Parkin TB (1994) Defining soil quality for a sustainable environment. Published by Soil Science Society of America and American Society of Agronomy 35
Hasebe A, Kanajava S, Takai Y (1985) Microbial biomass in paddy soil: II microbial biomass C measured by Jenkinsons fumigation method. Soil Sci Plant Nutr 34:349–359
Maji AK, Reddy GPO, Sarkar D (2010) Degraded and wastelands of India: status and special distribution. Directorate of information and publications of agriculture. Indian Council of Agricultural Research, New Delhi, p 158
Oldeman LR (1994) The global extent of soil degradation. In: Greenland DJ, Szabolcs I (eds) Soil resilience and sustainable land use. CAB International, Wallingford, pp 99–118
Pimentel D (2000) Genetically modified crops and the agroecosystem: comments on "Genetically modified crops: risks and promise" by Gordon Conway. Conserv Ecol 4(1):10
Sharda VN, Dogra P, Prakash C (2010) Assessment of production losses due to water erosion in rainfed areas of India. J Soil Water Conserv 65:79–91
Verma SK, Akhilesh S, Tomar PS (2016) Management of Chambal ravines for food and environment security. Technical Bulletin No. RVSKVV/61/2016. Rajmata Vijyaraje Scindia Krishi Vishwa Vidyalaya, Gwalior

# Soil, Landscape and Nutrient Management of Ravine Areas for Enhancing Crop Productivity and Livelihood Security

6

P. Dey, S. K. Verma, K. Ramesh, J. Somasundaram, and M. Vassanda Coumar

**Abstract**

Ravine formed from intricate network of gullies because of localized physical degradation due to surface runoff affecting the friable unconsolidated material in the formation of perceptible channels resulting in undulating terrain with fragile ecosystem. Over-exploitation of this vast tract of existing ravine lands coupled with improper management practices has led to deterioration of soil health and poses threat to adjoining productive agricultural lands. Under ravine landscapes, soil undergoes various changes due to accelerated erosional processes. With rapidly shrinking per capita availability of land, there is a growing need for restoration of these areas for productive land uses. Maintenance of soil organic carbon is one of the most important factors for aggregate stability, soil structural durability and nutrient availability in ravine areas. Deficiency of nutrients and poor water retention are two major causative factors of stagnation in crop productivity in ravine lands. In this chapter, an attempt has been made to consolidate various best nutrient management practices for soils under ravine region for increasing crop productivity and improving soil health. Also, there is a need to harness and manage the indigenous technical knowledge and fine-tune them to suit the modern needs. Overall, the twin aspect of devising strategies for leveraging resources to tackle the challenge of enhancing soil health and carbon sequestration will help in combating climate change without compromising economic development in ravine areas. The future R&D strategies for maintaining soil health, crop productivity and environmental sustainability in degraded ravine land have also been included.

P. Dey (✉) · K. Ramesh · J. Somasundaram · M. Vassanda Coumar
ICAR-Indian Institute of Soil Science, Nabibagh, Bhopal, Madhya Pradesh, India

S. K. Verma
Department of Soil Science & Agricultural Chemistry, RVSKVV, Vishwa Vidyalaya, Gwalior, Madhya Pradesh, India





J. C. Dagar, A. K. Singh (eds.), *Ravine Lands: Greening for Livelihood and Environmental Security*, https://doi.org/10.1007/978-981-10-8043-2_6

## 6.1 Soil and Landscape of Ravine Areas

The word 'ravine' denotes gullied land containing systems of gullies running parallel to each other and entering a nearby river flowing much lower than the surrounding tablelands. The gullies are formed because of localized surface runoff affecting the friable unconsolidated material in the formation of perceptible channels resulting in undulating terrain. The gullies are the first stage of excessive land dissection followed by their networking which leads to the development of ravenous land (Anderson et al. 2001). These gullied lands comprise of numerous small catchments along the river affecting millions of hectares of productive land. Ravines are the extreme examples of physical degradation due to water erosion. Wind and water are the major agents responsible for the land degradation, due to which about 5334 million tons of soil is lost (16 Mg $ha^{-1}$) annually (Mandal et al. 2008). Out of a total of 20.17% of geographical area categorized as wastelands in India, 0.65% is contributed by ravenous area (Mythili and Goedecke 2016). The ravine areas are the most fragile ecosystem and subjected to various kinds of natural resource losses and threat to biodiversity. These lands ingress in about 2 km surrounding area of river systems and their tributaries; ravine lands and adjoining marginal lands are predominantly treated as either wastelands or managed as low input farming systems (Singh et al. 2002).

With rapidly shrinking per capita availability of land, there is a growing need for restoration of these areas for productive land uses. Different land uses and land management options influence soil degradation or aggradation process; however all such relationships are specific to edaphic and climatic conditions. The Chambal ravine regions are one of the best examples for physical soil degradation due to water erosion. Under these landscapes, soil undergoes various changes due to accelerated erosional processes. Land use-induced erosion process affects the soil physical and chemical properties especially soil aggregates, soil organic carbon (SOC) and nutrient availability (Foster et al. 2003; Han et al. 2010). The SOC plays a crucial role in sustaining soil quality, crop production and environmental quality (Doren and Parkin 1994). Maintenance of SOC is especially important for nutrient availability and soil structural stability. Stability of aggregate is the measure of the structural stability of soils (Six et al. 2000). Aggregate distribution helps us to correlate with various factors such as erosion, infiltration rate, evaporation, etc. Favourable soil aggregation is important to improve soil fertility and quality with emphasis on SOC sequestration, increasing agronomic productivity, enhancing porosity and decreasing erodibility. Any reduction in soil aggregate stability is a powerful early indicator of the onset of land degradation. Several studies have been reported on impact of different land uses on soil properties under arable lands (Singh 1999; Joshi 2002).

Conversion of native forests and pristine soils to cultivation is usually accompanied by decline in SOC and deterioration of soil structure. The aggregate dynamics vary among different crops, crop rotations and cover crops (Jarecki and Lal 2003). Vegetation cover can influence soil aggregate stability because of contribution of organic matter through litter and plant root turnover (Angers and Caron 1998), root exudates and rhizo-deposition (Gyssel et al. 2005). Soil degradation is generally accelerated when perennial tree crops are replaced by annual row crops, primarily due to loss of canopy cover and organic matter as well as increased soil disturbance from tillage operations. Apart from that the potential of these trees for agroforestry systems to increase nutrient stocks on soils appears to be important in terms of maintaining or improving soil chemical properties and organic matter and protect the soil surface as related to the processes of litter or fine root production, decomposition and soil organic matter transformation (Datta and Singh 2007). Loss of nutrients by leaching, erosion and runoff can be minimized by tree-derived mulch, litter or foliage (Lal 1984; Sanchez 1987), and an annual crop or cover of legumes or other vegetation can provide an effective soil cover to arrest soil erosion (Sanchez 1995). Jha et al. (2010) reported that after 25 years of plantation of forest species in the ravine land, soil organic carbon content increased by more than twofold and mean weight diameter of aggregates by two- to sixfold in comparison to reference site (uncultivated ravenous land). Significant reduction in soil bulk density (4–18%) and increase in steady-state infiltration rate (1.5–2 times) were recorded under the forest land use system. Conversion of ravine to forest system resulted in significant improvement in moisture retention capacity of soil. Conversion of ravine landform to agriculture adversely affected the soil bulk density and infiltration rate.

The loss in crop productivity caused by land degradation is of great significance as it has direct impact on food security of the country. The stagnation in crop productivity in ravine lands has been found due to deficiency of nutrients, as they play an important role in physiological processes in crop plants which are directly linked with yield parameters. In addition to nutrient deficiency, vegetation in these ravine regions also suffers from unfavourable conditions such as poor in soil fertility, low in SOC, soil erosion, moisture stress and biotic interference. The inclement weather conditions coupled with very high summer temperature further aggravate the problem and make farming uneconomical. In such situations, less water and nutrient demanding technologies hold a good promise to sustain the productivity and provide alternative source of income to the farmers. However, increased yield through intensive farming with high-yielding crop varieties, use of inorganic fertilizers free from micronutrients as filler material, limited use of organic manures and restricted use of crop residue are the major factors contributing towards deficiency of micronutrients (Mathur et al. 2006). Soil properties have been continuously influenced by the management practices and land uses, in which the latter one has been identified as profound influence on the soil properties especially on soil aggregates and soil organic carbon (SOC).

Soil degradation is generally accelerated when perennial crops are replaced by annual row crops, primarily due to increased soil disturbance from tillage operations (Sharma and Behera 2010). Over-exploitation of this vast tract of existing ravine lands coupled with improper management practices led to deterioration of soil health and poses threat to nearby productive lands. Moreover, very scanty information is available on nutrient management for improving crop productivity and livelihood improvement under ravine land. Therefore, maintaining soil productivity through conservation measures and suitable crop and nutrient management holds key for ravine land sustainability.

## 6.2 The Ravines: Non-arable Wastelands

The National Wasteland Development Board (NWDB) defined wasteland as 'that land which is degraded and is presently lying unutilized (except as current fallow) due to different constraints'. Wastelands primarily consist of culturable and unculturable wastelands. Culturable wastelands are those lands which have potential for the development of vegetative cover or may be reclaimed at later stage. It includes gullies and/or ravenous lands, undulating upland surface, waterlogged and marshy areas, salt-affected land, shifting cultivation area, degraded forest land, sandy area, mining and industrial wasteland, strip land, pasture and grazing land. Unculturable wastelands are those lands which cannot be developed for vegetative cover. These lands are barren and cannot be put to significant uses. However, some of the areas of such lands can be converted into pasture land. It includes barren rocky area, steep slopes and snow-covered glacial areas.

Ravine lands are highly degraded dry lands, and the word 'ravine' means a deep gorge which represents the last stage of water erosion. Ravines are the systems of gullies running almost parallel to each other and draining into a river after a short distance with the development of deep gorges. A paper presented on 'Genesis of ravines of the lower Chambal valley, India' in an international congress defined ravines as a channel of ephemeral flow, denuded and guided essentially by the process of rejuvenated streams and having steep sides and head scarps with a width and depth always greater than a gully. A ravine is generally a fluvial sloe landform of relatively steep (cross-sectional) sides, on the order of 20–70% in gradient. They manifest highly degraded landforms as discernible through satellite imageries. Neglect of these ecosystems leads to total degradation of the area. These form natural drainage system in the river basins and an inevitable necessity for a landscape. Although these are found here and there in the length and breadth of the country, they became severe in the northern flood plains due to weak geomorphology, erodible nature of the alluvium, bare steep slopes open to intense rainfall, overgrazing and deforestation.

Chambal river is a perennial river and carries large quantity of water and sediment load to transportation. Chambal origin is considered from the Janapav valley at Malwa ridge surrounded by Manpur reserve forest, Madhya Pradesh, India. Chambal river has Deccan trap basalt as river bedded rock, Mesozoic age formation with dendritic drainage pattern at Malwa region (Kaushik and Ghosh 2015). Chambal river origin lies at an elevation of 840.6 m with latitude 22°27′36.5″N and longitude 75°41′14.4″E. Physiographically the area to the North of Chambal is characterized by moderately dissected plateau and formed undulating topography. However, a clear-cut distinction between gully and ravine has been drawn, based on their morphological and genetic characteristics by the geologists. Ravine formation begins along river sides and encroaches upon the catchment area by headward growth. Active gully systems commonly develop in unconsolidated materials due to changing patterns of land use and associated change in catchments hydrology. Monitoring gully development has provided important information on processes, rates and geomorphologic controls of gully initiation and its growth. Gully erosion due to river channel trenching is a problem that threatens vast tracts of the world's agricultural land. The damage is greatest in the alluvial plains of the semiarid and arid zones and most serious where it threatens precarious subsistence agricultural systems. The ravines are characterized by dominance of thorny vegetation and are classified as northern tropical ravine thorn forests (6B/C2) (Champion and Seth 1968).

## 6.3 Ravine Zones

Ravines and gullies universally in various parts of the world and India are not an exception. However, the largest ravine incidence is noticed in central and northern India, viz. Madhya Pradesh, Uttar Pradesh and parts of Rajasthan. The National Remote Sensing Agency, Hyderabad, utilizing the Landsat data, has estimated nearly 4 million hectares of ravine land in India. Notable ravine intrusions could be noticed to occur along the banks of Beas in Punjab; Chambal in Madhya Pradesh, Rajasthan and Uttar Pradesh; Kalisind, Banas, Morel and Gambhir in Rajasthan; Yamuna in Uttar Pradesh; and Mahi, Sabarmati, Narmada and Tapti in Gujarat. Ravines are the worst form of land degradation and characterized by edaphic, topographic and climatic adversities. A rough estimate suggests that every year ravine area expands at a rate of 8000 ha year$^{-1}$ (Sharma et al. 1980). The climate, topography and soils of the ravine region are typically harsh and thus adversely affect the composition and growth of the vegetation.

Ravines are formed when the upper layer of vegetal cover is not strong enough and the roots are unable to hold and bind the soil together. Constant rainfall erodes the soil and washes away the crust of the earth due to which, the water flow turns into drains, creating cracks. In due course of time, these cracks are further eroded and become large ravines. The devastating effect of raindrops striking bare soil by detachment of soil particles is the principal factor for soil erosion. In the initial stages of rainfall, splashing effect of the raindrops detaches the soil particles. But as the rainfall continues, on the bare soil, a sheet of water begins transporting loose soil down the slope. As the water accumulates in low areas, it takes shape into a stream, the deeper it becomes, the faster it flows, and the higher its erosive power becomes, cutting gullies while flowing down. Gullies erode entire soil profiles and often cut into unconsolidated material beneath the soil (Luna 2006).

### 6.3.1 Driving Energy for Ravines

(a) *Population and poverty*: Population pressure on land is very high because percentage of arable land to total area is much higher than the global average. Due to scanty resources, these ecosystems inhabit very sparse population with poor developmental activities.
(b) *Climate change and natural disasters*: Climate change is now accepted as a reality and has caused heavy damage to the ravines. Recurrent floods accelerate ravine formation at alarming rate in these regions.

(c) *Overgrazing and livestock*: Overgrazing is yet another determinant for ravine formation.
(d) *Inappropriate land use and management*: Land capability classification should be strictly adhered for successful rehabilitation of ravine lands. Inappropriate land management with faulty land management practices severely spoils the landforms and results in denuded ravine/fully formation.

### 6.3.2 SWOT Analysis of Ravine Ecosystems for Crop Production

SWOT analysis is a very simple but effective tool of analysis employed in management systems which help in rejuvenation and rehabilitation for policy formulation for sustainable utilization of ravine ecosystems in the country. It helps to analyse the inherent strengths of ravine systems with due credit to weaknesses and visualizes the future on the basis of present status. The strengths of the sector are the result of successful strategies of the past, and the opportunities can be reaped with blend of new and old strategies. The weaknesses and threats are the hurdles for development which may be due to past mistakes and newly emerging trends.

Abundant water availability is an inherent *strength* of ravine systems which helps to carry out nutrient management strategies for the establishment of the vegetation in these areas. It is to be noted that ravine is a separate entity and not just a simple network of gullies. However, the inherent *weakness* is that the heavy soil erosion due to floods impedes any development activity. The associated soil constraints, viz. zero topsoil, poor organic matter, salinity, waterlogging and low fertility, need to be addressed properly. Further the natural processes leading to ravines, like climate and geology, can be engineered only to a limited extent. The attractive *opportunity* rests with the establishment of pasture and tree-based farming systems for their rejuvenation. The government may restrict exploitation activity in these areas to decelerate further degradation of the ravines. For example, government of India has declared Chambal ravines in India as national Chambal sanctuary status for the upper and lower sector for protection against further degradation. The genesis of ravines has not been studied in detail till now which in due course may provide new opportunities for their development. The unauthorized mining in these ecosystems poses a severe *threat* to the development in this region.

## 6.4 Ravine Rehabilitation

There are many approaches to land rehabilitation through vegetation and management. These depend on severity of damage to the land resource, the goals of rehabilitation and availability of resources for repairing the damage. Improvement of degraded lands implies restoration or reclamation or rehabilitation. The IUCN guidelines for the restoration of degraded ecosystems define restoration as process bringing a degraded ecosystem or landscape back to the same prescribed, productive condition in short process of revitalization (Lamb 1988). Ecological restoration concentrates on processes such as persistence of species through natural recruitment and survival, functioning food webs and systems-wide nutrient conservation via relationships among plants, animals and the detritivore community. Reclamation on the other hand is referred as means that the site will be similar in ecological functioning after disturbance and will be habitable by similar but not necessarily the same organisms. Rehabilitation means that the land will be made useful but with different land use and usually with different species. The most intensive rehabilitation is sometimes termed reclamation, because the damage is so severe that soils must be replaced and landscapes may have to be reshaped (Ang 1994; Bradshaw 1997; Singh et al. 2002). An important goal of ecological rehabilitation is to accelerate natural succession to increase biological productivity, reduce rates of soil erosion, increase soil fertility and increase biotic control over biogeochemical fluxes within the recovering ecosystem (Mishra and Rath Monalisa 2013).

Degraded lands can be suitably reclaimed for agriculture or some alternate uses through afforestation and agroforestry. Both these measures are not only helpful in meeting the diverse needs of fuel, fodder, timber, etc. but are also able to improve environment condition, biodiversity and socio-economic conditions of the stakeholders. The climate, topography and soils of the ravine region are typically harsh and thus adversely affect the composition and growth of the vegetation. Rehabilitation of such degraded areas requires systematic and scientific approach which includes proper survey, choice of species and techniques for establishment of plant species. Rehabilitation of ravine lands involves treatment of table and marginal lands contributing runoff to the gullies and proper gullies/ravines on watershed basis. The soils of the ravenous regions are poor in fertility and physical conditions (Dagar and Mall 1980) as the top fertile layer is torn and the soil profile remained naked leaving behind rocky bottom with poor water retention. The land feature of ravines is rugged and characterized by three distinct parts, namely, ravine top or hump, slope and bottom/bed. In general, the depth of ravines ranges from 1 to 20 m but may reach up to 55 m.

Most of the ravenous lands are not fit for cultivation of agricultural crops and reclassified as nonarable lands. Crop cultivation in these lands is not desirable as it leads to accelerated erosion thus adding to their fast degradation. The best scientific land use for these lands is to place them under permanent vegetation involving forest and fruit trees along with forage grasses and energy plantations. Proper soil and water conservation structures need to be erected, in order to check the uninterrupted flow of runoff into the ravines. Peripheral bunds and permanent drop structures like spillway, brushwood check dams, gabions, live vegetative check dams and trenching were found suitable to rehabilitate the ravines (Chaturvedi et al. 2014). After successful checking of the ravine extension, the most important measure for reducing the risk of degradation of catchments of gullies and marginal lands along the ravines is the introduction of trees and grasses in the ravine lands. Hence, it requires an integrated approach of using gullies according to land capability classes, soil and water conservation measures and putting land under permanent vegetation cover involving afforestation, agroforestry, horticulture, pasture and energy plantations.

### 6.4.1 Techniques for Rejuvenation of Ravine Lands

Agro-techniques are those sustainable to improve the farming situation under the aberrant situation like flood/soil erosion and are helpful in restricting/preventing land degradation and/or ravine formation. Considering the weaknesses of the system for successful rehabilitation, enhancing the soil health of the ecosystems is the chief architect for successful crop production. Adding organic matter every year is perhaps the most important way to protect the soils from the onslaught of erosion damage. Regular additions of organic matter improve soil structure, enhance water and nutrient holding capacity, protect soils from erosion and compaction and support a healthy community of soil organisms. There are several other practices which not only increase organic matter in the soils but also helpful in preventing ravine formation. Alternate land use planning includes legume-based forage production and silvopasture; agroforestry may be the other focal point for improving present farming situation and restricting deterioration of ravines in the future.

### 6.4.2 Revegetation of Ravine Lands

Vegetation is a very important component of the ravines land management. The vegetation serves as a good indicator of erosion in the ravines. A ravine or portion of a ravine, with good vegetation that contains large trees and an abundance of grasslands, has only minimal erosion. The ravines are characterized by dominance of thorny vegetation. Champion and Seth (1968) have classified the natural vegetation of ravines as northern tropical ravine thorn forests. In India, classification of gullies has been made which is linked with treatment of gullies for different land uses like agricultural and nonagricultural ones. It considers the influence of soil characteristics and gully dimensions on suitability of reclaiming gullies for cropping, horticulture, forestry or growing grasses, etc. Deep and narrow gullies are recommended to be put under permanent vegetation of grasses and trees for resource conservation measures. The predominant vegetation in the degraded ravenous land comprises vilayati babul (*Prosopis juliflora*) followed by babul (*Acacia nilotica*), thorny legumes, euphorbias and dwarf grass species (*Cenchrus ciliaris*). There is occasional occurrence of neem (*Azadirachta indica*), papdi (*Holoptelea integrifolia*), shisham (*Dalbergia sissoo*), karanj (*Pongamia pinnata*) and chonkra (*Prosopis cineraria*). A study conducted by Jha et al. (2010) provided practical information regarding rehabilitation potential of different tree species (*Prosopis juliflora*, *Leucaena leucocephala*, *Acacia nilotica* and *Azadirachta indica*) that could be used for restoration of Ustifluvent soil susceptible to gully erosion in the semiarid region. The species *P. juliflora* promoted the most desirable changes in soil physical and water transmission properties. However, it becomes undesirable due to its poor fodder and timber value and its prolific nature of spreading and engulfing nearby agricultural fields. Hence, other species like *A. nilotica* and *A. indica* could be a better choice for restoring and rehabilitation of degraded ravine lands in the riparian zone along the river Yamuna.

Srinivasan et al. (1970) evaluated the performance of six promising grass species on sandy-loam soils in an overgrazed gullied area of W. Central India. *Andropogon ischaemum* (>3000 kg DM $ha^{-1}$) and two cv. of *Cenchrus ciliaris* (5451 DM kg $ha^{-1}$) gave the highest forage yields. *Dichanthium annulatum* and *Amphilophis* [*Bothriochloa*] *glabra* gave moderately high yields. *Panicum antidotale* was the least promising species in terms of yield and tillering capacity; its performance also deteriorated rapidly with age. All species were sufficiently succulent and palatable for use in pasture development.

Recently, importance of bamboo plant (*Dendrocalamus strictus)* has been identified for protective and productive use in ravine lands as an effective means of resource conservation by several authors. Bamboo acts as good soil binder owing to its dry and hardy nature, peculiar dense clump formation and extensive interlocking fibrous root systems, natural capacity to regenerate through its rhizomes which play an important role in preventing erosion, increasing water holding capacity and nutrient cycling under gully beds. In comparison to deciduous and coniferous plantations, bamboo plants are hydrologically best suited to degraded ravine lands. Bamboo plantation in the gully beds helps to improve groundwater level, land productivity, ecological parameters, microclimate and socio-economic conditions of livelihood of ravine land. Pande et al. (2012) also found bamboo plantation economically viable in ravines and suggested policy measures to development agencies and finance institutions for large greening of ravines in the country. A study was carried out at Central Soil and Water Conservation Research and Training Institute (CSWCRTI), Research Centre, Agra, on hydrologic and economic evaluation of bamboo plantations in gullied lands. After 4 years of bamboo plantation, permeability of soil and drainage capacity of soil were improved as a result runoff was reduced from 9.65% to 1.81% and soil loss from 4.27 to 0.60 Mg $ha^{-1}$ $year^{-1}$. Under bamboo plantation, carbon stock build-up in the ravine land has been improved and extensively quantified and reported (Singh et al. 2014). Also, the B:C ratio was around 1.89 with the best rate of return (19.3%) in Yamuna ravines through bamboo plantation.

## 6.5 Nutrient Status of Ravenous Lands

The soils of the ravenous regions are poor in fertility and physical conditions (Dagar and Mall 1980) as almost all the soil profiles are eroded leaving behind rocky, sandy or stiff clay with poor water retention or water-absorbing power. The process of soil formation is also very slow since little interaction takes place due to non-availability of water and vegetation roots; sometimes the lime develops into *kankar* pan (deposits of calcium carbonate) which further prevents the roots of the plants to penetrate and get water from deeper layers. The ravine soils vary in pH from 7.5 to 8.0. In the zones of lighter soils in Uttar Pradesh, Madhya Pradesh and Gujarat, the water holding capacity is low. Soils of these regions are deep alluvial deposits having sandy to sandy-loam texture. The soils of Chambal ravine regions are less deep alluviums as compared to those in Yamuna and Mahi ravine region in Uttar Pradesh, Kshipra river (Madhya Pradesh) and Gujarat. They have silty clay loam to clay texture and calcium carbonate content increases with increasing depth. These soils are prone to waterlogging when wet and to compaction when dry. The infiltration rates in these soils are low to moderate and soils are highly erodible in the upper layers.

In the recent past, micronutrient status of the cultivated soil is found to be towards deficient range for several micronutrients impeding successful crop production. The situation is alarming in the ravine lands due to severe soil erosion and topsoil loss. Somasundaram et al. (2009) studied micronutrient status of soils representing eight different land use systems in Chambal ravines. The DTPA-extractable micronutrients in soils showed wide variation under different land uses, and its content was in the order Mn>Fe>Zn>Cu. Mean value of DTPA Mn, Zn, Fe and Cu in surface soils varied from 6.98 to 23.85, 0.44 to 1.41, 2.14 to 11.59 and 0.23 to 2.88 mg $kg^{-1}$, whereas, in subsurface soils it varied from 5.01 to 23.03, 0.44 to 1.29, 1.49 to 10.24 and 0.22 to 2.58 mg $kg^{-1}$. Micronutrient content was relatively low in ravine lands as compared to other land uses. Iron (Fe) and zinc (Zn) deficiencies are the major disorders in agricultural fields (marginal land) along the Chambal ravine systems (Table 6.1).

Another study conducted by Somasundaram et al. (2012) has shown that ravenous lands had low clay content. The erosion degree has also varied from the under tablelands (no erosion or slight erosion; up to 25% of 'A' horizon is lost) to (severe to very severe erosion—'A' horizon is lost and 50% of B horizon is exposed) under ravine lands. The soils under perennial vegetations and cultivated areas had higher clay content (37.80–42.0%), whereas ravine lands recorded the lowest clay content (20.2–22.85%). Proper soil conservation measures along with additions of organic manures helps in reducing soil erosion and improvement of soil properties in ravenous lands.

## 6.6 Nutrient Management Approaches

Nutrient management in India over time has specific syndrome which can be summarized by abuse of nitrogen and disuse of potassium and generally coupled with overuse of phosphorus. It suggests inherent flaws in fertilizer application practices adopted by farmers that probably promote imbalance in nutrient applications. The problem of decreasing partial factor productivity of fertilizers can be arrested by addressing complex nutrient interaction. The DAP centric agriculture that is in vogue in many parts of India today is perhaps the main suspect which is causing problem of negative nutrient interaction of phosphate with micronutrients like Zn, Fe and Cu besides hampering the nutrient ratios of Ca+Mg/K ratio.

Blanket fertilizer recommendations prescribed for large areas, generally advocated by different state governments, are based on crop responses over large areas without considering the spatial and temporal variability of soil in terms of plant nutrient supplying capacity. Due to less complexity in arriving at plant nutrient application, these recommendations find advocacy with line departments. The other

**Table 6.1** Physicochemical properties of soils under different land use systems under distinct landscapes

| Land use | Depth (cm) | pH | EC (dS $m^{-1}$) | OrgC (%) | CEC [c mol ($p^+$) $kg^{-1}$] | Clay (%) | Available nutrients (kg $ha^{-1}$) | | |
|---|---|---|---|---|---|---|---|---|---|
| | | | | | | | N | P | K |
| *A. Table landscape* | | | | | | | | | |
| Rainfed cropping (single crop) | 0–15 | 7.7 | 0.30 | 0.36 | 31.70 | 40.8 | 280 | 14.62 | 405 |
| | 15–30 | 7.8 | 0.39 | 0.37 | 31.45 | 41.2 | 266 | 13.62 | 385 |
| Irrigated cropping (double crop) | 0–15 | 7.8 | 0.48 | 0.57 | 28.77 | 42.0 | 275 | 15.51 | 435 |
| | 15–30 | 7.8 | 0.41 | 0.39 | 28.60 | 41.6 | 252 | 13.47 | 475 |
| Agri-horti system | 0–15 | 7.7 | 0.52 | 0.42 | 21.15 | 38.4 | 350 | 18.13 | 395 |
| | 15–30 | 7.4 | 0.57 | 0.37 | 26.10 | 39.6 | 308 | 14.80 | 356 |
| Grassland | 0–15 | 7.3 | 0.26 | 0.63 | 18.55 | 38.3 | 392 | 24.42 | 484 |
| | 15–30 | 7.4 | 0.17 | 0.46 | 21.10 | 36.9 | 350 | 21.67 | 425 |
| *Leucaena leucocephala* | 0–15 | 7.1 | 0.29 | 0.72 | 24.40 | 37.8 | 392 | 24.31 | 425 |
| | 15–30 | 7.2 | 0.19 | 0.63 | 24.80 | 37.6 | 381 | 18.51 | 357 |
| *Soymida febrifuga* | 0–15 | 7.4 | 0.34 | 0.85 | 21.10 | 40.0 | 393 | 23.87 | 474 |
| | 15–30 | 7.4 | 0.30 | 0.70 | 26.00 | 38.8 | 315 | 19.40 | 474 |
| *B. Ravenous landscape* | | | | | | | | | |
| Ravine (mixed forest) | 0–15 | 7.9 | 0.43 | 0.77 | 16.90 | 20.2 | 258 | 20.52 | 430 |
| Top | 15–30 | 7.9 | 0.52 | 0.65 | 15.15 | 21.8 | 240 | 19.10 | 375 |
| Ravines (*Acacia senegal*) | 0–15 | 8.0 | 0.43 | 0.70 | 13.00 | 22.9 | 294 | 20.30 | 557 |
| | 15–30 | 8.0 | 0.47 | 0.61 | 15.25 | 23.5 | 275 | 18.20 | 430 |

Source: Somasundaram et al. (2009)

approach of fertilizer application is based on own/peer perception linked with fertilizer price, commodity price, availability of water and fertilizers, access to market and risk perception (abiotic and biotic). Such type of fertilizer application usually leads to imbalanced use of plant nutrients. Peer perception especially that of progressive farmers is also considered for adoption of plant nutrient prescription without considering farmers' own resource endowment. The third approach with two broad subsets is soil test-based fertilizer recommendation which is based on analysis of representative soil samples, correlation and calibration and finally fertilizer recommendation.

In one subset, site-specific nutrient management (SSNM), which is basically the plant-based approach, was developed in the 1990s by IRRI in collaboration with national partners across Asia to address serious limitations arising from blanket fertilizer recommendation for large areas in Asia. Another important subset, the targeted yield approach, wherein Ramamoorthy et al. (1967) established the theoretical basis and experimental proof for the fact that Liebig's law of the minimum operates equally well for N, P and K. This forms the basis for fertilizer application for targeted yields (Truog 1960). Among the various methods of fertilizer recommendation, the one based on yield targeting is unique in the sense that this method not only indicates soil test-based fertilizer dose but also the level of yield the farmer can hope to achieve if good agronomic practices are followed in raising the crop. The differentiation of significant multiple regression equations provides a basis for soil test-based fertilizer recommendation for maximum yield, profit and return per rupee investment on fertilizer. The resultant fertilizer adjustment equations have been tested in follow-up and frontline demonstrations conducted in different parts of the country. In these trials soil test-based rates of fertilizer application helped to obtain higher response ratios and benefit: cost ratios over a wide range of agroecological regions (Dey 2015a; Dey and Srivastava 2013; Singh et al. 2015). Targeted yield concept strikes a balance between 'fertilizing the crop' and 'fertilizing the soil'. The procedure provides a scientific basis for balanced fertilization and balance between applied nutrients and soil available nutrients.

In the targeted yield approach, it is assumed that there is a linear relationship between grain yield and nutrient uptake by the crop, and for obtaining a specific yield, a definite amount of nutrients is taken up by the plant. Once this requirement is known for a given yield level, the fertilizer needed can be estimated taking into consideration the contribution from soil available nutrients. More than 5000 frontline demonstrations have been organized by different centres of AICRP (STCR) in farmers' fields of different states to demonstrate the developed STCR technology. Front-line demonstrations conducted at farmers' field clearly brought out the superiority of STCR-IPNS fertilizer recommendations for different crops over blanket recommendation and farmer's practice in terms of higher yields with higher BCR/net returns. Front-line demonstrations conducted under Tribal Sub Plan (TSP) in Assam, Bihar, Chhattisgarh, Gujarat, Himachal Pradesh, Jammu and Kashmir, Jharkhand, Karnataka, Kerala, Madhya Pradesh, Maharashtra, Manipur, Odisha, Rajasthan, Tamil Nadu, Telangana, Uttar Pradesh and West Bengal with tribal farmers also clearly brought out the superiority of STCR-based fertilizer recommendation for different crops over blanket recommendation and farmer's practice and that even farmers with very little knowledge of modern agriculture could achieve yield target by practising STCR technology within ±10% variations of the target set (Dey 2015b, 2016).

Further, there is a wealth of knowledge for a range of measures that can help in developing agrotechnologies to overcome climate vulnerabilities. Research works from plateau region clearly demonstrate that indigenous people and their knowledge are central to the adaptive changes for sustainable agriculture using available natural resources essential to face the world's changing climate (Dey and Sarkar 2011). In one such practice for direct seeded rice, cow dung is powdered and mixed thoroughly with the soil after broadcasting of rice seeds and then planking (levelling the land surface after cultivation with wood log) is done. This process helps in better mineralization of nitrogen from manure for uptake by young plants, since nitrogen mineralization is essentially a microbial-mediated process and with powdering and mixing, the process occur faster. Research has shown that such practice helps better nutrient uptake by the young plants due to better mineralization of nitrogen from the manure under improved water holding capacity of the soil.

### 6.6.1 The Integrated Nutrient Management (INM) for Ravine Land Management

The basic concept underlying the integrated nutrient management system (INMS), nevertheless, remains the maintenance and possible improvement of soil fertility for sustained crop productivity on long-term basis and to reduce inorganic (fertilizer) input cost for ravenous lands. Since soils of ravine lands are nutrient poor with low organic matter, INM approach would be useful for building soil productivity and improving crop yields. INM is an approach for maintenance or adjustment of soil fertility in relation to plant nutrient supply at an optimum level for sustaining desired crop productivity. This is done through optimization of the benefits from all sources of plant nutrients in an integrated manner. Maintenance of soil health with maximum use of organic manure and recycling of organic waste along with chemical fertilizers and biofertilizers in an integrated manner are environmentally sound and cost-effective alternatives. Build-up of soil organic matter is an important strategy to rehabilitation of degraded wasteland (Lal and Bruce 1999). Soil organic matter is an essential but transient component that controls several physical, chemical and biological properties of soil (Carter 1996).

The three main components of INMS are:

1. Maintain or enhance ravine soil productivity through a balanced use of fertilizers combined with organic and biological sources of plant nutrients to increase soil organic matter.
2. Improve the stock of plant nutrients in the soils as the soils are hungry and thirsty.
3. Improve the efficiency of plant nutrients, thus, limiting losses to the environment as the soil health is very poor under ravenous lands.

Land degradation manifests itself chiefly in the form of water erosion as in the case of ravines, followed by wind erosion and biophysical and chemical deterioration. Soil is a very important base indicator for crop production. It integrates a variety of important processes involving vegetation growth, overland flow of water, infiltration, land use and management. Therefore, arresting the soil erosion and soil fertility decline including reduction in soil organic carbon, deterioration of physical properties, change in soil nutrient stock and build-up of toxic substances are necessary. Waterlogging, salinity and sodicity, acidity, soil pollution, loss of vegetation, sedimentation or burial of soils and exposure of stoniness/rockiness are some associated factors. Several studies have shown that in most of the regions, there is a net negative balance of nutrients and a gradual depletion of the organic matter content of soil, which could otherwise protect the soil from erosion. An improvement in the biophysical environment of the ravine areas could help to rebuild the soil productivity. Integrated nutrient management inherently has the organic matter component which is the prime requirement for arresting the soil degradation and enhancing soil health in ravine ecosystems. Thus, integrated nutrient management (INM) refers to the maintenance of soil fertility and of plant nutrient supply at an optimum level for sustaining the desired productivity through optimization of the benefits from all possible sources of organic, inorganic and biological components in an integrated manner.

The concept of integrated nutrient management (INM) takes into consideration the nutrient cycle involving soils, crops and livestock, nutrient deficiencies, organic recycling, conjunctive use of organic manures and mineral fertilizers and biological nitrogen fixing. In order to ensure that crop productivity increases in an efficient and sustainable manner without sacrificing soil productivity of future generations, we must keep the following into consideration:

- Balanced application of appropriate fertilizer is a major component of INM.
- Fertilizers need to be applied at the required level for optimal growth based on crop requirement and agroclimatic considerations:
  - Overapplication of nutrient neither helps in greater nutrient uptake nor significantly increases higher yield.
  - Underapplication can retard the crop growth and resulted in lower yields.
- Salinity affects the growth and yield of most of the crops. Yet, at a given salinity level, yield can be increased through balanced nutrient management. Recycling of organic waste from urban to rural areas is a potential and largely untapped source of nutrients for crop and farm.
- Municipal sewage sludge contains about 3.3% N, 2.3% P and 0.3% K. This urban waste needs to be treated carefully because it may contain toxic metals, parasites, etc.
- Use of green manures/organic manures.
- Organic manure increases the WHC of soils, as result the conductivity of soil solution decreases.
- Ploughing and levelling of the land.
- It helps in infiltration and percolation rate; therefore, salts leached down to the lower levels.

### 6.6.2 Integrated Nutrient Management (INM) for Improving Crop Productivity and Soil Health in Ravine Land

Integrated nutrient management including manures and biofertilizers, e.g. *Azospirillum*, *Rhizobium*, blue-green algae, phosphorus-solubilizing bacterium, mycorrhizae and VAM fungi in addition to chemical fertilizers (Rao and Venkateswarlu 1987), enhanced soil cover through vegetation and overall biomass production. An increase in soil organic matter/soil organic carbon acts as cementing agent for the stability of soil aggregates (Masri et al. 1996) which is a must for ravine lands and improves total nitrogen content (Harris 1995) to counteract poor soil fertility and vis-à-vis stability of soils. These practices in combination reduce the impact of beating action of rainfall expediting ravine formation through soil erosion. A long-term experiment was conducted in water-starved area of Ranchi on Typic Rhodustalfs where FYM application at 2.5 Mg $ha^{-1}$ along with fertilizers significantly enhanced the soil fertility to a considerable extent in legume- and rice-based cropping sequence (Singh et al. 2009b). This analogy applies to ravines also. Applying fertilizers through irrigation water, particularly through the drip system, termed as fertigation, provides the most effective way of supplying nutrients to the plant roots and enhancing nutrient use efficiency.

In experiments conducted at RVSKVV, Gwalior indicated that various inorganic and organic practices exert significant impact on wheat yield under ravenous lands. Inorganic and organic practices significantly increased grain/straw/biological yield of wheat over farmers' practice of application as 75% RDF+2.5 Mg FYM $ha^{-1}$ + PSB+ $ZnSO_4$ at 25 kg $ha^{-1}$, which was statistically identical to 150% RDF (Table 6.2).

The study revealed that combined application of organic and inorganic sources needs to be followed to satisfy the nutrient requirement of crops cultivated under ravenous lands. Combined application of different production factors (irrigation, fertilizer, farmyard manure and plant protection) under high input management recorded the highest increase in growth, yield and yield attributes of Indian mustard (*Brassica juncea*) in reclaimed ravine land of river Yamuna at Agra. Mustard yield increased by 58% with one irrigation given at branching, 82% with two irrigations given at branching and flowering and 101% with three irrigations given at branching, flowering and pod filling. Yield further increased by 118% with additional application of 60 kg N $ha^{-1}$ and 135% with further additional application of 10 Mg farmyard manure $ha^{-1}$ over low input management (control). The highest yield (2268 kg $ha^{-1}$) and the highest additional net returns (Rs 8823 $ha^{-1}$) were recorded under high input management. Another experiment conducted on ravine top on research farm of CSWCRTI, Central, Vasad, Gujarat, revealed that application of FYM at 10 Mg $ha^{-1}$ $year^{-1}$+N:P:K at 100:60:40 showed high gain during favourable weather condition in terms of better pearl millet crop growth parameters, higher yield, lower runoff, lower soil loss and build-up of SOC (Kumar et al. 2014).

Further, integrated nutrient management assumes centre stage for ravine lands since among the soil physical properties, aggregate stability (Thakuria et al. 2009; Bossuyt et al. 2001), infiltration and water transmission properties (Franzluebbers 2002) are the most sensitive to INM practices and their improvement certainly

**Table 6.2** Effect INM practices on yield (kg $ha^{-1}$) of wheat

| Treatments | Grain yield | Straw yield | Biological yield | Harvest index | Nutrient use efficiency (%) | | |
|---|---|---|---|---|---|---|---|
| | | | | | N | P | K |
| Farmer practices | 1915 | 2244 | 4159 | 46.04 | – | – | – |
| 100% RDF | 3306 | 3561 | 6867 | 48.14 | 41.74 | 8.33 | 22.08 |
| 150% RDF | 3385 | 4001 | 7386 | 45.82 | 48.05 | 9.39 | 31.17 |
| NPK application on the basis of STCR equation | 2922 | 2942 | 5864 | 49.82 | 34.48 | 6.48 | 20.92 |
| 50% RDF+5 Mg FYM $ha^{-1}$+PSB+ $ZnSO_4$ at 25 kg $ha^{-1}$ | 2936 | 3013 | 5949 | 49.35 | 63.21 | 11.65 | 35.41 |
| 75% RDF+2.5 Mg FYM $ha^{-1}$+ PSB+ $ZnSO_4$ at 25 kg $ha^{-1}$ | 3984 | 4454 | 8438 | 47.21 | 45.84 | 9.61 | 28.95 |
| FYM at 10 Mg $ha^{-1}$+PSB+*Azotobacter* | 2369 | 2413 | 4782 | 49.53 | 67.68 | 23.75 | 23.90 |
| LSD ($p \leq 0.05$) | 49.6 | 134.71 | | | 17.10 | 3.63 | 7.50 |

*RDF* recommended doses of fertilizers, *FYM* farmyard manure, *PSB* phosphorus-solubilizing bacteria

**Table 6.3** Integrated nutrient management in degraded ravine land on bulk density ($Mg\ m^{-3}$), mean weight diameter (MWD, mm) and moisture content (%)

| Tr. no. | Treatments | Bulk density ($Mg\ m^{-3}$) | MWD (mm) | Moisture (%, wt/wt) |
|---|---|---|---|---|
| T 1 | Farmer practices | 1.34 | 0.31 | 12.60 |
| T 2 | 100% RDF | 1.47 | 0.33 | 14.74 |
| T 3 | 150% RDF | 1.52 | 0.48 | 15.76 |
| T 4 | NPK application on the basis of STCR equation | 1.47 | 0.43 | 14.72 |
| T 5 | 50% RDF+5 Mg FYM $ha^{-1}$+ PSB+ $ZnSO_4$ at 25 kg $ha^{-1}$ | 1.42 | 0.42 | 14.16 |
| T 6 | 75% RDF+2.5 Mg FYM $ha^{-1}$+ PSB+ $ZnSO_4$ at 25 kg $ha^{-1}$ | 1.42 | 0.50 | 18.24 |
| T 7 | FYM at 10 Mg $ha^{-1}$+PSB+*Azotobacter* | 1.39 | 0.48 | 15.69 |
| | LSD ($p \leq 0.05$) | NS | 0.07 | 1.96 |

*RDF* recommended doses of fertilizers, *FYM* farmyard manure, *PSB* phosphorus-solubilizing bacteria

**Table 6.4** Integrated nutrient management in degraded ravine land on soil biological properties

| Tr. no. | Treatments | SMBC (μgC $g^{-1}$) | FDA (μg $g^{-1}\ h^{-1}$) | DHA (μg $g^{-1}$ TPF $g^{-1}\ h^{-1}$) |
|---|---|---|---|---|
| T 1 | Farmer practices | 51.71 | 5.83 | 12.43 |
| T 2 | 100% RDF | 54.85 | 6.09 | 26.83 |
| T 3 | 150% RDF | 63.79 | 8.59 | 36.29 |
| T 4 | NPK application on the basis of STCR equation | 56.27 | 7.22 | 28.73 |
| T 5 | 50% RDF+5 Mg FYM $ha^{-1}$+ PSB+ $ZnSO_4$ at 25 kg $ha^{-1}$ | 62.92 | 8.93 | 45.49 |
| T 6 | 75% RDF+2.5 Mg FYM $ha^{-1}$+ PSB+ $ZnSO_4$ at 25 kg $ha^{-1}$ | 66.05 | 11.24 | 59.46 |
| T 7 | FYM at 10 ton $ha^{-1}$+PSB+*Azotobacter* | 65.46 | 7.26 | 47.03 |
| | SEm± | 1.84 | 0.31 | 1.03 |
| | LSD ($p \leq 0.05$) | 5.67 | 0.94 | 3.19 |

*RDF* recommended doses of fertilizers, *FYM* farmyard manure, *PSB* phosphorus-solubilizing bacteria

amends the ravine soil to a large extent with improvement in soil health. In ravenous land, application of chemical fertilizer at 75% RDF along with 2.5 Mg FYM $ha^{-1}$ plus PSB and ZnSO4 at 25 kg $ha^{-1}$ improved significantly mean weight diameter (MWD) and moisture content over farmers' practices or inorganic treatment (Table 6.3). Also integrated nutrient management supply in ravines significantly improved biological properties of soil as a measure of soil microbial biomass carbon (SMBC), dehydrogenase activity (DHA) and fluorescein diacetate (FDA) over conventional fertilizer application (Table 6.4).

## 6.7 Conclusions

Even in the best farming practices, only 40–50% of the applied fertilizers are used by the crops, and the balance finds their way into the aquatic environment through runoff. The soils of the ravenous regions are poor in fertility and physically degraded in condition. Therefore, integrated nutrient management / practices involving soils, crops and livestock, nutrient deficiencies, organic recycling, conjunctive use of organic manures and mineral fertilizers are crucial practices to sustain productivity in degraded soils. Moreover, in ravine land, farmyard manure is applied in small quantity prior to plantation, but the quantity is too small to replace the nutrients lost through soil erosion. Hence, only chemical fertilizers are used for replacement of the last nutrients. In fact the availability of large quantities of farmyard manure for INM is considered as a major hurdle in promoting INM programs on mega-scales. Therefore, use of locally available manures such as green leaf/crop residue such as mustard and pigeon pea may offer great promises in these areas. By and large, it appears to be promising that the integrated use of chemical fertilizers and other organic nutrient sources in a judicious manner is the most viable and practicable strategy to rejuvenate the denuded ravine ecosystems.

## 6.8 Way Forward

Future strategies for maintaining soil health, crop productivity and environmental sustainability in degraded ravine land may include:

- STCR-based nutrient management for ravine areas is warranted as it has provided a strong foundation for pragmatic policy formulation towards fertilizer recommendations based on yield targets for better nutrient use efficiency as well as sustaining soil and crop productivity.
- Long-term strategies required to ensure adequate soil P and K supply in the ravenous land so that the crop growth is not limited.
- Appropriate integration of several organic components along with mineral fertilizers to restore soil nutrients in the degraded ravine land through INM practices.
- Comprehensive and concerted effort is needed in utilization of locally available components of INM involving rational and appropriate use of fertilizers and organics.

- Agrotechnologies for maximizing input use efficiency must form an integral part of the INM package.
- Adaptive research trials on large scale have to be undertaken to assess the INM technology with respect to agronomic productivity, ecological compatibility, economic profitability and social acceptability is necessary.
- Research on secondary and micronutrient status and for crop production in ravenous land is very limited, and an INM technology with secondary/micronutrients may provide land sustainability for crop production.

## References

Anderson JR, Ernest EH, John TR, Richard EW (2001) A land use and land cover classification system for use with remote sensor data. Geological Survey Professional Paper 964, A revision of the land use classification system as presented in U.S. Geological Survey Circular 671

Ang LH (1994) Problems and prospects of afforestation on sandy tin tailings in peninsular. Malaysia J Trop For Sci 7:87–105

Angers DA, Caron J (1998) Plant-induced changes in soil structure: processes and feedbacks. Biogeochem 45:55–72

Bossuyt H, Denef K, Six J, Frey SD, Merchx R, Paustian K (2001) Influence of microbial populations and residue quality on aggregate stability. Appl Soil Ecol 16:195–208

Bradshaw AD (1997) The importance of soil ecology in restoration science. In: Urbanska KM, Webb NR, Edwards PJ (eds) Restoration ecology and sustainable development. Cambridge University Press, Cambridge, pp 33–64

Carter MR (1996) Analysis of soil organic matter storage in agro-ecosystems. In: Carter MR, Stewart BA (eds) Structure and organic matter storage in agricultural soils. CRC Press, Boca Raton, pp 3–14

Champion HG, Seth SK (1968) A revised survey of forest types of India. Controller of Publn, New Delhi

Chaturvedi OP, Kaushal R, Tomar JMS, Prandiyal AK, Panwar P (2014) Agroforestry for wasteland rehabilitation: mined, ravine, and degraded watershed areas. In: Dagar JC, Singh AK, Arunachalam A (eds) Agroforestry systems in India: livelihood security & ecosystem services, advances in agroforestry, vol 10. Springer, New Delhi, pp 233–271

Dagar JC, Mall LP (1980) Studies on the vegetation of ravines and banks of the river Kshipra and its tributaries. J Indian Bot Soc 59:234–245

Datta M, Singh NP (2007) Growth characteristics of multipurpose tree species, crop productivity and soil properties in agroforestry systems under subtropical humid climate in India. J For Res 18(4):261–270

Dey P (2015a) Targeted yield approach of fertiliser recommendation for sustaining crop yield and maintaining soil health. JNKVV Res J 49(3):338–346

Dey P (2015b) Soil test crop response: what can be learnt? In: Jat ML, Majumdar K, McDonald A, Sikka AK, Paroda RS (eds) Book of extended summaries. National dialogue on efficient nutrient management for improving soil health, September 28–29, 2015. TAAS, ICAR, CIMMYT, IPNI, CSISA, FAI, New Delhi, pp 25–31

Dey P (2016) Soil health management. In: Katyal JC, Chaudhari SK, Divedi BS, Biswas DR, Rattan RK, Majumdar K (eds) Soil health: concept, status and monitoring. Bulletin of the Indian Society of Soil Science, vol 30, pp. 79–97

Dey P, Sarkar AK (2011) Revisiting indigenous farming knowledge of Jharkhand (India) for conservation of natural resources and combating climate change. Indian J Trad Knowl 10(1):71–79

Dey P, Srivastava S (2013) Site specific nutrient management with STCR approach. In: IISS contribution in frontier areas of soil research. Indian Inst Soil Sci, Bhopal, pp 259–270
Doran JW, Parkin TB (1994) Defining andAssessing soil quality. Defining soil quality for a sustainable environment. Soil Sci Soc Am Madison, WI, USA, 3–12
Foster DF, Swanson J, Aber I, Burke N, Brokaw D, Tilman KA (2003) The importance of land use legacies to ecology and conservation. Bio Sci 53:77–88
Franzluebbers AJ (2002) Water infiltration and soil structure related to organic matter and its stratification with depth. Soil Tillage Res 66:97–105
Gyssels G, Poesen J, Bochet E, Li Y (2005) Impact of plant roots on the resistance of soils to erosion by water: a review. Prog Physi Geogr 29:189–217
Han KH, Ha SG, Jang BC (2010) Aggregate stability and soil carbon storage as affected by different land use practices. In: Proceedings of international workshop on evaluation and sustainable management of soil carbon sequestration in Asian countries. Bogor Sept 28–29, pp 113–124
Harris HC (1995a) Long-term trials on soil and crop management at ICARDA. Adv Soil Sci 19:447–469
Jarecki M, Lal R (2003) Crop management for soil carbon sequestration. Crit Rev Plant Sci 22:1–32
Jha P, Mohapatra KP, Dubey SK (2010) Impact of land use on physico-chemical and hydrological properties of ustifluvent soils in riparian zone of river Yamuna. India Agroforest Syst 80:437–445
Joshi BK (2002) Effect of land use and elevation on soil properties of Betagad watershed – a case study from Indian central Himalaya. J Soil Water Cons 1(2&3):109–117
Kaushik P, Ghosh P (2015) Geomorphic evolution of Chambal river origin in Madhya Pradesh using remote sensing and GIS. Int J Adv Remote Sens GIS 4(1):1130–1141
Kumar G, Kurothe RS, Brajendra VAK, Rao BK, Pande VC (2014) Effect of farmyard manure and fertilizer application on crop yield, runoff and soil erosion and soil organic carbon under rainfed pearl millet (*Pennisetum glaucum*). Indian J Agric Sci 84(7):816–823
Lal R (1984) Soil erosion from tropical arable lands and its control. Adv Agron 37:183–248
Lal R, Bruce JP (1999) The potential of world cropland soils to sequester C and mitigate the greenhouse effect. Environ Sci Pol 2(2):177–185
Lamb D (1988) IUCN guidelines for restoration of degraded ecosystems. draft- world conservation union (IUCN), Gland
Luna RK (2006) Plantation forestry in India. Offset Printers and Publishers, Dehradun, pp 361–386
Mandal D, Sharda VN, Tiwari A (2008) Soil loss tolerance limits in planning conservation measures in different districts of India. CSW CRTI, Dehradun, p 112
Masri Z, Rayn J, Pala M (1996) Specific effects of a cereal – legume rotation on soil physical properties under Mediterranean type of condition. (Abstracts) International Conference on Land Degradation, June 10–14, Adan, Turkey, pp.51
Mathur GM, Deo R, Yadav BS (2006) Status of zinc in irrigated north-west plain soils of Rajasthan. J Indian Soc Soil Sci 53:359–361
Mishra CSK, Monalisa R (2013) Wasteland reclamation: novel approaches. Ecoscan 3:99–105
Mythili G, Goedecke J (2016) Economics of land degradation in India. In: Nkonya E et al (eds) Economics of land degradation and improvement – a global assessment for sustainable development, pp 431–470
Pande VC, Kurothe RS, Rao BK, Kumar G, Parandiyal AK, Singh AK, Kumar A (2012) Economic analysis of bamboo plantation in three major ravine systems of India. Agric Econ Res Rev 25(1):63–73
Ramamoorthy B, Narasimham RL, Dinesh RS (1967) Indian Farming 17(5):43–45
Rao AV, Venkateshwarlu B (1987) Nitrogenase activity of pearl millet-Azospirillum association in relation to the availability of organic carbon in the root exudates. Proc Indian Acad Sci 97:337
Sanchez PA (1987) Soil productivity and sustainability in agroforestry systems. In: Steppler HA, Nair PKR (eds) Agroforestry: a decade of development. ICRAF, Nairobi, pp 1–345
Sanchez A (1995) Science in agroforestry. Agrofor Syst 30:5–55

Sharma AR, Behera UK (2010) Green leaf manuring with prunings of Leucaena leucocephala for nitrogen economy and improved productivity of maize (*Zea mays*)–wheat (*Triticum aestivum*) cropping system. Nutr Cycl Agroecosyst 86:39–52

Sharma AK, Pradhan IP, Nema JP, Tejwani KG (1980) 25 years research on soil and water conservation in ravine lands of Gujarat. CS&WCRTI, Research Centre, Vasad

Singh KA (1999) Impact of various land uses on soil properties of sloping hills in Sikkim, Eastern Himalaya. Ind J Soil Cons 27(1):70–73

Singh RK, Singh SK, Tarafdar JC (2009) Influence of cropping sequence an management on soil organic carbon and nutrient status of Typic Rhodustalfs. J Indian Soc Soil Sci 56:174–181

Singh SK, Baser BL, Shyampura RL (2002) Chemical composition and charge behaviour of smectites in Vertisols of Rajasthan. J Indian Soc Soil Sci 50:106–111

Singh AK, Kala S, Dubey SK, Rao BK, Gaur ML, Mohapatra KP, Prasad B (2014) Evaluation of bamboo based conservation measures for rehabilitation of degraded Yamuna ravines. Indian J Soil Conserv 42(1):80–84

Six J, Paustian K, Elliott ET, Combrink C (2000) Soil structure and soil organic matter I. Distribution of aggregate size classes and aggregate associated carbon. Soil Sci Soc Am J 64(2):681–689

Somasundaram J, Singh RK, Parandiyal AK, Prasad SN (2009) Micronutrient status of soils under different land use systems in Chambal Ravines. J Indian Soc Soil Sci 57(3):307–312

Somasundaram J, Singh RK, Ali S, Sethy BK, Singh S, Lakaria BL, Chaudhary RS, Singh RK, Sinha NK (2012) Soil aggregates and other properties as influenced by different long term land uses under table landscape topography of Chambal region, Rajasthan, India. Indian J Soil Conserv 40(3):212–217

Srinivasan Y, Venna B, Thakur JY, Bonde WC (1970) Some grasses for pasture development in ravine lands. Indian J Agric Sci 40(3):268–272

Singh SR, Kundu DK, Tripathi MK, Dey P, Saha AR, Kumar M, Singh I, Mahapatra BS (2015) Impact of balanced fertilization on nutrient acquisition, fibre yield of jute and soil quality in New Gangetic alluvial soils of India. Appl Soil Ecol 92:24–34

Thakuria D, Talukdar NC, Goswami C, Hazarika S, Kalita MC, Bending GD (2009) Evaluation of rice-legume-rice cropping system on grain yield, nutrient uptake, nitrogen fixation, and chemical, physical, and biological properties of soil. Biol Fertil Soils 45:237–251

Truog E (1960) Fifty years of soil testing. Transactions of 7th International Congress of Soil Science, vol III, Commission IV, Paper No. 7, pp 46–53

# Some Practices of Gully Rehabilitation in Central Brazil

7

Silvio Carlos Rodrigues

**Abstract**

The recovery of eroded areas in the form of gullies in a Cerrado environment (Brazilian savanna) was the object of studies carried out at the Experimental Farm of the Federal University of Uberlândia, Brazil. The experiments were carried out seeking to use low-cost procedures and materials founded near the affected areas. Vegetative, edaphic, and mechanical techniques were used. The results show that after 11 years of implementation of the techniques, it could be noticed that changes in the soil cover and an onset of ecological succession are under way in the area. The best responses were achieved with the use of palisades to contain sediment exit, which managed to retain much of the sediment produced on the gully walls. Also, good results were achieved with the planting of herbaceous and shrub species, which partially covered the eroded soil. The encouraging results were also obtained about the construction of a fence to prevent the entry of livestock from neighboring farms and for fire protection keeping grass and weeds mowed around the fence. This approach got better results in the upper part of the water table, providing an increase in soil protection and consequent reduction of surface runoff, which also began to be controlled through terracing and construction of containment basins.

## 7.1 Introduction

The morphoclimatic domain of the Cerrado belongs to the category of savanna areas and occupies a prominent place among the tropical environments of the world, both for its ecological particularities and for its productive potential. The

S. C. Rodrigues (✉)
Geomorphology Lab, Federal University of Uberlândia, Uberlândia, Brazil
e-mail: silgel@ufu.br



J. C. Dagar, A. K. Singh (eds.), *Ravine Lands: Greening for Livelihood and Environmental Security*, https://doi.org/10.1007/978-981-10-8043-2_7

areas covered by these environments began to be investigated in the nineteenth century with the expeditions of von Martius, 1817–1820; Saint Hilaire, 1816–1822; and especially Warning, in 1892 (cited by Rodrigues 2014). Further studies on the Cerrado characteristics only began in the 1940s (Rawitscher 1948; Salgado-Labouriau 1961; Ferri 1969). Agricultural research gained pace in 1975, with the creation of the Embrapa (Brazilian Agricultural Research) System. From that time, the agricultural occupation of the Cerrado has grown, together with an accelerated deforestation. This has rapidly aggravated soil degradation and erosion, with the associated anthropic activity. The Cerrado area of Brazil has seasonal tropical climate. Annual rainfall is 1500 mm, higher than other savanna areas in Africa and Australia. The dry season extends during 5 or 6 months in most of the region and as much as 1 month longer in some areas.

The Cerrado soils are of low fertility and acidic with high levels of aluminum. Dystrophic soils cover approximately 90% of the total area. Most of it is relatively flat but with good drainage. The vegetation, with herbaceous species of 50–70 cm in height, is somewhat dense for savannah with a discontinuous arboreal canopy and shrubby elements characterized by twisted branches, thick bark, and large leathery leaves. Isolated areas with pronounced environmental deficiencies (sandy, litholic, or hydromorphic soils) create physiognomies with less biomass. Other areas, with ample supplies of water or amenable soils, present the characteristics of tropical forest. This biome includes the headwaters of several major Brazilian and South American watersheds which provide 80% of the country's energy matrix by way of large hydroelectric power plants (Oliveira et al. 2015). The intense exploration of the Cerrado, however, is leading to soil and water degradation, especially because of inadequate management and soil erosion.

The purpose of this study is to analyze the results of research conducted in a degraded area from which gravel had been removed for highway construction. The area was subsequently abandoned without recovery of the soil. Natural processes resulted, producing a unique environment referred to as *voçoroca* in Portuguese. This type of gully develops rapidly in weak, sandy soils that have little vegetation to hold the soil. If they are not contained in the early stages, they rapidly reach an advanced stage of degradation and can become enormous in size. The recovery of gullies is difficult. In most cases, great quantities of manpower and financial resources are required. Stop gap interventionist practices, performed to control or, at least, delay effects of degeneration, apply alternative methods but are still expensive.

## 7.2 Study Area

The study area is located on the right side of the Glória Stream, in the county of Uberlândia (State of Minas Gerais), between the geographic coordinates of 18° 58′19 "S and 48° 12′31" W Greenwich, in an altitude between 830 and 860 m above sea level (Fig. 7.1). The local geological sequence is composed by basalt (Serra Geral Formation) under sandstone (Marília Formation) and a Cenozoic sedimentary cover. The superficial material has a sandy-clay-loam to sandy-loam

**Fig. 7.1** Location of the study site

texture, resulting from the alteration of sandstone and alluvial materials (Rodrigues 2002). The soils around the gully, augmented by the extraction of gravel, were classified by Silva (2010) as oxisol in the upper part of the slope and inceptisol near the valley bottom. Locally the relief is composed, predominantly, of convex and tabular forms.

The soil at the site has an exposed Bw horizon of a typical oxisol, with a sandy loam texture. Natural soils in this watershed have huge variation because of their composition and the state of the site's conservation. The regional climate is characterized by well-defined seasons of rain in summer (from November to March) and dry winters (from May to September), controlled by continental and Atlantic air masses (Vrieling et al. 2007). Since 2006, management practices have been applied in the study area to stabilize the erosion using physical barriers (palisades) as reported by Silva (2010), Biulchi (2012), Pereira Júnior (2013), and replanting by Serato and Rodrigues (2010), Machado et al. (2014), and Confessor et al. (2016). The initial signs of stabilization in the area are visible, verifying the importance of these practices and the efficiency of the methods used.

Several factors have contributed to the emergence of the gully. One is directly related to anthropic activities. During the extraction of gravel, the natural land cover and the top soil were scraped away, leaving exposed sandy material, friable, with low levels of aggregation, compacted, with very little organic matter and low capacity for water infiltration (average of 0.8 mm per min) (Table 7.1).

**Table 7.1** Soil physical and texture characteristics of the study area (Silva 2010)

| Total density | Particle density | Porosity | Sand | Silt | Clay |
|---|---|---|---|---|---|
| $g\ cm^{-3}$ | | % | $g\ kg^{-1}$ | | |
| 1.49 | 2.67 | 44.07 | 886.0 | 19.5 | 94.5 |

**Table 7.2** Chemical properties of soil in the study area (Silva 2010)

| pH | P | K | Ca | Mg | Al | SB | t | T | V | m | OM |
|---|---|---|---|---|---|---|---|---|---|---|---|
| $H_2O$ | $mg\ dm^{-3}$ | | $cmol_c\ dm^{-3}$ | | | | | | % | | $gkg^{-1}$ |
| 4.9 | 4.2 | 0.02 | 0.25 | 0.1 | 0.5 | 0,37 | 0.87 | 1,5 | 25 | 55 | 5 |

The low fertility at the site is common in tropical soils and complicates efforts to revegetate and to stabilize erosion (Table 7.2).

While identifying characteristics of the erosion process in the region of the study, Bezerra et al. (2012a) reported that when ravines and gullies reach down to the layers of sandstone of Marilia Formation, there is a progression and expansion of the side walls. Another implication of this geological arrangement is the occurrence of water exudation at the different layers of soil.

## 7.3 Conservation Practices

Because of environmental variation and many types of soil degradation, several intervention techniques have been attempted to determine the best possible result. The conservation practices used have been vegetative, edaphic, and mechanical. Each of these acts on a specific part of the problem so that it is necessary, in most cases, to use more than one practice (Bertoni and Lombardi Neto 1990). Soil conservation is thus a complex of practices and tasks that are correlated, complete, and interdependent.

Vegetative practices are used to replace vegetation in areas where there is only bare soil, low fertility, or the presence of stony surfaces. The first procedure is to create a base for the fixation of roots. This was completed a mix of surface material with cattle dung as source of organic matter, from outside areas. A selection of leguminous and herbaceous plants was tested. Those most adapted to the reconstructed surface were planted and their growth monitored. The intention of this procedure was the protection against varied forms of erosion. The dense vegetation, with extensive root systems, would thus be most useful for the objective. Techniques of revegetation were used in the present study to establish perennial herbage of large and arboreal size. To this end, seedlings were planted in the area surrounding the gully, to protect the slopes, reduce the superficial flow, and retain as much as possible of the slope area. The seedlings stabilize the soil around the gully, securing the friable soil and providing a cover to avoid direct contact with rainfall, minimizing the splash effect; the roots also assist in retaining material by holding the soil avoiding its downstream channel loss.

Mechanical practices support the other practices and are applied concomitantly, to produce an environment conducive for the emergence of vegetation and its development. These practices have a somewhat higher cost due to the necessary machinery, but it is possible to implement some alternative methods to minimize costs. The construction of terraces and contour curves aids in the dispersion of concentrated rainfall as well as its retention (Bezerra et al. 2012b). They are thus very efficient in the control of gullies. It is recommended that terraces and curves be arranged at an average distance of 30 m from the head of the gully (Silva 2010).

As an additional practice, for more effective stabilization of the gully, the water flows at the site were diverted. In the high part of the slope, a group of terraces were constructed to decrease the flow downslope. The retention of the water flow reduces the loss of sediment enabling fixation of the plants in the lower parts of the gully. Retention basins were constructed at the end of each terrace. Trenches were constructed and filled with mulch, as additional barriers to the water flow.

The study area, in general, is in a state of intense degradation. The removal of vegetation has left the soil exposed causing it to become compacted by direct exposure and the impact of rain. The compacted soil reduces the infiltration so that the surface flow is intense and drains toward the main channel of the gully. The practices applied sought to stabilize the erosive processes in the gully and its surrounding landscape. The first action, in 2006, of the project was to build a fence to prevent the entry of livestock from neighboring farms and for fire protection keeping grass and weeds mowed around it. This action made it possible for the scarce grass to produce a seed bank for later regeneration. In the last 11 years, the area has been isolated and demonstrated vigorous regeneration, especially in areas of degraded Cerrado vegetation. The presentation of results is divided per the three types of intervention applied in the project.

### 7.3.1 Terrace Construction and Revegetation Techniques

Following the construction of terraces and contour curves at the head of the gully, attention was given to the lateral areas. These areas are in constant change; erosive processes are accelerated due to the surface flow which transports plant and soil material, contributing to the collapse of the slopes. The solution was to construct barriers, bags filled with soil. Soil analysis revealed very low fertility of the area, which suggested the need for chemical fertilization and addition of organic matter. This effort was associated with the planting of leguminous species (*Arachis pintoi* and *Macrotyloma axillare*) and grasses (*Brachiaria decumbens*, *Brachiaria humidicola*, and *Pennisetum purpureum*). Some of these results are visible in Figs. 7.2 and 7.3.

Along with these practices, vertical mulching of stones was applied at the top of the slope in 2007 (Fig. 7.4). After that the vegetation could appear and establish. The results are apparent in Fig. 7.5, when the gully channels have stabilized.

**Fig. 7.2** Stabilizing the gully head: (**a**) gully head collapse; (**b**) first step of vegetation growth; (**c**) second stage of rehabilitation with vegetation growth outside the gully

**Fig. 7.3** The stages of gully (**a**) before and (**b**) after the revegetation

**Fig. 7.4** Bare gully (left); and the same after mulching implementation in 2007

## 7.3.2 Construction of Palisades in the Channels of Erosion

The inside channel of the gully has a perennial spring, the water of which does not currently cause major damage, but on days of intense rainfall, the streams are concentrated toward the main channel of the gully, increasing considerably the

**Fig. 7.5** Mulching area in 2007 (left) and a view of the site in 2014 (right) after establishment of vegetation and stabilizing of gully channels

**Fig. 7.6** Water exfiltration on the gully head

erosive power along of the channel Figs. 7.6 and 7.7. This directly influences the deepening of the gully as well as the width of the channel.

The solution proposed for this situation was the construction of palisades, walls made from bamboo stakes and tree trunks to retain the sediment flow generated during the year and especially during intense rainfall. These palisades were 1 m in height, positioned at the bottom of the gully, and associated with leguminous shrubs-trunks in the temporary channels (Fig. 7.8).

The construction of the palisades in the interior of the gully yielded important results in terms of the implementation of sediment containment techniques. It was learned that when initiating the construction of large palisades in the bottom of gullies, the difficulties of reaching the location as well as the implementation of

**Fig. 7.7** Gully floor after a rainfall

**Fig. 7.8** Barrier built in the secondary channel of the gully. Notice the amount of sediment retained in the bottom of the channel

techniques must be taken into consideration. Gullies are places of difficult access. They have unstable slopes and collapses in the walls are frequent. One of the main problems was the difficulty of fixing the palisades to the instable walls. If the palisades are not firm at their bases, over time there may be partial or total collapse. Thus, the procedure was more effective in situations where the area above the palisade was filled in quickly after the first rains. At places where palisades formed small pools, instability was greater, and palisades as well as gully walls collapsed more easily. In the areas where the channels were temporary and the soil is humid during most of the year, the implementation of palisades was more effective. They promoted the invagination and the growth of vegetation, followed by a slow succession of herbaceous and shrub species (Fig. 7.9). Although the process has been slow, the vegetation that has settled in the area has provided greater protection to the surface of the soil. Over time it is expected that the ecological succession will generate biomass and the fixation of shrub species common in the humid environments of the Cerrado.

### 7.3.3 Recuperation of Slopes Above the Gully Head

Above the area of the gully head, some procedures were implemented with the aim of reducing or eliminating the direct accumulation of upstream surface flows with flows from the lower parts of the terrain. The objective was to retain water and/or

**Fig. 7.9** Palisade area after implementation (left) at initial stage and (right) after 10 years in 2017

**Fig. 7.10** Head of gully in 2008 (left) and same area in 2017 (right). Note that with the stabilization of the more intense erosive processes, vegetation could start establishing in the bottom of the gully

divert it safely to other locations, avoiding damage not only in terms of erosion but also benefit the surrounding area. The total area of the research was isolated with the construction of a fence to prevent grazing and trampling by neighboring cattle, as well as to prevent the practice of agricultural activities or extraction as had occurred previously, with the extraction of gravel. The practices adopted in the upper part of the gully not only improved the site but also benefitted the whole downstream area, as the retention of the surface water flows promoted a more stable environment in the lateral and lower parts of the gully. It became possible for vegetation to develop not only outside but also inside the gully (Fig. 7.10).

## 7.4 Conclusions

All the measures adopted in the upper part of the gully directly affected the lower parts. It was thus apparent that gullies must be treated as integrated systems that receive direct influence from their parts as well as from the surrounding areas. Although some aspects of the project did not yield immediate success, they provided favorable conditions for the stabilization of the area. Even the barriers that deteriorated with time, improved conditions for the appearance of vegetation. This vegetation subsequently began to retain sediment and protect the soil, in this way, substituting the barriers in function. Even the vegetation that did not establish ended up providing better conditions for the vigorous growth of other species, contributing to the continuous improvement of the environment, toward its successful recovery.

The recuperation of degraded areas requires time and money but alternative measures can help to reduce costs. In the present research such measures proved their effectiveness over a period of 11 years. Favorable conditions were created for the establishment of native plants, promoting the stabilization of the gully and the surrounding area.

**Acknowledgment** We are grateful to CAPES, FAPEMIG, and CNPq, who financed the researches that resulted in the methodologies and results presented in this chapter. We also want to remember all LAGES colleagues, master and doctoral students, who have contributed their efforts to the development of field and laboratory research over the last few years.

## References

Bertoni J, Lombardi Neto F (1990) Conservação do solo. Ícone ed., São Paulo, p 355

Bezerra JFR, Guerra AJT, Rodrigues SC (2012a) Relações entre potencial matricial no solo e cobertura vegetal em uma estação experimental, Uberlândia – MG. Soc nat 24(1):103–114. https://doi.org/10.1590/S1982-45132012000100009

Bezerra JFR, Guerra AJT, Rodrigues SC (2012b) Variação da umidade superficial do solo em parcelas experimentais com uso de geotêxteis biodegradáveis, Uberlândia–MG. Rev Bras Geomorf 13(1):93–100. https://doi.org/10.20502/rbg.v13i1.346

Biulchi DF (2012) Uso de Leguminosas Arbóreas no Controle da Evolução de Voçoroca no Domínio do Cerrado. Ph.D. thesis, Instituto de Geografia – Universidade Federal de Uberlândia, Uberlândia, Brazil

Confessor JG, Machado DFT, Rodrigues SC (2016) Procedimento de revegetação irrigada por carneiro hidráulico em área degradada por voçorocamento. Rev Sapiênc Soc Sab Prat Educ 5(2):112–126

Ferri MG (1969) Plantas do Brasil – Espécies do Cerrado. Edgard Blücher Ed. Ltda, São Paulo, p 239

Machado DFT, Confessor JG, Rodrigues SC (2014) Initial process of degraded area recovery. Through physical interventions and utilization of Leguminous spec. Cad Geo 24(1):42–54. https://doi.org/10.5752/P.2318-2962.2014v24nespp42

Oliveira PTS, Nearing MA, Wendland E (2015) Orders of magnitude increase in soil erosion associated with land use change from native to cultivated vegetation in a Brazilian savannah environment. Earth Surf Process Landforms 40:1524–1532. https://doi.org/10.1002/esp.3738

Pereira Júnior RA (2013) Uso de barreiras físicas em voçoroca e seus efeitos na retenção de sedimentos: um estudo de caso. Master Thesis. Instituto de Geografia, Universidade Federal de Uberlândia, Uberlândia, Brazil

Rawitscher F (1948) The water economy of the vegetation of the "campos cerrados" in southern Brazil. J Ecol 36:237–268

Rodrigues SC (2002) Impacts of human activity on landscapes in Central Brazil: a case study in the Araguari watershed. Aust Geogr Stud 40:167–178. https://doi.org/10.1111/1467-8470.00172

Rodrigues SC (2014) Degradação dos Solos no Cerrado. In: Guerra JTA, Jorge MCO (eds) Degradação dos Solos no Brasil. Bertrand Brasil, Rio de Janeiro, pp 51–85

Salgado-Labouriau ML (1961) Pollen grains of plants of the "Cerrado" I. An Acad Brasil Cienc 33(1):119–130

Serato DS, Rodrigues SC (2010) Avaliação e recuperação da área degradada (voçoroca) no interior da Fazenda Experimental do Glória no município de Uberlândia (MG). Bol Goiano Geogr 30(2):29–42. https://doi.org/10.5216/bgg.v30i2.13788

Silva AH (2010) Medidas físicas e biológicas com potencial para uso em recuperação de voçoroca no município de Uberlândia – MG. Master thesis, Instituto de Geografia, Universidade Federal de Uberlândia, Uberlândia, Brazil

Vrieling A, Rodrigues SC, Bartholomeus H, Sterk, G. (2007) Automatic identification of erosion gullies with ASTER imagery in the Brazilian Cerrados. Intern J Rem Sens 28(12):2723–2728. https://doi.org/10.1080/01431160600857469

# Gully Erosion and Land Degradation in Brazil: A Case Study from São Luís Municipality, Maranhão State

8

Antonio Jose Teixeira Guerra, Michael Augustine Fullen, José Fernando Rodrigues Bezerra, and Maria do Carmo Oliveira Jorge

**Abstract**

Soil erosion and land degradation are global problems and pose major problems in Brazil, which is characterized by a great diversity of soil types. This chapter discusses gully erosion and land degradation in Brazil, outlining a case study from São Luis City, Maranhão State. This diversity is related to a wide variety of landforms, climates, parent materials, vegetation cover and biota and is influenced by land use and management. Therefore, a brief description of different types of land degradation, mainly gully erosion and mass movements, is presented, and illustrative photographs are presented. These hazards affect both urban and rural areas within the extensive national territory (8,547,403 km$^2$). Consequently, these problems have serious environmental impacts and pose socio-economic challenges.

## 8.1 Introduction

Soil erosion and land degradation are global problems and pose major problems in Brazil. The hazard affects both urban and rural areas within the extensive national territory (8,547,403 km$^2$). In turn, these problems have serious environmental impacts and pose socio-economic challenges. It is important that soils are conserved for the present and future generations. Although erosion is a natural phenomenon,

A. J. T. Guerra (✉) · Maria do Carmo Oliveira Jorge
Department of Geography, Federal University of Rio de Janeiro, Rio de Janeiro, Brazil

M. A. Fullen
Faculty of Science and Engineering, University of Wolverhampton, Wolverhampton, UK

J. F. R. Bezerra
Department of Geography, State University of Maranhão, Maranhão, Brazil





J. C. Dagar, A. K. Singh (eds.), *Ravine Lands: Greening for Livelihood and Environmental Security*, https://doi.org/10.1007/978-981-10-8043-2_8

often human activity accelerates erosion processes throughout Brazil (Guerra et al. 2014). Gully erosion has been recognized throughout history as a major land degradation process and, in many cases, has been directly linked to unsustainable land management (Castillo and Gómes 2016). Therefore, since the early twentieth century, an increasing number of publications have described gully erosion in terms of its importance, contributory factors, associated geomorphological processes and its detrimental effects (Poesen 2011; Castillo and Gómes 2016). Gullies are present in most countries, and they have specific features in each of them, depending on several factors. Nevertheless, each country has its own word to characterize this erosion landform. Consequently, *a gully* is known as a *cárcava* in Spain, *ravine* in France, *lavaka* in Madagascar, *wadi* in Arabic, *donga* in South Africa, *voçoroca* in Brazil and *barranco* in Argentina (Castillo and Gómes 2016).

Soil erosion on US agricultural soils causes the loss of an average of 30 Mg $ha^{-1}$ $year^{-1}$; some eight times greater than rates of soil formation. A survey suggested the situation in Brazil is often worse, reaching 60 Mg $ha^{-1}$ $year^{-1}$ in southeastern Brazil (EMBRAPA -Brazilian Agricultural Research Corporation 2002). Particularly serious soil erosion is associated with six regions. These are northwestern *Paraná* State; the Central Plateau, in the Central Western Region; Western *Sao Paulo State*; the *Paraíba do Sul* middle drainage basin, in *Rio de Janeiro* State; and *Campanha Gaúcha* in *Rio Grande do Sul* State and *Triângulo Mineiro*, in western Minas Gerais State (EMBRAPA 2002; Guerra et al. 2014).

As per Goudie and Boardman (2010), the major areas of intense erosion are associated with both human and natural factors. Boardman (2006) suggested the following countries and regions are global erosion 'hotspots': The Loess Plateau of China, Ethiopia, Swaziland and Lesotho, the Andes, South and East Asia, the Mediterranean basin, Iceland, Madagascar, the Himalayas, the Sahel of West Africa, the Caribbean and Central America. We propose Brazil is also an erosion 'hotspot' (Guerra et al. 2014, 2017).

Brazil is characterized by a great diversity of soil types, corresponding to a wide variety of landforms, climates, parent materials, vegetation cover and biota. Diverse pedogenic processes thus facilitate the pedogenesis of multiple soil types. Related to this great diversity, Brazil has potentialities and limitations in terms of land use. This diversity is enhanced by regional differences in terms of settlement, land use and development, which may cause different erosion processes, including gully erosion, being responsible for land degradation (EMBRAPA 2002; Guerra et al. 2014, 2017).

Selby (1993) classified soil erosion as a geomorphological process which occurs on hillslopes, carried out by flowing water and splash processes. Selby termed this "erosion on hillslopes by raindrops and flowing water" and outlined the role of water in removing and transporting sediments, which he described as *wash*, a term adopted by many authors (Gerrard 1992; Evans 1993; Goudie and Viles 1997; Poesen et al. 2006; Goudie and Boardman 2010; Guerra et al. 2014, 2015). Although both soil erosion and mass movements are two forms of land degradation, and humans play important roles in these geomorphological processes, they present different modes of occurrence and diverse morphological features and, consequently, different ways of being identified and monitored (Varnes 1978; Small and Clark 1982; Abrahams 1986; Brunsden 1988; Gerrard 1992; Evans 1993; Selby 1993;

Guerra 1994; Goudie and Viles 1997; Favis-Mortlock and Guerra 1999; Fullen and Catt 2004; Crozier and Glade 2005; Morgan 2005; Van Beek et al. 2008; VanWesten et al. 2008; Goudie and Boardman 2010; Vente et al. 2011; Boardman and Favis-Mortlock 2013; Kanungo and Sharmas 2014; Oluwagbenga and Orimoogunje 2014; Guerra et al. 2015; Monsieurs et al. 2015; Vanmaercke et al. 2016). Nevertheless, the best way to avoid both forms of land degradation is acting preventively, which means understanding the risks of soil erosion and/or mass movements, to avoid them (Guerra et al. 2017). This chapter discusses gully erosion and land degradation in Brazil, outlining a case study from São Luis City, Maranhão State.

## 8.2 Gully Erosion and Land Degradation in Brazil

Analysis of land degradation in Brazil shows that gully erosion and mass movements are major geomorphological processes, influenced by varied climates, geomorphology, pedogenic processes and land use and management. Photographs obtained from several Brazilian experts help to illustrate these features. Related to this great diversity, Brazil has potentialities and limitations, in terms of land use and can be divided into the following five regions:

1. The Northern (or Amazon) Region, which is characterized by plains and low plateaux, a tropical climate, perennial high temperatures, high humidity, deep and highly weathered soils with high acidity and low natural fertility. These soils have low production potential, especially when they are poorly managed.
2. The North-Eastern Region has diverse climates, which vary from humid and warm, near the coast, to semiarid (warm and dry) in the interior. There is a semi-humid transition zone between these two areas. Generally, soils have medium to high natural fertility and are usually shallow, due to relatively low weathering rates (EMBRAPA 2002; Guerra et al. 2014).
3. The Central West Region mainly consists of a vast plateau, called the Brazilian Central Plateau. The Plateau has formed due to prolonged erosive processes over geological time. Regional characteristics include the tropical climate, with two distinct seasons, one humid and the other dry (both of approximately 6 months per year), and extensive areas with deep, well-drained, acid soils of low natural fertility. However, these soils have potential and can be improved with applications of lime and other organic and chemical fertilizers. Once these treatments have been conducted, the region generally has a favourable relief for mechanized agriculture.
4. The South-Eastern Region is characterized by plateaux and mountain ranges, which reach ≤2000 m altitude. The climate is tropical, with hot summers in the lowlands and more temperate conditions on the mountain ranges and plateaux. Soils are generally deep, well developed and have low natural fertility (EMBRAPA 2002).
5. The Southern Region has many soils developed on varied parent materials, including basic igneous, metamorphic and a diverse range of sedimentary rocks. Thus, landscapes are very diverse. The climate is sub-tropical with very well-defined seasons. The predominantly fertile soils have high potential for agro-pastoral use (EMBRAPA 2002; Guerra et al. 2014).

**Fig. 8.1** Gully erosion in Ipameri Municipality, Goiás State, on pasture on a sandy loam soil (Photo by Silvio Carlos Rodrigues)

Land degradation occurs throughout Brazil, being more dramatic in some regions. Therefore, a general description of gully erosion and land degradation is outlined here, for different States and regions within Brazil. In the *Cerrado* (Central Western Region), the settlement process has specific characteristics, which have caused different management issues (Rodrigues 2014). Within the *Cerrado* land degradation mainly takes the form of soil erosion (Figs. 8.1 and 8.2), desertification, loss of soil productive capacity, and modification of the low-order fluvial systems.

Land degradation has been active in Rio Grande do Sul State for many years. Here land degradation has specific characteristics, especially in terms of its historical context, where land appropriation without adopting suitable soil conservation practices can cause gully erosion (Figs. 8.3 and 8.4) (Verdum et al. 2014). The State has diverse geological formations, climates and vegetation types, which contribute to soil diversity. Vegetation clearance, slash-and-burn agriculture and intensive use of heavy agricultural machinery are the main causes of soil erosion in the State (Verdum et al. 2014).

Land degradation on the northern coast of São Paulo State is due to land use and vegetation clearance throughout its economic history, and these have caused gully erosion (Jorge 2014) (Figs. 8.5 and 8.6). Some municipalities currently face serious problems in terms of economic and environmental crises. The natural environmental system is very fragile; thus, settlement restrictions are essential, especially on steep slopes (Jorge 2014).

Soil erosion in semiarid Brazil has been surveyed by several researchers, mainly Correa et al. (2014), who discussed relationships between soil classes and land

**Fig. 8.2** Gully erosion in Uberlândia Municipality, Minas Gerais State, on pasture on a sandy loam soil. Pipes are evident on the gully wall (Photo by Silvio Carlos Rodrigues)

**Fig. 8.3** Rill and gully erosion on Quaternary sediments in São Francisco de Assis Municipality, Rio Grande do Sul State (Photo by Roberto Verdum)

**Fig. 8.4** Gully erosion on Quaternary sediments, on a sandy loam soil, due to >100 mm of rain in 24 h. São Francisco de Assis Municipality, Rio Grande do Sul State (Photo by Roberto Verdum)

degradation (Figs. 8.7 and 8.8). For their assessment, they recognized two broad soil types: those related to semiarid conditions (i.e. entisols, vertisols, ultisols and aridisols) and soils unrelated to current semiarid conditions (oxisols and alfisols). Inceptisols are present on slopes, and, when vegetation clearance is accompanied by land mismanagement, rill and gully erosion occurs (Correa et al. 2014).

Water erosion by rills and gullies is a major problem in the fragile environment of Amazonia (Albuquerque and Vieira 2014) (Figs. 8.9 and 8.10). There is a need for improved land use and management practices aimed at the maintenance of hydro-geomorphic mechanisms through human interventions. These include talus cuts for house and road construction; sand, clay and stone quarries for civil construction; and the need for new sites to expand urban areas.

Rill and gully erosion has long been a problem in Paraná State (Figs. 8.11 and 8.12). Soil erosion occurs mainly in north-western Paraná State, associated with natural landscape characteristics, soil properties and high rainfall index values (Santos and Westphalen 2014). Severe soil erosion is also related to the settlement in this part of the State, which started in 1930 with the establishment of coffee plantations. Vegetation clearance for plantations and urban growth have contributed to increased surface runoff and gully erosion.

Soil degradation in Rio de Janeiro State is related to soil loss, mainly due to soil erosion and mass movements (Soares da Silva and Botelho 2014). In turn, this causes river siltation, which is one of the main problems of the rivers in the State. Gully erosion is evident in several parts of Rio de Janeiro State, mainly due to agricultural activities and cattle ranges without conservation practices. These can lead

**Fig. 8.5** Gully erosion on a sandy loam soil in Ubatuba Municipality in São Paulo State (Photo by Maria do Carmo Oliveira Jorge)

to severe land degradation in several rural areas which are naturally fragile and have low soil fertility (Figs. 8.13 and 8.14). Urban soils, especially in mountainous areas, are prone to mass movements, causing the death of thousands of people over the three last decades (Figs. 8.15, 8.16, and 8.17) (Graeff et al. 2012). This is particularly a problem in Petrópolis Municipality.

## 8.3 Gully Erosion in São Luís Municipality

Socio-environmental problems have intensified due to accelerated urbanization without regard to environmental constraints. Therefore, erosion processes become more intense in urban areas in Brazil, mainly where soils are friable and unconsolidated. With this intensification, there is a need to produce research publications

**Fig. 8.6** Rill erosion on a compacted trail in Ubatuba Municipality. The high soil bulk density values (typically ~1.6 g $cm^{-3}$) are caused by trampling by the many tourists who use the trail to access the beaches (Photo by Maria do Carmo Oliveira Jorge)

(articles, book chapters, books, MSc dissertations and PhD theses) to analyse the factors controlling erosion processes and diminish and ultimately prevent their catastrophic effects (Mendonça 2006; Bezerra and Guerra 2007; Bezerra et al. 2009; Bezerra 2011; Guerra et al. 2014).

São Luís (2°19′09″–2°51′00″S, 44°01′16′–44°19′37″W) covers 831.7 $km^2$ of central-western Maranhão Island and has a total population of 1,011,943 (IBGE 2010) (Fig. 8.18). There are erosion features throughout the City, and gully erosion is especially problematic within the Bacanga drainage basin in central north-western Maranhão Island. Severe gully erosion poses serious risks to the people who live around these gullies. These include the gullies of Coeduc, Batatã, Gapara, Itaqui, Maracanã, Posto, Sacavém, Torre and Vila Maranhão.

The study area has conditions conducive to soil erosion, with an annual rainfall average over the past 30 years of >2000 mm, with a well-defined rainy and dry season. Most intense rains fall in summer (January–March), with daily amounts

**Fig. 8.7** Gully monitoring on calcic soils, Sierra da Capivara, south-east Piauí State (Photo by the 'Northeast of Brazil Quaternary Study Group')

>180 mm on occasions. The intense convectional storms on the northern part of Maranhão State have strong influences on all Maranhão Island. These rains influence the onset and evolution of gully erosion in both time and space.

Soils are erodible, being formed by the weathering of *Itapecuru* and *Barreiras* Formation, mainly formed of sandstones, clay deposits and unconsolidated sediments. Soils within São Luis Municipality are mainly alfisols, followed by entisols, which are very erodible because they usually have a sandy/silty texture and are friable and unconsolidated (Bezerra 2011).

Within the nine monitored gullies, 2.78% of the area of soils is clayey silt, 5.55% are sands, 8.33% are sandy loams, 25.00% silt and 55.55% silt loams (texture size ranges: sand (2000–53 μm), silt (53–2 μm) clay (<2 μm)). These soils are eroded by water during the rainy season and by wind during the dry season. Considering that >50% of the gully soils belong to the silt loam class, this explains the rapid retreat of gully heads during the rainy season (Bezerra and Guerra 2007; Bezerra 2011). In some cases, rain-splashed silts and clays seal topsoils causing soil sealing, creating crusts and decreasing infiltration rates, thus accelerating runoff erosion, especially on surfaces with no or sparse vegetation cover. Locally, the relief is formed by tabular and sub-tabular plateaux, with dissected hillslopes (Feitosa 1996; Maranhão 1998; Bezerra 2011). Human activities are accelerating erosion processes (Figs. 8.19 and

**Fig. 8.8** Gullies (actually rills) on Pleistocene colluvium under low-income houses, Pilar Municipality, Paraíba State (Photo by the 'Northeast of Brazil Quaternary Study Group')

**Fig. 8.9** Severe gully erosion on sandy loam soils in Itacoatiara Municipality, Amazonas State (Photo by Antonio Fábio Sabbá Guimarães Vieira)

**Fig. 8.10** Severe gully erosion on an archaeological site in Nova Cidade District, Manaus Municipality, Amazonas State (Photo by Fábio Sabbá Guimarães Vieira)

**Fig. 8.11** Gully erosion in Paraná City, Paraná State. The soil texture is sandy loam, and the main land use is unplanned urban residential development (Photo by Leonardo José Cordeiro Santos)

**Fig. 8.12** Gully erosion in Loanda Municipality, Paraná State. The soil texture is sandy loam and it is an urban residential area (Photo by Leonardo José Cordeiro Santos)

**Fig. 8.13** Gully erosion on pasture on an oxisol in Silva Jardim Municipality, Rio de Janeiro State (Photo by Antonio Soares da Silva)

**Fig. 8.14** Gully erosion on an abandoned mining area in Santo Antonio de Pádua Municipality, Rio de Janeiro State (Photo by Antonio Soares da Silva)

8.20). This is mainly due to poor soil management, vegetation clearance, disorganized urbanization and construction quarries extracting laterite, fine sand, silt and clay.

The settlement process of Maranhão Island, since 1612 to the present, has undergone evolutionary phases. A very important phase is related to an industrial epoch in the second half of the eighteenth century, when many factories located in São Luís City. Another phase of rapid growth was between 1950 and 1960, due to the migration of families from the central part of the North-Eastern Region, because of the great droughts of the 1950s. A third phase of urban growth occurred between 1970 and 1990, due to industrial activities (Cordeiro et al. 2013; Ferreira and Santos 2015). There are multiple factors causing soil erosion, including heavy rainstorms, friable soils, wind and inadequate soil management. Most erosion features occur on built areas, especially those with disorganized settlement.

Sacavém and Salina gullies are located on the central-western part of the study area in Sacavém District (Figs. 8.21 and 8.22). These features are situated on the north bank of the River *Bicas*, within the Bacanga drainage basin. The area shows some outcrops of the Barreiras Formation, with steep slopes and sparse vegetation cover. The main land use is urban and quarry exploitation (mainly sand for house construction), which is conducted by many people, including residents. Field monitoring has shown that the gully heads are endangering the stability of the

**Fig. 8.15** Gully erosion in a rural area in Petrópolis Municipality, Rio de Janeiro State (Photo by Antonio Jose Teixeira Guerra)

**Fig. 8.16** Mass movement scar in Petrópolis Municipality (January 2001). Fifty people died due to this catastrophe (Photo by Antonio Jose Teixeira Guerra)

**Fig. 8.17** Landslide scar beneath a condemned house in the Quitandinha District of Petrópolis Municipality, Rio de Janeiro State (Photo by Antonio Jose Teixeira Guerra)

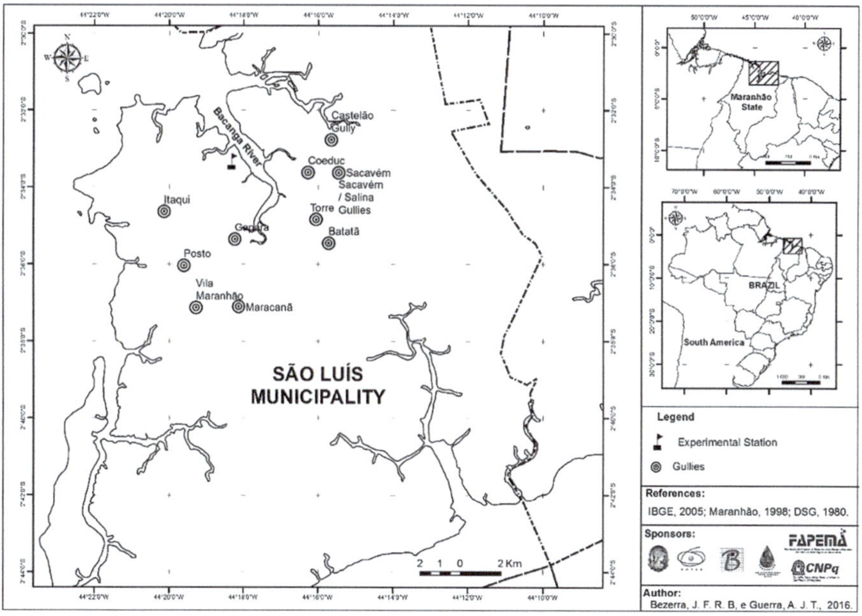


**Fig. 8.18** Maranhão Island, São Luis City, showing the location of the gullies and experimental station

**Fig. 8.19** Gully wall in Bacanga drainage basin, within the urban area of São Luís Municipality. The soil texture is silt loam (Photo by Jose Fernando Rodrigues Bezerra)

**Fig. 8.20** Araçagi gully, near the beach in São Luis City (Photo by Jose Fernando Rodrigues Bezerra)

**Fig. 8.21** Gully wall in Sacavém being recuperated with the application of Buriti palm-leaf geotextiles, together with grass seeds and NPK fertilizers. The association of geomorphology with other areas of knowledge, such as bioengineering, may offer tools to decrease soil erosion in São Luís and comparable areas (Photo by José Fernando Rodrigues Bezerra)

ELETRONORTE (Northern Brazil Central Electric Company) electricity pylon and surrounding houses.

Urban areas are usually highly disorganized, including absent or inadequate sewage and drainage systems, unpaved roads and sparse vegetation cover, which in combination can promote severe soil erosion. This class corresponds to the area around Salina and Sacavém gullies, which causes potential environmental risks to local people. On the intensely urbanized areas, impermeable soils impede water infiltration, which contributes erosive runoff, especially on the areas with little or no vegetation cover. Another economic land use is quarries. In the Sacavém gullies, this class is represented by mineral quarries (mainly sand). Constant quarrying causes the collapse of adjacent gully walls, accelerating both gully evolution and risks to residents (Fig. 8.23). Both land use classes are intimately related to the occurrence and acceleration of gully erosion. Gully monitoring shows most gully head retreat at *Castelão* (Fig. 8.24), *Sacavém* and *Salina* (Fig. 8.25).

**Fig. 8.22** The sand quarry in Sacavém gully accelerates gully evolution (Photo by Jose Fernando Rodrigues Bezerra)

**Fig. 8.23** Gully evolution on sandy loam soils and sediments in Bacanga drainage basin in São Luis City (Photo by Jose Fernando Rodrigues Bezerra)

**Fig. 8.24** Gully erosion in Castelão District. The main soil texture is silt loam, and the main land use is for leisure (walking, cycling and football), which are causing severe soil erosion (Photo by Jose Fernando Rodrigues Bezerra)

## 8.4 Conclusions

Soil erosion is the product of complex interactions between rainfall regime, soil properties, slope characteristics, vegetation cover and land use and management. Their interaction often produces excessive erosion rates in Brazil. Understanding soil erosion as a geomorphological process is an essential step towards developing effective soil conservation strategies. There is a close association between soil classes in Brazil and soil erodibility. Field measurements suggest erosion rates often far exceed tolerable levels and thus impair the ability of soil systems to sustainably produce crops.

This chapter has discussed processes of gully erosion and land degradation in Brazil, particularly using São Luis City (Maranhão State) as a case study. The hazards affect both urban and rural areas within the extensive national territory (8,547,403 $km^2$). Consequently, these problems have serious environmental impacts and pose socio-economic challenges. It is important that the soils are conserved, for the present and future generations. Quantitative data on measured soil erosion rates are sparse in Brazil. However, the few available studies show total soil loss is often >50 Mg $ha^{-1}$ $year^{-1}$ and can be >100 Mg $ha^{-1}$ $year^{-1}$. Thus, Brazil can be considered

**Fig. 8.25** Sacavém gully, where the main soil texture is sandy loam. This gully experiences severe erosion and deposition, due to sand quarrying activities (Photo by Jose Fernando Rodrigues Bezerra)

one of the global erosion 'hotspots', and more research is needed to better understand this environmental problem and consequently take remedial measures.

In São Luis Municipality, gully erosion is enhanced by the action of intense rains, acting in combination with vegetation clearance and the erodible texture of the sandy and sandy loam soils. Intense human action, with inadequate soil use and irregular settlement, also plays very important roles in gully inception and evolution.

## References

Abrahams AD (1986) Hillslope processes. Allen and Unwin, London

Albuquerque ARC, Vieira AFSG (2014) Erosão dos solos na Amazônia. In: Guerra AJT, Jorge MCO (eds) Degradação dos Solos no Brasil. Bertrand Brasil Press, Rio de Janeiro, pp 223–259

Beek V, Cammeraat E, Andreu V, Mickovski SB, Dorren (2008) Hillslope processes: mass wasting, slope stability and erosion. In: Norris JE, Stokes A, Mickovski SB, Cammeraat E, van Beek R, Nicoll BC, Achim A (eds) Slope stability and erosion control: ecotechnological solutions. Slope stability and erosion control. Springer, Amsterdam, pp 17–64

Bezerra JFR (2011) Geomorfologia e Reabilitação de Áreas Degradadas por Erosão com Técnicas de Bioengenharia de Solos na Bacia do Rio Bacanga, São Luís – MA. Tese de Doutorado. Rio de Janeiro: Universidade Federal do Rio de Janeiro, Programa de Pós- Graduação em Geografia

Bezerra JFR, Guerra AJT (2007) Geomorfologia e reabilitação de áreas degradadas por erosão com técnicas de bioengenharia. XII Simpósio Brasileiro de Geografia Física Aplicada. Natal. 09–13 julho 2007

Bezerra JFR, Guerra AJT, Lima LDM, Mendonça JKS (2009) Geomorfologia aplicada à reabilitação de áreas degradadas por voçorocamento utilizando geotêxteis produzidos com fibra do buriti na área urbana de São Luís – MA. 8 Simpósio Nacional de Controle de Erosão dos Solos, São Paulo. 22 março-2 abril 2009

Boardman J (2006) Soil erosion science: reflections on the limitations of current approaches. Catena 68:73–86

Boardman J, Favis-Mortlock DT (2013) The significance of drilling date and crop cover with reference to soil erosion by water, with implications for mitigating erosion on agricultural land in South East England. Soil Use Manag 29:1–8

Brunsden D (1988) Slope instability, planning and geomorphology in the United Kingdom. In: Hooke JM (ed) Geomorphology in environmental planning. Wiley, Chichester, pp 105–119

Castillo C, Gómes JA (2016) A century of gully erosion research: urgency, complexity and study approaches. Earth Sci Rev 160:300–319

Cordeiro EC, Burnett CFL, Ferreira AJA (2013) A efetivação da Grande São Luís: do que se trata e como foi tratada. In: dos Santos PC, de Mattos Júnior JS (eds) Desenvolvimento sócio espacial: novos olhares, vol 1, 1st edn. EDUEMA, São Luís, pp 111–133

Correa ACB, Souza JOP, Cavalcanti LCS (2014) Solos do ambiente semiárido brasileiro: erosão e degradação a partir de uma perspectiva geomorfológica. In: Guerra AJT, Jorge MCO (eds) Degradação dos Solos no Brasil. Bertrand Brasil Press, Rio de Janeiro, pp 127–169

Crozier MJ, Glade T (2005) Landslides hazard and risk: issues, concepts and approach. In: Glade T, Anderson M, Crozier M (eds) Landslide hazard and risk. Wiley, New York, pp 1–41

EMBRAPA (2002) Uso agrícola dos solos brasileiros. Embrapa Solos, Rio de Janeiro

Evans R (1993) Sensitivity of the British landscape to erosion. In: Thomas DSG, Allison RJ (eds) Landscape sensitivity. Wiley, New York, pp 189–210

Favis-Mortlock DT, Guerra AJT (1999) The implications of general circulation model estimates of rainfall for future erosion: a case study from Brazil. Catena 37:329–354

Feitosa AC (1996) Dinâmica dos processos geomorfológicos da área costeira a nordeste da ilha do Maranhão. Tese de Doutorado. Rio Claro: IGCE – Cp – UNESP

Ferreira AA, Santo SLCA (2015) Formação Socioambiental do Estado do Maranhão. In: Lígia T, Simonian L, Baptista ER (eds) Formação Socioambiental da Amazônia, vol 1, 1st edn. NAEA., 2015, Belém, pp 249–317

Fullen MA, Catt JA (2004) Soil management: problems and solutions. Edward Arnold, London

Gerrard J (1992) Soil geomorphology: an integration of pedology and geomorphology. Chapman and Hall, London

Goudie AS, Boardman J (2010) Soil erosion. In: Alcántara-Ayala I, Goudie A (eds) Geomorphological hazards and disaster prevention. Cambridge University Press, Cambridge, pp 177–188

Goudie A, Viles H (1997) The earth transformed: an introduction to human impacts on the environment. Blackwell, Oxford

Graeff O, Guerra AJT, Jorge MCO (2012) Catastrophic floods and landslides in Brazil. Geogr Rev 26:38–41

Guerra AJT (1994) The effect of organic matter content on soil in simulated rainfall experiments in West Sussex, UK. Soil Use Manag 10:60–64

Guerra AJT, Fullen MA, Jorge MCO, Alexandre ST (2014) Soil erosion and conservation in Brazil. Anu Inst Geocências – UFRJ 37:81–91

Guerra AJT, Bezerra JFR, Fullen MA, Mendonça JKS, Jorge MCO (2015) The effects of biological geotextiles on gully stabilization in São Luís, Brazil. Nat Hazards 75:2625–2636

Guerra AJT, Fullen MA, Jorge MCO, Bezzera JFR, Shokr SM (2017) Slope processes, mass movements and soil erosion: a review. Pedosphere 27(1):27–41

IBGE (2010) Censo Demográfico do Instituto Brasileiro de Geografia e Estatística, Rio de Janeiro

Jorge MCO (2014) Degradação dos Solos no Litoral Norte Paulista. In: Guerra AJT, Jorge MCO (eds) Degradação dos Solos no Brasil. Bertrand Brasil Press, Rio de Janeiro, pp 171–221

Kanungo DP, Sharmas S (2014) Rainfall thresholds for prediction of shallow landslides around Chamoli-Joshimath Region, Garhwal Himalayas, India. Landslides 11:629–638

Maranhão (1998) Secretaria de Estado do Meio Ambiente e Recursos Hídricos. Diagnóstico ambiental da microrregião da aglomeração urbana de São Luís e dos Municípios de Alcântara, Bacabeira e Rosário. São Luís

Mendonça JKS (2006) Uso sustentável de espécies de palmeiras da APA da Baixada Maranhense para controle e recuperação de áreas degradadas por erosão. Dissertação de Mestrado. São Luís: Universidade Federal do Maranhão, Departamento de Oceanografia e Limnologia, Programa de Pós-Graduação em Sustentabilidade de Ecossistemas. São Luís

Monsieurs E, Dessie M, Adgo E, Poesen J, Deckers J, Verhoest N, Nyssen J (2015) Seasonal surface drainage of sloping farmland: a review of its hydro-geomorphic impacts. Land Degrad Dev 26:35–44

Morgan RPC (2005) Soil erosion and conservation. Blackwell, Oxford

Oluwagbenga O, Orimoogunje I (2014) Forest cover change and land use dynamics in Oluwa forest reserve, southwestern Nigeria. J Landsc Ecol 7:25–44

Poesen J (2011) Challenges in gully erosion research. Landform Anal 17:5–9

Poesen J, Vanwalleghem T, De Vente J, Knapen A, Verstraeten G, Martınez-Casasnovas JA (2006) Gully erosion in Europe. In: Boardman J, Poesen J (eds) Soil erosion in Europe. Wiley, Chichester, pp 515–536

Rodrigues SC (2014) Degradação dos solos no Cerrado. In: Guerra AJT, Jorge MCO (eds) Degradação dos Solos no Brasil. Bertrand Brasil Press, Rio de Janeiro, pp 51–85

Santos LJC, Westphalen LA (2014) Erosão dos solos no noroeste do Paraná. In: Guerra AJT, Jorge MCO (eds) Degradação dos Solos no Brasil. Bertrand Brasil Press, Rio de Janeiro, pp 293–317

Selby MJ (1993) Hillslope materials and processes, 2nd edn. Oxford University Press, Oxford

Small RJ, Clark MJ (1982) Slopes and weathering. Cambridge University Press, Cambridge

Soares da Silva A, Botelho RGM (2014) Degradação dos solos no estado do Rio de Janeiro. In: Guerra AJT, Jorge MCO (eds) Degradação dos Solos no Brasil. Brazil, pp 261–292

Van Westen CJ, Castellanos Abella EA, Sekhar LK (2008) Spatial data for landslide susceptibility, hazards and vulnerability assessment: an overview. Eng Geol 102:112–131

Vanmaercke M, Poesen J, Mele BV, Demuzere M, Bruynseels A, Golosov V, Bezerra JFR, Bolysov S, Dvinskih A, Frankl A, Fuseina Y, Guerra AJT, Haregeweyn N, Ionita I, Imwangana FM, Moeyersons J, Moshe I, Saman AN, Niacsu L, Nyssen J, Otsuki Y, Radoane M, Rysin I, Ryzhov YV, Yermolaev O (2016) How fast do gully headcuts retreat? Earth Sci Rev 154:336–355

Varnes DJ (1978) Slope movements: types and processes. In: Schuster RL, Krizek RJ (eds) Landslide analysis and control. Transportation Research Board Special Report, vol 176. National Academy of Sciences, Washington, DC, pp 11–33

Vente J, Verduyn R, Verstraeten G, Vanmaercke M, Poesen J (2011) Factors controlling sediment yield at the catchment scale in NW Mediterranean geoecosystems. J Soils Sediments 11:690–707

Verdum R, Streck E, Vieira LFS (2014) Degradação dos Solos no Rio Grande do Sul. In: Guerra AJT, Jorge MCO (eds) Degradação dos Solos no Brasil. Bertrand Brasil Press, Rio de Janeiro, pp 87–125

# Technological Interventions for Managing Ravine Lands for Livelihood and Environmental Security

9

A. K. Sikka, P. K. Mishra, R. K. Singh, B. Krishna Rao, and Adlul Islam

**Abstract**

Rehabilitation of degraded ravine lands is immensely important to conserve land and water resources for sustaining and improving production and productivity and to counter impacts of climate change. There is a need to effectively implement improved package of practices along with soil and water conservation measures in marginal lands around the ravines. It is estimated that these practices may result in 9–28% improvement in cropping intensity and 20–66% increase in current yield levels with an overall improvement of 118–280% increase in net returns through increased crop production. If the shallow ravine lands are brought under scientific cultivation of high-value as well as low-value fruit trees, it may augment the supply of fruits in the daily diet of people in India. These horticulture-based systems are cost-effective with high benefit-cost ratio. In deep ravines, silvopastoral systems are viable technology for stabilization and productive utilization. Ravine area development will not only contribute to bridging the growing demand and supply gap for food, fodder, and fuel but also help restore livelihood security for resource constraint communities and mitigate climate change impacts, besides providing improved ecosystems services. Apart from

A. K. Sikka (✉)
IWMI – India Representative (Formerly DDG (NRM)), International Water Management Institute, New Delhi, India
e-mail: A.Sikka@cgiar.org

P. K. Mishra
Indian Institute of Soil & Water Conservation, Dehradun, Uttrakhand, India

R. K. Singh
Indian Institute of Soil & Water Conservation, Research Centre, Kota, Rajasthan, India

B. Krishna Rao
Indian Institute of Soil & Water Conservation, Research Centre, Vasad, Gujarat, India

A. Islam
Natural Resource Management Division, ICAR, New Delhi, India



J. C. Dagar, A. K. Singh (eds.), *Ravine Lands: Greening for Livelihood and Environmental Security*, https://doi.org/10.1007/978-981-10-8043-2_9

scientific and technological interventions for reclamation and productive use of ravine land, the support and involvement of the local farmers and local community in planning, project formulation, and implementation of different government programs and reform of local land tenure and social arrangements are equally important for the success of ravine reclamation programs.

## 9.1 Introduction

Soil and water are the two vital resources for agricultural development and sustaining life on the earth. About 52% of the Indian population depends on agriculture for their livelihood. India presently supports 18% of the world's human and 15% livestock population with only 2.4% of the land mass and 4.2% fresh water resource. With increase in population from 361 million in 1951 to 1.21 billion in 2011, there is tremendous demographic pressure on finite land and water resources. This has resulted in sharp decrease in per capita availability of agriculture land in India from 0.48 ha in 1951 to 0.13 ha in 2011 and projected to decrease to 0.08 ha in 2035. Similarly, per capita annual water availability has declined from 5177 $m^3$ in 1951 to 1614 $m^3$ in 2011 and further projected to be around 1154 $m^3$ by 2050 (Jain 2011).

Land is a valuable and fixed natural resource base and its proper use is essential for ecological balance. Inadequate use of natural resources aggravates the problem of land degradation (Bhushan and Yadav 2000). The problem is severe in the alluvial plains of the semiarid region and most serious in subsistence-oriented agriculture system in the developing countries (Yadav and Bhushan 2002). Land degradation is one of the most serious global challenges to sustainability of agriculture. As per the latest estimates (NAAS 2010) based on harmonized database, around 120.4 million ha (36.5% of the total geographical area) of the country is affected by various kinds of land degradation comprising of water erosion (82.6 million ha), wind erosion (12.0 million ha), chemical degradation (24.8 million ha), and physical degradation (1.0 million ha). Out of total degraded area, 104.2 million ha is arable land. Erosion-induced loss in crop production in rain-fed areas alone under major cereal, oilseed, and pulse crops has been estimated as 13.4 million Mg (~16%), which in economic terms is equivalent to ₹162.8 billion (Sharda et al. 2010). Besides, over 5.3 billion Mg of soil is lost annually through water erosion resulting in a loss of ~8 million Mg of plant nutrients (NPK).

The ravine lands are one of the most degraded and vulnerable ecosystems with severe on-site and off-site implications. A ravine system is a network of gullies developed by ephemeral streams with steep banks and a nearly vertical gully head and is deeper than 0.3 m (Fig. 9.1). Apart from land and soil quality degradation in gullied and inter-gullied areas and declined productivity for food, fuel, fodder, and biomass, these sites are major sediment-producing hot spots, hence increasing the risk for flooding and sedimentation. The subsistence farming practiced by resource constraint communities in ravine areas contributes to progressive degradation of these fragile ecosystems. Several natural and man-made factors have been

**Fig. 9.1** Gully head extending in tablelands

responsible for ravine formation. Erratic, short-duration, and high-intensity rainfall, erodible nature of soil, weak geology of alluvium, steep slopes and uneven terrain, faulty agricultural practices, illicit cutting of trees and bushes, and overgrazing are some of the factors responsible for formation of ravines. To arrest this negative trend and to bring gullied/ravinous land back to their productive capabilities, adoption of appropriate and innovative technologies for ravine land management needs to be planned on a scientific basis with participatory watershed management approach.

## 9.2 Extent of Ravine Land in India

Generally, ravine lands have developed in India along the major river systems on highly productive deep alluvial soils. The National Commission on Agriculture (GoI 1976) estimated 3.67 million hectares of ravine lands in India which constitute 1.12% of total geographical area of the country. Out of 3.67 million hectares of ravine lands in India, 2.36 million hectares (64%) are spread over in the states of Uttar Pradesh, Madhya Pradesh, Bihar, Rajasthan, and Gujarat (Dhruva Narayan1993). A rough estimate suggests that about 8000 ha are added to these ravines annually (Sharma et al. 1980). As per the estimate based on satellite data (NRSC 2011), 7412.03 $km^2$ area is under gullied and ravinous land in India, with three states of Rajasthan, Madhya Pradesh, and Uttar Pradesh accounting for about 56% of the total gullied and ravinous land (Table 9.1). A recent ravine area delineation effort by Indian Institute of Soil and Water Conservation (IISWC) indicated about 1.04 M ha in the ravine area of Gujarat (0.110 M ha), Rajasthan (0.274 M ha), Madhya Pradesh (0.312 M ha), and Uttar Pradesh (0.340 M ha). Under unmanaged conditions rate of ravine extension ranges from 0.6 to 1.0 m per year and is greater

**Table 9.1** State-wise distribution of gullied and/or ravinous area (km²) in India

| | 2005–2006 | | | 2008–2009 | | |
|---|---|---|---|---|---|---|
| | Gullied and/or ravinous land (medium[a]), km² | Gullied and/or ravinous land (deep/very deep ravine[b]), km² | Total gullied and/or ravinous land, km² | Gullied and/or ravinous land (medium) | Gullied and/or ravinous land (deep/very deep ravine), km² | Total gullied and/or ravinous land, km² |
| Andhra Pradesh | 405.48 | 2.89 | 408.37 | 381.81 | 3.00 | 384.81 |
| Bihar | 71.83 | 0.00 | 71.83 | 71.44 | 0.00 | 71.44 |
| Chhattisgarh | 142.90 | 0.00 | 142.90 | 74.68 | 15.50 | 90.18 |
| Delhi | 0.72 | 6.12 | 6.84 | 0.73 | 6.18 | 6.91 |
| Gujarat | 398.46 | 1.70 | 400.16 | 337.74 | 0.93 | 338.67 |
| Haryana | 0.00 | 0.96 | 0.96 | 1.22 | 0.96 | 2.18 |
| Himachal Pradesh | 170.23 | 4.52 | 174.75 | 167.79 | 14.37 | 182.16 |
| Jammu and Kashmir | 423.14 | 553.24 | 976.38 | 226.24 | 503.09 | 729.33 |
| Jharkhand | 106.14 | 0.00 | 106.14 | 191.47 | 0.00 | 191.47 |
| Karnataka | 127.11 | 0.00 | 127.11 | 67.97 | 0.00 | 67.97 |
| Madhya Pradesh | 1493.69 | 8.37 | 1502.06 | 1444.79 | 8.47 | 1453.26 |
| Maharashtra | 547.03 | 0.00 | 547.03 | 510.89 | 0.00 | 510.89 |
| Orissa | 671.19 | 0.00 | 671.19 | 431.29 | 0.00 | 431.29 |
| Punjab | 82.12 | 0.00 | 82.12 | 79.49 | 0.00 | 79.49 |
| Rajasthan | 1020.17 | 864.75 | 1884.92 | 1087.52 | 437.49 | 1525.01 |
| Tamil Nadu | 107.97 | 0.91 | 108.88 | 125.29 | 0.91 | 126.20 |
| Tripura | 0.00 | 0.00 | 0.00 | 0.60 | 0.00 | 0.60 |
| Uttar Pradesh | 1216.48 | 264.63 | 1481.11 | 923.99 | 274.59 | 1198.58 |
| West Bengal | 20.56 | 0.58 | 21.14 | 20.66 | 0.58 | 21.24 |
| Union Territory | 0.26 | 6.12 | 6.38 | 0.35 | 0.00 | 0.35 |
| **Total value** | **7005.48** | **1714.79** | **8720.27** | **6145.96** | **1266.07** | **7412.03** |

Source: NRSC (2011)
[a]Medium ravines: depth ranging from 2.5 to 5.0 m
[b]Deep ravines: depth of more than 5 m

on sandy soil than on the clay loam soils. The average annual loss of nutrients from these lands due to soil erosion has been estimated at 5.37–8.4 million tons (Mg) resulting into loss of production due to non-reclamation of ravines.

The Yamuna-Chambal ravine zone is the largest and most severely degraded area. The ravines flank the Yamuna river for nearly 250 km and in Agra and Etawah attain a depth of more than 80 m. Nearly 3,89,000 ha are affected along the Yamuna in southern Uttar Pradesh. The Chambal ravines flank the river Chambal in a 10 km wide belt, which extends southward from the Yamuna confluence to 480 km to the town of Kota in Rajasthan. Ravines also affect basins of several Chambal tributaries, viz., Mej, Morel, Kali Sindh, etc. Altogether, about 5000 km$^2$ area is affected. In Gujarat, ravine belt covers 5,00,000 ha and extends from the southern bank of the Tapti, banks of the Narmada, Watrak, Sabarmati, and Mahi basins. Besides these river basins, ravines are also found in Chota Nagpur, Mahanadi and upper Sone Valley, Indo-Gangetic plains, Siwaliks and Bhabar tract, and Western Himalayas even up to the Kashmir Valley (Dhruva Narayan 1993). The ravine lands warrant a comprehensive and long-term solution of the problem not only for resource sustainability but also for environmental remediation.

## 9.3 Classification of Ravines

The reclamation strategy for a ravine land largely depends on degree of terrain deformation and soil quality. Based on degree of terrain deformation and hazards involved, ravine lands are categorized into six reclaimability classes (Table 9.2). Shallow gullies up to 3.0 m depth are classified as class 1 and 2 lands and can be used for cultivation of seasonal crops with moderate land leveling works. Class 3

**Table 9.2** Gully reclaimability classes

| Class | Texture | Bed width (m) | Gully depth (m) | Side slope (%) | Climate |
|---|---|---|---|---|---|
| 1 | sicl, cl, l,<br>sl, sil, scl | >18/<6<br>($w_3$) ($w_1$) | <1.5<br>($gd_1$) | <5% ($S_1$) | Humid climate with well-distributed rainfall or having perennial irrigation |
| 2 | sicl, cl. sl,<br>sil, scl. | >6–18<br>($W_2$) | 1.5–3.0<br>($gd_2$) | 5–10% ($S_2$) | Humid climate with occasional dry spells or dry-cum-wet irrigation |
| 3 | sc, sic, c,ls | >18 ($W_3$) | 3–6 ($gd_3$) | 10–15%($S_3$) | Subhumid climate or situated at tail end of an irrigation system where water is occasionally not adequately available |
| 4 | Any texture excluding sand and with gravel | Any width | 6–9(gd4) | 10–15% ($S_3$) | Semiarid climate (without irrigation) with enough rainfall to sustain hardy horticultural plants |
| | Reclaimable classes decrease in priority for reclamation for agriculture from Class 1 to 3. Class 4 may be put under horticulture | | | | |
| 5 | Reclaimable classes 1–4 having the following hazards | | | | |
| | (a) Waterlogging/salinity has developed due to irrigation system | | | | |
| | (b) Back flow and flooding from a nearby stream or river. Hence, the gully need not be reclaimed for agriculture. It may be put under any suitable grass or tree spp. | | | | |
| 6 | Any texture including gravelly, etc. | Any width | >9 (gd 5) | 15% irregular (S4) | Semiarid with long dry spells and arid |
| | Gully humps may also be included in this class | | | | |

lands have deeper and narrower gullies and have more limitations for cultivation. Class 4 land has severe limitations for reclamation in the form of soil texture or gully dimensions. Steeper side slopes and deeper and narrower gully beds are the constraints for reclamation of these lands for cultivation. Such gullies may be put under horticulture crops. Class 5 and 6 lands are not suitable for high-value plantations or cultivation as these lands either are prone to seasonal backflows from a nearby river, have developed waterlogging and salinity due to irrigation, or have severe terrain deformation. These lands need to be retired for perennial plantations with suitable tree or grass species. Grass cutting for stall feeding or cut and carry system may be preferred over direct grazing, while fuel trees may be exploited on rotation basis.

Reclamation of ravine land needs location-specific technological packages after careful biophysical analysis following the above classification rating and shall be blended with locally preferred land use systems to accommodate need and preferences of the beneficiary households. A ravine reclamation and rehabilitation project need to be planned with the following broad objectives: to (1) ensure protection to tableland along the ravine areas by arresting the process of gully head extension, (2) reclaim degraded lands to optimum production levels and restore ecological balance through synergistic harmonization of community needs and production potential of these ecosystems, and (3) sensitization and empowerment of communities for strengthening self-sustaining livelihood support systems. The following section describes some of technological packages developed by ICAR-Indian Institute of Soil and Water Conservation and its regional centers for reclamation and productive utilization of ravine lands.

## 9.4 Ravine Reclamation

Generally, ravines are associated with alluvial deep soils; therefore, land leveling for agriculture is a preferred activity for reclaiming shallow ravines. However, the cost of land leveling is a deciding factor in medium-deep and deep ravine systems. Other measures of reclaiming the ravine systems include (a) checking the runoff and safe disposal of surplus runoff water into the ravines from the adjacent marginal land by constructing marginal bund, (b) storage of runoff/sediment in gullies and gully bed stabilization with earthen gully plugs or *bori-bund* or composite check dam at specific horizontal interval, and (c) utilization of groundwater potential through drip system. The shallow and medium ravines can be converted to terraced land and put under agricultural/agroforestry land use, while deep and very deep ravines may be put under permanent vegetative cover with adoption of suitable soil and water conservation practices. Medium-deep and deep gully systems which can't be economically reclaimed for cultivation shall be put under silvopastoral systems. Severe grazing and illicit felling are major constraints in managing these lands; therefore, closure against biotic interferences is a prerequisite for revegetating these lands.

### 9.4.1 Management of Shallow Ravines

About 33% of the ravine systems are shallow ravines (<1 m deep) and can be safely reclaimed for cultivation. Reclamation of shallow ravines includes (a) planning and establishing marginal or peripheral bund at the downstream periphery of shallow ravines, (b) leveling and slope smoothening on the upstream to marginal bunds for promoting in situ rainwater conservation and improving water and nutrient use efficiency of cropping systems, and (c) construction of earthen check dams in the drainage channel at regular interval equipped with a pipe or small masonry spillway to facilitate gully bed stabilization for raising economically useful crops in the gullied lands. Check dams are reported to be the most effective measure to prevent soil loss from the gully mouth in the Kunwari River, which is a major tributary of the Chambal where gully erosion is a prominent feature (Pani 2016).

#### 9.4.1.1 Improved Crop Production from Shallow Ravines

Reclamation of shallow ravine has shown tangible benefit by improved value of land and crop yield improvement by 200% due to improved moisture level. The intangible benefits include conservation of soil, water, moisture, and nutrients and protection of land. In a ravinous watershed located in southeastern Rajasthan, bunding alone and bunding with leveling increased crop production by 46% and 112%, respectively, over control (Table 9.3) (Singh et al. 2005). Mustard cultivation recovered 76% of the bunding cost and 64% of the bunding + leveling cost in the first year only after imposing land treatments (Table 9.4). This indicates that in ravine areas

**Table 9.3** Crop productivity (2 years average) in Sorghum grain equivalent (kg $ha^{-1}$) yield as influenced by shallow ravine reclamation

| Crops | Untreated ravine | Graded bunding alone | Bunded and leveled | Crop average |
|---|---|---|---|---|
| Chickpea | 620 | 1023 | 1290 | 978 |
| Mustard | 510 | 964 | 1528 | 1001 |
| Sorghum | 655 | 849 | 1241 | 915 |
| Soybean | 985 | 1269 | 1813 | 1355 |
| Mean | 692 | 1026 | 1468 | |
| LSD ($p \leq 0.05$) for Crop = 191, Treatment = 191 | | | | |

Source: Singh et al. (2005)

**Table 9.4** Tangible benefits of shallow ravine reclamation

| | Crop production (Seed yield kg $ha^{-1}$) | | | Additional return (₹ $ha^{-1}$) | |
|---|---|---|---|---|---|
| Land treatment | Control | Bunded | Bunded and leveled | Bunded | Bunded and leveled |
| Mustard | 453 | 857 | 1358 | 3636 | 8145 |
| Chickpea | 451 | 744 | 938 | 3223 | 4383 |
| Soybean | 788 | 1015 | 1450 | 2270 | 6620 |
| Sorghum | 655 | 849 | 1241 | 1552 | 4688 |
| Land treatment cost were 4750 ₹ $ha^{-1}$ for bunding and 12,800 ₹ $ha^{-1}$ for bunding with leveling | | | | | |

Source: Singh et al. (2005)

appropriate conservation measures coupled with suitable land use planning could result in convincing tangible benefit on short-terms basis, apart from their protective and long-term intangible benefits.

## 9.4.2 Management of Marginal Lands Around Ravines

The major crops grown in the ravine region are soybean, sorghum, pearl millet in *kharif* and wheat, barley, mustard, and coriander during *rabi*. Packages of better suited crops and varieties, optimum conservation measures, improved irrigation methods, and optimum nutrient management and integrated disease and pest management technologies have been evaluated and recommended by the centers of IISWC for ravine regions. Several studies have shown that scientific and judicious management of ravine lands may lead to 9–28% improvement in cropping intensity and 20–66% increase in current yield levels with an overall improvement of 118–280% increase in net returns through increased crop production (Rao et al. 2013, 2015). Some of the production systems in marginal lands around the ravines are described below.

### 9.4.2.1 Stubble Mulch Farming Tillage System

Stubble mulch farming tillage technique in combination with intercropping of one long-duration (pigeon pea or castor) and another short-duration (pearl millet or cowpea) crop is recommended in the marginal lands at the periphery of the ravines to enhance the productivity and resource conservation. Sowing of crops on contours and application of pearl millet straw (5–7.5 cm length) at the rate of 2 Mg $ha^{-1}$ as a mulch/protective cover to soil surface resulted in increase in crop yield by 42–178% and reduction in runoff by 60% and soil loss by 70% in rain-fed crops in the marginal lands around the periphery of ravines.

### 9.4.2.2 Vegetative Filters in Crop Fields and Water Courses in Marginal Lands of Ravines

Vegetative filter strips of 1–2 m width with combination of *Eulaliopsis binata* and *Dichanthium annulatum* grasses at spacing of 45 m are recommended for the marginal lands around the periphery of the ravines and in watercourses to prevent runoff and soil losses from crop fields, thereby reducing the sedimentation of downstream water bodies. Vegetative filters in marginal lands around the ravines are able to reduce runoff and soil and nutrient losses by 20%, 65%, and 75%, respectively, and increase the crop yields by 20% (Rao et al. 2013). *Dichanthium annulatum* grass in waterways reduces the soil loss up to 80% and sediment concentration in the runoff water by six times (a reduction from 4.2 to 0.65 g $l^{-1}$) and reduce the sedimentation of downstream reservoirs (Rao et al. 2015).

#### 9.4.2.3 Drumstick with Green Gram Followed by Fennel as Profitable System in Marginal Lands of Mahi Ravine

Drumstick (*Moringa oleifera*), a multipurpose tree along with *kharif* legume green gram (*Phaseolus radiatus*) followed by *rabi* spice crop fennel (*Foeniculum vulgare*), has been found profitable with a net return of ₹52,928 $ha^{-1}$ from drumstick + green gram + fennel as compared to traditional cropping system of tobacco (*Nicotiana tobacum*) in the marginal lands of Mahi ravines.

#### 9.4.2.4 Three-tier Systems of Rainwater Management in Ravines of Bundelkhand

The study conducted on rainwater management with three-tier (contour bund + dead furrows + harvesting of excess rainwater for lifesaving irrigation) system in Jalaun district of Bundelkhand, Uttar Pradesh, showed increase in productivity, profitability, and water use efficiency in ravine area through double cropping system (Singh 2011). The three-tier system comprised of collection of rainfall in cropland with contour bund until no harmful effect to rainy season crops, collection of excess water (after storage in contour bund) in dead furrows, and utilization of this harvested rainwater during moisture stress period through seepage system of rainwater utilization. After the second step, the excess rainwater is transferred to brick masonry check dams, constructed in lowest position of the catchments. The stored water is utilized for lifesaving irrigations during dry spells of monsoon season as well as winter season experimental crops. The remaining excess water after impounding in water storage structures is discharged into the nearby drainage system.

In double cropping system of sesamum-wheat, significantly higher sesamum seed yield of 448 kg $ha^{-1}$ under three-tier system of rainwater management was obtained as compared to other systems of moisture management. Wheat grown after sesamum under three-tier system of rainwater harvesting gave significantly higher grain yield of 4560 kg $ha^{-1}$ in comparison with other moisture management practices. Three-tier system of rainwater management resulted in significantly higher black gram yield of 700 kg $ha^{-1}$as compared to the yield recorded in mono-tier, two-tier, and conventional systems of soil moisture management under black gram-wheat cropping system. Similarly, highest wheat yield of 4618 kg $ha^{-1}$ was also found in three-tier system of rainwater management after harvesting of black gram, which was significantly superior over other soil moisture management practices. The significantly higher wheat equivalent yield of 7248 kg $ha^{-1}$ was found in sesamum-wheat cropping system compared with wheat equivalent yield of 7068 kg $ha^{-1}$ recorded in black gram – wheat cropping system under three-tier system of rainwater management. The highest net return of ₹41,510 $ha^{-1}$ was obtained from sesamum-wheat cropping system, which was significantly superior to the net return received from black gram-wheat cropping system (₹41,045 $ha^{-1}$) under three-tier system of rainwater management (Singh 2011).

**Table 9.5** Economic potential of *Ziziphus mauritiana-Aloe vera* plantation at Yamuna ravines

| | *Ziziphus* with *Aloe vera* | | Sole ziziphus |
|---|---|---|---|
| Particulars | Aloe vera | Ber | |
| Yield of economic part (Mg $ha^{-1}$ $year^{-1}$) | 30[a] | 7.5–8.0 | 7.5–8.0 |
| Cost of cultivation (₹)[b] | 20,000 | 12,500 | 12,500 |
| Gross income (₹) | 60,000 | 48,000 | 48,000 |
| Net income (₹) | 40,000 | 35,000 | 35,000 |
| Net income from the total system (₹ $year^{-1}$) | 75,000 | | 35,000 |

Source: Mahapatra and Jha (2009)

[a]Yield of *Aloe vera* is based on three harvesting in a 4-year life cycle

[b]Cost of cultivation includes field preparation, fertilizer, planting material, and plant protection measures (Base Year 2008)

#### 9.4.2.5 Ber with *Aloe vera* Cultivation System for Yamuna Ravines

*Ber* (*Ziziphus mauritiana*) is one of the important fruits of the hot, semiarid ravinous tracts of the Indo-Gangetic region. The cultivars of *ber* suitable for this region include Banarasi, Ponda-safeda and Gola. *Aloe vera* (vernacular *Gwarpatha* or *Ghritkumari*) is a rich source of polysaccharides. This xeric plant is an important medicinal crop having very high cosmo-therapeutic value. The crop is suitable for degraded soil of ravine region around the river Yamuna. The crop is successfully grown between the tree interspaces of the mature and established *ber* orchards. The *Aloe vera* is a hardy plant and shows resistance to moisture and nutrient stresses and can withstand salinity and alkalinity. With a modest initial investment, return is accrued after a gestation period of 1.5 year (Table 9.5) (Mahapatra and Jha 2009). Therefore, this technology would be more suitable for Yamuna and Chambal ravines in the districts of Agra, Mainpuri, Etawah, Morena, Bhind, and Dholpur of UP, MP and Rajasthan.

### 9.4.3 Management of Medium and Deep Ravines

More careful planning is required for ravines deeper than 3 m. A successful ravine rehabilitation program for medium and deep ravine should focus on stabilization of (a) gully head, (b) side slopes, (c) gully bed (drainage line treatment for flattening the gully beds), and (d) vegetative establishment. The three ravine centers of IISWC located at Agra, Kota, and Vasad have evaluated suitable conservation measures for stabilization of gully head, side slopes, and gully beds and recommended suitable design specifications. Once erosion process is arrested and gully systems are stabilized, these lands can be used to contribute to livelihood and environmental security for local inhabitants. Some of the recommended land use systems for these lands are listed here.

**Fig. 9.2** Staggered contour-trenching

#### 9.4.3.1 Staggered Trenching with Gooseberry-Based Production System in Chambal Ravines

Staggered trenching is a practice of excavating trenches across the land slope directly below one another in alternate rows that breaks slope length and increases opportunity time for retention of runoff (Fig. 9.2). Staggered contour trenching (SCT) density of 417 trench per ha at 75% runoff trapping potential has been found to be the most protective, productive, and economical conservation measure for the medium and deep ravines of the Chambal (Sethy et al. 2012). This technology conserves on an average 85% runoff and reduced 9.1 Mg $ha^{-1}$ $year^{-1}$ soil loss over no trench. The average productivity and benefit-cost ratio of gooseberry-fruit equivalent improved by 118% and 99%, respectively, over the no trench (Sethy et al. 2012). The staggered contour trenching system is suitable under rain-fed condition in medium and deep ravinous lands along the bank of the major rivers like Chambal, Yamuna, Kali Sindh, and Mahi and their tributaries in the states of Rajasthan, Uttar Pradesh, Madhya Pradesh, and Gujarat. In medium and deep ravine lands for supporting the horti-pastoral land use system, staggered contour trenches of 417 with 75% of runoff trapping were recommended for higher production and optimal resource conservation (Ali et al. 2017).

#### 9.4.3.2 Horticultural Production Systems for Medium-Deep Ravines

Ravines with multidirectional slopes, where cost of leveling or terracing may be high, shall preferably be utilized for production of underutilized or hardy fruits. The establishment of fruit-based system (when irrigation facilities are available) is quite

**Fig. 9.3** Fruit-based system in Chambal ravine areas

profitable preposition for such lands (Fig. 9.3). However, provisions for occasional irrigation by installing drip irrigation would be required for establishing horticultural land use systems. Since ravine lands are located in the vicinity of river systems, developing micro-irrigation facility shall be feasible in most instances. Hardy fruit species like gooseberry (*Emblica officinalis*), ber (*Ziziphus mauritiana*), pomegranate (*Punica granatum*), wood apple (*Feronia limonia*), fig (*Ficus carica*), karonda (*Carissa carandas*), bael (*Aegle marmelos*), lasoda (*Cordia myxa*), karil (*Capparis decidua*), tamarind (*Tamarindus indica*), phalsa (*Grewia asiatica*), jamun (*Syzygium cumini*), and guava (*Psidium guajava*) can be grown with ease in shallow ravines with provision of micro-irrigation system like drip. For planting fruit trees on multidirectional slopes, half slanting pits (usually 1 $m^3$) are dug up with minor leveling of the planting area and leaving interspaces unleveled. To improve the moisture availability in sloping lands, half-moon-shaped catchment is developed near planted tree. Micro-catchment shaping is carried out with mild slope on the upstream of planted seedling, leading the rainwater to the roots of planted seedlings and making a crescent-shaped bund (half-moon) on the downstream side of pit at 0.75–1.0 m from the tree stem. The horticultural land use systems with higher benefit-cost ratio and adequate marketing support have been found cost-effective in reclaimed ravine lands (Table 9.6).

#### 9.4.3.3 Agri-horticultural Systems

As suggested above, fruit trees such as *Citrus* spp., *Emblica officinalis*, *Psidium guajava*, *Punica granatum*, *Ziziphus mauritiana*, *Aegle marmelos*, *Cordia myxa* and *Carissa carandas* are frequently used for growing along with field crops in reclaimed shallow ravine lands/marginal lands along the ravines, as they are compatible with most field crops and can very well withstand the low-resource

**Table 9.6** Economic return and benefit-cost ratio of horticultural production system from reclaimed ravines

| Name of the species | Bael | Lemon | Karonda | Lasoda |
|---|---|---|---|---|
| Spacing | 6 × 6 m | 4 × 4 m | 2 × 2 m | 6 × 6 m |
| Varieties | NA5 | Kagzi | American red | Puskar |
| Number of trees per ha | 278 | 625 | 2500 | 278 |
| Average yield from 5th year (kg $ha^{-1}$) | 2780 | 6250 | 2500 | 1668 |
| Average yield -6–10 year (kg $ha^{-1}$) | 11,120 | 7500 | 8750 | 5560 |
| Average yield -11–15 year (kg $ha^{-1}$) | 16,680 | 9375 | 10,000 | 12,510 |
| Average yield -16–20 year (kg $ha^{-1}$) | 12,510 | 6250 | 7500 | 11,120 |
| Benefit-cost ratio | 2.83 | 2.89 | 2.62 | 2.25 |

conditions. Gooseberry *cv* NA7 is raised at a spacing of 8 × 8 m, and cultivation of field crops is taken up in the interspaces of the tree rows. Kagzi lime (*Citrus limon*) is raised at a spacing of 5 × 5 m, and cultivation of field crops is taken up in the interspaces of the tree rows. Guava *cv* L-49/Allahabad *safeda* is raised at a spacing of 6 × 6 m, and cultivation of field crops is taken up in the interspaces of the tree rows. Field crops like green gram (IPM 02-3), soybean (JS 9560), and sesame (RT 46) are raised during *kharif* season in the interspaces of planted trees with regular cultivation practices.

#### 9.4.3.4 Horti-pastoral Systems

The system involves growing of fruit trees and grasses in combination as per the suitability of the species to a particular site. Such system sustains the requirement of rural masses for fodder through leaves, grasses, and fruits. Suitable combinations for ravines are bael-karad grass (*Aegle marmelos-Dichanthium annulatum*), bael-dhaman grass (*A. marmelos-Cenchrus ciliaris*), gooseberry-dhaman (*Emblica officinalis-C. ciliaris*), ber-karad (*Z. mauritiana-D. annulatum*), and bordi-dhaman (*Z. nummularia-C. ciliaris*). Results of growing different grass species between productive and unproductive *ber* trees with varying degree of successes have been reported. Ber-based pastoral studies on sandy rangelands of Rajasthan revealed that *Cenchrus ciliaris-Ziziphus mauritiana* system produced 1.2 Mg $ha^{-1}$ forage and did not affect the fruit yield (Sharma and Diwakar 1989). Sharma et al. (1994) reported that the productivity of wasteland in Western Rajasthan could be increased by planting top feed species *Z. nummularia* in combination with grasses such as *C. ciliaris*. In ravine lands, land use significantly influences various physiochemical properties (Agnihotri and Yadhav 1995; Agnihotri et al. 2007). After 25 years of plantation of forest species in Yamuna ravines, soil organic carbon content increased by more than twofold and mean weight diameter of aggregates by two- to sixfold in comparison with reference site (Jha et al. 2010).

#### 9.4.3.5 Silvopastoral Systems

Ravine lands with severe terrain deformation such as medium-deep gullies and the ravine lands which are under community ownership/village panchayat lands should preferably be put under silvopastoral land use systems. Severe grazing and illicit

**Table 9.7** Site suitability and production potential of suitable tree and grass species

| Name of species | Planting location | Productivity ($Mg\ ha^{-1}$) | Useful products |
|---|---|---|---|
| **Suitable tree species** | | | |
| *Acacia nilotica* | Hump top, slope, and ravine beds | 20–25 | Fodder, fuel, timber |
| *Acacia tortilis* | Hump top, slope, and ravine beds | 40–60 | Fuel, fodder and small timber |
| *Azadirachta indica* | Marginal lands, hump top and beds | 40–60 | Fodder, fuel medicinal oil, and timber |
| *Balanites aegyptiaca* | Hump top and slopes | 30 | Fuel wood, medicine, fodder |
| *Prosopis juliflora* | Hump top, slope, and ravine beds | 60–90 | Charcoal, fuel wood, pods for feed and fencing materials |
| *Tamarix dioica* | Swampy areas on gully bottom | 20 | Reclamation of saline soils, fuel |
| *Soymida febrifuga* | Marginal lands and hump top | 20 | Fodder, fuel and light timber |
| *Eucalyptus tereticornis* | Marginal lands, swampy areas on gully bottom | 26–37 | Poles and fuel wood |
| *Albizia lebbeck* | Marginal lands, hump top | 85–90 | Fodder and fuel |
| **Suitable grasses for ravine top and slopes** | | | |
| *Dichanthium annulatum* | Degraded land, ravine hump, and slope | 6–7 | Fodder grass, cut before flowering for use as hay and silage. Dry grass splits used for mat preparation |
| *Cenchrus ciliaris* | Hump top, slope, and ravine beds | 6–7 | Excellent for pasture in hot, dry areas, fodder, hay, silage soil conservation, folk remedies for kidney pain, tumors, sores and wounds |
| *Brachiaria mutica* | Marginal lands, ravine hump slope, and beds | 8–9 | Nutritive green fodder, rapid spread by stolons, excellent soil binder, silage and hay making |
| *Cenchrus setigerus* | Ravine hump and slopes | 5–6 | Valuable stand over feed in low-rainfall areas. Some value for erosion control in arid and semiarid land |
| *Panicum antidotale* | Ravine top, slope | 4–6 | Highly nutritious green fodder, plant smoke used to treat smallpox and throat infection |

felling are major contributing factors for degradation of these lands. Closure of ravine area against biotic interferences is a prerequisite for revegetating these lands. Out of the species tried, best performance was observed by *Eucalyptus tereticornis*, *E. camaldulensis*, *Acacia nilotica*, and *Acacia tortilis* (Table 9.7). The ravine lands confined largely to the states of Uttar Pradesh, Madhya Pradesh, Rajasthan, and Gujarat could be developed as silvopastoral systems by introducing different grasses

(*Cenchrus* spp., *Panicum antidotale*, *Pennisetum pedicellatum*, and *Dichanthium annulatum*).

As has been discussed in Chap. 1 of this book, Dagar and Mall (1980) and Dagar and Pathak (2005) have expressed that if we stop grazing in ravine areas, natural grasses will appear and establish along with regenerating woody vegetation and protect the soil from erosion. Grasses such as *Sehima nervosum*, *Dichanthium annulatum*, *Bothriochloa pertusa*, *Chrysopogon fulvus*, *Cynodon dactylon*, and *Cenchrus ciliaris* tolerate biotic pressure and get established easily. *Acacia nilotica*, *A. eburnea*, *Balanites aegyptiaca*, *Dichrostachys cinerea*, *Capparis decidua*, and *Clerodendrum phlomidis* form excellent natural silvopastoral system in ravine lands. Even cultivated forages may be introduced.

#### 9.4.3.6 Bamboo-based Production System

Researchers have established the importance of bamboo plant as an effective means for natural resource conservation (Lawler 1993; Yanhui and Yongmin 1995). Ben-zhi et al. (2005) reported that soil erosion was low in bamboo plot (178 kg $ha^{-1}$) as against other forest plantation types. The high stem flow amount and funneling ratio of bamboo plants in comparison with deciduous and coniferous plants makes better rainfall absorption and hydrologically best suited plantation in degraded ravine lands (Rao et al. 2012a). It acts as a barrier and filter strip for overland flow, thus promoting soil deposition and slowing down water flow speed along riverbanks and in deforested areas. *Dendrocalamus strictus* which covers about 45% of the area under bamboo in India has been demonstrated to be very useful species for economic utilization of ravines (Fig. 9.4).

**Fig. 9.4** Cultivation of bamboo for better economic gains and gully stabilization

After extensive investigation on the production and protection value of *Dendrocalamus strictus* for Chambal, Mahi, and Yamuna ravines, Indian Institute of Soil and Water Conservation's Research Centers in Kota, Vasad and Agra have recommended it for planting in medium and deep ravines. It is estimated that out of 3 million ha area of ravines, approximately 1 million ha area of deep ravines are potentially suitable for planting of *Dendrocalamus strictus*. Annually 4000 culms of this bamboo per ha can be harvested by planting in ravines (Rao et al. 2012b). The benefit-cost ratio (1.98) of plantation in ravines is lucrative with an economic return of 19.3% over a 20-year period (Pande et al. 2012). The intangible benefits of carbon sequestration and prevention of soil erosion are the added benefits. The Yamuna ravine lands under the bamboo increased the soil pH, organic carbon and no runoff (Singh et al. 2014). The economic analysis on bamboo revealed that a cash outflow of ₹48,000 $ha^{-1}$ from seventh year onwards (Singh et al. 2015) was obtained. This has implications for resource conservation and production system sustainability. In addition, benefits accrued to society in terms of value of nutrient saved through soil conservation and incremental soil carbon buildup.

## 9.5 Livelihood and Environmental Security

Generally, the ravine areas have 15–32% arable lands and 60–85% nonarable lands. Under the current scenario, it is estimated that scientific and judicious management of ravine land would increase 10–50% of existing arable lands, develop irrigation facilitation for its 30–60% arable lands, and improve 9–28% cropping intensity and 20–66% of current yield levels with an overall increase of 118–280% in the net returns through increased crop production. Tangible and intangible benefits which were recorded from a typical rain-fed ravinous watershed located at BadaKhera village in Bundi District of southeastern Rajasthan are summarized in Tables 9.8 and 9.9.

About 20% area of this watershed is tableland with multidirectional slopes (2–10%), and the gully network spreads in about 80% of the downstream watershed area. The highest order of stream was 4 and drainage density of study area was found to be 14.86 km per $km^2$. Out of 682.5 ha of total area of watershed, private agricultural lands occupy 378.9 ha, and the remaining 303.6 ha were community lands. Reclamation of shallow ravine has improved value of land and improved crop yield up to 200% due to improved moisture regime. The intangible benefits include conservation of soil, water, moisture, and nutrients and protection of land. Overall benefit-cost ratio of the entire project worked out to be 1.54:1, and that is quite favorable considering problem complexity encountered by the farmers in ravinous region. It was also observed that strengthening village-level institutions and promoting social fencing could further improve the results.

It is now well recognized that for sustainable development of degraded lands, involvement of people (landless and beneficiaries) is very essential. Successful ravine reclamation requires the support and involvement of the local cultivator and

**Table 9.8** Economic returns from different interventions

| Intervention | Direct benefits | Indirect benefits |
|---|---|---|
| Bunding and leveling | Increased crop yield (89% increase in mustard observed) | Improvement in land quality, ground water recharge |
| Spillways | Increased crop yield through increased moisture regime near structures (24.6 ha) | Improvement in land quality, ground water recharge |
| Gully control structures | Increased area under cultivation (9.12 ha); increased fuel and fodder production | Improvement in land quality, ground water recharge |
| Water harvesting structures | Increased water availability for irrigation and drinking for cattle | Ground water recharge |
| Crop demonstrations | Increased crop yields | Sustainability in agriculture production |
| Horticulture plantations | Increased fruit production from 5th year onward | Sustainability, stability, and diversification in agriculture production |
| Afforestation works | Increased fuel and fodder production | Erosion control, microclimate improvement |

**Table 9.9** Intangible benefits from different interventions

| S no | Interventions | Benefits |
|---|---|---|
| A | Water conservation measures, leveling, bunding, check dams/spillway | Increased *in situ* moisture conservation; groundwater recharge; improvement in water quality; retained soil fertility because of reduction in sediment yield; increased vegetation or regeneration; drought mitigation |
| B | Agri-horticultural system | Additional regular employment; ensuring food security by stabilizing income; risk minimization under frequent drought occurrence; changes in microclimate in area due to increased vegetation |
| C | Water harvesting structures (village pond, percolation tank, farm pond) | Checking migration of livestock due to availability of drinking water for livestock; increased irrigated area; groundwater recharge |
| D | Social awareness camps/skill development trainings | Awareness creation for natural resource conservation; women empowerment; increased social status; improved quality of life (health, education, etc.) |

the community and also reform of local land tenure and social arrangements. In the recent decades, efforts have been made to institutionalize the organization of community and beneficiaries and ensuring their involvement in the project planning, formulation, implementation, and maintenance (Bhan 2013). Pani (2016) emphasized the need to bring pixels closer to people – make people central to a bottom-up approach for land-use monitoring and evaluation with effective use of remote sensing and geographic information systems (GIS) techniques to solve some of the key problems faced in the ravine region, such as poor land records.

## 9.6 Government Policies and Programs for Management of Degraded Land

Problems of land degradation and soil erosion in India were duly acknowledged by policy makers in the "First Five-Year Plan" itself, and various soil conservation programs were initiated with provisions of suitable legislations and regulations to prevent land degradation and promote conservation practices. Again in 1976, the National Commission on Agriculture accorded higher priority to the ravine problem. Over the period, various programs have been launched in the country to address the issue of erosion such as Strengthening of State Land Use Boards (SLUBS); National Land Use and Conservation Board (NLCB); Soil Conservation Training Centre, DVC Hazaribagh; Centrally Sponsored Scheme of Soil Conservation for Enhancing Productivity of Degraded Lands in the Catchments of River Valley Projects; and Centrally Sponsored Scheme of Soil Conservation in the Catchments of Flood Prone Rivers. These programs helped to check the land degradation problem to some extent. However, success stories are scattered through the length and breadth of the landscape. Realizing the importance and urgency to protect soil and water resources, Government of India established a chain of Soil Conservation Research, Demonstration and Training Centers during first and second 5-year plans (1954–1957). Subsequently, these centers became the constituents of Indian Council of Agricultural Research (ICAR) and now functioning as Regional Centers of Indian Institute of Soil and Water Conservation (IISWC). Over a period of 60 years, these centers have been engaged in various aspects of characterization of ravines, technology development for rehabilitation and productive utilization of ravine lands, and dissemination of these technologies through regular training and outreach activities.

The Government of India has implemented several programs, namely, National Watershed Development Project for Rainfed Areas (NWDPRA), Soil Conservation in the Catchments of River Valley Project and Flood Prone River (RVP&FPR), Integrated Wastelands Development Program (IWDP), Drought Prone Areas Program (DPAP), and Desert Development Program (DDP) across the country. These programs have been amalgamated into a single modified program called Integrated Watershed Management Program (IWMP) with effect from February 26, 2009. The major activities taken up under IWMP inter alia include ridge area treatment, drainage line treatment, soil and moisture conservation, rainwater harvesting, nursery raising, afforestation, horticulture, pasture development, etc. About 57.61 million ha area has been developed under various watershed development programs of Ministry of Agriculture and Ministry of Rural Development since its inception up to 2011–2012. The Integrated Watershed Management Program (IWMP) has now been amalgamated as the Watershed Component of the *Pradhan-Mantri Krishi Sinchayee Yojana* (PMKSY) with effect from 2015 to 2016.

## 9.7 Summary

With increasing demographic pressure on natural resource base and far-reaching on-site and off-site ecological and economic consequences of land degradation, reclaiming every inch of the land for sustainable productive utilization has emerged as a priority. Reclamation of ravine lands is a complex task, and it requires adequate technological, financial, and policy support. There are time-tested technological solutions and regionally differentiated recommendations for management of ravine lands. There is a need to sensitize beneficiary groups, developmental agencies, and policy makers for the need of sustainable developmental potential of ravine lands. Nevertheless, the success of the ravine reclamation program would be closely linked with careful microlevel planning, adoption of economically viable and site-specific production systems, effective community mobilization, and efficient monitoring and policy support. Ravine area development will not only contribute to bridging the growing demand and supply gap for food, fodder, fuel, fiber, and woody produces, but it will also help in restoring livelihood security for resource constraint communities.

## References

Agnihotri RC, Yadav RC (1995) Effects of different land uses on infiltration in ustifluvent soil susceptible to gully erosion. Hydrol Sci J 40(3):395–406

Agnihotri RC, Yadav RC, Jha P (2007) Erodibility of an ustifluvent soil in relation to ecological settings of riparian zone of river Yamuna, India. Indian J Soil Conserv 35(3):226–229

Ali S, Sethy BK, Singh RK, Parandiyal AK, Kumar A (2017) Quantification of hydrologic response of staggered contour trenching for horti-pastoral land use system in small ravine watersheds: a paired watershed approach. Land Degrad Dev 28(4):1237–1252

Ben-zhi Z, Mao-yi F, Jin-zhong X, Xiao-sheng Y, Zheng-cai L (2005) Ecological functions of bamboo forest: research and application. J For Res 16(2):143–147

Bhan S (2013) Land degradation and integrated watershed management in India. Int Soil Water Conserv Res 1(1):49–57

Bhushan LS, Yadav RC (2000) Management of degraded lands with special reference to gullied and ravines. Indian Farming August 27–29

Dagar JC, Mall LP (1980) Studies on the vegetation of ravines and banks of the river Kshipra and its tributaries. J Indian Bot Soc 59:234–245

Dagar JC, Pathak PS (2005) Grassland dynamics and their management. Range Manag Agroforestry 26(1):7–31

Dhruva Narayan VV (1993) Soil and water conservation research in India. Indian Council of Agricultural Research, Pusa, p 454

GoI (1976) Report of the National Commission on Agriculture 1976. National Commission on Agriculture, Government of India, Ministry of Agriculture and Irrigation, New Delhi

Jain SK (2011) Population rise and growing water scarcity in India – revised estimates and required initiatives. Curr Sci 101(3):271–276

Jha P, Mohapatra KP, Dubey SK (2010) Impact of land use on physico-chemical and hydrological properties of ustifluvent soils in riparian zone of river Yamuna, India. Agrofor Syst 80:437–445. https://doi.org/10.1007/s10457-010-9338-3

Lawler DM (1993) The measurement of river bank erosion and lateral channel change: a review. Earth Surf Process Landforms 18:777–821

Mohapatra KP, Jha P (2009) *Aloe vera* cultivation in interspaces for supplementing productivity of ber orchards in Reclaimed Yamuna ravines. Tech. Brochure. ICAR-IISWC, Dehra Dun

NAAS (2010) Degraded and wastelands of India: status and spatial distribution. National Academy of Agricultural Sciences (NAAS), New Delhi

NRSC (2011) Wastelands Atlas of India 2011 (Change analysis based on multi-temporal satellite data of 2005–06 and 2008–09). Land Use and Cover Monitoring Division (LRUMG), Remote Sensing & GIS Applications Area, National Remote Sensing Centre, Indian Space Research Organisation, Department of Space, Govt. of India, Hyderabad

Pande VC, Kurothe RS, Rao BK, Gopal K, Parandiyal AK, Singh AK, Ashok K (2012) Economic analysis of bamboo plantation in three major ravine systems of India. Agric Econ Res Rev 25(1):63–73

Pani P (2016) Controlling gully erosion: an analysis of land reclamation processes in Chambal Valley, India. Dev Prac 26(8):1047–1059. https://doi.org/10.1080/09614524.2016.1228831

Rao BK, Kurothe RS, Singh AK, Parandiyal AK, Pande VC, Kumar G (2012a) Bamboo Plantation Based Technological Interventions for Reclamation and Productive Utilization of Ravine Lands, CSWCRTI, Technical Bulletin No. T-62/V-4: 30p

Rao BK, Kurothe RS, Pande VC, Gopal K (2012b) Throughfall and stemflow measurement in bamboo (*Dendrocalmus strictus*) plantation. Indian J Soil Conserv 40(1):60–64

Rao BK, Vishwakarma AK, Pande VC, Kurothe RS, Mishra PK, Kumar G, Baraiya MJ (2013) Vegetative filters for reducing the sediment loads from agricultural fields. Proceedings of All India Seminar on "Recent Advances in Watershed Development Programme" (AISRAWDEP), September 5 6, 2013 at Ahmednagar, Maharastra

Rao BK, Mishra PK, Kurothe RS, Pande VC, Gopal K (2015) Effectiveness of *Dichanthium annulatum* in watercourses for reducing sediment delivery from agricultural watersheds. Clean – Soil, Air, Water 43(5):710–716. https://doi.org/10.1002/clen.201400265

Sethy BK, Parandiyal AK, Ali S, Kumar A, Singh RK (2012) Developed staggered contour trenching (SCT) technology for horti-pastoral land use in medium deep ravines for Chambal ravine system Tech. Brochure. ICAR-IISWC, Dehra Dun

Sharda VN, Dogra P, Prakash C (2010) Assessment of production losses due to water erosion in rainfed areas of India. J Soil Water Conserv 50(2):79–91

Sharma SK, Diwakar GD (1989) Economic evaluation of horti-pastoral system on arid region of western Rajasthan. Indian J Range Manag 10(2):119–122

Sharma AK, Pradhan IP, Nema JP, Tejwani KG (1980) 25 years research on soil & water conservation in ravine lands of Gujarat. CS&WCR&TI, Research Centre, Vasad

Sharma SK, Singh RK, Tewari JC, Burman U (1994) Silvipastoral studies in arid and semi-arid degraded lands of Western Rajasthan. In: Agroforestry systems for degraded lands, vol II, pp 749–754

Singh RA (2011) Rain water management with three tier systems in ravines affected area of Bundelkhand. Indian J Soil Conserv 39(1):14–19

Singh RK, Prasad SN, Shakir A, Kumar A, Singh KD, Prasad A, Singh SV, Parandiyal AK (2005) On-farm evaluation of conservation measures to performance of rainfed crops in semi-arid region. Indian J Soil Conserv 33(2):141–143

Singh AK, Kala S, Dubey SK, Rao BK, Gaur ML, Mohapatra KP, Prasad B (2014) Evaluation of bamboo based conservation measures for rehabilitation of degraded Yamuna ravines. Indian J Soil Conserv 42(1):80–84

Singh AK, Kala S, Dubey SK, Pande VC, Rao BK, Sharma KK, Mahapatra KP (2015) Technology for rehabilitation of yamuna ravines – cost effective practices to conserve natural resources through bamboo plantation. Curr Sci 108(8):1527–1533

Yadav RC, Bhushan LS (2002) Conservation of gullies in susceptible riparian areas of alluvial soil regions. Land Degrad Dev 13:201–219. https://doi.org/10.1002/ldr.493

Yanhui W, Yongmin L (1995) Hydrological characteristics of a moso-bamboo (*Phyllostachys pubescens*) forest in South China. Hydrol Process 9(7):797–808

# Formation, Control, and Management of Deep (Chambal) Ravines: An Energy Concept

10

V. S. Tomar and S. K. Verma

**Abstract**

The bank of Chambal River is as high as 30–40 m in Chambal division of Madhya Pradesh, and soils are alluvial in nature with high infiltration rate (>1.2 m $day^{-1}$), which increases with depth in the profile. The aggregation is very poor, low organic matter further causes poor binding capacity, and the particle size is also very fine that makes water convenient to reach lower reason and subsurface water flow which in turn creates very hydrodynamic pressure and hydraulic gradient causing deep ravine formation right at head point. The major issue that needs attention is checking of gully advancement as it is advancing day by day by converting about 8000 ha tableland into ravine every year. The indigenous and innovative technologies identified are suitable to fight against this issue and are helpful to control soil, water, and nutrient losses and further advancement to engulf good productive tableland. Many options like system modules are identified as very good option for eco-friendly, feasible, economical, and remunerative use of ravine lands.

## 10.1 Introduction

Theoretically ravine formation takes place by runoff surface water that fails to penetrate soil profile but instead will tend to accrue at the surface or flow over it in the form of sheet erosion and finally converted into shallow or deep ravines. The bank of Chambal River is as high as 30–40 m in Chambal division of Madhya Pradesh, and soils are alluvial in nature with high infiltration rate (>1.2 m $day^{-1}$), which increases with depth in the profile. The aggregation is very poor, low organic matter

V. S. Tomar · S. K. Verma (✉)
Department of Soil Science & Agricultural Chemistry, RVSKVV, Vishwa Vidyalaya, Gwalior, Madhya Pradesh, India





J. C. Dagar, A. K. Singh (eds.), *Ravine Lands: Greening for Livelihood and Environmental Security*, https://doi.org/10.1007/978-981-10-8043-2_10

further causes poor binding capacity, and the particle size is also very fine that makes water convenient to reach lower reason and subsurface water flow that causes deep ravine formation right at head point.

When the water is supplied from above to the soil surface, whether by precipitation or irrigation, it typically penetrates the surface and is absorbed into successively deeper layers of the profile. At times, however, a portion of the arriving water may itself be partitioned between the amount that returns to the atmosphere by direct evaporation from the soil or by the extraction and transpiration of plants and the amount that continues to seep downward and eventually recharges the groundwater reservoir. The motion of any inhomogeneous solution in a porous body brings about another process that differs from diffusion in its mechanism but that tends to produce an analogous or synergetic tendency to mix and eventually to even out the differences in concentration or composition among different proportions of the flowing solution. This process, which at times predominates over diffusion, is called hydrodynamic dispersion. It results from the microscopic nonuniformity of flow velocity in the soil's conducting pores. Since water moves faster in wide pores than in narrow ones and faster in the center of each pore than along its wall, some portion of the flowing solution moves ahead while other portion lags.

The groundwater system is a three-dimensional flow field; therefore, it is important to understand how the vertical components of groundwater movement affect the interaction of groundwater and surface water. A vertical section of a flow field indicates how potential energy is distributed beneath the water table in the groundwater system and how the energy distribution can be used to determine vertical components of flow near a surface water body. The term hydraulic head, which is the sum of elevation and water pressure divided by the weight density of water, is used to describe potential energy in groundwater flow systems. Water that infiltrates at land surface moves vertically downward to the water table to become groundwater. The groundwater then moves both vertically and laterally within the groundwater system. Movement is downward and lateral on the side, mostly lateral in the center, and lateral and upward on the other side of the profile. Soil parameters responsible for extension of ravine formation and management aspects of ravines have been discussed in this chapter.

## 10.2 Land Physiography

The Chambal is an interstate river and part of its catchment lies in Rajasthan and Uttar Pradesh. Chambal originates from the northern slopes of the Vindhyan mountains, about 14.4 km southwest of Mhow, district Indore, Madhya Pradesh (MP), at an elevation of 854 m above MSL. At first, it flows in a northerly direction in Madhya Pradesh for a length of about 320 km, and after passing by the historic fort of Chaurasigarh, it flows northeast for about 286 km through Rajasthan. Later, it forms the border between Madhya Pradesh and Rajasthan along Morena district for a length of 216 km continuing in the same direction; it then forms the boundary between Madhya Pradesh and Uttar Pradesh (UP) along Bhind district for about 112 km before finally joining the river Yamuna in UP after flowing for another

**Table 10.1** Drop in the slope of the Chambal riverbed

| Distance from the source (km) | Slope/drop in riverbed (m $km^{-1}$) |
|---|---|
| 0–16 | 20 |
| 230–246 | 2 |
| 386–430 | 1.6 |

64 km. The drop in the slope of riverbed is gentle except in three reaches, two in the upper reaches before the Chaurasigarh gorge and the third below it (Table 10.1).

In the third reach, the drop in the river slope is 1.6 m per km. Formation of ravines is maximum in this reach. This situation shows that riverbed has a very gentle slope, whereas the tablelands from where tributaries originate are at a much higher level. The differences in the two elevations really initiate the process of active erosion which is aided by the climate and soil type and other related parameters. The vertical fall between bank and the riverbed is 10–15 m at places and ranges up to 20 m. The vertical drop of the tributaries is a highly influencing factor for gully and ravine development. In general, Chambal River is in the degradation stage, and so depth of the riverbed is likely to increase with time to achieve a dynamic equilibrium with the vertical drops.

The factors phenomenon of gully erosion tries to expand in the lateral direction all along river courses. Gullies and ravines flank for about 2 km along the banks. The districts of Morena (including Sheopur) and Bhind suffer very badly due to its erosion-prone soil characteristic (sandy loam with increase in sand with depth, poor organic carbon content, low binding capacity, high infiltration rate, poor aggregation, and high dispersion ratio). In the rainy season, even under drizzling condition, the rainwater enters to the profile; after attaining the level of riverbed, it flows horizontally toward river as subsurface flow; after moving to some distance, its hydrodynamic velocity carries sand with flow; near to river it becomes successful to remove sand and makes hollow space; and the whole profile collapsed and makes a deep ravine. In the formation of Chambal ravines, the traditional way of ravine formation, i.e., sheet→ rill→ gully→ ravine, does not occur, and it develops due to subsurface flow only.

## 10.3 Soil Water Properties

### 10.3.1 Equation for Transient Water Flow

The water conservation equation relates water fluxes, storage changes, and sources or sink of water. When it is combined with the Buckingham-Darcy flux equation (Darcy 1856), an equation may be derived to predict water content or matric potential in soil during transient flow. It can be assumed for simplicity that the flow is vertical and that no plant roots are present ($r_w = 0$); the equation becomes

$$\frac{\partial \theta}{\partial t} = \partial / \partial z \left[ K(h) \left( \frac{\partial h}{\partial z} + 1 \right) \right]$$

This equation may not be solved in the form it is in, because it contains two unknown functions ($\theta$ and $h$) and only one equation. This difficulty may overcome by using the water characteristics or matric potential-water content function $h(\theta)$ to eliminate either $\theta$ or $h$ from equation.

Darcy's law by itself is sufficient to describe steady or stationary, flow process, in which the flux remains constant and equal throughout the conducting medium (and, hence, the potential and gradient at each point remain constant in time). Unsteady or transient, flow processes, in which the magnitude and possibly even the direction of flux and potential gradient vary in time, require the inclusion of an additional law, namely, the law of conservation of matter. The mass conservation law, expressed in the equation of continuity, states that if the rate of inflow into the volume element exceeds inflow, storage must be decreasing.

Consider first the simplest case, one-dimensional flow, with $q_x$ as the flux in the direction of $x$. As per Darcy's law in horizontal flow, the solution becomes as

$$\frac{\partial\theta}{\partial t} = \partial/\partial x\left[K\left(\frac{\partial Hp}{\partial x}\right)\right]$$

where K is the hydraulic conductivity of the saturated soil (the saturated conductivity) and $H_p$ is the pressure head.

While in vertical flow, $\partial z/\partial z = 1$ therefore, for this case,

$$\frac{\partial\theta}{\partial t} = \partial/\partial x\left[K\left(\frac{\partial Hp}{\partial x}+1\right)\right]$$

In an isotropic soil that is also homogeneous, the K values of all points are equal.

Differential equations, who's the solutions give the water content as a function of time and position in a wetting or drying system, are obtained by substitution of the appropriate flux in Darcy's equation. For horizontal one-dimensional flow in the x direction

$$\frac{\partial\theta}{\partial t} = \partial/\partial x\left[D(\theta)\left(\frac{\partial\theta}{\partial x}\right)\right]$$

It should be noted that for saturated water flow, water content $\theta$ remains constant with time and therefore, $\partial\theta/\partial t = 0$, with flux written in terms of hydraulic head H.

The more case in which $k$ is not uniform in all direction due to soil anisotropy is discussed in detail by Childs (1969), the equation reduces to

$$\frac{\partial\theta}{\partial t} = k\left[\partial 2\,Hx/\partial x^2\right]$$

Which is known as Laplace equation for horizontal flow and it has numerous important uses in solving equation of saturated flow?

This clearly indicates that the vertical flow after reaching to certain depth, higher buildup potential forces to flow toward the direction where gradient is abruptly very high on same value of distance $x$ and flow start at a high volume and it carries out

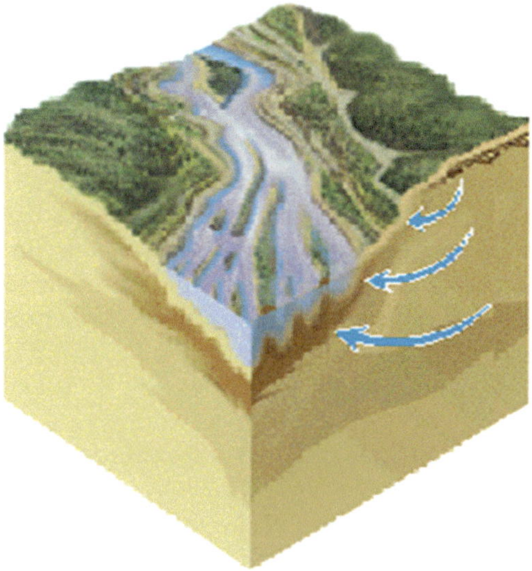

**Fig. 10.1** Pictorial view of water -flow in a ravine area

all dispersed material (mostly sand) and makes a pithole on terminal point of channel. During monsoon, the action remains and continues for a longer period that converts pithole in a small tunnel. The profile over the tunnel gets slipped down when it becomes wet as there are very poor aggregation and organic matter and no presence of roots.

The presence of plant roots or any kind of manipulation in soil isotropy in a direction, i.e., horizontally, water flow in the whole system can be modified, and the direction of flow can be diverted easily to vertical way rather than horizontal (Fig. 10.1).

## 10.4 Mechanism of Ravine Development on Chambal River

The mechanism of ravine development on Chambal River is quite different as compared to any other river of the area. The formation of Chambal ravines does not follow the ways described in theories, and it is always deep (more than 10 m) in nature right from adjoining tableland. The data shows that the Chambal ravines are still advancing at the rate of about 800 m $year^{-1}$. The major role in its formation is played by soil water energy balance and accelerated by profile's soil texture, elevation difference between riverbed and adjoining tableland, low particle binding energy, and high water transmission properties. Sometimes elevation difference from riverbed to tableland is more than 30 m. The soil profile is alluvial in nature, and sand content increases with depth (reaches 95% at a depth of 6 m) with very high water intake rate (more than 1.25 m $day^{-1}$). The soil is very poor in organic matter content (less than 0.2%) and particle binding energy (less than −40 mV), and this causes very high dispersion in fine sand particles with the moving water in saturated and transient water flow. The infiltrating water achieves a very high hydrodynamic energy after reaching a certain depth, and it further accelerated with depth which causes displacement of soil along with horizontally passing waters. The water column near to vertically exposed profile has horizontally a very high hydraulic head drop, and dispersed phase of sandy texture causes pithole formation in adjoining area of recently formed ravine which makes a path for its further advancement. The formed pithole further becomes big to bigger in continuously dazzling water and makes a larger cavity at lower depth, and the whole profile collapses, and dispersed soil gets lost with flowing subsurface water. The continuous horizontal subsurface water flow with relatively higher rate and energy accelerates the formation of deep ravine.

### 10.4.1 Stages of Ravine Development

Normally four stages of ravine development are generally recognized by geomorphologists (Fig. 10.2).

1. *Pothole stage*

   In this initial stage, parallel potholes are formed along the river below the flat ground by the mutual action of water and soil particles.
2. *Tunneling stage*

   A further stage in the development is the deepening of potholes and formation of a tunnel from the potholes to the adjoining ravine. With continuous corrosion, these tunnels enlarge in size.

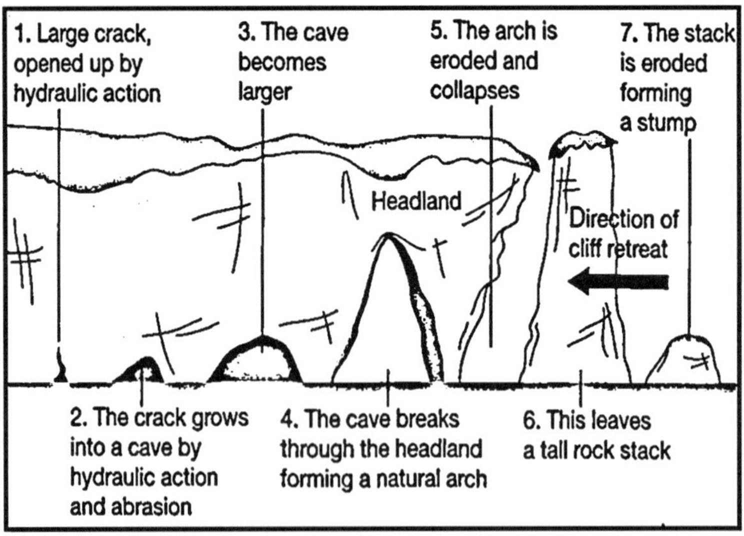


**Fig. 10.2** Schematic diagram of ravine development stages

**Fig. 10.3** Ravine formation process at Chambal River

3. *Collapsing stage*

   In this stage, the roofs of tunnels collapse when rain is dazzling and the profile gets wet and slides in the water channel and is carried away to the river; therefore formation of ravines takes place as shown in Fig. 10.3 up on the surface.
4. *Recession stage*

The side and head scarps of ravines recede continuously. With the continuous recession of side scarps, alluvial bluffs are disconnected and thus begin to shatter into conical blocks. In this last stage, maximum width of ravine bottoms takes place.

### 10.4.2 Major Catalysts That Contributes in Deep Ravine Formation

There are following factors which contribute in formation of deep ravines:

*Elevation differences*: There is about 30–50 m elevation difference on tableland at bank of Chambal River and riverbed; this is one of the very effective reasons for formation of deep ravines.

*Soil texture and related properties*: Soils on the bank of Chambal in Sheopur, Morena, and Bhind district of Madhya Pradesh are alluvial in nature, and sand fraction increases with depth (reaches ≥90% at 25 m depth) that makes soil highly permeable with very high intake capacity.

- The water-dispersible aggregates show that the soil of ravine/study area has very weak binding properties. The study shows that the percentage of silt and clay fractions of soil are more in disperse condition (dispersion ratio of 0.80–0.90) in all soil profiles and depths.
- More aggregates (50–70%) are in finer size (>125 μ) fraction, and only very less (8–14%) are in coarser (>1 mm) size.
- The recorded value of zeta potential (mv) showed a value of only −40 to −45 that indicates a very poor presence of colloidal fraction.
- The average value of particle size (radius in nm) of the particles below 0.2 mm showed a value of only 150–300 that indicated that most of the particles are in very fine in size.

*Hydraulic properties*: High infiltration rate (˃5 cm $h^{-1}$) at tableland around river facilitates downward water movement in these areas particularly when rainfall less than its infiltration is one of the potential reasons to create a hydraulic pressure and facilitates subsurface water flow that carries away sands by making a wide gap.

*High hydrodynamic pressure*: The rainwater that infiltrates downward rapidly causes underground flow toward high potential gradients because of open space at very short (few cm horizontal) distance and makes sufficient pressure to move the sand layer existing there along with water stream flowing toward riverbed.

*Dispersed soil condition*: The soil particles are highly dispersed and remain arranged when it is dry, but when it becomes moist in dazzling rains, the whole profile gets slipped due to presence of tunnels created by sand removal from subsurface streams.

## 10.5 Classification of Chambal Ravines

The ravines in Madhya Pradesh are classified in four categories (Gupta and Prajpati 1983), i.e., shallow (up to 1.5 m deep), medium (1.5–5 m deep), deep (5–10 m deep), and very deep (>10 m deep) ravines. The ravines of Chambal River are deep to very deep type of ravines, which are mostly in Sheopur, Morena, and Bhind districts of Madhya Pradesh, while shallow and medium ravines are existing in Kunwari, Sindh, Sank, and Asan River in Gwalior Morena and Datia districts.

## 10.6 Soil Conservation Strategies for Control Measures of Chambal Ravines

The technologies developed by various researchers are not suitable for Chambal ravines due to its different nature, posing a serious challenge to check its further advancement and gobbling some other new villages. Some new approaches have been started for various stages through a project undertaken recently funded by ICAR. The following steps have been initiated.

### 10.6.1 Management of Upper Reaches of Ravines

The major issue that needs attention of worker is checking of gully advancement as it is advancing day by day by converting about 8000 hectare tableland into ravines every year. Some work has been initiated in the project which includes native as well as innovative.

#### 10.6.1.1 Indigenous Technology

After discussion with thousands of villagers, the fact emerged that peoples are trying since hundreds of years to check the advancement and save their villages, but it never worked; permanent solution only got partial success in delaying the situation for some years. Among many methods, the shaping of advancing gullies (Figs. 10.4 and 10.5) and planting of shelterbelts of trees are quite impressive. Shaping was done manually and on individual basis by making the edge in slanting condition (say 120° rather than 90° angle) and planting of grass *Saccharum munja* as vegetative barriers on the slopes. These two methods are found very effective if carried out on community basis and with the help of modern machineries.

**Fig. 10.4** Location of vegetative barrier for protection of village settlement at Aisah

**Fig. 10.5** Vegetative barrier which protects the village from advancement of ravines

Childs (1969) equation clearly indicates any increase in the distance is going to decrease

$$\frac{\partial \theta}{\partial t} = k\left[\partial 2\, Hx / \partial x^2\right]$$

the gradient in horizontal direction; hence increase in $x$ with increasing depth will reduce the chances of vertical flow. The vegetative covers on the slopes will again reduce the chances of horizontal flow by increased soil particle binding with the help of roots and it exudes.

**Soil Conservation and village(Aisah) protection from Advancing Ravines (A Success Story)**

The villagers of village Aisah (Tehsil Ambah, district Morena in Madhya Pradesh, India) practiced various soil conservation measures to save the village from converting into ravines. The Aisah was shifted approximately 1.5 km from its original location due to advancement of ravines, and this is an ongoing process. The villagers planted the *Acacia nilotica* (Babool) trees at the ravine formation site. The strip of 20 m plantation approximately 70 m in length acted as vegetative barrier. Figure 10.4 shows the vegetative barrier (encircled green), and Fig. 10.5 depicts the vegetative barrier which protects the village from advancement of ravine.

#### 10.6.1.2 Innovative Techniques

Two methods, i.e., terracing at ravine advancing point (Fig. 10.6c, d) and slopping the vertical fall (>2: 1, Fig. 10.6c), have been introduced under the project of Niche Area of Excellence, and it may give a permanent solution to the problem. A strong vegetative barrier at suitable intervals is provided on slopes of terrace and diagonal surface of slanting face of the slopes.

**Fig. 10.6** Different options for checking ravine advancement: (**a**) hedge plantation at *top*, (**b**) mechanical structures to check runoff, (**c**) changed slope, and (**d**) multistep leveling (Photos by SK Verma)

### 10.6.2 Management of Gullies

Many options have been introduced by the project to manage the gullies and conserve the natural resources; it includes masonry structures, gabions, submersible dams, earthen dams (Fig. 10.7), plugging of gullies, and temporary bunds (*bori bandhan*). All the measures normally got failed due to leakage from side or gushing of water stream from below the structures. The submersible dam has some edges over other structures provided it has been raised at least from a depth of 1.00 to 1.25 m. A modified form of submersible dam plus gabion was introduced to check runoff and reduce the cost, and it is working satisfactorily.

*Management of deep ravines*: Deep ravines can manage by introducing submersible dams with sufficient arms and putting earthen dams over it. These structures are also useful in making water harvesting/recharging tanks that will contribute in raising groundwater.

*Multistep leveling system or zero loss technology*: A very innovative, simple, and practical methodology for partial reclamation of deep Chambal ravine has been developed. This technology involves use of heavy earth machine (Hitachi 110 or 220) only for cutting of top edge and uses this soil for chocking of drainage gullies at specified distance with appropriate strength. Construction of peripheral bunds

Missionary structures *Bori-bandhan*

Gabion Improvised gabion

**Fig. 10.7** Different structures (masonry, *bori bandhan*, gabion, and improvised gabions) for soil water conservation (Photos by SK Verma)

and partial modification of slope at every piece of land is mandatory. The reclaimed flat land is usable for crop cultivation (35–50%), modified slopes can be used for medicinal/fruit trees/grasses/silvopastoral system (30–45%), and plugged deep gullies are good for water storage. This is the cheapest and practical methodology to make ravines usable. The reclamation cost varies only from ₹35,000 to ₹40,000 depending on depth of ravines. The system is eco-friendly and remunerative and takes care of all aspects of sustainability. Figure 10.8 depicts some of the technological interventions followed in the project.

## 10.7 Monitoring Soil and Nutrient Losses (N, P and K) Through Mechanical and Agronomical Soil and Water Conservation Measures

Adaptation of various options either innovative or indigenous technology helps to control soil, water, and nutrient losses (Tables 10.2 and 10.3). Ravine soils generally suffer from very huge natural resource losses, and the soil loss (Angima et al. 2003) may go more than 25 Mg ha$^{-1}$year$^{-1}$ along with ˃80% of rainfall as runoff (Table 10.2), and they loss huge quantity of nutrients (Table 10.3) every year with a tune of 356 kg ha$^{-1}$ (N, P and K). These losses can be minimized even to zero by adopting various kinds of management systems.

A. View of partially reclaimed valley

B. Water harvesting tank in valley

C. Silvo-pastoral system on side of ridge

D. Crops on flat land and plantation on slopes

**Fig. 10.8** Multistep leveling system (zero loss technology) (Photos by SK Verma)

**Table 10.2** Erosion losses computed and actually recorded under different systems (Verma et al. 2016)

| Module | Computed | | Recorded | |
|---|---|---|---|---|
| | Silt loss (Mg $ha^{-1}year^{-1}$) | Runoff (cm) | Silt loss (Mg $ha^{-1}year^{-1}$) | Runoff (cm) |
| $M_0$ (diversified cropping system) | 18.24 | 32.42 | 28.7 | 30.8 |
| $M_1$ (agri-horticulture module) | 07.54 | 26.94 | 16.2 | 25.6 |
| $M_2$ (horti-medicinal/pastoral modules) | 02.87 | 25.60 | 24.0 | 25.8 |
| $M_3$ (silvi- medicinal module) | 14.37 | 25.21 | 18.6 | 25.9 |
| $M_4$ (silvi- medicinal module) | 11.50 | 25.90 | 14.8 | 24.6 |
| $M_5$ (silvopastoral module) | 11.50 | 26.37 | 08.4 | 24.2 |
| MSL (multistep leveling) | Nil | Nil | Nil | Nil |

**Table 10.3** Estimated losses (kg $ha^{-1}$) of major nutrients (N, P and K) through runoff under different systems adopted for ravine lands (Verma et al. 2016)

| Modules | N | $P_2O_5$ | K | Total |
|---|---|---|---|---|
| $M_0$ (diversified cropping system) | 88.97 | 8.61 | 258.30 | 355.88 |
| $M_1$ (agri-horticulture module) | 49.36 | 7. 56 | 145.80 | 202.72 |
| $M_2$ (horti-medicinal/pastoral modules) | 74.40 | 7.20 | 216.00 | 297.60 |
| $M_3$ (silvi- medicinal module) | 67.66 | 5.58 | 167.40 | 240.64 |
| $M_4$ (silvi- medicinal module) | 45.88 | 4.44 | 133.20 | 183.52 |
| $M_5$ (silvopastoral module) | 26.04 | 2.52 | 75.60 | 104.16 |
| MSL (multistep leveling) | Nil | Nil | Nil | Nil |

## 10.8 Suggested Crop/Tree Plants Identified for Ravines with Different Degrees of Slopes

1. **Steep slope land (>6% slope) with eroded soil**

   Five different modules, viz., horti-pastoral/silvopastoral system (grasses+ legumes/legume shrubs)/medicinal modules, were selected for growing in ravine lands. The plantation was done at geometry of 3 × 3 m with provision of five rows of grass (at row to row distance of 50 cm) between plants rows.

2. **Systems modules for deep ravine management**

   The works carried out at Chambal ravine identified some very potential system modules to minimize soil and water losses from such lands; they are:

   – *$M_1$ Diversified cropping system modules*

   (a) Pearl millet – taramira – fodder pearl millet
   (b) Green gram/black gram – mustard – cucurbits
   (c) Sesbania – mustard – watermelon
   (d) Pigeon pea + cluster bean intercropping – muskmelon

   – *$M_2$ Agri-horti modules*

   (a) *Emblica officinalis* (gooseberry) + (pearl millet-*Sesamum*)
   (b) *Ziziphus mauritiana* (ber) + (pigeon pea and cluster bean – onion)
   (c) *Moringa oleifera* (drumstick) tree + (zinger-taramira)

   – *$M_3$ Horti-medicinal/pastoral modules*

   (a) *Ziziphus mauritiana* (ber) + *Stylosanthes scabra*
   (b) *Emblica officinalis* (gooseberry) + *Chrysopogon fulvus*
   (c) *Ziziphus mauritiana* (ber) + *Cymbopogon* + *Ocimum*

   – *$M_4$ Silvi-medicinal module*

   (a) *Acacia catechu* tree with (*Cymbopogon* grass + *Aloe vera*)
   (b) *Pongamia pinnata* tree + (satawar (*Asparagus*) + lemon grass)
   (c) *Aegle marmelos* fruit tree + (castor + turmeric)
   (d) *Salvadora oleoides* tree + (pigeon pea-chandrasur)

   – *$M_5$ Silvopastoral module*

   (a) *Azadirachta indica* tree + guinea grasses (*Panicum maximum*)
   (b) *Pongamia pinnata* (Karanj) tree + Napier grass
   (c) *Prosopis cineraria* tree + perennial grasses (marvel grass)
   (d) *Ailanthus excelsa* tree + para grass
   (e) *Albizia lebbeck* + perennial grasses

3. **Sloppy land (3–6% slope) with shallow/medium soil:** Agronomical measures, i.e., strip/contour planting against slope (grass strips between crop strips for assuring low soil losses):
   - Green gram
   - Pearl millet
   - Pigeon pea-cluster bean intercropping
4. **Mild slope land (>1–3% slope) with medium/deep soil**
   - Green gram-taramira (*Eruca sativa*) cropping systems
   - Pearl millet-mustard cropping systems
5. **Flat land (1% or fewer slopes) with deep soil:** Suitable for intensive cropping system, i.e.:
   - Sesame-taramira
   - Pearl millet-mustard
   - Green gram-wheat

Five different modules, viz., diversified cropping, agroforestry, agro-horticultural, silvi-medicinal, and silvopastoral systems, were tested for its suitability and tried to categorize on its application basis for protection, conservation, and production of system:

- *For soil and water conservation*:
  - Silvopastoral system>silvi-medicinal>agroforestry>agro-horticultural> cropping systems
- *For economical purpose:*
  - Cropping system>agro-horticultural>agroforestry>silvi-medicinal>silvo-pastoral
- *Most feasible system:*
  - Agro-horticultural>agroforestry>silvopastoral>silvi-medicinal system>cropping systems
- *Eco-friendly system:*
  - Silvopastoral>silvi-medicinal>agro-horticultural>agroforestry>cropping system

## References

Angima SD, Angima DE, Stott MK, Neill O, Ong CK, Weesies GA (2003) Soil erosion prediction using RUSLE for central Kenyan highland conditions. Agric Ecosyst Environ 97:295–308

Childs EC (1969) Physical basis of soil water phenomena, introduction. Book Depart, Shepherdstown

Darcy H (1856) Less fontaines publiques de la ville de Dijon. Dalmont, Paris

Gupta RK, Prajpati MC (1983) Reclamation and use of ravine lands. Desert Resour Technol (Jodhpur) 1:221–262

Verma SK, Akhilesh S, Tomar PS (2016) Management of Chambal ravine for food and environment security: technical bulletin no. RVSKVV/61/2016. Rajmata Vijaya Scindia Krishi Vishvavidyalaya, Gwalior

# Management of Chambal Ravines for Income Enhancement

11

G. P. Verma, Y. P. Singh, Anil Kumar Singh, and S. K. Verma

**Abstract**

Deep to very deep ravines in Chambal valley represent the worst form of land degradation by runoff water. Initial splash erosion caused by raindrop goes unnoticed, but subsequent formation of rills and gullies was too conspicuous to be missed, and this was the stage of erosion to be tackled through well-established and tested technology evolved by research for rainwater management on watershed basis. But policymakers and planners thought soil and water conservation to be one-time operation involving mostly mechanical structures. Accordingly, budget and targets (hectarage) were fixed. Soil conservationists constructed field bunds and property bunds in the name of contour/graded bunds and leveled the ravine land without measures to hold the loose soil in situ; tanks were constructed at sites having no catchments, and structures of inappropriate type and designs were constructed at improper sites. Consequently, the land degradation went from bad to worst. A network of gullies changed into network of deep and very deep gullies still marching ahead unabated as deep and large ravines, despite of spending millions of rupees again on leveling, bunding, and aerial spray of seeds for afforestation. The fact is that foolproof technology for reclamation and control of very deep ravines is just not available nor can it be evolved by conducting research on well-managed

G. P. Verma
Department of Soil Science, Jawaharlal Nehru, Krishi Vishva Vidayalaya, Jabalpur, Madhya Pradesh, India

Y. P. Singh (✉)
AICRP-IWM, R. V. S. Krishi Vishwa Vidyalaya-ZARS, Morena, India

A. K. Singh
RVS Krishi Vishwa Vidyalaya, R. V. S., Gwalior, India
e-mail: vcrvsaugwa@mp.gov.in

S. K. Verma
Department of Soil Science & Agricultural Chemistry, RVSKVV, Vishwa Vidyalaya, Gwalior, Madhya Pradesh, India



J. C. Dagar, A. K. Singh (eds.), *Ravine Lands: Greening for Livelihood and Environmental Security*, https://doi.org/10.1007/978-981-10-8043-2_11

research station farms. To be purposeful, the research must be conducted right on the land under very deep ravines. Thanks to efforts made by RVS. Krishi Vishwa Vidyalaya for getting some funds from Indian Council of Agriculture Research, New Delhi, for conducting research under NICRA Project at Aisah village district Morena, whose almost all land is under deep ravines and has therefore been shifted to a new site. Very good beginning was made during the last 2 years of operation of project to demonstrate how the inaccessible land with deserted look can be changed into greenery of economic plants yielding vegetables, fruits, food grains, fiber, fuel, and raw material for valuable medicines. However, the technology to make the soil stay in situ rather than in transit for conservation farming on deep ravines is still to be perfected. The performance of various types of land shaping tried to control the advance of deep gullies and reclaim these could not be confirmed within the limited period of 3–4 years. Hence, there is a need to conduct further research on (1) to confirm various observations already made, (2) to implement a strong program of hydrometeorological monitoring to have locality-specific information for precise designing of various structures, (3) to test anicut system for gully control and ravine reclamation in the upper most area where general land slope and the slope of main gully bed are >3%, and (4) to work out the manurial requirement of various crops to be grown on soil after land shaping.

## 11.1 Introduction

Land and water resources are a valuable gift of the nature to a nation, and prosperity of the nation depends on how well these resources are managed. Inappropriate management of land and rainwater results in water erosion of varying degree depending upon the erodibility of soil and erosivity of rainfall. Water erosion can be so mild as to lose negligible quantity of only fine particles from cultivated fields or so severe that it results in the formation of a network of gully, sub-gully, sub-sub-gully, and so on – ravines rendering the cultivated fertile land unsuitable for crop production, i.e., complete degradation of once very fertile agricultural land. Consequently, the ravine land is classed as wasteland. It is not only a loss of nonrenewable land resource but leads to other processes destructive to national economy, such as floods in rivers, siltation of water reservoirs and consequent loss in their storage capacity, and damage to railway lines, roads, and other public properties. Ravines particularly those along river Chambal encroach upon inhabited villages destroying the residential houses and dividing a village into many parts and thereby making the life of their inhabitants miserable. Ultimately these villages must be shifted to new sites to avoid loss of lives. Besides these economic losses, an important socio-economic problem created by Chambal ravines in Madhya Pradesh, Uttar Pradesh, and Rajasthan states of India is providing shelter to dacoits and consequently making the area dacoit-infested. The Chambal region of Madhya Pradesh has provided huge number of warriors to the nation and is blessed with very fertile soil and good quality of water. The region is a major producer of oilseeds, pulses, and the milk. But unfortunately, the precious land now continues to be converted into ravines. The need to manage the ravine land to control ravine formation and to reclaim those already formed cannot be overemphasized. It is necessary not only for improving ecology and socio-economic condition of the people but also for the survival of present and future generations (Tomar et al. 2015). Therefore, an effort has been made to assess the situation of ravines of Chambal valley and review the efforts made so far to reclaim these.

## 11.2 How Are Ravines Formed?

It is well established that water erosion is initiated through three consecutive processes, viz., (1) dispersion of soil particles by impact of raindrop (splash erosion), (2) movement of fine particles with runoff particularly on flat (slope <1%) land with less permeable soil (sheet erosion), and (3) formation of rills by concentrated runoff. Tejwani et al. (1975) observed that gullies owe their origin to rill erosion on erodible soils by runoff water. Even on apparently level land but with highly erodible soils, rill formation on surface takes place due to slight depressions in the field and consequent concentration of runoff to some extent. With intense rains, much quantity of concentrated runoff with many times higher abrasive power runs through these rills taking away large chunks of soil along with it. Thus, rills are converted into gullies, and these get deeper and wider with time due to scouring of gully bed at first and then falling away of both banks. The problem does not stop here as gully head also starts falling and the gully extends length wise as well. Extension in length of gullies is usually faster than in width, and this is how the gully advanced into new area toward the ridge, i.e., in opposite direction of flow of runoff. Not only this, a gully gives outside gully (sub-gully) which in turn forms tertiary gully (sub-sub-gully), and thus land destruction continues unabated. How fast and to what extent all this happens depends on erodibility of the soil and erosivity of rainfall.

On highly erodible soils, a network of shallow gullies is produced at a fast rate in the form of shallow ravines to start with and then medium and deep ravines later on. The process continues to form deeper and deeper ravines with time. As already stated, depth of ravines goes on increasing year after year till its bed level is the same as that of a natural stream into which the ravines discharge the runoff water carried by them. Thus, the depth of ravines is ultimately governed by the bed level of the rivulet or the river, the banks of which have been converted into ravines. Once the bed level of ravines is the same as that of the river, there is no further deepening of ravines with time, and thus the depth of ravines gets stabilized. But by this time, on both the banks of river, a huge scenario of ravines is created. That is, whole of the land on both sides of river is covered by deep to very deep ravines (depending upon soil type), and the land surface is almost lost completely and remains intact (somewhat) as narrow and pointed vertices of the steep banks of ravines in the whole area. While ravines may be stabilized near natural watercourse after attaining certain depth, the gully heads of ravines may be still active eating into the fertile cultivated land, if no measures are taken to control the advancement of the gullies. The process of rain-induced erosion is illustrated in Fig. 11.1.

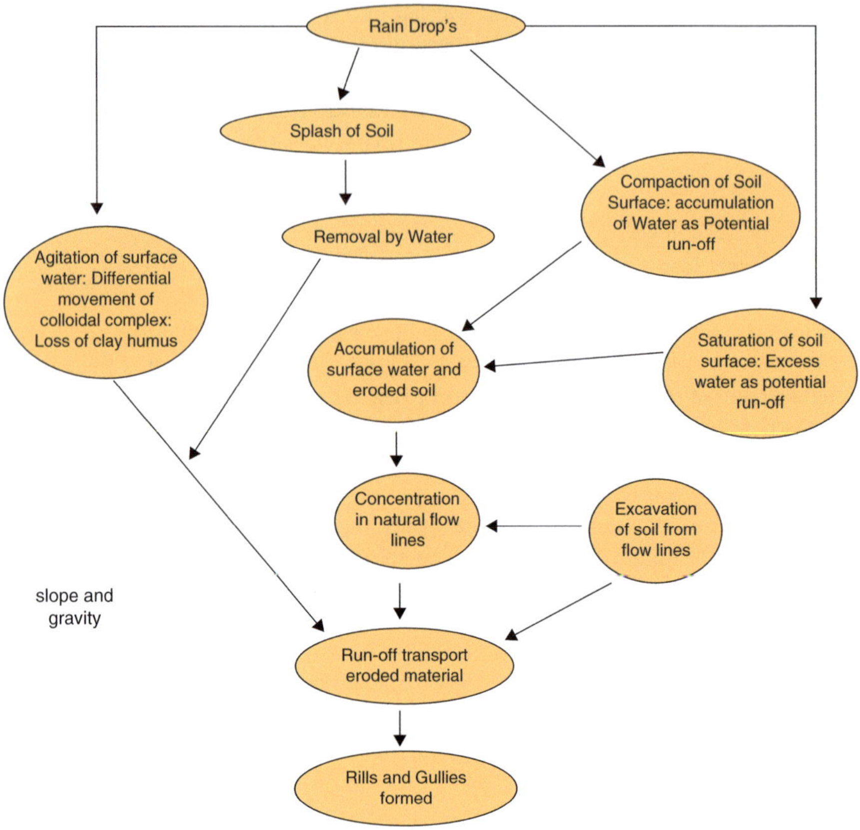


**Fig. 11.1** Process of rain-induced erosion

### 11.2.1 Other Views

Ravine ecosystems occur worldwide and the phenomenon of ravine evolution has been discussed in detail. Some workers related it to the climate change causing modification to ground cover, concentration of runoff, and changes to the level of groundwater tables. Others believed that it was caused by human activities which induced the concentration of surface flow. The arguments of both groups were so well documented that reviewers had to conclude that both explanations could be right (Haigh 1984). Most of the scientists agree that ravine formation was due to inappropriate management of rainwater. Further, according to Sobolev (1948), Ollier and Brown (1971), Tejwani et al. (1975), and Schwab et al. (1981), the following five steps were involved in the development of a ravine system: (1) origin of a rill by runoff water; (2) conversion of the rill into a small gully; (3) enlargement of gully, depth- and widthwise; (4) development of a sub-gully on main gully; and (5) repetition of the process on main gully and sub-gully till gullies change into a ravine system. However, Sharma (1979, 1980) opined that ravine originates as a shallow hole created by subterranean solution of calcium carbonate nodules. But this idea is not widely accepted because ravines are formed even on soils which have no calcium carbonate nodules in soil profile. In India, ravine evolution does not appear to be related to climate too, unlike the opinion of scientists in the west. The average annual rainfall varies widely in various important ravine zones of India. In Chambal-Yamuna zone, it varies from 800 to 1300 mm, while in Gujarat-Rajasthan zone, it is 500–750 mm, in sub-Himalayan zone 1125–1255 mm, and in Chota Nagpur Plateau 1125–1500 mm (Sharma 1980). Roy and Mishra (1969) have made a distinction between the gully and the ravine. Both gully and ravine originate as a rill, and the casual agency is "runoff" water – a part of rainwater, the quantum of which is a function of catchment (land slope and texture, permeability, and dispersion coefficient of soil). Now, whether a gully stays as a gully or gets converted into a ravine depends on fragility and depth of soil, erosivity of rainfall, human interference, and time lapsed after initiation of rill formation. The maximum depth of a gully/ravine is governed by the bed level of the natural stream (*nala*, rivulet or river) into which it discharges the runoff. The gully/ravine remains active, i.e., bed erosion, etc. continue till its bed level is in equilibrium with bed level of the natural stream. Further, if drainage lines of a micro-catchment of the ravine land are closely examined, its main gully and sub-gullies can be identified.

## 11.3 Advancement of Chambal Ravines

The ravines on the bank of Chambal river where it flows through alluvial tract are very deep (depth up to 70 m) all along their length which may be 3 km or more. The advancement of gully head cuts of these ravines toward the ridge takes place in the form of deep gullies only through a phenomenon characteristic of these ravines. That is, such ravines do not march ahead through slumping of gully heads by runoff water flowing on the land surface as in the case of shallow and medium ravines. On the other hand, because of high permeability of the soil and occurrence of a sandy/sandy loam horizon at a depth of 6–8 m or more from the surface of normal cultivated land, an underground water current of rainwater permeated through soil starts flowing toward the deep/very deep gully and carries soil particles with it creating a small cavity in the side (or bank) of the gully. The cavity grows with time due to underground water current, and soon a crack appears on the land surface about a meter or so away from the gully head. In due course of time, the crack widens, and ultimately the whole chunk of soil falls into the gully and gets washed away by runoff (Verma et al. 2012). Thus, the gully advances by a meter or more during the entire rainy season depending upon its duration and the intensity of rainfall.

At some spots, the cavity stays below the upper layer of soil till the end of rainy season, and the chunk of soil above the cavity falls off in the next rainy season leading to advancement of the deep gully only then. This hypothesis of gully advancement is supported by limited work conducted abroad. For example, Manendez-Duarte et al. (2007) studied relations between relief factors, type of substratum, and the development of gully erosion and reported that (1) there are robust correlations between morphometry, relief, and erosion intensity and (2) gullying is strongly influenced by the presence of a deep regolith, and gully development is associated with land sliding process which is ultimately controlled by groundwater movement through a porous regolith. Regional uplift and deforestation are the most likely causes of channel incision and gullying. Bergonse and Reis (2016) worked on large gully systems in tributary basins of the Tagus river (Portugal) and found that gully systems initiated as bank gullies in response to an incision by the drainage network in the basins and then evolved as a function of upslope contributing area developing progressively more sinuous outlines because of head cut bifurcation. The pre-erosion topographic form was shown to have only a marginal effect with results suggesting that gully systems have grown essentially through mass movements in head cuts and side walls influenced by subsurface water.

To conclude, rills develop into gullies by the abrasive force of runoff water, and gullies get deeper and wider depending upon the soil depth, resulting into formation of shallow, medium, and deep gullies on both alluvial and black soils. Further on highly permeable and very deep soils with a sandy layer at a depth of 6–8 m, the surface runoff appears as underground water and starts getting collected in the porous (sandy) layer. But if this layer meets or opens into a very deep (>10 m depth) gully, the underground/subterranean water starts flowing laterally through the porous (sandy) layer (exposed to air in very deep gully) taking away with it the soil mass and thus creating a cavity in the side wall of the gully. The cavity enlarges with time. When the cavity is big enough (both in diameter and length), the soil layer up to surface and above the cavity becomes instable, and a crack appears on the soil surface. Soon this whole chunk of soil falls off resulting in advancement of very deep gully as very deep gully only. The underground current of water cannot be stopped by afforestation or by gully plugs constructed in the deep/very deep gully for reclamation purpose.

## 11.4 Classification of Ravines

Ravines have been categorized in various classes based on different criteria. The most common classification is that based on dimensions of ravines.

### 11.4.1 Based on Depth

Different workers have suggested different standards of the depth for classification of the ravines they have studied, as shown in Table 11.1.

**Table 11.1** Classification of ravines based on depth

| Class | Bhulyan (1967) | Seth et al. (1969) | Tejwani (1974–1975) | Gupta and Prajapati (1983) |
|---|---|---|---|---|
| Very shallow | – | Depth <1.5 m<br>Bed width up to 3.0 m | – | – |
| Shallow | Depth <1 m | Depth <1.5 m<br>Bed width >3.0 m | Depth <1 m; bed width up to 18 m | Depth <1.5 m |
| Medium | Depth 1–5 m | Depth 1.5–5.0 m<br>Bed width >3.0 m | Depth 3–9 m<br>Bed width 18 m<br>Side slope 8–15% | Depth 1.5–5.0 m |
| Deep | Depth >5 m | – | Depth 3.0–9.0 m<br>Bed width >18.0 m | Depth 5.0–10.0 m |
| Very deep | – | – | Depth >9.0 m<br>Side slope steep to vertical | Depth >10 m |

### 11.4.2 Classification of Ravines Based on Reclaimability

Bali and Karale (1977) divided ravines into five classes – Classes I–V – considering their depth; width; bed slope; hard pan or rock, especially on the inter-gully land; and groundwater level.

Class I: Very good ravine land that can be reclaimed readily with minor reclamation measures. These include shallow ravines, width 30 m or more, no calcareous layer within 1.5 m depth.

Class II: Good ravine land that can be reclaimed with minor operations requiring more dozing work than Class I. Bed slope 5–10%.

Class III: Moderate land with medium depth ravines, bed slope 10–15%, calcium carbonate hard pan or bed rock within 0.5–1.0 m depth from surface.

Class IV: Poor land with medium to deep gullies of narrow width, hard pan or bed rock within 0.5 m depth, very costly to reclaim or may not be suitable for cultivation after reclamation.

Class V: Land unsuitable for reclamation and can be developed for forestry/grassland.

The classification seems to be based on extent of leveling effort required for reclamation. Leveling of ravine land for its reclamation has already been proved to be most expensive yet not lasting permanently. Also the land with deep ravines is unsuitable even for reclamation for alternate land use, e.g., agroforestry or silvopastoral systems.

Kandrika and Dwivedi (2013) measured bed width and depth of ravines successfully using cartosat-1 (an Indian earth observations satellite) stereo image with

2.5 m spatial resolution. They delineated three categories of ravines, namely, shallow (<3 m deep and <18 m bed width), medium deep (3–9 m deep and >18 m bed width), and deep (>9 m deep) for their reclamation.

### 11.4.3 Classification of Ravines According to Shape (Cross Section)

Based on shape of the ravines, these can be divided into two classes, viz., U-shaped and V-shaped. U-shaped ravines are formed by runoff water on alluvial soils (entisols) which are very fragile (having very high rate of soil erosion) throughout their whole depth. V-shaped ravines are formed on black soils (vertisols and associated soils) which are not so erodible, particularly in lower layers. That is, in vertisols, surface soil is easily eroded and washed away, while lower layers resist soil erosion. However, very deep ravines (depth greater than 30 m) on Chambal banks tend to be V-shaped. It is quite probable that initially the surface runoff had carved U-shaped medium/deep gullies, but later on subsurface current became active as soon as the depth of gully was greater than the depth at which sandy/sandy loam layer existed in the soil profile.

## 11.5 Extent of Chambal Ravines

India has four major areas of severe ravine erosion and the largest is Yamuna-Chambal ravine zone (Sharma 1980). Ravines flank river Chambal all along its length from its origin to its confluence with Yamuna river. Chambal is an interstate river originating in Madhya Pradesh (MP) and flowing through MP, Rajasthan, and Uttar Pradesh (UP). Accordingly, parts of its catchment lie in Rajasthan and UP besides MP in which major part of its catchment lies. Chambal, a major tributary of river Yamuna, emerges from Singar Chori peak of northern slopes of Vindhyan ranges – a spot about 14.4 km southwest of Mhow town of Indore district at an elevation of 854 m above mean sea level (MSL). Initially, it flows northward in MP for 320 km, and after passing by the historic fort of Chaurasigarh, it flows northeast for about 286 km through Rajasthan. Continuing in the same direction (northeast) further, it forms the boundary between Rajasthan and MP along Sheopur and Morena districts for a length of 216 km. Later, it travels eastward and southeast as a boundary between MP and UP along Bhind district for about 112 km. Finally, after flowing for 64 km in UP, it joins the river Yamuna. Thus, Chambal valley extends through MP, Rajasthan, and UP states, and the river Chambal with its tributaries forms the infamous Chambal ravines. The area under Chambal ravines in the three states is shown in Table 11.2. The total area under ravines formed by Chambal and its tributaries is 482.7 thousand ha of which 193.4 thousand ha is in MP, 210.3 thousand ha in Rajasthan and 79.0 thousand ha in UP.

Chambal ravines represent the most severe land degradation and, therefore, can be rightly called "cancer of the land" as they have so far challenged all efforts of

**Table 11.2** Extent of Chambal ravines in MP, Rajasthan, and UP (river-wise in 000 ha)

| Madhya Pradesh | | Rajasthan | | Uttar Pradesh | |
|---|---|---|---|---|---|
| River | Ravine area | River | Ravine area | River | Ravine area |
| Chambal | 95.0 | Chambal | 130.3 | Chambal | 79.0 |
| Asan | 18.2 | Kali Sindh | 40.0 | | |
| Sank | 2.3 | Parwati | 40.0 | | |
| Seep | 13.6 | | | | |
| Parwati | 57.2 | | | | |
| Kuno | 7.1 | | | | |
| Total | 193.4 | Total | 210.3 | Total | 79.0 |

Source: Action Aid, India (1996)

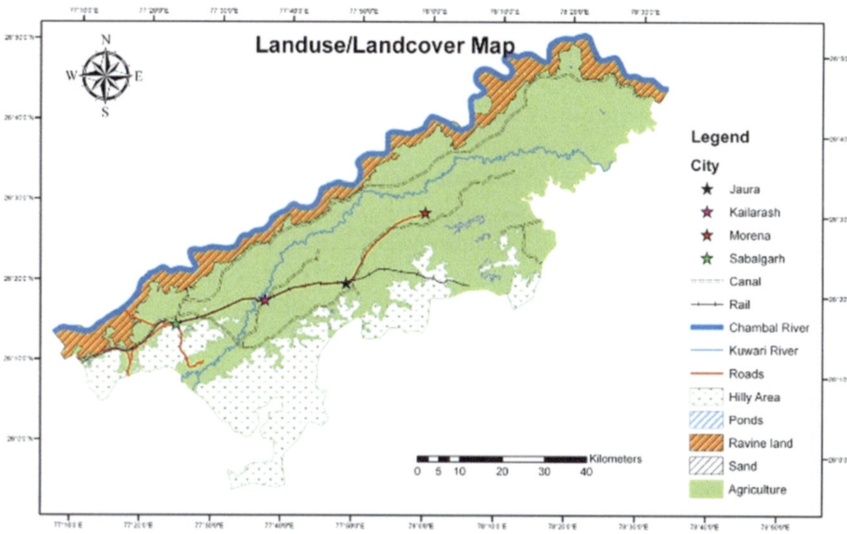


**Fig. 11.2** Chambal ravines in Chambal division (MP)

their control and reclamation. The severity varies according to (1) nature and properties of the soil occurring in the catchment areas of river Chambal and its tributaries and (2) configuration of the bed of Chambal river from its origin to the point of joining the river Yamuna (Fig. 11.2).

Chambal valley is covered by two prominent soil types, viz., black soils (vertisol and associated soils) and alluvial soils (entisols). The vertisols are not so erodible and only V-shaped gullies are formed by runoff water. But alluvial soils of the Chambal valley are highly fragile and have very high rate of soil erosion due to their high dispersion coefficient. On alluvial soils, a network of gullies is formed rapidly, as already explained, and deep gullies are U-shaped with broad base and almost vertical banks, but very deep gullies, formed on the bank of river Chambal where it flows through alluvial soil, are V-shaped with almost vertical sides. The configuration of bed of river Chambal and its tributaries, i.e., their bed slope and bed elevations above MSL, governs the degree and extent of erosion caused by them. The bed level controls the maximum depth of ravines formed by its tributaries, while the bed slope decides the velocity of flow of the river/tributary and the consequent erosive power. The steeper the slope of bed, the greater the degree of bed corrosion, and because of this, the deeper the ravines formed. An abrupt fall in altitude between the main Vindhyan hill range, situated in southwest and western part of MP, and the course of Chambal river in northern and northeastern directions together with alluvial deposits in between results in heavy soil erosion in Chambal valley. The bed level of river Chambal controls the maximum depth of ravines formed by it and its tributaries.

## 11.6 Configuration of River Bed of Chambal

To understand the magnitude and severity of ravines in Chambal valley, it is necessary to know the cross section of the bed of Chambal river from its origin to its point of merger with Yamuna river (Table 11.3).

As seen from Table 11.3, the point of origin of Chambal river is at an elevation of 854 m above MSL, and at a point about 16 km away from the origin, the elevation of river bed is approximately 532.48 m above MSL. This represents the biggest drop (321.52 m) in the bed level of the river over its length of 16.1 km. Consequently, the bed slope is also much steeper (about 2.0%) over the stretch as compared to remaining part of its run. The second noticeable drop (32.2 m) in the river bed is over 230.2–246.3 km stretch of the river, i.e., again over 16.1 km length, there is a drop of 32.2 m which means a bed slope of about 0.2%. Both these drops in the river bed are in the upper reaches before the Chaurasigarh gorge. The third comparatively big drop in the river bed is below this gorge over 386.4–429.9 km stretch, and it is quite mild – a drop of only 69.17 m over 43.5 km length of the river, i.e., a bed slope of 0.16%. Except in these three stretches, the river bed slope is quite gentle. Further, Chambal river bed is not stable in these three stretches; the river bed is likely to degrade further to achieve a dynamic equilibrium with the vertical drops. After first stretch, i.e., beyond 16.1 km length, the elevations of the river bed are much lower as compared to those of tableland from where tributaries of the river Chambal

**Table 11.3** Cross section of Chambal river bed

| Distance from origin (km) | Elevation (m above MSL) | Drop in bed (m) | Slope of bed | |
|---|---|---|---|---|
| | | | m $km^{-1}$ | Percent |
| Up to 16.1 | 854.00–532.48 | 321.52 | 19.97 | 2.00 |
| 16.1–230.2 | 532.48–265.00 | 267.48 | 1.24 | 0.12 |
| 230.2–246.3 | 265.00–232.80 | 32.20 | 2.00 | 0.20 |
| 246.3–386.4 | 232.80–138.93 | 93.87 | 0.67 | 0.07 |
| 386.4–429.9 | 138.93–69.76 | 69.17 | 1.59 | 0.16 |
| 429.9–606.0 | 69.76–32.56 | 37.20 | 0.20 | 0.02 |
| 606.0–998.0 | 32.56–0.00 | 32.56 | 0.08 | 0.01 |

originate. The vertical fall between the riverbank and the river bed varies from 10 to 20 m over the black soil region of the Chambal valley, and that explains why gullies/ravines are comparatively less deep in this region as the vertical drop of the tributaries is a highly influencing factor for gully and ravine development. On the other hand, the vertical fall between the riverbank and the river bed is much higher after the river enters the alluvial soil zone, i.e., beyond 600 km of its initial run. This high fall associated with fragile nature of alluvial soil results in severe ravines formation as deep as 70 m and degrading the land completely into sharp triangular vertices. There is no intact land left between vertices of two adjacent and very deep gullies.

## 11.7 Scenario of Chambal Ravines

The landscape of Chambal ravines includes scenarios of different types of ravines formed in watersheds of various tributaries or river Chambal joining it on either side, as well as the ravines (deep) formed along the Chambal river itself. From Kota to Dholpur, the Bundi-Ranthambore line of hills and, beyond Dholpur, the ravines of Chhoti Parvati form the northern boundary of the Chambal ravines. The southern boundary consists of the watersheds of tributaries of Chambal, viz., Kali Sindh, Parvati, Kuno, and Seep. Beyond these rivers, the southern boundary is well marked by the ravines of Kunwari river. As already stated Chambal River has its initial run in MP on steep land to start with and then on gradual slope land entering the Bundi district of Rajasthan through a gorge. Heading toward northeast, the river is joined by its tributaries on both sides. These tributaries carve out an intricate network of ravines on both sides of river Chambal which itself also forms ravines along its banks. The ravines of Chambal river and major tributaries, viz., Talera, Ghorapachar, and Kural joining the main river on its left bank, form a narrow, elongated belt along the piedmont of the northern Bundi hills. On the right bank, the river Chambal is joined by its tributaries, viz., Kali Sindh, Parvati, and Kuno rivers, which along with main river form lot of gullies as their basin is covered by black soil (medium in depth) and elevation of Chambal river bed is also not too much below the general level of the adjoining catchment areas of the tributaries. Consequently, the ravines of this triangular area are characteristically narrow, V-shaped, parallel to each other, and 3–4 m deep and have stabilized with time, though these appear to be at comparatively younger stage of development. Further down, after the confluence of the Banas and Seep rivers with Chambal near Pali at Rameshwar, it is seen that with increasing depth of alluvium, the depth of the ravines increases considerably, the shape of ravines changes from V-shape to U-shape from Pali to Dholpur. On the left bank of the Chambal, there is a terrain of wastelands stretching over the southern part popularly known as Dangland. From Dholpur downward, the ravines are more stark and devoid of vegetation. They are characteristically U-shaped with vertical walls and flat bottoms. The depth varies from 30 to 60 m. This dissected tract consists of long strips between the Chambal and the Yamuna and the Chambal and the Kunwari and tributaries like Asan. They seem to erode each other's banks cutting into the adjacent areas. This erosion has resulted in the formation of numerous platforms, knolls, and plain bluffs. On the right bank of Chambal also, after Kuno river basin, a belt of ravines nearly 3–4 km wide exists up to Morena, like that on the left bank already described. Further eastward, these ravines almost merge with the ravines of Yamuna and Kunwari and make a bewildering network of ravines. The Chambal-Yamuna confluence is joined by Kunwari, Sindh, and Pahuj resulting into a big junction of five rivers, called locally as Pachnada, near the border of Bhind and Etawah districts. The entire countryside in Bhind and Etawah district where Chambal has largely dissected the land into irregular shapes is made of steep ridges, deep trenches, and low sloping hills and thus divided itself into many streams. During rainy season, these streams become flooded and are difficult to cross over. The areas neighboring the junction of these streams with Yamuna and Kunwari are wild and untamed yet picturesque and present a unique view. As far as the eye can see, there are labyrinths of dissected ravines and some valleys green with Acacia trees.

## 11.8 Recurring Spread of Ravines

Formation of ravines is a continuous process and it is faster and more severe on the alluvial soil of Chambal valley. The soil is very susceptible to erosion as it has sandy loam texture and a very high dispersion coefficient –0.8 to 0.9 (Verma et al. 2016). The soil cannot stand the impact of raindrop and starts running ahead the runoff water (Verma 2003). The march of ravines continues and in Chambal Division alone 809 ha land is converted into ravines every year, rendering the adjoining productive land unfit for cultivation. A study by Chambal Command Area Development Authority, Gwalior, revealed that in Chambal Division, ravines covered 22.83 thousand ha in 1943–1944 and 31.07 thousand ha in 1975–1976 an increase of 82,400 ha in 32 years, i.e., 2575 ha per year. They further brought out that during the earlier period of 1943–1950, the growth of ravines was 1380 ha year$^{-1}$, where as in the later period of 1950–1951 to 1975–1976, it increased to 2130 ha year$^{-1}$. Singh and Rao (1996) used remote sensing techniques and reported that the area under gullies and ravines in 66 villages of Morena district increased to 218.695 km$^2$ (21,869 ha) in 1994 from 130.15 km$^2$ (13,015 ha) in 1969, i.e., an increase of 88.545 km$^2$ (8854 ha) in 25 years. This amounts to a land loss of 354 ha year$^{-1}$ or more than 5 ha year$^{-1}$ in each village. A survey conducted along the Chambal river by Rajasthan Forest Department in 1969 estimated that ingress rate for the river catchment was 0.869% over a period of 20 years (Singh 2014). The ravines attain depths ranging from 5 to 70 m. The ravines are not only destroying agricultural land but gobbling up village after village forcing inhabitants of the village to shift to new sites. The process is continuing unabated year after year, and it is estimated that ravines have affected 948 villages in Bhind and Morena districts. The ravines are eating into the social life of the villages Nayakpura, Rubara, Ajitpura, Khandoli, Jaghina, Rithona, Mahuwa, Sessanio, and Gaushpur. In Porsa block, Ratanbasai village has split into eight segments.

The streets and roads have been destroyed, and it takes a tough walk of 3 km to cover all segments of the old village. Aisah village in Morena district has been already gobbled up completely by Chambal ravines and shifted to a new site on Morena – Porsa road. Mrigpura village in the same district is being devoured by ravines so rapidly that all the land around it has turned into deep pits and the village stands divided into several parts. The inhabitants survive on rainfed cultivation on flat bottoms of U-shaped ravines and by rearing livestock. Poverty, social injustice, and menace of dacoits have really exacerbated their problems (Tomar et al. 2015).

## 11.9 Earlier Efforts on Ravine Reclamation

The seriousness of the problem of ravine formation and their spread in Chambal valley area was duly realized during early years of the twentieth century by rulers of erstwhile Gwalior State and subsequently not only by Madhya Bharat State (now part of MP State) Government and the national government but also by international agencies such as the World Bank, European Economic Communities (EEC) now called European Union (EU), and so on. Consequently, many efforts at various levels were made to reclaim the ravines and put them to agricultural use.

As early as 1919, the ruler of the erstwhile Gwalior State appointed a commission with the following terms of reference: (1) to suggest ways and means to arrest further extension of ravines and (2) to suggest means for making available fodder, fuel, and timber for agricultural purposes to such villagers who had no access to forest nearby. Later, the Government of India invited Dr. Schuhart, an American expert on soil conservation, to suggest solution of the problem. He visited ravine-affected areas in 1945 and had suggested contour bunding starting from the ridge, control on grazing, and afforestation for soil conservation in the area. Based on these recommendations, numerous attempts were made to develop Chambal valley by reclaiming the ravines during the last few decades. Rightly the objective was to stop the large-scale land degradation and to restore the degraded land to some vitality and use for biomass production.

After independence, the Government of MP executed several ravine reclamation projects for Chambal ravines, viz., Chhonda project (1955–1956), Bagchini project (1955–1956), Nayakpura project (1956–1957), Deori Hingona project (1959–1965), Jawasa project (1962–1970), and Dimani-Chandpur project (1967–1968). Under these programs an area of 9080 ha was treated from the First Five-Year Plan up to 1970–1971, comprising 3100 ha of ravine reclamation and 5980 ha of ravine afforestation at a cost of INR 4.423 million. The Government of India has invested heavily on measures to control soil erosion. Vast sums of money have been allocated to soil conservation in various 5-year plans during 1969–1990. The budget for soil conservation for this period was INR 16 billion (Kerr and Sanghi 1993). In 1971, ravine control scheme started under central sector protecting 13,500 ha tableland and leveling 3000 ha land, at a cost of INR 20.12 million up to 1981.

Later, Chambal Multipurpose Hydel Project for Chambal valley was started as a joint venture of the Government of the Madhya Bharat (now part of MP) and

**Table 11.4** Earlier ravine reclamation work (Action Aid, India 1996)

| S. No. | Scheme | Sector | Period | Area covered (ha) | Expenditure (INR million) |
|---|---|---|---|---|---|
| 1. | Ravine reclamation | State | 1955–1973 | 1240 | 3.020 |
| 2. | Ravine reclamation | Central | 1970–1984 | 1748 | 10.215 |
| 3. | Ravine erosion control (CADA) | State and World Bank | 1976–1987 | 13,500 | 20.120 |
| 4. | Dacoity-prone area development program (UP and MP) | Central | 1988–1992 | 8005 | 36.244 |

Rajasthan to exploit the irrigation potential of the river Chambal. Accordingly, besides construction of dams and network of canals to irrigate 28.3 thousand ha in Chambal Division, rectangular fields were created by large-scale land leveling using bulldozers on the part of command area under ravines. But soon there were problems particularly with rectangular fields, and the World Bank supported Command Area Development Project was taken up in 1975 with a financial credit of INR 24 million. But the results were again not worth the investment (Anonymous 1997). In 1980, a World Bank Project was launched in Chambal Command Area, covering 82,000 ha land under aerial seeding of tree species. Unfortunately, the seed did not stay on slope and gathered on the bed of deep ravines. The result was that not a single plant grew on slopes but a thick forest of *Prosopis juliflora* developed on the bed of ravines. Being thorny plants, the thick vegetation pushed the wild blue bulls (locally called *neel gai*) to farmers' fields ruining their crops. During 1988–1992, ravine reclamation was undertaken in MP and UP as per Government of India's instructions under Dacoity Prone Area Development Program. Construction of 476 km of peripheral bund along with tableland treatment of 5229 ha and shallow ravine treatment of 27,776 ha was undertaken at a cost of INR 36.244 million in MP (Table 11.4).

During 1987–1994, European Union (EU) funded a project entitled "Integrated Watershed Management Program in Ravine Areas of Chambal and Yamuna Catchments" on 48,000 ha land in Agra and Etawa districts at a cost of about INR 800 million. Very good work on land development (leveling/bench terracing), introduction of improved and high-yielding varieties of crops, development of roads, and introduction of tube-well irrigation was undertaken, and immediate results were startling. But the approach was very expensive. Leveling of land to create rectangular fields with pipe outlets for runoff from field to field worked very well initially but gave up after some years due to lack of maintenance, the requirement of which was rather high. Alternatively, if an effort was made to manage the runoff/rainwater on watershed basis by converting main gullies into grassed waterways, allowing rainwater to run from field to field over grassed field bunds/appropriate cheap outlets and finally into stabilized waterway through appropriate wasteweirs, the rainwater surplus arrangement would have been much cheaper yet more stable and needing less maintenance as observed by Monitoring and Evaluation Team of Water and Power Consultancy Services (WAPCOS) (Verma 1999a, b).

### 11.9.1 Why Were Earlier Efforts Not Fruitful?

While the projects on ravine reclamation undertaken in MP state were of pioneering nature in the country, the benefits derived from these were not commensurate with the investment. In all these projects, either the priorities for multi-sectoral development were not provided, or even if provided, these were predominantly devoted to development in one or two selected sectors. The scope of work and the nature of funding in most cases were too limited to have any multi-sectoral impact. Invariably, all these projects were treated as purely government schemes because the soil conservation service of the State Government was the main player and executioner of the project and people's participation was more of a rhetoric rather than reality. Other reasons for the failure of these projects were:

1. Watershed approach for simultaneous treatment of tableland, peripheral land, and the ravine land was completely lacking.
2. Emphasis was given only on reclamation of ravines and not on prevention of formation of ravines by managing the runoff on watershed basis.
3. Whatever technology was evolved by scientists could not be transferred to farmer's fields as such because farmers had no trust in soil conservation service. Also, too much emphasis was laid on dozing for leveling and bunding for making rectangular fields for cultivation. There were breaches in field bunds as no appropriate provision was made for surplus runoff to get out of field safely (without causing erosion).
4. Appropriate and adequate size structures were not made perhaps due to high cost and lack of basic information on rates of runoff, rainfall intensity, etc. In absence of such information, over-designing was necessary but not followed perhaps because of cost factors. The well-built masonry structures were still there but surplus runoff did not pass through them. Instead, the bund breached on either side of the structure and made a way for the runoff. At some places, runoff tunneled through under the masonry crest wall, leaving the structure intact but redundant. The probable reasons for such failures could be (i) undersize of structures, (ii) inappropriate location of these, and (iii) lack of integration of vegetative cover and mechanical structures.
5. Ravine afforestation was not always undertaken in the same watershed/micro-watershed where other measures for ravine reclamation were undertaken. This showed lack of coordination between different line departments, viz., soil conservation and forest department.

### 11.9.2 Watershed-based Management of Ravine Lands

#### 11.9.2.1 Management of Rainwater

Since formation of ravines is a function of uncontrolled runoff, the management of rainwater on watershed basis is the only appropriate approach for the control and reclamation of ravines (Verma 1983). Watershed-based management of land and water resources for gully control and consequent enhanced and stable agricultural production has already been demonstrated on farmers' fields at Indore, MP (Anonymous 1979; Verma 1982). This experience of black soils cannot be transferred to alluvial soils for ravine control and reclamation as such because alluvial soils particularly in Chambal valley have much higher dispersion coefficient and pitching of earthen bunds/banks with boulders is not successful unlike black soils. Also, the problem of caving under the structures is serious. Bhushan et al. (1992) rightly recommended construction of peripheral and contour bunds on light-textured soils for ravine control. However, their suggestion for bench terracing and leveling to reclaim very small, i.e., very shallow (up to 1 m deep), small, i.e., shallow (<3 m deep), and medium gullies (3–9 m deep) is not right because bench terracing and leveling are too expensive and therefore not needed for very shallow and shallow gullies. Also, ravines are not widely spaced gullies everywhere but a network of gullies, and the inter-gully spaces are too narrow strips of land to construct bench terraces particularly in case of very deep ravines. Again, Bhushan and Yadav (2000) divided watershed with ravine land into four zones, viz., (1) zone of conservation and control, (2 and 3) zones of conservation and reclamation, and (4) zone of backwater flow. Apparently, they did not consider the whole watershed as a single unit and failed to suggest measures for management of rainwater on watershed basis to control formation of ravines as well as their reclamation. Obviously, what is needed is foolproof technology based on local experience, transferred to farmers' fields with their cooperation and active involvement (Verma 1987). Land and water management works must be planned for the entire watershed starting from the ridge and including the tableland which usually exists just below the ridge and above the ravine land.

#### 11.9.2.2 Tableland

Usually it is flat or gradually sloping and thus runoff water may not cause serious erosion. However, to increase the productivity of such land, efforts must be made for in situ water conservation and more profitable use of the conserved water. For this, field bunding (in alluvial soils) is necessary. If the average rainfall of the area is less and infiltration rate of the soil high, field outlets may not be necessary as most of the rainwater will seep into groundwater quickly. In case of high-intensity rain, the runoff may flow over these grassed field bunds (15–20 cm high). If bunds are constructed afresh, compaction of bunds and planting of grass may be done, or natural grass may be allowed taking over to create a vegetative cover on the bunds. Bunding will prevent the sheet and rill erosion by interfering with free flow of runoff, particularly if the field is cropped during rainy season.

#### 11.9.2.3 Peripheral/Water Diversion Bund

Between the tableland and the ravine land, a peripheral or water diversion bund must be constructed to lead the runoff from tableland to a safe place (water reservoir, farm pond, or a stable grassed waterway depending upon the location) . Peripheral bund is essentially a graded bund the cross section of which will depend upon the runoff to be tackled and the type of soil. The bund should be covered with vegetation like grasses of soil conservation, agave (*Agave americana*), munj (*Saccharum munja*), kans (*Saccharum spontaneum*), vetiver (*Vetiveria zizanioides*), etc. on the upper side (the side along which water will flow) and the utility trees on the outer side (side facing ravine land) (Singh and Verma 2016). Suitable deep-rooted tree species like babul (*Acacia nilotica*), shisham (*Dalbergia sissoo*), or fodder-cum-fuel species like subabool (*Leucaena leucocephala*) or fruit trees like plum (*Prunus domestica*), guava (*Psidium guajava*), aonla (*Emblica officinalis*), jamun (*Syzygium cumini*), etc. can be planted.

### 11.9.3 Shallow and Medium Gullies/Ravines

For ravine control and reclamation of shallow and medium gullies/ravines, the following steps must be taken:

(a) Identification of main gully.
(b) Its stabilization by bund across the general slope and with a drop structure (gabion or masonry structure) of appropriate size. Such bunds will be repeated at different points on the main gully and perpendicular to it (Anonymous 1979). The center of the structure should be on the flow line of the gully, and its foundation should be at least 30 cm below the flow line.
(c) Smoothening of land on both sides of the gully (between two bunds) to reduce the slope of land below 3% on black soil and below 1% on fragile alluvial soil (on which erosion can be checked by good crop cover in rainy season with crop rows sown across the slope and adopting other conservation practices). Leveling of hummocks and isolated steep banks of secondary/tertiary ravines is to be done with minimum disturbance to the topsoil.
(d) Planting of *munj* (*Saccharum munja*) grass or vetiver (*V. zizanioides*) on the bunds to stabilize these.
(e) Construction of runoff collection ponds/tanks at suitable sites.

The technology is based on the following assumptions: (1) surface runoff causes formation of shallow and medium gullies, (2) division of micro-watershed into small segments by construction of contour/graded bunds reduces the amount and velocity of runoff to be diverted along the bund to the stabilized (main) gully, (3) reduced velocity does not enable runoff to form rills, and (4) stabilization of main gully through drop structures raises the bed level and thus reverts the process of bed erosion.

### 11.9.4 Treatment of Deep Ravines

Lower parts of the watershed have deep and broad (U-shaped) main gullies on alluvial soil and V-shaped gullies on black soil. The steps (b) to (e) as above can be continued to raise the level of the bed of main gully and grading the land on both sides to the periphery of the watershed to reduce the slope to less than 3% on black soil and to less than 1% on fragile alluvial soil. If broad U-shaped gullies are formed wide apart, i.e., if inter-gully space is sufficient, terracing of banks of the gully can be done.

### 11.9.5 Treatment of Very Deep Ravines

The reclamation and control of very deep ravines formed in central and lower Chambal valley where the river Chambal flows through alluvial soil tract is very difficult and expensive. As a matter of fact, no proven technology is available for it, and whatever technology is recommended by research workers so far does not work. Earthen gully plugs and afforestation of deep ravines has been recommended and tried in the past without much success. The idea of reclaiming deep/very deep ravines by just afforestation does not work because of too steep and deep sides of gullies and very narrow and pointed tops (Fig. 11.3.). Moreover, these are

**Fig. 11.3.** Very deep Chambal ravines

inaccessible. In a World Bank funded ravine reclamation project, aerial spray of seeds on ravine land was undertaken with disastrous results.

## 11.10 Recent Efforts of Reclamation of Ravine Lands

During past two decades, several efforts have been made for reclamation of ravine lands. Some of these have been described in brief here.

### 11.10.1 Anicuts and Afforestation

In 2007–2008, the forest department of MP Government undertook a project on reclamation of Chambal ravines through anicuts and afforestation. It was an innovative approach by a team consisting of foresters, local farmers, and village leaders. The construction of anicuts and afforestation was taken up in forest land of Bindwa village, district Morena, and of Himmat Pur village, Bhind district. An impact study of the project led to the facts that (1) farmers' land in the adjoining area was protected against formation of ravines, (2) productivity of various crops was improved many times, and (3) project works paid heavy dividends and benefit/cost ratio was 5.89:1 for Bindwa village land and 7.31:1 for Himmat Pur village. Hence, anicut construction is quite promising as a measure for management of very deep Chambal ravines (Singh et al. 2016).

### 11.10.2 NICRA Project

Some research work to manage very deep ravines for food and environment security was initiated under ICAR-funded National Initiative on Climate Resilient Agriculture (NICRA) project entitled "Identification of efficient integrated modules for sustainable management of ravines (Chambal) and carbon sequestration for climate resilience" at Aisah village, district Morena, MP. A heavy emphasis was laid on plantation of several species of native trees, fruit trees, and medicinal plants (the details of this experiment have been discussed by Singh et al. in this book). But the ravine land under deep ravines was inaccessible, and therefore dozing a path to the main gully and limited shaping to enable plantation of trees was necessary. A critical evaluation of various works executed under the project for reclamation and control of very deep ravines was as follows:

#### 11.10.2.1 Greening of Ravines

Lot of greenery was created in the area by plantation of fast-growing species and by rejuvenation of natural vegetation due to protection of the area. But the growth of trees, grass, etc. did not stop the march of deep ravines into fertile land.

#### 11.10.2.2 Steep Sides and Pointed Tops

A great problem in case of deep and very deep ravines was a network of these with no level land at all, i.e., side gullies and tertiary gullies were also as deep as main gully with steep sides and pointed tops. Also, the deep gullies and side deep gullies did not advance through cutting action of surface runoff, as in case of medium and shallow ravines. The advancement of shallow and medium gullies could be easily controlled by diverting the runoff through construction of a water diversion bund of appropriate size and design.

#### 11.10.2.3 Earthen Gully Plugs

Earthen gully plugs were tried knocking down the tops of banks of gullies, but due to high permeability of soil, runoff flowed through original flow lines, making holes in the bund (gully plug) at its base though it was quite broad at the base.

#### 11.10.2.4 Checking of Gully Advance

For this shaping of gully sides to reduce the slope of sides, i.e., making them slant (90° angle changed to 120°) and planting of munj (*Saccharum munja*), grass on the slanting sides was undertaken. The villagers had tried this technique long back but it did not solve the problem permanently. It rather delayed the advance.

#### 11.10.2.5 Shaping of Gully-Side Tops

Construction of terraces by knocking down the pointed tops of ravine banks using JCB machine (CAT Backhoe Loader 424B2) - to create level land and less slopping sides (>2:1) protected by strong vegetative barriers at suitable intervals. The performance is to be watched over years (Verma et al. 2016), but project period of 3 years was too short for this.

#### 11.10.2.6 Extent of Land Shaping

Ideal extent of land shaping was to cut about one third to half top part of ravine bank and pack the loose soil on the bed of gully. This created a comparatively broader plate form in place of a pointed top which can be used for growing a crop or fruit trees, i.e., crop alone or agri-horti- or silvopastoral system to stabilize the plate form and the soil filled part. Because of the loose soil, it will not be stable in spite of compaction and will be washed away with rainwater unless masonry/gabion structure is constructed at the end of gully or at its confluence with main gully. One more precaution to be taken here is that the foundation of the structure is at least 30 cm deep below the original flow line, so that there is no under-caving below the structure by runoff water.

#### 11.10.2.7 Multi-step Leveling System Technology

The top was cut using heavy earth-moving machine (Hithachi 110 or 220) so as to completely block the gully. Construction of peripheral bank and partial modification of slope at every leveled piece of land (step) is mandatory. The reclaimed flat land (35–50%) is suitable for crop production, side slopes (less slopping, 30–45%) for silvo- or horti-pastoral system, and plugged deep gullies for water storage. The cost of reclamation varies from INR 35,000 to INR 40,000 per ha (Verma et al.

**Fig. 11.4** Gabion structure with sediment

2016). The biggest weakness of the technology is poor runoff management. Complete leveling of ravine land failed for the same reason. Also, no rainwater could be stored against even strong earthen bunds attempted earlier in the project.

#### 11.10.2.8 Performance of Gabions

Gabions as drop structures have worked very successfully in stabilizing the medium and deep gullies on black soils (Anonymous 1979) as well as on alluvial soils (Verma 2003, 2009) where bed corrosion was caused by surface runoff (Fig. 11.4). But these were not successful in very deep gullies/ravines (formed on alluvial soil) where bed corrosion/gully head advance was through subterranean flow or underground current of water. The gabion structure seems to be too porous (with bigger voids) to check the flow of underground water. Alternatively, a masonry wall (part of crest wall of masonry drop structure) could serve as a dyke against the undercurrent. Accordingly, the gabion was modified to have a masonry foundation cum shoulder and a gabion upper part (upper crest wall). This modified structure worked successfully during the first rainy season but could not be observed for its performance later as the project was over. Thus, confirmation of this result is necessary. Also, in light alluvial soils, gabions should be constructed of dressed boulders instead of irregular-shaped boulders, the use of which creates too big voids to hold back silt (of runoff).

#### 11.10.2.9 Steep Bed of Main Gully

A serious problem faced in stabilization of a gully is slope of the bed of the gully. The bed slope is usually higher (>3%) starting from gully head. If bed slope is 3%, a gabion/masonry drop structure will have to be constructed every 33.3 m, i.e., three structures in 100 m length of the gully, and this will be too expensive.

Alternatively, an anicut against the slope and across the main gully can be constructed through whole width of the micro-watershed. Anicut is an earthen bund so constructed as to hold all the runoff and silt against it, and therefore, no surplus arrangement (construction of wasteweir) is necessary. However, in high rainfall areas, if the quantity of runoff expected is high, wasteweir in the form of a composite gabion is provided over the main flow line (main gully). The spacing of anicuts is decided to take into consideration the slope of the bed of main gully and the height of anicut which in turn depends on the size of anicut, i.e., the height, length, and breadth of the anicut are determined by the quantity of runoff to be stored against it. The height of anicut can be taken as the "vertical interval" to calculate the spacing of anicuts. Also, earthen bunds (anicuts) constructed in alluvial soils cannot serve as submersible check dams even after pitching these very well with dressed boulders fixed in cement over earthen dam. Pitching earthen dams with boulders is very successful in black soil region but not in alluvial soil tract (Fig. 11.5).

**Fig. 11.5** Anicut structure

#### 11.10.2.10 Appropriate Vegetative Cover

Above all, appropriate vegetative cover must be established to protect all the treated land against erosion which could result in formation of rills, washes, and gullies even on level land (due to high erodibility of ravine land/soil).

#### 11.10.2.11 Runoff Collection and Recycling

Runoff collection and its recycling and its use for recharge of underground water is a very important component of ravine reclamation and control technology. Runoff collection can be done in several ways depending upon the site in watershed. Runoff will get stored temporarily even against peripheral bund, contour/graded bunds, and submersible check dams, the construction of which will be necessary for the control of ravine formation. Also, at suitable sites farm ponds or tanks can be constructed to store runoff (Fig. 11.6). If at any site, construction of a peripheral bund is not feasible, construction of a bund along the boundary of two farmers both having ravine land along the boundary will be very appropriate to hold silt and water upstream the bund. The bund must be of an appropriate cross section according to the size of catchment above bund. Also, it should be compacted and protected by planting *munj* (*Saccharum munja*) grass, agave (*Agave americana*), etc. Further, it must be provided with a wasteweir of appropriate type and size to dispose excess of runoff. An important point for the site of wasteweir of tank was that it must be located above the natural flow line and its foundation should be deep enough to avoid undercaving below the wasteweir.

**Fig. 11.6** Runoff collection tank

#### 11.10.2.12 Appropriate Land Use Plans

Along with execution of land and water management works in a watershed, sustainable livelihood systems must be devised for different capability classes of land in the watershed. Not only more remunerative crop plans have to be introduced, but also the crop-based farming system has to be replaced by mixed farming of any type suiting different farmers. Cropping with dairy animals, cropping with poultry, goat farming, piggery, cropping with fisheries (where runoff collection in a farm pond/tank has been done) or water chestnut-singhara (*Eleocharis dulcis*) cultivation, and development of silvopastoral system on a slopping eroded soil planting improved grasses, shrubs, and medicinal plants like *Aloe vera* and legumes, agri-horticulture system, and beekeeping suiting the farmers' resources must be introduced.

To summarize, the strategy to manage very deep Chambal ravines must be a combination of limited but precise land shaping, construction of appropriate mechanical structures (anicuts, masonry and gabion drop structures, bunds, wasteweirs, etc.), and creation of vegetative cover. Structures for runoff collection at appropriate sites must be built. Also, appropriate land use plan for biomass production is a must.

## References

Action Aid India (1996) Chambal ravines-a glimpse of the scenario. Paper presented at the "Workshop on Chambal Ravines" held at Gwalior on Feb. 6–7, 1996

Anonymous (1979) Gully control. Indo-UK Dry Farming Project (ICAR), JNKVV (now RVSKVV) Campus Collage of Agriculture, Indore (M.P)

Anonymous (1997) Chambal Command Area Development Agency, Ninth Five Year Plan (1997–2002), Chairman, CADA, Gwalior, M.P

Bali YP, Karale RL (1977) Reclamability classification of ravines for agriculture. Soil Cons Digest 5:40–47

Bhulyan S (1967) Survey of ravine lands in Rajasthan. Proceedings 11th Silvi-culture conference. Forest Research Institute, Dehradun, U.P. (now Uttarakhand)

Bhushan LS, Yadav RC (2000) Management of waste lands with special reference to ravines. Indian Farming 50(5):22–31

Bhushan LS, Yadav RC, Prajapati MC (1992) Rehabilitation of ravines. Indian Farming 42(9):17–28

Bergonse F, Reis E (2016) Controlling factors of the size and location of large gully systems: a regression- based exploration using reconstructed pre-erosion topography. Catena 147:621–631

Gupta RK, Prajapati MC (1983) Reclamation and use of ravine lands. Desert Resour Technol (Jodhpur) 1:221–262

Haigh MJ (1984) Ravine erosion and reclamation in India. Geoforum 15(4):543–546

Kandrika S, Dwivedi RS (2013) Reclamative grouping of ravines using Cartosat-1 PAN stereo data. J Indian Soc Remote Sensing 41(3):731–737

Kerr JM, Sanghi NK (1993) Indigenous soil and water conservation in India's semi-arid tropics. Appl Manag 3:255–289

Manendez-Duarte R, Marquing J, Fernandez-Manendez S, Santos R (2007) Incised channels and gully erosion in Northern Iberi Peninsula: controls and geomorphic setting. Catena 71:267–278

Ollier CD, Brown MJF (1971) Erosion of a young volcano in New Guinea. Z Geomorphol 15:12–28

Roy K, Mishra PK (1969) Formation of Chambal ravines. Indian Forester 95(3):24–28

Schwab GO, Frevert TWT, Barnes KK (1981) Soil and water conservation engineering, 3rd edn. Willey, New York

Seth SP, Bhatnagar RK, Chauhan SS (1969) Reclamability classification and nature of ravines of Chambal command areas. J Soil Water Cons India 17:39–44

Sharma HS (1979) The physiography of the lower Chambal valley and its agricultural development: a study in applied geomorphology. Concept, New Delhi
Sharma HS (1980) Ravine erosion in India. Concept, New Delhi
Singh R (2014) Characterization, rehabilitation and sustainable management of Chambal ravines. Lecture delivered in Training on "Management of soil health and Degraded Lands for sustainable Agriculture" held at RVSKVV, Gwalior, Feb. 14 to March 6, 2014
Singh S, Rao KK (1996) Land use pattern changes due to ravines using remote sensing techniques in Morena Distt. M.P. Paper presented at the "Workshop on Chambal Ravines" held at Gwalior on Feb. 6–7 1996
Singh YP, Verma GP (2016) Watershed based ravine reclamation. Global ravine conference on managing ravines for food and livelihood security. Organized at RVSKVV, Gwalior on March 7–10, 2016
Singh YP, Sinha SB, Singh AK, Verma GP (2016) Management of "Chambal ravines" through anicut and afforestation: an impact assessment in Madhya Pradesh, India. Global ravine conference on Managing ravines for food and livelihood security organized at RVSKVV, Gwalior on March 7–10, 2016
Sobolev SS (1948) Development of erosion process in the European USSR and their control (in Russian). Izvestiya an SSSR, Moscow
Tejwani KG (1974–75) Classification and reclamation of gullied lands. J Soil Water Conserv India 24–25:29–40
Tejwani KG, Gupta SK, Mathur HN (1975) Soil and water conservation research (1956–71) Indian Council of Agricultural Research, New Delhi
Tomar VS, Verma GP, Singh YP (2015) Land degradation in India-strategies to manage Chambal ravines in Madhya Pradesh. JNKVV Res J 49:366–373
Verma GP (1982) Land and water management in semi-arid rainfed areas. Review of Soil Research in India, Indian Society of Soil Science, New Delhi
Verma GP (1983) Watershed Management Practices for the Control of Erosion and Water-logging under Rainfed Condition. Lecture delivered at the "Training on Management of Black soils under Rainfed conditions" organized under the auspices of NARP at JNKVV Campus Indore, M.P. from Sept. 1 to 10, 1983
Verma GP (1987) Farmer's Participation in Watershed Management, Experience of Indo-UK Dry farming Project, Indore, M.P., India. Paper presented at the International workshop on "Farmers and agricultural research: complementary methods" held at the Institute of Development Studies, University of Sussex, Brighton, U.K., July 26–31, 1987
Verma GP (1999a) Monitoring and evaluation report of water and power consultancy services (WAPCOS) evaluation team for project "integrated watershed management Programme on Ravinous land of Chambal and Yamuna catchments". Central Water and Power Commission, New Delhi
Verma GP (1999b) Ravine soils and their management. Paper presented at the workshop on Land resources of M.P. held at Bhopal on Aug. 17, 1999 under the auspices of Department of Agriculture, M.P. Govt. and N. B.S.S. and L.U.P. Nagpur
Verma GP (2003) Final Report submitted to Depeartment of Science and Technology GO, New Delhi on the project "Watershed-based Development of Ravine land" Centre of Sustainable Agricultural and Rural Development, C.P. Colony, Morar, Gwalior
Verma GP, Singh YP, Dubey SK (2012) Watershed based reclamation and control of Chambal ravines. Conservation of natural resources for food and environmental security In: Dubey et al. (eds) 2912, pp 329–341
Verma SK, Singh A, Tomar PS (2016) Management of chambal ravines for food and environment security. Rajmata Vijayaraje Scindia Krishi Vishwa Vidyalaya, Gwalior, M.P

# Reclamation of Ravine Lands for Higher Production

# 12

M. L. Soni, V. Subbulakshmi, P. S. Renjith, J. C. Dagar, and N. D. Yadava

**Abstract**

Ravine lands are spread over an area of about 4.3 million hectare in India, but their largest incidence (about 72%) is found along the rivers Yamuna, Chambal, Mahi, and Sabarmati and their tributaries in Uttar Pradesh, Madhya Pradesh, Rajasthan and Gujarat states. These are the worst form of land degradation and formed when the vegetal cover is not strong enough to hold and bind the soil together from being carried away by the flowing rainwater. Loss of land due to soil erosion and uncertain crop yield from eroded soil increases the risk to sustain the livelihood of inhabitants of ravine-affected area. According to an estimate, India loses about 3 million tons of food grain due to the formation of ravine lands. These losses can be minimised through adoption of scientific techniques to reclaim and check their further extension. For rehabilitation and controlling the further extension of these lands, integration of soil-water conservation measures with suitable agroforestry systems using fruit trees/shrubs, forage grasses and energy plantations seems viable option. Many species of fruit trees, shrubs, grasses, energy plantations and medicinal use have been identified for their environmental suitability in ravine lands. Using appropriate engineering technology and selecting appropriate species, these lands can improve the livelihood of local inhabitants through production of fodder for animals, fuelwood for energy requirement and food crops and fruits for human consumption. Tree-based technologies have additional environmental benefits. Growing trees with annual crops or pasture grasses help in retaining soil moisture by reducing runoff and simultaneously reduce nutrient leaching losses through root and mycorrhizal

M. L. Soni (✉) · V. Subbulakshmi · P. S. Renjith · N. D. Yadava
ICAR-Central Arid Zone Research Institute, Regional Research Station, Bikaner, Rajasthan, India

J. C. Dagar
Indian Council of Agricultural Sciences, Krishi Anusandhan Bhavan-II, Pusa, New Delhi, India





J. C. Dagar, A. K. Singh (eds.), *Ravine Lands: Greening for Livelihood and Environmental Security*, https://doi.org/10.1007/978-981-10-8043-2_12

systems, regulate biodiversity, improve soil quality and sequester huge amounts of carbon in their biomass. This chapter describes the formation of ravine lands, the ways to reclaim and sustain them through integrated approach of soil-water conservation technologies and agroforestry systems with the emphasis to improve the economic condition and livelihood of the farmers, conserve natural resources and provide ecosystem services under climate change scenario.

## 12.1 Introduction

Ravines are formed due to surface runoff, deforestation, overgrazing, bad tillage practices and mismanagement of soils susceptible to water erosion (Ali 1974; Tejwani et al. 1975; Dagar and Mall 1980; Prajapati et al. 1982). This erosion susceptibility is partly due to the intensity and concentration of rainfall during the monsoon and partly due to the easily erodible parent material highly susceptible to water erosion and responsible for the ravines development (Narain et al. 1979). According to a report of the National Commission on Agriculture (1976), the indiscriminate land-use practices are other reasons for ravine development. Pani (2016) demonstrated the ravines of Chambal with steep slopes and narrow widths. Such ravines are found where the subsoil is more resistant to erosion as compared to the topsoil. The shape of ravines converts into a U shape, where the thickness of alluvium is high. Pani and Mohapatra (2001) observed three different types of ravines in Chambal valley following multiple criteria of shape, depth, side slope and head slope. These were: young (narrow base width, with steep side slopes), mature (wide base width with moderate side slopes) and old (very wide base width with gentle side slopes).

Existence of ravines is not new, but their rapid spread of course is a recent phenomenon. Land degradation by ravines affects large area in India. The faulty agricultural practices convert about 8000 ha of land into ravines every year (Sharma et al. 1980), and approximately 0.5% of the ravine catchment area is being eroded. According to the latest land degradation data in India, about 82.57 million hectares (m ha) of land is degraded due to water erosion, which constitutes about 68% of the total degraded lands of India (120.72 m ha) (CSWCRTI 2012a, b). The ravine lands are spread over an estimated area of 3.97 m ha along the rivers Yamuna, Chambal, Mahi, and Sabarmati and their tributaries (Ministry of Agriculture 1984; Singh et al. 2015) and constitute about 1.12% of total geographical area of 328 m ha in India (National Commission on Agriculture 1976). The study conducted by National Remote Sensing Agency (NRSA) based on the Landsat data has estimated nearly 4.32 m ha of ravine land (Kar et al. 2009). Although ravines occur all over India, but they are largely spread (about 72%) in Uttar Pradesh, Madhya Pradesh, Rajasthan and Gujarat. A review of gully erosion and ravine spreading rates of lower Chambal valley shows that both the ravenous and the marginal lands have increased in recent years. This chapter describes the production constraints and reclamation of ravine lands through integrated approaches of soil-water conservation technologies and agroforestry systems and the ways to improve the income of the farmers, conserve natural resources and provide ecosystem services under climate change scenario.

## 12.2 Production Constraints in Ravine Lands

The poor management of ravine lands poses potential threat to nearby productive agricultural lands. Ravine erosion not only damages agricultural lands but also destroys roads, railway tracks, habitations and other public properties and, thus, adversely affecting rural ecology. According to National Commission on Agriculture, the ravine lands not only create problems where they exist but also degrade the adjacent arable lands affecting production potential in a major way. According to an estimate, about 3 million tons of food grain are lost every year due to the formation of ravine lands in India (Das 1985). The annual damage to the tablelands accounts for ₹40 million (GoI 1972). This happens because of the poor investment in management of these systems. Hence, there is a need to give proper attention and focus to reclaim these degraded lands and protect the nearby arable and nonarable lands from further degradation.

## 12.3 Need for Conservation Measures

Soil erosion is the major threat in ravine areas. Loss of land due to soil erosion and uncertain crop yield from eroded soil increases the risk to sustain livelihood of the inhabitants of that area. As the topsoil erodes, the infiltration rates and water availability become limited for crop production. The rainwater does not penetrate into the compact subsoil as rapidly as the loose topsoil. This leads to loss of water as runoff and limit its availability for crop production. Under such circumstances, soil conservation measures, and adoption of agroforestry systems using indigenous species seems viable option. Increasing vegetative cover through agroforestry systems for the reclamation of ravine lands will conserve the land and water resources, enhance the flora and fauna population, restore ecosystem and mitigate the impacts of climate change.

There are many classical examples where the farmers of ravine-affected area shifted their pattern of livelihood from crops alone to agroforestry or animal husbandry-based systems and improved their livelihood. Shifting for better option and to secure livelihoods to meet basic needs of food is a natural phenomenon. People change their livelihood strategies as an adaptive response to changes in their environment. Some livelihood flourish, while other diminishes, and this ebb and flow are the result of a changing livelihood context. In ravine-affected Piprai village of Morena district of Madhya Pradesh (India), the villagers shifted their livelihood from crop-based to animal husbandry-based systems. Thomas et al. (2011) identified 60 floral and 27 faunal species of economic importance from these ravines. Out of 60 species, 21 provided foods, 17 fodders, 19 fuelwood, and 21 medicines, and 4 were of other commercial importance. The villagers stretched out to harness the biodiversity of the area to make their food secure and living sustainable.

## 12.4 Management Options

Ravenous lands are not fit for growing agricultural crops because they lead to accelerated erosion and fast degradation. Planting fruit and forest trees along with shrubs/forage grasses and energy plantations with appropriate soil and water conservation measures are the best scientific use of these lands. Suitable soil and water conservation structures need to be erected to check the uninterrupted flow of runoff into the ravines. Peripheral bunds, permanent drop structures like spillway, brushwood check dams, gabions, live vegetative check dams and trenching were found suitable to rehabilitate the ravines (Chaturvedi et al. 2014). After successful checking of the ravine extension, the trees/shrubs/grasses are introduced as alley, boundary plantation or sole tree plantation so as to reduce the possibilities of gully formation in the future. The tree/shrubs/grasses should have the capabilities to withstand stress, fast growing, easy to establish, multipurpose uses, fibrous roots and potential to ameliorate the soil.

## 12.5 Potential of Agroforestry in Ravine Lands

Agroforestry plays an essential role in reducing vulnerability, enhancing resilience of farming systems and buffering households against climate-change-related risks (Dhyani and Handa 2014). At present agroforestry meets about 50% fuelwood demand, 2/3 small timber, 70–80% raw material for plywood, 60% raw material for paper pulp and 9–11% of the livestock fodder demand, additionally to its environmental benefits (NRCAF 2013). It is estimated that by 2050, an increase in 1.5 times fodder, two times food grains and fuelwood, three times timber production and a three-fold increase in production of biodiesel will be required to meet future demand of growing population (Dhyani et al. 2013). Agroforestry may contribute considerably to meet out this demand and provide ecological services (Table 12.1). There are ample evidences to show that the agroforestry may contribute considerably in recycling of the nutrients, improvement of soil fertility, overall (biomass) productivity and carbon sequestration as compared to annual cropping system (Dhyani et al. 2009).

At present agroforestry is reported to occupy 25.32 m ha area in India and has a further scope of its expansion in 28.0 m ha of fallow lands, degraded pastures, groves, salt-affected and ravine lands. Thus, a total of 53.32 m ha area of the country may potentially be brought under agroforestry in future, which can make agroforestry a significant land-use activity, next to agriculture (140.86 m ha) and forestry (69.63 m ha) (Dhyani et al. 2013).

Agroforestry interventions are of prime importance to increase the production and productivity of degraded ravine lands. It will meet out the desired forest cover of 33% and mitigate the impacts of climate change. Most agroforestry systems help to conserve land and water resources and improve soil conditions in many ways (Nair

**Table 12.1** Total domestic demand for various commodities

| Items | 2010–2011 | Projected for 2025 | Projected for 2050 | Contribution from agroforestry in 2050 |
|---|---|---|---|---|
| Food grains (million Mg[b]) | 218.20 | 320.00 | 457.1 | 41.14[a] |
| Fruits (million Mg) | 71.20 | 106.00 | 305.3 | 47.74[a] |
| Fodder (million Mg) | 1061.00 | 1170.00 | 1545.0 | 154.50 |
| Fuelwood (million Mg) | 308.00 | 479.00 | 629.0 | 308.00 |
| Timber (million Mg) | 120.00 | 171.00 | 347.0 | 295.00 |
| Biodiesel (million Mg) required for 20% blending of diesel | 12.94 | 22.21 | 37.92 | 30.34 |
| Area (m ha) required for tree born oilseeds (TBOs) | 12.32 | 15.86 | 21.67 | 17.34 |

Source: CAFRI (2015)

[a]Food grains/fruits production from systematic agroforestry systems, viz. agri-silviculture/agri-horticulture only considered

[b]Mg denotes megagram

2008). It is a standout amongst the most reasonable alternative land-use systems for degraded lands for maximum sustainable productivity while preserving the environment. In many countries forest departments are integrating agroforestry programs with conventional silvicultural practices as people started realising that agroforestry is a practical solution for environmental protection and low-cost alternative for food production also (Swaminathan 1987). Many aspects of past development have shown the acceptance of agroforestry system for land management and its applicability to both farm and forest (Nair 1993) and so are possible for ravine lands also.

## 12.6 Interventions Required for Rehabilitation of Ravine Lands

Depending upon the extent of degradation and agroecological characteristics of the land, different types of reclamation strategies have been suggested in the context of ravine-affected areas. According to land capability class of VI and VII, the most economic use of ravine land is the permanent vegetation. The deep ravines may be stabilised completely by closing the land for grazing and undertaking the regular afforestation program supported by engineering measures such as check dams, ponds and runoff control structures. Later, the program of farm forestry, grass land and pasture development may be undertaken. Closing the ravine lands for grazing will allow the fast recovery of vegetation. In ravine tract of Vasad (Gujarat), simple closure resulted in 35% increase in tree cover in 5 years (Tejwani et al. 1975). Similar observations are recorded at Kota in the Chambal ravine region (Verma et al. 1969). (for more details see Chap. 3 written by JC Dagar in this volume).

Pani et al. (2011) and Pani (2016) suggested the ways through which ravine land is being reclaimed. These includes (1) stopping the growth of ravines by putting boulders (mainly sandstones), bushes, straws and other crop residues on gully heads and (2) creating check dams and cemented walls on gully heads (bringing lost agricultural land back to agriculture by land levelling and afforestation). A simple plan for reclamation of ravine is as follows:

- Survey, investigation of land and planning
- Treatment of tableland
- Protection of marginal land for raising crops
- Main ravine land reclamation
- Ravine stabilisation through biological measures.

### 12.6.1 Survey and Investigation Regarding Land Use Planning

A very comprehensive soil and land-use survey is required before proceeding for ravine reclamation. The general plan consists of classification of the ravine area, land use and proposed ravine reclamation measures and needs to be developed from the relevant aerial photographs (scale 1:6000) on watershed basis consisting of the following points:

- Recognition of flat lands and peripheral lands where ordinary soil conservation measures would be needed to intercept runoff
- Demarcation of areas where intensive soil working would be involved in bench terracing and major land levelling
- The selection of ravine waterways and protecting them by gully head control and gradient control structures for safe disposal of runoff from the relevant watershed
- Selection of sites for tube wells and tentative alignment of distribution system
- Location of ravine stabilisation structures

The next step is the field work required, which consists of planning of terraces, calculation of earth work involved in each terrace and preparation of beneficiary-wise cost estimates.

### 12.6.2 Treatment of Table/Level Land

The treatment of tableland is a vital element in ravine reclamation program. It aims at systematic handling of the runoff water from top of the watershed so that the benefits of soil and water conservation work are evenly distributed and further spread of ravines is checked. The objective of tableland treatment is to reclaim ravine lands for agricultural purposes (Fig. 12.1). The table/level lands are cultivated using non-erosive crop sequence. Under rainfed conditions, measures like contour bund farming and contour ridge and furrow system help conserving and holding much of rainfall where it occurs. Depending upon the rainfall, soil, topography and cropping system contour bunding should be done. For light rainfall areas with 0.5% slope, contour bunding is suitable. In high rainfall areas with heavy soils

**Fig. 12.1** Use of levelled land for growing arable crops (Photo by SK Verma)

of low permeability, graded channel terrace should be adopted in preference to contour bunding. The development of systematic waterways will lead to better runoff disposal. No ravine reclamation can be done safely unless the whole watershed is treated and reclaimed from top to bottom.

Integrated watershed management programs (IWMP) are being implemented in India on a massive scale since 1991, which is most sustainable multipurpose strategy. A review of more than 300 integrated watershed management projects indicated that in majority of them, total crop production increased by 50–123% (Joshi et al. 2005). Water-harvesting technologies resulted in 50–156% increase in irrigated area under different schemes, which increased average cropping intensity by 64% (NAAS 2009). Apart from increasing agricultural productivity, these projects helped the stakeholders in generating employment, and about 47% of degraded lands have been treated for rehabilitation (Sharda et al. 2008).

### 12.6.3 Protection of Marginal Land for Raising Crops

This is the most vulnerable strip lying in between the tableland and gully head. Unless protected, gully or ravine heads fast extend due to runoff causing waterfall erosion. Good tableland soon gets engulfed, eventually to be turned into a ravine. Tejwani and Bhardwaj (1982) reported that closure of the eroded areas for grazing reduced the soil loss from 3.3 to 0.6 Mg $ha^{-1}$ at Vasad, Kota and Agra. They further reported that contour bunds at 90–190 cm vertical interval with 0.9–1.3 $m^2$ cross section area were most effective for proper retention and disposal of rainwater in alluvial soils. For black soils with less than 4% slope at Kota, graded bunds of 0.7–1.05 $m^2$ in cross section and 0.5–1.0% grade were adequate. In Bellary black soils, graded bunds of 0.87 $m^2$ cross section and vertical interval of 0.6–0.76 m and grades of 0.1–0.2% were found suitable (Chittaranjan and Patnaik 1980). In all these examples, the bunds need to be planted with grass like *Dichanthium annulatum*, *Andropogon ischaemum* and *Cenchrus ciliaris* for stabilising the bunds.

Ravine reclamation work mostly consists of bench terracing supported by field outlets for safe disposal of runoff water. From the past experiences, it is observed that 0.20 ha terraces (size = 50 × 40 m) are most economic. Bench terrace operations consist of the following steps:

1. Cutting of earth
2. Transporting towards the fill and compaction
3. Construction of risers (bunds)
4. Dressing and consolidation of whole work
5. Construction of field outlets and runoff control structures
6. Reconsolidation of work after the rain
7. Construction of irrigation structures

Although in ravine reclamation work, usually cutting of 1 m depth of soil is involved, but quite frequently 2–3 m height is also required for this purpose. Dressing of bunds of the prescribed specifications is done by manpower. After a rainy season, the uneven settlement of breaches must be attended immediately after the rains when optimum soil moisture is available. This operation helps in stabilising the newly developed bench terraces (also see Chapter by Akhlesh Singh and Verma in this book).

### 12.6.4 Soil Conservation Engineering Structures to Reclaim Ravine Lands

The lower region of the ravine system which comprises mostly medium to deep ravines is stabilised by soil conservation engineering structures installed at the strategic points to ensure ravine stabilisation and to prevent further growth of ravines (Fig. 12.2). The soil conservation engineering structures include gully plugs, check dams, drop spillways, drop inlets, chutes, etc.

#### 12.6.4.1 Gully Plugs

To stabilise the gully bed and to check further expansion of gullies, gully plugs are provided at vertical intervals between 2 and 3 m and length between 30 and 40 m strip, wherever drainage density is extremely high and every individual watershed has been reduced to 3–5 ha. Gully plugs are economical and usually made without

**Fig. 12.2** Engineering measures to check runoff in deep ravines (Photo by SK Verma)

any outlets. This helps in stabilising the gully bed, thus checking the phenomenon of gully extension.

#### 12.6.4.2 Check Dams

These are small, temporary structure constructed across gullies, aligning perpendicularly to the direction of water flow in order to offset soil loss by reducing the energy of runoff. A check dam placed in gully interrupts and minimises the runoff velocity, thereby increasing infiltration (Agoramoorthy et al. 2008). Check dams are constructed after detailed survey and hydrological analysis of the ravine area. Construction of bunds and check dams in the ravines captures water and accumulates silt by checking runoff. At ravenous areas of Vasad (Gujarat), runoff reduced from 35% to 27% of the rainfall, and available water increased because of soil-water conservation measures. This eventually increased the agricultural production of the area (Singh and Dayal 1975). Check dams can be constructed by using different local materials:

- *Log check dams*
- To make these structures, logs and posts are placed across the gully. Planks, heavy boards, slabs and poles can also be used to construct this type of check dams. They are used to stabilise gully heads and to hold the soil material carried by water in the gully. They are constructed along the small emerging gullies.
- *Loose stone check dams*

- They are constructed with the help of small rocks. They control channel erosion along the gully bed and stop waterfall erosion by stabilising gully heads. This type of check dams is constructed to check the runoff from the branches of a gully network.
- *Boulder check dams*
- These types of check dams are made with boulders to stabilise the main gully channels of continuous gullies. It is constructed to control channel erosion and stabilise gully heads in all regions.
- *Gabion check dams*
- Gabion check dam is a flexible, permeable structures constructed in gullies by filling of stones in wire mesh cage. It decreases the average upstream slope by creating a sedimentation bench. Gabion check dam limits the velocity of water thereby increases water infiltration and reduces the soil loss in upstream and sedimentation in reservoirs.

Small gullies (up to 3 m deep and 8 m or more wide) may be reclaimed by levelling the bed and construction of diversion-cum-check bunds (1.5 m cross section) at 30–45 m intervals and providing grass ramps and composite check dam at the end of the gully. Medium gullies (3–9 m deep, 18 m or wider with side slope 8–15%) may be reclaimed by levelling the bed and constructing a series of composite earth and brick masonry check dams at vertical intervals of 1.2 m and terracing the side slopes. Larger gullies are reclaimed by masonry check provided with aprons on lower side and their wing wall properly embedded in stream sides.

#### 12.6.4.3 *Bori* Bunds

*Bori* bund is a series of earthen check dams constructed in degraded gullies. In order to maintain the stable slope, the earth-filled polythene/jute bags (called *bori*) are arranged as the more number of rows at bottom, the less number of rows at the middle and one row at the top. After that they are covered with soil on top and other sides also. The top *boris* from one end of the bund are removed and kept at upstream and downstream to serve as outlet. They are stabilised by growing suitable grasses in two rows on each *bori* bund, one each at upstream and downstream edge. In between the check dams, general plantations are raised in staggered manner (CSWCRTI 2012a).

In an experiment conducted on three-tier rainwater management system, viz. conventional/mono-tier (contour bund), two-tier (contour bund + dead furrows) and three-tier (contour bund + dead furrows + harvesting of excess rainwater for lifesaving irrigation) systems in ravines of Bundelkhand, Uttar Pradesh, Singh (2011) reported significantly higher seed yield of sesamum-wheat and black gram-wheat cropping system under three-tier system of rainwater management over other systems.

#### 12.6.4.4 Continuous/Staggered Contour Trenching

Management of belowground interactions is most important in agroforestry systems particularly in dry climates, semiarid tropics and on infertile soils (Rao et al. 2004). Trees suppress productivity of field crops depending upon the type of crop and above- and belowground growth behaviour of trees. In semiarid environment, pruning tree roots may help in reducing the competition for belowground resources (Korwar and Radder 1994) particularly when crops are grown with trees in agroforestry systems (Wajja-Musukwe et al. 2008).

In ravine beds, staggered contour trenches (0.5 m width × 0.5 m depth × 2 m length) are excavated across the slope at 4 × 4 m spacing. Bunds are made by heaping the excavated soil on downstream side of the trench to retain the moisture before rain starts. Seedlings are planted at the centre of the trench (CSWCRTI 2012a). The continuous or staggered trenches and contour furrow helps in *in situ* rainwater harvesting (Mishra and Tripathi 2013). Staggered contour trenches are effective in intercepting the incident rainfall, preventing gully extension, controlling soil erosion, accelerating vegetation establishment, encouraging natural regeneration of species, increasing biomass production and improving soil fertility of degraded lands. (Bhushan et al. 1986). Sharda et al. (1982) reported that contour trenching completely checked the runoff from 25% to almost nil and facilitated deep soil-water storage (Singh 2009). Samra et al. (2004) observed that survival of oak seedlings planted in contour trenches was 20% more than planted on pits in Sainji watershed in Mussoorie. In Chambal ravines watershed area, staggered trenching for trapping 75% runoff doubled the productivity of 6-year-old horti-pastoral system of goose berry (*Emblica officinalis*) + *Cenchrus ciliaris* (Sethy et al. 2010). Creation of 'V'-shaped structure with 5% slope between the two trees is a recommended practice to facilitate the movement of water near the root zone of trees and improving the survival, growth and production of fruit trees. Runoff yield can further be increased by covering the upper part of 'V'-shaped catchment with polythene sheets.

### 12.6.5 Ravine Stabilisation Through Biological Measures

While reclaiming ravines with vegetative (biological) measures like raising fodder reserves, fuel-fodder plantation or industrial plantation or orchards, stabilisation of ravines against erosion is sought through gully plugging followed by putting vegetative/biological barriers. Vegetative barriers are cost effective as compared to mechanical measures on mild slopes. Grasses such as *Cenchrus ciliaris*, *Dichanthium annulatum*, *Cynodon dactylon*, *Cymbopogon nardus*, *C. martinii*, *C. fluxuosa*, *Heteropogon contortus*, *Chrysopogon montanus*, *Saccharum spontaneum*, *S. munja*, and *Eragrostis curvula* and legumes like *Tephrosia candida*, *Clitoria ternatea*, *Desmodium* spp. can very well be grown in the ravines. Chandra Prakash et al. (1999) enlisted 34 grasses with potential use as vegetative barrier (live bunds) in different agroecological regions of India. Raising vegetative barrier with grasses increased maize and wheat yields by 23–40% and 10–20%, respectively, in addition to 0.6–1.7 Mg $ha^{-1}$ $year^{-1}$ grass yield (Ghosh 2010). Hedge rows of trees such as *Leucaena leucocephala* and *Gliricidia sepium* are found very effective in controlling soil erosion when planted across slope. *Eucalyptus tereticornis* and Bhabar grass (*Eulaliopsis binata*) planted in Shiwaliks were found quite remunerative and effective to control soil erosion (Sharda and Venkateswarlu 2007).

## 12.7 Agroforestry Interventions

Adoption and maintenance of engineering structures for ravine reclamation involve capital investment and community involvement. Therefore, it has been suggested to restore ravine lands to permanent vegetation through introduction of trees and grasses (Tejwani et al. 1975). After proper treatment, the small and medium gullies can be put under crop cultivation, and the deep and narrow gullies are put under permanent vegetation of grasses and trees (Tejwani et al. 1975).

The soils of ravenous regions are poor in physical properties, viz. infiltration and water retention capacity (Dagar and Mall 1980), as there remains only rocky, sandy or stiff clay due to severe erosion. They face moisture stress during most part of the year. Hence, their rehabilitation requires an integrated approach of soil-water conservation measures with suitable location-specific agroforestry systems. The technologies of water harvesting developed in India resulted in 50–156% increase in irrigated area under different schemes, which increased average cropping intensity by 64% (NAAS 2009). Apart from increasing agricultural productivity, these projects helped the stakeholders in generating employment, and about 47% of degraded lands have been treated for rehabilitation (Sharda et al. 2008). After controlling the erosion through the engineering measures, the incorporation of agroforestry may be given top priority. Location-specific agroforestry systems offer a cost-effective and sustainable method to combat the degradation of wastelands.

Different agroforestry practices have been applied successfully in reclamation of ravines. In Yamuna ravines at Agra, bamboo-based agroforestry system produced 30–33 harvestable culms every 3 years after proper establishment and produced an average yield of 4000 poles $ha^{-1}$ (Prajapati et al. 1993). Adoption of silvopastoral system with multipurpose trees is also a viable option for economic utilisation of ravines as advocated in this book by many authors.

### 12.7.1 Choice of Species

The tree and grass species (Table 12.2) selected should have ability to withstand stress, fast growing, easy to establish, multipurpose uses, fibrous roots and potential to ameliorate the soil. Production of different tree species in ravines has shown that *Acacia nilotica* is the most promising fuelwood tree species. The suitable grasses for gully stabilisation are *Dichanthium annulatum*, *Cenchrus ciliaris* and *Sehima nervosum* (Chaturvedi et al. 2011). In Doon valley with 4% slope, three live bunds of Guinea (*Panicum maximum*), Bhabar (*Eulaliopsis binata*) and Vetiver (*Vetiveria zizanioides)* grass reduced the runoff by more than 18% and soil loss by more than 78% as compared to cultivated fallow (Bharadwaj 1990–1991). Species of *Cymbopogon* have successfully been utilised to stabilise bunds in Morena of Chambal ravines (Fig. 12.3).

**Table 12.2** Choice of species for ravine lands

| Species | Planting location | Uses |
|---|---|---|
| *Acacia catechu* | Gully slopes and humps | Fuelwood, fodder and minor forest products (MFP) |
| *Acacia leucophloea* | Gully slopes | Fodder and fuelwood |
| *Acacia nilotica* | Hump top, slope and ravine beds | Fodder, fuelwood and small timber |
| *Acacia tortilis* | Hump top, slope and ravine beds | Fuelwood and fodder |
| *Ailanthus excelsa* | Gully slopes and humps | Fodder, fuelwood and timber |
| *Albizia lebbeck* | Hump top | Fodder and fuelwood |
| *Azadirachta indica* | Hump top and beds fodder | Fuelwood, pesticides and timber |
| *Balanites aegyptiaca* | Hump top and slope | Fuelwood and MFP |
| [a]*Brachiaria mutica* | Hump top, slope and ravine beds | Fodder, soil binder and silage |
| [a]*Cenchrus ciliaris* | Hump top, slope and ravine beds | Fodder, hay, silage and soil conservation |
| *Dalbergia sissoo* | Hump top and slope | Fuelwood and small timber |
| *Dendrocalamus strictus* | Gully bottom | Poles and fuelwood |
| [a]*Dichanthium annulatum* | Hump top and slope | Fodder and mat |
| *Eucalyptus tereticornis* | Gully bottom | Poles and fuelwood |
| *Gmelina arborea* | Ravine beds | Small timber |
| *Leucaena leucocephala* | Gully slopes and humps | Fodder and fuelwood |
| [a]*Panicum antidotale* | Ravine top and slope | Fodder and soil binder |
| *Prosopis juliflora* | Hump top, slope and ravine beds | Charcoal, fuelwood and fencing |
| *Soymida febrifuga* | Hump top | Fodder, fuel and small timber |
| *Tamarix dioica* | Swampy areas in gully bottom | Reclamation of saline soils and fuelwood |
| *Ziziphus* sp | Hump top and slope | Fruit and fuelwood |

Source: Uthappa et al. (2016); modified from Sikka et al. (2016)
[a]Grasses

Bamboo (*Dendrocalamus strictus*) is highly promising for soil conservation and productive utilisation of gullies. The bamboo species are hydrologically best suited to ravine lands (Rao et al. 2012). Bamboo roots play an effective role in increasing infiltration, reducing runoff and protecting soil from further extension of gully beds (Singh et al. 2015). Some legume species, viz. *Macroptilium atropurpureum*, *Clitoria ternatea*, *Atylosia scarabaeoides*, *Stylosanthes hamata*, *Rhynchosia capitata*, *Alysicarpus* spp. and *Stizolobium deeringianum* can also be planted in ravine areas which improve the soil and provide fodder also.

**Fig. 12.3** Stabilisation of bunds by using *Cymbopogon* grass in Chambal ravines (Photo courtesy SK Verma)

### 12.7.2 Planting and Aftercare

After removing bushes in the ravine lands, the soil should be loosened by a deep ploughing with a disc or country plough to improve soil physical conditions for proper aeration and moisture conservation. Pits of recommended size should be made and filled with 1–3 kg of FYM (farmyard manure). Tall, 1-year-old nursery-raised seedlings with ball of earth should be planted in the pit or trench made for planting. Planting is recommended in the early part of the rainy season having sufficient soil moisture. In the initial 2 years after planting, lifesaving irrigations with suitable irrigation intervals need to be provided for their establishment. Special care is wanted to protect the seedlings from biotic and abiotic stresses. Weeding and application of manures at regular intervals will help the plants to grow stronger in ravine lands (CSWCRTI 2012a). Organic manures and fertilisers should be applied during winter or dry period as per requirement of tree species.

### 12.7.3 Different Agroforestry Systems for Ravine Lands

The following agroforestry systems would be the best options for rehabilitation of degraded ravines.

#### 12.7.3.1 Agri-silvicultural System

In this system, agronomic crops are grown together with trees/shrubs on the same unit of land in such a manner that the yield of crops are not affected by the tree/shrub species and the overall productivity of system is improved. Reduction in crop productivity can be minimised by proper silvicultural practices, viz. pruning, thinning, etc. Apart from conservation of soil and water, the trees provide additional income in the form of fodder, fuelwood, fruits, etc. In ravine areas of Kota (Rajasthan), higher yield, land equivalent ratio and soil organic carbon were recorded in *Leucaena*-based alley cropping systems. Lopped *Leucaena* leaves with 50% recommended dose of N through urea registered highest productivity of sorghum + pigeon pea intercropping system and improved the soil nutrient status. Effect of boundary plantations of different tree species on intercrops was also evaluated in ravine areas of Kota. The results showed that castor and pigeon pea grew well with *Acacia nilotica* followed by *Azadirachta indica* and *Albizia lebbeck* (Dhyani et al. 2007).

#### 12.7.3.2 Agri-horticultural System

This system involves growing fruit trees in combination with arable crops. Under agroforestry systems, fruit trees are the first preference of farmers because of short gestation period, risk cover and regular income to meet household needs and aesthetic value. Depending on the suitability of environment, various fruit-based agroforestry systems can successfully be raised in humps and gully beds (Verma et al. 1986). Lemon (*Citrus limon*), mango (*Mangifera indica*), ber (*Ziziphus mauritiana*) and goose berry (*Emblica officinalis*) can successfully be included in agri-horti systems. In Bundelkhand regions, goose berry-based agri-horti system is highly profitable and sustainable as the fruit tree is a regular bearer and the associated crops are less affected and require low inputs and management (Dwivedi et al. 2007). In agri-horti system fruit trees are normally grown at wider spacing, and crops like wheat, sorghum, sesamum, black gram and pigeon pea are grown as intercrops. Pruning at nonbearing stage of horticulture trees in agri-horti systems also provides additional income in the form of leaf fodder and fuelwood.

Various root management options were tried to increase fruit yield in ber-based agri-horti systems (ber + pearlmillet-wheat) in alluvial soils of Yamuna ravines (Table 12.3). Root management practices, viz. trenching, bottomless bitumen drums polythene lined pits, etc., checked the tree roots extension in cropping zone and improved crop yield over without root management practices. The highest ber yield (7347 kg $ha^{-1}$) was recorded in bottomless drums treatment. This study revealed that growing pearl millet-wheat in ber plantations established in bottomless bitumen drums is not only beneficial in producing higher and sustained crop yields, but it also produces significantly higher fruit yield, tree volume and fuelwood (Om Prakash et al. 2011).

#### 12.7.3.3 Agri-horti-silvicultural Systems

In this system, arable crops are grown with fruit trees and multipurpose tree species (MPTs) like subabool (*Leucaena leucocephala*), drumstick (*Moringa oleifera*) and

**Table 12.3** Grain yield (kg $ha^{-1}$) of pearl millet and wheat and fruit yield (kg $ha^{-1}$) in ber-based agri-horti systems in Yamuna ravines

| | Pearl millet | | | | Wheat | | | | Ber fruit | | | |
|---|---|---|---|---|---|---|---|---|---|---|---|---|
| Treatments | 1998 | 1999 | 2000 | 2001 | 1998 | 1999 | 2000 | 2001 | 1998 | 1999 | 2000 | 2001 |
| $T_1$ | 1562 | 2490 | 2015 | 2575 | 3866 | 4116 | 4040 | 3850 | 7000 | 5900 | 8688 | 7800 |
| $T_2$ | 1525 | 2440 | 1585 | 2560 | 3558 | 3953 | 3820 | 3663 | 5282 | 5760 | 7545 | 7100 |
| $T_3$ | 1475 | 2408 | 1725 | 2300 | 3641 | 3906 | 3715 | 3706 | 5180 | 4900 | 7616 | 7180 |
| $T_4$ | 1408 | 2144 | 1510 | 2270 | 3453 | 3603 | 3450 | 3444 | 6440 | 5240 | 8171 | 7460 |
| $T_5$ | 1680 | 2652 | 2140 | 2656 | 4150 | 4409 | 4265 | 4150 | – | – | – | – |
| LSD ($p \leq 0.05$) | 169 | 212 | 77 | 60 | 201 | 284 | 90 | 77 | 12.48 | 8.88 | 13.20 | 9.18 |

$T_1$ – tree planted in bottomless used bitumen drums of 90 cm height and 25 cm radius

$T_2$ – tree planted in polythene (thick black, 120 μ) lined pits (up to 90 cm)

$T_3$ – trench (40 cm deep and 30 cm wide) in two sides of tree

$T_4$ – tree without root management

$T_5$ – sole cropping system of pearl millet-wheat

Source: Om Prakash et al. (2011)

*Acacia* spp. These MPTs provide fodder, fuelwood and fruits and protect the fruit trees and crops from adverse climate and improve the soil through nitrogen fixation. The system is very profitable if irrigation (lifesaving) facilities are available.

#### 12.7.3.4 Silvopastoral System

The ravine lands could sustainably be developed through silvopastoral system (Fig. 12.4). Dagar and Pathak (2005) suggested that in highly eroded habitats local trees, perennial grasses and legumes such as *Stylosanthes*, *Alysicarpus*, etc. may be introduced with due care. High-value medicinal species such as *Aloe vera*, *Ocimum americanum*, *O. sanctum*, *Withania somnifera*, *Adhatoda vasica*, *Barleria prionitis*, *Solanum xanthocarpum*, etc. and biofuels such as *Euphorbia antisyphilitica*, *Jatropha curcas* and *Pongamia pinnata* (at the bottom of ravines) can easily be blended in these habitats. The animals are either stall- fed or managed under rotational grazing. Sustainable forage production may be achieved by pruning and lopping the side branches of trees. Experiment conducted to check the degradation of ravine lands through suitable silvopastoral systems showed that growing *Cenchrus ciliaris* with *Acacia nilotica* in the gully bottom and *Acacia tortilis* in gully top was economical in the Yamuna ravines of Agra. The other tree species, viz. *A. eburnea*, *A. leucophloea*, *A. catechu*, *Azadirachta indica*, *Albizia lebbeck*, *Balanites roxburghii*, *Butea monosperma*, *Dalbergia sissoo*, *Dendrocalamus strictus*, *Dichrostachys cineria*, *Eucalyptus* spp., *Feronia limonia*, *Pongamia pinnata*, *Prosopis juliflora* and *Ziziphus mauritiana*, have been found to adapt easily in the

**Fig. 12.4** A view of ravine stabilisation through silvopastoral system

ravines of river Yamuna at Agra and Kshipra at Ujjain. Among grasses, *Dichanthium annulatum*, *Cenchrus ciliaris*, *Bothriochloa pertusa*, *Chrysopogon fulvus*, *Themeda triandra*, *Heteropogon contortus*, *Sehima nervosum*, *Tragus biflorus*, *Iseilema laxum*, *Cynodon dactylon* and *Saccharum munja* flourish well in ravine lands. Trees such as *Eucalyptus tereticornis*, *Prosopis juliflora* and *Tamarix dioca* and grass *Brachiaria mutica* were suitable for saline riverbank. *Cenchrus ciliaris* interplanted with *Dendrocalamus strictus* is the best combination for gully stabilisation and to get economic return in the Mahi ravines (Dhyani et al. 2007). Jha et al. (2010) suggested the inclusion of *P. juliflora*, *L.leucocephala*, *A. nilotica* and *A. indica* as a better choice for reclamation of degraded ravine lands in the riparian zone.

For stabilising the degraded ravine lands and improving their productivity, CSWCRTI, Research Centre located at Vasad in Gujarat, developed a bamboo + Anjan grass (*Cenchrus ciliaris*)-based silvopasture system. Planting of bamboo on the gully beds and Anjan grass on the side slopes and interspaces of gully bed proved a good choice for economic utilisation of gullied lands, provides livelihood support and gave high returns. In addition to this, the technology checked water erosion and prevented soil and nutrient loss from ravines. Planting grasses protected the ravine slopes and produced green fodder to the tune of 7.1 Mg $ha^{-1}year^{-1}$ on the slopes and 10 Mg $ha^{-1}year^{-1}$ from interspaces. The production of bamboo and grasses fetched an income of INR 6000–INR 27,000 $ha^{-1}$ $year^{-1}$ and INR 3000–INR 6000 $ha^{-1}$ in 5 years period, respectively. In addition to economic benefits, the bamboo-based silvopastural system absorbed about 80% of rainfall and reduced soil loss from 20 Mg $ha^{-1}$ $year^{-1}$ (prior to plantation) to less than 1 Mg $ha^{-1}$ $year^{-1}$ because of soil-binding effects of bamboo and grass roots (ICAR 2017). In the long run, reclamation of ravine lands help in conservation of natural resources and carbon sequestration. This technology has been taken up by state agriculture department, forest department and other agencies for its further extension among the farmers for improving their livelihood. Using this technology, about thousand hectares of community and wastelands in Mahi river stretch in Gujarat have been reclaimed by various agencies.

#### 12.7.3.5 Agri-silvopastoral Systems

To minimise the risk of degradation of gullies, introduction of trees in cropping lands and perennial grasses on field bunds are important options. Such systems where trees, grasses and crops are grown together are termed as agri-silvopastoral system. Depending on slope, extent of problem and needs, the plantations may be done in the form of alley, boundary plantation or scattered trees. All along the boundary facing gullies, hedges of suitable species such as *Carissa carandas* supported by perennial grasses must be placed. The overall economic returns of such systems in ravine habitats have been found better as compared to sole cropping (Prasad 1994a, b; Parandiyal et al. 2006).

#### 12.7.3.6 Horti-silvopastoral System

In this system, horticultural trees, viz. ber (*Z. mauritiana*), karonda (*Carissa carandas*), pomegranate (*Punica granatum*), etc., are grown for fruit purpose, and MPTs, viz. species of *Leucaena*, *Moringa*, *Dalbergia*, *Bombax*, etc., and grasses such as *Cenchrus ciliaris* and *Dichanthium annulatum* are grown for their multipurpose uses. The grasses, viz. *Cenchrus*, *Chrysopogon* and *Vetivera*, and legumes like *Stylosanthes* can be grown with trees for soil conservation purpose and to provide fodder to the livestock. In Chambal ravines, planting goose berry on gully humps, bamboo on gully bed and buffel grass in interspace with trenching treatment increased goose berry fruits from 1.31 to 6.61 Mg $ha^{-1}$ and grass yield from 7.71 to 9.91 Mg $ha^{-1}$ in treatment involving trenching as compared to no trenching treatment (Sethy et al. 2010).

#### 12.7.3.7 Multi-tier Agroforestry System

It is a system of raising plant species of different heights with agricultural crops on the same piece of land. The system has minimum three tree layers of different heights, viz. timber trees, horticultural crops, herbaceous plants, etc. This type of agroforestry system can be followed to minimise the risk of land degradation in ravine lands during droughts and floods and get extra income. Multi-tier system produces maximum dry matter per unit area by intercepting maximum solar radiation and by efficient utilisation of land, water and space. The study conducted in Gujarat indicated that the sapota (*Achras zapota*)-*Jatropha curcas*-based three-tier agroforestry system with herbal medicinal crops, namely, basil (*Ocimum sanctum*), kalmegh (*Andrographis paniculata*) and mint (*Mentha arvensis*), gave higher economic returns as compared to sole crop.

## 12.8 Effect of Agroforestry on Natural Resource Conservation in Ravine Lands

The ravine areas are one of the most vulnerable ecosystems and exposed to multifarious natural resource losses and threat to biodiversity. Ravine area development will not only contribute in bridging the gap between growing demand and supply of food, fodder and fuel but also help in restoring livelihood security for resource constraint communities (Sikka et al. 2016). Agroforestry systems provide unique opportunity to bring together the different components of farming system which in turn will help in better management of natural resources. The outputs from an agroforestry system generally exceed that of an arable cropping system or sole plantations and facilitate economic usage of environmental resources. In addition, they maintain and improve soil health in long term. Agroforestry systems afford many environmental benefits and have the potential to contribute significantly in improving or maintaining soil quality of the region, enhancing flora and fauna, increasing vegetative cover and sequestering huge amounts of carbon in their biomass.

### 12.8.1 Reducing Runoff and Improving Soil Moisture Retention

Different agroforestry systems have been successful in reclamation of ravines and conservation of resources. Given enough time, agroforestry systems can bring significant improvement in soil properties of the ravine areas. Agroforestry systems protect soil against runoff, erosion and from surface compaction (Young et al. 1986). They improve moisture retention capacity and soil available water (Saha et al. 2007). Study conducted in catchment area of Chambal ravines at Kota showed that the silvopasture system of *Acacia nilotica* + *Dichanthium annulatum* and sole grass of *Dichanthium annulatum* decreased the runoff to 5.8% and 2.6%, respectively, as compared to 14.7% in agricultural catchments. The soil loss was reduced to 1.26 and 0.62 Mg $ha^{-1}$ in silvopasture and pasture system as compared to 3.0 Mg $ha^{-1}$ from agricultural catchments. Dosskey (2002) observed that the agroforestry technologies like riparian forest buffers are very effective in decreasing the flow of runoff and stabilise streams. Reduction of flow speed helps water to infiltrate in soil and allows for natural filtration of wastes or chemicals and thus improves groundwater quality.

The importance of bamboo plantation as an effective means for natural resources conservation and soil erosion control has been well documented in different ravines of India and abroad (Lawler 1993; YanHui and Yongmin 1995; Sharma et al. 1980; Kurothe and Nambiar 2001, Deng et al. 2003; Sharda et al. 1982; Zhou et al. 2005). Bamboo-based agroforestry models, viz. bamboo + ginger, bamboo + soybean and bamboo + turmeric, bamboo + pigeon pea, have successfully been established in managing degraded lands (Jha and Lalnunmawia 2004; Sheshadri 1985; Shanmughavel and Francis 2001). The study conducted in Mahi ravines of India showed that bamboo-based vegetation was quite promising in terms of controlling runoff, soil loss and economic returns. Farmers can earn a net income of Rs. 8000–10,000 per hectare with bamboo plantation from seventh year onward (Kurothe and Nambiar 2001; Kurothe et al. 2012; Pande et al. 2012).

In Yamuna ravines of Agra, staggered plantation of two rows of bamboo as vegetative barrier reduced runoff from 9.65% to 1.81% and soil loss from 4.2 to 0.6 Mg $ha^{-1}$ $year^{-1}$ (Table 12.4, Singh et al. 2015). In Mahi ravines of Vasad (Gujarat), the watershed with bamboo-based silvopasture system showed more than 80% of rainwater utilisation which is either used by the plants or percolated deep to recharge the groundwater. Due to reduction in runoff, soil loss was decreased to <1 Mg $ha^{-1}$ $year^{-1}$ as compared to 20 Mg $ha^{-1}$ $year^{-1}$ from degraded ravines. Reduction in runoff

**Table 12.4** Runoff and soil loss under bamboo plantation in the Yamuna ravines

| Year | Seasonal rainfall (mm) | Runoff from bamboo planted area (mm) | Runoff (%) | Soil loss (Mg $ha^{-1}$ $year^{-1}$) |
|---|---|---|---|---|
| 2010 | 456 | 44.0 | 9.65 | 4.27 |
| 2011 | 226 | 6.04 | 2.67 | 0.66 |
| 2012 | 531 | 14.50 | 2.73 | 0.78 |
| 2013 | 494 | 8.96 | 1.81 | 0.60 |

Source: Singh et al. (2015)

and soil loss due to bamboo plantation was attributed to the extensive fibrous root system of bamboo to hold the soil. The fibrous rooting system of bamboo increases permeability and water penetration. The dense foliage protects the soil against beating action of rainwater. Further, the habit of bamboo to produce new culms from underground rhizomes allows its harvesting without disturbing the soil (Ben-zhi et al. 2005). Bamboo plants conserve soil moisture and mitigate the adverse effects of drought on flora and fauna (Sharma et al. 1992). Pande et al. (2012) found that the bamboo plantation is economically viable in ravines. He suggested policy measures to development agencies for greening of ravines in large area.

### 12.8.2 Checking Soil and Nutrient Loss

Soil and nutrient losses are reduced due to the adoption of agroforestry systems. Silvopastoral systems have been advocated most ideal for ravine lands and highly erodible soils (Prajapati et al. 1989; Dagar 1995, 2015; Chaturvedi et al. 2014). Grewal (1993) found minimum soil loss in silvopastoral system (Table 12.5). Research conducted on bamboo plantations in ravine lands showed that the soil loss is decreased considerably and silt retention increased due to the growth of a dense vegetative cover in the gully beds (Singh et al. 2015). In another study by Mishra and Rao (2012), lowest runoff and about 50–67% reduction of nutrient loss were seen in the system of small earthen check dams.

Tiwari and Kurothe (2006) evaluated the effect of six vegetative barriers of *Dichanthium annulatum*, *Cenchrus ciliaris*, *Vetiveria zizanioides*, *Eulaliopsis binata*, *Sacharum munja* and *Aloe barbadensis* on soil and nutrients losses in reclaimed Mahi ravines with 2% slope on down cultivation of pigeon pea and reported that vegetative barriers reduced the runoff by 19.7–50.1% and soil loss by 51.1–80.3% over control. The highest runoff (65.0%) and soil loss (13.7 Mg $ha^{-1}$) were recorded under control, and lower runoff (32.4%) and soil loss (2.7 Mg $ha^{-1}$) were recorded under *Dichanthium annulatum* as compared to other treatments. Nutrients losses followed the trend of soil loss. Nitrogen loss under different vegetative barriers ranged from 5.9 to 17.4 kg $ha^{-1}$, whereas P and K losses ranged

**Table 12.5** Soil and nutrient loss from different and land-use systems in Shivalik

| Land-use systems | Soil loss (Mg $ha^{-1}$) | Runoff (%) | Nutrient loss (%) N | Nutrient loss (%) K |
|---|---|---|---|---|
| Eucalyptus-bhabar grass | 0.07 | 0.05 | 0.46 | 0.90 |
| *Acacia catechu*-forage grass | 0.24 | 2.00 | 6.97 | 0.52 |
| *Leucaena*-napier grass | 0.28 | 4.40 | 6.60 | 1.20 |
| Teak-*Leucaena*-bhabar | 0.43 | 3.30 | 2.08 | 0.55 |
| *Eucalyptus*-*Leucaena*-turmeric | 0.59 | 2.60 | 2.47 | 0.73 |
| Poplar-*Leucaena*-bhabar | 1.54 | 4.80 | 5.90 | 1.10 |
| Sesamum-rape seed | 2.69 | 20.50 | 42.50 | 3.00 |
| Cultivated fallow | 5.65 | 23.00 | 51.30 | 5.00 |

Source: Grewal (1993)

1.5–2.4 kg $ha^{-1}$ and 4.0–5.3 kg $ha^{-1}$, respectively. The total system productivity in terms of pigeon pea grain equivalent yield was in the order of *V. zizanioides* (1.88 Mg $ha^{-1}$), *S. munja* (1.83 Mg $ha^{-1}$), *D. annulatum* (1.78 Mg $ha^{-1}$), *Eulaliopsis binata* (1.62 Mg$ha^{-1}$) and *C. ciliaris* (1.41 Mg $ha^{-1}$) as compared to control (1.70 Mg $ha^{-1}$).

### 12.8.3 Improving Soil Properties

Other major benefits of agroforestry are soil improvement through physical, chemical and biological properties. Growing trees in conjunction with annual crops or pastures improve soil fertility by fixing nitrogen from the air, recycle the nutrient from deeper soil layers to soil surface through litter fall and its decomposition and improving organic carbon status of soil (Nair et al. 1999). The effect of bamboo plantation on improved soil fertility in terms of nitrogen, phosphorus, potassium and organic carbon has been shown by Singh et al. (2014). They showed that the value of nutrient saved through bamboo plantation was Rs. 2126–5555 $ha^{-1}$. Mixed bamboo stands exhibit elevated amounts of advantageous soil nutrients and superior soil qualities including soil porosity, aeration and bulk density as compared to monoculture stands (Zheng and Hong 1998; Zhang et al. 2010). Growing bamboo trees in agroforestry systems enriches soil fertility and microbial and soil enzyme activity (Zhihua et al. 2013).

Inclusion of grasses in the agroforestry systems further improves the soil structure and soil organic matter in these habitats. Trees in such systems help retain moisture by reducing runoff and simultaneously reduce nutrient leaching losses through root and mycorrhizal systems. Reduction in runoff also reduces intensity of flooding in downstream areas and helps maintain the system functions, which also regulate biodiversity. Study conducted at ICAR Research Complex for NEH Region at Umiam, Meghalaya showed that the tree species, viz. *Pinus kesiya*, *Alnus nepalensis*, *Parkia roxburghii*, *Michelia oblonga* and *Gmelina arborea* lowered bulk density, increased soil organic carbon and porosity, increased water stable aggregates (>0.25 mm) and decreased soil erodibility as compared to the natural fallow (Saha et al. 2007).

Jha et al. (2010) evaluated the impact of plantation of different tree species (*Prosopis juliflora*, *Leucaena leucocephala*, *Acacia nilotica* and *Azadirachta indica*) on selected physico-chemical and hydrological properties of soil and compared with uncultivated ravenous wasteland zone along the river Yamuna. After 25 years of plantation, there was twofold increase in soil organic carbon and two- to sixfold increase in mean weight diameter of aggregates, 4–18% reduction in soil bulk density and 1.5- to twofold increase in steady-state infiltration rate as compared to the reference site. This proves the potential of agroforestry systems for restoration of soil health.

**Table 12.6** Comparison of carbon stock in bamboo and tree forest ecosystems (Mg C $ha^{-1}$)

| Parts/forest | Arbour and shrub | Litter | In soil | Total |
|---|---|---|---|---|
| Moso bamboo in Lin'an (medium intensity management) | 34.20 | 0.66 | 71.48 | 106.34 |
| Chinese fir at 15th year | 53.60 | 3.43 | 93.16 | 203.79 |
| Moso bamboo in Yong'an (medium intensity management) | 61.30 | 3.01 | 197.36 | 261.67 |
| Deciduous broad-leaved forest | 47.75 | 5.85 | 208.90 | 262.50 |

Source: Venkatesh et al. (2005)

### 12.8.4 Biomass Production and Carbon Sequestration

Agroforestry contribute significantly to ecosystem protection through biomass production and carbon sequestration. Production of 4.5 Mg $ha^{-1}$ of air dry grass + firewood from silvopastoral systems in degraded ravine lands proved the effectiveness of such perennial systems for productive utilisation of degraded ravine lands (Sharda and Venkateswarlu 2007).

A long-term study (Prajapati et al. 1993) to assess the fuel and fodder production in Yamuna ravines revealed that after 15 years of planting, the fuelwood production was 28.7 Mg $ha^{-1}$ in *Acacia tortilis* and 27 Mg $ha^{-1}$ in *A. nilotica* raised at 3 × 3 m spacing. The biomass ranged between 7.6 Mg $ha^{-1}$ at ravine top (5 × 5 m spacing) and 34.4 Mg $ha^{-1}$ at ravine bottom (3 × 3 m spacing) in *A. nilotica*, while the corresponding yields of *A. tortilis* were 11.5 and 30.2 Mg $ha^{-1}$, respectively. The mean annual pasture yield ranged from 1.5 Mg $ha^{-1}$ $year^{-1}$ at 3 × 3 m spacing to 1.8 Mg $ha^{-1}$ $year^{-1}$ at 5 × 5 m spacing under *A. nilotica* and 1.8–2.1 Mg $ha^{-1}$ $year^{-1}$, respectively, under *A. tortilis* at similar spacing. The top feed production from respective spacing was 3.8 and 5.2 Mg $ha^{-1}$ in *A. nilotica* and 3.8 and 3.1 Mg $ha^{-1}$ in *A. tortilis* at the age of 14 years. In these trees carbon sequestered is about 46–48% of total biomass.

In another experiment, Parandiyal et al. (2006) studied the impact of *A. nilotica*, *Azadirachta indica* and *Albizia lebbeck* trees as boundary plantations grown in the marginal lands near ravines and found that the yield reduction near trees was compensated well from the income generated by selling the fuelwood produced by trees. They also found that the cereal crops are usually more compatible in agroforestry systems as compared to leguminous crops in rainfed situation. Under alley cropping system, sorghum was found more compatible with *Leucaena* as compared to pigeon pea (Prasad 1994a).

Bamboo-based agroforestry system is another hope for biomass production, income generation and C-sequestration in ravine areas. Venkatesh et al. (2005) identified *Dendrocalamus giganteus*, *D. hookerii* and *Bambusa nutans* as the superior bamboo species in terms of biomass production and improving fertility status of acid soils in the NEH region of India. Table 12.6 shows that bamboo will sequester prominent amount of carbon, if proper management is practiced. In another study, Tariyal et al. (2013) assessed the total carbon stock and carbon sequestration potential of four bamboo species and found the highest carbon stock in *Dendrocalamus strictus* (381.50 Mg $ha^{-1}$) and the lowest carbon stock in *Bambusa vulgaris* (160.11 Mg $ha^{-1}$). However, the rate of carbon sequestration was maximum in *B. balcooa* (99.81 Mg $ha^{-1}$ $year^{-1}$) and minimum in *B. vulgaris* (57.77 Mg $ha^{-1}$ $year^{-1}$).

Yadava and Thokchom (2015) assessed the carbon sequestration of bamboo and other tree species and observed that fast-growing trees absorb carbon dioxide more quickly as compared to slow-growing ones. Bamboo can sequester up to 62 Mg $CO_2$ $ha^{-1}$ $year^{-1}$, whereas other young forest can sequester 15 Mg $CO_2$ $ha^{-1}$ $year^{-1}$.

## 12.9 Effect of Agroforestry Systems on Socio-economic Conditions in Ravine Lands

Agroforestry systems on ravine lands not only checks soil erosion but also generate employment and fetch substantial economic remunerations to the farming community or the region. Agroforestry systems provide the economic, marketable and high-value products of medicinal importance, fruits, fuelwood, fodder, gum, resin and high-grade timber. They reduce the chance of total crop failure and ensure increments in farm incomes due to improved and sustained productivity, especially in problem lands such as ravines. These systems can cater to the needs of local masses and help in improving the rural living standards through sustained employment, enhanced incomes and improved nutrition and health due to increased quality and diversity of food outputs. National Wastelands Development Council is promoting the restoration of large areas of nonproductive land by agroforestry approaches.

Economic assessment of bamboo plantation under different soil conditions was done by Pande et al. (2012) in Yamuna, Chambal and Mahi ravine systems. The analysis suggested a cash outflow ranging from INR 30,550 to INR 48,000 $ha^{-1}$. The soil conservation practices benefitted INR 2125–INR 5555 $ha^{-1}$ to the society in terms of the value of nutrient and INR 41,000 $ha^{-1}$ for soil carbon build-up. The study suggested that high cost of establishment of agroforestry systems can be met by converging with the land-based schemes of central and state governments like MGNREGA. Bamboo and *Anjan* grass-based silvopasture system has been developed by CSWCRTI, Dehradun which checks water erosion, prevents soil loss and provides high returns from the ravines. In ravines of Yamuna and Chambal, *Jamunapari* breed of goats and sheep were reared on the biomass of grasses planted with trees, shrubs and bamboos (Tiwari 2008), which reduced the fodder cost and improved the economic status of people in these areas. Drumstick (*Moringa oleifera*)-based agri-horticultural systems was tested on the reclaimed ravine lands of Mahi River in the Gujarat. The idea was to replace the tobacco (*Nicotiana tabacum*) monocropping system. It was observed that the drumstick (*Moringa oleifera*)-based agri-horticultural systems was as remunerative as tobacco. It also provided environmental services in terms of higher soil carbon built up, saving of irrigation water and nutrients. The estimate of the nutrient saving and soil carbon built up was valued at Rs. 657 and Rs. 3696 $ha^{-1}$, respectively (Pande et al. 2014).

## References

Agoramoorthy G, Chaudhary S, Hsu MJ (2008) The check dam route to mitigate India's water shortages. Nat Resour J 48:565–583

Ali (1974) Halt the menace of sloppy lands. Indian Farmers Dig 7(9):31–34

Ben-zhi Z, Mao-yi F, Jin-zhong X, Xiao-sheng Y, Li Z-c (2005) Ecological functions of bamboo forest: research and application. J For Res 16(2):143–147

Bharadwaj SP (1990–1991) Annual Report. CSWCRTI, Dehradun, pp.41–42

Bhushan LS, Tiwari AK, Mishra PR (1986) Contour trenching – an effective measure to control erosion on sloping lands. In: Proc. National Symposium on problems of arid and semi- arid regions and their remedial measures, Feb. 08–09, 1986, Jodhpur, pp 83–85

CAFRI (2015) Vision 2050. Central Agroforestry Research Institute, Jhansi

Chaturvedi S, Kaushal R, Tewari S, Dhyani VC (2011) Agroforestry: an integrated land management option for fragile ecosystem. J Funct Environ Bot 1(1):45–53

Chaturvedi OP, Kaushal R, Tomar JMS, Prandiyal AK, Panwar P (2014) Agroforestry for wasteland rehabilitation: mined, ravine, and degraded watershed areas. In: Dagar JC, Singh AK, Arunachalam A (eds) Agroforestry systems in India: livelihood security and ecosystem services, advances in agroforestry vol 10. Springer, New Delhi, pp 233–272

Chittaranjan S, Patnaik US (1980) Hydrology and engineering. In: Research on soil and water conservation in semi-arid deep black soils. CSWCRTI Research Centre, Bellary, pp 48–76

CSWCRTI (2012a) Final project report (2007–2012) on hydrologic and economic evaluation of bamboo plantations in gullied lands under major ravine systems of India. Central Soil & Water Conservation Research & Training Institute, Dehradun

CSWCRTI (2012b) Annual report 2011–2012. Central Soil and Water Conservation Research and Training Institute, Dehradun

Dagar JC (1995) Agroforestry systems for Andamans and Nicobar Islands in India. Int Tree Crops J 8(2–3):107–128

Dagar JC (2015) Agroforestry for restoration, conservation and resilience of waste/degraded lands: challenges and opportunities. In: Dhyani SK, Newaj R, Alam B, Dev I (eds) Agroforestry: present status and way forward. Biotech Books, New Delhi, pp 323–362

Dagar JC, Mall LP (1980) Studies on the vegetation of ravines and banks of the river Kshipra and its tributaries. J Indian Bot Soc 59:234–245

Dagar JC, Pathak PS (2005) Grassland dynamics and their management. Range Manag Agrofor 26(1):7–31

Das DC (1985) Problem of soil erosion and land degradation in India. Proceedings of the National Seminar on Soil Conservation and Watershed Management, New Delhi

Deng Y, Peng Y, Wang Y, Yang Z (2003) Soil conservation efficiency of different models converting from steep farming slope plots in yaan, Schian, China. Int J Sediment Res 18(4):340–345

Dhyani SK, Handa AK (2014) Agroforestry in India: current scenario. Indian Farming 63(11):6–8

Dhyani SK, Samra JS, Ajit, Handa AK, Uma (2007) Forestry to support increased agricultural production: focus on employment generation and rural development. Agric Econ Res Rev 20:179–202

Dhyani SK, Newaj R, Sharma AR (2009) Agroforestry: its relationship with agronomy, challenges and opportunities. Indian J Agron 54(3):70–87

Dhyani SK, Handa AK, Uma (2013) Area under agroforestry in India: an assessment for present status and future perspective. Indian J Agrofor 15(1):1–11

Dosskey MG (2002) Setting priorities for research on pollution reduction functions of agricultural buffers. Environ Manag 30:641–650

Dwivedi RP, Tewari RK, Kareemulla K, Chaturvedi OP, Rai P (2007) Agri-horticultural system for household livelihood – a case study. Indian Res J Ext Edu 7(1):22–26

Ghosh BN (2010) Vegetative barriers for erosion control in Western Himalayan region. Technology Brochure, CSWCRTI, Dehradun, p 8

GoI (1972) Ravine reclamation program for dacoits infested area of Uttar Pradesh, Madhya Pradesh and Rajasthan. Report of working group on ravine reclamation, Ministry of Home Affairs, Government of India (GoI), New Delhi

Grewal SS (1993) Agroforestry systems for soil and water conservation in Shivalik. In: Agroforestry in 2000 AD for semi-arid and arid tropics, NRC for Agroforestry, Jhansi, India, pp 82–85

ICAR (2017) http://www.icar.org.in/en/node/5101 [Accessed on 20/05/2017]

Jha LK, Lalnunmawia F (2004) Agroforestry with bamboo and ginger to rehabilitate degraded areas in north east India. J Bamboo Rattan 2(2):103–109

Jha P, Mohapatra KP, Dubey SK (2010) Impact of land use on physico-chemical and hydrological properties of ustifluvent soils in riparian zone of river Yamuna, India. Agrofor Syst 80:437–445. https://doi.org/10.1007/s10457-010-9338-3

Joshi PK, Jha AK, Wani SP, Joshi L, Shiyani RL (2005) Comprehensive assessment of watershed management in agriculture: meta-analysis to assess impact of watershed program and people's participation. Research Report 8. ICRISAT and Asia Development Bank, pp 21

Kar A, Moharana PC, Raina P, Kumar M, Soni ML, Santra P, Ajai, Arya AS, Dhinwa PS (2009) Desertification and its control measures. In: Kar A, Garg BK, Singh MP, Kathju S (eds) Trends in arid zone research in India. Central Arid Zone Research Institute, Jodhpur, pp 1–47

Korwar GR, Radder GD (1994) Influence of root pruning and cutting interval of *Leucaena* hedgerows on performance of alley cropped rabi sorghum. Agrofor Syst 25:95–109

Kurothe RS, Nambiar KTN (2001) Eco-system restoration and sustainable production in degraded Mahi ravines and its effect on hydrology and sedimentation. In: Proc. National Conference on resource conservation and watershed management (RCWM-2001), May 23–25, 2001, Dehradun

Kurothe RS, Gaur ML, Rao BK, Parandiyal AK, Singh AK (2012) Conservation and production potentials of bamboo in ravine lands. CSWCRTI., ISBN 978-81-924172-1-9, Dehradun, p 160

Lawler DM (1993) The measurement of river bank erosion & lateral channel change: a review. Earth Surf Process Landf 18:777–821

Ministry of Agriculture (1984) Report of the working group on reclamation and development of ravines for formulation of the five-year plan. Government of India, New Delhi

Mishra PK, Rao BK (2012) Hydrologic and economic evaluation of bamboo plantations in gullied lands under major ravine systems of India. Final Project Report of National Bamboo Mission Sponsored Research & Development Project (2007–2012). National Bamboo Mission, Department of Agriculture & Cooperation, Ministry of Agri., Govt. of India, Krishi Bhawan, New Delhi. Available at http://nbm.nic.in/Reports/Report-4.pdf

Mishra PK, Tripathi KP (2013) Soil and water conservation research for land management in India. Indian J Dryland Agric Res Dev 28(1):01–18

NAAS (2009) State of Indian Agriculture. National Academy of Agricultural Sciences (NAAS), New Delhi, p 256

Nair PKR (1993) An introduction to agroforestry. Kluwer Academic publisher with cooperation ICRAF Dordrecht, London

Nair PKR (2008) Agroecosystem management in the 21st century: it is time for a paradigm shift. J Trop Agric 46:1–12. Florida, USA

Nair PKR, Buresh RJ, Mugendi DN, Latt CR (1999) Nutrient cycling in tropical agroforestry systems: myths and science. In: Buck LE, Lassoie JP, Fernandes EC (eds) Agroforestry in sustainable agricultural system. CRC Press, Boca Raton, pp 1–31

Narain P, Verma B, Singhal AK (1979) Studies of soils of Chambal with regard to their erodibility. Ind J Soil Conserv 7(2):37–42

National Commission on Agriculture (1976) Report of the National Commission on Agriculture, Part 5, Resource Development. Government of India, New Delhi, pp 107–322

NRCAF (2013) Vision 2050. National Research Centre for Agroforestry, Jhansi

Pande VC, Kurothe RS, Rao BK, Kumar G, Parandiyal AK, Singh AK, Kumar A (2012) Economic analysis of bamboo plantation in three major ravine systems of India. Agric Econ Res Rev 25(1):63–73

Pande VC, Kurothe RS, Singh HB, Tiwari SP (2014) A case study on agri-horticulture production system for income and environmental security in Mahi ravine lands of Gujarat. In: Verma KS, Panwar P, Kaushal R, Chuhan S, Chander J, Chandel RS (eds) ISTS-IUFRO Conference on "Sustainable Resource Management for Climate Change Mitigation and Social Security" March 13–15, 2014. pp. 30

Pani P (2016) Controlling gully erosion: an analysis of land reclamation processes in Chambal Valley, India. Dev Pract 26(8):1047–1059

Pani P, Mohapatra SN (2001) Delineation and monitoring of gullied and ravenous lands in a part of lower Chambal valley, India, using Remote Sensing and GIS – a research paper published in the proceedings of ACRS 2001 – 22nd Asian conference on remote sensing, 5–9 November 2001, Singapore, vol 1, pp 671–675

Pani P, Mishra DK, Mohapatra SN (2011) Land degradation and livelihoods in semiarid India: a study of farmers' perceptions in Chambal Valley. Asian Profile 39(5):505–519

Parandiyal AK, Kumar A, Prasad A, Singh KD (2006) Studies of tree crop association under boundary plantation system in southern Rajasthan. Paper in National Symposium for livelihood security, environmental protection and biofuel production. NRC Agroforestry, Jhansi, December 16–18, 2006

Prajapati MC, Joshie P, Rathore BL, Dubey LN (1982) Surface water management for grass and tree land development in ravenous watersheds – a case study. Ind J Soil Conserv 10(2/3):30–38

Prajapati MC, Nambiar KTN, Singh JP, Malhotra BM, Sharda VN (1989) Effect of goat grazing on vegetation, runoff and soil loss in a stabilized watershed vis-à-vis animal growth. Indian J Soil Conserv 17(1):9–16

Prajapati MC, Tiwari AK, Nambiar KTN, Puri DN, Singh JP, Malhotra BM (1993) Fuel and fodder production in Yamuna ravines at Agra. Ind J Soil Conserv 21(3):8–13

Prakash C, Raizada A, Samra JS, Sastry G (1999) Vegetative barriers for resource conservation. Bulletin No. T-41/D-29. CSWCRTI, Dehradun

Prakash O, Dubey RK, Dubey SK, Deshwal JS (2011) Evaluation of Ber (*Zizyphus mauritiana* Lamk.) based agri-horti systems under various tree root management practices on reclaimed Yamuna ravines. Ind J Soil Conserv 39(1):59–62

Prasad A (1994a) Alley cropping in *Leucaena leucocephala* with field crops for sustained productivity. In: Proceedings international conference on sustainable development of degraded lands through agroforestry in Asia and Pacific, New Delhi, vol I, pp 471–478

Prasad A (1994b) Shelterbelt of *Leucaena* and its effect on crop yield in rainfed agroforestry system. In: Proceedings international conference on sustainable development of degraded lands through agroforestry in Asia and Pacific, New Delhi, vol I, pp 499–506

Rao MR, Schroth G, Williams SE, Namirembe S, Schaller M, Wilson J (2004) Managing belowground interactions in agro-ecosystems. In: van Noordwijk M, Cadisch G, Ong CK (eds) Below-ground interactions in tropical agro-ecosystems: concepts and models with multiple plant components. CABI Publishing, Wallingford, pp 309–328

Rao BK, Kurothe RS, Singh AK, Parandiyal AK, Pande VC, Kumar G (2012). Bamboo plantation based technological interventions for reclamation and productive utilization of ravine lands. CSWCRTI, T-62/V-4, pp 30

Saha R, Tomar JMS, Gosh PK (2007) Evaluation and selection of multipurpose tree for comparing soil hydro-physical behaviour under hilly ecosystem of north east India. Agrofor Syst 69:239–247

Samra JS, Chandra Prakash Sastry G, Raizada A (2004) Technical manual on contour trenching for rehabilitation of degraded lands. Central Soil and Water Conservation Research and Training Institute, Dehradun, p 60

Sethy BK, Parandiyal AK, Shakir A, Ashok K, Singh RK (2010) Cost effective conservation measures for management of medium and deep ravenous lands. Annual Report. CSWCRTI, Dehradun, pp 73–74

Shanmughavel P, Francis K (2001) Intercropping trials of four crops in plantations. J Bamboo Rattan 1(1):3–9

Sharda VN, Venkateswarlu B (2007) Crop diversification and alternate land use systems in watershed management. In: Best-bet options for integrated watershed management. Proceedings of the comprehensive assessment of watershed programs in India, 25–27 July 2007, ICRISAT, Patancheru, pp 111–128

Sharda VN, Bhushan LS, Raghuveer (1982) Hydrological behaviour of ravinous watersheds under different land uses. In: International symposium on hydrological aspects of mountainous watersheds. University of Roorkee, Roorkee, pp 14–18

Sharda VN, Juyal GP, Naik BS (2008) Watershed development in India: status and perspective. Allied Printers, Dehradun, p 219

Sharma AK, Pradhan IP, Nema JP, Tejwani KG (1980) 25 years research on soil & water conservation in ravine lands of Gujarat. CSWCRTI Research Centre, Vasad

Sharma BD, Hore DK, Pandey G, Wadhwa BM (1992) Genetic resources of bamboos in the NE region of India. Indian J For 15(1):44–51

Sheshadri P (1985) Intercropping of bamboo (*D. strictus*) with soybean- an agroforestry study. Ph.D thesis, Tamil Nadu Agricultural University, Coimbatore, p 480

Sikka AK, Mishra PK, Singh RK, Krishna BK, Islam A (2016) Management of ravines for food, livelihood and environment security. In: Souvenir-Global ravine conference on managing ravines for food and livelihood security. RVSKVV, Gwalior, India, March 7–10, 2016, pp 1–14

Singh G (2009) Soil water dynamics, growth of *Dendrocalamus strictus* and herbage productivity influenced by rainwater harvesting in Aravalli hills of Rajasthan. For Ecol Manag 258:2519–2528

Singh RA (2011) Rain water management with three tier systems in ravines affected area of Bundelkhand. Ind J Soil Conserv 39(1):14–19

Singh A, Dayal R (1975) Development of water resource in ravine utilization -a case study. Soil Conserv Dig 3(1):26–30

Singh AK, Kala S, Dubey SK, Rao BK, Gaur ML, Mohapatra KP, Prasad B (2014) Evaluation of bamboo based conservation measures for rehabilitation of degraded Yamuna ravines. Ind J Soil Conserv 42(1):80–84

Singh AK, Kala S, Dubey SK, Pande VC, Rao BK, Sharma KK, Mohapatra KP (2015) Technology for rehabilitation of Yamuna ravines – cost-effective practices to conserve natural resources through bamboo plantation. Curr Sci 108(8):1527–1533

Swaminathan MS (1987) The promise of agroforestry for ecological and nutritional security. In: Steppler HW, Nair PKR (eds) Agroforestry: a decade of development. ICRAF, Nairobi, p 30

Tariyal K, Upadhyay A, Tewari S, Melkania U (2013) Plant and soil carbon stock and carbon sequestration potential in four major bamboo species of North India. J Adv Lab Res Biol 4(3):90–98

Tejwani KG, Bhardwaj SP (1982) Soil and water conservation research in India, Part II, Proceedings of the 12th International Congress of Soil Science, New Delhi, pp 608–21

Tejwani KG, Gupta SK, Mathur HN (1975) Soil and water conservation research 1956–71. Indian Council of Agricultural Research, New Delhi

Thomas M, Sahu P, Shrivastava A, Hussain Z (2011) Biodiversity and livelihood options of people in Chambal ravine of Morena district, Madhya Pradesh, India. J Trop For 27(3):40–56

Tiwari SK (2008) Agro-forestry. Available at http://nsdl.niscair.res.in/jspui/bitstream/123456789/656/1/revised%20agroforestry.pdf

Tiwari SP, Kurothe RS (2006) Effect of vegetative barriers on soil and nutrients losses at 2% slope on agricultural lands of reclaimed Mahi ravines. Ind J Soil Conserv 34(1):37–41

Uthappa AR, Chavan SB, Singh M, Sridhar KB, Dev I, Ram A, Sathish BN, Kumar M, Dwivedi RP, Singh R, Singh RK, Singh R, Tewari RK, Handa AK, Chaturvedi OP (2016) Tree diversity in ravines and their rehabilitation through agroforestry interventions in Bundelkhand Region of India. Indian J Agrofor 18(1):77–83

Venkatesh MS, Bhatt BP, Kumar K, Majumdar B, Singh K (2005) Soil properties influenced by some important edible bamboo species in the north-eastern Himalayan region, India. J Bamboo Rattan 4(3):221–230

Verma B, Singh B, Saraf NH, Monappa K (1969) Suitability and economics of grasses for reclamation and stabilization of Mahi ravines in Gujarat. Indian For 95(1):33–44

Verma B, Chinamani S, Bhola SN, Rao DH, Prasad SN, Prakash C (1986) Twenty-five years of research on soil and water conservation in ravine lands of Rajasthan. Central Soil and Water Conservation Research and Training Institute Publication, Research Centre, Kota, pp 1–215

Wajja-Musukwe TN, Wilson J, Sprent J, Ong CK, Deans DJ, Okorio J (2008) Tree growth and management in Ugandan agroforestry systems: effect of root pruning on tree growth and crop yield. Tree Physiol 28:233–242

Yadava PS, Thokchom A (2015) Carbon sequestration potential of the forests of North Eastern India. In: Goel M, Sudhakar M, Shahi RV (eds) Carbon capture, storage and utilization: a possible climate change solution for energy industry. TERI Publishers, New Delhi, pp 149–165

YanHui W, Yongmin L (1995) Hydrological characteristics of a moso-bamboo (*Phyllostachys pubescens*) forest in South China. Hydrol Process 9(7):797–808

Young A, Cheatle RJ, Muraya P (1986) The potential of agroforestry for soil conservation. Part III. Soil changes under agroforestry (SCUAF): a predictive model. ICRAF Working Paper 44. ICRAF, Nairobi

Zhang CS, Xie GD, Fan SH, Zhen L (2010) Variation in vegetation structure and soil properties and the relation between understory plants and environmental variables under different *Phyllostachys pubescens* forests in Southeastern China. Environ Manag 45:779–792. https://doi.org/10.1007/BF02857909

Zheng YS, Hong W (1998) Management of *Phyllostachys pubescens* stand Xiamen. Xiamen University Publishing House

Zhihua T, Lihua C, Xinxiao Y, Yushan Z (2013) Effect of bamboo plantation on rhizosphere soil enzyme and microbial activities in coastal ecosystem. J Food Agric Environ 11(3&4):2333–2338

# Soil and Water Conservation in Ravinous Watersheds: Case Studies from Uttar Pradesh in India

13

Suraj Bhan and Sanjay Arora

**Abstract**

Ravines and gullies occur all over India especially in states of Madhya Pradesh and Uttar Pradesh. The state of Uttar Pradesh has taken the lead role in reclamation of ravine lands starting from 1884 to till recently, but most of the project activities gradually approached toward the requirement of reclamation based on integrated farming system approach on watershed basis. Some projects have resulted in cumulative gain in terms of area reclaimed, people benefited, and knowledge gained which can be fruitfully utilized for future ravine reclamation planning. However, the reason of the not much benefit from the earlier projects was due to failure of stable use of reclaimed land for productive purposes. Still about 75% of the ravine land needs to be effectively reclaimed for productive use. In this chapter, an attempt has been made to identify the gaps and suggest approaches and components for effective reclamation of ravine lands on watershed basis. The reclamation should aim for improvement of socio-economic conditions of ravine areas indicating the scope of providing sustainable livelihood to the people of the area. The region has potential for diversified crops including highly remunerative medicinal and aromatic crops. Introduction of these crops calls for an integrated bio-industrial approach for locally processing the produce that would provide additional income to farmers and generate employment.

S. Bhan (✉)
Soil Conservation Society of India, National Societies Block A/G-4, National Agricultural Science Centre (NASC) Complex, New Delhi, India

S. Arora
ICAR-Central Soil Salinity Research Institute, Regional Research Station, Lucknow, India





J. C. Dagar, A. K. Singh (eds.), *Ravine Lands: Greening for Livelihood and Environmental Security*, https://doi.org/10.1007/978-981-10-8043-2_13

## 13.1 Introduction

Uttar Pradesh (UP) is one of the states of India having the largest population dependent on agriculture. It has varying types of soil and topography, but the major agricultural area constitutes the part of Indo – Gangetic terrain and the great Indian plenteous. Large number of rivers flow through this plain and act as major drainage system. The banks of these rivers are generally sloping and eroded. Where the river cuts a channel very deep from its bank, the runoff from the tableland having to negotiate a large vertical fall in short distances, a situation is created for rapid development of gullies along the riverbanks, resulting in what is commonly called "ravines." Once a ravine is formed, it grows by the phenomenon of saturation and slip of its head and sides. Further, in broader sense ravines are a system of gullies created by water from top land to the deep rivers below and running parallel to each other. Development of gullies is the last phase in the process of soil erosion by the agency of water starting from the beating of raindrops on the bare soil resulting in "splash erosion," the movement of water and washing away of the soil in shallow depths over a large surface call "sheet erosion," development of rills along the micro-depressions on the land surface causing "rill erosion," and the whole process ending ultimately in gully formation. Gullies or ravines are nature's waterways having a distinct watershed consisting of gently sloping tablelands, the steeply sloping peripheral lands, and the ravinous land proper at the confluence with the river (Dhar 2006). Since the formation and development of ravines are due to soil erosion, the remedy lies in adopting comprehensive soil conservation measures on a watershed basis. This would control and stabilize the growth of fast spreading ravines and consequently help in reclaiming the ravine land for gainful utilization.

To study the technology of the ravine reclamation, a long-term pilot project was taken up in the year 1969–1970. It was observed that ravines which are less than 5–6 m deep can economically be bench terraced and provided irrigation facilities with underground distribution system for intensive agricultural purposes. Deep ravines may be stabilized by check dams, natural vegetation, and plantation with fencing at a nominal cost. The result of the pilot project has encouraged further work in the possibilities of reclaiming the ravinous land on large scale.

A working group on ravine reclamation was set up by the Government of India which published its report in October 1972 (Anonymous 1972). Apart from a survey of the ravine problem, it has worked out a tentative plan for reclamation of 0.73 million hectares of ravinous land of Uttar Pradesh at a total cost of ₹6.08 billion. To meet the above expenditure, a phasing was made to reclaim the entire area by the end of this century. The economic feasibility for reclaiming the ravine area indicate that bankable projects could be taken up with the economically viable schemes which would start giving results within 3–4 years after the reclamation. It is in this context that the present scheme for the World Bank is proposed for the districts which have ravinous area on extensive scale.

Investigations on topography maps and aerial photographs revealed that about 1.23 million hectares of land is covered with ravines in Uttar Pradesh (UP). The problem is not static, but it is extending and spreading every year adding to its share some more area from adjoining tableland. In case the ravines grow and engulf the good flat land at present rate, it is estimated that the entire watershed would become ravinous in about a century and with the present rate of deterioration of soil fertility and top soil, the state will be losing about Rs. 30 million per year. It is, therefore, necessary to control the spread of ravines, as well as to reclaim the ravinous area wherever practicable.

The two aspects of the program are complementary as desired results will not be achieved in the absence of one or the other. An effective program will have to tackle the problem in an integrated manner that the upper portion of the watershed where the runoff drains out water is put under effective control measures and the lower part of the watershed, as it flows toward the river basin, is reclaimed through soil conservation measures for agriculture, horticulture, afforestation, and pasture lands. It has been estimated that the reclaimed areas will not only add to the cultivated area of the state but will yield an additional income of ₹1000 million every year by the way of additional food, fuel, timber, and fodder grasses. The program will also provide additional avenues of work for engineers, agriculturists, horticulturists, foresters, and skilled and unskilled workers of the area.

## 13.2 Historical Background

The first attempt in the direction of soil and water conservation in this state was started in the ravines of Etawah district by its collector late Fisher, who in the year 1884 persuaded some local landlords to hand over to the State Government about 1200 ha of ravinous land. The treatment of this area was taken up in the same year, and plants of various species were planted, and the area came out as a very good forest, and the experiment carried out gave valuable indication for the possibilities of successful afforestation in the ravines. A distinctive method adopted here was the construction of large water storage structures to improve moisture contents and also control the spread of ravines. Due to conservation of moisture, the forest created by Fisher became a classical example of successful ravine afforestation in the future (Fig. 13.1), and the National Agriculture Commission adopted intensive productive forestry operations under irrigated conditions.

In 1947, the State Government set up a land management block with the objective of checking further erosion and reclamation of ravines in badly eroded areas, particularly the ravines of Jamuna and Chambal rivers. In Etawah district, the ravinous areas cover 61,683 ha area which is 14.2% of the geographical area of the district (Rao et al. 2005). In the year 1948, a pilot project for ravine reclamation was taken up in Dalelnagar area of Etawah district. Under this project, the methodology adopted was construction of check dams for siltation of the upstream areas for ultimate formation of agricultural lands. Another project was taken up in P.D.P. Etawah for ravine reclamation in an area of 75 acres (30 ha) and was attempted through extension methods. Although these individual projects can be considered a success in their way, they did not lead to large-scale ravine land development through people's own efforts.

**Fig. 13.1** Afforested ravine area, converted to real forest

A conference was held in Agra in the year 1951 for making up large-scale afforestation work on the borders of UP and Rajasthan to arrest the spread of Rajasthan desert. The scheme consisted of reclamation of wasteland by wind breaks and shelter belts. The aim was also to conserve moisture along the dry tracts which were devoid of vegetative growth and to provide fuel and fodder for the village population and cattle. Under this scheme, some ravine afforestation work was also undertaken. The scheme was, however, later abandoned due to the difficulties of obtaining land transfer. The State Government of UP decided in 1964 that the ravines should be controlled by protecting the adjoining tableland so that further advancement of gullies may be checked. The area down below was put under afforestation work. As a result of this decision, the soil conservation wing of Agriculture Department set up ten soil conservation units for safeguarding of tablelands and treatment of peripheral lands. These units are still working but due to the financial difficulties, their activities are very much limited. In 1967, a national policy on ravine reclamation was formulated. The objective of the national policy was to arrest their further advancement and save the valuable tablelands. It was also aimed to utilize the ravine land for productive purposes like raising of cultivated crops, grasses, horticulture, forestry, or industrial raw materials depending upon land capability, techno-economic feasibility, and social setting in an area. These recommendations necessitated the development of an integrated approach for treatment of ravines on watershed basis.

In the year 1969–1970, a pilot project was taken up to treat the ravinous area from top to bottom. This required a classification of ravines as per their depth. It was proposed to have a ravine control program on the tableland, and a part of peripheral land and deep ravines was to be developed for intensive agriculture through bench terracing and leveling for irrigated agriculture. Consequently, two projects were started one each in Kanpur and Rae Bareli districts. It was found that the watershed approach and a comprehensive treatment of the area costed about ₹3000–4000 per hectare on an average. The overall area under cultivation in ravinous watershed increased and so was the cropping intensity and per hectare production. Thus, three-fold benefits from ravine reclamation work encouraged to take up the work on a larger scale.

Land is a precious natural resource. The existence of mankind depends on land resources. Its unscientific use for immediate gains without considering the long-term sustainability leads to land degradation. India is endowed with a rich diversity of natural resources. However, the need for food and nutritional security is increasing due to expanding population. The present population of India is expected to stabilize at about 1.7 billion by 2050. This would result in significant reduction of per capita availability of land, water, forest, and other forms of natural resources. The annual food grain requirement is likely to reach 350 million Mg by the year 2050 which is to be met out from the shrinking land resources. Therefore, quality of land needs to be maintained to enhance and sustain higher productivity, and land degradation rates have to be effectively controlled.

About 53% of the total geographical area of the country suffers from different kinds of land degradation, and UP accounts for 9%. About 4.0 million ha (11.03%) area of the country is suffering from ravine problem, out of which UP shares 1.230 million ha (31% of ravine land of the country) spread over in 25 districts of the State. Out of these districts, five districts, namely, Agra, Banda, Etawah, Hamirpur, and Jalaun, contribute 0.70 million ha (62%) and are extensively in the catchments of Yamuna and Chambal rivers and their tributaries (Bhushan and Sharma 1997).

Ravines are a system of gullies running more or less parallel to each other and draining into a major river or its tributary with development of deep gorges and riverine terraces (*kachars*) at the end. Development of a gully is the last phase in the process of soil erosion by the agency of water starting from the beating of raindrops on bare soil resulting in "splash erosion," the movement of water in shallow depths over a large surface causing "sheet erosion," development of rills along the micro-depressions on the land surface causing "rill erosion," and the whole process ending in gully formation ultimately (Fig. 13.2).

Gullies or ravines are nature's waterways having a distinct watershed consisting of gently sloping tablelands, the severely sloping peripheral lands, and the ravines proper from the ravine head to their mouth at the confluence with the river. Since the ravines are formed by the process of soil erosion and they may be in a state of continuing active growth, the remedy lies in adopting comprehensive soil conservation approach on a watershed basis.

About 74% of landholdings of the state belong to marginal farmers. About 16% of the holdings are small farmers. The most significant fact is that about three quarters of total holdings in UP account for only 21% of the total cultivable land and more than 50% of the farmers have holdings less than 0.5 ha each. Experience has shown that strategies and program for recovery of degraded lands in UP has been limited and sporadic. This is a very vital area of economic significance for the state where nearly 80% of people live in rural areas and depend on land for sustenance.

**Fig. 13.2** Gully formation

The need of the hour is a concerned drive to recover ravine lands for sustainable uses and to prevent existing agricultural and other productive lands from suffering depletion or further degradation (Anonymous 2003). Many strategies and policy thrust in this direction called for which would include investments, infrastructure, technology inputs, soil and water conservation, participative mobilization of people, stakeholder involvement, and close coordination between different departments, institutions, agencies, and rural community groups. In this chapter the possible strategies for development of ravines, based on past experiences of research and development along with the new technologies, have been discussed. In this regard the scope and effectiveness of faming system approach for rehabilitation of ravine lands have also been addressed.

## 13.3 Extent of Ravines, Their Classification, and Land Use in India

In India about 3.97 million ha area has been estimated to be affected by ravines are given in Table 13.1.

To determine the proper land use for ravinous land, the relevant factors like ravine depth, width, side slopes, stage of stabilization, watershed area shape, length, irrigation, availability, special hazard like flooding, soils, land ownership, present land use, and other physical and socio-economic factors must be studied and interpreted into ravine reclamability classification. The ravines which can be economically reclaimed into agricultural land should be identified giving due consideration to the alternate use of scarce machinery, equipment, and economic resources which may have to be invested into such lands for reclamation. In general, shallow and wide ravines where irrigation is available should be considered for agriculture.

**Table 13.1** Extent of ravines in India

| State | Rivers and their tributaries | Area[a] (000 ha) Tejwani et al. (1975) | Area[b] (000 ha) (IAB, 1990; Chaturvedi et al. 2014) |
|---|---|---|---|
| Uttar Pradesh | Yamuna and tributaries, Chambal, Gomati, Betwa, and *Kholas* of Ganges | 123 | 1230 |
| Madhya Pradesh | Chambal, Assan, Kali Sindh, Kshipra, etc. | 600 | 683 |
| Rajasthan | Chambal, Parvati, Kali Sindh, Mej, Mangali, Banas, Parwan, Alnia, Talera, Chora-Pachar, Morel, Banas, Banganga, and Sabie | 340 | 452 |
| Gujarat | Sabarmati, Mahisagar, Watrak, Meshwa, Banas, Narmada, Dhadhar, Saraswati, Rupem, Tapi, Kim, Purna, Ambika, Mindota, Khari, Par, Auranga, Kotak | 280 | 400 |
| Maharashtra | Ratnagiri, Godavari, etc. | 20 | 020 |
| Punjab and Haryana | Ghaggar, Satluj, and Beas | 120 | 120 |
| Bihar | – | 600 | 600 |
| Tamil Nadu | Cauvery, Pennar | 60 | 060 |
| West Bengal | Kangsabati | 104 | 104 |
| Odisha | Salandi, Baitarani, Brahmani, Mahanadi, etc. | – | 113 |
| Himalaya foothills (including Assam and Himachal Pradesh) | | – | 193 |
| Total | | 2247 | 3975 |

[a]Mentioned by Tejwani et al. (1975) and [b]Indian Agriculture in Brief (1990) cited by Chaturvedi et al. (2014)

The other ravinous areas should be retired to permanent vegetation with closure and aids to vegetative growth where necessary. Regular afforestation for fuel, timber, or industrial raw materials may also be considered in selected blocks where land can be procured and other conditions are favorable. For afforestation purposes, complete control of land should be transferred to the Forest Department. The National Forest Policy also points attention toward the need for checking "the erosion progressing a pace along the tree-fewer banks of the great rivers leading to ravine formation and on stretches of undulating waste lands depriving the adjoining fields of their fertility." In fixing the proportion of forest area, the National Policy resolution further states, "In the plains, where the ground is flat and erosion is normally not a serious factor, the proportion to be attained should be placed at 20% and in view of the pressure of agriculture, effort at the extension of tree-lands should be concentrated on river banks and other convenient places not suitable for agriculture." The ravine lands, with protective afforestation done now, can serve as a conserved land resource to be drawn upon for intensive use in the future.

**Table 13.2** Category-wise distribution of ravine areas

| Sl. No. | Category | Area (in 000 ha) | Percentage |
|---|---|---|---|
| 1 | Marginal land | 150.3 | 37.8 |
| 2 | Shallow ravines | 108.1 | 27.2 |
| 3 | Medium ravines | 84.7 | 21.3 |
| 4 | Deep ravines | 54.4 | 13.7 |
| | Total | 397.5 | 100 |

Ravine area includes marginal lands, shallow ravines, medium ravines, and deep ravines. It is estimated that about 37.8% area falls under marginal lands, 27.2% area under shallow ravines, 21.3% area under medium ravines, and about 13.7% area under deep ravines. As per above category-wise ravine land distribution is given in Table 13.2.

Indian Photo-interpretation Institute, Dehradun, has classified the ravines in the following classes:

- Flat uplands gently sloping or tablelands
- Peripheral lands around ravines
- Shallow ravines – depth up to 2.5 m, measured in the middle of second-order ravines
- Medium ravines – depth 2.5–5.0 m measured as above
- Deep ravines – depth >5 m measured as above
- Wide valley bottoms – width >16 m, slope >5%
- Younger river terraces or *kachar* land
- Hilly areas

In UP, most of the deep ravines and valley bottom, etc. are under the possession of Forest Departments as reserve forests.

### 13.3.1 Extent of Ravines in UP

In UP, the total ravine area is 1.23 million ha spread over 25 districts (Table 13.3), of which 5 districts (Agra, Banda, Etawah, Hamirpur, and Jalaun) contributes about 62%. The ravine lands are distributed in the catchment of 14 rivers (Table 13.4) of which Yamuna and Betwa contribute about 60% of the total land. In these areas the people are very poor and have adapted to unsocial activities and end up forming groups of dacoits. The natural physical inhospitable features of the area (Fig. 13.3) also support these activities. It is fact that four districts (Agra, Banda, Hamirpur, Jalaun) alone contribute about 0.70 million ha (62%) of the area and most of the area possess very depleted natural vegetation (Fig. 13.4). District-wise ravine lands in Uttar Pradesh and river-wise total ravine land are given in Tables 13.3 and 13.4.

**Table 13.3** District-wise ravine lands in Uttar Pradesh

| District | Ravine area (000 ha) |
|---|---|
| Agra | 17.6 |
| Aligarh | 1.4 |
| Allahabad | 1.6 |
| Banda | 1.2 |
| Barabanki | 0.9 |
| Bulandshahr | 0.5 |
| Etawah | 15.8 |
| Farrukhabad | 4.0 |
| Fatehpur | 3.1 |
| Hamirpur | 12.9 |
| Jalaun | 11.4 |
| Jaunpur | 1.3 |
| Jhansi | 0.1 |
| Kanpur | 5.5 |
| Lucknow | 0.2 |
| Mainpuri | 5.9 |
| Mathura | 1.4 |
| Meerut | 0.6 |
| Mirzapur | 7.6 |
| Muzaffarnagar | 3.0 |
| Pratapgarh | 1.9 |
| Raebareli | 0.9 |
| Shahjahanpur | 0.8 |
| Sultanpur | 1.8 |
| Varanasi | 0.8 |
| Total | 123.0 |

**Table 13.4** Total ravine land in Uttar Pradesh (river wise)

| Rivers | Ravine area (000 ha) |
|---|---|
| Yamuna | 38.9 |
| Chambal | 7.9 |
| Kali | 8.9 |
| Ken | 2.0 |
| Gomti | 3.2 |
| Sengar | 5.1 |
| Betwa | 22.7 |
| Dhasan | 5.1 |
| Pahuj | 5.3 |
| Sai | 2.9 |
| Ganga | 3.5 |
| Son | 7.7 |
| Kaveri | 5.0 |
| Sindh | 4.8 |
| Total | 123.0 |

**Fig. 13.3** Ravine lands deprived with any vegetation

**Fig. 13.4** Course of river having depleted natural vegetation and formation of ravines

## 13.4 Generic Features and Constraints of Ravines Areas

### 13.4.1 Soils and Geology

Soils of ravine are generally deep alluvial with extensive network of gullies running parallel to each other along riverbanks in all the ravine area of the state. The texture of soil varies from sandy loam to clay loam in texture and neutral to mildly alkaline in reaction. The peripheral lands are free from carbonates and pH varied from 6.5 to

7.5. The soils are moderately well drained and *kankar* (calcareous) layer is observed throughout the profile depth. In some ravinous areas, dark clay-textured soils are also found. The parent material is basaltic rocks which are broken into small fragments on exposure to rain. Hardpans are also found in certain districts at varying depths, which seriously inhibit afforestation work especially if the rotations are longer than 10 years. *Kachar* soils, i.e., soil on the riverine terraces at the mouth of ravines, are medium-textured and moderately well-drained and have neutral to mildly alkaline reactions. The soils are severely eroded on steep slopes exposing *kankar* nodules. Due to excessive runoff, the soil is almost dry throughout the year. The calcareous nature of soil, moisture stress, and uncontrolled grazing has ruined the lands with vital vegetal cover.

### 13.4.2 Water Resources

Ravine systems occur at the lower end of the watershed and therefore provide sites for the good storage for irrigation in the downstream reclaimed areas. The average rainfall is 750 mm. Water storage structures would greatly help the development of horticulture and forest plantation as experienced around Kanpur by reclamation of ravines. The location of the ravines again provides good opportunities for lift irrigation from the river below. However, lift irrigation is economical only when the heads are low. There are successful tube wells having a discharge about 2 cus (cubic foot per sec), located just above the ravine lands, and having a command of about 100 ha.

### 13.4.3 Vegetation

The ravine lands are generally devoid of a good vegetative cover. The ravine area in Uttar Pradesh has poorly grown and stunted stands of tree species such as *Acacia leucophloea* (reonja), *Prosopis cineraria* (gheonlear), *Azadirachta indica* (neem), *Holoptelea integrifolia* (chirol), *Acacia catechu* (khair), *Balanites aegyptiaca* (ingua), and *Flacourtia indica* (katai); shrubs such as *Capparis decidua* (karil), *Carissa carandas* (karaunda), *Ziziphus nummularia* (jharberi), and *Dichrostachys cinerea* (kiari); and grasses *Desmostachya bipinnata* (dab), *Apluda mutica* (parwa), and *Bothriochloa pertusa* (phulaica) are prominent. For detailed vegetation, please see Chap. 3 of this book by JC Dagar.

### 13.4.4 Agriculture in Ravine Lands

Because of nondevelopment of irrigation facilities, the rain-fed agriculture is practiced in ravine lands. Coarse cereals, pulses, and oilseeds are generally grown without proper fertilizers. Due to the availability of high-yielding varieties, some improvement in crop productivity is seen, but the practice is still traditional one. The major crops with its average productivity during *kharif* are pearl millet (300 kg

$ha^{-1}$), pigeon pea (300 kg $ha^{-1}$), sorghum (450 kg $ha^{-1}$), and black gram (200 kg $ha^{-1}$), while during *rabi* season, major crops are gram, mustard, barley, and pea, which produce 400, 300, 600, and 300 kg $ha^{-1}$, respectively.

### 13.4.5 Ownership and Socio-economic Backdrop

The ownership is an important factor which has bearing on land use pattern. In UP the ravine lands are generally privately owned (60%) and the rest as village community and government land. The privately owned lands are being cultivated by stakeholders with their own perceptions, risk factors, and resource availability. The community and government lands are afflicted with the tragedy of commons. Use of community land for cattle grazing and illicit felling of trees also produces devastating degradation.

The inhabitants of ravine region have poor socio-economic conditions as is evident from:

- Small landholding.
- 68% of total farm families holding is below 0.1 ha.
- 20% families owning 1–2 ha.
- 10% of the families owned land in range of 2–5 ha.
- 2% holdings have 5 ha or above.
- A large family size of 10 members per family.
- Nearly 70% land is rain-fed, and 30% is irrigated through groundwater resources
- requiring greater investment as well as operational cost.
- Fertilizer use is practically nonexistent.
- Lack of communication facilities.
- High percentage of illiteracy.
- Poor health of animals with low productivity due to fodder shortage.
- Low risk-bearing capacity.

### 13.4.6 Constraints

There are several bottlenecks which make it difficult to rehabilitate these lands. Some of these include the following: ravines are affected by severe soil erosion, and in most of cases "A" horizon of soil has been lost to such an extent that the soil fertility has become very poor; the development of water resources – surface and groundwater – is inadequate with the result that the rain-fed farming is only a practice of cultivation; biodiversity is poor due to soil erosion, water scarcity, scrub forests, and scanty vegetation; the inaccessibility and remote situations in ravines have created law and order problem to the state; and blue bulls are a menace to crops and vegetation.

**Table 13.5** Progress of state sector schemes of ravine reclamation (in '000 ha)

| Plan/years | Ravine reclamation for agriculture | Ravine afforestation | Total | Expenditure ₹ in million |
|---|---|---|---|---|
| First plan (1951–1956) | – | 02.81 | 02.81 | 2.073 |
| Second plan (1956–1961) | – | 04.41 | 04.41 | 2.892 |
| Third plan (1961–1966) | 07.28 | 04.89 | 12.17 | 3.909 |
| 1966–1969 | 02.22 | 01.64 | 03.86 | 10.990 |
| 1969–1970 | 17.09 | 05.00 | 22.09 | 5.282 |
| 1970–1971 | 12.00 | 05.00 | 17.00 | 4.702 |
| Grand total | 38.59 | 23.75 | 62.34 | 29.848 |

### 13.4.7 Reclamation Efforts

The Agriculture Department set up ten soil conservation subdivisions for reclamation of ravine lands by bunding of tablelands and treatment of peripheral lands. Up to the years 1970–1971, 38,590 ha of tablelands and peripheral lands were treated, and 23,750 ha of ravine lands were afforested, with an expenditure of ₹29.848 million (Table 13.5). Table 13.6 provides a summary of overall previous efforts of ravine reclamation made by the Government of Uttar Pradesh state through their own funds and through the Government of India and the European Commission and NABARD-sponsored schemes. Initially only some interventions like contour bunding and land leveling were attempted, but gradually it took the shape of integrated watershed development approach. In the process, it has addressed about 32,141 million ha area through various measures and activities like agroforestry, afforestation, and animal husbandry development. The corresponding amount of expenditure was in the tune of ₹12,31 million. Through all these efforts, both the people and officials have started understanding the effects of all these efforts.

### 13.4.8 Economic Evaluation of the Program

The first benefit of the ravine control and reclamation plan would be prevention of damage to the 8000 ha of good tablelands valued at ₹40 million, which is occurring every year in the country. This will be the benefit of the perspective plan considering all the ravine area in the country. For 0.33 million hectares of the ravine land to be tackled in the 7-year plan alone, the rate of their retrogression into the tablelands would bc 1650 ha annually valued at ₹8.25 million. Apart from the damage prevented, the direct benefits of the 7-year plan would be ₹2.337 million for an investment of ₹2836.2 million for all the three states. This works out to a net return of

**Table 13.6** Summary of previous ravine development efforts

| S. No. | Year (scheme ID) | Funding | Subsidy | Locations | Achievements (ha) cost (million₹)# | Treatments |
|---|---|---|---|---|---|---|
| 1 | 1884 | State | Nil | Etawah | 1200 ha | Afforestation in ravine lands |
| 2 | 1945 | Government of India | Nil | Nil | Called American experts for formulation of strategy, program, and action plan | |
| 3 | 1965–1966 to 1977–1978 (state ravine reclamation scheme) | State Government | 25% to all beneficiary | Eight districts | 129,165 ha | CB and LL |
| 4 | 1969–1970 to 1988–1989 (PP on ravine reclamation) | 100% Government of India | 100% on all works | Four districts | 53,782 | CB, LL and CD |
| 5 | 1987–1988 to 1991–1992 (CSS in dacoit-infested districts) | 100% Government of India (50% grant 25% loan) | 100% to small and marginal farmers | Nine districts | 25,885 (113.351) | CB, LL, PB and CD |
| 6 | 1987–1988 to 1992–1993 (IWD in ravinous catchment of Yamuna and Chambal river valley with EEC assistance) | 100% EEC | 100% to all works | Two districts (Agra and Etawah) | 63,898 (for all activities) (819.30) | CB for tableland<br>PB and L smoothening ravine land<br>Terracing – reclamation<br>Plantations<br>Ag. production<br>Maintenance of old works<br>Irrigation<br>Link road<br>Etc. |

(continued)

**Table 13.6** (continued)

| S. No. | Year (scheme ID) | Funding | Subsidy | Locations | Achievements (ha) cost (million₹)# | Treatments |
|---|---|---|---|---|---|---|
| 7 | 1997–1998 to 1999–19200 (IWD in Ravinous ecosystem of Yamuna, NABARD assistance) | NABARD 90% loan 10% Government of UP | 1005 to small and marginal | Three districts (Agra, Firozabad, and Etawah) | 41,501 (222.619) | All physical activities excluding afforestation |
| 8 | 2000 Feb. to Dec. 2004 (ravine stabilization with EEC assistance) | 80% EEC 20% Government of UP | 100% to small and marginal farmers | Six districts | 6981 ha/other activities (76.254) | Land development<br>Animal husbandry development<br>Ag. extension |
| 9 | Total | | | | 321,412 ha (1231.524[a]) | |

*PP* pilot project; *LL* land leveling; *CSS* centrally sponsored schemes; *PB* peripheral bunding; *EEC* European Economic Communities; *CD* check dam; *IWD* integrated watershed development; *CB* contour bunding; *NABARD* National Bank for Agriculture and Rural Development
[a]Excluding Sl. No. 1 and 2. #Values in parenthesis is cost in₹)

11.2% per annum on the investment. For each of the three states of Uttar Pradesh, Madhya Pradesh, and Rajasthan, the annual return on the investment of ₹945.4 million would be ₹77.9 million, the percentage return remaining the same, i.e., 11.2%.

## 13.5 Proposed Approach and Policy Options for Ravine Area Reclamation

The "Integrated Development" approach of the ravine area is proposed to be taken on "Watershed" basis. These areas are rain-fed. Hence it is suggested that one need to follow the "Common Guidelines" developed by the National Rainfed Area Authority (NRAA) in this regard (Anonymous 2008). This has also the approval of Planning Commission and Ministries. The guiding philosophies of this guideline include equity and gender sensitivity, decentralization, facilitating agencies, centrality of community participation, capacity building and technology transfer, and organizational restructuring.

Now these are to be appropriately translated into action frame as the possible structure of the guidelines. The general technical approach to the development and integrated farming system models is discussed in later sections. This approach is in line with the Ravine Reclamation Policy of Ministry of Agriculture, Government of

India, 1970. The essential part of the Policy includes work to be done on a watershed basis; shallow and medium ravines to be reclaimed for cultivation and restored to the villagers; irrigation from tube wells or pumping from the river for the ravine land reclaimed for agriculture which was made mandatory; and deep ravines to be afforested with fruit, fodder, fuelwood, timber, and other species wanted by the villagers.

### 13.5.1 General Approach for Development

#### 13.5.1.1 Protection of Marginal Lands

Proper protection and treatment of marginal lands which includes tableland of slope <3% would be taken for enhancement of agricultural production. Following measure would be taken to treat marginal land depending upon the site requirement.

(a) Contour bunding
(b) Peripheral bunding
(c) Land smoothening leveling with provision of safe water disposal structure
(d) Construction of farm ponds
(e) Crop management demonstration in the treated area
(f) Horticulture development
(g) Agroforestry
(h) Pasture development

#### 13.5.1.2 Protection of Shallow Ravines

Through treatments of shallow ravines including peripheral lands, i.e., constituting the sloping belt between tableland and the actual initiation of ravine lands, additional area could be brought under cultivation. Following measures would be taken for reclamation of shallow ravines:

(a) Leveling/Terracing
(b) Construction of grass waterways for safe disposal of runoff
(c) Gully head control structures
(d) Loose boulder/earthen structures in upper reaches with vegetative support
(e) Gully check dams/percolation tanks in lower reaches of the gully
(f) Earthen check dams/percolation tanks in lower reach of the gully
(g) Water harvesting structures/farm ponds
(h) Agroforestry
(i) Horticulture
(j) Afforestation
(k) Pasture development
(l) Crop management demonstration in developed areas

#### 13.5.1.3 Stabilization of Medium Deep Ravines

Reclamation of these areas is difficult because of:

(a) Deep ravine areas are not accessible for physical treatment. In most of the places, it would require aerial seeding which is very expensive.
(b) Most of the deep ravines are having dense vegetation because these areas are less affected by grazing problems.
(c) Structural measures if proposed would of high dimensions and expensive.
(d) With the treatment of upper reaches, the erosion problem in deep ravines can be stabilized to a large extent as the silt load and runoff from upstream aggravates the erosion problem in deep ravines.

Apart from the above reclamation work in the area, the following site development activities could be taken up: (i) construction of link road and (ii) livestock development under this program along with fodder production demonstration which can be carried out on small plots of farmer's land (0.01 ha). In reclaimed area, crop production demonstration would be carried out under which selected farmers would be provided with high-yielding seeds, fertilizers, and pesticides. A package of ₹1000 (1US$ = ₹67) would be given for demonstration plot of small size (0.1 ha).

### 13.5.2 Policy Options and Related Issues

#### 13.5.2.1 Survey and Planning

As the formation of ravine is a dynamic, the exiting information of ravines may be updated regularly based on ground observations. Wherever the facilities of aerial photography exist, it should be interpreted to find out the extent of ravines, the depth, width, length, vegetation cover, and other physical factors characterizing a particular ravinous system. A basin-wise master plan for tackling the entire ravine problem should be prepared and suitably phased into Five-Year Plan periods indicating implications of finance (including foreign exchange machinery and trained personnel). Within the framework of the master plan, watershed work plans may then be prepared on the basis of surveys, integrating measures under all technical disciplines like engineering, forestry, agriculture, etc.

#### 13.5.2.2 Ravine Stabilization

Ravines grow by waterfall erosion and retrogression of the head, erosion on the sides, and degradation of the bed by high velocity runoff waters from the tributary watershed. In order to stabilize the ravines, therefore, soil conservation measures like bunding, terracing, and waterways are necessary in the tablelands and the peripheral land. Runoff should be diverted off the numerous ravine heads into a particularly chosen waterway ravine which can be protected with head control and bed gradient control structures. Such engineering structures should, however, be supplemented with vegetative measures like grass establishment and protective afforestation at all sites with erosion hazard. Irrespective of the use to which the already ravine land may be put in due course, the works of ravine stabilization to stop their further growth into the valuable tableland above must constitute the minimum priority program.

#### 13.5.2.3 Land Rights and Transfer Issues

The ravine land is under the ownership of individuals, village Panchayat and the Government (mostly revenue department). In order to enable survey, planning, control, and utilization measures on a scientific basis of watershed planning, it will become necessary to obtain cooperation of the local people, owners, and local village communities. Methods of persuasion should be adopted, and provisions of existing soil conservation legislation already enacted in the states of UP, Madhya Pradesh, Rajasthan, and Gujarat may be tried for all ravine control and reclamation operations. If, however, reallotment of reclaimed land to progressive farmers on a self-contained colonization basis is attempted, special legislation to enable land take-over from various parties and redistribution may become necessary. Special legislative provisions may also be necessary for transfer of individual and village common lands to the Forest Department for creation of large-scale afforestation works. The need of all the concerned states in the neighborhood should be kept in view while taking such measures.

#### 13.5.2.4 Social Problems

The problem of ravines is not only an erosion problem but also a development problem of utilizing the nation's land resources. The area bas posed a big social problem also because a number of dacoit gangs have sprung up making use of the naturally available hideouts in the ravines. The dacoit problem is also related to the economic situation in the area where the available land is being continuously depleted by the encroachment of ravines. The wholesale leveling of ravines to wipe out the hideouts of the dacoits cannot be considered as an economically feasible proposition. Measures to stabilize the ravines for selective agriculture and general grassland and forest development would not necessarily fit in with the requirements of law and order enforcement which will have to be tackled independently of the ravine land stabilization and utilization policy.

## 13.6 Integrated Farming System Models for Sustainable Development of Ravines

Farming systems refer to deliberate raising of crops, forest and fruit trees, and mushroom and animals including fisheries, piggery, and duck farming and sericulture on a given unit of land to increase the productivity and profitability, to upgrade natural resource base, and to achieve overall improvement of the environment through conservation of natural resources (Subha Reddy 2008). These also try to provide livelihood support to the people living in the area. Integrated farming system refers to farming which introduces a change in farming techniques for maximum economic return through appropriate combinations of agri-horti-agroforestry-animal husbandry mix of activities best suited for the location and people's satisfaction and also conserving the natural resources.

Integrated farming system has the advantages of increasing economic yield per unit area per unit time, profitability, and sustainability; provides balanced nutritious food for the farmers, pollution-free environment, and opportunity for effective recycling of one product as input to other component, money round the year; solves the energy, fodder fuel, and timber crisis; avoids degradation of forest and enhances the employment generation; increases input use efficiency; and finally improves the livelihood of the farming community. In these processes both top tablelands and sloping lands are addressed for stabilization from erosion and put into appropriate production systems. The stabilization process includes all appropriate measures as suggested earlier. The valley is put under appropriate conservation measures such as gully land control, grassed waterways, check dams, etc. These water bodies then form the sources of lifesaving irrigation and also help in increasing groundwater of the area. Thus, whole activities are incorporated as integrated farming system.

### 13.6.1 Suggestive Farming System Models

#### 13.6.1.1 Crop-based Farming Systems

In this system animals are raised on agricultural wastes, and animal power is used for agricultural operations, and voids are used as manure and fuel. In rice-based production system at Orissa, in situ conservation of rainwater by optimum wear height, conserving excess water in the refuges constructed at the downstream of rice field, and rearing of fish in the refuges in the medium land enhanced the total productivity. This system recorded the highest net returns with benefit to cost (BC) ratio (2.78) as compared to the growing of rice alone. The average fish yield in this system was 1107 kg $ha^{-1}$ in 6 months. The cropping intensity was increased from 100% to 131%. Under cereal-based production system, the cropping pattern with 35.93% of food grains, 25.71% of pulses, 20.7% of oilseeds, 17.3% of commercial crops, and 1.17% of fodder crops in total holdings of small farmers stabilizes the farm income at Dharwad. The animal component (poultry bird) helped the farmers during the drought year to stabilize the farm income as compared to crop alone. At Hyderabad, a marginal farmer having 0.5 ha grown cotton (bunny) and maize + pigeon pea systems recorded the BC ratio of 3.47 and 4.43, respectively.

Agri-sheep farming with ten lambs and growing crops and use of farm by-products in 1 ha of marginal lands gave the net returns of ₹27,500 $ha^{-1}$ as compared to growing cotton alone at Warangal in Andhra Pradesh (₹8700 $ha^{-1}$) under cotton-based production system. Poultry-based integrated farming system alone generated 245 man-days in a year.

In oilseed-based production system, groundnut cake is the most important by-product used as protein source and cattle feed at Anantapur. Groundnut haulm farming system offers gainful employment in rain-fed areas. In this scarce rainfall zone of Andhra Pradesh, the highest net returns were recorded with farming with the poultry.

#### 13.6.1.2 Agroforestry-based Farming System

Perennial grass components, besides imparting stability to crop production in arid areas, also act as vegetative filter strips for prevention of wind and water erosion

(Narayan et al. 1995). Moreover, the grass component improves the soil organic matter and starts giving production from the establishment year onward. Growing of grasses and legumes reveals that moth bean (*Vigna aconitifolia*) and green gram (*Vigna radiata*) in the ratio of 2:1 with grasses are quite promising. These crops, besides giving full yield of grass component, give the bonus yield of pulses (130–280 kg $ha^{-1}$) and fodder (350–600 kg $ha^{-1}$).

The grass component in agri-pasture and silvopasture systems was more profitable than the arable farming. The agri-silviculture system is recommended for land capability class IV with annual rainfall of 750 mm. A large number of tree-crop combinations, particularly of $N_2$, fixing trees with sorghum, groundnut, castor, and pulses were evaluated in alfisols and vertisols. Short-duration dryland crops such as pearl millet (*Pennisetum typhoides*) and black gram (*Vigna mungo*) combined with widely spaced tree rows have been found compatible in semiarid tropical areas.

Management of ravines and gullied land is done by adopting suitable mechanical and biological measures. Very small (depth less than 1 m), small (depth 1–3 m), medium (depth 3–9 m with bed width of 18 m), and deep gullies (9 m depth and large bed width) can be reclaimed into level bench terraces by cutting the soil from humps and side slopes and filling the same into adjacent gullies and depressions. Gullied lands, however, can be reclaimed by using suitable tree, bushes, and grass species. The overall aim of adopting mechanical measures and vegetation is to achieve more production without wasting runoff water and deterioration of land. Nevertheless, agroforestry land use provides ample opportunities for conservation of natural resources like soil and water. Vegetative cover of trees and grasses minimizes the loss of sediment and runoff and maximizes infiltration of water into the soil instead of going waste to the gullies. The combined application of mechanical and vegetative measures helps in moisture conservation and increasing the profile moisture for crop production, forage production, and horticultural crops.

#### 13.6.1.3 Livestock-based Production System

The livestock farming system in rain-fed agriculture is complex and generally based on traditional socio-economic considerations. An understanding of production factors (livestock, capital, feed, land, and labor) and processors (description, diagnoses, technology design, testing, and extension) that effect animal production is prerequisite for livestock integration. The productivity of livestock in farming systems in rain-fed agriculture can be improved by increased fodder production as an intercrop with cereals, relay and alley cropping as well as forage production on bunds. Improving the feeding value of stover by chopping, soaking with water, urea treatment, strategic supplementation of concentrate, urea-molasses-mineral block for enhanced utilization, improvement in productivity of grasses quantum and distribution decides the effective growing season and it becomes critical in selecting cropping systems for a given region. In degraded ravine lands, establishment of fodder banks in areas where surplus fodder can be made available, artificial insemination with semen-accredited bulls, removing low-grade animals through castration, and adoption of preventive measures like vaccination and deworming through health camps are some beneficial effects which will enhance the productivity and income of the farmers.

Experiment conducted with five land use systems during 2001–2006 at Umiam, Meghalaya, revealed that alternate land use systems like dairy-based, silvopastoral, and agri-horti-silvo-pastoral systems are better than shifting cultivation. Due to shifting cultivation, the total water yield (494 mm), base flow (360 mm), and runoff were highest (134 mm) inducing soil erosion. The soil pH, organic carbon, and cation exchange capacity were also better in the four alternate land use systems than shifting cultivation. Agri-horti-silvo-pastoral system involving crop production in the bottom 33% bench terraces, horticultural crops in the middle one third portions with contour bunding, and silviculture in the top portion was found sustainable in terms of hydrology, soil pH, and economics. Almost the entire rainfall (>99%) was absorbed in situ in the agri-horti-silvo-pastoral system; thus the runoff (0.84 mm per year) and peak flow rate was lowest (0.60 mm $hr^{-1}$) restricting soil erosion. Agri-horti-silvo-pastoral land use worked out to be the most profitable farming with an output/input ratio 2.15, followed by dairy-based land use system (1.69). A similar experiment was conducted in different watersheds with 30–40% hill slopes under land capability class VIIe at Barapani, Meghalaya (Satapathy and Sarangi 2010). Each of the microwatersheds was put to land use as per its physiological suitability (soil depth and percent slope) and to cover a wide range of land use options, viz., agri-pastoral system for small/marginal farmers having family laborers, agri-horti-silvo-pastoral system with pig as a livestock component, and silvopastoral system to generate supplemental income from the land unsuitable for agriculture and horticulture. A combination of physical (complete terracing under agropastoral, bottom portion terracing in agri-horti-silvo-pastoral and dairy farming system, half-moon terracing for fruit trees) and biological conservation measures (growing rice-bean-guinea-grass-broom grass and fruit trees) was followed as per need of the different land use systems. The results of the study indicated that integration of various enterprises by resource-poor farmers on small size of landholding is viable due to better utilization of land, water, input, and output resources with livestock rearing as subsidiary sources of income compared to arable farming alone.

## 13.7 Monitoring and Evaluation of Ravine Reclamation Project

Monitoring and evaluation of the project plays a very important integral part of the project. It is through these activities the project's planned progresses are monitored and then its performance is evaluated. Finally, based on this evaluation, the project impact is analyzed. Again, for monitoring and evaluations, one needs to develop the key monitoring indicators. These indicators are then monitored over space and time. Table 13.7 provides the key monitoring indicators along with the parameters and sample observations.

**Table 13.7** Key performance monitoring indicators for evaluation of project benefits

| Components of development | Key performance indicators | Monitoring and evaluation | Critical assumptions |
|---|---|---|---|
| Sustainable reclamation of ravine land and crop production | i. Area of ravine land reclaimed | Report from independent and monitoring agencies | Timely allocation of land |
| | ii. Land use | Reports provided by RSAC | LUGs, SICs formed expeditiously |
| | iii. Soil fertility | Bank review missions | Smooth interdepartmental coordination |
| | iv. Cropping pattern and cropping intensity | | Improved rain-fed farming |
| | v. Crops productivity | | |
| | vi. Diversification | | |
| | vii. Improved quality of biodiversity | | |
| | viii. Market value of reclaimed land | | |
| | ix. People's participation in development process | | |
| Socio-economic upliftment with improvement in income and quality of life | i. Household income | Progress reports by independent monitoring agencies | Convergence of other schemes and programs |
| | ii. Health and education | UPBSN review reports | Basic rural infrastructure provided |
| | iii. Standard of living improved | Bank review missions | Marketing facilities become available |
| | iv. Women empowerment | | |
| | v. Equity in benefits distribution | | Availability of technical services |
| Improved livestock development | i. Breed of cows and buffaloes | Reports from independent monitoring agencies | Timely supply of improved bulls/the buffaloes and male and female goats |
| | ii. Milk production | Reports provided by livestock development department (LDD) | Smooth coordination between UPBSN and LDD |
| | iii. Grazing pressure | Bank review missions | Nutritious feed and fodder availability |

(continued)

**Table 13.7** (continued)

| Components of development | Key performance indicators | Monitoring and evaluation | Critical assumptions |
|---|---|---|---|
| Runoff and silt monitoring | i. Runoff | Report from independent monitoring agencies | Timely establishment of SMS |
| | ii. Soil loss | Review reports of UPBSN | Timely collection of data and analysis |
| | iii. Nutrient loss quality | Bank review mission | |
| | iv. Water conservation | | |
| | v. Irrigation | | |
| | vi. Groundwater levels | | |

### 13.7.1 Impact of Agroforestry on Runoff and Soil Loss

The role of trees and grasses on erosion control is one of the most widely acclaimed reasons for including them on farmlands that are prone to erosion hazards. The tree and grass reduce erosion by enriching and binding the weak surface soil, providing protection against erosion with canopy and leaf litter, and impeding the velocity and erosive ability of surface runoff by stem, surface roots, and litter. Thus, trees and grasses play an important role in soil and water conservation, especially, in ravine region. Hydrological evaluation studies conducted on ravine lands clearly brought out that various agroforestry systems were found effective enough in reducing the runoff and soil loss. Least runoff and soil loss was recorded under permanent cover of grasses. Likewise, silvopastoral system also gave lower values of runoff (4.9% of rainfall) and soil loss (1.06 Mg $ha^{-1}$) than agricultural land use (runoff 14.2% and soil loss 2.79 Mg $ha^{-1}$). Among different agricultural crops, legumes (green gram and cowpea) were more effective in reducing erosion losses (runoff, soil loss, and splash) from agricultural fields as these crops quickly develop their canopy and provide a better cover on the ground than the cereal crops.

The trees improve the long-term physical condition of the soil for plant growth. The soils are improved greatly through continuous incorporation of organic residues and their litter. The soil properties and available nutrient status under different land uses in Yamuna ravines clearly showed better maintenance of organic carbon and availability of nitrogen, phosphorus, and micronutrients under horticulture and forest land uses than agriculture. Finding of low concentration of calcium carbonate in soils under horticulture and forest land use may be due to higher organic carbon content in these soils, which might have helped in dissolving the calcium carbonate and consequent precipitation in soil profile.

### 13.7.2 Sustained Production Through Agroforestry

Crop production in ravine region is risky and yields are unstable. Agroforestry systems provide considerable security against the risk of failure of agricultural crops during drought years. Introduction of grasses and trees in cultivated lands is a measure which improves crop production through improved soil and atmosphere environment. Alley cropping is quite popular for ravine lands because, from this system, not only soil erosion is checked but also the production of agricultural crops is increased and farmer gets fuel and fodder without any adverse effect on crops. Apart from this, soil fertility and profile moisture are also enhanced.

The combination of trees and perennial grasses is an accepted land use for deep ravines (capability class VI to VII) which are uncultivable for agricultural crops. It provides a complete protection and successive buildup of the endangered land resources besides sustainable production of fuel, fodder, and a number of minor forest products. Deep ravines and gullied land can be managed through fuel/fodder plantations. Species like *Acacia* and *Prosopis* and grasses such as *Dicanthium* and *Cenchrus* can be planted on humps and gully slopes. Some species of bamboo also have been found beneficial and economic in gully beds.

## 13.8 Concluding Suggestions

A review of ravine reclamation in India reveals that Uttar Pradesh has taken the lead role since 1884 for reclamation of ravine lands in the country. Most of the activities under different soil conservation projects gradually approached toward the requirement of reclamation on integrated farming system approach on watershed basis. Giving all due regards to the past efforts as per official records, so far only about 25% of the ravine lands have been attempted for reclamation. Hardly any assessment of stabilized reclaimed areas is available. The reasons of the failure of stable use of reclaimed land in productive purposes have not been clearly identified. These reasons for failures are very important for future work plan to address the leftover 75% of the ravine lands.

The poor socio-economic conditions of ravine areas call for immediate actions. The scope of providing sustainable livelihood to the people of the area through reclamation of the ravine areas is great under the given agroecological conditions and with people's participation. It has potential of giving diversified crop outputs along with medicinal and aromatic crops. This would call for an integrated bio-industrial approach for locally processing the outputs, and these are expected to provide farmers more return and also will widen the scope employing more people. Hence, a pragmatic view of the ravine development is very important. In this regard the following suggestions will be important for the consideration of the policy implementation:

- A synthesis study must be conducted to evaluate the specific and overall effects of all previous ravine reclamation projects. The reasons of failure and success must be identified, and from there a way forward must be worked out for future action.

- For detailed studies of ravines, multidisciplinary projects must be undertaken involving ecologists, geologists, soil scientists, engineers, agroforesters, and socio-economists.
- By application of modern technological tool, we must work out the exact area falling under different types of ravines.
- The future ravine reclamation programs must be based on integrated participatory and watershed approach.

## References


Anonymous (1972) Working group report on "Ravine Reclamation"

Anonymous (2003) Guidelines on rehabilitation of gullied and ravine lands. Department of Land Resources, Ministry of Rural Development, Govt. of India, New Delhi

Anonymous (2008) Common guidelines for watershed development projects. Govt. of India, National Rainfed Area Authority, Ministry of Agriculture, Govt. of India, New Delhi

Bhushan LS, Sharma AK (1997) Management of ravine watershed. Central Soil & Water Conservation Research & Training Institute, Research Centre, Agra

Chaturvedi OP, Kaushal R, Tomar JMS, Prandiyal AK, Panwar P (2014) Agroforestry for wasteland rehabilitation: mined, ravine, and degraded wasteland areas. In: Dagar JC, Singh AK, Arunachalam A (eds) Agroforestry systems in India: livelihood security and ecosystem services, advances in agroforestry, vol 10. Springer, New Delhi, pp 233–272

Dhar TN (2006) Sustainability or bust: reclaiming degraded lands in U.P. Society for Himalayan Environment Rehabilitation & People's Action, Lucknow, pp 1–164

IAB (1990) Indian agriculture in brief. Ministry of Agric, New Delhi

Narayan D, Nitant HC, Bhushan LS (1995) Agroforestry as a sustainable land management system for ravines and gullied areas. J Soil Water Conserv 39:26–31

Rao RVS, Mahapatra SK, Verma TP, Lal T (2005) Use of remote sensing data to delineate ravinous areas of Etawah District, Uttar Pradesh. J Soil Water Conserv 3(3&4):125–130

Reddy S (2008) Integrated farming system models for sustaining rural livelihood in rainfed areas. Training Lecture, Winter School, CRIDA, Hyderabad

Satapathy KK, Sarangi SK (2010) Long term effect of alternate land use systems on hydrology, soil properties and economics in fragile hill ecosystem. J Soil Water Conserv 9(1):46

Tejwani KG, Gupta SK, Mathur HN (1975) Classification, control and reclamation of ravine lands. In: soil and water conservation research (1956–1971). ICAR Pub, New Delhi, pp 254–310

# Agroforestry Potential for Higher Productivity from Degraded Ravine Watersheds

14

O. P. Chaturvedi, J. C. Dagar, A. K. Handa, Rajesh Kaushal, and V. C. Pandey

**Abstract**

Ravines are considered among one of the most degraded wastelands. The ravenous watersheds can be managed through agroforestry systems successfully, which in turn provide food, timber, fuelwood, biofuels, fodder, feed and medicinal and industrial non-timber products and also help in conserving and rehabilitating ecosystems. It is now widely accepted that the future of livelihood and environmental security will depend upon the attention paid to conservation, sustainable development and management of natural resources and reclamation of degraded lands including ravines. If judiciously managed, agroforestry offers affordable alternative for managing the degraded resources for higher productivity and environmental safety. The suitable agroforestry technologies of planting forest and fruit tree species along with grasses and remunerative crops can reclaim ravine watersheds and help in generating income for resource-poor farmers. Agroforestry in ravine watersheds may be considered a low-hanging fruit and not only helps in land reclamation but also ensures environmental and livelihood security of different stakeholders.

O. P. Chaturvedi (✉) · A. K. Handa
ICAR-Central Agroforestry Research Institute, Jhansi, Uttar Pradesh, India

J. C. Dagar
Indian Council of Agricultural Sciences, Krishi Anusandhan Bhavan-II, Pusa, New Delhi, India

R. Kaushal
ICAR-Indian Institute of Soil and Water Conservation, Dehradun, India

V. C. Pandey
ICAR-Indian Institute of Soil and Water Conservation, Research Centre Vasad, Dehradun, India





J. C. Dagar, A. K. Singh (eds.), *Ravine Lands: Greening for Livelihood and Environmental Security*, https://doi.org/10.1007/978-981-10-8043-2_14

## 14.1 Introduction

In India, water-induced erosion is the most serious type of soil degradation, and it accounts for nearly 65% of land degradation; and it is estimated that soil loss of 5334 million Mg occurs annually, which works out to be 16.35 Mg $ha^{-1}$ per annum (Dhuruva Narayana and Ram-Babu 1983). This not only degrades soil resources at site but along with runoff soil particles chocks the water courses and reaches reservoirs as sediments. Samra and Narain (1998) reported that about 29% of the total eroded sediment is permanently lost to the sea, 61% is redistributed and the remaining 10% is deposited in the reservoirs, resulting in loss of their storage capacity. Many engineering-based measures have been evolved for soil conservation in different watersheds of the country, but most of these in isolation are not suitable for reclamation and management of ravine lands. Agroforestry offers affordable alternative in place of expensive conventional conservation measures involving forest and fruit trees, forage grasses and other useful species for providing variety of products to meet various requirements of the people. These products are considered insurance against risks caused due to weather aberrations, controlling erosion hazards and ensuring sustainable production of the land on a long-term basis. Agroforestry is considered to be a low-hanging fruit especially in ravines, where we need a perennial cover. The main components of agroforestry systems are trees and shrubs, crops, pastures and livestock, together with the environmental factors of climate, soils and landforms. Applying agroforestry techniques for ravine lands and the rest of the degraded catchment, both food and high income can be generated, while at the same time ravines are rehabilitated for better social and environmental gains. Woody perennials in agroforestry enhance soil fertility, reduce erosion, improve water quality, enhance biodiversity, increase aesthetics and sequester carbon. The role of agroforestry in management of ravines and degraded watersheds is described in this Chapter.

## 14.2 Water Conservation through Integrated Watershed Management with Agroforestry Approach

In India, hydrological investigations have been carried out primarily on experimental watersheds and micro-watersheds with different land uses including agroforestry. Dhuruva Narayana (2002) compiled information on different aspects of different land uses such as Shola forests describing vegetation of natural versus man-made forests. Other land uses mentioned included tree plantation in ravenous watersheds in Shivaliks cultivating grasses with *Eucalyptus tereticornis*, Doon Valley under tree-crop experimental watersheds and another on experimental watersheds involving trees in Meghalaya. Trees play a significant role in reducing the amount of the rainfall reaching the soil due to canopy interception and in moisture conservation. Maintenance of high ground cover formed by combinations of trees, herbaceous vegetation, litter and twigs during the period of rains significantly reduces the soil erosion. Tomar et al. (1992) observed that when 4 Mg $ha^{-1}$ of mulch (consisting of fine leaves and herbaceous plants) was applied to wheat crop, soil moisture levels were higher under mulch treatments at all stages of growth as compared to un-mulched conditions; and the effectiveness in conserving water was increasingly proportional to the speed of decomposition of the mulches. This helped in moisture conservation and checking soil erosion on mild slope.

Studies on interception of rainfall under a natural oak forest at Agarakhal in the outer Himalayas revealed that maximum interception was recorded under low rainfall groups (< 10 mm) and it decreased progressively in the medium (10–50 mm), high (50–100 mm) and very high (>100 mm) rainfall groups. However, the terminal velocity of the raindrops and consequently splash on the forest floor (having leaf litter) were considerably reduced (Dhyani et al. 2006b). Similarly, from a comparative study of stemflow, through fall and interception in a natural forest and plantation in the Nilgiris, maximum interception of rainfall was recorded in Shola forest (33.8%), followed by the plantations of *Acacia mearnsii* (25.2%) and *Eucalyptus globulus* (21.9%). In Shivaliks, the interception of rainfall by the crowns in *Dalbergia sissoo* was more than *E.* hybrid; and there was a gradual increase in interception *in D. sissoo* with age but not so in *Eucalyptus* because of the difference in development of crown.

It was stated by Young (1997) that the effects of tree-crop mixture on the water balance are cumulative from year to year or between season to season in bimodal climates. If trees cause soil moisture depletion, this goes on during rain deficit months and becomes quite critical till equilibrium is established after commencing of rains. On the other hand, if the water is conserved, the water saved can be carried forward through dry periods in the lower soil horizons, particularly in deep alluvial soils, as is the case in Shivaliks. This has clearly been documented by Grewal et al. (1994) in northern India having highly variable monsoon rainfall, where *Eucalyptus tereticornis* plantation used 29% more rainfall than sesamum (*Sesamum indicum*) and rapeseed (*Brassica napus*) crops but also retained more by reducing runoff, so that the soil water content in the soil at the end of the trial was the same.

The integrated watershed management (IWM) is the key factor in soil and water conservation. Agroforestry combined with conventional conservation measures in a catchment area including where ravines are formed plays an important role in social, economic and environmental sustainability along with livelihood factor in a watershed area. The projects taken up by the Central Soil & Water Conservation Research & Training Institute in Sukhomajri, Bunga and Relmajra have shown significance of these social values due to the successful implementation of conservation programs during watershed management (Samra 2006). Besides improvement in agroforestry components (crops, grasses and trees) in the catchment, groundwater recharge and rise in water table due to IWM have been spectacular in different regions of the country as has been summarized by Samra and Narain (1998) in Table 14.1.

Ram-Babu et al. (1997) evaluated 32 watersheds representing ten major agroecological regions in the country and found that the productivity of arable land increased by 420–1540 kg $ha^{-1}$ with an average of 780 kg $ha^{-1}$ and that of nonarable land by 1.0–8.5 Mg $ha^{-1}$ (average 5.5 Mg $ha^{-1}$). There was sustained availability of fodder grasses throughout the year for rearing of livestock; therefore, the milk production

**Table 14.1** Effect of integrated watershed management strategies on groundwater recharge in different regions of India

| Watershed | Surface storage capacity created (ha-m) | Observed rise in groundwater table (m)[a] |
|---|---|---|
| Bazar-Ganiyar (Haryana) | 79.0 | 2.0 |
| Behdala (HP) | 18.0 | 1.0 |
| Bunga (Haryana) | 60.0 | 1.8 |
| Chhajawa (Rajasthan) | 20.0 | 2.0 |
| Chinnatekur (AP) | 5.6 | 0.8 |
| GR Halli (Karnataka) | 6.8 | 1.5 |
| Joladarasi (Karnataka) | 4.0 | 0.2 |
| Siha (Haryana) | 42.2 | 2.0 |

Source: Samra and Narain (1998)

[a]Difference between pre-project and post-project water table

increased by three times on sustained basis after taking conservation measures in the watersheds. Soil loss also decreased by 10–80% (depending upon management) and consequently the runoff by 2–42%, leading to better in situ moisture conservation. These programs were found to be successful and were economically sound with benefit-cost ratio varying from 1.10 to 2.94 and internal rate of return 12.3–37% depending on the status of management.

During the last one decade, several studies have been initiated under the integrated watershed development (IWD) program in Bundelkhand region implementing in situ and ex situ interventions. The landscape in the region is undulating suffering from acute soil erosion problems and at times even gullies are formed. The most common in situ interventions were land levelling in agricultural field after bunding and contour bunding. The crops and fruit trees were cultivated across the slope, which helped in harvesting surface runoff, allowing more water to percolate, and disposing excess runoff safely from the fields. Excess runoff was harvested in water-harvesting structures of different capacities raised in watershed areas, depending upon availability of funds and also on the needs of farmers and availability of runoff water. In Garhkundar-Dabar watershed, field bunding was carried out in 52 ha land area (20% of agricultural land), and contour cultivation was promoted in the rest of the agricultural land (Singh et al. 2016). This practice created an opportunity to accumulate surface runoff along the contour line and protected soil from erosion. Building check dams and low-cost gully control measures on the stream network as ex situ measures reduced peak discharge and runoff velocity and harvested a substantial amount of runoff in watershed (Fig. 14.1) and increased groundwater recharge.

Not only rainwater was harvested in these structures, which was later used for irrigation, but at the same time, these structures trapped sediment which protected the river ecosystem. Singh et al. (2016) have reported that in total nine check dams having storage capacity between 1000 and 6500 $m^3$, 150 low-cost gully control

**Fig. 14.1** Rainwater harvesting structures in Bundelkhand region

structures (called gabions locally) of 30–100 $m^3$ capacity and 15 number of drainage structures for safe disposal of excess water from agricultural fields were constructed in their study to develop 35,000 $m^3$ of storage space (~40 $m^3$ $ha^{-1}$) in watershed. The water in the check dams could be used directly for irrigation purpose and simultaneously facilitated artificial groundwater recharge. The IWD interventions changed the hydrological components in the study area as evapotranspiration increased from 58% to 68% and runoff reduced from 35% to 25%. In these studies, the groundwater recharge and baseflow enhanced from 7% to 11% and 1.2% to 7.2%, respectively, of rainfall received in monsoon as compared to no intervention stage. During initial years of studies also, soil loss from the treated watershed was 45–70% lower than untreated one, while with passage of time the efficiency will increase to a greater extent.

Besides soil and water conservation measures, agroforestry-based interventions were studied in these watersheds. Fruit trees such as gooseberry (*Emblica officinalis*), ber (*Ziziphus mauritiana*), guava (*Psidium guajava*) and lemon (*Citrus limon*) were cultivated along with several food and fodder crops. Forest trees such as teak (*Tectona grandis*) and kumat (*Acacia senegal*) trees were grown as boundary plantation. To increase water use efficiency, pressurized irrigation (e.g. drip irrigation) was used. These studies further indicated that the integrated water management system is more efficient than any system in isolation. It was evident from the results reported by Singh et al. (2016), which showed that the cropping intensity in treated watersheds increased from 69% to 207% while the productivity of different crops enhanced by 20–60%. There was significant increase in farmers' income, who saved both water and energy. The average economic water productivity increased from ₹2–5 to ₹5.0 $m^{-3}$, and average income increased more than double from ₹11,500 to ₹27,500 $ha^{-1}$ $year^{-1}$.

## 14.3 Agroforestry for Soil and Moisture Conservation

Soil erosion is considered to be one of the important factors for soil degradation. It is a well-established fact now that agroforestry interventions have the potential to increase infiltration rate of rainwater, check the runoff, and consequently reduce soil erosion. Vegetation starts playing its role just at the stage of falling of raindrop on soil surface. Trees, annual crops and grasses play a significant role through canopy, surface litter and root systems. They act as a physical barrier against running surface water (runoff) and reduce the kinetic energy and break the impact of raindrops which decrease the erosive capacity (Young 1997; Andreu et al. 1998). More details about the raindrop and other erosion processes are given in Chap. 2 of this book.

The ground vegetation acts as a sponge and absorbs large quantities of water during the rainy season. Agroforestry systems influence the availability and utilization of available moisture in the field. The vegetation takes care of both excess water due to high-intensity rainfall and lack of water due to extended drought periods through addition of soil organic matter which helps in absorbing a higher amount of water without causing surface runoff and improving water absorption capacity during extended drought period (Young 1997). It is a well-known fact that in dry season, deep root systems of the trees explore a larger soil volume in the soil profile in search of water and nutrients from different soil depths. Shading by tree canopy also helps in reducing evapotranspiration, thus indirectly helping in conserving of soil moisture. This fact was observed by Neumann et al. (1989) who observed that the potential evaporation in the middle of the field having agroforestry system was reduced on an average by 36%. In Costa Rica, Bronstein (1984) reported higher soil moisture content under *Erythrina poeppigiana* tree than in open fields during dry season. The soil moisture contents in the top 0–15 cm layer in a field having agroforestry systems were measured and found that both during dry and wet seasons, the moisture content in agroforestry fields was generally higher than that of control, i.e. the crop field without tree component.

Stem and roots also trap sediments and stop their downslope movement and facilitate water infiltration. Litter and other organic residue (mulch) change soil physical, chemical and biological properties. The litter fall improves the structure of the soil and hence produces water-stable aggregates which help in reducing surface runoff volume. Soil biota influences soil properties through formation of stable aggregates, development of organo-mineral complexes by improving macroporosity and continuity of pores from surface to the subsoil which ultimately increase the water transmission and reduce runoff and check gully formation and extension.

Deep tree roots further stabilize slopes and help prevent shallow landslides. Young (1997) and Dagar (Chap. 3 of this book) have explored the role of vegetation in checking soil erosion and ravine formation.

Many agroforestry systems have been established on highly eroded land which have shown their efficiency in reducing soil erosion and ameliorating the land in terms of improving organic carbon and other physical properties of soil such as infiltration rate and conservation of moisture. For example, while comparing some agroforestry models for eroded lands in the Shivaliks, Grewal (1993) found that under *Eucalyptus* hybrid-*Eulaliopsis binata* (*bhabbar* grass) system, the soil loss was negligible (0.07 Mg $ha^{-1}$) followed by *Acacia catechu-Pennisetum purpureum* (napier grass) (0.24 Mg $ha^{-1}$) and *Tectona grandis-Leucaena leucocephala-Eulaliopsis binata* (0.43 Mg $ha^{-1}$) in comparison to 5.65 Mg $ha^{-1}$ under cultivated fallow and 2.69 Mg $ha^{-1}$ under arable crops such as *Sesamum indicum-Brassica campestris* system. Besides soil loss, runoff and nutrient losses were also lower under agroforestry systems as compared to arable crops and fallow land.

In another study in Shivaliks, Yadav et al. (2005) compared four gooseberry (*Emblica officinalis*) fruit tree-based agroforestry systems, viz. sole gooseberry (no grass), gooseberry + *Chrysopogon fulvus* (*Dholu* grass), gooseberry + *Pennisetum purpureum* (hybrid napier var. NB-21 grass) and gooseberry + *Cajanus cajan* (perennial pigeon pea var. ICP-8094). These systems were compared for their run-off and soil conservation potentials. The runoff on an average, over a period of 10 years, was 8.0, 13.1 and 18.6% for gooseberry + *Chrysopogon fulvus*, gooseberry + hybrid napier and gooseberry + perennial pigeon pea, respectively, over control (pure gooseberry). The runoff increased during post-bearing years. The effectiveness of different intercrops in reducing soil loss was in the order of *Chrysopogon* > napier> pigeon pea with, respectively, 81%, 56% and 25% reduction over sole gooseberry. It was because the root system of *C. fulvus* was fibrous and efficient in holding the soil. Contrary to runoff, sediment losses were less during post-bearing stages. In another study undertaken in Akola, Maharashtra, under All India Coordinated Research Project in highly eroded land, *Leucaena leucocephala* barrier maintained as hedgerow significantly reduced soil loss from 1.75 to 0.80 Mg $ha^{-1}$ and increased cottonseed yield from 990 to 1160 kg $ha^{-1}$ when grown as intercrop with *Leucaena* (AICRP 1994).

Further, in watersheds of north-western outer Himalayas in Doon Valley having 1740 mm annual rainfall, contour-paired rows of *Leucaena* hedge, *Leucaena* and *Eucalyptus* trees together and 0.75-m-wide planted grass (*Panicum maximum*) barrier at 1.0 m interval in maize (*Zea mays* arable crops) were compared for their soil conservation efficiency. It was found that all these systems brought down runoff from 16% to 43% of rainfall and soil loss from 7 to 21 Mg $ha^{-1}$ $year^{-1}$ (Narain et al. 1992; Narain and Grewal 1994) as depicted in Table 14.2. The sediment deposition along the hedge and tree rows increased considerably with consequent reduction in soil loss. Application of *Leucaena* leaf mulch before or at harvesting of maize increased moisture conservation and yield of following wheat crop because of nitrogen fixation.

**Table 14.2** Effect of paired rows of barrier hedges, grass strips and trees on runoff and soil loss in maize at 4% slope in Doon Valley

| Treatment | Runoff (%) | Soil loss (Mg $ha^{-1}$ $year^{-1}$) |
|---|---|---|
| Maize on contour | 40.0 | 21.0 |
| *Leucaena* hedge | 21.3 | 12.1 |
| *Panicum* (0.75 m wide) | 36.7 | 7.0 |
| *Eulaliopsis* (0.75 m wide) | 42.7 | 10.0 |
| *Vetiveria* (0.75 m wide) | 39.6 | 8.1 |
| *Leucaena* trees (6–8 years) | 20.4 | 8.4 |
| *Eucalyptus* trees (6–8 years) | 16.3 | 5.8 |
| Average of agroforestry land uses | 30.0 | 8.7 |

Source: Narain et al. (1992) and Narain and Grewal (1994)

Further, in another study in watershed of Doon Valley, Dhyani et al. (2006a) reported that raised forests of fast-growing species of *Eucalyptus camaldulensis* and *E. grandis* resulted in reduction in runoff by 28% and the peak rate of discharge by 73% as compared to the natural brushwood vegetation, which was in depleted condition due to biotic interference and high erosion. The favourable influence was attributed mainly to the dense and the luxuriant undergrowth under the light crown of *Eucalyptus* trees. They also mentioned that under different kinds of vegetation such as degraded Shola forest, raised forest of blue gum (*Eucalyptus globulus*) and black wattle (*Acacia mearnsii*), total soil nitrogen, organic carbon, and exchangeable K, Ca and Mg contents increased which in turn helped in more infiltration of rainwater, hence low runoff and erosion. The soil erosion under these species was almost negligible. Yadav et al. (2011) observed that in Vasad ravine areas, *Acacia nilotica*, *Dendrocalamus strictus* and *Eucalyptus* hybrid plantations produced dry litter to the extent of 8.1, 6.0 and 1.8 Mg $ha^{-1}$ $year^{-1}$, and the runoff was almost negligible.

Dhyani et al. (2006a) compared soil loss in the Nilgiri Hills on 16% sloping land, under vegetation cover of different species. The runoff and soil loss data observed for 5 years showed that although all the treatments gave very low runoff, the arborescent vegetation provided higher runoff as compared to well-protected grassland vegetation. The highest runoff (1.27%) of the total precipitation was observed from the blue gum tree (*Eucalyptus globulus*) cover and 1.08% from *Rhus*- and *Chromolaena* (*Eupatorium*) *odoratum*-dominated brushwood cover and almost nil from grassland cover.

It has been proven worldwide in humid, subhumid and sub-tropical climate that alley cropping involving *Leucaena leucocephala*, *Gliricidia sepium*, *Cassia siamea*, *Morus alba* and *Pithecellobium dulce* hedges, when plated across slope in highly sloping land up to 30% slope, along with fodder grasses and legumes as intercrops are effective for mulching, which in turn conserves moisture and helps in erosion control. The alley cropping with different leguminous trees has been found efficient for erosion control on sloping lands (including ravines) and improves moisture and nutrient availability for increased crop production maintaining soil health. Dagar

**Fig. 14.2** Tea cultivation along with *Grevillea robusta* on sloping land

(1995), Kumar and Kunhamu (2011) and Dagar et al. (2014) reported the role of alley cropping and other vegetation in controlling runoff and soil erosion and crop production on sloping land in a tropical climate of Bay Islands and other coastal ecologies such as Kerala. *Leucaena leucocephala* and *Gliricidia sepium* have been widely used also on bunds for this purpose. In hilly humid regions of Kerala, tea (*Thea chinensis*) cultivation along with *Grevillea robusta* (Fig. 14.2) is very popular and successful in erosion control and economic production.

Trees in agroforestry, besides checking erosion also ameliorate soil in terms of increasing organic matter, fixing nitrogen, and nutrient uptake and recycling through litter fall. They improve soil physical properties primarily due to increased soil organic matter and activities in root zone through biological activities by soil micro- and macrofauna (Young 1997; Nair and Latt 1997). Growing of nitrogen-fixing alley crops on the field bunds or across slopes helps in fixation of atmospheric nitrogen, and it reduces the leaching losses of mineral nitrogen. Saha et al. (2012) reported that tree root system also mobilizes phosphorus, potassium and other trace elements. Such improvement in soil properties has a direct bearing on long-term sustainability, productivity and soil quality through increased porosity, higher rainwater infiltration and greater aggregate stability (Campbell et al. 1994). This quality also helps in checking soil erosion in watershed areas.

Long-term studies (Vishwanathan et al. 1999; Samra et al. 1999b) were conducted on silvopastoral systems in the highly degraded ravenous areas and boulder riverbed lands. The results of 14-year studies have indicated the potential of this system for biomass production and improving soil fertility involving *Dalbergia*

**Table 14.3** Effect of bioengineering and vegetation measures on landslide measures (based on 30 years of observations)

| Description | Before treatment | After treatment |
|---|---|---|
| Runoff (mm) | 55 | 38 |
| Dry weather flow (days) | 100 | 250 |
| Sediment load (Mg $ha^{-1}$ $year^{-1}$) | 320 | 5.5 |
| Vegetation cover (%) | <5 | >95 |
| Nala bed slope (%) | | |
| Lower reach | 12 | 7 |
| Middle reach | 23 | 14 |
| Upper reach | 54 | 44 |
| Toe cutting | Severe | Nil |

Source: Dhyani et al. (2006b)

*sissoo* + *Chrysopogon fulvus*, *Acacia catechu* + *Eulaliopsis binata*, *Eucalyptus* hybrid + *C. fulvus* and *Leucaena* + *Panicum maximum*. The soil erosion under the system was almost negligible. Similar observations were also made by Saha et al. (2007, 2010) who planted various multipurpose tree species such as *Pinus kesiya*, *Alnus nepalensis*, *Parkia roxburghii*, *Michelia oblonga* and *Gmelina arborea* on hilly terrain of north-east India and observed soil physical behaviour by these species for quite a long time. Among the tree species tested, *P. kesiya*, *M. oblonga* and *Alnus nepalensis* were found more suitable as bio-ameliorants and increased average soil organic C by 96.2%, porosity by 10.9%, aggregate stability by 24.0% and available soil moisture by 33.2% and simultaneously reduced bulk density and erosion ratio by 15.9% and 39.5%, respectively.

In hilly areas landslide and torrents result in mass erosion. Agroforestry along with mechanical measures has been found to be effective for their rehabilitation. Nalotanala catchment on Dehradun-Mussoorie road was successfully stabilized by bioengineering measures including agroforestry interventions (Sastry et al. 1981). Planting of shrubs such as *Ipomoea carnea* and *Vitex negundo* and grasses such as napier (*Pennisetum purpureum*) and *Arundo donax* with trees such as *Erythrina suberosa*, *Dalbergia sissoo*, *Pueraria hirsuta* and *Acacia catechu* was quite effective, and the entire area was completely rehabilitated with vegetation cover within 10 years. Based on 30 years of observation, Dhyani et al. (2006b) reported the results of these studies and stated that the sediment load was 320 Mg $ha^{-1}$ $year^{-1}$ before treatment, which reduced to 5.5 Mg $ha^{-1}$ $year^{-1}$ with the improvement of vegetation cover from <5% to >95%; and the dry weather flow, which used to last hardly for 100 days after the cessation of monsoon, increased to 250 days (Table 14.3) after the treatment.

The grass species play an important role in establishing the sloped banks of torrents. Among suitable species *Aristida cynantha*, *Chrysopogon fulvus*, *Arundo donax*, *Pennisetum purpureum* (napier) and *Cynodon plectostachys* (giant star grass) are important. The repelling type of spurs was found most suitable for the quick reclamation of concave banks. *Salmalia malabarica*, *Cordia myxa* and *Broussonetia papyrifera* gave better performance for wooden posts, while grass *P. purpureum* was adjudged the best for stabilizing fresh deposits of silt near the spur and other structures. For the utilization and stabilization of the reclaimed areas behind the banks, fuel-cum-fodder plantations of *D. sissoo* and grass *Chrysopogon fulvus* were found quite suitable. Vishwanathan and Joshi (1994) reported that boulder torrent beds falling under class V land use could be utilized for fuel and fodder purposes by planting tree and grass combination of *D. sissoo* + *C. fulvus* and *Acacia catechu* + *C. fulvus*, which could produce 64–71 Mg $ha^{-1}$ fuelwood and over 50 Mg $ha^{-1}$ air dry grass in 17 years. The oven-dry fodder yield from *C. fulvus* was 1.07 Mg $ha^{-1}$, while for industrial grass *Eulaliopsis binata*, it was 2.17 Mg $ha^{-1}$ with about 400 kg $ha^{-1}$ lopped fodder leaves of tree *Bauhinia purpurea* making the system economically viable and comparable in productivity to maize-wheat cropping on fertile land.

Samra et al. (1995) reported the results of small watershed of size 59 ha at Relmajra in Shivaliks. The catchment area was treated with several conservation measures. It was observed that catchment conservation and channel bioremediation in raised bed level of streams reduced the bed slope and with passage of time changed the channel morphology and retained 2.54 ha-m (38,100 Mg) of silt in 850-m-long section within 2 years. This amounted to saving of de-siltation cost of reservoir of ₹1.9 million at ₹50 per Mg. In this area, the runoff was reduced from 21.3% to 6.8% in a spell of 6 years with an average rainfall of 1000 mm per annum. Average soil erosion rate of 623 Mg $ha^{-1}$ came down to 164 Mg $ha^{-1}$ during the first 3-year period, that is, 7.4% reduction, which will be more with passage of time. In total, there was tremendous improvement in biomass productivity, biodiversity, wildlife, land levelling index and soil organic carbon.

Sharda et al. (1998) reported that when this watershed area of 59 ha consisting of natural forest was brought under blue gum (*Eucalyptus globules*) plantation, it reduced total runoff and base flow by 16% and 15%, respectively, during first rotation of 10 years. The effect of coppiced growth for 10 years was still higher and reduced total and base flow by 25% and 27%, respectively. Still better conservation of resource has been realized by Samra (2006) in small scale watersheds of 500–1000 ha area. It was reported that in a well-managed silvopastoral system of ravenous region, the runoff was reduced from 40% (before implementation of watershed project) to 13% after implementation of the project and the soil loss reduced from 30 Mg $ha^{-1}$ per annum to 2 Mg $ha^{-1}$ per annum.

## 14.4 Impact of Agroforestry for Carbon Sequestration in Eroded Ecologies

In recent times rise in temperature and greenhouse gases is causing change in climate which is considered one of the greatest environmental threats as it is affecting the functioning of different ecosystems. Due to rainfall aberrations, most of the watersheds along rivers are facing enhanced soil erosion as a consequence of which the area under gullies and ravines is increasing at fast rate. Lal (1995) has estimated that 1.14 Pg C $year^{-1}$ (Pg = Petagram; 1 Pg = $10^9$ Mg) is emitted into the atmosphere by water erosion. Despite its global significance, erosion-induced carbon emission into the atmosphere remains misunderstood and an unquantified component of the global carbon budget. Further, Lal (2003) while dealing with the role of soil erosion in influencing carbon budget explained that soil erosion is a four-stage process involving detachment, breakdown, transport/redistribution and deposition of sediments. The soil organic carbon (SOC) pool is influenced during all four stages.

It is presumed that a combination of mineralization and C export by erosion processes causes a severe depletion of the SOC pool on eroded soil as compared with uneroded or slightly eroded soils. The depletion is many times higher in gully and ravine erosion sites. Depending on the delivery ratio of the soil particles or the fraction of the sediment delivered to the river system, gross erosion by water is estimated to be 75 billion Mg, of which 15–20 billion Mg are considered to be transported by the rivers into the aquatic ecosystems and eventually into the ocean (Lal 2003). Assuming a delivery ratio of 10% and SOC content of 2–3%, it was emphasized that the amount of total C displaced by erosion process on the earth may be 4.0–6.0 Pg $year^{-1}$. Thus, soil erosion has a strong impact on the global C cycle. Therefore, the adoption of soil conservation-effective measures may reduce the risks of C emission and sequester C in soil and biota.

We know the fact that agroforestry systems are the most efficient land use for sequestering carbon, particularly in eroded habitats. The practice of eco-restoration of degraded ravine lands through agroforestry systems has great potential for maintaining and increasing the land-based carbon sinks. Agroforestry can store carbon through enhancing build-up of soil organic matter and by biomass production especially in woody components that can pool more $CO_2$ from the air and can play an important role in reducing vulnerability and increasing resilience of farming systems. Sequestering carbon through tree-based systems is now being considered as an attractive economic opportunity for carbon trading. ICRAF has projected that the carbon market may exceed US$1 trillion by 2025; therefore, significant funds could potentially be available to finance sustainable rural development and adaptation to climate change (Ajit et al. 2013). In several studies the potential of agroforestry systems for carbon sequestration has been explored in different agroclimatic zones of India. Potential of soil carbon sequestration rate through restoration of water erosion-related degradation has been estimated to be 80–120 kg $ha^{-1}$ $year^{-1}$, while total SOC sequestration potential has been estimated to be 2.6–3.9 Tg $year^{-1}$ (Lal 2006). It was further stated that the total potential of soil carbon sequestration in India is 92 Tg C $year^{-1}$ (1 Tg = $10^6$ Mg) or about 30% of the fossil fuel emission. In addition to that, the improvement of soil quality can help in production of 6–12 million extra food grains per year. Thus, using judicious agroforestry technologies, the ravine lands can be used as potential sink for carbon sequestration, and trading C credits can provide additional income to farmers.

## 14.5 Agroforestry for Biodiversity Conservation

As discussed in Chap. 3 in this book that due to extraordinary anthropogenic interferences, the ravine lands are degraded ecologies but are rich in both animal and plant biodiversity. In many studies it has been proved that most of the agroforestry systems are diversity-enhancing land use systems (Atta Krah et al. 2004; Jose 2009). Agroforestry (AF) systems help in enhancing and conserving biodiversity in ravine-affected areas by the following means:

1. The area is to be protected from grazing and other biotic interferences before developing AF systems.
2. Providing habitats and resources for a wide range of plant and animal species including micro- and macrofauna.
3. Creating congenial environment for moisture conservation and establishment of diverse species.
4. Preserving and multiplying the germplasm of sensitive and economically important species.
5. Reducing the rates of conversion of natural habitat by providing a more productive, sustainable alternative to traditional agricultural systems that may involve clearing natural habitats.
6. Domestication of diversified species of high economic value.
7. Regulating the environment.
8. Providing other ecosystem services such as erosion control and water recharge, thereby preventing the degradation and loss of surrounding habitat.

Samra et al. (1999a) observed the impact of regulating grazing and anthropogenic interference on a degraded ravine land over 24 years by analysing floristic diversity and community composition as site quality indicator. There was tremendous improvement in the abundance and density of grasses and trees at the three topographic situations as a function of years of protection.

Dagar (Chap. 3 of this book) has given an extensive account on impact of grazing on vegetation composition and biodiversity change in ravine lands. Vegetation cover is considered as one of the important input parameters in most of the models predicting runoff, erosion and environmental impacts (Samra 2006). Frequent trampling and raking of grazing beyond their carrying capacity promoted erosion of resources. An exponential negative relationship was also observed by Sharma (2004) between infiltration rate and stock density of animals and depicted by the expression: ($Y = 203.4\ X^{-0.23}$), where Y depicted as infiltration rate and X as number of grazing animals per ha. However, Samra (2006) found that moderate grazing produced better biodiversity and vegetative cover as compared to heavy and no grazing (Table 14.4).

Dispersal and seed germination of many species improve after passing through the animal digestive tract. For example, seeds of *Acacia nilotica* and *Prosopis juliflora* germinate faster when passed through the digestive tract of goats, which browse upon the pods of these species. Proper management of grasslands and

**Table 14.4** Effect of different summer grazing intensities on species richness and diversity

| | Grazing intensity | | |
|---|---|---|---|
| Indices | Heavy | Moderate | No grazing |
| No. of total species | 15 | 41 | 27 |
| No. of abundant species | 11 | 24 | 14 |
| No. of very abundant species | 7 | 20 | 5 |
| H (diversity index) | 2.26 | 3.01 | 2.61 |
| Evenness of species distribution | 0.95 | 0.78 | 0.60 |

Source: Modified from Samra (2006)

grazing according to their carrying capacity will maintain and improve qualities of natural resources and also in checking soil erosion in ravine lands.

Traditional agroforestry systems such as home gardens in tropics, boundary plantations, multi-storeyed plantation-based systems, jhum cultivation, Khejri (*Prosopis cineraria*)-based and *Faidherbia albida*-based system in dry regions of India and Africa, respectively, and silvopastoral systems in varied ecological zones are valuable compromise between biodiversity conservation and profitable use of natural resources (Leaky 1998). Contrary to traditional agroforestry systems, modern agroforestry systems and technologies (alley cropping, bio-drainage, woodlots, improved fallows, fodder banks, windbreaks and shelterbelts) are generally developed using only a few selected tree species – often in mono-tree species systems, usually with preferred characteristics such as high-yielding, fast-growing and nitrogen-fixing trees and arboreal structure. These systems aim at planting the trees in regular pattern and, therefore, mimic the temperate models where monoculture is more prevalent (Cooper et al. 1996). These systems maximize ecosystem processes and structural complexity rather than increasing number of species, and results in low diversity make the system vulnerable to insect or diseases. These systems, however, can be useful in maintaining biotic connectivity and can improve soil faunal and avian diversity to some extent particularly in ravine areas.

## 14.6 Agroforestry for Fodder Production from Ravine Lands

During lean period and scarcity of fodder, trees and shrubs are often lopped or pruned and used as leaf fodder in arid and semiarid and hill regions. The phenomenon is popularly known as top feed. The leaf fodder yield depends on nature of species, initial age of the plant, lopping intensity and interval as well as agroclimatic conditions. Fodder from trees is mainly available from two parts, viz. leaf twigs and pods. This forage is usually rich in proteins, vitamins and minerals but in general low in phosphorus and crude fibre. The importance of top feeds increases with the severity of drought and progression of drought season. Other uses such as for live fencing or boundary plantations or alley cropping are complimentary. Besides using for specific purpose, these are also used as fuelwood or mulch and also as fodder. The top-feed trees are considered very important in vegetation stabilization and sustained productivity of rangelands and also play an important role as windbreaks and by providing shade for the grazing animals.

In ravine lands, local people take their herds for grazing and also lop the trees for fodder. The foliage of *Balanites roxburghii*, *Feronia limonia*, *Cordia rothii*, *Ziziphus nummularia*, *Z. mauritiana* and many leguminous trees such as *Prosopis cineraria*, *Acacia nilotica*, *Pithecellobium dulce*, *Dalbergia sissoo* and *Leucaena leucocephala* are very rich in nutrients and protein. These are regularly fed to the animals, specially to goats and sheep. Camel browses on these trees and likes the foliage of *Salvadora oleoides* and *S. persica*. In dry regions, particularly in Rajasthan, the pods of *Prosopis juliflora* are collected and are mixed in large amount in animal feed. *P. cineraria* and *Z. nummularia* are the lifeline of people in dry regions and are lopped and also dried for fodder during scarcity. Other important fodder trees include *Albizia lebbeck*, *Acacia leucophloea*, *A. eburnea*, *A. senegal*, *Ailanthus excelsa*, *Azadirachta indica*, and *Ficus* spp. for the arid and semiarid regions and *Grewia optiva*, *Morus alba*, *Celtis australis*, etc. for the hilly regions. Thus, tree fodder provides enough nutrients and can serve very well as a green fodder supplement. Besides providing green fodder, such leaves are also conserved in the form of hay and silage to supplement feed during scarcity periods. These are important constituents of the silvopastoral system in ravine areas.

Silvopastoral systems are the lifeline of the people living in ravine ecologies. These may be strengthened using trees such as *Dalbergia sissoo*, *Dichrostachys cinerea*, *Azadirachta indica*, *Acacia nilotica*, *A. tortilis*, *Albizia lebbeck*, *A. amara*, *L. leucocephala*, *Cassia siamea*, *Prosopis juliflora*, *Anogeissus pendula* and *Hardwickia binata*. Grass species like *Bothriochloa intermedia*, *Cenchrus ciliaris*, *C. setigerus* and *Dichanthium annulatum* perform better than other species. Solanki (1998) reported that on the top and slopes of ravines 2.8–3.0 Mg $ha^{-1}$ dry forage and at bottom 6.3 Mg $ha^{-1}$ forage could be obtained. After 6 years, the trees could be lopped for fodder, which produced on an average 2.5 Mg $ha^{-1}$ fodder every year. The economic analysis of this system gave a high benefit-cost ratio of 1.42 in a 12-year rotation.

## 14.7 Biofuel Species Suitable for Eroded Lands

To meet the growing needs of energy in the country, cultivation of biofuel and bioenergy plantations is generally advocated particularly in degraded soils. Ravines are one of the most suitable areas for cultivation of these plantations. The potential tree borne oilseeds (TBOs) holding promise for biofuel and suitable for growing in ravine areas include *Jatropha curcas*, *Pongamia pinnata*, *Simarouba glauca*, *Azadirachta indica* and *Madhuca indica*. It has been found that TBOs can successfully be intercropped with annual crops such as pearl millet, cluster bean cowpea, sesame, sunflower, French bean, black gram, green gram and groundnut at various places. Castor (*Ricinus communis*) is another interesting plant, which can successfully be grown in beds of ravine lands in isolation or with other crops. It is currently cultivated on about 700,000 hectares mostly in Gujarat and Andhra Pradesh under rainfed conditions. The average seed production of castor is 1250 kg $ha^{-1}$, and the yield in terms of oil varies from 350 to 650 kg of oil per hectare when no maintenance is applied to the crop (www.castoroil.in). The biodiesel can be obtained by transesterification of castor oil using either ethanol or methanol as the transesterification agent (Shrirame et al. 2011).

*Euphorbia antisyphilitica* commonly known as candelilla wax plant is another succulent laticiferous potential hydrocarbon yielding petro-crop having potential for degraded lands. It yields 8–10% biomass utilized as biofuel. It can be grown successfully on degraded sandy and calcareous soils in arid and semiarid regions. The crop, when irrigated with saline water, produced ~ 23 Mg $ha^{-1}$ dry biomass in 2 years. It has been found to be a low nutrient demand crop as it required only 16 and 40 kg $ha^{-1}$ of phosphorus and nitrogen, respectively, for optimum biomass production. It is also low-water requiring and produced 17.5 and 15.25 Mg $ha^{-1}$ dry biomass with saline water (12 dS $m^{-1}$) irrigation at Diw-CPE ratio of 0.1 and 0.2, respectively, as compared to 10.9 Mg $ha^{-1}$ under rainfed condition (Dagar et al. 2012). For large-scale cultivation, it can be grown successfully on marginal calcareous and sandy soils of dry regions including ravine lands, and one or two life-saving irrigations with saline water will give optimum biomass.

## 14.8 Bioenergy Plantations Suitable for Eroded Lands

In rural areas, 70–80% energy comes from biomass of trees and shrubs mostly found growing in degraded lands. Due to the agroforestry initiatives, large amount of wood is now being produced from outside the conventional forestlands. As the landholdings are shrinking and to meet the food requirement for ever-increasing population, we need to grow more food crops on arable land and can hardly afford to grow energy plantations on these lands. Therefore, degraded lands such as affected by ravines and other wastelands are the most suitable for this purpose. Small and marginal farmers are now providing domestically produced wood products through short rotation agroforestry practices (Dhyani et al. 2006a). *Prosopis juliflora* is the major source of fuelwood used in rural areas throughout the country. *Prosopis* biomass is now also used extensively for the boilers of the power generation plants in Andhra Pradesh. The fuelwood potential of indigenous and exotic tree species such as *Acacia nilotica*, *A. auriculiformis*, *A. tortilis*, *A. senegal*, *Azadirachta indica*, *Casuarina equisetifolia*, *Dalbergia sissoo*, *Eucalyptus camaldulensis*, *E. tereticornis*, *Prosopis juliflora*, *P. cineraria* and *Ziziphus mauritiana* with their calorific values ranging from 17.3 to 20.8 MJ $kg^{-1}$ has been assessed as remarkable. Most of these can successfully be grown in ravine lands. Species such as *Casuarina equisetifolia*, *Prosopis juliflora*, *Leucaena leucocephala* and *Calliandra calothyrsus* have been prominent due to their potential for providing wood energy at the highest efficiency, in shorter rotation, and due to their high adaptability to diverse habitats including ravine lands and dry climates. Some other bushes such as *Capparis decidua*, *Salvadora oleoides*, *Tamarix articulata*, *Balanites roxburghii*, *Acacia eburnea*, *A. nilotica*, *A. jacquemontii*, *Tecomella undulata*, *Prosopis cineraria*, *Ziziphus nummularia*, etc. found growing natural in ravines are harvested and used as fuelwood by local people. Agroforestry plantations on community land and as live fence on farm boundaries have immense potential in this regard. The crop residues obtained from cotton, mustard, pigeon pea (*Cajanus cajan*) and discs prepared from cow dung are extensively used in rural households as bioenergy.

## 14.9 Potential Silvopastoral Systems for Ravine Lands

Silvopastoral systems are the most promising alternate land use which integrate multipurpose trees, shrubs, legumes and grasses, mostly on nonarable, degraded ravine lands for optimizing land productivity (Fig. 14.3). A judiciously managed silvopastoral system not only gives livelihood security to the village folks through their cattle management but also helps in conservation of vegetation, soil and nutrients and provides timber and firewood on a sustainable basis. Trees in silvopastoral systems tolerate extreme soil and climatic conditions, whereas the grasses provide good ground cover to check soil erosion. The system provides resilience by ensuring continued and sustainable multiple outputs such as forage, fuel, fibre and industrial raw material, besides other positive environmental effects including carbon build-up (Puri and Nair 2004). These systems are helpful in enhancing livestock and crop/forage productivity in ravine lands. The results obtained during the past four decades indicate that fodder and fuelwood production could be increased

**Fig. 14.3** Silvopastoral system with moderate grazing

severalfolds from the existing levels through proper management of these wastelands (Ghosh et al. 2015; Vijay et al. 2015; Dagar 2017).

In a long-term study in Bundelkhand region, Rai et al. (1999, 2001) reported productivity enhancement in eroded lands through tree-grass components in various silvopastoral systems involving trees such as *Acacia nilotica* var. *cupressiformis*, *Albizia lebbeck*, *A. procera* and *Hardwickia binata*. *Albizia amara*-based silvopastoral system produced 8 Mg $ha^{-1}$ $year^{-1}$ fodder followed by *Acacia tortilis*-based system producing 7.6 Mg $ha^{-1}$ $year^{-1}$ forage. Sharma et al. (1994) observed that in highly degraded arid lands when natural vegetation degraded due to overgrazing was replaced with managed silvopastoral system consisting of *Cenchrus ciliaris* grass and the top feed tree species *Ziziphus nummularia* and *Grewia tanax*. Similarly, Prajapati et al. (1993) improved a ravine-dominated area in Yamuna catchment and obtained 1.5–2.0 Mg $ha^{-1}$ $year^{-1}$ mean yield of grass forage and top feed yield of 4.2 Mg $ha^{-1}$ $year^{-1}$ from a 14-year-old *Acacia nilotica* and *Acacia tortalis* + *Cenchrus ciliaris* silvopastoral system. Chauhan et al. (1993) established a sustainable silvopastoral system along 30–60% slopes comprising nitrogen-fixing tree *Alnus nepalensis*, pineapple and forage crops like *Panicum maximum* or *Setaria sphacelata* coupled with *Stylosanthes guianensis* in 1:1 ratio. Protein-rich forage yield of 13.5 Mg $ha^{-1}$ was obtained from the combination of leguminous *Stylosanthes* and *Setaria* grass. In addition to this, 2.3 Mg $ha^{-1}$ litter from *Alnus nepalensis* and 4000 fruits $ha^{-1}$ from pineapple were obtained. This system also restored the fertility of soils. Chaturvedi and Khan (2009) in their socio-economic studies reported that silvopastoral system on an average cycle of 10 years can generate 120 man-days $ha^{-1}$ employment annually. A separate chapter on the subject by Ghosh et al. has been included in this book.

## 14.10 Technological Interventions for Rehabilitation of Ravine Lands

The conventional technologies for ravine reclamation are usually engineering-based, which are exceptionally costly and difficult to adopt by an individual farmer. Based on international deliberations, Sharma et al. (2005) and Indian Society of Soil Science (2006) brought out proceedings dealing with strategies and future challenges for improving the health of degraded lands and watersheds for enhancing agricultural production. Samra (1997) and Samra and Narain (1998) reviewed the work related to soil and water conservation and reclamation of ravine lands in India. Most of the work is related to contour bunds, gully plugs and land modifications. Most of the peripheral bunds are stabilized by sodding with grasses such as *Dichanthium annulatum, Cenchris ciliaris* and *Panicum antidotale*. The excess run-off is safely disposed off through grassed ramps or piped outlets. In the Chambal ravines, where the soil is relatively fine and the rainfall is heavier, graded bunds in association with grassed waterways and chute drop structures for the safe disposal of water have been found better than the contour bunds.

For exploration of various soil erosion parameters and indices, in India still the universal soil loss equation (USLE) is relevant and continues to be the most practical and widely used erosion-prediction tool. It states:

$$A = RKLSCP$$

where $A$ is average annual soil loss, R is rainfall erosivity, K is soil erodibility, L is slope length, S is slope steepness, C is crop cover and P is conservation practice. The parameters used in USLE have been determined for different crops in different regions and also for agroforestry land uses (Tables 14.5 and 14.6). It has been revealed that the integrated development of watersheds has reduced the runoff or moderated flooding of downstream areas and improved in situ moisture conservation for increased biomass production (Table 14.7).

**Table 14.5** Rainfall erosivity (R) and soil erodibility (K) factors for different regions

| USLE-related parameters | Dehra Dun | Chandigarh | Agra | Kota | Vasad |
|---|---|---|---|---|---|
| Soil type | Silty loam | Sandy loam | Sandy loam | Clay loam | Sandy loam |
| Annual rainfall (mm) | 1740 | 1100 | 704 | 784 | 826 |
| Rain factor-R | 1048 | 569 | 277 | 354 | 648 |
| K-factor | 0.124 | – | 0.070 | 0.106 | 0.13 |

Source: Samra and Narain (1998) Crop management factor (C) for different crops and topographic (LS) factor for Bellary and Vasad have been reported by Samra and Narain (1998) elsewhere

**Table 14.6** Crop management (C) factor for different agroforestry land uses at Dehra Dun

| Sequence I | C-factor | Sequence II | C-factor | Average-C |
|---|---|---|---|---|
| Sole *Leucaena* | 0.01 | Sole *Leucaena* | 0.008 | 0.009 |
| Sole *Eucalyptus* | 0.008 | Sole *Eucalyptus* | 0.007 | 0.008 |
| Grass + *Leucaena* | 0.109 | Turmeric + *Leucaena* | 0.242 | |
| Grass + *Eucalyptus* | 0.003 | Turmeric + *Eucalyptus* | 0.219 | |
| Maize-wheat + *Leucaena* | 0.257 | Maize-wheat + *Leucaena* | 0.304 | 0.278 |
| Maize-wheat + *Eucalyptus* | 0.160 | Maize-wheat + *Eucalyptus* | 0.225 | 0.189 |
| Grass | o.112 (o.o48)[a] | Turmeric + *Leucaena* hedge | 0.246 | |
| Maize | 0.425 | Maize + *Leucaena* hedge | 0.377 | |

Source: Samra and Narain (1998)
[a]Excluding establishment period

**Table 14.7** Impact of integrated watershed management practices on runoff and soil loss

| | Runoff (percent of rainfall) | | Soil loss (Mg ha$^{-1}$) | |
|---|---|---|---|---|
| Watershed | Pretreatment | Post-treatment | Pretreatment | Post-treatment |
| Fakot (UP) | 42 | 14.2 | 11.9 | 2.5 |
| GR Halli (Karnataka) | 14 | 1.3 | 3.5 | 1.0 |
| Behdala (HP) | 30 | 15.0 | 12.0 | 8.0 |
| Joladarasi (Karnataka) | 20 | 7.0 | 12.0 | 2.3 |
| Una (HP) | 30 | 20.0 | 12.0 | 10.0 |

Source: Samra and Narain (1998)

**Table 14.8** Sediment deposition along vegetative barriers in soil conservation plots at Dehra Dun

| Vegetative barrier | No. of years in existence | Sediment deposited (Mg $ha^{-1}$ $year^{-1}$) | Average deposition (Mg $ha^{-1}$ $year^{-1}$) | Soil loss (Mg $ha^{-1}$ $year^{-1}$) |
|---|---|---|---|---|
| *Leucaena* hedge in turmeric field | 3 | 47.31 | 15.77 | 7.57 |
| *Leucaena* hedge in maize field | 3 | 183.98 | 61.33 | 12.09 |
| *Leucaena* trees in maize field | 9 | 256.54 | 28.50 | 8.82 |
| *Eucalyptus* trees in maize field | 9 | 185.58 | 20.62 | 5.80 |
| *Leucaena* trees in turmeric field | 9 | 90.06 | 10.01 | 6.78 |
| *Eucalyptus* trees in turmeric field | 104 | 11.52 | 7.12 | NA |

Source: Narain et al. (1998)

Arora and Mohan (1996) reported that under rainfed conditions of Dehra Dun, agri-horticultural agroforestry system of peach (*Prunus persica*) and turmeric (*Curcuma longa*) was found to be highly remunerative and gave highest returns as compared to other systems. In general, growing of lemon grass (*Cymbopogon flexuosus*) as intercrop lowered the nutrient status of peach leaves, especially Zn, N and P compared with turmeric intercrop. The grass has higher nutrition absorbing capacity. Samra et al. (1995) reported the results of other fruit-based systems. They found that kinnow (*Citrus reticulata*)-turmeric agroforestry system in V-shaped micro-catchment with mulberry (*Morus alba*) on field bunds produced 4.34 Mg $ha^{-1}$ kinnow fruits, 1.11 Mg $ha^{-1}$ turmeric spice from interspaces and 2.24 Mg $ha^{-1}$ cane for basket making from mulberry twigs, along with 2.16 Mg $ha^{-1}$ wood and 0.69 Mg $ha^{-1}$ leaves for sericulture. In another system, plantation of kinnow at 4 × 4 m spacing and bhabbar (*Eulaliopsis binata*) grass at 50 × 50 cm as intercrop after minor levelling at Relmajra watershed provided early returns to the farmers, and the system was found highly remunerative and viable.

It has now been recognized that engineering methods alone are not much effective for controlling soil erosion, particularly in most difficult situations like ravine lands; rather integrated approaches involving vegetation means for controlling erosion are most effective. During the last two decades, the role of agroforestry-based models has gained momentum in controlling erosion and ravine extension. Narain et al. (1998) have found that agroforestry barriers are quite effective in controlling runoff and soil loss and in managing several watersheds. In many studies, the reduction in crop yield was fully compensated by the tree biomass reduction in crop yield due to trees which can be minimized to greater extent by shifting the management of trees to contour hedgerows. These not only check runoff and soil loss on highly sloping lands but also contribute in moisture conservation through mulch, soil amelioration through litter and fine root decomposition and fixing nitrogen in soil when the hedgerow is of leguminous trees. The pruned biomass of hedges is used as fodder or mulch and green manure and sticks for fuelwood. The total sediment deposition along hedgerows (3-year period) and tree rows (9-year period) was found to range from 184 to 256 Mg $ha^{-1}$, equivalent to 15–20 mm soil depth (Table 14.8) showing the efficiency of the system.

Verma et al. (2015) have reported that the extension and deepening of gully beds can successfully be prevented with gully plugs of various materials such as brushwood live hedges, earth, sandbags, brick masonry and boulders. These measures were applied successfully in Chambal ravines near Morena in Madhya Pradesh. From experiences it could be inferred that a composite check dam of earth and brick masonry (spillway portion) is necessary for larger catchments. It is an established fact that the soils of the ravine areas are poor in fertility and physical conditions, as almost all the soil profiles are eroded leaving behind rocky, sandy or stiff clay with poor water retention or water-absorbing power. At times, the entire profile is found to be eroded up to the extent of rocks below. The process of soil formation is very slow or negligible because of such drastic situations. Sometimes, in the soil profile, the lime develops into *kankar* (calcareous) pan which further prevents the roots of the plants to penetrate for getting water from deeper layers.

Ravine lands can be suitably reclaimed by putting them under tree cultivation following suitable technologies. Depending on the problems and needs of the area, suitable tree species may be introduced as alley, boundary plantation or scarred tree plantation in the field. Medium and shallow gullies can be utilized under silvopastoral and horti-pastoral systems and deep gullies under sole tree plantations or along perennial grasses preferably using native species, which have adapted to erosion stress since time immemorial. These have the capacity of regeneration from the rootstock or exposed root buds due to soil erosion. Tree species such as *Acacia nilotica*, *A. eburnean*, *A. leucophloea*, *Ailanthus excelsa*, *Albizia lebbeck*, *Azadirachta indica*, *Dalbergia sissoo*, *Dendrocalamus strictus*, *Eucalyptus camaldulensis*, *E. citrodora*, *E. tereticornis*, *Pongamia pinnata*, *Emblica officinalis*, *Salmalia malabarica* and *Tectona grand*is are found quite promising for ravines in Gujarat and also for Chambal ravines. *Butea monosperma*, *Dalbergia sissoo*, *Acacia eburnea*, *A. nilotica*, *Feronia limonia* and several perennial grasses constitute natural savannah vegetation on top of ravines of Chambal and tributaries like Kshipra river. For Yamuna ravines, *D. sissoo*, *A. nilotica*, *A. indica*, *A. lebbeck*, *P. pinnata*, *Holoptelea integrifolia*, *D. strictus* and *Eucalyptus* hybrid have been reported to be most suitable species. In Kota, *D. strictus* and *Salix tetrasperma* in gully beds and *A. nilotica* on the top of the ravines were reported very suitable. Prajapati et al. (1993) established a successful silvopastoral system at Agra in Chambal ravines by planting trees of *Acacia nilotica* and *Acacia tortilis* at 3 × 3 m spacing on top, slopes and bottom of ravines along with *Cenchrus ciliaris* grass. They could harvest about 1.5–2.0 Mg $ha^{-1}$ $year^{-1}$ forage from grass and 3.8–5.2 Mg $ha^{-1}$ top feed from the trees annually, and about 28 Mg $ha^{-1}$ fuelwood from each species was obtained.

Bamboo and Anjan grass (*Cenchrus ciliaris*)-based silvopasture system has been evolved and found successful for enhancing productivity of ravines at Vasad in

Anand district of Gujarat. The technology has been found suitable for gullied lands of Ahmedabad, Anand, Bharuch, Banaskantha, Kheda, Mehsana, Panchmahals, Sabarkantha, Surat Vadodara and Valsad districts. It involves cultivation of *Dendrocalamus strictus* bamboo in gully beds at 5 m × 5 m spacing (staggered). New slips of grass *Cenchrus ciliaris* are planted on slopes as well as in the interspaces of bamboo planted in gully bed. Staggered contour trenches (60 cm × 60 cm × 1.8 m length) are dug out in ravine bed across the slope.

Pandey et al. (2012) reported that following the technology, the gullies were stabilized completely showing negligible runoff and soil loss during rainy season. Moreover, on an average 7–10 Mg $ha^{-1}$ fresh forage was harvested annually from stabilized gullies. After 7 years of plantation, 30% of the total bamboo culms (per clump ~1000 bamboo poles $ha^{-1}$ $year^{-1}$) could be harvested every year giving suitable income to the stakeholders. It has been stated that over a production period of 20 years, the benefit cost ratio of 1.85: 1 was obtained. Further, it was estimated that more than 80% of total rainfall was either utilized by the plant or percolated deep to recharge the groundwater. Due to low runoff, soil loss was reduced to less than 1 Mg $ha^{-1}$ $year^{-1}$ compared to 10 Mg $ha^{-1}$ $year^{-1}$ from degraded ravines prior to the plantation.

Further, Sethy et al. (2011) established forest and fruit trees-based silvopastoral system at Kota in Rajasthan for the ravine watersheds. The gooseberry fruit trees were grown on ravine humps, bamboo on gully beds and *Cenchrus ciliaris* grass as interspace forage crop. Soil moisture was conserved by trapping runoff through trenches numbering 278 $ha^{-1}$. The fruit yield was found to be 6.61 Mg $ha^{-1}$ in treatment involving trenching as compared to only 1.31 Mg $ha^{-1}$ in control (without any trench). The production of grass ranged from 7.71 to 9.91 Mg $ha^{-1}$. The average number of harvestable clumps in bamboo ranged from 8 to 16 under different treatments. The interesting observation was that the trenching helped in moisture conservation up to 87% and reduction in runoff 78% and reduction in soil loss was 90%. The improvement in production potential was found to be 218.7%. This clearly shows the success of bioremediation measures of controlling soil erosion and ravine formation. Further it can easily be concluded that silvopastoral system involving forest trees like bamboo and fruit trees like gooseberry along with suitable grasses is most ideal for checking erosion in ravines, stabilizing gullies, conserving rainwater, sequestering carbon, enhancing biodiversity and sustaining productivity for livelihood security.

## 14.11 Conclusions

Besides conventional engineering measures for controlling soil erosion due to water and extension of ravines, agroforestry is the viable option for controlling soil erosion and reclamation of ravine lands. Sustainable development through scientific agroforestry interventions has a great scope and potential for productivity enhancement and efficient conservation of the resources through reducing soil and water erosion, soil enrichment, microclimate amelioration, air- and water-quality improvement, carbon sequestration and biodiversity conservation. Lot of work has been done in the past which has laid strong scientific foundation in accepting the claim of agroforestry for resource conservation including in most unfavourable situations like ravine lands. However, still there is need to highlight the intangible benefits of agroforestry which are generally overlooked while considering cost and benefit of any agroforestry system. Integrated watershed development approach involving agroforestry technologies such as silvopastoral system and alley cropping will help in stabilizing gullies and ensuring livelihood and environmental security.

## References

AICRP (1994). Annual report of the all India coordinated research project on dryland agriculture, main centre, Akola (Maharashtra), India

Ajit, Dhyani SK, Ramnewaj, Handa AK, Prasad R, Alam B, Rizvi RH, Gupta G, Pandey KK, Jain A, Uma (2013) Modeling analysis of potential carbon sequestration under existing agroforestry systems in three districts of Indo-Gangetic plains in India. Agrofor Syst 87(5):1129–1146

Andreu V, Rubio JL, Cerni R (1998) Effects of Mediterranean shrub cover on water erosion. J Soil Water Conserv 53:112–120

Arora YK, Mohan SC (1996) Identification of companion crops for fruit orchard, Annual Report. CSWCRTI, Dehra Dun, pp 47–48

Atta-Krah K, Kindt R, Skilton JN, Amaral W (2004) Managing biological and genetic diversity in tropical agroforestry. Agrofor Syst 61:183–194

Babu R, Dhyani BL, Agarwal MC, Samra JS (1997) Economic evaluation of watershed management projects, Bull No. T-33/D-23. CSWCRTI, Dehra Dun

Bronstein GE (1984) Production de uno pesture de *Cynodon plecostachyus* asoucda con arboles de *Cordia alliodora* con arboles de *Erthryna poeppegriana* of Sin arbolts. Tesis Maq. Sci. CATIE, Turrialba, Costa Rica

Campbell BM, Frost P, King JA, Mawanza M, Mhlanga L (1994) The influence of trees on soil fertility on two contrasting semi-arid soil types at Matopos, Zimbabwe. Agrofor Syst 28:159–172

Chaturvedi OP, Khan IA (2009) Agroforestry research perspectives and potential. In: Chaturvedi OP, Venkatesh A, Dwivedi RP, Singh R, Dhyani SK (eds) Agroforestry: natural resource sustainability, livelihood and climate moderation. Satish Serial Publishing House, Delhi, pp 27–42

Chauhan DS, Dhyani SK, Desai AR (1993) Productivity potential of grasses in association with *Alnus nepalensis* and pine apple under silvi-horti-pastoral system of agroforestry in Meghalaya. Indian J Dryland Agric Res Dev 8:60–64

Cooper PJM, Leakey RRB, Rao MR, Reynolds L (1996) Agroforestry and the mitigation of land degradation in the humid and sub-humid tropics of Africa. Exp Agric 32:235–290

Dagar (1995) Agroforestry systems for Andamans & Nicobar Islands in India. Int Tree Crops J 8(2–3):107–128

Dagar JC (2017) Potentials for fodder production in degraded lands. In: Ghosh PK, Mohanta SK, Singh JB, Vijay D, Kumar RV, Yadav VK, Sunil K (eds) Approaches towards fodder security in India. Studera Press, New Delhi, pp 333–364

Dagar JC, Yadav RK, Sharif A (2012) *Euphorbia antisyphilitica*: a potential petro-crop for degraded calcareous soils and saline water irrigation in dry regions of India. J Soil Salinity Water Qual 4(2):86–91

Dagar JC, Pandey CB, Chaturvedi CS (2014) Agroforestry: a way forward for sustaining fragile coastal and island agro-ecosystems. In: Dagar JC, Singh AK, Arunachalam A (eds) Agroforestry systems in India: livelihood security & ecosystem services, Advances in Agroforestry, vol 10. Springer Publishers, New Delhi, pp 185–232

Dhuruva Narayana VV (2002) Soil and water conservation research in India. ICAR, New Delhi, p 444
Dhuruva Narayana VV, Ram-Babu (1983) Estimation of soil erosion in India. J Irrig Drain Eng 109(4):419–434
Dhyani SK, Kareemulla K, Handa AK (2006a) Agroforestry potential and returns across agro-climatic zones. NRCAF, Jhansi
Dhyani SK, Sharda VN, Samra JS (2006b) Sustenance of soil, water and environmental quality through agroforestry: India's 25 years experiences and a future perspective. In: Proc international conference on soil, water and environmental quality-issues and strategies. Indian Society of Soil Science, New Delhi, pp 146–154
Ghosh PK, Mahanta SK, Singh JB Pathak PS (2015) Grassland: a global resource perspective. (Proceedings of XXIII international grassland congress, 2015 held in New Delhi by Range Management Society of India, Jhansi, India)
Grewal SS (1993) Agroforestry systems for soil and water conservation in Shivalik. In: Agroforestry in 2000 AD for semi-arid and arid tropics, NRCAF, Jhansi, India, pp 82–85
Grewal SS, Juneja ML, Singh K (1994) Rainwater conservation and use by an energy plantation of *Eucalyptus tereticornis* and rainfed field crops in north India. J Trop For Sci 6:408–421
Indian Society of Soil Science (2006) Proceedings International Conference on soil, water and environmental quality-issues and strategies. Indian Society of Soil Science, New Delhi, p 484
Shibu J (2009) Agroforestry for ecosystem services and environmental benefits: an overview. Agrofor Syst 76(1):1–10
Kumar BM, Kunhamu TK (eds) (2011) Quarter century of agroforestry research in Kerala: a compendium of research publication. Kerala Agriucltural University, Thrissur, p 404
Lal R (1995) Global soil erosion by water and carbon dynamics. In: Lal R, Kimble JM, Levine E, Stewart BA (eds) Soils and global change. CRC/Lewis Publishers, Boca Raton, pp 131–141
Lal R (2003) Soil erosion and global carbon budget. Environ Int 29:437–450
Lal R (2006) Carbon sequestration and climate change with special reference to India. In: Proceedings international conference on soil, water and environmental quality-issues and strategies. Indian Society of Soil Science, New Delhi, pp 295–302
Leakey RRB (1998) Agroforestry for biodiversity in farming systems. In: Collins W, Qualset C (eds) The importance of biodiversity in agroecosystems. Lewis Publishers, New York, pp 127–145
Nair PKR, Latt CR (1997) Directions in tropical agroforestry research. Agrofor Syst 38:1–249
Narain P, Grewal SS (1994) Agroforestry for soil and water conservation in the western Himalayan region of India. I. Runoff, soil and nutrient losses. Agrofor Syst 39(2):175–189
Narain P, Singh G, Joshie P (1992) Technological needs of vegetative land protection measures. Proc 7th ISCO Conf, Sydney Australia, pp 638–643
Narain P, Singh RK, Sindhwal NS, Joshi P (1998) Agroforestry for soil and water conservation in the western Himalayan valley region of India. 2. Crop and tree production. Agrofor Syst 39:191–203
Neumann IF, Pietrowicz P, Reifsnyder WS, Darnhofer TD (1989) Light and water availability in the fields with and without trees. An example from Nyabisindu in Rwanda meteorology and agroforestry. In: Proceedings of an international workshop on the application of meteorology to agroforestry systems planning and management, Nairobi 9–13 February 1987, pp 401–406
Pandey VC, Kurothe RS, Rao BK, Kumar G, Parandiyal AK, Singh AK, Kumar A (2012) Economic analysis of bamboo plantation in three major ravine systems of India. Agri Eco Res Rev 25(1):49–59
Prajapati MC, Nambiar KTN, Puri DN, Singh JP, Malhotra BM (1993) Fuel and fodder production in Yamuna ravines at Agra. Indian J Soil Conserv 21(3):8–13
Puri S, Nair PKR (2004) Agroforestry research for development in India: 25 years of experiences of a national program. Agrofor Syst 61:437–452
Rai P, Solanki KR, Rao GR (1999) Silvipasture research in India – a review. Indian J Agroforest 1:107–120

Rai P, Yadav RS, Solanki KR, Rao RS (2001) Growth and pruned biomass production of multipurpose tree species in silvipastoral system on degraded lands in semiarid region of Uttar Pradesh, India. Forest Tree Livelihood 11:347–364

Saha R, Tomar JMS, Ghosh PK (2007) Evaluation and selection of multipurpose tree for improving soil hydro-physical behaviour under hilly eco-system of north east India. Agrofor Syst 69(3):239–247

Saha R, Ghosh PK, Mishra VK, Majumdar B, Tomar JMS (2010) Can agroforestry be a resource conservation tool to maintain soil health in the fragile ecosystem of north-east India? Outlook Agric 39(3):191–196

Saha SR, Chaudhary RS, Somasundaram J (2012) Soil health management under hill agroecosystem of North East India. Appl Environ Soil Sci 2012:9. https://doi.org/10.1155/2012/696174

Samra JS (1997) Status of research on watershed management. Presented at 173rd meeting of governing body of ICAR, March 1997

Samra JS (2006) Participatory watershed management for improved soil and water quality. In: Proc international conference on soil, water and environmental quality-issues and strategies. Indian Society of Soil Science, New Delhi, pp 22–29

Samra JS, Narain P (1998) Soil and water conservation. In: Singh GB, Sharma BR (eds) 50 years of natural resource management research. ICAR, New Delhi, pp 145–176

Samra JS, Bansal AK, Sikka SP, Mittal SP, Agnihotri Y (1995) Resource conservation through watershed management in Shiwalik foothills, Bulletin T-28/C-7. CSWCRTI, Research Centre, Chandigarh

Samra JS, Raizada A, Prajapati MC, Bhushan LS (1999a) Floristic diversity and community composition in Yamuna ravines. In: Impact assessment of long-term production. Bulletin No T-41/A-2. Central Soil & Water Conservation Research & Training Institute, Dehra Dun

Samra JS, Vishwanathan MK, Sharma AR (1999b) Biomass production of trees and grasses in a silvipasure system on marginal lands of Doon Valley in north-west India, 2. Performance of grass species. Agrofor Syst 46:197–212

Sastry G, Mathur HN, Tejwani KG (1981) Landslide control in north-western outer Himalayas, Bulletin No R8/D6. CSWCRTI, Dehra Dun

Sethy BK, Parandiyal AK, Ali S, Kumar A, Singh RK (2011) Cost effective conservation measures for management of medium and deep ravenous lands, Annual Report. CSWCRTI, Dehradun, pp 73–74

Sharda VN, Samraj P, Samra JS, Lakshman V (1998) Hydrological behaviour of first generation coppiced blue gum plantations in Nilgiri sub-watershed. J Hydrol 211:50–60

Sharma PD (2004) Managing natural resources in the Indian Himalayas. J Indian Soc Soil Sci 52(4):314–331

Sharma SK, Singh RS, Tiwari JC, Burman U (1994) Silvopastoral studies in arid and semiarid degraded lands of western Rajasthan. In: Singh P, Pathak PS, Roy M (eds) Agroforestry systems for degraded lands. Oxford & IBH, New Delhi, pp 749–754

Sharma BR, Samra JS, Scott CA, Wani SP (eds) (2005) Watershed management challenges: improving productivity, resources and livelihoods. International Water Management Institute, Colombo. xiv + 336 p

Shrirame HY, Panwar NL, Bamniya B (2011) Biodiesel from castor oil-a green energy option. Low Carbon Econ 2:1–16

Ramesh S, Chaturvedi OP, Tewari RK, Dwivedi RP, Rizvi RH (2016) Drought proofing and rural livelihood security through watershed and agroforestry interventions in Bundelkhand region. (A case study of Garhkundar-Dabar Watershed). ICAR-Central Agroforestry Research Institute, Jhansi, p 22

Solanki KR (1998) Agroforestry. In: Singh GB, Sharma BR (eds) 50 years of natural resource management research. ICAR, New Delhi, pp 447–476

Tomar VPS, Narain P, Dadhwal KS (1992) Effect of perennial mulches on moisture conservation and soil-building properties through agroforestry. Agrofor Syst 19:47–70

Verma SK, Trivedi SK, Singh A, Tomar AS (eds) (2015) Management of degraded lands and soil health. Biotech Books, New Delhi, p 249

Vijay D, Srivastava MK, Gupta CK, Malviya DR, Roy MM, Mahanta SK, Singh JB, Maity A, Ghosh PK (2015) Sustainable use of grassland resources for forage production, biodiversity and environmental protection. Proceedings 23rd international grassland Congress (Keynote Lectures), Range Management Society of India, Jhansi, India, p 383

Vishwanathan MK, Joshi P (1994) Economic utilization of boulder riverbed land for fodder, fiber and fuel, Annual Report. CSWCRTI, Dehra Dun, p 31

Vishwanathan MK, Samra JS, Sharma AK (1999) Biomass production of trees and grasses in a silvipasture system of marginal lands of Doon Valley in north-west India. 1. Performance of tree species. Agrofor Syst 46:181–196

Yadav RP, Prasad R, Aggarwal RK, Agnihotri Y, Samra JS (2005) Aonla (*Embilica officinalis*) based silvi-pastoral systems for soil and water conservation in degraded lands of Shivalik foothills. Indian J Soil Conserv 33(3):207–212

Yadav TH, Smith GM, Warden K, Peter YT (2011) Facilitating the wider use of agroforestry for development in India. Dev Pract 11:414–438

Young A (1997) Agroforestry for soil management, 2nd edn. C.A.B. International, Wallingford, p 320

# Forest and Fruit Trees-based Agroforestry Systems for Productive Utilization of Ravine Lands

15

A. K. Parandiyal, R. Kaushal, and O. P. Chaturvedi

**Abstract**

Among the various forms of land degradation, ravines are the worst manifestation of terrain deformation by water. The land degradation due to ravines is a major problem along several river systems in thc alluvial zones. These may extend up to 2–3 km in to the tablelands, and vast areas go out of cultivation due to the process of degradation. It has been estimated that about 3.7 million hectares of land has been affected by gully erosion along several river systems in the alluvial zones in India. Since land resources are finite, requisite measures are required to reclaim degraded and wastelands so that areas going out of cultivation due to social and economic reasons are replenished by reclaiming these lands and by arresting further loss of production potential. Agroforestry as a traditional land-use adaptation may potentially support livelihood improvement through simultaneous production of food, fruit, fodder and firewood as well as mitigation of the impact of land degradation. Rehabilitation of ravine lands involves treatment of table- and marginal lands contributing runoff to the gullies and economic utilization of gullied ravines. It requires an integrated approach of using gullies per land capability classes, soil and water conservation measures and putting land under permanent vegetation cover involving afforestation, agroforestry, horticulture, pastures and energy plantations. Selection of species of trees/shrubs/grasses is important in ravine rehabilitation efforts.

A. K. Parandiyal (✉)
ICAR-Indian Institute of Soil and Water Conservation, Research Centre, Agra, Uttar Pradesh, India

R. Kaushal
ICAR-Indian Institute of Soil and Water Conservation, Dehradun, India

O. P. Chaturvedi
ICAR-Central Agroforestry Research Institute, Jhansi, Uttar Pradesh, India



J. C. Dagar, A. K. Singh (eds.), *Ravine Lands: Greening for Livelihood and Environmental Security*, https://doi.org/10.1007/978-981-10-8043-2_15

## 15.1 Introduction

Land, a nonrenewable resource, is the key for all primary production systems. In the entire world, total area of degraded land exceeds 1.9 billion hectares, which is more than the total territory of arable land. The largest areas of degraded land are in Asia (748 million hectares), Africa (495 million hectares) and Latin America (305 million hectares). Over the years, India's landmass has also suffered from different types of degradations. Degradation of land is caused by biotic and abiotic disturbances. Land degradation, like climate change, is an anthropogenic-induced process and poses the biggest threat to sustainable livelihood security of the farming communities across the country. The degraded land and wasteland in India cover 120.4 million hectare (Mha) out of which water erosion is the predominant degradation class covering around 93 Mha of area affecting more than 28% of total geographical area (TGA) of the country (ICAR 2010). Among the various forms of land degradation by water, ravines are the worst manifestation of land degradation by water. The land degradation due to ravines is a major problem along several river systems in the alluvial zones in India. Since land resources are finite, requisite measures are required to reclaim degraded land and wastelands so that areas going out of cultivation due to social and economic reasons are replenished by reclaiming these lands and by arresting further loss of production potential. Agroforestry as a traditional land-use adaptation may potentially support livelihood improvement through simultaneous production of food, fodder and firewood as well as mitigation of the impact of land degradation.

## 15.2 Land Degradation Due to Ravines

Ravines are the system of gullies running almost parallel to each other and draining in to a river after a short distance with the development of deep gorges. These may extend up to 2–3 km in to the tablelands, and vast areas go out of cultivation due to the process of degradation. It has been estimated that ~4 Mha of land has been affected by gully erosion along several river systems in the alluvial zones in India (Fig. 15.1). The major problem areas are in Uttar Pradesh, Madhya Pradesh, Rajasthan, Gujarat and Bihar (Dhruvnarayan 1993). Sharma (1980) classified the areas under ravines in five major zones based on morphological characters, erosion rates and intensity of ravines and suggested that a gully initiates along the animal trails, roads and paths on the agricultural uplands. On the other hand, ravine is a function of river action and begins along the river and encroaches upon the catchment area by headward growth. Another view is that both gullies and ravines are the result of surface runoff mismanagement (Prajapati et al. 1982), especially deforestation, overgrazing and ill-considered tillage in an environment which is particularly susceptible to erosion (Kaul 1962; Tejwani 1968). The indiscriminate land-use practice leading to a disturbance of the hydrological balance is one of the reasons. The major variables that affect upland soil erosion by water in the catchment of

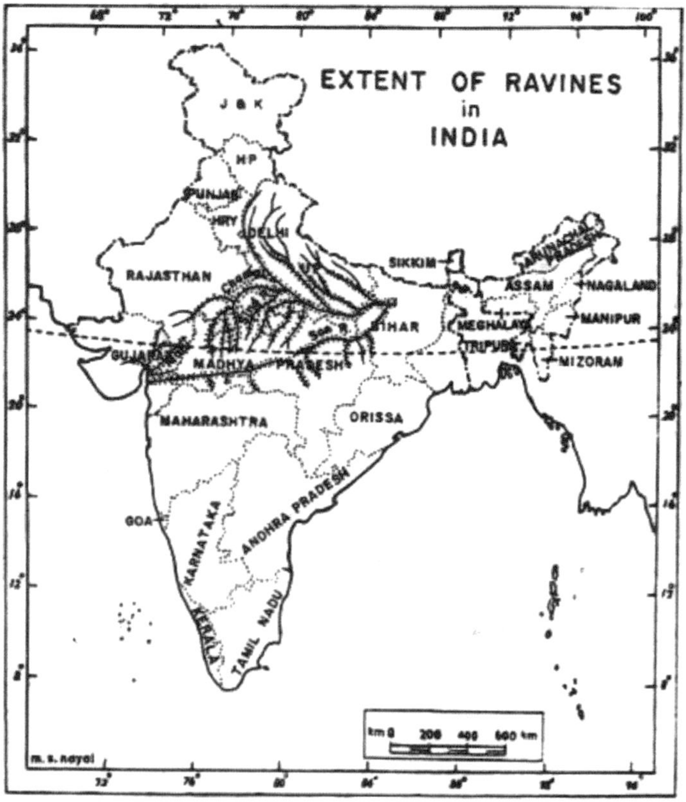


**Fig. 15.1** Distribution of ravines in India (Source: Sharda and Kurthe 2012)

ravines are climate, inherent soil characteristics, topography, land use and runoff management practices.

If left unattended the ravine lands tend to ingress in to the adjacent tablelands by headward extension, and vast areas of fertile lands go out of cultivation. Studies on the rate of ravine extension in Rajasthan, Gujarat and Uttar Pradesh have revealed that the area under ravines is extending incessantly. Ravine extension rates vary from 0.73% to 18.7% per annum (Prakash et al. 1987; Prakash et al. 1990; Tiwari et al. 1987; Nema et al. 1982) for the Chambal, Yamuna and Mahi ravine systems. Singh et al. (1966) considering a very conservative estimate of ravine extension, at 0.175% per annum, warned that Rajasthan may lose 13,750 ha of cultivated land in 25 years. Khan (1972) reported that ravines in Madhya Pradesh were increasing at 2.5% per annum, thus engulfing about 809 ha of land every year. Thus, it is rightly said that extending ravines marks the footprint of death for the cultivator.

## 15.3 Constraints for Development of Vegetation in Ravines

In nature, the ravine lands present severe constraints for development of vegetation. The major constraints are as follows:

### 15.3.1 Highly Undulating Topography

'Ravine' means a deep gorge and represents the last stage of water erosion. The land feature of ravines is rugged and characterized by three distinct parts, namely, (a) ravine top or hump, irregular in size and shape having multidirectional slope and highly eroded soils, (b) ravine slope which are steep and devoid of moisture and (c) ravine bottom/bed, narrow and difficult to cultivate. In general, the depth of ravines ranges from 1 to 20 m but may reach up to 55 m (Roy and Mishra 1969). The humps of ravines face high moisture stress as these are exposed from all sides and are thus highly desiccated. Ravine slopes are too steep to allow any infiltration of moisture during rains. In the ravine beds, moisture availability is better, but light availability may be a limiting factor. Generally, ravines of the Chambal River region are 'V' shaped where the subsoil is comparatively resistant to erosion than the topsoil, whereas 'U'-shaped ravines are found where thickness of alluvium increases and the soils are light textured. Such types of ravine are found along the rivers of Yamuna and Mahi. Barring the marginal and shallow gullied lands, which can be reclaimed economically and can be utilized for agriculture, most of the ravine lands are not fit for cultivation of agricultural crops and are classified as nonarable lands. Crop cultivation in these lands is not desirable as it leads to accelerated erosion, thus, adding to their fast degradation. The best scientific land use for these lands is to place them under permanent vegetation involving horticulture, horti-pasture, silvopastoral, energy plantation and plantation for timber or other forest produce.

### 15.3.2 Unfriendly Climate

Ravine regions face extremes of temperature. While the temperatures may be higher than 48°C during the hottest period, the coldest month may frequently witness a low temperature of 3°C. Low moisture and high evaporation (ET) characterize the climate of this region. The rainfall pattern is highly erratic as more than 90% of annual rainfall is received during June to September in intense storms which results in excess runoff due to poor permeability of soils and no adoption of resource conservation measures, and the rest of the 9 months face high moisture constraint. Relative humidity may be as low as 7% and evaporation very high during summers. The potential ET is more than double of the annual precipitation. The climate of the Chambal ravine region in Rajasthan and Madhya Pradesh is semiarid to subhumid (Agro-climatic zone IV) with annual average rainfall of 600–800 mm, mostly received from July to October in intense storms. The year has three distinct seasons: Hot and dry summer (April to June), rainy monsoon (July to October) and cold and dry winter (November to March). Almost similar rainfall pattern exists in Yamuna ravine region in Uttar Pradesh and Mahi ravine region in Gujarat (Bhushan and Saxena 1984; Singh et al. 1972, 1976; Prajapati et al. 1977; Prakash and Rao 1986).

### 15.3.3 Low Soil Fertility

The soils of the ravinous regions are poor in fertility and physical conditions. In the gullies the topsoils are eroded, and the exposed subsoils are low in organic matter content. In the zones of lighter soils in Uttar Pradesh and Gujarat, the water-holding capacity is low. Soils of these regions are deep alluvial deposits having sandy to sandy loam texture. In these zones, U-shaped ravines occur. The soils of Chambal ravine regions are less deep alluviums as compared to those in Yamuna and Mahi ravine region in Uttar Pradesh and Gujarat. These have silty clay loam to clay texture, and calcium carbonate content in these increases with increasing depth. Sometimes the lime develops into *kankar* (calcareous) pan which further prevents the roots of the plants to penetrate and get water from deeper layers. These soils are prone to waterlogging when wet and to compaction when dry. The infiltration rates are low to moderate, and soils are highly erodible in the upper layers. As the soils in lower layers are resistant to erosion as compared to soil, in these zones, V-shaped gullies are formed with narrow beds. The ravine soils vary in pH from 7.5 to 8.0. These soils are prone to waterlogging when wet and to compaction when dry.

### 15.3.4 High Biotic Pressure

Ravine lands are owned by resource-poor people. These lands experience extreme biotic pressure of domestic animals, which are low yielding and are managed under free-grazing system. Ravine lands are also exposed to high biotic pressure of wildlife including blue bull, *sambar*, rabbits and wild *langurs*. As productivity of land is low, the exposure to high biotic pressure leads to disappearance of utilizable vegetation.

### 15.3.5 Uneconomical Natural Vegetation

The ravines are characterized by dominance of natural thorny vegetation. The dominance of unpalatable vegetation enhances the biotic pressure on the planted vegetation, thus reducing the establishment and survival of economically utilizable vegetation. Extremes of climatic and edaphic conditions result in low productivity in adjoining marginal lands also. Champion and Seth (1968) have classified the natural vegetation of ravines under type 6BC2, i.e. northern tropical ravine thorn forests. The typical vegetation consists of low-growing type of coarse grasses like *Aristida royaliana*, *Aristida adscensionis*, *Eremopogon foveolatus*, *Heteropogon contortus*, *Themeda triandra*, *Heylandia latebrosa*, *Eragrostis* spp. and *Eleusine* spp. and rarely occurring *Dichanthium annulatum* among grasses and bushy shrubs of *Ziziphus nummularia, Capparis zeylanica, Capparis decidua, Crotalaria burhia, Cassia tora, Tephrosia purpurea, Xanthium strumarium, Calotropis procera, Clerodendrum phlomidis, Grewia pilosa, Grewia tenax, Lantana camara*, etc. along with xerophytic trees like *Dichrostachys cinerea, Acacia senegal, Acacia leucophloea, Acacia nilotica, Prosopis juliflora, Balanites aegyptiaca*, etc. Sajwan (1975, 1976), Prajapati (1995), and Balaji and Nitant (2002) have described the floristic wealth of Yamuna ravines. Dagar and Mall (1980) described the vegetation ecology of Kshipra ravines in detail. Parandiyal et al. (2000) studied the floristic diversity of Chambal ravines. The details of vegetation are given in Chap. 3 of this book.

## 15.4 Scope of Forest and Fruit Trees-based Agroforestry Systems in Ravines

Ravine lands shall be protected from further degradation to ensure that unrestricted expansion of ravine does not engulf the adjoining lands. The optimum utilization of ravine lands is also essential for exploiting their available potential for providing various goods and services and to achieve the various national objectives like reduction in regional inequalities and poverty, creation of income-generating and productive employment opportunities and improvement of ecological balance with a systematic and strategic approach for holistic development of these degraded areas. We can do little to alter climate, soil or topography, but we can do a lot to determine land use and manage runoff. Since conservation practices are designed to use land judiciously and manage runoff properly, they are our best opportunity to control soil erosion by water. In the year 1976, the National Commission on Agriculture estimated an annual loss of ₹1570 million worth of food grains, fodder and firewood production in the absence of judicious management of the ravine infested lands. Considering the average inflation rate of 8.15% per annum, this loss could be in the tune of ₹30,923 million in the year 2015. The shallow ravines can be economically reclaimed for agriculture if resources for irrigation can be created. The medium and deep ravines which cannot be economically/ecologically reclaimed for agriculture, due to vulnerability to accelerated erosion and rapid land degradation during tillage, can augment supplies of fruits, fuelwood, fodder industrial timber, bamboo and other minor forest produce if utilized judiciously.

In this situation, tree-based farming offers excellent opportunities because of several advantages. The major advantages of tree-based farming are the ability of tree species to withstand harsh agroecological conditions such as moisture stress and tolerance to various degrees of salinity. Trees can establish on poor soils and improve the soil productivity through deep root system, recharging of groundwater, recycling of mineral nutrients from lower profiles of the ground, contribution of organic matter and improvement of microclimate by providing shade and checking the wind velocity.

**Table 15.1** Percentage of household and average MPCE by primary source of energy for cooking in India 2007–2008

| | Rural | | Urban | |
|---|---|---|---|---|
| Primary source of energy used for cooking | % of households | Average MPCE (₹) | % of households | Average MPCE (₹) |
| Fuelwood and chips | 77.6 | 707 | 20.1 | 780 |
| LPG | 9.1 | 1389 | 61.8 | 1764 |
| Dung cake | 7.4 | 713 | 1.4 | 801 |
| Kerosene | 0.6 | 945 | 7.6 | 1109 |
| No cooking arrangement | 1.7 | 2654 | 5.7 | 3033 |
| All[a] | 100 | 772 | 100 | 1472 |

Source: NSSO (2010)

[a]Includes sources (3.96% in rural and 3.5% in urban areas) other than reported in col.1 MPCE stands for monthly per capita expenditure

## 15.4.1 Augmenting Fuelwood Supply in Rural India

In India, the land under cultivation increased from 118 million ha in 1951 to 142 million ha by 1987, which is almost consistent since then, rather decreased due to developmental activities. This expansion was accompanied by diversion of about 4.5 million ha of forestland to agriculture. This resulted in over-exploitation of remaining forests for supplies of fuelwood, fodder and other forest resources by increasing population. The household sector is one of the largest users of energy in India, accounting for about 30% of final energy consumption, excluding energy used for transport (Reddy 2003). The rural population in India relies heavily on traditional biomass-based fuels (fuelwood, crop residues and animal dung) for meeting the energy needs. As per the results of the Household Consumer Expenditure Survey conducted by National Sample Survey Organization (NSSO) in the year 2007–2008 (Table 15.1) in rural India, over 77% of households in the country continued to depend on firewood and chips for cooking, with only 9% using LPG. In urban India, 62% of households used LPG as major fuel for cooking, and 20% of households used firewood and chips (NSSO 2010).

The rural areas having no forests usually compensate fuel needs by increasing the share of agricultural residues, dung cake, and kerosene. The use of cow dung for fuel usually leads to reduced soil fertility due to unavailability of farmyard manure for agriculture. In the current scenario, the medium and deep ravine lands (accounting for approximately 2.45 million ha in the country) are the potential areas which can be utilized for developing energy plantations under silvopastoral systems. These areas have the potential for raising biomass for meeting requirements of household energy especially energy for cooking. The fuelwood species may be raised as sole plantation in the form of energy plantation or under silvopastoral systems.

### 15.4.2 Supplementing Fodder Supply in Rural India

The demand for feed and fodder for the livestock sector has also been rising steadily for the past six decades (post-independence period) with multifold increase in the livestock population in the country. India has one of the largest livestock populations in the world (512.1 million) with 57% of the world's buffalo and 16% of cattle as per livestock data of 2012. Apart from this, number of poultry in India is 729.2 million. The livestock sector alone contributes ₹5,37,5350 million (during 2012–2013) or nearly 25.6% of total value of output in the agriculture, fishing and forestry sector. The overall contribution of the livestock sector in total GDP is nearly 4.11% at 2012–2013 prices. The egg production in India was 69.73 billion during 2012–2013 with the per capita availability of 57 eggs per annum. The meat production was 5.95 million Mg during 2012–2013. India ranks first in milk production, followed by the USA, China, Pakistan and Brazil, accounting for 18.5% of world production, achieving an annual output of 146.3 million Mg during 2014–2015 (Nddb 2016) as compared to 137.69 million Mg during 2013–2014 recording a growth of 6.26%, whereas the Food and Agriculture Organization (FAO) has reported a 3.1% increase in world milk production from 765 million Mg in 2013 to 789 million Mg in 2014. The per capita availability of milk in India has increased from 176 g per day in 1990–1991 to 322 g per day by 2014–2015. It is more than the world average of 294 g per day during 2013. This represents a sustained growth in availability of milk and milk products for the growing population. Dairying has become an important secondary source of income for millions of rural households engaged in agriculture (GOI 2016). Still the average per animal production of milk in India is far below the milk production of countries like Israel and the USA. In Israel, the annual milk yield of a cow may be up to 10.5 Mg, while the average daily production of an Indian cow ranges from 2.36 to 7.02 kg per cow per day. This is approximately 10% of the milk output of dairy animals in the developed countries. The demand for milk in India in 2020 has been estimated as 135–156 million Mg (Delgado et al. 2001; Rao and Birthal 2008).

One of the notable characteristics of India's livestock sector is that almost its entire feed requirement is met from crop residues and by-products; grasses, weeds and tree leaves gathered from cultivated and uncultivated lands and grazing on common lands and harvested fields. In the animal production system, important factors deciding the productivity include breed, animal health management and feed/nutrition. Land allocation to cultivation of green fodder crops is limited and has hardly ever exceeded 5% of the gross cropped area (GoI 2009). Hence, the supply of feed has always remained short of normative requirement (GoI 1976; Singh and Mujumdar 1992; Ramachandra et al. 2007), restricting realization of the true production potential of livestock. For example, the actual milk yield of bovines is reported to be 26–51% below the attainable yield under field conditions (Birthal and Jha 2005), which otherwise could have been realized with better feeding, breeding and disease management, and feed scarcity is the main limiting factor for improving livestock productivity. The demand-supply estimates for green and dry fodder in India for 1995 to 2025 by Tenth Five Year Plan Document, Government of India, suggested that there will be a deficit of up to 61.1% in green fodder, 21.9% in

**Table 15.2** Scenario of feed and fodder requirement and availability in India (million Mg)

| Year | Supply | | Demand | | Deficit as % of demand (actual demands) | |
|---|---|---|---|---|---|---|
| | Green | Dry | Green | Dry | Green | Dry |
| 1995 | 379.3 | 421 | 947 | 526 | 59.95 (568) | 19.95 (105) |
| 2000 | 384.5 | 428 | 988 | 549 | 61.10 (604) | 21.93 (121) |
| 2005 | 389.9 | 443 | 1025 | 569 | 61.96 (635) | 22.08 (126) |
| 2010 | 395.2 | 451 | 1061 | 589 | 62.76 (666) | 23.46 (138) |
| 2015 | 400.6 | 466 | 1097 | 609 | 63.50 (696) | 23.56 (143) |
| 2020 | 405.9 | 473 | 1134 | 630 | 64.21 (728) | 24.81 (157) |
| 2025 | 411.3 | 488 | 1170 | 650 | 64.87 (759) | 24.92 (162) |

Source: Based on Tenth Five Year Plan Document, Government of India. Figures in parentheses indicate the deficit in percentage

dry crop residues and 64% in feeds. Supply and demand scenario of forage and roughage is presented in Table 15.2.

The current scenario calls for optimum utilization of all nonarable/degraded lands, especially ravine lands, for augmenting fodder supplies wherever possible by placing them under silvopastoral land uses. Ravines are suitable areas where the fodder production, including tree fodder production, under different systems can be taken up on a large scale. Several multipurpose tree species can produce leaf fodder and are amenable to lopping. These tree fodders may be palatable and are largely free from the anti-quality factors for use as feed for livestock.

### 15.4.3 Potential for Fruit Production in Ravines

Shallow and medium ravines are the potential areas for fruit production under agri-horticultural and horti-pastoral systems. India is the second largest producer of fruits in the world. The annual production of fruits in India during 2013–2014 was 88.98 million Mg from a 7.23 million ha area (NHB 2015). Still the per capita availability of fruit in India remained <20 g per day. The Indian Council of Medical Research in India has recommended a minimum daily dietary input of 80 g fruits for an adult. Thus there is a huge gap in ideal nutritional demand and supply of fruits in the country. As the productivity of fruits in India has barely increased from 10.03 to 12.33 Mg $ha^{-1}$ during the past two decades, there is a low possibility of further improvement in productivity of fruits in India. As increasing population has already put lot of pressure on arable lands for meeting the food demands, thus further expansion of area under fruits at the cost of food production is highly improbable. Therefore, we need to utilize the degraded lands which cannot be placed under agricultural land use due to vulnerability to further degradation or due to inherent limitation for producing agricultural crops. Shallow ravines can be easily placed under agri-horticultural systems with some investment on irrigation resources.

As suggested earlier, the best scientific land use for medium and deep ravines, which are prone to further degradation if placed under continuous tillage, due to ecological or hydrological instability, is to place them under permanent vegetation. Raising hardy and underutilized fruit species either as sole crop or under a horti-pastoral system is desirable in medium deep ravines. The fruit production may lead to higher economic benefits and regular income for a long period as the fruit production systems have easy marketability of the produce, scope for combining food crops and other arable crops with fruit production and farmers have scope for shifting from low-input, uneconomic agricultural production to assured sustainable farming systems.

## 15.5 Major Agroforestry Systems for Ravine Rehabilitation

The major agroforestry systems for ravine rehabilitation include systems that can be adopted in the marginal lands adjacent to ravines and systems for reclaimed shallow, medium and deep ravines.

### 15.5.1 Agroforestry Systems for Marginal Lands Adjacent to Gullied Lands

Marginal lands adjacent to gullies of ravines form the catchment of gullies and contribute to excess runoff to gullies. Management of these lands is important to arrest the ingress of extending gullies in to these lands. Agroforestry practices can be used as important tools for reducing runoff and soil loss from agricultural watersheds. Normally, interest in the use of agroforestry practices and contour grass strips for various environmental benefits relates to their potential to increase infiltration, reduce runoff and reduce non-point source pollution. Multi-strata systems, combined with litter cover and dense root systems, hold runoff when it first reaches the surface and subsequently promote infiltration. In an agroforestry system, the vegetation plays two major roles: (i) the fine root system holds soil in place, reducing susceptibility to erosion, and (ii) plant stems act as barrier to runoff water, thus decreasing the flow velocity and consequently enhancing sedimentation. Agroforestry systems used in these marginal lands may be agri-horticulture or agri-silviculture systems.

#### 15.5.1.1 Agri-horticultural Systems for Marginal Lands

In an agri-horticulture system, usually fruit trees are grown in rows (Figs. 15.2, and 15.3), and intercrops are cultivated on both arable crops and grasses for optimum utilization of the land during initial establishment of fruit plantations. The fruit trees are also found as scattered in the agriculture fields or may be introduced on field boundary of the fields. Several fruit trees have been evaluated for marginal lands adjacent to ravines by the three research centres of the Indian Institute of Soil and Water Conservation working on ravines of Chambal river system, Mahi river system and Yamuna river system which cover the major ravine lands in the country.

**Fig. 15.2** Bael (*Aegle marmelos*) fruit trees in agricultural fields adjacent to Chambal ravines

**Fig. 15.3** Inter-cultivation of wheat in *Ziziphus mauritiana* plantation

The fruit species found suitable for productive utilization of marginal lands of ravines include sapota (*Achras zapota*), mango (*Mangifera indica*), *kagzi* lemon (*Citrus aurantifolia*), kinnow mandarin (*Citrus reticulata*), *Ber* (*Ziziphus mauritiana*), guava (*Psidium guajava*), gooseberry (*Emblica officinalis*), *Bael patra* (*Aegle marmelos*), *Lasoda* (*Cordia dichotoma*) and *Karonda* (*Carissa carandas*) as they are compatible with most field crops and can withstand the low-resource conditions well. Large-scale adoption of any agri-horticultural system requires knowledge of marketing avenues and dedicated management for profitable disposal of produce. Unlike food crops, the market of fruits is more specialized and depends largely on postharvest processing, speedy disposal of produce, infrastructure for rapid transportation, etc. These aspects are needed to be taken care of while adopting any agri-horticultural system in ravines, which may not have very good connectivity to major markets of the country. There are some important case studies in Chambal and Yamuna ravines:

(a) *In Chambal ravines:* Studies conducted at IISWC Research Centre, Kota, found that gooseberry-, *kagzi* lemon- and guava-based agri-horticulture land uses are suitable for Chambal ravine region (Parandiyal et al. 2005). These can be suitably applied in marginal lands along the ravines and reclaimed shallow ravines in the Mahi and Yamuna ravine systems also. Gooseberry (*cv* NA 7) is raised at a spacing of 10 × 10 m, and cultivation of field crops is taken up in the interspaces of the tree rows. Kagzi lemon is raised at a spacing of 6 × 6 m, and cultivation of field crops is taken up in the interspaces of the tree rows. Guava (*cv* L-49, Allahabad *safeda*) is raised at a spacing of 8 × 8 m, and cultivation of field crops such as green gram (K 851), soybean (JS 335) and sesame (RT 46) is raised in the interspaces of planted trees with regular cultivation practices.

    The cost of developing the fruit plantation varied from ₹22,780 per ha (with intercultural cost for a 5-year period) in gooseberry to ₹26,450 in guava and ₹28,230 in lemon at prices of 2013. The average returns from the fruit trees were ₹84,000 per ha per annum in gooseberry, ₹120,000 (at 2013 prices) per ha per annum in guava and ₹90,000 from lemon on full maturity with fruiting starting from the fourth year of planting in these species. The yield of soybean grown in association with different fruit trees ranged from 933 kg with guava to 981 kg with lemon in the initial stages of plantation. The yield of *Sesamum* ranged from 583 kg with guava to 614 kg with lemon trees in the initial stages of plantation, and the yield of green gram ranged from 773 kg with guava to 794 kg with lemon trees in the initial stages of plantation.

(b) *In Yamuna ravines:* Kumar and Srivastava (1998) in a study of suitable intercrops in Yamuna ravines in Agra reported better vegetative growth in kinnow mandarin as compared to *mosambi* irrespective of intercrops grown. In this study, both plantations were planted at a spacing of 5 × 5 m. Spinach, fenugreek, mustard and potato were grown as intercrops in *rabi* (winter), and watermelon, lady's finger, brinjal and *lobia* in *zaid*; and green gram in *kharif* season. Maximum fruit yield (3828 kg $ha^{-1}$) in *mosambi* was obtained with fenugreek as the intercrop followed by spinach. The maximum yield of intercrops grown in both the fruit crops was recorded for spinach (16,000 and 8400 kg $ha^{-1}$) and minimum for potato (10,400 and 5600 kg $ha^{-1}$). In another study, Balaji et al. (2002) assessed the performance of gooseberry fruit trees and of intercrops in reclaimed ravines in class II and class IV lands. Maximum (1352 kg $ha^{-1}$) yield of *rabi* mustard was obtained in sunhemp green manured plots. Maximum

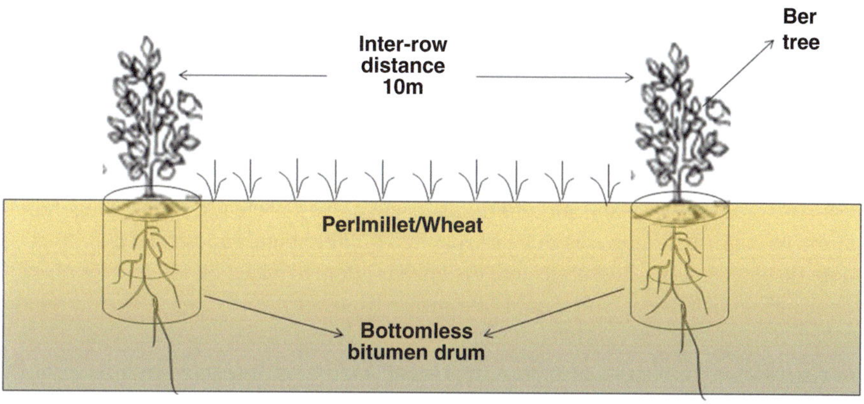


**Fig. 15.4** Representation of bottomless bitumen drum for planting of Ber (*Ziziphus mauritiana*) fruit trees

moisture conservation (15.8 cm in 100 cm profile at 32 days after sowing) was recorded in sunhemp green manured plots.

#### 15.5.1.2 Root Management Options

Low and erratic rainfall coupled with dry hot summer makes crop cultivation a risky proposition in marginal lands adjacent to ravines. Agri-horticultural system having a fruit crop tolerant to these stress conditions opens up opportunity for the farmer to get more assured income. However, the yield of field crops is considerably affected when they are grown in association with tree crops. Competition in the root zone for moisture and nutrients are the major reasons for such antagonistic effects. A suitable root management practice would be able to minimize the adverse effects of tree crops on the field crop. Five tree root management options, viz. $T_1$, tree planted in bottomless used bitumen drums of 90 cm height and 25 cm radius (Fig.15.4); $T_2$, tree planted in polythene (120 μ thick) lined pits; $T_3$, trenches (30 cm wide and 40 cm deep) in two sides of trees; $T_4$, trees without root management and $T_5$, crop without *Ber* (*Ziziphus mauritiana*) trees, were evaluated in Ber + pearl millet-wheat sequence in semiarid Indo-Gangetic alluvium soils of Yamuna ravines from 1988 to 2001 (Prakash and Kumar 2001; Prakash et al. 2001). Trees planted in bottomless bitumen drums produced significantly higher fruit yield (7347 kg ha$^{-1}$), tree volume (35.1 m$^3$) and firewood (2180 kg ha$^{-1}$). All tree root management options improved crop yield over $T_4$ which was equivalent to 13.76%, 7.48% and 7.28% in wheat and 17.89%, 10.64% and 7.86% in pearl millet under $T_1$, $T_2$ and $T_3$, respectively. However, tree root management options (except $T_1$) decreased fruit yield than T4 (6827 kg ha$^{-1}$). Among tree root management options, $T_1$ produced significantly higher grain yield of pearl millet (2161 kg ha$^{-1}$) and wheat (3968 kg ha$^{-1}$) and produced minimum reduction in pearl millet (5.30%) and wheat (6.48%) grain yields than sole cropping.

#### 15.5.1.3 Boundary Plantation of Multipurpose Trees in Marginal Lands

Multipurpose trees are traditionally planted or retained along the field boundary in the semiarid tropics and more so in the lands adjacent to ravines. Apart from providing fuel and fodder, these trees are maintained as assets to provide revenues at maturity. The trees raised under this system shall be amenable to lopping with less-spreading crowns and preferably shall have deep-rooted systems. Depending on the width of field, the tree rows are planted on field bunds constructed at 30 m or more distance apart. The tree-to-tree distance is adopted based on the species chosen for planting. The trees which could be grown along the boundary in ravine region included *Acacia nilotica, Acacia leucophloea, Leucaena leucocephala, Azadirachta indica, Albizia lebbeck, Soymida febrifuga, Salvadora oleoides* and *Tecomella undulata.* These trees are compatible for use in boundary plantation with most of the crops grown in ravine regions. For example, castor (*var* Aruna, GAUCH- 1, GCH-4) or pigeon pea (*var* Prabhat, IPCL-151) were found compatible for these boundary plantation systems.

#### 15.5.1.4 Case Studies

In a study of boundary plantation system in marginal lands of Chambal ravines at Kota, Parandiyal et al. (2008) reported better returns in a system comprising of boundary plantation of *Acacia nilotica, Azadirachta indica* and *Albizia lebbeck* in association with field crops of castor and pigeon pea. The average yield of castor in marginal lands of ravine region was obtained to be 600–1000 kg per ha, and yield of pigeon pea was 700–1000 kg per ha. The average reduction in castor production in association with trees varies from 4% in association with *Acacia nilotica* to 5.4% with *Albizia lebbeck.* In 2014, the price of pigeon pea was ₹43,500 per Mg, and market price of castor was ₹42,000 per Mg. The returns from fuelwood production by trees vary from ₹25,260 in *Acacia nilotica* to ₹30,240 per ha in *Albizia lebbeck* in a rotation of 12 years. Production of fodder at 3–5 Mg $ha^{-1}$ $year^{-1}$, bio-insecticides, timber on tree maturity if trees are not harvested early, yield of by-products like tannins, etc. and soil amelioration due to addition of soil nitrogen are added advantages.

**Case Study 2**

Singh and Pande (2010) reported the results of a study on drumstick (*Moringa oleifera*)-based plantation system developed for marginal lands in ravine regions of Gujarat. They reported the net returns of ₹31,079 $ha^{-1}$ (at 2009–2010 prices) from drumstick + green gram-fennel cropping system. In this system seedlings of drumstick (*Moringa oleifera*) *cv* PKM- 1, raised from seeds in the nursery, were planted at plant-to-plant distance of 4 m and row-to-row distance of 16 m. Green gram (*Phaseolus radiatus*) was grown during *kharif* season as a rain-fed crop in an alley cropping system in between two rows of drumstick (Fig. 15.5).

Fennel (*Foeniculum vulgare*) was grown in alley cropping system in between two rows of drumstick after harvesting of green gram. Drumstick is a rain-fed horticultural crop and does not require additional irrigation water for normal growth and development. Green gram, a rain-fed cover crop grown in *kharif*, reduces soil

**Fig. 15.5** Green gram with *Moringa oleifera*

erosion. Growing of green gram-fennel would therefore save about 70% water which may be made available for other crops. Drumstick plant is a highly valuable plant with exceptional nutritional properties. The leaves are highly nutritious and a significant source of beta-carotene, vitamin C, protein, iron and potassium. The fresh leaves are cooked and used like spinach, and the dried ones are crushed into a powder and used in soups and sauces. It has been used successfully to combat malnutrition among infants and women of childbearing age. The leaves, roots, seed, bark, fruit, flowers and immature pods act as cardiac and circulatory stimulants and possess antitumor, antipyretic, antiepileptic, anti-inflammatory, antiulcer, antispasmodic, antihypertensive, antioxidant, antidiabetic, antibacterial and antifungal activities. Pruned leaves are nutritious and used as fodder. Air-dried stems and branches are used as fuelwood.

### 15.5.2 Agroforestry Systems for Medium and Deep Ravine Lands

The medium and deep ravines are unsuitable for tilling for agricultural crop cultivation, and most of the times, these cannot be reclaimed economically by terracing and/or levelling. Even if these medium and deep ravines are levelled with substantial cost, the hydrological stability of reclaimed land remains questionable. Thus, the most suitable scientific land use for these medium and deep ravine lands is placing these under permanent vegetation where periodic tilling of land is not required. Silvopastoral land use, involving cultivation of multipurpose trees with interplanting of palatable grasses, is an important production system for economic utilization of

**Fig. 15.6** Neem (*Azadirachta indica*)-*Dichanthium annulatum* grass-based silvopastoral system in agricultural fields adjacent to Chambal ravines

medium and deep ravines (Fig. 15.6). It aims at augmenting production of fuelwood while utilizing the interspaces of planted trees for production of fodder grasses.

Fuelwood tree species such as *Acacia nilotica, Acacia leucophloea, Ailanthus excelsa, Bauhinia variegata, Leucaena leucocephala, Azadirachta indica, Albizia lebbeck, Soymida febrifuga, Salvadora oleoides, Tecomella undulata, Pongamia pinnata, Melia azedarach, Erythrina indica, Holoptelea integrifolia* and *Dalbergia sissoo* are suitable for use in deep ravines. Planting of locally available promising grasses like *Dichanthium annulatum, Cenchrus ciliaris, Cenchrus setigerus, Sehima nervosum* (rat's tail grass), *Panicum maximum* (Guinea grass) and *Chloris gayana* (Rhodes grass) can successfully be cultivated in interspaces of the multipurpose trees.

#### 15.5.2.1 Case Studies

Sethy et al. (2011) evaluated the production potential of goose berry based hortipastoral system of ravine lands for sustainable production at IISWC Research Centre, Kota, from 2006 to 2011 under different runoff management regimes. Soils of the study site were very deep clay loam in texture, were low in organic carbon and had available nitrogen and phosphorus but medium to high in potash with a slightly alkaline pH. The climate is semiarid tropical with an average annual rainfall of 789 mm, mostly received during July to September, while almost 8–9 months are dry. Four ravinous micro-watersheds ($RW_1$ = 0.4 ha, $RW_2$ = 1.4 ha, $RW_3$ = 1.1 ha and $RW_4$ = 1.0 ha) were treated with four different trenching densities (i.e. 0%, 25%, 50% and 75% runoff trapping potentials), i.e. no trenches, 139 trenches $ha^{-1}$, 278 trenches $ha^{-1}$and 417 trenches $ha^{-1}$with uniform size of 3.0 mm × 0.6 × 0.45 m. A uniform land-use management system applied in all the micro-watersheds included fruit tree gooseberry (*Emblica officinalis*) on ravine humps, bamboo on gully beds and interspaces with *Cenchrus ciliaris* grass. The numbers of plants were kept in 278 $ha^{-1}$ with spacing of grass slips at 0.5 × 0.5 m. After 6 years of planting in ravines, the total productivity of ravinous lands was observed as 3.90, 4.52, 6.31 and 8.53 Mg $ha^{-1}$ $year^{-1}$, fruit equivalent under $RW_1$, $RW_2$, $RW_3$ and $RW_4$, respectively. While the production of fruit ranged from 1.31 Mg $ha^{-1}$ in control (no trenches) to 6.61 Mg $ha^{-1}$ in treatment involving trenching for trapping 75% runoff, the production of grass ranged from 7.71 to 9.91 Mg $ha^{-1}$. The average number of harvestable clumps in bamboo ranged from 8 to 16 under different treatments. The cost and return economics indicated that treatments $RW_4$, $RW_3$, $RW_2$ and $RW_1$ started yielding positive return at the end of third, fourth, fifth and sixth years, respectively. The trenching at 75% runoff trapping potential density proved superior to others in terms of moisture conservation (87.3%), reduction in runoff (78.23%) and soil loss (90%) and improved production potential (218.7%) of ravinous lands under medium and deep ravinous lands. Hence 75% runoff trapping potential trenching density is recommended for the region.

### 15.5.3 Evaluation of Underutilized Fruit Species for Suitability in Chambal Ravines

In another study carried out from 2006 to 2014 for evaluation of underutilized fruit species for suitability in Chambal ravines, four underutilized tree species, namely, *Aegle marmelos* (cv. N.B. 5), *Cordia myxa* (cv. *Puskar*), *Annona squamosa* (cv. *Mammoth*) and *Carissa carandas* (cv. *American Red*), were evaluated under a horti-pastoral system (Meena et al. 2015). The study was initiated with the aim of generating information on suitability and production potential of underutilized fruit species in ravinous degraded lands as well as the efficacy of half-moon- shaped micro-catchment shaping treatment with and without interplanting of grasses. The experiment was conducted with four fruit species and four interspaces management, viz. clean tilled, clean tilled with micro-catchment shaping, micro-catchment shaping with *Cenchrus ciliaris* and micro-catchment shaping with *Dichanthium annulatum*. The fruit species were planted with Bael, *Aegle marmelos* (8 × 8 m); Lasoda, *Cordia myxa* (6 × 6 m); custard apple, *Annona squamosa* (4 × 4 m) and Karonda, *Carissa carandas* (2 × 2 m). Both the grasses *Cenchrus ciliaris* and *Dichanthium annulatum* were planted with 30 × 30 cm spacing in interspaces of trees as per treatments. The treatments were imposed on the hump top of the ravine area. *Carissa carandas* recorded 100% survival followed by *Cordia myxa* (97%), *Aegle marmelos* (96%) and *Annona squamosa* (46%) after 2 years of planting. The maximum height in all four species was recorded under clean tilled with half-moon shaping treatment followed by clean tilled, micro-catchment shaping with *D. annulatum* and micro-catchment shaping with *C. ciliaris* grass, respectively. In the initial fruiting stage, the highest yield was recorded in Karonda (3.3 Mg $ha^{-1}$) because of early fruiting and highest number of plants accommodated in per unit area in all treatments followed by Bael (3.0 Mg $ha^{-1}$), Lasoda (1.5 Mg $ha^{-1}$) and custard apple (0.7 Mg $ha^{-1}$), respectively. After 8 years of planting, the average grass cover of *Cenchrus ciliaris* was 78%, while *Dichanthium annulatum* recorded 71% average grass cover in their respective treatments. The cover of both grasses was lowest (58% in *Dichanthium annulatum* and 65% in *Cenchrus ciliaris*) when planted with *Lasoda* due to its spreading type of growth behaviour, and it was highest (75% in *Dichanthium annulatum* and 86% in *Cenchrus ciliaris*) with custard apple. *Cenchrus ciliaris* provided higher average annual grass yield (8.31 Mg $ha^{-1}$dry wt.) than *D. annulatum* (8.02 Mg $ha^{-1}$dry weight) for a 3-year period. Thus, *Carissa carandas*-, *Cordia myxa*- and *Aegle marmelos*-based pastoral horti-pastoral systems are considered successful in Chambal ravines.

### 15.5.4 Fuel and Fodder Production Under Silvopastural System in Yamuna Ravines at Agra

Prajapati et al. (1993) in a 15-year study at research centre in Agra assessed the fuel and fodder production potential of *Acacia nilotica* and *Acacia tortilis* in Yamuna ravines. Soils of the study site were sandy loam to loamy sand and alluvial in nature with moderate permeability. The quantum of '*kankar*' inter-calcination was higher on ravine top than on ravine slopes and beds. The climate is a dry extreme type with an average annual rainfall of 755.3 mm, mostly received during July to September, while almost 8–9 months are dry. The minimum temperature during January falls to 0°C, and the highest temperature during May shoots to 48°C. The recurrence interval of floods and droughts is 10 and 5 years, respectively. The plantations of *Acacia nilotica* and *Acacia tortilis* were raised by planting seedlings at 3 × 3 m, 4 × 4 m and 5 × 5 m spacing. The plot size was 0.3 ha, covering top, slopes and bottom of ravines in 6.7 ha catchment which was mechanically stabilized with 20 gully plugs. The interspaces were sown with *Cenchrus ciliaris* grass in furrows 1 m apart. The spacing did not affect the growth of trees, and it was observed that even 3 × 3 m spacing appeared wide for a rotation of 15 years. Topographic location was significant for growth of both species with highest growth in both species on ravine bed followed by ravine sides and ravine top. After 15 years of planting, the fuelwood yield was found to be maximum (28.7 Mg $ha^{-1}$) in *Acacia tortilis* when raised at 3 × 3 m spacing. The fuelwood yield ranged between 7.6 Mg $ha^{-1}$ at ravine top with 5 × 5 m spacing and 34.4 mg $ha^{-1}$at ravine bottom with 3 × 3 m spacing in *Acacia nilotica*, while the corresponding yields of *Acacia tortilis* were 11.5 Mg $ha^{-1}$ at ravine top with 5 × 5 m spacing to 30.2 Mg $ha^{-1}$ at ravine bottom with 3 × 3 m spacing. The mean annual pasturage yield ranged from 1.52 Mg $ha^{-1}$ $year^{-1}$ at 3 × 3 m spacing to 1.8 Mg $ha^{-1}$ $year^{-1}$ at 5 × 5 m spacing under *Acacia nilotica*, while it was 1.8–2.06 Mg $ha^{-1}$ $year^{-1}$, respectively under *Acacia tortilis* at similar spacing. The top feed production in *Acacia nilotica* ranged between 3.8 Mg $ha^{-1}$ with 3 × 3 m spacing and 5.2 Mg $ha^{-1}$ with 5 × 5 m spacing at 14 years of age, while it was 3.8–3.1 Mg $ha^{-1}$, respectively, in *Acacia tortilis*.

## 15.6 Selection of Land Use and Species for Ravine Lands

Several factors and criteria need to be carefully considered before deciding the appropriate land use and suitable tree species for use in ravine region (Table 15.3). Important factors include proximity of land to gullies, state of gully stabilization,

**Table 15.3** Criteria for selection of land use and species in ravines

| Topographic location | Soil type | Resource availability | Recommended land use | Tree/shrub species[a] | Soil working |
|---|---|---|---|---|---|
| Marginal lands along ravines | Heavy | Supplemental irrigation available | Agri-horticulture | *Emblica, Psidium, Citrus, Aegle, Punica, Carissa, Cordia, Feronia* | Pits 1 × 1 × 1 m |
| Marginal lands along ravines and reclaimed shallow ravines | Heavy | Supplemental irrigation available for crops | Agroforestry boundary plantation | *Acacia, Leucaena, Prosopis cineraria, Azadirachta, Albizia, Balanites, Jatropha* | Pits 0.5 × 0.5 × 0.5 m |
| Marginal lands along ravines | Light | Supplemental irrigation available | Agri-horticulture | *Emblica, Psidium, Citrus, Aegle, Punica, Carissa, Cordia, Ziziphus* | Pits 1 × 1 × 1 m |
| Marginal lands along ravines and reclaimed shallow ravines | Light | Supplemental irrigation available for crops | Agroforestry boundary plantation | *Dalbergia, Acacia, Prosopis, Leucaena, Melia, Bombax, Ailanthus, Azadirachta, Salvadora, Tecomella* | Pits 0.5 × 0.5 × 0.5 m |
| Medium and deep ravines, ravine top gently sloping | Heavy | Rain-fed life- saving irrigation available | Horti-pasture | *Emblica, Psidium, Citrus, Aegle, Punica, Carissa* + grasses *Dichanthium, Cenchrus, Panicum* | Pits 1 × 1 × 1 m; staggered trenches 0.5 × 0.5 × 2 m |

(continued)

**Table 15.3** (continued)

| Topographic location | Soil type | Resource availability | Recommended land use | Tree/shrub species[a] | Soil working |
|---|---|---|---|---|---|
| Medium and deep ravines, ravine top gently sloping | Light | Rain-fed life-saving irrigation available | Horti-pasture | *Emblica, Psidium, Citrus, Aegle, Punica, Carissa, Ziziphus, Cordia*+ Grasses: *Dichanthium, Cenchrus, Brachiaria, Stylosanthes* (legume) | --do-- |
| Medium and deep ravines slopes <33% | Heavy | Rain fed | Silviculture/ grass production | *Azadirachta, Acacia, Jatropha* + grasses | --do-- |
| Medium and deep ravines slopes <33% | Light | Rain fed | Silviculture/ grass production | ….do--- | --do-- |
| Ravine slopes >33% | Heavy or light | Rain fed | Grass production | Grasses | Continuous contour furrows 10 cm deep |
| Medium and deep ravines beds | Heavy | Rain fed | Silvopasture/ bamboo cultivation | Bamboo, *A. nilotica, P. cineraria, Leucaena* + grasses | Pits 0.5 × 0.5 × 0.5 m |
| Medium and deep ravines beds | Light | Rain fed | Silvopasture/ bamboo cultivation | Bamboo, *Dalbergia*, Bamboo, *A.nilotica, P. cineraria, Leucaena, Neem, Melia, Bombax, Ailanthus*, Neem + grasses | --do-- |
| Waterlogged ravine beds | – | – | Fuel/fodder production | *Tamarix dioica, Pongamia* pinnata + *Brachiaria* grass para grass | --do-- |

[a]For full names of species, see the text.

topography of the land, texture of soil and availability of resources especially the irrigation facilities. Another important factor which needs careful consideration is extent of biotic interference prevalent at the proposed site.

The proximity to gullies and the size of gullies and catchment of gullies influence the vulnerability of the land to degradation due to water erosion. The state of gully stabilization affects the choice of species for productive and protective roles in ravines. Stable gullies with low vulnerability can easily be placed under trees, shrubs and grasses with high economic value which may need periodic inter-culture. In case of areas adjacent to active gullies, the species which quickly establish with low input and low inter-culture and have low resource requirement need to be selected as these areas are more prone to accelerated erosion if exposed to regular and periodic inter-culture. The topography of the land concerned affects the choice of land use and species. Ravine humps which are exposed to sun from all sides usually face high moisture stress and desiccation and need to be placed under xeric species.

Ravine beds may have accumulated silt deposited due to runoff water flowing from upper catchment, resulting in better availability of nutrients and soil moisture and better establishment and growth of mesophytic species. Thus, these can be placed under tree, shrub and grass species of higher economic value. Though ravine lands are usually located in the proximity of some river systems, these are usually owned by resource-poor people. If farmers can afford to develop facilities for life-saving supplemental irrigation, options for agri-horticulture and agriculture with boundary plantation of valuable multipurpose trees, which may need periodic irrigation during a dry spell, may be considered for plantations, particularly fruit trees of dry regions.

## 15.7 Conclusion

Ravines are too unfriendly for cultivation of high-value crops, but these are gaining importance with progressively declining per capita land availability in the country. Agroforestry technologies provide ample opportunities for economic utilization of the nonarable ravine lands along with marginal and reclaimed ravine lands while arresting their further degradation. Moisture conservation techniques like staggered trenches improve survival and growth of planted vegetation in the ravines. After protection from grazing, silvopastoral or fruit-based agroforestry systems involving species of dry areas are the most appropriate options for rehabilitation of ravine lands.

## References

Balaji B, Nitant HC (2002) Phyto-sociological studies of Yamuna ravines. Range Manag Agroforest 23(2):110–114

Balaji B, Nitant HC, Prakash OM (2002) Aonla bass agroforestry system for moisture conservation and soil productivity in degraded ravine lands

Bhushan LS, Saxena SC (1984) Rainfall erosion index for Agra. Indian J Soil Conserv 12(2 & 3):24–29

Birthal PS, Jha AK (2005) Economic losses due to various constraints in dairy production in India. Indian J Animal Sci 75:1476–1480

Champion HG, Seth SK (1968) A revised survey of forest types of India. Controller of Publication, New Delhi

Dagar JC, Mall LP (1980) Studies on the vegetation of ravines and banks of the river Kshipra and its tributaries. J Bot Soc 59:234–245

Delgado C, Rosegrant MW, Meijer S (2001) Livestock to 2020: the revolution continues. Paper presented at the Annual Meeting of the International Trade Research Consortium (IARTC), Auckland, New Zealand, 18–19 January. Available at http://www.ilri.org/ILRI_Dev/miscpdf/delgado.pdf

Dhruvnarayan VV (1993) Soil and water conservation research in India. Indian Council of Agricultural Research, New Delhi, p 449

GoI (1976) Report of the National Commission on Agriculture. Part 3, demand and supply. Controller of Publications, New Delhi

GoI (2009) Year-wise area under crops—all India. Available at: http://dacnet.nic.in/eands/LUS-2006-07/Summary/tb3.13.pdf

GOI (2016) Press Information Bureau, Government of India, Ministry of Finance, 26-February-2016 12:07 IST (http://pib.nic.in/newsite/printrelease.aspx?relid=136849)

ICAR (2010) Degraded and wastelands of India status and spatial distribution. http://www.icar.org.in/files/Degraded-and-Wastelands.pdf

Kaul ON (1962) Forage yield from Chambal ravines. Indian For 88(11):832–836

Khan MA (1972) Ravines and their afforestation along Chambal river in Madhya Pradesh. J Soil Conserv India 20 & 21:8–13

Kumar A, Srivastava AK (1998) Intercropping in mosambi and kinnow mandarin. Annual Report (Research Centre). Central Soil and Water Conservation Research and training Institute, Research Centre, Agra, pp 40–41

Meena HR, Parandiyal AK, Kumar A, Meena GL (2015) Evaluation of different underutilized fruit species with varying interspaces management in Chambal ravines. Annual report. IISWC, Dehradun, pp 44–45

Nddb (2016) (http://www.nddb.org/information/stats/milkprodindia)

Nema JP, Kamannavar HR, Shah KA (1982) Studies on extension of ravines – Vasad. Annual report. CSWCRTI, Dehradun

NHB (2015) Indian horticulture database – 2014 National Horticulture Board, Ministry of Agriculture, Government of India, www.nhb.gov.in

NSSO (2010) Household consumer expenditure in India, 2007–2008, National Sample Survey Organization (NSSO) 64th Round, Ministry of Statistics and Programme Implementation Government of India

Parandiyal AK, Samara JS, Singh KD, Ratan S, Rathore BL (2000) Floristic diversity of Chambal ravines under varying levels of protection. Indian J Soil Conserv 28(2):160–166

Parandiyal AK, Ashok K, Arjun P (2005) Evaluation of different field crops under rainfed agri-horticultural system in South-Eastern Rajasthan. Paper presented in National Seminar on Commercialization of Horticulture in Non-Traditional Areas held at Central Institute for Arid Horticulture, Bikaner on February 05–06, 2005

Parandiyal AK, Ashok K, Arjun P, Singh KD, Prasad SN (2008) Effect of direction of tree planting on yield of rainfed crops under boundary plantation in medium black soils of south eastern Rajasthan. Indian J Agric Sci 78(7):618–621

Prajapati MC (1995) Floristic wealth of Yamuna ravines. Indian J Soil Conserv 23(1):4–61

Prajapati MC, Agarwal MC, Bhaskar KS (1977) Rainfall features and agricultural droughts at Agra. Ann Arid Zone 16(2):176–184

Prajapati MC, Joshi P, Rathore BL, Dubey LN (1982) Surface water management for grass and tree land development in ravinous watersheds – a case study. Indian J Soil Conserv 10(2 & 3):30–38

Prajapati MC, Nambiar KTN, Puri DN, Singh JP, Malhotra BM (1993) Fuel and fodder production inYamuna ravines at Agra. Indian Journal of Soil Conservation. 21(3):8–13

Prakash OM, Kumar D (2001) Tree root management for their productivity in Yamuna ravinous lands under semiarid conditions. Paper presented in National Conference on Resource Conservation and Watershed Management, May 23–25, at CSWCRTI, Dehradun
Prakash C, Rao DH (1986) Frequency analysis of rainfall data for crop planning – Kota. Indian J Soil Conserv 14(2):23–26
Prakash C, Rao DH, Kurothe RS (1987) Ravine extension in twenty years – a case study, Abst: National Symposium on Land and Water Management in ravines. CSWCRTI, Research Centre, Agra. March 19–22, 1987:6
Prakash C Rao DH, Kurothe RS (1990) Encroachment of ravines in table lands along Chambal river in semiarid region of Kota (Rajasthan) Post Symposium. In: Proceedings of international symposium on "Water erosion, Sedimentation and Resource conservation". Dehradun, Oct.09–13, 1990:47–50
Ramachandra KS, Taneja VK, Sampath KT, Anandan S, Angadi UB (2007) Livestock feed resources in different agro-ecosystems of India: availability, requirement and their management. National Institute of Animal Nutrition and Physiology, Bangalore
Rao P, Birthal (2008) Livestock in mixed farming systems in South Asia. International Crop Research Institute for Semiarid Tropics, Patancheru
Reddy SB (2003) Overcoming the energy efficiency gap in India's residential sector. Energy Policy 31(11):1117
Roy K, Mishra P (1969) Formation of Chambal ravines. Indian For 95(3):160–164
Sajwan SS (1975) Ground flora and their distribution in forest plantation in Jamuna ravines. Indian For 101(5):269–275
Sajwan SS (1976) Dominant grasses and their associates in ravinous tracts of Jamuna. Soil Conserv Digest 4(1):26–35
Sethy et al. (2011) evaluated the production potential of goose berry based hortipastoral system of ravine lands for sustainable production at IISWC Research Centre, Kota, from 2006 to 2011 under different runoff management regimes.
Sharda VN, Kurthe RS (2012) Introduction-bamboo cultivation in ravines. In: Kurothe RS, Gaur ML, Krishna Rao B, Parandiyal, A.K. , Singh AK (eds) Conservation and production potential of bamboo in ravine lands. Central Soil & Water Conservation Research & Training Institute, Dehradun
Sharma HS (1980) Ravine erosion in India. Concept Publishing Co., New Delhi, pp 1–96
Singh P, Mujumdar AB (1992) Current status of feed and forage management of livestock in India. Agric Situat India 47(5):375–382
Singh HB, Pande VC (2010) Drumstick based agri-horti system for replacing tobacco in central Gujarat, Ext. bulletin pub. Central Soil & Water Conservation Research & Training Institute (ICAR), Research Centre, Vasad
Singh G, Mishra PR, Bullian S (1966) Ravine ravage must be controlled. Indian Farming XVI(3):21–22
Singh A, Shah CM, Dayal R (1972) Point rainfall analysis of Soil Conservation Research Centre, Vasad (Gujrat). Indian For 98(9):514–519
Singh A, Shah CM, Kamannavar HK (1976) Rainfall erosivity analysis of Soil Conservation Research Centre, Vasad (Gujrat). Indian For 102(2):126–132
Tejwani KG (1968) Classification of ravine lands. Proc. Symp. on reclamation and use of wastelands in India. New Delhi, 161p
Tiwari AK, Bhushan LS, Raghuvir (1987) Rate of extension of ravines in Agra region, Abst: National Symposium on Land and Water Management in ravines. CSWCRTI, Research Centre, Agra. March 19–22, 1987:5

# Silvopastoral System: A Key Component for Rehabilitation of Ravine Lands

16

P. K. Ghosh and S. K. Mahanta

**Abstract**

The land degradation due to ravines and gullies is quite extensive in India, largely occurring in states of Uttar Pradesh, Madhya Pradesh, Bihar, Gujarat and Rajasthan. The soils of ravine lands are usually poor in quality and physical conditions. These are also not good for cultivation being stony, sandy or firm clay with lower ability to retain moisture. The climate, topography and soils of the ravine regions are characteristically ruthless and thus badly affect the composition and growth of different kinds of plant species. The ravines are characterized by the presence of higher proportion of thorny vegetation. Rehabilitation of such degraded lands requires an appropriate approach based on science and technology. It must include survey, identification and selection of appropriate plant species and also techniques of their establishments in the field. Indeed, it requires an integrated approach for rehabilitation of gullies based on quality of land, adoptable soil and water conservation measures and development of permanent vegetation cover involving trees and pastures. Grasses, shrubs and trees which are fast growing, can easily be established and have fibrous roots are suitable for ravine lands. Tree species such as *Acacia nilotica*, *A. eburnea*, *A. catechu*, *Butea monosperma* and *Dalbergia sissoo* are preferred for gully slopes/humps, while *Dalbergia sissoo*, *Tectona grandis*, *Dendrocalamus strictus*, etc. are planted in the ravine beds. Studies on forage production under silvopastoral system indicated that it is possible to get 3.5–4.0 Mg $ha^{-1}$ of dry forage annually without having an adverse effect on growth of tree species, besides positive impact on runoff and soil loss. Hence, degraded ravine lands need to be put under productive use through silvopastoral systems for economic benefits as well as environmental protection.

P. K. Ghosh (✉) · S. K. Mahanta
ICAR-Indian Grassland and Fodder Research Institute, Jhansi, India





J. C. Dagar, A. K. Singh (eds.), *Ravine Lands: Greening for Livelihood and Environmental Security*, https://doi.org/10.1007/978-981-10-8043-2_16

## 16.1 Introduction

Ravines are described as channels of short-lived flow, denuded and guided basically by the process of rejuvenated streams, and have steep sides and head scarps with a width and depth always greater than a gully. The usual depth of different types of gullies, i.e. small, medium and large, varies from <1 m, 1–5 m and 5–10 m, respectively, while the depth of ravines is more than 10 m. Ravine channels are widespread in the alluvial deposits, and they are the most prominent topographic features along majority of the rivers in India (Hajgh 1984). They may spread up to 2–3 km surrounding the river resulting in land degradation and keep vast areas unsuitable for cultivation. In terms of size, ravines are usually smaller than valleys but larger than gullies. Literally the 'ravine' indicates a deep gorge and kind of water erosion which has reached to its final stage. Ravines are the systems of gullies, which operate usually parallel to each other for a short distance and then drain into a river with the formation of deep gorges. These are developed when the upper layer of vegetal cover becomes weak and the roots are unable to hold and bring together the soil. Continuous raining erodes the upper layer of soil and removes the crust of earth; the water flow then turns into drains and creates cracks. With the passage of time, these cracks are further eroded and converted into large ravines. Thus, stream bank erosion is a major cause of land degradation, leading to deteriorated drainage systems, which ultimately govern natural calamities in terms of floods, and nonpoint source pollution in ravine lands. However, in general ravine lands represent the kind of lands which are degraded and unsuitable for cultivation and remain unused. Efforts have been made to utilize these lands in the recent past to produce food, fodder, timber and fuels and also for conservation of natural resources. Development of such degraded lands requires an appropriate approach based on science and technology and should include survey, selection of species and establishment of plant species in the fields (Chaturvedi et al. 2014). In fact, it requires an integrated approach of using gullies based on quality of land, soil and water conservation measures and development of permanent vegetation cover on the land involving trees and pastures.

Ravine lands are found in many areas/countries and have many names. They are also reported to be present in Zimbabwe, Central Russia, Madagascar and different parts of south-west America where they are linked with the name of *arroyo*. Gully erosion due to river channel trenching is a big problem throughout the world, and it threatens the vast tract of agricultural land. The damage is maximum in the alluvial plains of semiarid and arid zones, although it is a more serious problem in developing world where it is threatening precarious subsistence-oriented agricultural systems. In India, the ravines are found along the rivers of Yamuna, Gomati, Ganga and their tributaries in Uttar Pradesh; Chambal, Assan, Kalishindh, Kshipra and their tributaries in Madhya Pradesh; Chambal, Banganga, Sabi and their tributaries in Rajasthan; and Mahisagar, Narmada, Sabarmati, Watrak, Saraswati, Dhadhar and others in Gujarat (Tejwani et al. 1975). Besides these, substantial ravine erosion is also observed in Chota Nagpur, Bihar, along river Mahanadi and Upper Son Valley of Mirzapur, all on the southern fringes of the Gangetic plain and the Siwalik and Bhabar tract of the western Himalayan foothills (Sharma 1980). Thus, land degradation due to ravines and gullies is extensive. The National Commission on Agriculture

(1976) estimated that 3.67 million ha is under ravines. But the National Remote Sensing Agency, based on the Landsat data, has estimated recently around 4 m ha of ravine lands. The largest occurrence was observed in states like Uttar Pradesh, Madhya Pradesh, Bihar and parts of Rajasthan.

## 16.2 Ravine Vegetation

The land features of ravines are uneven and bumpy and categorized by three different parts, namely, ravine top, being irregular in size and shape, slope and bottom/bed. The deepness of ravines usually ranges from 1 to 20 m, but sometimes it can extend up to 55 m. In India, 'V'-shaped ravines are observed along Chambal river where the subsoil is relatively resistant to erosion than the topsoil. Ravines of 'U' shaped are observed in Yamuna and Mahi rivers, where the soils are light textured and depth of alluvium increases with the passage of time (Dagar and Mall 1980; Dagar 1995). The soils of ravine lands are poor in quality, and physical conditions are also not good since the soils are being eroded continuously keeping them rocky, sandy or stiff clay with poor ability to absorb or retain water (Dagar and Mall 1980). The rate of soil formation is also very slow since there were little interactions due to non-availability of water and vegetation roots. These soils are susceptible to water logging when there is adequate moisture and to compaction when there is scarcity of moisture. The water infiltration rates in these ravine soils are low to moderate, and such soils are prone to erosion, particularly the upper layers.

In general, the climate, topography and soils of the ravine region are typically ruthless and thus harshly affect the composition and growth of the vegetation. The ravines are characterized by the presence of higher proportions of thorny vegetations. Indeed, the natural vegetation of ravines has been branded as northern tropical ravine thorn forests (Champion and Seth 1968). The typical ravine vegetation is consisted of low-growing type of coarse grasses like *Aristida adscensionis*, *Eremopogon foveolatus*, *Heteropogon contortus*, *Themeda triandra*, *Eragrostis* spp., *Eleusine* spp., *Dichanthium annulatum*, etc., and among grasses, bushy shrubs are predominant, which include *Ziziphus nummularia.*, *Capparis zeylanica*, *C. decidua*, *Crotalaria burhia*, *Cassia tora*, *Tephrosia purpurea*, *Xanthium strumarium*, *Calotropis procera*, *Clerodendrum phlomidis*, *Grewia pilosa*, *G. tenax*, *Lantana camara*, etc., and stress-resistant trees like *Dichrostachys cinerea*, *Acacia senegal*, *A. leucophloea*, *A. nilotica*, *Prosopis juliflora*, *Balanites aegyptiaca*, etc. are common (Dagar and Mall 1980; Prajapati 1995; Parandiyal et al. 2000). Grasses which are hardy like *Sehima nervosum*, *Heteropogon contortus*, *Dichanthium annulatum*, *Chrysopogon fulvus* and *Cynodon dactylon* and legumes like *Alysicarpus rugosus* and *Rhynchosia minima* can withstand biotic pressure, and they are usually found in ravine vegetation. But grasses like *Themeda triandra*, *Iseilema laxum*, *I. prostratum* and *Apluda mutica* and legumes like *Indigofera glandulosa*, *Rhynchosia capitata* and *Alysicarpus longifolius* are found when animal grazing is withdrawn and the area is protected with fencing (Dagar and Mall 1980; Dagar and Singh 1980; Dagar 2001).

The predominant vegetation of Yamuna ravines is reported to be comprised of vilayati babul (*Prosopis juliflora*), deshi babul (*Acacia nilotica*), thorny legumes,

Euphorbias and different dwarf grass species. Sometimes, shrub/tree species like neem (*Azadirachta indica*), papdi (*Holoptelea integrifolia*), shisham (*Dalbergia sissoo*), karanj (*Pongamia glabra*) and chonkra (*Prosopis cineraria*) were also observed (Singh et al. 2015). For detailed vegetation of ravines, see Chap. 3 in this book.

## 16.3 Silvopastoral Systems in Ravines

When gully reclaimability ratings (Table 16.1) are considered, gullies of classes 5 and 6 should be put under perennial vegetation of trees and grasses producing fuel and forages for animals (Verma and Bhushan 1986). In fact, except the marginal and shallow gullied lands, which can only be rehabilitated cost-effectively and used for agricultural crop production, majority of the ravine lands are not fit for agricultural purposes and are categorized as nonarable lands. Moreover, agricultural crop production in such lands is also not advisable as it will lead to faster erosion causing further land degradation.

The most appropriate and logical land use for such uncultivable lands is to put them under permanent vegetation of trees and forages, although vegetation in ravine lands suffers from different kinds of unfavourable conditions like nutrient deficiency, moisture stress and biotic interference. The adverse weather conditions coupled with very high summer temperature again exaggerate the problem and make crop farming uneconomical. In such situations, technologies having less

**Table 16.1** Gully reclaim ability ratings and classifications (Chaturvedi et al. 2014)

| Rating | Remarks |
|---|---|
| Class 1 | Shallow gullies need only little levelling work for reclamation for crop farming. They have favourable quality soil for crop cultivation. The beds are spacious enough to convert them into agricultural fields but require minor constructing activities |
| Class 2 | Gullies under this class need moderate levelling work for reclamation towards crop farming. These gullies are 1.5–3.0 m deep, and sides are more sloping than what are being observed in gullies under Class 1. Maintenance of such reclaimed fields requires comparatively more care and little expensive |
| Class 3 | It has more limitations than Class 2. Soil qualities are lighter or heavier than loam soils, and the gullies are deeper, creating more problems for reclamation and then maintenance of reclaimed gullies for crop cultivations |
| Class 4 | They have severe limitations for reclamation for crop farming when considered in terms of soil quality or gully sizes. Steeper slopes of the sides and more deepness of gully make it a borderline case for reclamation for crop cultivation. Such gullies should be considered for agri-horticultural purposes after making narrow terraces |
| Class 5 | This category includes all the gullies mentioned under Classes 1–4, but they cannot be reclaimed for crop farming as they are susceptible to seasonal backflows from a nearby river or suffer from water logging and salinity problems, due to some other adverse factors. Such gullies should be put under perennial vegetation of silvopastoral systems, comprising of suitable fuel and fodder trees and grasses |
| Class 6 | Gullies with more severe limitations than Class 4, when considered in terms of soil quality and gully features like depth and side slopes, are brought under this category. They should be considered for perennial vegetations only |

demand for moisture and nutrients like silvopastoral systems are quite promising to sustain the productivity as well as farm income and can be exploited as alternative source of farmer's livelihood.

Silvopasture is a very old and traditional practice and consists of integration of woody preferably fodder plants with pastures to take advantage of the soil fertility improvement by trees. The silvopastoral systems are usually defined as growing ideal combination of grasses, legumes and trees for optimizing land productivity, conserving plants, soil and nutrients and producing forage, timber and firewood on a sustainable basis. This includes replantation, substitution or intervention in the existing vegetation by desirable species. Recently, the silvopastoral system is practiced to optimize, rather than maximize, production from the same unit of land through integrated management. In fact, silvopasture is as an agroforestry practice, which is specifically designed and managed for the production of trees, tree products, forage and livestock and most importantly in semiarid and arid regions (Klopfenstein et al. 2008). Animal draft, milk, milk products, meat, wool, hides, skins, etc. provide a wide canvas of livelihood inputs from silvopastoral system. Trees and bushes provide green fodder during dry season when grasses are not available. These systems are more suited to degraded lands like ravines and for areas facing acute shortage of fodder. It has been observed that suitable species for silvopasture for ravine lands vary based on climate and soil types (Table 16.2).

But highly degraded and poor-quality soils along with extremes of temperature and scarcity of moisture hold back the survival and growth of trees/shrubs in ravine lands. Hence, selection of the species should be based on:

- Ecological aptness of tree species
- The compatibility of tree-pasture-soil interactions
- Favourable structural designs of trees
- Money making value of tree species
- Farmer's/stakeholder's perception or preference for the tree species

**Table 16.2** Suitable trees, shrubs, grasses and legumes for ravines

| Planting site/soil type | Trees/shrubs | Grasses/legumes |
|---|---|---|
| Salt-affected ravines | *Leucaena leucocephala*, *Acacia nilotica*, *Dendrocalamus strictus*, *Dalbergia sissoo*, *Albizia lebbeck*, *Prosopis juliflora*, *Terminalia arjuna*, *Azadirachta indica*, *Dichrostachys cinerea*, etc. | *Chrysopogon fulvus*, *Cenchrus ciliaris*, *Cenchrus setigerus*, *Pennisetum pedicellatum*, *Saccharum spontaneum*, *Dichanthium annulatum*, *Phaseolus atropurpureus*, *Stylosanthes* sp., *Panicum antidotale*, *Panicum maximum*, *Pennisetum purpureum*, *Brachiaria mutica*, etc. |
| Ravine side slopes/tops | *Acacia catechu*, *Dalbergia sissoo*, *Dichrostachys cinerea*, *Leucaena leucocephala*, *Acacia nilotica*, *Acacia tortilis*, *Acacia senegal*, *Albizia lebbeck*, *Azadirachta indica*, etc. | |
| Ravine beds | *Pongamia pinnata*, *Tectona grandis*, *Eucalyptus*, *Dendrocalamus strictus*, *Morus alba*, *Broussonetia papyrifera*, *Gmelina arborea*, *Ipomoea carnea*, etc. | |

**Table 16.3** Promising silvopastoral models for degraded/ravine lands under semiarid condition (Roy et al. 2005)

| Tree | Pasture | |
|---|---|---|
| | Grass | Legume |
| *Acacia nilotica* | *Cenchrus ciliaris* | *Stylosanthes hamata, S. scabra* |
| | *Dichanthium annulatum* | *S. hamata* |
| *Acacia tortilis* | *Cenchrus ciliaris, C. setigerus* | *S. hamata, S. scabra* |
| | *Dichanthium annulatum* | *S. hamata* |
| *Azadirachta indica* | *Cenchrus ciliaris* | *S. hamata* |
| | *Dichanthium annulatum* | *S. hamata* |
| *Albizia amara* | *Cenchrus ciliaris, C. setigerus* | *Macroptilium atropurpureum, S. hamata* |
| | *Dichanthium annulatum* | *S. hamata* |
| | *Panicum maximum* | *S. hamata* |
| *Albizia lebbeck* | *Cenchrus ciliaris, C. setigerus* | *Clitoria ternatea, S. hamata* |
| | *Chrysopogon fulvus* | *Clitoria ternatea, S. hamata* |
| | *Sehima nervosum* | *Clitoria ternatea, S. hamata* |
| *Albizia procera* | *Cenchrus ciliaris, C. setigerus* | *Clitoria ternatea, S. hamata* |
| *Dalbergia sissoo* | *Cenchrus ciliaris* | *S. hamata, S. scabra* |
| | *Dichanthium annulatum* | *S. hamata, S. scabra* |
| *Leucaena leucocephala* | *Cenchrus ciliaris, C. setigerus* | *S. hamata, S. scabra* |
| | *Dichanthium annulatum* | *S. hamata, S. scabra* |
| | *Panicum maximum* | *S. hamata* |
| *Ziziphus mauritiana* | *Cenchrus ciliaris* | *S. hamata* |

Grasses, shrubs and trees which are fast growing and easy to establish and have fibrous roots to withstand different stresses associated with ravine lands are appropriate for silvopastoral systems (Table 16.3). It means that the plant species chosen should have the capability to tolerate adverse conditions and offer quick protection against the erosion of soils, while the forage (grasses/legumes) component should be a perennial crop with following characteristics:

- Suitable for animal grazing
- Capability to withstand heavy utilization
- Well-matched with the site (soil, temperature, precipitation)
- Ability to grow under partial shade and moisture stress
- Producing more under intensive management

### 16.3.1 Steps for Developing Silvopastoral System

**Protection from biotic agencies:** First step of developing any system in degraded ravines is to protect the area from biotic interferences such as over-grazing and deforestation activities. Ravine lands need to be protected through fencing. The protection from biotic agencies helps in rapid establishment of grasses and trees, besides promoting better biomass yields from vegetation. Along with fencing, some conservation measures for eroded soils like check dam construction and runoff disposal should also be adopted. By adopting simple measure of protection over 30 years, the gully systems of different rivers like Yamuna, Chambal and Mahi have been reported to be stabilized and converted into naturally growing dense forests of trees, shrubs and grasses. It has also prevented runoff and soil loss and increased infiltration as well as enriched biodiversity including wildlife (Chinnamani 1986). Studies conducted in ravines of Chambal and Yamuna rivers revealed that natural ravines supported rearing of four goats $ha^{-1}$ without any erosion problems when controlled grazing system was followed (Prajapati et al. 1989).

- *Selection of tree/shrub species*: Selection of tree is of crucial significance, and there are many factors and forces (as mentioned above) that govern the choice of species for such activities. In Indian conditions, tree species like *Acacia catechu* and *Dalbergia sissoo* are ideal for gully slopes and humps since they have dormant period during the dry summer months and thus can easily overcome the moisture deficiency in the soils, while *Dalbergia sissoo*, *Pongamia pinnata*, *Tectona grandis*, *Dendrocalamus strictus*, etc. are planted in the beds of a ravine where moisture availability is comparatively more than the top or side slopes of a ravine. While selecting species for cultivation in ravine lands, preference must be given to those local species which are already growing in natural conditions there and have regenerating capacity (Dagar and Mall 1980).
- *Establishment of tree seedlings*: The techniques for establishing trees in silvopastoral systems are similar to those followed in tree plantation program elsewhere, like seed collection, raising of nursery stock, soil working, transplanting, cultural operations, disease and pest control, etc. These operations vary from site to site. Initially seedlings are raised in a nursery, and then they are planted in pits of 45 × 45 × 45 cm size at the top and bottom of a ravine. In ravines with mild slopes, half slanting pits are made for planting, but in case of steep side slopes, contour furrows are made for seeding and raising of plants. However, insecticides should be used in low rainfall areas to prevent termite attack and protection of seedlings. Sometimes mulching of soils with available weeds/shrubs is also beneficial in conservation of moisture.
- *Selection of shade tolerant pasture species*: With the growth of tree canopy, the understory pasture species experience varying degrees of shade conditions. The tolerance of pasture species to shade conditions, in such a situation, becomes important for their growth and thus overall productivity of silvopasture. Various grass and legume species, viz. guinea grass and stylo, have been reported to perform well under shade. Introduction of legumes is also recommended, which, in addition to improve the nutritive value of grasses, act as an excellent ground cover, eliminate weeds and improve soil fertility. For development of pastures, land preparation activities should be initiated at the onset of monsoon. Pelleted grass seeds should be sown in 10 cm deep furrows usually dug along the contours. However, for quick establishment of grasses, rooted slips are planted at a

spacing of 50 × 50 cm. Erecting water spreaders are also required in ravines with mild slopes to ensure better distribution of water to planted grasses (Prajapati et al. 1982). When grasses are planted on ravine top, they usually become better compatible with trees as the availability of light is better compared to those plated at ravine bottom.

### 16.3.2 Forage Production and Soil Conservation

Studies on forage production under silvopastoral systems have shown that it is possible to get substantial quantity of forage and fuelwood from ravine lands. In one study, silvopastoral system was established having *Acacia nilotica* tree with *Cenchrus ciliaris* and *Dichanthium annulatum* grasses in ravines around Agra (Singh and Puri 1975). During the end of 5-year period, *C. ciliaris* had peak yield of 3.93 Mg DM $ha^{-1}$ $year^{-1}$ in 18 × 18 m tree spacing treatment, while *Dichanthium annulatum* had peak yield of 3.55 Mg DM $ha^{-1}$ $year^{-1}$ in 9 × 9 m tree spacing treatment. The subsequent evaluation of the same plantation revealed that there was a decline in percent composition of *Dichanthium annulatum* declined with passage of time due to shade, and the grass establishment was observed only in depressions and gully bottoms which constituted only 10% of ravine area. However, the yield of *Cenchrus ciliaris* increased under tree canopies, preferring closer tree spacing having cool and shade conditions. Planting of *Cenchrus ciliaris* and *C. setigerus* grasses in ravine lands yielded 5–7 Mg $ha^{-1}$ air-dry fodder or 18–20 Mg $ha^{-1}$ of green fodder per annum. Besides it reduced runoff and soil loss significantly to the tune of 6–10 Mg $ha^{-1}$ per annum (Sharda et al. 1982). It also led to improvement in soil quality with addition of organic matter. However, mixed vegetation of trees and grasses was superior in reducing the runoff and soil loss in ravenous watersheds. In another study from a silvopastoral system, dry forage yield of 3.5–4.0 Mg $ha^{-1}$ was obtained annually without having an adverse effect on growth of tree species.

In a 14-year-old silvopastoral system of *Acacia nilotica* and *A. tortalis* trees along with grass *Cenchrus ciliaris* in Yamuna ravines, dry forage biomass of 1.5–2 Mg $ha^{-1}$ and 4.2 Mg $ha^{-1}$ of top feed was obtained annually (Prajapati et al. 1993).

Ravines of Chambal at Kota, when covered under silvopastoral system with treatments like *Acacia* + *Dichanthium annulatum* and *Dichanthium annulatum* alone, generated 5.8% and 2.6% of runoff and 1.26 and 0.62 Mg $ha^{-1}$ of soil loss in respective treatments as compared to 14.7% of runoff and 3 Mg $ha^{-1}$ of soil loss from agricultural catchment. It also yielded 4.5 Mg $ha^{-1}$ of air-dry grass + firewood indicating the role of grasses and trees as an alternative land use for protection and productive utilization of such ravine lands (Sharda and Venkateswarlu 2007).

A 4-year study (2010–2013) conducted by Singh et al. (2015) on bamboo (*Dendrocalamus strictus*) plantation on gully beds of Yamuna ravines also indicated considerable decrease in soil loss and brought changes in moisture and silt retention (Table 16.4). This is also attributed to the establishment of a dense vegetative cover in the gully beds. Bamboo plant roots played an important role by providing protection to soil in gully beds and reduced runoff by increasing infiltration of soil.

**Table 16.4** Runoff and soil loss under bamboo plantation in Yamuna ravines (Singh et al. 2015)

| Year | Seasonal rainfall (mm) | Runoff from planted area (mm) | Runoff (%) | Soil loss (Mg $ha^{-1}$ $year^{-1}$) |
|---|---|---|---|---|
| 2010 | 456 | 44 | 9.65 | 4.27 |
| 2011 | 226 | 6.04 | 2.67 | 0.66 |
| 2012 | 531 | 14.50 | 2.73 | 0.78 |
| 2013 | 494 | 8.96 | 1.81 | 0.60 |

**Table 16.5** Performance of goats as influenced by different models of pastures in Yamuna ravines

| Attributes | *Zizyphus* sp.-based silvopasture | *C. ciliaris* pasture | Natural pasture[a] |
|---|---|---|---|
| Dry matter intake (g)/W $^{0.75}$ | 102.5 ± 13.7 | 93.18 ± 10.5 | 72.17 ± 6.2 |
| Crude protein intake(g)/W $^{0.75}$ | 9.91 ± 3.25 | 9.77 ± 2.92 | 7.25 ± 0.56 |
| Dry matter digestibility (%) | 65.61 ± 2.95 | 58.09 ± 2.97 | 52.71 ± 1.71 |
| Crude protein digestibility (%) | 57.08 ± 5.10 | 58.71 ± 4.88 | 50.87 ± 2.49 |
| Average daily body weight gain (g) | 65.13 ± 5.71 | 29.60 ± 2.60 | 9.21 ± 4.70 |

[a]Comprised of *Saccharum munja*, *Saccharum spontaneum*, *Dichrostachys bipinnata*, *C. ciliaris*, *C. setigerus* and other seasonal forbs

Tripathi et al. (2015) evaluated the performances of yearling Barbari male goats through grazing trials in *Ziziphus-Cenchrus ciliaris* (sown) silvopastoral, *C. ciliaris* sole pasture and natural pasture of Yamuna ravines. The performance of silvopastoral system was better than sown pasture alone and natural pasture after protection (Table 16.5).

The Regional Research Centre of Central Soil and Water Conservation Research and Training Institute in Vasad (Gujarat) has developed a silvopastoral system involving bamboo and anjan (*Cenchrus ciliaris*) grass to improve the productivity of ravine lands (Fig. 16.1). Ravine lands, which remained unproductive earlier, were successfully converted into productive lands by planting bamboo, a fast-growing species, on the gully beds and anjan grass (*Cenchrus ciliaris*) on the side slopes and the interspaces of gully bed for profitable utilization of gullied lands. This technology gave high economic returns and also reduced water erosion and thereby helped in protection of soil.

The results showed that the reclamation of ravine lands improved farm income and helped in the conservation of natural resources and enhancing carbon sequestration in the system. The cost of planting bamboo and grass root slips in ravine was about INR 22,000 per hectare. Planting of grass protected the ravine slopes and reduced soil loss and yielded 7.1 Mg $ha^{-1}$ $year^{-1}$ of green fodder. The grass yield from interspaces of bamboo planted on bed portion was about 10 Mg $ha^{-1}$ $year^{-1}$ during first 5 years generating an income of INR 3000–6000 per ha during 5-year period at Vasad (ICAR 2016). The bamboo planted on the ravine bed could generate up to INR 6000–27,000 $ha^{-1}$ $year^{-1}$. In India, many organizations such as the Anand-based Foundation for Ecological Security, Gujarat State Watershed Management

**Fig. 16.1** Bamboo- and anjan grass-based silvopasture in ravine lands

Agency, Gujarat State Land Development Corporation, Forest and Agricultural Departments and other user agencies have adopted this proven technology for reclamation and stabilizing the degraded ravine lands in different states. It has improved the productive utilization of those ravine lands and contributed substantially to livelihood of different stakeholders and farmers (ICAR 2016).

## 16.4 Conclusion

In developing countries like India with ever increasing human and livestock population, the requirement for forage is increasing continuously. This has put pressure on different types of waste and unproductive lands and warrants putting these wastelands into productive use for sustained and regular availability of forage. Such unproductive lands, in general and ravine wastelands in particular, have the potential to supply the required forage and fuelwood. Hence, ravine wastelands must be put into productive use through silvopastoral systems to attain economic returns as well as environmental protection. Indeed, planting of different kinds of fodder and fuelwood trees and grasses on ravine lands provides a huge scope to mitigate the shortage of fuel and forage for rural households and becomes a supplementary source of regular income to the farmers, besides preserving these lands against further deterioration. In the recent past, efforts have led to development and adoption of better technologies in ravine lands which have considerable impact on production and protection. But in the future, the following issues need to be addressed/considered for better soil health and production from ravine lands:

- Ravines/gullies need to be treated in accordance with gully reclaimability classifications.
- Development and providing the quality planting materials/forage seeds and animal breeds for developing productive silvopastoral systems.
- Exploring wild native fodder trees and shrubs suitable for different types of ravine lands through collections, improvement and domestication.
- Development of site-specific silvopastoral models for optimizing production, carbon sequestration and biodiversity conservation.

- Capacity building of the stakeholders and enhancing transfer of technology.
- Fine-tuning of ravine land rehabilitation technologies already developed, keeping in mind variability in ravine lands.
- Policy initiatives for supporting farmers through suitable cheap loans, insuring their crops and animals against drought, diseases and pests (both in crops and animals), natural calamities and all kind of damages including tree damage, creating market facilities and facilitating carbon credit against the carbon sequestered in the system.

## References

Champion HG, Seth SK (1968) A revised survey of forest types of India. Controller of Publication, New Delhi

Chaturvedi OP, Kaushal R, Tomar JMS, Prandiyal AK, Panwar P (2014) Agroforestry for wasteland rehabilitation: mined, ravine and degraded watershed areas. In: Dagar JC, Singh AK, Arunachalam A (eds) Agroforestry systems in India: livelihood security & ecosystem services, advances in agroforestry. Springer, New Delhi, pp 233–271

Chinnamani S (1986) Forestry, social forestry and agroforestry in ravine reclamation. Lecture delivered in summer institute on wasteland in arid and semi-arid zones and technologies for their improved utilization. ICAR-CAZRI, Jodhpur

Dagar JC (1995) Soil conservation values of some grasses and forbs. Proc Indian Nat Sci Acad B 61:163–170

Dagar JC (2001) Biological spectrum and successional trends in Kshipra watershed areas. Indian J For 24:351–356

Dagar JC, Mall LP (1980) Studies on the vegetation of ravines and banks of the river Kshipra and its tributaries. J Indian Bot Soc 59:234–245

Dagar JC, Singh VP (1980) The ecological studies of the vegetation of the river Kshipra and its tributaries. Biologia 26:43–62

Hajgh MJ (1984) Ravine erosion and reclamation in India. Geoforum 15:543–561

ICAR (2016) Bamboo and anjan grass based silvipasture system enhanced the productivity of ravines. http://www.icar.org.in/en/node/5101

Klopfenstein TJ, Erickson GE, Bremer VR (2008) Use of distillers by-products in the beef cattle feeding industry. J Anim Sci 86:1223–1231

Parandiyal AK, Samara JS, Singh KD, Ratan S, Rathore BL (2000) Floristic diversity of Chambal ravines under varying levels of protection. Indian J Soil Conserv 28:160–166

Prajapati MC (1995) Floristic wealth of Yamuna ravines. Indian J Soil Conserv 23:4–61

Prajapati MC, Joshi P, Rathore BL, Dubey LN (1982) Surface water management for grass and tree land development in ravinous watersheds: a case study. Indian J Soil Conserv 10:30–38

Prajapati MC, Tiwari AK, Nambiar KTN, Singh JP, Malhotra BM, Sharda VN (1989) Effect of goat grazing on vegetation, runoff and soil loss in a stabilized ravine watershed vis-à-vis animal growth. Indian J Soil Conserv 17:9–16

Prajapati MC, Nambiar KTN, Puri DN, Singh JP, Malhotra BM (1993) Fuel and fodder production in Yamuna ravines at Agra. Indian J Soil Conserv 21:8–13

Roy MM, Kushwaha D, Gupta SK, Singh M (2005) Lessons learnt in promoting MPTS on wastelands of farmers in semiarid rainfed regions. Forests, Trees Livelihoods 15:383–388

Sharda VN, Venkateswarlu B (2007) Crop diversification and alternate land use systems in watershed management. In: Proceedings of comprehensive assessment of watershed programs (best-bet options for integrated watershed management) in India, ICRISAT, Patancheru, 25–27 July, 2007

Sharda VN, Bhushan LS, Raghuvir (1982) Hydrological behaviour of ravenous watersheds under different land uses. In: Proceeding international symposium on hydrological aspects of mountainous watersheds, Roorkee, 4–6 November, 1982

Sharma HS (1980) Ravine erosion in India. Concept Publishing Co., New Delhi, pp 1–96

Singh JP, Puri DN (1975) Economic utilization of ravines for fuel-cum-fodder. Annual Report, Central Soil and Water Conservation Research and Training Institute, Dehradun, pp 39–40

Singh AK, Kala S, Dubey SK, Pande VC, Rao BK, Sharma KK, Mahapatra KP (2015) Technology for rehabilitation of Yamuna ravines – cost-effective practices to conserve natural resources through bamboo plantation. Curr Sci 108:1527–1533

Tejwani KG, Gupta SK, Mathur HN (1975) Classification, control and reclamation of ravine lands. In: Soil and water conservation research (1956–1971). Indian Council of Agricultural Research Publication, New Delhi, pp 254–310

Tripathi P, Tripathi MK, Chaudhary UB, Ravindra K (2015) Comparison of enriched silvipasture and Cenchrus ciliaris pasture with natural pasture in ravenous soils for goat production under rainfed conditions (paper ID: 1044), International Grassland Congress, New Delhi, 20–24 November, 2015

Verma B, Bhushan LS (1986) Management of gullied watersheds. Indian J Soil Conserv 14:105–110

# Bamboo-based Bioengineering Interventions for Rehabilitation of Ravines

17

B. Krishna Rao, V. C. Pande, R. S. Kurothe, A. K. Singh, and A. K. Parandiyal

**Abstract**

In India, the ravines are occupied by about 4 million hectares. Rehabilitation of degraded ravine lands with various kinds of vegetation is immensely important to increase forest cover and to meet the requirements of ever-increasing population in the country. Bamboo is considered to be one of the potential species suitable for ravine lands. It is one of the fastest-growing plants having extensive fibrous root system, connected rhizome system, dense foliage, and yielding sufficient amount of leaf litter. A research and development project entitled "Hydrologic and economic evaluation of bamboo plantations in gullied lands under major ravine systems (Mahi, Chambal, Yamuna) of India" was implemented during the period 2008–2015, with financial assistance by National Bamboo Mission, Ministry of Agriculture, Govt. of India. A part of this chapter is based on the final report of the project submitted to National Bamboo Mission. Several experiments were conducted to study the effect of bamboo-based interventions for sustainable natural resource management in degraded gully lands. The high stemflow amount and funneling ratio of bamboo plants in comparison to deciduous and coniferous plants make it better rainfall absorbing and hydrologically best-suited plantation in degraded ravine lands. Annual return from the bamboo-based interventions for gully beds was worked to be ₹88,780, ₹70,000, and ₹63,910 in "*bori-bund* reinforced with bamboo," "trenching," and "bamboo live check dam" treatments, respectively, with benefit-cost ratio of 2.09, 2.05,

The chapter is based on final report of the project submitted to National Bamboo Mission, Ministry of Agriculture, Govt. of India, by the authors; available on *nbm.nic.in/Reports/Report-4.pdf*.

B. Krishna Rao (✉) · V. C. Pande · R. S. Kurothe
ICAR-Indian Institute of Soil & Water Conservation Research Centre, Vasad, Anand, Gujarat, India

A. K. Singh · A. K. Parandiyal
ICAR-Indian Institute of Soil and Water Conservation, Research Centre, Agra, Uttar Pradesh, India





J. C. Dagar, A. K. Singh (eds.), *Ravine Lands: Greening for Livelihood and Environmental Security*, https://doi.org/10.1007/978-981-10-8043-2_17

and 1.96, respectively. It is recommended that these interventions can be implemented in degraded ravine lands.

## 17.1 Introduction

Ravines constitute a network of gullies, which form one of the worst kinds of land degradation processes. These cover ~4 million ha land, which is about 1.12% of total geographical area of the country (NCA 1976). Very extensive degradation of land has occurred along some of the major river systems of the country, and in many states this degradation is in the form of deep gullies. The National Commission on Agriculture has repeatedly indicated that gullied lands not only create problems where they exist but are also the root causes for degradation of adjacent arable lands affecting production potential in a major way. There are several major areas of severe ravine erosion in India. These degraded ravines can be safely put under permanent vegetation for their sustainable use to meet the requirements of fodder, fuelwood, and in areas of shallow gullies food crops can also be grown. Rehabilitation of ravine lands with various kinds of vegetation is immensely important to increasing the desired level of forest cover as targeted in National Forest Policy, Government of India. The increase in vegetation in degraded ravine lands enhances the flora and fauna and thereby restoring natural ecosystem to counter impacts of climate change and to conserve land and water resources for sustain and improves the production and productivity of degraded ravine lands. Bamboo has been identified as one of the most potential species for protection of soil from erosion and productive utilization of these lands.

India has one of the richest bamboo resources in the world, second only to China in bamboo production. More than 100 vital species of bamboo have been reported to be found in India, which grow naturally at heights ranging from sea level to over 3500 m in varied habitats (Tewari 1992). Bamboo is known to be one of the fastest-growing plant species with a growth rate ranging from 30 to 100 cm per day in favorable growing season. Bamboo species can reach to a height of 36 m with diameter ranging from 1 to 30 cm (United Nations 1972). It grows so fast that a culm can reach its full height just within 2–3 months period. Based on research it has already been established that its plantation is quite efficient in conserving of natural resources. Simon and Collison (2002) while exploring runoff mechanism on riparian lands revealed that under bamboo plantation, the concentration of runoff was reduced considerably, and the cover also checked the offside damage at valley bottom. Lawler (1993) and Yanhui and Yongmin (1995) also reported the positive utility of bamboo plantation for enhancing natural resource conservation including soil and moisture conservation.

Sharma et al. (1980) and Kurothe and Nambiar (2001) demonstrated successfully the utility of bamboo plantation as effective cover for soil and water conservation measures in Mahi ravines. They reported that bamboo-based vegetation was quite promising both in terms of controlling runoff and soil loss and generating economic returns. In one well-conducted study, it was observed that from 7th year onward, a net income of ₹8000–10,000 was easily obtained without any risk (Kurothe and Nambiar 2001; Kurothe et al. 2012; Pande et al. 2012). Comparing the soil erosion rate under different vegetation cover, Zhou et al. (2005) reported that the soil erosion was very low in bamboo plot as against other types of forest plantations. Lu et al. (2007) and Rao et al. (2012a) observed that the bamboo plants had higher stemflow in comparison to other deciduous and coniferous forest tress, due to their special canopy characteristics. Deng et al. (2003) studied five different conversion models of vegetation including of bamboo comparing their soil conservation efficiency. Out of the five models, bamboo grooves intercropped with fodder grass were more effective in controlling soil erosion. The parameters such as extensive fibrous root system, connected rhizome system in soil, comparatively dense foliage, and huge leaf litter production on the soil surface, and fast-growing nature are some important valuable characteristic features of bamboo for controlling runoff and soil erosion. The comparatively dense foliage protects the soil against beating by raindrops, hence checking splash and sheet erosion at initial stage itself. Further, its habit of producing new culms from underground rhizomes allows pole harvesting without disturbing the soil (Ben-zhi et al. 2005). In India, different bamboo-based agroforestry models have been worked out by many workers. For example, the systems bamboo + soybean by Sheshadri (1985), bamboo + turmeric and bamboo + pigeon pea by Shanmughavel and Francis (2001), and bamboo + ginger by Jha and Lalnunmawia (2004) were successfully established in managing the different types of degraded lands. Christanty et al. (1996) reported that bamboo as sole plantation and in agroforestry mode has tremendous restoring ability.

National Bamboo Mission, Ministry of Agriculture, Government of India, has clearly emphasized on the conservation value of bamboo plantations. Keeping these considerations in view, a research and development project entitled "Hydrologic and economic evaluation of bamboo plantations in gullied lands under major ravine systems" was carried at Central Soil and Water Conservation Research and Training Institute, Research centers at Vasad under Mahi ravines, Kota under Chambal ravines, and Agra under Yamuna ravines during the years from 2007 to 2015 with financial assistance from National Bamboo Mission, Government of India. The study was primarily focused on utilizing and judging overall performances of bamboo plantations for controlling soil erosion on beds and banks of gullies and streams under real field conditions. Under this project research and development activities at three specific locations, namely, Mahi ravines at Vasad (Gujarat), Chambal ravines at Kota (Rajasthan), and Yamuna ravines at Agra (UP) encompassing about 15 ha gullied land at each location were selected. This project was implemented in two locations of each ravine system. The results and success of the same have been reported here in brief (*nbm.nic.in/Reports/Report-4.pdf*).

## 17.2 Suitability Bamboo Species for Ravine Lands

*Dendrocalamus strictus* (Manvel or Solid bamboo) (Fig. 17.1) and *Bambusa bambos* (Katas or Thorny bamboo) (Fig. 17.2) are important species for semiarid regions of India. Besides these, *D. hamiltonii* is also found in small patches in semiarid regions of Uttar Pradesh and Madhya Pradesh. No other species has been reported to be suitable for ravine lands.

Most of the bamboos are considered to be hungry and thirsty plants and require moisture and nutrients for establishment. Therefore, they need to be grown in the perfect place with fertile soil and place where suitable and timely rainfall occurs. They need to be provided both fertilizers and water regularly for establishment and best growth. As ravine lands are characteristically devoid of fertility and are mostly moisture deficient, only the hardier species can be grown in these conditions, and comparatively it takes longer for these to be established and proper growth. *Dendrocalamus strictus* is the most widely planted and successful species for growing in the ravines. It is considered to be the hardest of all Indian bamboos; hence it thrives even in areas which are under successive period of severe drought. Further, it is also a frost-hardy species. In the less harsh areas of ravine region where moisture deficit is not too prominent, *Bambusa bambos* can also be planted.

Bamboos such as species of *Dendrocalamus* grown in ravines are usually clump-forming types which give propagules in peripheral region from rhizomes. After germination, it produces a grass-like seedling in the first year. The plumule,

**Fig. 17.1** Clump of *Bambusa bambos*

**Fig. 17.2** Clump of *Dendrocalamus strictus*

which appears as a conical bud covered by sheathing scaly leaves, develops rapidly into a thin wiry stem bearing single leaves, which alternate at the nodes and the leaf bases cover the stem. Fibrous roots form at the base of the young shoot and successive pointed buds appear on the rhizome. These buds form short rhizomes which curve upward to produce an aerial shoot. The process of rhizome and shoot production continues for several years, and with the previous year's shoots dying down, new rhizomes grow deeper and new shoots appear each succeeding year. Usually *D. strictus* requires 12–13 years under natural forest conditions and 6 years in artificially established stands, to form a mature clump. In ravines, one third of the old mature culms per clump are harvested every year for natural regeneration starting from 7th year of growth. Because of extensive rhizome-root system and accumulation of leaf mulch, bamboo serves as an efficient agent in preventing runoff and soil erosion, conserving moisture in reinforcement of embankments and drainage channels.

## 17.3 Performance in Ravine Lands

One experiment was conducted in ravines of Mahi river where 2-year-old bamboo seedlings were planted in contour staggered trenches (60 × 60 cm × 1.8 m length) at spacing of 3 × 3 m. The seedlings were planted at the center of the trench after filling the excavated soil up to 30 cm depth, and equal volume of soil was refilled after

**Table 17.1** Growth parameters of bamboo in Mahi ravines

| Growth parameters | Age of bamboo (years) | | | | | | | |
|---|---|---|---|---|---|---|---|---|
| | 4 | 5 | 6 | 7 | 8 | 9 | 10 | 11 |
| Clump circumferences (m) | 1.1 | 1.2 | 1.4 | 1.6 | 1.9 | 2.2 | 2.4 | 2.7 |
| Culm height (m) | 4.2 | 4.8 | 5.0 | 5.7 | 6.2 | 7.2 | 7.7 | 8.7 |
| Culm diameter (cm) | 2.0 | 2.1 | 2.5 | 2.6 | 2.7 | 3.1 | 3.3 | 3.7 |
| Old culms (nos) | 4.6 | 6.0 | 6.6 | 10.1 | 13.0 | 13.5 | 16.0 | 20.0 |
| New culms (nos) | 2.1 | 2.6 | 2.6 | 3.1 | 3.6 | 3.1 | 4.0 | 4.0 |
| New culms to old culm (%) | 55.2 | 43.3 | 34.7 | 30.5 | 27.4 | 23.1 | 25.0 | 20.0 |

Source: Adopted from Dhruva Narayana (1993)

**Table 17.2** Comparison of bamboo clumps harvested at regular and irregular intervals in Mahi ravines

| Clump class | Range of culms in a class | Regular harvesting | | Irregular harvesting | |
|---|---|---|---|---|---|
| | | New culms | Old culms | New culms | Old culms |
| I | 1–5 | 1.3 | 2.7 | 0.9 | 4.4 |
| II | 6–10 | 2.1 | 6.4 | 2.0 | 6.4 |
| III | 11–15 | 2.8 | 10.3 | 2.7 | 10.0 |
| IV | 16–20 | 3.9 | 13.9 | 3.6 | 13.8 |
| V | 21–25 | 4.3 | 19.4 | 3.9 | 18.5 |
| VI | 26–30 | 6.2 | 22.4 | 4.5 | 22.4 |
| VII | 31–35 | 8.3 | 27.2 | 5.6 | 27.1 |
| VIII | 36–40 | 10.5 | 27.6 | 5.5 | 32.0 |
| IX | 41–45 | 7.4 | 35.0 | 6.0 | 37.0 |
| X | 46–50 | 11.0 | 37.0 | – | – |
| XI | 51–55 | 13.5 | 38.5 | – | – |
| XII | 56–60 | 11.7 | 46.3 | – | – |
| XIII | 61–65 | 17.0 | 47.0 | – | – |
| XIV | 66–70 | 14.5 | 52.5 | – | – |

Source: Adopted from Dhruva Narayana (1993)

planting, i.e., total refilling being one third of the excavated soil. The growth parameter of bamboo such as culm height and diameter improved with the age of plantation (Table 17.1).

Effect of regular and irregular harvesting on production potential was examined by comparing bamboo clumps harvested at 3-year regular felling cycle and the clumps harvested after 11th year of plantation (Table 17.2). It was revealed that up to 20 old culms/clump, the regeneration rate of new culms was not different in the two cases.

Another study was conducted planting *D. strictus* at a space of 6 × 6 m distance in the deep ravines located on the right bank of river Mahisagar at village Vasad in Kheda district of Gujarat. Planting was done on contours in staggered manner. Various felling cycles were tried for economic exploitation of bamboos. Harvesting of old culms representing 30% of the total number of culms every year gave the maximum production. The annual average yield per clump obtained

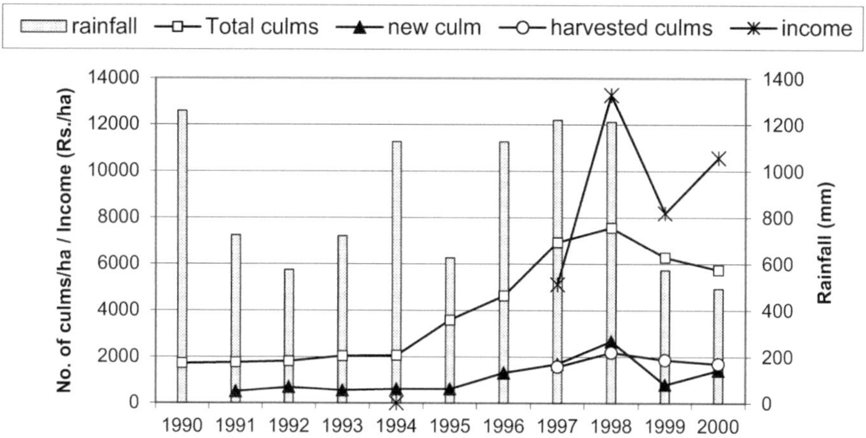


**Fig. 17.3** Bamboo yield and income over the period of observation

varied from 6.21 to 11.69. This harvesting intensity is, therefore, being adopted in the harvest of bamboo plantation. Kurothe and Nambia (2001) reported results of a bamboo plantation raised in 5 × 5 m spacing (staggered) on ravine bed of the Mahi ravines (Fig. 17.3). The growth varied during different periods because of variation in rainfall.

### 17.3.1 Performance of Bamboo Under Different Technological Interventions

Rao et al. (2012b) tested the performance of bamboo with different soil and moisture conservation interventions (Figs. 17.4 and 17.5). The survival under different treatments varied from 40% to 80% in different ravine systems. The highest survival (up to 80%) was recorded in gully beds treated with bamboo plantation with supportive staggered trenches followed by bamboo plantation with earthen check dams. Bamboo saplings planted during 2007–2008 were measured for different growth parameters in the year 2013–2014. The average height was maximum (353 cm) under earthen gully plug treatment, followed by staggered trenches (330 cm) in Mahi ravines (Table 17.3). Similar trend was observed for clump diameter.

The growth parameters of bamboo at Chambal ravines are presented in Table 17.4. The better growth performance was recorded in bamboo plantation with earthen gully plugs followed by trenches. It was observed in Yamuna ravines that plantation of rows of bamboo in the upstream and downstream side of small earthen gully plugs had a significantly higher culm height (126.20 cm), culm diameter (7.56 mm), crown size (72.86 cm), and maximum of 12 culms compared to others (Table 17.5). The better growth in bamboo plantation with conservations measures was mainly due to absorption of rainfall thereby increasing in moisture holding status.

**Fig. 17.4** Bamboo plantation with live supportive staggered trenches

**Fig. 17.5** Bamboo plantation with (left) live check dam and (right) earthen check dam

**Table 17.3** Bamboo growth under various treatments—Mahi ravines

| Treatment | Average height (cm) | Average clump diameter (cm) | Number of clumps | Average collar diameter (cm) |
|---|---|---|---|---|
| Bamboo + trench | 330 | 150.7 | 12 | 2.4 |
| Bamboo as live | 281 | 111.4 | 9 | 1.94 |
| Check dam | | | | |
| Bamboo + earthen gully plug | 353 | 154.1 | 15 | 2.72 |

**Table 17.4** Bamboo growth under various treatments in Chambal ravines

| Treatment | Average height (cm) | Number of culms | Average collar diameter (cm) |
|---|---|---|---|
| Bamboo + trench | 340 | 7 | 2.7 |
| Bamboo as live | 310 | 6 | 2.6 |
| Check dam | | | |
| Bamboo + earthen gully plug | 360 | 7.5 | 2.8 |

**Table 17.5** Bamboo growth under various treatments in Yamuna ravines

| Treatment | Av. height (cm) | Average culm diameter (cm) | Average crown size (cm) | No. of culms/ clump |
|---|---|---|---|---|
| Bamboo + trench | 108.7 | 7.9 | 66.8 | 9.7 |
| Bamboo as live | 99.2 | 6.2 | 49.1 | 6.3 |
| Check dam | | | | |
| Bamboo + earthen gully plug | 126.2 | 7.6 | 72.8 | 12 |

**Table 17.6** Silt deposition (cm) behind bamboo plantation at different positions

| Treatment | Starting | Middle | End |
|---|---|---|---|
| Average depth of silt deposition (cm) | | | |
| Bamboo plantation with supportive staggered trenches | 7.2 | 4.7 | 6.9 |
| Bamboo plantation to act as live check dam | 3.8 | 2.1 | 3.3 |
| Bamboo plantation with earthen gully plugs | 10.3 | 7.3 | 8.6 |
| Control | 2.1 | 0.0 | 1.4 |

### 17.3.2 Comparison of Different Treatments in Silt Deposition and Gully Extension

The highest silt deposition was recorded in gully treated by bamboo plantation with earthen gully plugs (Table 17.6). The silt deposition was highest in the start of the gullies. Bamboo-based interventions for gully heads and bank stabilization gave net returns at the rate of ₹700 per 10 m length from seventh year onward. Gully head extension was observed up to 1.4 m in gully head without peripheral bund, and it was almost nil in gully head with peripheral bund supported by bamboo plantation. Gully bank extension was observed up to 0.4 m in gully bank without bamboo plantation, and it was negligible in gully bank with bamboo plantation.

### 17.3.3 Bamboo Economics Under Different Conservation Treatments

When different conservation techniques were employed in Mahi ravines, a total number of culms per clump were 15, 12, and 9 in *bori*-bund reinforced with bamboo, trenching, and bamboo live check dam treatments, respectively. Considering the recommended harvest cycle of 30% mature bamboo culms per clump, the harvestable bamboo culms per clump worked out to be 5, 4, and 3 in these treatments, respectively. Based on this, the available bamboos harvested for sale were 3126, 2500, and 2312 in respective treatments. The costs of *bori*-bund reinforced with bamboo, trenching, and bamboo live check dam treatments were estimated as ₹96,160 $ha^{-1}$, ₹37,500 $ha^{-1}$, and ₹57,588 $ha^{-1}$, respectively. The net annual return from these treatments, thus, varied from ₹63,910 to 88,780.

The economic indicators for bamboo plantation in different treatments revealed that for a production span of 20 years, the net present worth varies from ₹222,588 $ha^{-1}$ in bamboo live check dam treatment to ₹319,731 $ha^{-1}$ in *bori*-bund reinforced with bamboo treatment. Similarly, the benefit-cost ratio worked out to be 2.09, 2.05, and 1.96 in respective treatments. The internal rate of return revealed that bamboo performance in trenching gives best rate of return (20.2%) as compared to *bori*-bund reinforced with bamboo (19.7%) and bamboo live check dam (19.3%) treatments.

## 17.4 Impact of Bamboo on Hydrological Behavior in Ravines

Various hydrological processes such as rainfall, interception (throughfall, stemflow, interception loss), initial abstractions (depression storage, detention storage), infiltration, evaporation, runoff, etc. are influenced by bamboo plantations in ravines in various ways. These are discussed in brief here.

Throughfall and Stemflow

Rainfall intercepts with canopy, and part of it that passes through the canopy to reach the ground is known as throughfall. This process describes how wet leaves shed excess water onto the ground surface. When rainfall exceeds canopy storage capacity, some water is funneled as stemflow and delivered more efficiently to the soil around the root zone. Rates of throughfall are higher in areas of forest where the leaves are broad-leaved, because the flat leaves allow water to collect. Rates of throughfall are lower in coniferous forests as conifers can only hold individual droplets of water on their needles. Throughfall in bamboo can be measured by small rain gauges kept on the shade of the vegetation (Fig. 17.6).

**Fig. 17.6** Measurement of throughfall

**Fig. 17.7** Stemflow measurement

The low-cost plastic stemflow collar with attached rain gauge (Fig. 17.7) was successfully used to record the stemflow amount and intensity without involving much manpower and is found economical and simpler as compared to costly and sophisticated instruments. The throughfall varied from 43% to 72%, stemflow 7–22%, and interception losses varied from 12% to 50% of the total rainfall received at canopy level. The funneling ratio varies from 22 to 91. The high stemflow amount and funneling ratio of bamboo plants in comparison to deciduous and coniferous plants make them hydrologically best-suited plantation in degraded ravine lands (Rao et al. 2012a).

Infiltration

It is the entry of water into the soil surface. The measure of infiltration of water into the soil is an important indication concerning: the efficiency of irrigation and drainage, optimizing the availability of water for plants, improving the yield of crops, and minimizing erosion. The infiltration tests were conducted in the ravine lands treated with bamboo plantations. The observed data was fitted in Philip's equation. The infiltration rate was found to be low in bamboo plantation in degraded gully beds in comparison to humps and slopes (Fig. 17.8). This is mainly due to silt deposition due to bamboo plantations on the bed.

Runoff and Soil Loss

When rainfall occurs, after interception with canopy, partly it goes to soil as infiltration depending upon the condition of soil and vegetation. When excess, water flows on surface known as runoff and it carries with it soil particles as soil loss. In this particular case, the rainfall data were collected from meteorological observatory of the Central Soil and Water Conservation Research and Training Institute, Research Centre, Vasad. Runoff was recorded using stage level recorder installed with a weir. Runoff samples were collected periodically during storms for sediment concentration computations. Runoff generated by storms was retained upstream the check dam in most of the cases. However, higher intensity storms produced more runoff. Bamboo-based interventions retained most of the runoff, and negligible sediment

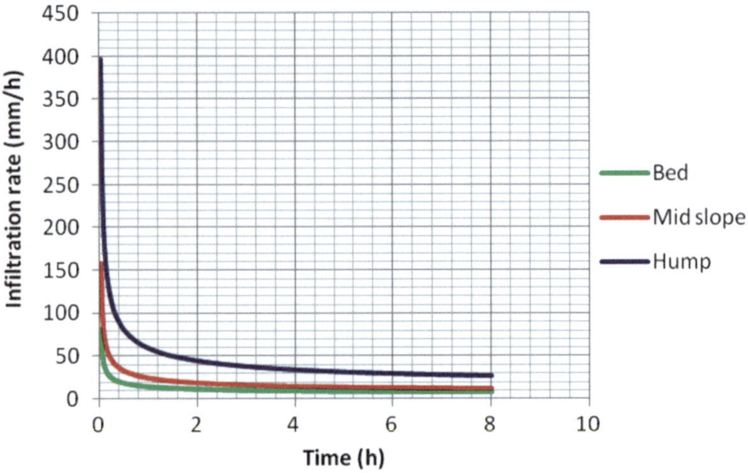


**Fig. 17.8** Infiltration rate of the ravine lands treated with bamboo plantations

yield was observed from this watershed. Similar untreated watershed produced about 20 Mg ha$^{-1}$ sediment which might be deposited in the river or major dams downstream and reducing their capacities (Kurothe and Nambiyar 2001).

The experiments were conducted at Khorwad village, and runoff from various bamboo-based interventions was measured. The average annual rainfall of the Khorwad was reported to be 870 mm. The total rainfall of Khorwad during study period was 980 mm and 840 mm in 2010 and 2011, respectively. Total rainy days were 32 and 34 in 2010 and 2011, respectively. The lowest runoff (79 mm and 67 mm in 2010 and 2011, respectively) and soil loss (5 and 4 Mg ha$^{-1}$ in 2010 and 2011, respectively) were recorded in W3 watershed, i.e., treatment T3 (bamboo plantation with earthen check dams made of sandbags) followed by W1 watershed (runoff 127 and 104 mm and soil loss 10 and 7 Mg ha$^{-1}$ in 2010 and 2011, respectively), i.e., treatment T1 (bamboo plantation with staggered trenches) in comparison to control watershed W4 (runoff 284 and 183 mm and soil loss 16 and 12 Mg ha$^{-1}$ in 2010 and 2011, respectively). Bamboo-based interventions absorb more than 80% of rainfall and reduced the soil and nutrient losses up to 70% (Rao et al. 2012b).

## 17.5 Soil Organic Carbon Enhancement Due to Bamboo Plantations

Soil organic carbon was estimated from soil samples taken from bamboo plantation sites after 20 years of plantation. Similar samples were taken from the fallow ravine lands. The analysis revealed that soil carbon buildup of 41 Mg ha$^{-1}$ was observed over a fallow ravine land. This carbon buildup sustains in the soil if plantation is maintained for a longer period, following the recommended harvest cycle of harvesting 30 % mature culms per clump. Imputing a value to this at a shadow carbon price of US $20 per Mg C (Atkinson et al. 2006) (₹value 50 = US $1) worked out to be ₹41,000 per ha, ranging between ₹10,250 per ha and ₹82,000 per ha at carbon price range of US $5 per Mg C and US $40 per Mg C, respectively.

## 17.6 Nutrient Saving by Bamboo Plantation

Soil conservation value was estimated in terms of nutrients saved from the loss to river in the downstream. The nitrogen, phosphorus, and potash content of soil in a degraded ravine land of this region ranged between 101–470 kg $ha^{-1}$, 24–95 kg $ha^{-1}$, and 216–470 kg $ha^{-1}$, respectively. This works as a lower and upper bound for the nutrients in the soil conserved because of bamboo plantation. In ravine lands, farm yard manure is applied by the farmers in small quantity prior to plantation, but the quantity is too small to replace the nutrients lost through soil erosion. Hence, only chemical fertilizers are used for replacement of nutrients lost. Further, nitrogen is closely related with carbon in the soil under plantation. Since, evaluation of soil carbon is done separately; the benefit of this nutrient was not summed up in the nutrient saved. The value of nutrients, thus, saved through bamboo plantation has been estimated to be ₹2126–5555 $ha^{-1}$.

## 17.7 Limitations and Policy Interventions for Bamboo in Ravine Lands

Despite the bamboo having potential for ravine lands in terms of productivity and economic profitability and availability of technical know-how in terms of its conservation value and capacity of conserving nutrient loss, the bamboo has not gained the desired momentum in reclaiming degraded lands in general and ravines in particular. The main limiting factors and required interventions in the promotion of the bamboo in ravines are presented here:

### 17.7.1 Lack of Awareness

Bamboo has been traditionally raised, to a limited extent, in village commons and homestead gardens. However, it remains a forest species in respect of industrial use and commercial plantations. There is a need for changing the 'forestry mindset' to the 'farming mind set' and creating awareness on the commercial viability and profitability of the species. In this regard, extensive awareness and capacity building programs are required at all level viz. farmers, extension workers, village level societies, tree growers' cooperative societies, etc. It is interesting to note that recently Government of India has brought a law that the bamboo is removed from the definition of a tree for the sake of farmers and now it will be treated non-tree species and will be out of forest gamut. The research and development (R&D) under the sector needs to be complimented with awareness creation, demonstration, and extension support for a multiplier effect. The self-help groups (SHGs), village forest committees (VFCs), Tree Grower's Cooperative Societies (TGCS), farmers clubs, etc. could be used as vehicle for such extension programs.

### 17.7.2 Policy for Leasing Revenue Degraded Ravine Lands for Forest Plantations

A large portion of degraded ravine lands, which could be profitably used for raising bamboo, are under the control of the Revenue Department/ Forest Department. Such lands are neither properly utilized nor leased away to the interested entrepreneurs for development. It is, therefore, desirable that the state governments evolve suitable policies for leasing of the revenue wastelands to the farmers/SHGs/entrepreneurs/industries for raising forest plantations for industrial use. Andhra Pradesh, Tamil Nadu, and Chattisgarh have come up with such policies for greening the wastelands. Other states may also follow to encourage the public-private partnership for greening the degraded ravine lands.

### 17.7.3 Need for Organized Supply Chain Management

Though a vast market exists for the bamboo sector, there is a need for an organized market with assured minimum price regime to encourage the small farmers to take up bamboo plantations on their private wastelands. There is a need to promote bamboo-based industries in states having major ravine areas.

### 17.7.4 Need for Incentives for Raising Bamboo Plantations

Cultivation of bamboo in degraded ravine lands and the agricultural land nearer to these is a new activity and may be given fillip through extending capital/interest subsidy to the farmers/entrepreneurs for raising bamboo plantations. The facilities of cheap loans and also insurance against natural calamities must be provided to the farmers growing bamboo. Bamboo-based industries must be installed in rural areas to facilitate better income to bamboo growers.

## 17.8 Conclusions

The studies conducted under National Bamboo Mission Sponsored Research and Development project on bamboo-based interventions in ravine reclamation proved to be quite promising in terms of generating technologies controlling runoff and soil loss, carbon sequestration, and economic returns. These techniques are very much useful for reclamation in ravine lands of various states. Despite of bamboo potential in degraded ravine lands in terms of controlling runoff and soil loss and economic returns, it has not been tapped to its fullest potential; hence wider and alternative utilities of bamboo are being "rediscovered" in India with increased attributes and potential toward ecological and soil conservation utilities in degraded ravine lands. Field visits, farmers' days, training programs, and workshops can be organized for sensitization and upscaling of these techniques among various stakeholders like policy makers, state and central government departments, funding agencies, NGOS, Tree Growers Cooperative Societies, and farmers. However, there is a need to integrate the various stakeholders at various levels (production to consumption) for evolving workable strategies to promote the bamboo-based interventions, which would not only be instrumental in addressing environmental concerns but also economic and livelihood security of habitants of these vulnerable lands.

**Acknowledgments** Authors are thankful to Dr. VN Sharda and Dr. PK Mishra former and present directors of Indian Institute of Soil and Water Conservation, respectively, for providing valuable guidance in this study. The authors are thankful to National Bamboo Mission, Ministry of Agriculture, Govt. of India for providing funds to this study. We are thankful to Dr. JC Dagar for his valuable input in finalizing this manuscript.

## References

Ben-zhi Z, Mao-yi F, Xie J-z, Yang X-s, Zheng-cai L (2005) Ecological functions of bamboo forest: research and application. J For Res 16(2):143–147

Christanty L, Mailly D, Kimmins JP (1996) "Without bamboo, the land dies": biomass, litterfall and soil organic dynamics of a Javanese bamboo talun – kebun system. For Ecol Manag 87:75–88

Deng Y, Peng Y, Wang Y, Yang Z (2003) Soil conservation efficiency of different models converting from steep farming slope plots in Ya'an, Sichuan, China. Int J Sediment Res 18(4):340–345

DhruvaNarayana VV (1993) Soil and water conservation research in India. Publications and Information Division, Indian Council of Agricultural Research, Krishi Anusandhan Bhavan, Pusa, New Delhi, p 449

Jha LK, Lalnunmawia F (2004) Agroforestry with bamboo and ginger to rehabilitate degraded areas in north east India. J Bamboo Rattan 2(2):103–109

Kurothe RS, Nambiar, KTN (2001) Eco-system restoration and sustainable production in degraded Mahi ravines and its effect on hydrology and sedimentation. In: Proc. National Conference on Resource Conservation and Watershed Management (RCWM-2001), May 23–25, 2001, Dehradun

Kurothe RS, Gaur ML, Rao BK, Parandiyal AK, Singh AK (2012) Conservation and production potentials of bamboo in ravine lands. CSWCRTI, Dehradun. 978-81-924172-1-9: 160pp

Lawler DM (1993) The measurement of river bank erosion & lateral channel change: a review. Earth Surf Process Landf 18:777–821

Lu SY, Liu CP, Hwang LS, Wang CH (2007) Hydrological characteristics of a making bamboo woodland in Central Taiwan. Taiwan J Sci 22(1):81–93

National Commission on Agriculture (1976) Report of National Commission on Agriculture, Part 5: Resource Development, Govt of India. New Delhi, pp 107–322

Pande VC, Kurothe RS, Rao BK, Gopal K, Parandiyal AK, Singh AK, Ashok K (2012) Economic analysis of bamboo plantation in three major ravine systems of India. Agric Econ Res Rev 25(1):63–73

Rao BK, Kurothe RS, Pande VC, Gopal K (2012a) Throughfall and stemflow measurement in bamboo (*Dendrocalamus strictus*) plantation. Indian J Soil Conserv 40(1):60–64

Rao BK, Kurothe RS, Singh AK, Parandiyal AK, Pande VC, Kumar G (2012b) Bamboo plantation based technological interventions for reclamation and productive utilization of ravine lands. CSWCRTI, Technical Bulletin No. T-62/V-4, 30pp

Seshadri P (1985) Intercropping of Bamboo *(D. strictus)* with soybean- an agroforestry study. Ph.D thesis, Tamil Nadu Agricultural University, Coimbatore, 480pp

Shanmughavel P, Francis K (2001) Intercropping trials of four crops in plantations. J Bamboo Rattan 1(1):3–9

Sharma AK, Pradhan IP, Nema JP, Tejwani KG (1980) 25 years research on soil & water conservation in ravine lands of Gujarat. CS&WCR&TI, Research Centre, Vasad

Simon A, Collison AJC (2002) Quantifying the mechanical and hydrologic effects of riparian vegetation on stream-bank stability. Earth Surf Process Landf 27:527–546

Tewari DN (1992) A monograph on bamboo. International Book Distributors, Dehra Dun, 498 pp

United Nations (1972) The use of bamboo and reeds in building construction. Department of Economic and Social Affairs, Document No. ST/SOA/113, p 95. United Nations, New York, USA

Yanhui W, Yongmin L (1995) Hydrological characteristics of a moso-bamboo (*Phyllostachys pubescens*) forest in South China. Hydrol Process 9(7):797–808

# *Leucaena leucocephala*: A Leguminous Tree Suitable for Eroded Habitats of Hawaiian Islands

18

Kazue L. Ishihara, Michael D. H. Honda, Ahmed Bageel, and Dulal Borthakur

**Abstract**

*Leucaena leucocephala* is divided into two subspecies: (i) *L. leucocephala* subsp. *leucocephala*, which is known as "common leucaena" or "koa haole," and (ii) *L. leucocephala* subsp. *glabrata*, which is known as "giant leucaena." Common leucaena is a small bushy shrub that forms a lot of seeds, because of which it can spread easily and is considered invasive. Giant leucaena, on the other hand, is a tree with large branches. It produces much less seeds and is not considered invasive. Giant leucaena can grow to become big trees of up to ~20 m in height, or it can be grown as a legume fodder by maintaining the plants as dwarf bushes through repeated harvest of foliage, up to ten times a year. The wood of giant leucaena can be used for timber, paper pulp, or biofuel production. In the Hawaiian Islands, many eroded habitats, such as roadsides, hill slopes, and barren lands, are generally occupied by common leucaena. Replacing common leucaena with giant leucaena in these locations may create new opportunities for cattle farming, animal feed production, and development of other related industry. Dr. James Brewbaker and his colleagues at the University of Hawaii developed many high-yielding cultivars of giant leucaena suitable for both fodder and wood production. Besides growing in Hawaii, these cultivars are also widely grown in many other tropical and subtropical countries.

K. L. Ishihara · M. D. H. Honda · A. Bageel · D. Borthakur (✉)
Department of Molecular Biosciences and Bioengineering, University of Hawaii at Manoa, Honolulu, HI, USA
e-mail: dulal@hawaii.edu





J. C. Dagar, A. K. Singh (eds.), *Ravine Lands: Greening for Livelihood and Environmental Security*, https://doi.org/10.1007/978-981-10-8043-2_18

## 18.1 Introduction

*Leucaena leucocephala* (leucaena) is a nitrogen-fixing tree legume that is widely distributed in the tropical and subtropical countries of the world. It is a small- to medium-sized tree in the tribe Mimosease, subfamily Mimosoideae of the family Leguminoase, originally native to Mexico, but now common throughout Africa, Australia, and the tropical and subtropical regions of Asia and the North and South Americas. It is commonly known as "koa haole" in Hawaii where it grows widely on mountain slopes and barren lands. In Hawaii, natural populations of leucaena are found in areas with the annual rainfall between 625 and 1600 mm and the elevation of up to 170 m above sea level in the higher rainfall windward sections or up to 330 m in the drier leeward locations (Dijkman 1950).

Leucaena growing in Hawaii can be divided into two groups: (i) common leucaena (*Leucaena leucocephala* subsp. *leucocephala*) and (ii) giant leucaena (*Leucaena leucocephala* subsp. *glabrata*) (Zarate 1984; Brewbaker 2010). The name "koa haole" is generally used to describe the common leucaena, which is a seedy shrub or small tree that originated in Southern Mexico or Guatemala (Wheeler and Brewbaker 1988). Morphological observations and limited isozyme studies in Hawaii did not reveal much variations among the populations of common leucaena suggesting that common leucaena is made up of a single genotype (Sun 1996; Brewbaker 2016). Because of the seediness property, common leucaena can spread easily and therefore, generally known as an invasive species (Fig. 18.1a, b). In sharp contrast to the shrubby form of common leucaena, the giant leucaena has an arboreal form with large glabrous branches (Fig. 18.1c, d). It produces much less seeds and is not considered as invasive. Because of the highly productive properties and its worldwide success as a long-lived and highly nutritious forage tree, the giant leucaena was known as the "miracle tree" during the 1970s and early 1980s (Shelton and Brewbaker 1994). The giant leucaena originated in the region of Zacatecas in Guerrero, Mexico, and was introduced to Hawaii in the 1960s (Brewbaker and Hylin 1965). *L. leucocephala* subsp. *glabrata* can be divided into two types according to branching pattern: Salvador type and Peru type (National Academy of Sciences 1977). The Salvador type is tall, high-yielding, and sparingly branched at the base; the Peru type is tall, high-yielding, and branches low down on the trunk. Generally, the Salvador type is more vigorous and matures earlier than the Peru type. The Peru type cultivars are generally more suitable and productive as forage than the Salvador type. The weedy and bushy *L. leucocephala* subsp. *leucocephala* is sometimes referred to as the "Hawaiian type," which is a misnomer since Hawaii is not its place of origin and it is abundant in many other places besides Hawaii.

**Fig. 18.1** Common and giant leucaena. (**a**) A photo showing a common leucaena plant growing near the curb of the access road to Sandy Beach in Honolulu; (**b**) a photo showing the seediness of common leucaena growing on a barren land in Hawaii Kai, Honolulu; (**c**) a photo showing two trees of giant leucaena in Waimanalo, Hawaii; and (**d**) a photo showing an 11-year-old seedless triploid leucaena (KX4-Hawaii) at the Waimanalo Research Station, University of Hawaii

## 18.2 Habitat

Leucaena is a hardy, fast-growing tree with high adaptability to a range of ecological conditions, although it prefers tropical climates. It is tolerant to full sun to moderate shade. Whenever heavy frost kills its aboveground growth, the crowns sprout vigorously again (Shelton and Brewbaker 1994). It grows well under varied soil conditions including dry phosphatic clay, sand, slit, and limestone (Mislevy et al. 1989; Mushaka and Maruzane 1998). It also has tolerance to acidic pH up to 5.0 (Brewbaker 2008), drought (Cavalcante and Perez 1995), and low levels of NaCl (~25 mM) (Anthraper and Dubois 2003). It can grow successfully even in relatively less fertile soils because of its ability to fix nitrogen in symbiosis with *Rhizobium* (Moawad and Bohlool 1984). Leucaena provides various environmental benefits including soil erosion control and improved soil fertility.

The natural distribution of leucaena indicates its preference for relatively steep and rocky slopes. With its deep root system, leucaena can support itself to grow in loose stands such as areas with thin and eroded soils or slopes. Its roots can pierce compact strata and break up impervious soils so that other plants can grow (Takahashi and Ripperton 1949; Dijkman 1950). Its few lateral roots also grow downward with the taproot. It was shown that the roots of leucaena trees grew over 2 m deep in 1 year and 5 m in 5 years even in young volcanic, well-weathered soil under humid conditions in which the plant is not required to grow down to reach deep water-carrying strata (Dijkman 1950).

The root system allows leucaena to grow in dry conditions because the plant can obtain water from deep strata that are not accessible to most other herbaceous plants (Shelton and Dalzell 2007). Leucaena thus has high adaptability to drought even for a long period of time. For example, during dry summer months in the University Farm or Poamoho Experimental Farm, Hawaii, it grew satisfactorily with only three irrigations, while irrigation was necessary every 10–14 days for alfalfa and napier grass to survive (Takahashi and Ripperton 1949).

The deep-rooted leucaena can also access nutrition from deeper strata. The plant acts as a "pump" of soil nutrition because the nutritive elements obtained by the plant eventually go back to the surface soil as it drops its leaves that readily decay (Dijkman 1950). Although there is no published report, the giant leucaena with its deep root system should be a suitable plant species for reabsorbing nitrate and phosphates in runoff water from crop fields to rivers and reaching our drinking water system. Not only that, it enhances soil fertility through nitrogen fixation (Dijkman 1950; Brewbaker et al. 1990). Nitrogen fixation promotes grass growth and strong ground cover, which helps prevention of soil erosion (Shelton and Dalzell 2007).

## 18.3 Uses of Leucaena

Leucaena is a multipurpose plant with various applications in industry, agriculture, and agroforestry (Shelton and Brewbaker 1994). Its young leaves, pods, and seeds are used not only as a fodder but also as vegetables for human consumption in Central

America, Indonesia, and Thailand (Bourges and Morales de León 1980; Mawardi 1985; National Academy of Science 1997; Hughes et al. 2007). Many cultivars of giant leucaena are grown as nutritious fodder for farm animals; they can be planted with narrow spacing of 0.5–1 m apart in rows 1–3 m apart and managed at low height (~3 m) by repeated harvest of foliage with cutting intervals ranging between 6 and 12 weeks during the growing season (Qamar et al. 2015). Due to its fast-growing nature and high biomass productivity, the potential of it as raw material for pulp and paper industries has gained attention (López et al. 2008; Pandey and Kumar 2013). The plant is also useful for soil remediation; recent studies show that it can effectively remove textile dyes or heavy metal contaminations from soils (Jayanthy et al. 2014; Juson et al. 2016). A combination of strong branches and deep root system makes it a suitable plant species for use as windbreaks. It can withstand strong wind without suffering much damage, although it may lose some leaves (National Academy of Sciences 1977). Soil erosion on slopes is a common natural phenomenon. Growing leucaena on such locations will be of a great benefit because it can grow vigorously and improve the soil quality for other plants to grow to prevent soil erosion. In Japan, its leaves are being used for the manufacture of herbal tea (Hongo et al. 1990). A leucaena company has been established to produce herbal tea and livestock feed in Okinawa in 1986 (Tawata et al. 2008).

## 18.4 Nutritional Quality of Leucaena Foliage

Leucaena is known as the "alfalfa of the tropics" because of the nutritional value and high palatability of its foliage. Young shoots have up to 31% protein on a dry weight basis, which decreased to 14% after 10 weeks (Tangendjaja et al. 1986). Garcia et al. (1996) reviewed 65 publications between 1946 and 1992 for nutritive value and forage productivity of leucaena and found that the medial concentration of crude protein was 29.2%. In the same study, they found that the medial concentrations of mimosine and tannin in the foliage were 4.3 and 1.01% of dry matter, respectively. Despite having high protein contents and high palatability, its foliage has two negative attributes; it has high amounts of a nonprotein amino acid mimosine, which is toxic to animals, and it has high amounts of condensed tannin, which has been identified as a major factor limiting the nutritive value of its foliage (Wheeler et al. 1995; Garcia et al. 1996; Osborne and McNeill 2001; Chanchay and Poosaran 2009). The mimosine contents of different parts of the shoot vary from 1% to 12%; old stems contain the lowest and the growing tips the highest amounts (Jones 1979). Young leaves contain the ~4.5% mimosine on a dry weight basis, which decreased to ~2% in 10-weak-old leaves (Tangendjaja et al. 1986). Chanchay and Poosaran (2009) also found 4.4% mimosine in the leaves. Soedarjo and Borthakur (1996a, b) determined that young leaves and pods contained as high as 6–10% mimosine on a dry weight basis. Condensed tannins are polyphenolic compounds that bind with soluble proteins and make them insoluble and indigestible. The leaves contain up to 37.6% total tannin, comprising both hydrolysable and condensed tannins

(Chanchay and Poosaran 2009). Soedarjo and Borthakur (1996b) found that young leaves of common leucaena contained only ~18% protein. They suggested that protein concentrations might have been overestimated in some of the earlier reports, because mimosine present in the samples interacts with the reagents for protein estimation, giving an overestimate of the protein content. They also described a more accurate method of determining soluble protein content in its tissues containing mimosine.

## 18.5 Role of Mimosine in Leucaena

It is not known why leucaena produces mimosine. Mimosine and its degradation product 3-hydroxy-4-pyridone (3H4P) have some antifungal properties (Tawata et al. 2008). Negi et al. (2014) estimated that if the total carbon and nitrogen used for production of mimosine were diverted for growth, the leucaena tree would have grown at least 20% larger. They also proposed that mimosine may provide a mechanism of drought resistance in it. According to this idea, during favorable weather conditions, when water and nutrients are available, leucaena synthesizes mimosine and accumulates in different parts of the plant, including the foliage. Under drought conditions, mimosine is degraded by the enzyme mimosinase, which is present in its chloroplast. They further suggested that during drought conditions, some chloroplast membrane may disintegrate, and mimosinase from the chloroplast stroma may come in contact with mimosine in the cytoplasm (Fig. 18.2).

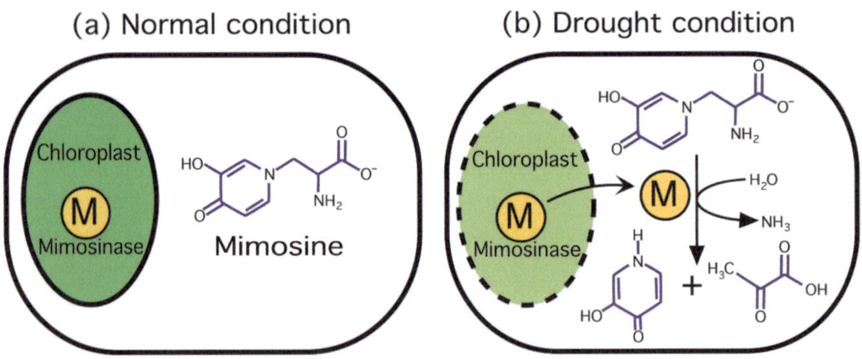


**Fig. 18.2** A possible role of mimosine in leucaena (**a**) In the leaves mimosine may be stored in the cell cytoplasm. A mimosine-degrading enzyme, mimosinase, is in the chloroplast, and thus mimosine and mimosinase are separated by membranes. (**b**) However, under drought conditions some chloroplast membrane may break, resulting in the release of mimosinase from the chloroplast to the cytoplasm, where mimosine is degraded by mimosinase. The degradation products of mimosine are used by the plant for survival during drought (Negi et al. 2014)

## 18.6 Water Use Efficiency and Erosion Control

A study in Queensland showed that leucaena pastures have the similar rooting depth and water use as the original native woodlands that covered the area before agricultural development, which helps to maintain hydrological balance by reducing deep drainage that can lead to dryland salinity (Poole 2003). Its pastures also have shown to have high water use efficiency compared with other pasture types; so planting leucaena across the slope will encourage water infiltration and reduces runoff and erosion (Dalzell et al. 2006). For instance, a pasture in Africa growing leucaena and *Setaria sphacelata* was shown to have reduced runoff and soil erosion loss (Roose and Ndayzigiye 1997). Once established, it shows excellent erosion control characteristics. In many leucaena pastures, little runoff is observed even after high-intensity rainfall, and water is generally clear and free from soil particles (Shelton and Dalzell 2007). Its role as hedgerow crop in alley cropping to control runoff and erosion has been discussed at many places in this book particularly in Chap. 3.

## 18.7 Drought Tolerance of Leucaena

Attributes of root competitiveness, such as taproot length, lateral root length density, mycorrhization, nodulation and nitrogen fixation, disease resistance, and flexibility in response to water and nutrient availability in the soil, are some of the important determinants of leucaena's success as a stress-tolerant tree legume in tropical and subtropical environments. Research at the International Livestock Research Institute in Kenya by Ezenwa and Atta-Krah (1992) noted that compared to *Gliricidia sepium*, leucaena seedlings grown in soils allocate more nutrient resources for growth of the taproot than on lateral roots until about 12 weeks. This may be one of the reasons why leucaena is naturally resistant to drought. It is a deep-rooted species and can extend its roots up to 5 m to exploit underground water (Brewbaker et al. 1972). In dry areas, it remains unaffected by drought if its deep roots can reach groundwater. It also shows voidance responses toward drought condition through leaflet folding during dry spells to prevent water loss and by dropping some leaves under severe drought conditions (Brewbaker 1987). Rao et al. (2008) reported that the net photosynthetic rate and transpiration rate decreased, and stomatal resistance increased in it in response to high water stress. They also observed that under high water stress, it maintained higher water potential ($\psi$) and proline content, indicating drought resistance. Similarly, Yige et al. (2012) studied the physiological responses of leucaena seedlings in leaf water content (LWC), chlorophyll content, and photosynthetic capacity to drought stress. They observed that the seedlings had a high ability to keep water as its LWC, which did not decrease significantly until 12-day drought. The chlorophyll content increased significantly but the chlorophyll (a/b) ratio decreased. Although water use efficiency increased, the rates of photosynthesis and transpiration and stomatal conductance of leaves significantly decreased due to drought. Thus, its seedlings could maintain a reasonable physiological activity by maintaining LWC, increasing chlorophyll content, and reducing photosynthetic activity under drought stress.

## 18.8 Susceptibility to Psyllids

Although leucaena has high adaptability to various biotic and abiotic stresses, there is one known serious pest, the leucaena psyllid (*Heteropsylla cubana*). It originated first in the Caribbean (McClay 1989) and was reported in 1983 to be a serious pest on common leucaena (Othman and Prine 1984), and it finally spread to Hawaii by April 1984 (Sorensson and Brewbaker 1984; Brewbaker 2016). In the same year, it dramatically defoliated leucaena on the island of Oahu and throughout the world. Psyllids cause damage on plant growth especially by feeding on juvenile leaflets and defoliating tip shoots (Funasaki et al. 1989). With a short life cycle of about 2 weeks and with the ability to lay up to 400 eggs in their lifetime, the pest grows exponentially and quickly infests plants especially in warm, moist conditions. Trials to test psyllid tolerance of *Leucaena* species were performed in multiple countries, including Thailand, Mexico, Philippines, and the USA, immediately after the psyllid outbreak (Brewbaker 2016). Also, efforts have been made to develop psyllid-resistant cultivars of giant leucaena by making crosses with other *Leucaena* species with higher resistance against the psyllid, such as *L. esculenta* and *L. pallida* (Brewbaker 2008) (see Sect. 18.13).

## 18.9 Leucaena-*Rhizobium* Symbiosis

In greenhouse experiments using the $^{15}N$-labelling method, leucaena was observed to have a consistently increasing pattern of nodulation, dry biomass accumulation, and nitrogen yield over a period of 16 months after planting (Kadiata et al. 1995). In field experiments, using the $^{15}N$ isotope dilution and the total nitrogen difference methods, leucaena K636 was found to fix consistently high levels of atmospheric nitrogen fixation even after third cuttings following 36 months of planting (Sanginga et al. 1989). Most of the *Rhizobium* strains that nodulate leucaena in Hawaii can degrade mimosine completely and use it as a source of carbon and nitrogen (Soedarjo et al. 1994; Soedarjo and Borthakur 1996a). TAL1145 was listed a competitive strain for nodulation of Leucaena in several reports (Moawad and Bohlool 1984; Somasegaran and Martin 1986; George et al. 1994). In competition experiments under field conditions using six indigenous *Rhizobium* strains on Leucaena grown in oxisol and mollisol soils in Hawaii, strain TAL1145 was found to be most competitive (Moawad and Bohlool 1984). Soedarjo and Borthakur (1998) constructed several mimosine-nondegrading ($Mid^-$) mutants of the mimosine-degrading ($Mid^+$) strain TAL1145 and used them in competition experiments with TAL1145 on it. The results of their experiments showed that the mimosine-degrading ability of strain TAL1145 provides a competitive advantage for nodulation of it which contains mimosine in all parts of the plant, including roots and root nodules. *Rhizobium tropici* strain CIAT899 also nodulates Leucaena (Martínez-Romero et al. 1991), although it does not degrade mimosine. By growing leucaena under hydroponic conditions, Soedarjo and Borthakur (1998) showed that some amount of mimosine is secreted in its root exudates. They proposed that mimosine in the rhizosphere binds with ferric iron ($Fe^{3+}$) to produce mimosine-$Fe^{3+}$ complex, which is then taken up by *Rhizobium* and used as a source of nutrients. *Rhizobium* also converts $Fe^{3+}$ into ferrous iron ($Fe^{2+}$) and releases excess $Fe^{2+}$ to the rhizosphere, where it is taken up by its roots (Fig. 18.3a). Soedarjo and Borthakur (1998) also showed that mimosine is present in the leucaena root nodule, where it is used as a source of carbon and nitrogen by resident nodule rhizobia that have not differentiated into the nitrogen-fixing bacteroid form (Fig. 18.3b). The genes for mimosine degradation from TAL1145 have been isolated and characterized (Fox and Borthakur 2001; Borthakur et al. 2003; Awaya et al. 2005, 2007).

## 18.10 Leucaena-Mycorrhiza Symbiosis

Leucaena root system can form symbiotic relationship with soil vascular-arbuscular mycorrhizal fungi (VAMF). Leucaena appears to be highly dependent on VAMF interactions for obtaining nutrients from low-nutrient soils. In phosphorus (P)-deficient soils, uptake of P was significantly greater in its plants containing VAMF interactions than ones without it (Yost and Fox 1979; Habte and Manjunath 1987). In studies comparing VAMF-inoculated and non-inoculated plant, inoculated plants had increased uptake of mobile and immobile nutrients such as P, potassium (K), calcium (Ca), and copper (Cu) from soil (Manjunath and Habte 1988; Huang et al. 1985). Leucaena grown in soils applied with rock phosphate and the VAMF *Glomus aggregatum* had significantly increased uptake of Cu, P, and zinc (Zn) (Manjunath et al. 1989). The increased uptake of nutrients due to VAMF interactions was correlated with increased plant growth. Leucaena inoculated with VAMF compared to non-inoculated had higher dry weight yields, increased plant growth, and greater root and leaf length and area (Huang et al. 1985; Manjunath et al. 1989; Habte and Manjunath 1987; Manjunath and Habte 1988; Osonubi et al. 1991). VAMF has been shown to increase nodulation by *Rhizobium* in leucaena (Manjunath et al. 1984; Purcino et al. 1986). Dual inoculation with VAMF and either *Rhizobium* or P-solubilizing bacteria has been shown to increase nutrient uptake, plant growth, nodulation, and root colonization in a synergistic way (Manjunath et al. 1984; Young 1990; Purcino et al. 1986). Increased VAMF interactions may also play a role in drought tolerance of leucaena. Under water-stressed conditions, leucaena inoculated with VAMF-stimulated P and N uptake (Osonubi et al. 1991). Experimental data collected by Huang et al. (1985) suggested that mycorrhizae enhanced water absorption through the VAMF root system. Increasing potential for nitrogen fixation and acquisition of soil nutrients and avoidance of drought stress may help to establish its root system under a wide range of environmental conditions. Experiments conducted by Brandon et al. (1997) confirmed the importance of VAMF on early seedling growth of it. Studies conducted by Puthur et al. (1998) showed that *in vitro* tissue-cultured micro-propagated plantlets transferred to VAMF-inoculated soils

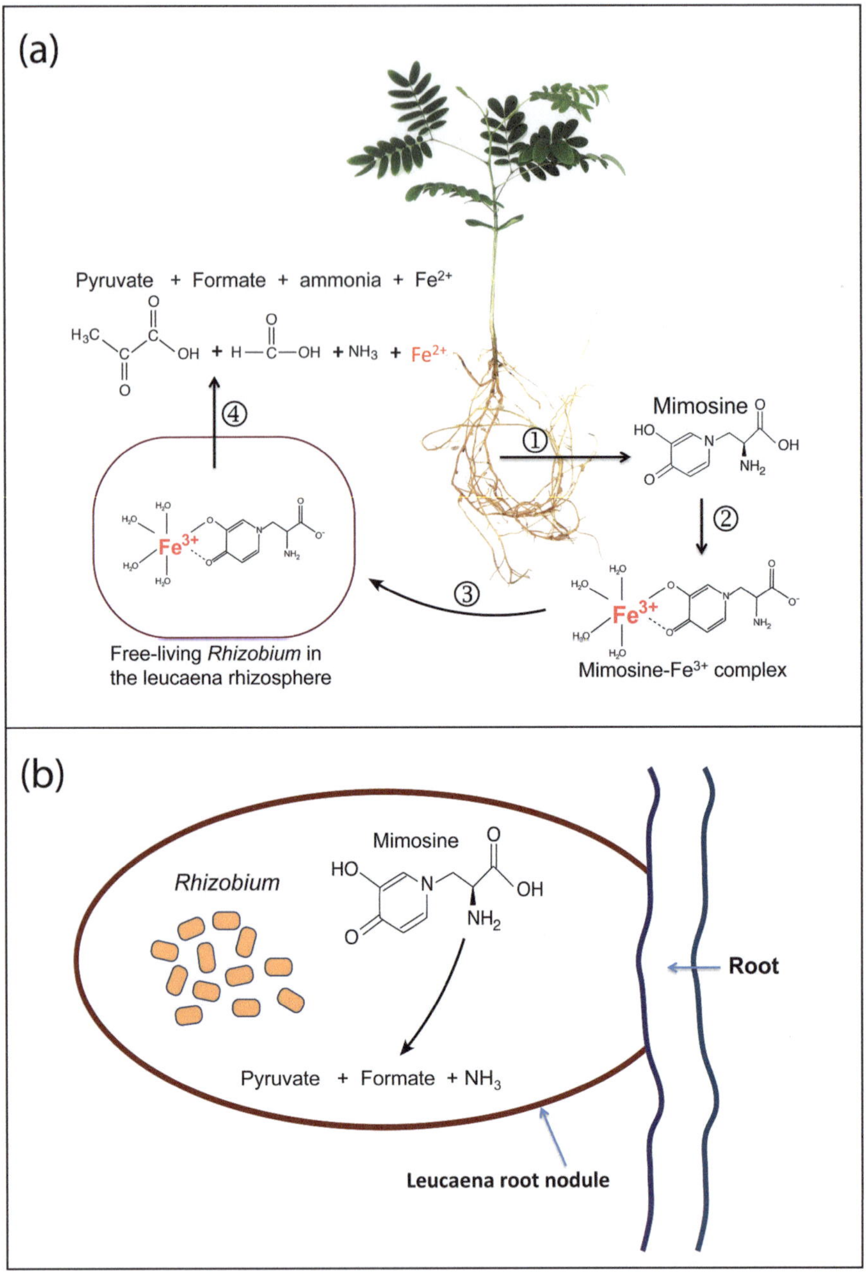


**Fig. 18.3** A possible role of mimosine in the leucaena rhizosphere. (**a**) some amount of mimosine is secreted by its roots to the rhizosphere (step 1); mimosine binds with $Fe^{3+}$ to form a mimosine-$Fe^{3+}$ complex (step 2), which is taken up by free-living rhizobia or other bacteria in the rhizosphere (step 3); rhizobia utilize mimosine by degrading it into pyruvate, formate, pyruvate, and ammonia (step 4). Rhizobia also converts $Fe^{3+}$ to $Fe^{2+}$ and release excess $Fe^{2+}$ in the rhizosphere, where it is taken up by the leucaena roots. **(b)** Rhizobium bacteria present in the root nodule of leucaena also convert mimosine into pyruvate, formate, and ammonia, which are incorporated into metabolic pathways of both bacteria and the plant.

had much better growth and establishment compared to plantlets transferred to non-inoculated soils. Plant genes that are induced because of inoculation by VAMF have been identified in several plants, including *Medicago truncatula*, although such genes have not been yet identified in leucaena. Leucaena transcriptome has been already sequenced and described (Ishihara et al. 2016); therefore, it might be interesting to identify genes that are upregulated and highly expressed because of VAMF inoculation in its root system.

## 18.11 Genetics of *Leucaena leucocephala*

The genus *Leucaena* includes 17 diploid and 5 tetraploid species growing throughout the world, including *L. leucocephala* (Hughes et al. 2007). *Leucaena* species have the basic chromosome number of $x = 26$ or $x = 28$. Most of them are diploid (2n = 2x) having 52 or 56 chromosomes, but there are five known tetraploid species (2n = 4x), including *L. leucocephala*, which is an allotetraploid with a total of 104 chromosomes.

Although chloroplast DNA analyses suggested *L. pulverulenta* or its close relative to be the maternal ancestor (Harris et al. 1994), Harris (1995) further speculated that *L. leucocephala* might have originated from hybridization of *L. pulverulenta* and *L. lanceolata* because they grow in the same geographic area where *L. leucocephala* is also found abundantly. *L. leucocephala* evolved through allopolyploidization following spontaneous hybridization between two diploid species, which was facilitated by human translocation of seed in south-central México over the last 10,000 years (Hughes et al. 2007). Based on the analysis of six SCAR-based nuclear loci, nrDNA ITS, and four cpDNA regions, Govindarajulu et al. (2011a) separated *L. lanceolata* into two distinct species – *L. lanceolata* and *L. cruziana.* Based on another study on polyploid *Leucaena* species using the same sequence-based method, they concluded that *L. pulverulenta* (2n = 56) and *L. cruziana* (2n = 52) are the maternal and paternal diploid progenitors, respectively, of *L. leucocephala* (Govindarajulu et al. 2011b). Accordingly, the chromosome number in *L. leucocephala* should have been 108 instead of 104; apparently, four chromosomes were lost following allopolyploidization. Using FACS flow cytometry, Govindarajulu et al. (2011b) estimated the genome size of *L. leucocephala* to be 2.52–2.93 pg/2C, which is equivalent to 1232–1433 MB per haploid genome.

## 18.12 Leucaena Germplasm Collection and Breeding in Hawaii

In 1962 Dr. James Brewbaker and his colleagues collected 347 accessions of common leucaena from different parts of the world. They also made six expeditions in Latin America collected ~500 accessions of giant leucaena. Some of these accessions grew to mature heights of about 45 ft in 4 years and have high-quality wood and fodder (Brewbaker 2010). Brewbaker and his students planted a duplicate trial of 72 accessions in Waimanalo, Hawaii, and successfully selected many outstanding giant leucaena lines, including K8, K28, K29, K67, and K72. The cultivar "Hawaiian Giant K8" was selected from the progenies of line K8 with emphasis on wood and forage yields (Brewbaker 2016). A few of the promising cultivars developed at the University of Hawaii are listed in Table 18.1. Dr. Brewbaker also introduced leucaena cultivars K5 and K500 in Hawaii from Australia, where they are known as Cunningham and CPI18614, respectively (Brewbaker 2016). K5 is a highly branching, Peru type cultivar of *L. leucocephala* subsp. *glabrata*, whereas K500 is a *L. leucocephala* subsp. *glabrata* Salvador type cultivar that originated from a cross made between the Salvador type and the more branched Peru type of *L. leucocephala* spp. *glabrata* in Australia. Furthermore, Dr. Brewbaker and his students developed many seedless triploid cultivars by making interspecific crosses between diploid and tetraploid species (see Sect. 18.13).

## 18.13 Development of Interspecies Hybrid Cultivars in Hawaii

To make the multipurpose plant more useful and manageable, many studies have been conducted to artificially develop leucaena hybrids with higher quality in Hawaii. The objectives of leucaena breeding in Hawaii include increasing biomass yield (Austin et al. 1997), psyllid resistance for lowlands, and cooler temperature tolerance for the upland tropics and subtropics.

Gonzalez et al. (1967) made interspecies crosses between *L. pulverulenta* (2n = 56) and common leucaena (2n = 104) with a goal to enhance forage quality considering that *L. pulverulenta* contained relatively lower content of mimosine. The very vigorous F1 hybrids had 2n = 80 with 26 bivalents and 28 univalents but were reasonably seedy. The chromosome numbers in advanced progenies obtained by selfing of the F1 hybrids ranged from 2n = 56 to 2n = 104, and the trees varied widely in sterility, arboreal vigor, and leaflet and pod traits. Thus, chromosome numbers and morphological characteristics showed high variations in segregating generations. Several seedless progenies that grew into trees were also obtained among the advanced generation progenies (Gonzalez et al. 1967). Sorensson and Brewbaker (1994) have shown that *Leucaena* species have high interspecific crossability with 77% of 118 combinations of 15 species producing viable seeds. These hybrids are often highly adaptable as they inherit characteristics from both parents. They made many interspecies crosses involving both diploid (2n = 52, 56) and tetraploid (2n = 104, 112) species of *Leucaena*. For these interspecies hybridization, they used six diploid species with 2n = 52 (*L. esculenta, L. lanceolata, L. macrophylla, L. trichandra, L. trichodes,* and *L. shannonii*), five diploid species with 2n = 56 (*L. collinsii, L. greggii, L. pulverulenta, L. retusa,* and *L. salvadorensis*), three tetraploid species with 2n = 104 (*L. leucocephala, L. diversifolia,* and *L. pallida*), and one tetraploid species with 2n = 112 (*L. confertiflora*). Ninety-nine hybrids were obtained from the 141 combinations that formed viable seeds (Sorensson 1993). All diploid species and *L. pallida* (2n = 104, tetraploid) were shown to be self-incompatible. All intercrosses of the four tetraploids were fully fertile, whereas the F1 hybrids obtained from a cross between diploids (2n = 52, 56)

**Table 18.1** Some promising cultivars of *L. leucocephala* subsp. *glabrata* developed at the Waimanalo Research Station, University of Hawaii

| Variety | Parent species | Characteristics | Reference/ Source |
|---|---|---|---|
| K636 or Tarramba | *L. leucocephala* subsp. *glabrata* accessions collected from Coahuila, Mexico | High forage yield; good seedling vigor; cold tolerant; low branching; highly psyllid resistant; less seedy, difficult for seed multiplication | Brewbaker (1987) |
| K8 or Hawaiian Giant | *L. leucocephala* subsp. *glabrata* accessions collected from Zacatecas in Mexico | Most widely gown leucaena cultivar in the world. It combines vegetative vigor with aggressive arboreal growth, high leaf, and wood yield. It is grown in the Philippines for charcoal and fuel, and in Hawaii as a fast-growing, deep-rooted, windbreak. When harvested for forage every 8–12 weeks, it has produced much higher yields than other common cultivars; susceptible to psyllid | Brewbaker (1975) |
| K28 | *L. leucocephala* subsp. *glabrata* accessions collected from El Salvador | Multipurpose cultivar; high foliage and wood yield; seedling vigor; tolerant to acidic soil; widely distributed and considered superior in wood yields to K8. Reported to perform marginally better than K636 in acid soils | Brewbaker and Hylin (1965) |
| K29 | *L. leucocephala* subsp. *glabrata* Salvador type accessions collected from Honduras | Multipurpose cultivar; high foliage and wood yield; seedling vigor, low seediness but full male fertility; difficult for seed multiplication | Brewbaker and Hylin (1965) |
| K67 | *L. leucocephala* subsp. *glabrata* Salvador type accessions collected from El Salvador | High foliage and wood yield; high seed production; seedling vigor; like K8 in foliage yield | Brewbaker et al. (1972) |
| K132 | *L. leucocephala* subsp. *glabrata* Peru type accessions collected from Zacatepec, Morelos, Mexico | Lower in branching; large pods favored as food source; high foliage and wood yield | Brewbaker (2016) |
| K584 | *L. leucocephala* subsp. *glabrata* accessions collected from Veracruz, Mexico | It has a combined stature of both Salvador and Peru type; giant leucaena with erect form like K636 but has more branching like the Peru type; highly psyllid resistant | Brewbaker (2016) |
| Wondergraze | K636 and K584 | Seedling vigor; shorter stature, basal branching, and bushier like the Peru type; like K636 for psyllid and cold tolerance; high forage yield; good forage quality and palatable | Brewbaker (2016) |

and tetraploids (2n = 104, 112) were mostly sterile (Sorensson and Brewbaker 1994). A vigorous interspecies hybrid variety, KX1, with impressive morphological characteristics, such as low-forking and arboreal vigor, was selected from the fertile progenies obtained from a cross between *L. diversifolia* (2n = 104) and *L. pallida* (2n = 104). Another interspecies hybrid variety, KX2, was selected and advanced from the progenies of a cross between *L. leucocephala* (2n = 104) and *L. pallida* (2n = 104). It was selected for its tolerance against the psyllid, and the resulting hybrid is self-sterile yet has high forage yield as well as high resistance to the psyllid insects and cold temperatures (Brewbaker 2008). Similarly, vigorous self-fertile progenies (KX3) were obtained from a cross between *L. diversifolia* (2n = 104) and *L. leucocephala* (2n = 104). The interspecies hybrid variety designated as "KX3-Hawaii" was derived from the KX3 progenies, followed by two cycles of recurrent selection at a midland station in Hawaii (700 m). The most distinguishing characteristics of this hybrid are its superior arboreal vigor and cold tolerance (Brewbaker 2016). The seedless triploid hybrid "KX4-Hawaii" was derived from an interspecies hybrid of *L. esculenta* (2n = 52) and *L. leucocephala* subsp. *glabrata* (2n = 104). The hybrid shows vigorous growth with arboreal plant type (Fig. 18.1d). It is a seedless triploid hybrid with high drought tolerance, wide environmental adaptability, and no known pests or diseases. It also grows rapidly to 15 m in 5 years with an open canopy, so it is attractive as an ornamental and shade tree (Brewbaker 2013). Due to the ease of interspecific crossing, *L. leucocephala* is useful to develop superior *Leucaena* hybrids with a higher potential for commercial use. Another seedless triploid hybrid, named KX5, was derived from a cross between *L. diversifolia* (2n = 104) and *L. pulverulenta* (2n = 56). It shows impressive arboreal vigor, tall, forking, and tolerance to psyllid, cold temperature condition, and low pH (Sorensson and Brewbaker 1994; Brewbaker 2016).

## 18.14 Conclusion

Thickets of "koa haole" or common leucaena are seen on the roadsides, hill slopes, and barren lands of the Hawaiian Islands, while trees of giant leucaena are generally limited to only few plantations. Cattle and poultry farms on the Hawaiian Islands, which are located ~4000 km away from the coast of California, must import a lot of animal feed from the US mainland. The high-yielding varieties of giant leucaena provide an opportunity to Hawaii for being self-sufficient with animal feed. Because of the nitrogen-fixing and fast-growing nature of leucaena, growing and maintaining plantations for animal feed should be profitable. The main obstacle for implementing this idea is a common misunderstanding among people that leucaena is "koa haole" (Brewbaker 2010). A strong extension program will be needed to explain to farmers the difference between the cultivated giant leucaena and the weedy common leucaena.

A major advantage of growing giant leucaena is that it can be managed as a legume fodder by maintaining the plants as dwarf bushes at a manageable height (~1–3 m) by repeated harvest of foliage, up to ten times a year. For high fodder yield, it can be grown at a close spacing of 1.0 m between rows and 0.5 m between trees. In this way, even the Salvador type of cultivars such as K636, K8, K28, K29, or K67 can be grown for fodder; repeated cuttings convert the trees into thick shrubby bushes with a lot of branches and foliage. The cultivars of Peru type, such as Wondergraze, K132, and K584, naturally form a lot of basal branches; so they are even easier to manage as bushy shrubs through repeated harvest of foliage. The presence of mimosine in the leucaena foliage also should not be a big concern, because ruminants grazing on it in Hawaii generally contain rumen bacteria that can degrade and detoxify mimosine (Jones and Megarrity 1986). Ruminants that do not carry such mimosine-degrading rumen bacteria can be inoculated with *Synergistes jonesii* (Alison et al. 1992). Alternatively, mimosine present in its foliage can be removed, easily and inexpensively (Soedarjo and Borthakur 1996a, b, Tawata et al. 2008).

The trees of the Salvador type of cultivars, if not harvested repeatedly for fodder, grow to become big trees, up to ~20 m in height within a few years. The wood can be used for timber, paper pulp, or biofuel production. They can be grown in the slopes of hills to prevent erosion and landslide. They can also be easily grown to convert barren eroded lands into beautiful forests within a few years. The cultivars of giant leucaena do not form too many seeds; therefore, they are not invasive. The triploid cultivars are sterile and they generally do not form seeds. Triploid leucaena shows hybrid vigor and they can be propagated from cuttings.

The giant leucaena may provide opportunities for new industries in the future. Besides supporting dairy and paper industry, the foliage can be used for extraction of mimosine, which is extensively used in biomedical research besides using as an iron-chelating drug. Another compound, 3-hydoxy-4-pyridone, which is used for research on iron chelation treatment, can be manufactured from mimosine, using a recombinant enzyme mimosinase or rhizomimosinase (Negi et al. 2013, 2014; Negi and Borthakur 2016). Recently, Nguyen et al. (2015) developed several amino alcohol derivatives of mimosine that can be used as pesticides. Researchers from the same laboratory also previously developed two propionate derivatives of mimosine that can be used as herbicides (Xuan et al. 2013). After extraction of mimosine from the leaves, the mimosine-free leaves can be dried to produce herbal tea (Tawata et al. 2008). Considering several key attributes of leucaena, such as nitrogen-fixing ability, drought tolerance, easy cultivation, and high protein content of foliage, it will be worthwhile to conduct more research on developing nutritious and palatable feed from it for all animals, including poultry, horses, pigs, and fishes.

**Acknowledgments** This work was supported by the USDA NIFA Hatch project HA05029-H, managed by CTAHR, University of Hawaii at Manoa, Honolulu. Authors would like to thank Dr. James Brewbaker for the useful discussion.

## References

Allison MJ, Mayberry WR, Mcweeney CS, Stahl DA (1992) *Synergistes jonesii*, gen. nov., sp. nov.: a rumen bacterium that degrades toxic pyridinediols. Syst Appl Microbiol 15:522–529

Anthraper A, Dubois JD (2003) The effect of NaCl on growth, $N_2$ fixation (acetylene reduction), and percentage total nitrogen in *Leucaena leucocephala* (Leguminosae) var. k-81. Am J Bot 90:683–692

Austin MT, Early RJ, Brewbaker JL, Sun W (1997) Yield, psyllid resistance, and phenolic concentration of leucaena in two environments in Hawaii. Agron J 89:507–515

Awaya JD, Fox PM, Borthakur D (2005) *pyd* genes of *Rhizobium* sp. strain TAL1145 are required for degradation of 3-hydroxy-4-pyridone, an aromatic intermediate in mimosine metabolism. J Bacteriol 187(13):4480–4487

Awaya JD, Walton C, Borthakur D (2007) The *pydA-pydB* gene produces an active dioxygenase-hydrolase that degrades 3-hydroxy-4-pyridone, an intermediate of mimosine metabolism. Appl Microbiol Biotechnol 75:583–588

Borthakur D, Soedarjo FPM, Webb DT (2003) The *mid* genes of *Rhizobium* sp. strain TAL1145 are required for degradation of mimosine into 3-hydroxy-4-pyridone and are inducible by mimosine. Microbiology 149:537–546

Bourges H, Morales de León JC (1980) The experience in Mexico on the utilization of non-conventional protein sources. Nutr Food Sci 2:277–287

Brandon NJ, Shelton HM, Peck DM (1997) Factors affecting the early growth of *Leucaena leucocephala*. 2. Importance of arbuscular mycorrhizal fungi, grass competition and phosphorus application on yield and nodulation of leucaena pots. Aust J Exp Agric 37:35–43

Brewbaker JL (1975) Registration of Hawaiian Giant K8 Leucaenal (Reg. No. 16). Crop Sci 15(6):885

Brewbaker JL (1987) *Leucaena*: a multipurpose tree genus for tropical agroforestry. In: Steppler HA, PKR N (eds) Agroforestry – a decade of development. International Council for Research in Agroforestry (ICRAF), Nairobi, pp 289–324

Brewbaker JL (2008) Registration of 'KX2-Hawaii', interspecific-hybrid leucaena. J Plant Registration 2:190–193

Brewbaker JL (2010) Leucaena is not Koa Haole. Hawaii For J 5:5–8

Brewbaker JL (2013) 'KX4-Hawaii', seedless interspecific hybrid Leucaena. Hortscience 48:1–2

Brewbaker JL (2016) Breeding Leucaena: tropical multipurpose leguminous tree. Plant Breed Rev 40:43–121

Brewbaker JL, Hylin JW (1965) Variations in mimosine content among *Leucaena* species and related Mimosaceae. Crop Sci 5:348–349

Brewbaker JL, Pluckett D, Gonzalez V (1972) Varietal variation and yield trials of *Leucaena leucocephala* (Koa haole) in Hawaii. Hawaii Agric Exp Station Bull 166:3–29

Brewbaker JL, Sorensson CT, Wheeler RW (1990) New tree crops from interspecific Leucaena hybrids. In: Janick J, Simon JE (eds) Advances in new crops. Timber Press, Portland, pp 283–289

Cavalcante ADMB, de Perez SCJGA (1995) Effects of water and salt stresses on germination of *Leucaena leucocephala* (Lam.) de Wit seeds. Pesqui Agropecuaria Bras 30:281–289

Chanchay N, Poosaran N (2009) The reduction of mimosine and tannin contents in leaves of *Leucaena leucocephala*. Asian J Food Agro-Ind:S137–S144

Chen Y, Chen F, Liu L, Zhu S (2012) Physiological responses of Leucaena leucocephala seedlings to drought stress. Procedia Eng 28:110–116

Dalzell SA, Shelton HM, Mullen BF, Lauren PH, McLaughlin KG (2006) Leucaena: a guide to establishment and management. Meat & Livestock Australia LTd, Sydney. isbn:1 7419 1013 7

Dijkman MJ (1950) Leucaena-A promising soil-erosion-control plant. Econ Bot 4:337

Ezenwa IV, Atta-Krah AN (1992) Early growth and nodulation in Leucaena and Gliricidia and the effects of pruning on biomass productivity. In: Mulongoy K, Gueye M, Spencer DSC (eds)

Biological nitrogen fixation and sustainability of tropical agriculture. International Institute of Tropical Agriculture, Ibadan, pp 171–178

Fox PM, Borthakur D (2001) Selection of several classes of mimosine-degradation-defective Tn3Ho*gus*-insertion mutants of *Rhizobium* sp. strain TAL1145 on the basis of mimosine-inducible GUS activity. Can J Microbiol 47:488–494

Funasaki GY, Lai PY, Nakahara LM (1989) Status of natural enemies of Heteropsylla cubana Crawford (Homoptera: psyllidae) in Hawaii, pp. 153–158 in Leuceana psyllid: problems and management. In: Proceedings of an International Workshop held January 16–21, Bogor

Garcia GW, Ferguson TU, Neckles FA, Archibald KAE (1996) The nutritive value and forage productivity of *Leucaena leucocephala*. Anim Feed Sci Technol 60:29–41

George MLC, Young JPW, Borthakur D (1994) Genetic characterization of *Rhizobium* sp. strain TAL1145 that nodulates tree legumes. Can J Microbiol 40:208–215

Gonzalez V, Brewbaker JL, Hamill DE (1967) Leucaena cytogenetics in relation to the breeding of low mimosine lines. Crop Sci 7(2):140–143

Govindarajulu R, Hughes CE, Bailey CD (2011a) Phylogenetic and population genetic assessment of diploid Leucaena reveal cryptic species diversity and patterns of allopatric divergent speciation. Am J Bot 98:2049–2063

Govindarajulu R, Hughes CE, Alexander PJ, Bailey CD (2011b) The complex evolutionary dynamics of ancient and recent polyploidy in *Leucaena* (Leguminosae; Mimosoideae). Am J Bot 98:2064–2076

Habte M, Manjunath A (1987) Soil solution phosphorus status and mycorrhizal dependency in *Leucaena leucocephala*. Appl Environ Microbiol 53:797–801

Harris SA, Hughes CE, Ingram R, Abbott RJ (1994) A phylogenetic analysis ofLeucaena (Leguminosae: mimosoideae). Plant Syst Evol 191(1–2):1–26

Harris SA (1995b) Systematics and randomly amplified polymorphic DNA in the genusLeucaena (Leguminosae, mimosoideae). Plant Syst Evol 197(1-4):195–208

Hongo F, Shiroma S, Kawashima Y, Sunagawa K, Kawamoto Y, Tawata S (1990) Effect of *Leucaena* tea on rats. Leucaena Res Rep 11:74–75

Huang RH, Smith WK, Yost RS (1985) Influence of vesicular-arbuscular mycorrhizae on growth, water relations, and leaf orientation in *Leucaena leucocephala* (Lam.) de Wit. New Phytol 99:375–378

Hughes CE, Govindarajulu R, Robertson A, Harris SA, Bailey CD (2007) Serendipitous backyard hybridization and the origin of crops. Proc Natl Acad Sci U S A 104:14389–14394

Ishihara KL, Honda MDH, Pham DT, Borthakur D (2016) Transcriptome analysis of *Leucaena leucocephala* and identification of highly expressed genes in roots and shoots. Transcriptomics 4:135

Jayanthy V, Geetha R, Rajendran R, Prabhavathi P, Karthik Sundaram S et al (2014) Phytoremediation of dye contaminated soil by *Leucaena leucocephala* (subabul) seed and growth assessment of *Vigna radiata* in the remediated soil. Saudi J Biol Sci 21:324–333

Jones RJ (1979) The value of *Leucaena leucocephala* as a feed for ruminants in the tropics. World Anim Rev 31:13–23

Jones RJ, Megarrity RG (1986) Successful transfer of DHP-degrading bacteria from Hawaiian goats to Australian ruminants to overcome the toxicity of *Leucaena*. Aust Vet J 63:259–262

Juson AEDS, Martinez MKM, Ching JA (2016) Accumulation and distribution of heavy metals in *Leucaena leucocephala* Lam. and *Bougainvillea spectabilis* Willd. plant systems. J Exp Biol Agric Sci 4:1–6

Kadiata BD, Mulongoy K, Isirimah NO (1995) Dynamics of nodulation, nitrogen fixation, nitrogen use and biomass yield over time in pot-grown *Leucaena leucocephala* (Lam.) de Wit. Biol Fertil Soils 20:163–168

López F, García MM, Yánez R, Tapias R, Fernández M et al (2008) Leucaena species valoration for biomass and paper production in 1 and 2 year harvest. Bioresour Technol 99:4846–4853

Manjunath A, Habte M (1988) Development of vesicular-arbuscular mycorrhizal infection and the uptake of immobile nutrients in *Leucaena leucocephala*. Plant Soil 106:97–103

Manjunath A, Bagyraj DJ, Gopala Gowda HS (1984) Dual inoculation with VA mycorrhiza and *Rhizobium* is beneficial to *Leucaena*. Plant Soil 78:445–448

Manjunath A, Hue NV, Habte M (1989) Response of *Leucaena leucocephala* to vesicular-arbuscular mycorrhizal colonization and rock phosphate fertilization in an oxisol. Plant Soil 114:127–133

Martínez-Romero E, Segovia L, Mercante FM, Franco AA, Graham P et al (1991) *Rhizobium tropici*, a novel species nodulating *Phaseolus vulgaris* L. beans and *Leucaena* sp. trees. Int J Syst Bacteriol 41:417–426

Mawardi AMA (1985) The general uses of lamtoro (*Leucaena*) in Indonesia. Leucaena Res Rep (EUA) 6:57–58

McClay AS (1989) Distribution of *Leucaena* psyllid and its natural enemies in Mexico: implications for biological control. In: Napompeth B, MacDicken KG (eds) Leucaena psyllid: problems and management, Proceedings of international workshop, Bogor, Indonesia; Winrock international and NFTA, Hawaii, p139–143

Mislevy P, Blue WG, Roessler CE (1989) Productivity of clay tailings from phosphate mining. I Biomass crops. J Environ Qual 18:95–100

Moawad H, Bohlool BB (1984) Competition among *Rhizobium* spp. for nodulation of *Leucaena leucocephala* in two tropical soils. Appl Environ Microbiol 48:5–9

Mushaka A, Maruzane D (1998) Performance of some multipurpose tree species: the forestry commission experience. Afr J Online 72:10–14

National Academy of Sciences (1977) Leucaena – promising tree and forage crop for the tropics. National Academy of Sciences, Washington, DC, p 121

Negi VS, Borthakur D (2016) Heterologous expression and characterization of mimosinase from *Leucaena leucocephala*. Methods Mol Biol 1405:59–77

Negi VS, Bingham J-P, Li QX, Borthakur D (2013) *midD*-encoded 'rhizomimosinase' from *Rhizobium* sp. strain TAL1145 is a C–N lyase that catabolizes L-mimosine into 3-hydroxy-4-pyridone, pyruvate and ammonia. Amino Acids 44:1537–1547

Negi VS, Bingham J-P, Li QX, Borthakur D (2014) A carbon-nitrogen lyase from *Leucaena leucocephala* catalyzes the first step of mimosine degradation. Plant Physiol 164:922–934

Nguyen BCQ, Chompoo J, Tawata S (2015) Insecticidal and nematicidal activities of novel mimosine derivatives. Molecules 20:16741–16756

Osborne NJT, McNeill DM (2001) Characterization of *Leucaena* condensed tannins by size and protein precipitation capacity. J Sci Food Agric 81:1113–1119

Osonubi O, Mulongoy K, Awotoye OO, Atayese MO, Okali DUU (1991) Effects of ectomycorrhizal and vesicular-arbuscular mycorrhizal fungi on drought tolerance of four leguminous woody seedlings. Plant Soil 136:131–143

Othman AB, Prine GM (1984) Leucaena accessions resistant to jumping plant lice. Leucaena Res Rep 5:86–87

Pandey VC, Kumar A (2013) *Leucaena leucocephala*: an underutilized plant for pulp and paper production. Genet Resour Crop Evol 60:1165–1171

Poole H (2003) Dryland salinity management in Central Queensland using *Leucaena leucocephala*. 4th year project, bachelor of encironmental science (natural resource science). University of Queensland, Brisbane

Purcino AAC, Lurlarp C, Lynd JQ (1986) Mycorrhiza and soil fertility effects with growth, nodulation and nitrogen fixation of *Leucaena* grown on a typic eutrustox. Commun Soil Sci Plant Anal 17:473–489

Puthur JT, Prasad KVSK, Sharmila P, Pardha Saradhi P (1998) Vesicular arbuscular mycorrhizal fungi improves establishment of micropropagated *Leucaena leucocephala* plantlets. Plant Cell Tissue Organ Cult 53:41–47

Qamar IA, Muhammad I, Louhaichi M (2015) Managing rangelands: promoting sustainable tree species *Leucaena leucocephala*: a versatile tree producing nutritious fodder for ruminants. Technical Report, ICARDA January 15, 2015. https://doi.org/10.13140/RG.2.1.2108.5920

Rao PB, Kaur A, Tewari A (2008) Drought resistance in seedlings of five important tree species in Tarai region of Uttarakhand. Int J Trop Ecol 49:43–52

Roose E, Ndayzigiye F (1997) Agroforestry, water and soil fertility management to fight erosion in tropical mountains of Rowana. Soil Technol 11:109–119

Sanginga N, Zapata F, Danso SKA, Bowen GD (1989) Effect of successive cutting on nodulation and nitrogen fixation of *Leucaena leucocephala* using 15N dilution and the difference methods. In: Proceedings of the Eleventh International Plant Nutrition Colloquium, Wageningen, The Netherlands, 30 July–4 August 1989

Shelton HM, Brewbaker JL (1994) *Leucaena leucocephala* – the most widely used forage tree legume. In: Gutteridge RC, Shelton HM (eds) Forage tree legumes in tropical agriculture. CAB International, London, pp 15–29

Shelton HM, Dalzell S (2007) Production, economic and environmental benefits of leucaena pastures. Trop Grasslands 41:174–190

Soedarjo M, Borthakur D (1996a) Mimosine produced by the tree-legume *Leucaena* provides growth advantages to some *Rhizobium* strains that utilize it as a source of carbon and nitrogen. Plant Soil 186:87–92

Soedarjo M, Borthakur D (1996b) Simple procedures to remove mimosine from young leaves, pods and seeds of *Leucaena leucocephala* used as food. Int J Food Sci Technol 31:97–103

Soedarjo M, Borthakur D (1998) Mimosine, a toxin produced by the tree-legume *Leucaena* provides a nodulation competition advantage to mimosine-degrading *Rhizobium* strains. Soil Biol Biochem 30:1605–1613

Soedarjo M, Hemscheidt TK, Borthakur D (1994) Mimosine, a toxin present in the tree legume *Leucaena*, induces a mimosine-degrading enzyme activity in some strains of *Rhizobium*. Appl Environ Microbiol 60:4268–4272

Somasegaran PS, Martin RB (1986) Symbiotic characteristics and *Rhizobium* requirements of a *Leucaena leucocephala* × *Leucaena diversifolia* hybrid and its parental genotypes. Appl Environ Microbiol 52:1422–1424

Sorensson CT (1993) Production and characterization of interspecific hybrids of the tropical tree *Leucaena* (Leguminosae: Mimosoideae). Dissertation, University of Hawaii, Honolulu

Sorensson CT, Brewbaker JL (1984) Newly introduced psyllid in Hawaii injurious to *Leucaena*. Leucaena Res Rep 5:91–93

Sorensson CT, Brewbaker JL (1994) Interspecific compatibility among 15 *Leucaena* species (Leguminosae: Mimosoideae) via artificial hybridizations. Am J Bot 81:240–247

Sun WG (1996) Genetic improvement of *Leucaena* and *Acacia koa*. Dissertation, University of Hawaii, Honolulu

Takahashi M, Ripperton JC (1949) *Koa haloe* (*Leucaena glauca*) Its establishment, culture and utilization as a forage crop. Bulletin 100 University of Hawaii Agricultural Experiment Station

Tangendjaja B, Lowry JB, Wills RBH (1986) Changes in mimosine, phenol, protein and fibre content of Leucaena leucocephala leaf during growth and development. Aust J Exp Agric 26:315–317

Tawata S, Fukuta M, Xuan TD, Farah Deba F (2008) Total utilization of tropical plants *Leucaena leucocephala* and *Alpinia zerumbet*. J Pestic Sci 33:40–43

Wheeler RA, Brewbaker JL (1988) Leucaena research for Hawaiian forest and rangeland applications. Trans West Sect WildLife Soc 24:94–97

Wheeler RA, Norton BW, Shelton HM (1995) Condensed tannins in *Leucaena* species and hybrids and implications for nutritive value. In: Shelton HM, Piggin CM, Brewbaker JL (eds) *Leucaena* – opportunities and limitations. ACIAR, Canberra, pp 112–118

Xuan TD, Tawata S, Khanh TD (2013) Herbicidal activity of mimosine and its derivatives. In: Price AJ, Kelton JA (eds) Herbicides-advances in research, agricultural and biological sciences. Intech. Chapter 15, Rijeka, pp 299–312

Yost RS, Fox RL (1979) Contribution of mycorrhizae to P nutrition of crops growing on an oxisol. Agron J 71:903–908

Young CC (1990) Effects of phosphorus-solubilizing bacteria and vesicular-arbuscular mycorrhizal fungi on the growth of tree species in subtropical-tropical soils. Soil Sci Plant Nutr 36:225–231

Zarate PS (1984) Taxonomic revision of the genus *Leucaena* from Mexico. Bul Intl Group Study Mimosoideae 12:24–34

J. L. Brewbaker, (1975) Registration of Hawaiian Giant K8 Leucaenal (Reg. No. 16). Crop Science 15 (6):885

Stephen A. Harris, (1995) Systematics and randomly amplified polymorphic DNA in the genusLeucaena (Leguminosae, Mimosoideae). Plant Systematics and Evolution 197 (1-4):195-208

# Ravines: Prospective Zone for Carbon Sequestration

# 19

J. Somasundaram, A. K. Parandiyal, Pramod Jha, Brij Lal Lakaria, R. K. Singh, B. L. Mina, S. Kala, and Shakir Ali

**Abstract**

Ravines are the extreme form of land degradation owing to water erosion and along with gullies of various sizes occupy about 10.37 million ha area in the country which could be utilized for sequestering carbon through improving vegetation densities/plantations. In fact, ravines are the most fragile ecosystems that have very low soil carbon content due to their light texture and poor aggregate stability. Thus, there is an urgent need to manage and restore these lands with suitable cultural and management practices. In fact, utilization of medium and deep ravine lands for regular cultivation always remains challenging; however, these ravines can be alternatively utilized for energy plantation, augmentation of fuel and fodder demands for local populace and production of hardy underutilized fruits and oil seed-bearing tree/shrub species. Ravine lands, which are economically unsuitable for agriculture, can be successfully stabilized by planting fast-growing species like bamboo on the gully beds and buffel grass/dhaman grass (*Cenchrus ciliaris)* on the side slopes and the interspaces of gully bed for economic utilization of gullied land. It was also evident that *Acacia nilotica*, *Bambusa* species and *Aegle marmelos* are highly suitable for ravine area and provide a substantial role in carbon sequestration under ravine landforms of Chambal and Yamuna river. Similarly, *Prosopis juliflora* and *Azadirachta indica* have greater ability for sustaining site productivity due to their greater leaf litterfall and fine root production

J. Somasundaram (✉) · P. Jha · B. L. Lakaria
ICAR-Indian Institute of Soil Science, Nabibagh, Bhopal, Madhya Pradesh, India

A. K. Parandiyal
ICAR-Indian Institute of Soil and Water Conservation, Research Centre, Agra, Uttar Pradesh, India

R. K. Singh · B. L. Mina · S. Kala · S. Ali
ICAR-Indian Institute of Soil and Water Conservation, Research Centre, Kota, Rajasthan, India



J. C. Dagar, A. K. Singh (eds.), *Ravine Lands: Greening for Livelihood and Environmental Security*, https://doi.org/10.1007/978-981-10-8043-2_19

under Yamuna ravines. In this chapter, an attempt has been made to address some of the issue of ravines vis-à-vis potential zone for carbon sequestration.

## 19.1 Introduction

The land degradation due to water is a major problem in India and affects more than 28% of total geographical area (TGA) of the country (ICAR 2010; Dhruvanarayana 1993). Among various forms of land degradation, water affects 10.37 million ha area under ravines and gully formation (GOI 2003; ICAR 2010), and this is a major problem along several river systems in the alluvial zones in India. Ravines are the worst form of land degradation due to water erosion and spread about 2 km vicinity of river systems and their adjoining tributaries. Ravine lands and adjoining marginal lands are mainly treated as either wastelands or managed as low-input farming systems. With rapidly shrinking per capita availability of land, there is growing need for restoration of these areas for productive land uses and for perennial vegetation to stabilize the ecosystems. Depletion of the soil organic matter/carbon (SOC) results in soil degradation, which may further jeopardize economic and ecological consequences, because the SOC serves as a sink of principal plant nutrients and governs other pedogenic functions. Soil carbon sequestration through the restoration of soil organic matter can further reverse land degradation and restore soil "health" through restoring soil biota and the array of associated ecological processes. The rapid increase in global carbon dioxide ($CO_2$) level in the atmosphere in the recent past and its potential for adversely impacting the world climate has raised concerns world over. Continued rise in atmospheric $CO_2$ concentration (~400 ppm) and its adverse effect on global climate make it necessary to explore all possible mitigation strategies. Among various strategies to reduce the atmospheric $CO_2$ concentration, enhancing the natural carbon sequestration capabilities of degraded or underproductive lands by vegetation/afforestation is the major thrust area. It is essential that lands with a degraded vegetation cover having lower carbon biomass density (the mass of carbon per unit area) than the potential for that site are identified as potential sites for improving as carbon sink. The Kyoto protocol proposed carbon (C) reduction through decreasing fossil fuel emission or accumulating C in vegetation and soil (Oelbermann et al. 2004). A total of 173.64 million ha degraded land have reported in India, in which degraded forest land covers about 19.50 million ha and 4.91 million ha land degraded due to shifting cultivation (MOA 1985). However, recent estimates indicated that ravines occupy 2.06 m ha in India and gully formation occurs in 8.31 million ha (35%) area (ICAR-NAAS 2010) (Fig. 19.1). Utilization of medium and deep ravine lands for recurring cultivation is always a challenging; however, these ravines can be potentially utilized for energy plantation, augmentation of fuel and fodder demands for local populace and production of hardy underutilized fruits and oil seed-bearing tree/shrub species. The degraded land under ravines is the most potential area which can more easily be used for enhanced carbon sequestration through improving vegetation densities. In this chapter, an attempt has been made to address some of the issue of ravines vis-à-vis potential zone for carbon sequestration.

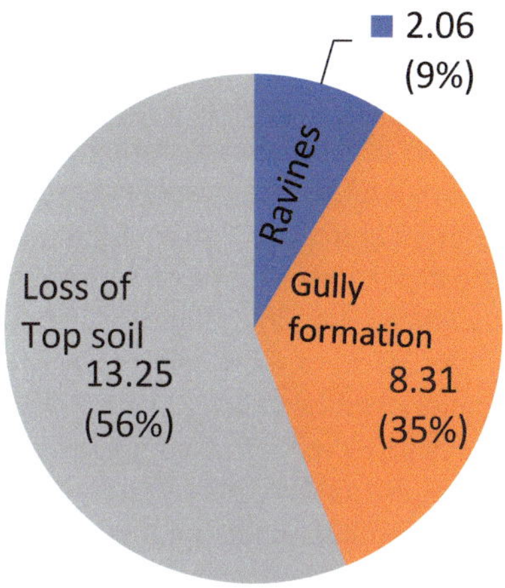


**Fig. 19.1** Forms of water erosion

## 19.2 Constraints of Ravines in Vegetation Development

Ravines are the system of gullies running almost parallel to each other and draining in to a river after a short distance with the development of deep gorges. If left unattended, these ravines tend to ingress into adjoining tablelands, and vast areas are becoming unproductive. The major problem areas are in Uttar Pradesh, Madhya Pradesh, Rajasthan, Gujarat and Bihar (Sharma 1980; Verma et al. 1986; ICAR-NAAS 2010). Many factors are responsible for the formation of ravine land such as erratic, short-duration and high-intensity rainfall; erodible nature of soil; multidirectional slopes; weak geology of alluvium; steep slopes and uneven terrain; faulty agricultural practices; and illicit cutting of trees and bushes.

In nature, ravine lands possess severe constraints for development of vegetation. The major constraints are highly undulating topography resulting high desiccation of land, unfriendly climate with long dry period and extremes of temperature during a year, highly eroded soils with low fertility status, low soil moisture availability and retention and high biotic pressure results in poor natural vegetation. Thus, in nature the ravines are characterized by dominance of thorny vegetation. The dominance of unpalatable natural vegetation enhances the biotic pressure on the planted vegetation, thus reducing the establishment and survival of economically utilizable vegetation. Indeed, extremes of climatic and edaphic conditions result in low productivity in adjoining marginal lands also. Therefore, except shallow ravines which are amenable to terracing and/or levelling and major part of ravines, i.e. medium and deep ravines, which are highly vulnerable to accelerated erosion and further degradation if placed under recurring tillage are not fit for cultivation for high-value agriculture or development of vegetation with high economic value. Deep and narrow gullies are stabilized by putting them under permanent vegetation of grasses and trees. In such cases, bamboo plantation for productive and protective utilization of such degraded lands is not only a profitable option for local stakeholders but also financially and economically viable policy option for different developmental agencies (Singh et al. 2015).

## 19.3 Scope of Carbon Sequestration in Ravine Lands

Soil organic matter is a key indicator for soil quality. The amount, diversity and activity of soil fauna and microorganisms are directly related to the organic matter. The biological changes due to organic matter in soil influence physical properties in term of aggregation; infiltration rate; available water capacity of the soil, as well as resistance against erosion by water and wind; and chemical properties in terms of dynamics and bioavailability of main plant nutrient elements to plants (Robert et al. 1996). Most of the world's soils used for agriculture have been depleted of organic matter over the years under the conventional farming systems of ploughing or hoeing before every crop, compared with their state under natural vegetation. This degradation process can be reversed through changes in land management practices. In many farmers' fields, in humid and subhumid climates as well as under irrigation, organic matter contents have been increasing rapidly after a change in land management procedures to conservation agriculture, including zero or minimum tillage and retention of residues on the soil surface. Even in semiarid conditions, the system is working but at lower rates of carbon sequestration.

Other management issue is the use of fertilizers and irrigation that have a direct impact on plant production and hence on soil organic matter (SOM) content. The rate of SOM mineralization depends mainly on temperature and oxygen availability (drainage), land use, cropping system and soil and crop management (Lal 1995). Carbon sequestration in agricultural soils through adopting conservation tillage coupled with other improved land management can be permanent if the farmers continue to use these practices. Ravines also have potential for carbon sequestration in degraded lands which can be restored by measures such as land levelling, plantation of locally dominating perennial trees and sowing strips of legumes in the spaces wherever possible and by needful fertilizer application. Also, higher production potential is interlocked in this ravinous uncultivated wasteland. To a large extent, the rates at which organic matter is provided to and cycles through riparian soils mediate their effectiveness as living filters between agricultural fields and surface waters. Litterfall and fine root production is a major pathway for carbon and nutrient cycling in forest ecosystems, and their turnover depends on a variety of factors such as species, age groups, canopy cover, weather conditions and biotic factors (Lodhiyal et al. 2002; Stewart and Frank 2008).The increased primary productivity would then initiate a virtuous cycle of better soil cover, greater root mass in the soil, increased soil porosity, reduction or elimination of runoff and erosion and improved moisture availability for the vegetation. This process results in increased amounts of organic matter in the soil, including in the deeper layers. Once the productivity of arable land and grazing land in these ravines is increased, they become more resilient against drought. This would reduce the pressure on forest areas and improve the chances of their preservation or resource-conserving management.

The fine roots of grasses and forest trees play a crucial role in the fluxes of energy and matter in the soil ecosystem and carries out the essential function of soil resource acquisition (Buyanovsky et al. 1987). Root decomposition is potentially important regulator of carbon and nutrient cycling (Ostertag and Hobbie 1999). The belowground inputs into tropical soils may be of great importance for maintaining the soil organic matter level because litter, in general, is rapidly mineralized to $CO_2$, whereas the root residues, which are in close contact with the mineral soil, have more chances to be stabilized by clay particles or sesquioxides (Oades 1988). While placing these lands under alternative land uses, those species may be given priorities which are capable of higher carbon sequestration.

## 19.4 Ravines: A Potential Zone for Carbon Sink

The study was conducted to assess the soil carbon status under four different land uses, viz. plantation of *Leucaena leucocephala* (10 years old) and *Soymida febrifuga* (10 years old) both planted on the sloping tablelands in the immediate vicinity of ravines, two vegetation types of mixed forest (20 years of natural vegetation) and plantation of *Acacia senegal* (12 years old) both in the ravine land. Representative samples were collected from four quadrants in each different land uses. Results of the study revealed that among the various vegetation types, ravines under mixed forest recorded greater accumulation of organic carbon (1.43% in top of ravine) followed by *Soymida febrifuga* (0.85%), *Leucaena leucocephala* (0.72%) and *Acacia senegal* (0.70% in top of ravine) plantations (Table 19.1). This was attributed due to addition of greater biomass through leaf litter from both trees and ground storey vegetation (Somasundaram et al. 2012).

Thus, bringing marginal lands and ravine land under perennial vegetation not only helps in sequestering organic carbon but also helps in moderating microclimate in surrounding area. In fact, litterfall and fine root biomass addition are the crucial agents of nutrient recycling in forest ecosystems under ravine lands. In this line, Jha et al. (2014) reported that fine roots of *Azadirachta indica* have greater carbon stabilization potential than other species such as *Jatropha curcas*, grass (*Cenchrus ciliaris*), *Leucaena leucocephala* and *Prosopis juliflora* in the Yamuna ravine region.

**Table 19.1** Soil organic carbon status under different land use systems

| Sl. No. | Land use history | Physiography | pH | OC (%) |
|---|---|---|---|---|
| 1. | 10-year-old *Leucaena leucocephala* plantation under marginal lands | Gently sloping (up to 1% slope), moderately well-drained tablelands in the immediate vicinity of ravines | 7.06 | 0.72 |
| 2. | 10-year-old *Soymida febrifuga* plantation under marginal lands | Gently sloping (up to 1% slope), moderately well-drained tablelands in the immediate vicinity of ravines | 7.44 | 0.85 |
| 3. | Over 20-year-old undisturbed forest of *Prosopis juliflora* and mixed shrubs | Ravine land with 6–9 m deep "V"-shaped and more than 18 m wide gully network is under influence of backflow from Chambal River | 7.46 | 1.43 |
| 4. | Over 12-year-old *Acacia senegal* plantation | | 8.03 | 0.70 |

Adopted from Somasundaram et al. (2012)

Similarly, Jha and Mohapatra (2009) examined the leaf litterfall, fine root mass, production and turnover rate in the upper soil (0–30 cm) under four locally dominant tree species such as *Leucaena leucocephala*, *Acacia nilotica*, *Azadirachta indica* and *Prosopis juliflora* in Yamuna ravine region of India. It was evident that *Prosopis juliflora* and *Azadirachta indica* had greater ability for sustaining site productivity due to their greater leaf litterfall and fine root production. *Prosopis juliflora* recorded maximum above- and belowground biomass and consequently enhanced nutrient fluxes in the degraded ravine lands. Notwithstanding the fact, due to its poor fodder and timber value and vast-spreading nature to adjoining agricultural fields, *P. juliflora* is considered undesirable. Therefore, other species like *Azadirachta indica* could be an alternative for restoring the degraded ravinous land.

Another study was conducted at ICAR-Indian Institute of Soil and Water Conservation, Research Centre, Kota, during 2011 to 2015 at on-farm and off-farm sites for assessing the carbon stock as influenced by different tree species in Chambal ravines and adjoining forest areas. At the on-farm site, two existing horti-pastoral systems, namely, bael + dhaman (*Aegle marmelos* + *Cenchrus ciliaris*) and Indian gooseberry (Anola) + dhaman (*Emblica officinalis* + *Cenchrus ciliaris*), and two silvopastoral systems, namely, desi babool + dhaman grass (*Acacia nilotica* + *Cenchrus ciliaris*) and karanj + dhaman (*Pongamia pinnata* + *Cenchrus ciliaris*), were studied. The study site represented the typical Chambal ravines and was in district Kota of Rajasthan. The another off-farm study sites of three natural forests consisted (i) Sahpur forest, (ii) Mamoni forest and (iii) Kaloni forests, located in Shahabad range – Baran division. All kind of vegetation was inventoried at both study sites through sample plots for preparing inventory map depicting top storey (trees), middle storey (shrubs) and ground flora (grasses + herbs). Periodic observations on the biomass growth and litter production characteristics of the vegetation were recorded. The vegetation in the on-farm site comprised of top storey of the tree species planted and ground flora dominated by *Cenchrus ciliaris* and annual herbs. Off-farm site in the natural forest consists of three different strata. The highest litter production was recorded in *Acacia nilotica* (6.23 Mg $ha^{-1}$) followed by *Aegle marmelos* (5.66 Mg $ha^{-1}$), while the lowest litter production was recorded in *Pongamia pinnata* (3.92 Mg $ha^{-1}$) in agroforestry systems. In the natural forest, the highest litter production was recorded in *Boswellia serrata* (9.02 Mg $ha^{-1}$), while the lowest litter production was recorded in *Acacia catechu* (1.62 Mg $ha^{-1}$). The rates of litter decomposition were highest in *Acacia nilotica*, *Acacia catechu* and *Emblica officinalis* (7 months for 99% decomposition), while it was lowest in *Terminalia belerica*, *Buchanania cochinchinensis* and *Diospyros melanoxylon* (11 months for 99% decomposition).

SOC stock was assessed under different land use systems, namely, degraded ravine lands, farmlands and reserved forest of Shahabad region. Among three land use systems, the SOC stocks were in the order of reserved forest > degraded ravine lands > farmlands (Fig. 19.2a–c). Under degraded ravine lands, about eight tree species were also assessed for SOC stocks; the data revealed that bamboo plantation registered the highest SOC stocks followed by gooseberry plantation. Similar trends were followed for $CO_2$ equivalent stock [Unpublished data of ICAR-IISWC,

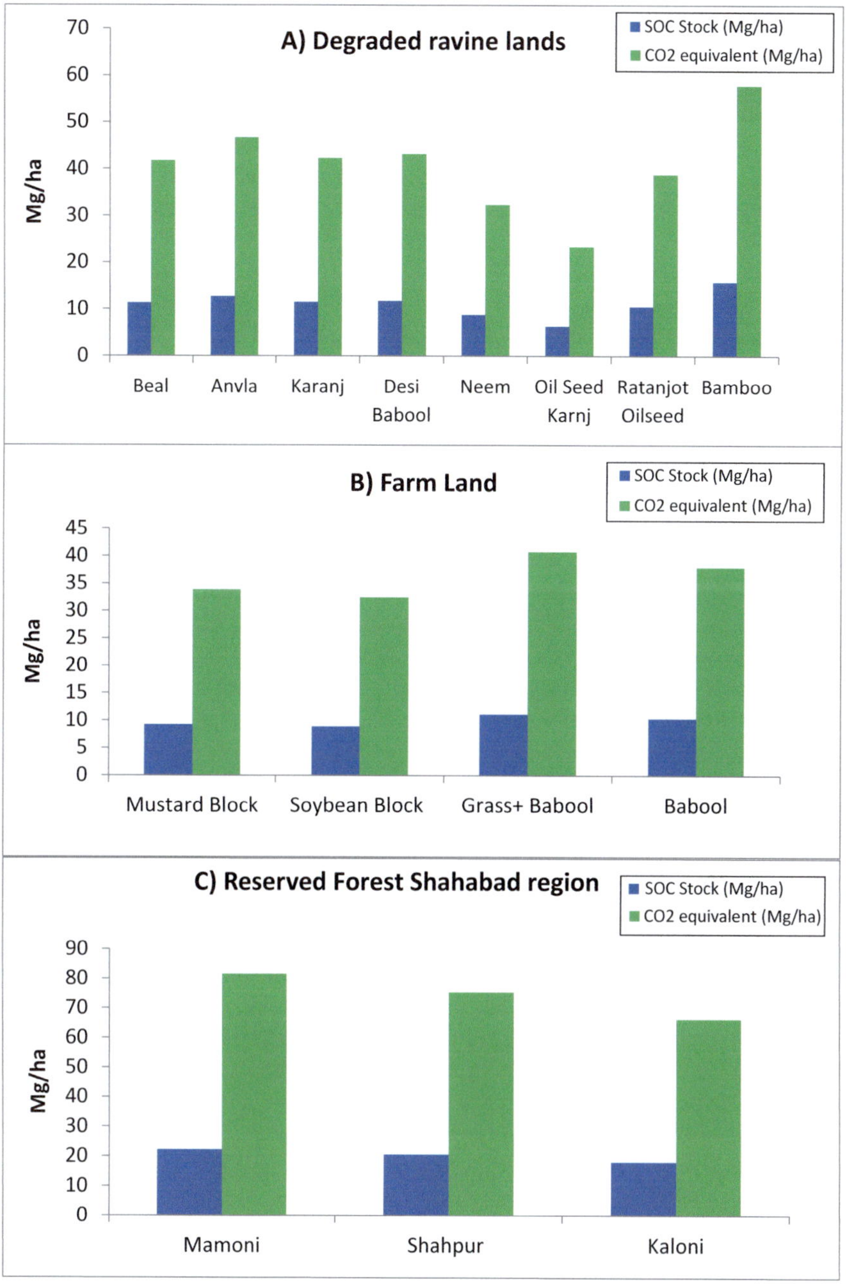


**Fig. 19.2** Carbon sequestration (Mg $ha^{-1}$) under different land use system at 0–15 cm depth (Adapted and Redrawn from Annual Report of ICAR-IISWC 2015–16; Parandiyal et al. 2016)

Research Centre, Kota; Parandiyal et al. (2016)); Annual Report (2015–2016)]. It is concluded from study that total biomass production and carbon stock were observed higher under *Acacia nilotica* followed by *Aegle marmelos*-based production systems. As a result, higher amount of $CO_2$ sequestration potential and soil nutrient build-up were estimated under *Acacia nilotica* followed by *Aegle marmelos.* Therefore, *A. nilotica* and *Aegle marmelos* species were highly suitable for ravinous area and provide a substantial role of carbon sequestration in Chambal and Yamuna ravines.

## 19.5 Carbon Sequestration Under Ravine Lands

Soil erosion by water is major soil degradation process, and globally more than 1 billion hectares' areas are affected. The soil loss generally varies from 1 to 10 Mg $ha^{-1}$ $year^{-1}$, and even in some cases, it may exceed 50 Mg $ha^{-1}$in cropped lands. These figures increase even by many folds during ravine formation. In fact, organic matter in the upper soil horizon is an important part of this soil loss. Moreover, measurement of exact carbon pool is difficult because of the heterogeneity in time and space. The global loss of carbon by erosion has been estimated to be about 150–1500 million Mg per year (Lal 1995; Gregorich et al. 1998). Measures to enhance the C sequestration in ravines include land levelling and terracing of high humps, trenching to arrest water flow to downstream and plantation of suitable trees. The soil cover increases the water infiltration rate by several hundred percent and prevents water evaporation; hence, soil moisture is increased. Especially in dry areas, soil cover has an important role in water economy. It also decreases the temperature, thus slowing the rate of OM mineralization. Thus, plantation of tree and grass species at the upper reaches is recommended where likely chances of soil erosion are more.

The increase in biomass involves both above-ground biomass and roots. Considerable progress could be made in this connection, especially as concerns grasslands and pasture, by selection of deep-rooting species and varieties. Crop residue management is another important method of sequestering C in soil and increasing the soil OM content. The positive effects of using crop residues to induce C sequestration have been estimated by Lal (1997) at 0.2 Pg C $year^{-1}$ with transformation of 15% of the total C (1.5 Pg C globally). Similarly, slope stabilization of bunds and terraces, contour or graded bunding for moisture conservation and erosion control, sodding of irrigation channels and bunds, construction of check dams wherever necessary, sowing of cover crops or mixed cropping at lower reaches with best package of practices, drainage line treatment with species like bamboos, *Ipomoea*, etc. These practices would help increase the biomass productivity in these lands, and that would help increase the C turnover in soil through leaf fall, root carbon mineralization, etc. In lower reaches where cultivation is possible, different cultural and management practices can be adopted that would enhance C sequestration. Minimum tillage operations would be sufficient to stimulate N release from SOM. Tillage practices have been causing the general decrease in OM of intensively cultivated soils in the past.

Adoption of conservation agricultural practices in these degraded lands through judicious use of crop residues covering about more than 30% of the soil surface helps in sustaining soil health (Lal 1997). Under conservation agriculture, 0.5–1.0 Mg C $ha^{-1}$ $year^{-1}$ can be sequestered in humid temperate conditions, 0.2–0.5 Mg C $ha^{-1}$ $year^{-1}$ in humid tropics and 0.1–0.2 Mg C $ha^{-1}$ $year^{-1}$ in semiarid zones (Lal 1999). But zero tillage with cover crops and/or green manures in a complex rotation patterns leads to large amounts of carbon sequestration.

Singh et al. have given details (in Chap. 20 of this book) of carbon storage in various agroforestry systems raised in Chambal ravines of Madhya Pradesh.

### 19.5.1 Bamboo, Carbon Sequestration and Ravine Reclamation

Permanent vegetation of grasses and trees is found effective in stabilizing in deep gullies landforms. Bamboos are generally grown in poor soil and are found effective for rehabilitation of degraded lands. This enables it to be used as a starting point in restoring degraded land. Adaptive capability, nutrient and water conservation of bamboos enables it as forerunner plants in combating land degradation. Because of the fast-growing nature and the dense foliage of bamboos, it can maintain the thick layer of litter. This litter layer maintains microclimate in the understory and soil moisture, the most important factors for the restoration of degraded lands. Table 19.2 shows that well-managed bamboo forests are likely to be a lower static carbon store as compared with other forest types (varying from 122 to 263 Mg C $ha^{-1}$).

Pande et al. (2012) carried out economic analysis using data from three major ravine systems, viz. Mahi, Chambal and Yamuna, to examine economic viability of plantation under different soil conditions. The analysis suggested a cash outflow ranging from ₹30,550 per ha to ₹48,000 per ha from the seventh year onwards to individual stakeholders in the region, in addition to the benefits accrued to society in terms of value of nutrient (₹2125–5555 per ha) saved through soil conservation and incremental soil carbon build-up (₹41,000 per ha) with the recommended harvest practice of harvesting one-third old culms per clump over the life of plantation. Carbon stock build-up under bamboo plantation has been extensively quantified. Carbon stock comprises living biomass and soil organic matter. While the former is depleted with the harvest of bamboo, the latter is retained in the soil till the bamboo

**Table 19.2** Comparison of carbon stock in bamboo and tree forest ecosystems (Mg C $ha^{-1}$)

| Forest/parts | Arbor and shrub | Litter | In soil | Total |
|---|---|---|---|---|
| Moso bamboo in Lin'an (medium intensity management) | 34.2 | 0.66 | 71.48 | 106.34 |
| Chinese fir at 15th year | 53.6 | 3.43 | 93.16 | 203.79 |
| Moso bamboo in Yong'an (medium intensity management) | 61.3 | 3.01 | 197.36 | 261.67 |
| Deciduous broad-leaved forest | 47.75 | 5.85 | 208.9 | 262.50 |

Source: Adapted from Mishra et al. (2014)

plantation lasts. The carbon build-up in soil under bamboo plantation was estimated with a scenario that the degraded ravine land with no plantation would not only have lost the soil nutrients but also lost the opportunity of building soil carbon under the plantation. Therefore, the intangible benefit of the carbon build-up justifies the utility of bamboo plantation in degraded ravines to national and international stakeholders and the policymakers.

Soil carbon build-up under plantation is valued using an estimate of the shadow price of carbon drawn from the climate change damage literature. This price reveals the present value of (future) damages caused by a ton of carbon (equivalent) emissions (Mendelsohn 2003). A value of US$ 20 per Mg C is used as the social cost of carbon. Sensitivity analysis is performed with estimates of US$ 5 per Mg C and US$ 40 per Mg C (Atkinson and Gundimeda 2006).

## 19.6 Conclusions

Carbon sequestration is a slow and long-term process that needs continuous inputs of biomass into the soil system. In fact, ravines are the most fragile ecosystems that have very low carbon content due to their light texture and poor aggregate stability. Thus, there is need to manage and restore these lands with suitable cultural and management practices including conservation measures such as surface smoothening, land levelling and bunding, etc. This ecosystem needs a surface cover against direct impact of raindrop that causes detachment of soil particles. Involving practices that enhance carbon input through above- and below-ground biomass help in improving soil organic matter and promote favourable changes in soil physical and chemical and biological properties. There is also a need for slope stabilization through locally adapted grasses and shrubs during rainy season. The lower regions of the ravines are capable enough to produce annual crops. These practices and measures not only hold promise in conserving this natural resource but also help in enhancing carbon input into the soil through crop residue, roots and leaf fall under different vegetation, which in turn would help increase the C turnover in soil. Over a period of time, this carbon input would help in restoring the soil carbon in these fragile ravine ecosystems.

## References

Annual Report (2015–2016). ICAR-IISWC, published by Director, ICAR-Indian Institute of Soil and Water Conservation, Dehradun, Uttrakhand, pp 48–49

Atkinson G, Gundimeda H (2006) Accounting for India's Forest Wealth. Working Paper No. 5/2006, Madras School of Economics, Chennai, 2006

Buyanovsky GA, Kucera CL, Wagner GH (1987) Comparative analyses of carbon dynamics in native and cultivated ecosystems. Ecology 68:2023–2031

Dhruvanarayana VV (1993) Soil and water conservation research in India. Directorate for Information and Publications of Agriculture, ICAR, KAB-II, Pusa, pp 1–454

GoI (2003) Rehabilitation of gullied and ravine lands. Technical brochure No-1/2003. Department of Land Resources, Ministry of Rural Development, Government of India (GoI), New Delhi, p 44

Gregorich EG, Greer KJ, Anderson DW, Liang BC (1998) Carbon distribution and losses: erosion and deposition effects. Soil Tillage Res 47:291–302
ICAR- NAAS (2010) Degraded and wastelands of India: status and spatial distribution., Directorate of Information and Publications of Agriculture, Indian Council of Agricultural Research, Krishi Anusandhan Bhavan I, Pusa, & National Academy of Sciences, New Delhi 110 012
Jha P, Mohapatra KP (2009) Leaf litterfall, fine root production and turnover in four major tree species of the semi-arid region of India. Plant Soil. https://doi.org/10.1007/s11104-009-0027-9
Jha P, Mohapatra KP, Dubey SK (2014) Fine roots carbon mineralization and soil carbon stabilization under major tree species of the semi-arid region of India. Natl Acad Sci Lett 37(5):413–418
Lal R (1995) Global soil erosion by water and carbon dynamics. In: Lal R, Kimble J, Levine E, Stewart BA (eds) Soil and global change. CRC Lewis Publishers, Boca Raton, FL, pp 131–142
Lal R (1997) Residue management, conservation tillage and soil restoration for mitigating greenhouse effect by $CO_2$ enrichment. Soil Tillage Res 43:81–107
Lal R (1999) Soil management and restoration for C sequestration to mitigate the greenhouse effect. Prog Environ Sci 1:307–326
Lodhiyal N, Lodhiyal LS, Pangtey YPS (2002) Structure and function of shisham forests in central Himalaya, India: dry matter dynamics. Ann Bot (Lond) 89:41–54
Mendelsohn R (2003) The social cost of carbon: an unfolding value. Paper presented to the Social Cost of Carbon Conference, Department of Environment, London
Mishra G, Giri K, Panday S, Kumar R, Bisht NS (2014) Bamboo: potential resource for eco-restoration of degraded lands. J Biol Earth Sci 4(2):B130–B136
MOA (1985) Indian agriculture in brief, 20th edn. Ministry of Agriculture (MOA) Government of India, New Delhi
Oades JM (1988) The retention of organic matter in soils. Biogeochemistry 5:35–70
Oelbermann M, Voroney RP, Gordon AM (2004) Carbon sequestration in tropical and temperate agroforestry systems: a review with examples from Costa Rica and southern Canada. Agric Ecosyst Environ 104:359–377
Ostertag R, Hobbie SE (1999) Early stages of root and leaf decomposition in Hawaiian forests: effects of nutrient availability. Oecologia 121:564–573
Pande VC, Kurothe RS, Rao BK, Kumar G, Parandiyal AK, Singh RK, Kumar A (2012) Economic analysis of bamboo plantation in three major ravine Systems of India. Agric Econ Res Rev 25(1):49–59
Parandiyal AK, Kala S, Singh RK, Mina BL (2016) Carbon stock under tree based production systems in the degraded Chambal ravines. In: National symposium on Agroforestry for environmental challenges, sustainable land use, biodiversity conservation and rural livelihood options (3–5 December 2016) at ICAR-CAFRI, Jhansi (U.P.) India
Robert EB, Brian RF, Peter AB (1996) Incorporation of nitrogen and carbon from spawning coho salmon into the trophic system of small streams: evidence from stable isotopes. Can J Fish Aquat Sci 53(1):164–173
Sharma HS (1980) Ravine erosion in India. Concept Publishing House, New Delhi, pp 1–90
Singh AK, Kala S, Dubey SK, Pande VC, Rao BK, Sharma KK, Mahapatra KP (2015) Technology for rehabilitation of Yamuna ravines – cost-effective practices to conserve natural resources through bamboo plantation. Curr Sci 108:1527–1533
Somasundaram J, Singh RK, Ali S, Sethy BK, Singh D, Lakaria BL, Chaudhary RS, Singh RK, Sinha NK (2012) Soil aggregates and other properties as influenced by different long term land uses under table landscape topography of Chambal region, Rajasthan, India. Indian J Soil Cons 40(3):212–217
Stewart AM, Frank AF (2008) Short sampling intervals reveal very rapid root turnover in a temperate grassland. Oecologia 157:453–458
Verma B, Chinamani S, Bhola SN, Rao DH, Prasad SN, Prakash C (1986) Twenty-five years of research on soil and water conservation in ravine lands of Rajasthan. Central Soil and Water Conservation Research and Training Institute Publication, Research Centre, Kota, pp 1–215

# Sequestering Carbon in Ravine-prone and Eroded Landscapes

20

Rattan Lal

**Abstract**

Ravine lands, the most severe form of water erosion leading to formation of badlands, occupy about 4 Mha (1.22%) of land area in India. These lands have strongly impacted economic, social, cultural, and environmental features of the region. This article describes the cause-effect relationship, suggests some preventive and control measures, and assesses the potential of soil carbon sequestration. Formation of ravine lands is triggered by tectonic activity, climate change, or drastic change in land use. In India, ravines are formed in the semiarid climate with short (3-month) monsoonal rains and a long dry season. On the basis of morphological features (depth, width, density, etc.), ravines can be categorized into shallow, medium, and deep. An effective restoration/rehabilitation necessitates characterization of ravines using satellite imagery and other remote sensing techniques validated by ground-truthing. For an effective control measure, runoff into the ravine land must be diverted using techniques of safe disposal and a perennial vegetation cover established. Deep ravines must also be afforested to stabilize further growth. Medium and shallow ravines can be bulldozed for agricultural forestry land use. Carbon sequestration in soil and biomass, with a technical potential of 2.6–5.3 TgC $year^{-1}$, can generate another income stream for farmers (US\$300–640 million $year^{-1}$) and promote the adoption of best management practices for both preventative and control measures.

R. Lal (✉)
Carbon Management and Sequestration Center, The Ohio State University, Columbus, OH, USA
e-mail: lal.1@osu.edu





J. C. Dagar, A. K. Singh (eds.), *Ravine Lands: Greening for Livelihood and Environmental Security*, https://doi.org/10.1007/978-981-10-8043-2_20

## 20.1 Introduction

The ravine erosion is among the major causes of land/soil degradation globally but especially in India. A ravine is a fluvial slope landform of steep side slope gradient (20–75%), in between a canyon and a valley, and formed through severe stream-cutting erosion (Morris 1992). Simply put, ravine comprises of a network of deep gullies characterized by steep gradient. A ravine may or may not have an ephemeral stream, except flash floods during and after a torrential rain. Because of a severe cutting of landscape, ravine formation leads to the development of "badlands" topography. Badlands are severely eroded terrain form where softer sedimentary rocks and clay-rich soils are strongly eroded by water, wind, or both. These lands are characterized by steep slopes, none or minimal vegetation cover, and a high drainage density (Bryan and Yair 1982; Torris and Bryan 1997; Parsons and Abraham 2009; Thomas 2011). The landscape is extremely dissected and is difficult to cross even on a horseback. Yet, the dissected landscape can be rather scenic with numerous contrasting colors of surface, subsoil, and the exposed regolith along with spectacular columns of soil (pedestals) protected by a rock or hard surface layer. Both ravine and gully are severe forms of water erosion caused by deep and concentrated flow. Gullies, U or V-shaped channels, are >30-cm deep formed on land of >5% slope gradient and prone to soil loss of >10 Mg $ha^{-1}$ $year^{-1}$. In comparison, ravines are extreme form of water erosion caused by deep and concentrated flow and high stream flow velocity. In general, an intricate network of parallel gullies forms ravines. In some cases, a gullied land degrades into a ravine-infested land because of the presence of unconsolidated and relatively loose and soft sediments (Mudgal 2005). The landscapes prone to ravine formation have slope gradient of >5%. The drainage density is high and the landscape may have multidirectional slopes. Terrain attributes leading to formation of ravines include steep slope, profile curvature, and concentrated/accumulated flow at high velocity. Both gully and ravine grow by head cutting.

Ravine and gully erosion together affect about ~1.22% of total land area of India of 329 Mha (Haigh 1984). Regions prone to ravine formation are along the margins of Gangetic Basin and also in the semiarid northwest regions of India. There are four primary regions prone to ravine formation: (i) Shivalik foothills in Punjab, (ii) the Yamuna-Chambal zone, (iii) Chota Nagpur zone, and (iv) Gujrat zone (Haigh 1984). A large proportion of the ravine-related badlands in India are located in the central Indo-Gangetic Plains (Joshi and Nagare 2013). Indeed, the Chambal Badlands in the Indo-Gangetic Plains are extensive and cover a large land area (Joshi and Nagare 2013; Chatterjee et al. 2009). Some reports have estimated that India has ~4 Mha of ravine or badlands (Chatterjee et al. 2009), of which 2.77 Mha are in Uttar Pradesh (Dwivedi and Ramana 2003) and about 0.32 Mha each in Madhya Pradesh, Rajasthan, and Gujrat (Mudgal 2005). About 20% of the total land area in Chambal district is affected by ravines. Ravine-prone badlands of Chambal are comparable in their magnitude and extent to that of the Dakota Badlands of the US. Ravines in the Chambal area have been in existence for $10^3$–$10^4$ years (Joshi 2014), and are characterized by an intricate and densely incised network of ravines in a 10-km-wide zone (Joshi 2014). The area affected by ravine land in India is reportedly expanding at the rate of 8000 ha per year (Dwivedi and Ramana 2003).

There are a few systems to classify ravines based on their morphological features (depth, width). In Madhya Pradesh, Gupta and Prajapati (1983) classified ravines into four categories based on their depth: (i) shallow, <1.5 m deep; (ii) medium, 1.5–5 m deep; (iii) deep, 5–10 m deep; and (iv) very deep, 5–10 m deep. A major cause of urgency with regard to the ravine development is the rapid rate by which ravines encroach and destroy farmland in its path (Singh and Dayal 1975). The average rate of expansion of ravines in Rajasthan has been reported at 500 ha year$^{-1}$or 1.75 ha year$^{-1}$for 1000 ha of existing ravine land (Singh et al. 1996; Tejwani et al. 1975). The rate of 1.93 ha for each 1000 ha is reported in Madhya Pradesh (Sharma 1980). All other factors remain the same; the rate of expansion is more on sandy than on clayey or loam soils (Sharma 1976).

Because of the extreme form of erosion and loss of the entire topsoil and most of the soil solum, the exposed regolith is devoid of soil organic carbon (SOC) and is low in nutrient reserves. Lack of vegetation cover, coupled with none or minimal input of biomass-C and high temperatures, the SOC stock is low and declining. Ravine-prone landscape has low SOC stocks compared with those of the un-eroded landscapes within the same biome. Therefore, these degraded and depleted badlands have a large technical potential of C sequestration through restoration and rehabilitation. A systematic and science-based approach to restoration of ravine lands has the potential to increase C stock in soil and in the below- and aboveground biomass. Increase in the terrestrial C stock of ravine/badlands has numerous benefits: improvements in soil quality/health, renewability and purification of water, increase in the above- and belowground biodiversity, increase in net primary productivity, and the overall improvement in the environment quality.

The objective of this review article is to describe processes, factors, and causes of ravine formation; explain evolutionary processes; assess reclaimability of ravines and recommend techniques of their reclamation; and evaluate the technical potential of soil/terrestrial C sequestration through restoration of badlands in India.

## 20.2 Evolution of Ravines and Factors Affecting Susceptibility

Ravines are erosional forms which are transformed from inter-rill → rill → gully →to ravine erosion under site-specific conditions of terrain, lithology, vegetation cover, land use, and climate. These lands mostly occur in semiarid climate with a pronounced dry season (Joshi 2014). Heavy monsoons for a short period and prolonged dry season weaken the alluvial sediments and transform into ravines. Ravines are formed by two types of concentrated flow at high velocities: (i) surficial and (ii) subterranean. The subterranean flow is caused by "piping" or the so-called tunnel flow that can occur in sodic soils. Tejwani et al. (1975) explained the evolutionary pathways of ravines primarily on the basis of the surficial runoff characteristics. In

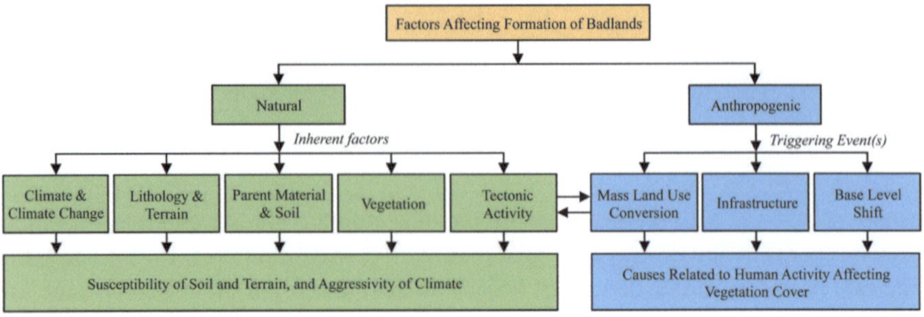


**Fig. 20.1** Principal controls of the processes, factors, and causes affecting trigger, rate, and extent of the area prone to the ravine erosion

addition, there is a wide range of interacting factors affecting the initiation (trigger point), rate of expansion, and the total land area prone to ravine formation (Fig. 20.1).

Two principal types of factors are inherent (natural) and anthropogenic (Gallart et al. 2013; Joshi and Nagare 2013). Important among inherent factors are the climate and the climate change, lithology (parent material) and the terrain (slope, gradient, aspect, shape, one vs. multidirectional), soil type, vegetation, and the percent vegetation cover and the frequency and magnitude of tectonic activity. Any drastic perturbations (e.g., climate change, tectonic activity) that can alter the terrain (i.e., mass movement) can trigger the ravine formation. There are numerous examples of the initiation of ravine because of tectonism (Rai et al. 1999; Mather et al. 2002; Harvey 2002; Larue 2005; Faulkner 2008). The trigger point can also be attributed to the climate change during the Pleistocene-Holocene era (Avni 2005; Joshi and Nagare 2013). Rapid and extensive land use change (i.e., deforestation of steep slopes) can also trigger the ravine formation (Gallart et al. 2013; Joshi and Nagare 2013). With regard to the lithological factors, the soil derived from unconsolidated material with low cohesion/cementation may lead to ravine formation (e.g., alluvial deposits in the Chambal Valley). Piping (i.e., tunnel erosion) and other lithological factors are of relatively lesser significance (Gallart et al. 2013). Soil dispersivity (e.g., slaking, erodibility), in relation to structure and dominance of exchangeable cations ($Na^+$ vs. $Ca^{+2}$, $Mg^{+2}$ or $K^+$), is an important attribute not only for the surface layer but also for the entire profile. Soil dispersivity of subsoil horizons is pertinent to pipe erosion and to expansion of the ravine network. Terrain characteristics which cause mass movement (i.e., landslide, mudflow) aggravate ravine formation. Vegetation cover is a major determinant of the runoff and sediment transport, and lack of vegetation cover aggravates the soil erosion both by wind and water.

While ravine lands primarily occur in semiarid climate, badland formation in the Chambal Valley may have coincided with the strong intensification of the SW monsoon at the end of the Last Glacial Maximum (Joshi 2014). In addition, evolution of the Chambal Badlands may also be attributed to the combined effect of strengthening of the SW monsoons and the block uplift by tectonic activity causing a sudden rejuvenation in relief and change in the course of the river (Joshi 2014). In recent past, Pani and Mohapatra (2001)) reported 17% increase in the ravine area between 1984 and 1998.

Peter Drucker (an American-Austrian educator) stated that "You can't manage what you can't measure" or "what gets measured gets improved." Yet, Edward Deming (an American management consultant) is quoted to state, "There are many things that cannot be measured and still must be managed." Therefore, measurement of the land area prone to ravine erosion, ravine type, and the rate of increase are important prerequisites to their control and restoration. In addition to the traditional techniques of ground survey, interferometric SAR is also being used for characterization of ravines as a function of their density, depth, and surface cover (Chatterjee et al. 2009). While traditional use of aerial photographs is a useful technique, the spectral responses of the ravined land are also measured by the space-borne multispectral landsat to delineate reclamative groupings of the land (Dwivedi and Ramana 2003). Kandrika and Dwivedi (2013) used the Cartosat-1, and India Earth Observatory Satellite, stereo images with 2.5 m spatial resolution in deriving morphometric information in ravines with the goal of reclamative grouping. Kandrika and Dwivedi used the anaglyph data to delineate different types of ravines on the basis of their depth and width: shallow (<3 m deep and <18 m wide), medium (3–9 m deep and >18 m wide), and deep (>9 m deep).

## 20.3 Reclamation of Ravine Lands

High population density and paucity of land resources necessitate reclamation of badlands for agriculture and other land uses. The immediate task is to restore 4 Mha of land affected by ravines in India and also nullify the current annual rate of growth of badlands estimated at 8000 ha year$^{-1}$ (Dwivedi and Ramana 2003). Thus, understanding the mechanisms of ravine formation is essential to identifying the appropriate techniques of restoration. Therefore, a clear understanding of the processes, factors, and causes of initiation and expansion of gullies into ravines is important.

The first step in designing a reclamation program is to delineate/demarcate reclamation groupings using remote sensing data (Dwivedi and Ramana 2003). Estimating dimensions and categories of ravines are important to identifying the reclamation measures and strategies of control.

Reclamation of ravine lands is a biophysical issue but also a socio-economic and a cultural challenge. Rather than agriculture, it may involve a new or a different land use. There is also a distinction between restoration, reclamation, and rehabilitation. The term restoration implies that land is to be returned to its former use. In comparison, reclamation implies that some new use of land is envisaged. The term rehabilitation is used when visual improvement is involved (Pani 2016; Bradshaw and Chadwick 1980).

In the context of controlling the biophysical processes, the runoff amount, rate, and velocity must be reduced. Once the ravine has been triggered, subsequent expansion must be controlled by installation of massive diversion ditches to divert the runoff away from the ravine head. Reclamation strategy would involve a series of steps to be considered in a well-planned manner:

### 20.3.1 Soil and Water Conservation in the Surrounding Uplands

Management of the surrounding land area (watershed that feeds water into the ravine land) is essential to reducing runoff and diverting any runoff away from the ravines. This would involve establishment of a perennial ground cover and installation of diversion channels. Any crop cultivation must be done by conservation agriculture (CA) with cover cropping and mulch farming (Lal 2015) in conjunction with strip cropping and contour hedges of perennials. Any grazing must be controlled and at a low stocking rate, in conjunction with establishment of pastures comprising of a mixture of native species of grass, legumes, and forage shrubs.

### 20.3.2 Restoring Deep Ravine Land

The goal is to establish vegetation cover with locally adaptive species which can quickly establish an effective canopy cover. The steep side slopes and the ravine bed land must be protected by establishing a perennial ground cover.

### 20.3.3 Shallow and Medium Ravine

Some of the shallow ravines can be leveled by land forming using bulldozers. However, the leveled land must be protected against rill/inter-rill erosion by establishing some fast-growing species.

### 20.3.4 Riparian Land Management

The riparian land must also be protected (Yadav and Bhushan 2002), preferably with a perennial vegetation cover. In addition to sedimentation control, vegetation cover within the riparian zone will also filter out pollutants, retain sediments, purify water, and sequester C in soil and the biomass.

The techniques of reclamation and restoration must be natural (biotic) measures used in conjunction with the engineering techniques of terracing, diversion channels, riprap, drop structures, gabions, etc. While being cost-effective (Singh et al. 2015), these measures must also be socially and culturally acceptable.

## 20.4 Soil and Terrestrial Carbon Sequestration

Most ravine lands and the surrounding areas are strongly depleted of their soil/ecosystem C stocks. Thus, upon reclamation/rehabilitation/restoration, these lands have a large potential of soil/ecosystem C sequestration. With a total land area of about 4 Mha, the rate of ecosystem C sequestration may be 0.25–0.5 Mg $ha^{-1}$ $year^{-1}$ for SOC and 0.5–2 Mg $ha^{-1}$ $year^{-1}$ for the biomass (root and shoot) C. Assuming that land use upon reclamation/rehabilitation/restoration is 25% perennial, 25% agroforestry, 48% croplands, and 2% riparian, the potential of soil and biomass C sequestration is given in Table 20.1.

The data in Table 20.1 show that total technical potential of C sequestration in reclaimed ravine land with adoption of BMPs is 2.6–5.3 TgC $year^{-1}$. With the societal value of C at $120 per MgC (Lal 2014), the terrestrial C sequestration in restored ravine land is worth US$300 to $640 million per year. Even if the market value of the terrestrial C sequestered is merely $150 million, this would be an appropriate incentive for land managers to adopt and pursue a judicious and sustainable use of restored ravine lands.

## 20.5 Ecological and Socio-economic Impact of Restoring Ravine Lands

Ravine lands adversely impact the socio-economic system of the region through destruction of the agricultural lands, physical infrastructure, and displacement of human and animal population. Ravines have disrupted lives and well-being of millions of people in the affected areas. Thus, prevention, control, and restoration of ravine lands have economic, social, cultural, political, and ecological ramifications. It is a five-win situation. It can have a strong and positive effect on all aspects of human life, along with improvements in aesthetic aspects of the land, enhancement of water quality, and renewability and improvement of the above- and belowground biodiversity.

**Table 20.1** Technical potential of C sequestered in soil and biomass with adoption of best management practices on reclaimed ravine lands in India

| | | Rate of C sequestration ($MgCha^{-1}$ $year^{-1}$) | | Total C sequestration ($TgCyr^{-1}$) | | |
|---|---|---|---|---|---|---|
| Land use | Area (Mha) | Soil | Biomass | Soil | Biomass | Total |
| Plantation forestry | 1 | 0.25–0.5 | 1–2 | 0.25–0.50 | 1.0–2.0 | 1.25–2.5 |
| Agroforestry | 1 | 0.2–0.4 | 0.5–1.5 | 0.20–0.40 | 0.5–1.5 | 0.7–1.9 |
| Cropland | 2 | 0.2–0.3 | – | 0.4–0.6 | 1 | 0.4–0.6 |
| Riparian land | 0.1 | 0.5–1.0 | 2–3 | 0.05–0.10 | 0.2–0.3 | – |
| Total | | | | 0.9–1.6 | 1.7–3.8 | 2.6–5.3 |

*Mg* megagram = $10^6$ g = 1 metric ton, *Tg* teragram = $10^{12}$ = 1 million metric ton

## 20.6 Conclusions

Ravine lands are the drastic and strong form of water erosion with severe adverse impacts on quality of soil and water, biodiversity, agronomic and economic productivity, and human well-being. Thus, all efforts must be made to prevent, control, and restore/rehabilitate and reclaim the ravine land. For adoption of effective prevention, control, and restoration measures, it is crucial to establish the cause-effect relationship so that the causes can be systematically and objectively addressed.

The ravine formation may be triggered by a tectonic activity with or without a drastic land use change (deforestation) leading to mass movement and landslides. Ravines occur in semiarid regions where a long dry season (8–9 months) is followed by intense monsoonal rains (3–4 months) causing flash floods with high-velocity concentrated flow.

The implementation of effective restoration measures begins with characterization of ravine (depth, width, density, side slopes, etc.) so that appropriate measures can be identified. Landsat imagery, aerial photography, and remote sensing techniques can be useful in ravine characterization.

The watershed surrounding the ravine must be protected through establishment of a perennial vegetation cover. If it is suitable for an arable land use, it must be managed by a CA system adopted in conjunction with mulch farming, cover cropping, contour hedgerows, etc. Any runoff must be diverted away from the ravine land.

Shallow and medium-depth ravine land can be bulldozed following the land-forming techniques. Subsequent land use must also involve CA, agroforestry, and other conservation-effective measures. Deep ravine lands must be afforested with native species adapted for the region. Riparian land must also have perennial vegetation cover.

The soil and ecosystem carbon sequestration can generate another income stream for farm managers and make restorative efforts a five-win option.

## References

Avni Y (2005) Gully incision as a key factor in desertification in an arid environment, the Negev highlands, Israel. Catena 1000:83–90

Bradshaw AD, Chadwick MJ (1980) The restoration of land: the ecology and reclamation of derelict and degraded land. Blackwell Scientific Publications, Oxford, p 317

Bryan R, Yair A (1982) Perspectives on studies of badland geomorphology. In: Bryan R, Yair A (eds) Badland geomorphology and piping. Geo Books, Norwich, pp 1–12

Chatterjee RS, Saha SK, Kumar S, Mathew S, Lakhera RC, Dadhwal VK (2009) Interferometric SAR for characterization of ravines as a function of their destiny, depth and surface cover. ISPRS J Photogram Remote Sens 64(5):472–481

Dayal R (1975) Halt the march of ravines. Intensiv Agric 13(4):20

Dwivedi RS, Ramana KV (2003) The delineation or reclamative groups of ravines in the Indo-Gangetic alluvial plains using IRS-1D LISS-III data. Int J Remote Sens 24(22):4347–4355

Faulkner H (2008) Connectivity as a crucial determinant of badland morphology and evolution. Geomorphology 100:91–103

Gallart F, Marignani M, Péres-Gallego N, Santi E, Maccherini S (2013) Thirty years of studies on badlands, from physical to vegetational approaches – a succinct review. Catena 106:4–11
Gupta RK, Prajapati MC (1983) Reclamation and use of ravine lands. Desert Resour Tech (Jodhpur) 1:221–262
Haigh MJ (1984) Ravine erosion and reclamation in India. Geoforum 15(4):543–531
Harvey AM (2002) Effective timescales of coupling within fluvial systems. Geomorphology 44:175–201
Joshi VU (2014) Landscapes and landforms of India. In: Kale VS (ed) World geomorphological landscapes. Springer, New York, pp 143–149
Joshi VU, Nagare VB (2013) Badland formation along the Pravara River, Western Deccan, India. Can neotectonics be the cause? Z Geomorphol 57(3):349–370
Kandrika S, Dwivedi RS (2013) Reclamative grouping of ravines using Cartosat-1 PAN stereo data. J In Soc Remote Sens 41(3):731–737
Lal R (2014) Societal value of soil carbon. J Soil Water Conserv 69:186A–192A
Lal R (2015) A system approach to conservation agriculture. J Soil Water Conserv 70(4):82A–88A
Larue JP (2005) The status of ravine-like incisions in the dry valleys of the Pays de Thelle (Paris basin, France). Geomorphology 68:242–256
Mather AE, Stokes M, Griffiths JS (2002) Quaternary landscape evolution: a framework for understanding contemporary erosion, southeast Spain. Land Degrad Dev 13:89–109
Morris CG (1992) Dictionary of information, science and technology. Academic Press, New York
Mudgal MK (2005) Socio-economic impact of ravine lands- a case study of River Chambal basin of state of Madhya Pradesh, India. Geophy Res Abstracts 7:00529
Pani (2016) Controlling gully erosion: an analysis of land reclamation processes in Chambal, India. Dev Pract 26(8):1047–1059
Pani P, Mohapatra SN (2001) Delineation and monitoring of gullied and ravine lands in a part of lower Chambal Valley, India using remote sensing and GIS. Proc ACRS Singapore 1:671–675
Parsons AJ, Abraham AD (2009) Geomorphology of desert environments, Second edn. Springer Science and Business Media, Berlin, p 831
Rai R, Maurya DM, Chamyal LS (1999) Tectonic control on distribution and evolution of ravines in the lower Mahi valley, Gujrat. J Geological Soc India. 53:669–6674
Sharma HS (1976) Morphology of ravines of the Morel basin, Rajsthan, India. Int Geogr 1:219–233
Sharma HS (1980) Ravine erosion in India. Concept, New Delhi
Singh A, Dayal R (1975) Development of water resources in ravine utilization-a case study. Soil Conservation Digest 3(1):26–30
Singh AK, Kala S, Dubey SK, Pande VC, Rao BK, Sharma KK, Mahapatra KP (2015) Technology for rehabilitation of Yamuna ravines- cost effective practices to conserve natural resources through bamboo plantation. Curr Sci 108:1527–1533
Singh JS, Singh KP, Jha AK (1996) Final Technical Report. Submitted to the Ministry of Coal, Government of India, p 116
Tejwani KG, Gupta SK, Mathur HN (1975) Soil and water conservation research 1956–1971. ICAR, New Delhi
Thomas D (2011) Arid zone geomorphology: process, form, and change in drylands, 3rd edn. Wiley, Hoboken, p 648
Torri S, Bryan R (1997) Micropiping processes and biancana evolution in southeast Tuscany, Italy. Geomorphology 20:219–235
Yadav RC, Bhushan LS (2002) Conservation of gullies in susceptible riparian areas of alluvial soil regions. Land Degrad Dev 13(3):201–219

# Greening of Ravines: A Strategy for Viable Production and Mitigation to Climate Change

# 21

Anil Kumar Singh, Y. P. Singh, and S. K. Verma

**Abstract**

Since formation of ravines is a function of uncontrolled runoff, the management of rainwater on watershed basis is the only strategy for the control and reclamation of ravines. Deep ravines pose the problems of inaccessibility, very steep slope of ravine banks, strong water flow lines, and high permeability of soil making limited land shaping unavoidable. After necessary land shaping in world infamous Chambal ravines covering an area of 304,300 ha in Madhya Pradesh, Rajasthan, and Uttar Pradesh states of India, strategy for viable production and mitigation of climate change was framed, and accordingly five different modules of plantation were tried, namely diversified cropping, agri-horticultural, horti-pastoral, silvi-medicinal, and silvopastoral, using land according to its suitability. The greatest carbon was produced under diversified cropping system followed by horti-pastoral, agri-horticultural, silvi-medicinal, and silvopastoral system during the first year, whereas maximum carbon production with agri-horticulture is followed by horti-pastoral, diversified cropping system, silvi-medicinal, and silvopastoral system from second year onward. Similarly, under crops' higher yield and carbon production with 150% of recommended dose of fertilizers were obtained. The highest carbon production among crops and plants was with *Moringa oleifera* followed by *Dalbergia sissoo* and pearl millet followed by

A. K. Singh (✉)
RVS Krishi Vishwa Vidyalaya, R. V. S., Gwalior, India
e-mail: vcrvsaugwa@mp.gov.in

Y. P. Singh
AICRP-IWM, R. V. S. Krishi Vishwa Vidyalaya-ZARS, Morena, India

S. K. Verma
Department of Soil Science & Agricultural Chemistry, RVSKVV, Vishwa Vidyalaya, Gwalior, Madhya Pradesh, India



J. C. Dagar, A. K. Singh (eds.), *Ravine Lands: Greening for Livelihood and Environmental Security*, https://doi.org/10.1007/978-981-10-8043-2_21

maize, respectively. The land and water management works must be planned for entire watershed starting from the ridge and including tableland. While technology for shallow and medium ravines has been worked out, afforestation of deep ravines is recommended.

## 21.1 Introduction

Indian agriculture is a gamble because of its dependency on climate and is highly prone to the risks due to uncertainty of rains and frequent drought, as two third of the agricultural land is rain-fed and even the irrigated system is dependent on monsoon. Greenhouse gases may rise by 25–90% by the year 2030 relative to 2000, and the earth may warm by 3 °C at the end of the century (IPCC 2007). Even with a temperature rise of 1–2.5 °C, the IPCC has predicted serious disastrous effects, including reduction in crop yields in shortage, tropical and subtropical areas leading to increased risk of food, spread of climate responsive diseases such as malaria, and an increased risk of extinction of 20–30% of all biodiversity present on earth. Climate change projection using PRECIS-A1B scenario for 2030 (2021–2050) indicates 11.6% increase in annual rainfall as compared to the baseline (1961–1990). However, this increased rainfall may also cause more floods and soil erosion. Goswami et al. (2006) in their analysis observed that extreme rainfall events of above 100 mm are increasing in frequency as compared to moderate events in Central India. The model also projected an average increase of 1.81–2.0 °C rise in daily maximum temperature and 2.0–2.4 °C rise in minimum temperature by 2030 whereas in the long term (2080), the overall impact of climate change would likely to receive relatively more rainfall in the future than today but in more intense storms and a generalized warming, increased incidences of drought, and uneven and untimely rains which would have strong implications for crop production and water resources management.

Among all greenhouse gases (GHG), $CO_2$ is of major importance. Between 2000 and 2011 the atmospheric concentration of $CO_2$ increased from 369 to 391.5 ppm (Coway and Tans 2012) and as on today it has exceeded 400 ppm. $CO_2$ emissions from industrial processes and fossil fuel combustion contributed about 78% of the total GHG. However, sequestering $CO_2$ from point sources or atmosphere through natural techniques (afforestation, reforestation, natural regeneration of forests, and adaptive agriculture) is more economically and ecologically sound in increasing the carbon storage capacity of the terrestrial ecosystems. IPCC (2007) also indicated in its special report that the conversion of wasteland and grassland to agroforestry has the best potential to soak up atmospheric $CO_2$ other than direct benefits.

Out of 328 million ha geographical area of the country, nearly 175 million ha is subjected to various types of degradation, and out of this about 100 million ha is severely eroded (Dhruvanarain and Babu 1983). An estimated area of approximately 4 million ha wasteland is in the form of ravines in the country (ICAR Report 2010). These landforms are almost totally devoid of fertility and hardly sustain any vegetation (Fig. 21.1).

**Fig. 21.1** Highly degraded ravenous area. Left, Chambal ravines; right, Yamuna ravines

The estimated average annual loss of production of food grains from this landmass is estimated at 3.0 million tons (Chaturvedi et al. 2016). The ravines are not static but are increasing in their spread every year. The average rate of encroachment into the tablelands is estimated at 8000 ha per annum (Anonymous 2003). Some of the constraints to rehabilitate the ravine lands may include:

i. Inaccessibility due to extremely dissected land surface made limited land shaping unavoidable.
ii. Highly steep slopes of ravine banks: too steep to hold seeds on them and steep ravine beds.
iii. Highly fragile and permeable surface soil generating an undercurrent of water running through sandy layer in soil profile at 2–3 m depth which unlike surface runoff cannot be diverted by storm drain/water diversion bund.
iv. Very strong flow lines of runoff along ravine beds which cause failure of gully plugs/earthen bunds.
v. Very low content of organic matter and colloidal complex resulting in high percentage of small (<125 μ) aggregates and thus soil very susceptible to water erosion even if the field has slight slope or no slope but slight depressions caused by cultivation/ploughing.
vi. Highly impoverished soil (by erosion) – quite unproductive.

Greening of these lands with forest and fruit trees, forages, and high-value crops such as medicinal crops will be the most efficient use of these resources, which will not only increase productivity but also sequester carbon and mitigate climate change. Based on long-term experimentation, some of these efforts have been highlighted in this chapter.

## 21.2 Land Use and Land Cover Map

Table 21.1 shows the status of land use and land cover of five districts (Bhind, Datia, Gwalior, Morena, and Sheopur) of Madhya Pradesh falling under Chambal catchment. It is divided into six major classes, viz., ravine/wasteland, water bodies, forest, fallow/barren, habitation, and agriculture (Fig. 21.2). The continuous increase in area of ravine was recorded during the period 1943–1944 to 1975–1976, but it

**Table 21.1** Area (ha) and percentage coverage under different LULC classes of northern part of Madhya Pradesh

| | Morena | | Bhind | | Sheopur | | Gwalior | | Datia | |
|---|---|---|---|---|---|---|---|---|---|---|
| Class name | Area | % | Area | % | Area | % | Area | % | Area | % |
| Ravine/wasteland | 72,144 | 14.4 | 19,532 | 4.4 | 18,848 | 2.8 | 13,900 | 3.0 | 3945 | 1.3 |
| Water body | 5810 | 1.4 | 953 | 0.5 | 2128 | 0.3 | 4144 | 0.9 | 1316 | 0.4 |
| Forest | 59,566 | 11.9 | 4596 | 1.0 | 310,000 | 46.5 | 112,100 | 24.5 | 29,344 | 9.9 |
| Fallow/barren/open | 180,421 | 35.2 | 215,569 | 48.4 | 179,095 | 26.9 | 123,769 | 27.1 | 64,899 | 21.9 |
| Habitation | 2127 | 0.4 | 2277 | 1.1 | 100.07 | 0.02 | 4045 | 0.9 | 586 | 0.2 |
| Agriculture | 181,305 | 36.2 | 202,446 | 45.4 | 156,136 | 23.4 | 198,586 | 43.5 | 195,721 | 66.2 |

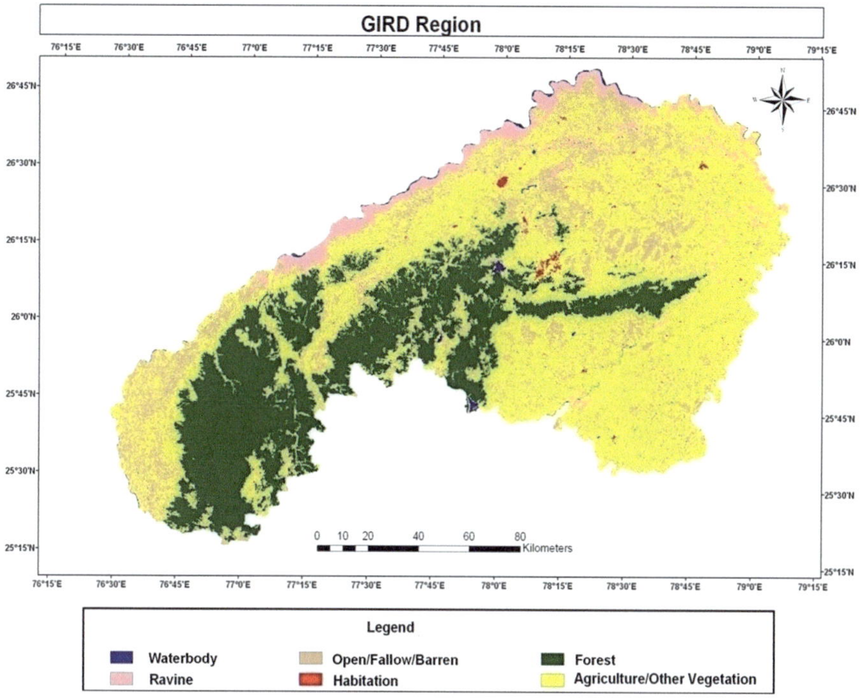


**Fig. 21.2** Land use and land cover map of study area (Gird region)

decreased drastically in recent years particularly last 3 years (2012–2014) due to decreasing land availability and escalating land prices farmers are forced to reclaim ravines on the riverbanks where the gullies are shallow or medium in nature (e.g., Bhind district where it is now only 19,000 ha instead of 119,000 ha).

## 21.3 Experimentation in Morena and Gwalior Areas (Case Studies)

### 21.3.1 Climatic of Study Areas

The challenges of diverse climate changes and climate variability do exist in ravine land areas, associated with general problems such as low ground water table, uneven and untimely distribution of rains, droughts and heat stresses, frost, and hailstorm. The Chambal region experiences subtropical climate where hot winds during summer flow for a great part of the day, and night temperature remains high. The hottest months are May and June (temperature ranges from 44 to 50 °C). The winter commences in October, and the month of December and January are the coldest, the minimum temperatures touching the freezing point at night. Winter rains are erratic

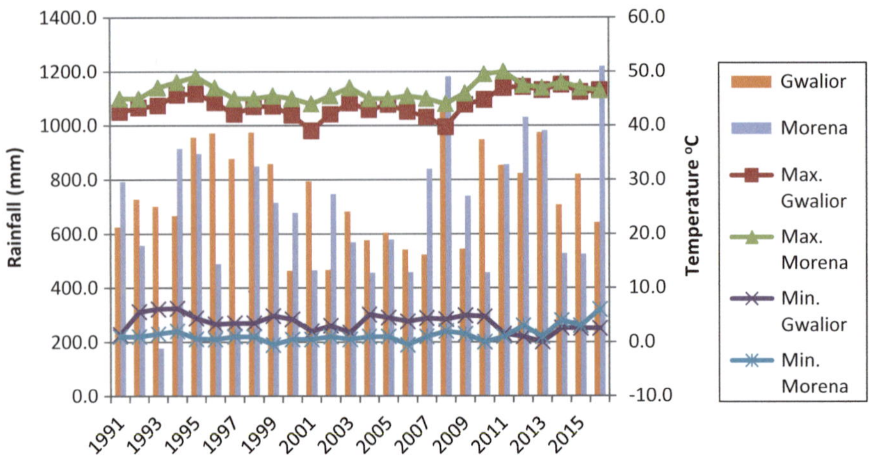


**Fig. 21.3** Weather data (1991–2015) of Gwalior and Morena

and irregular. The average annual rainfall of the Chambal divisions is 891.4 mm, and a major portion is received in July, August, and September. The weather remains sultry and humid. The long-term temperature and rainfall data of 25 years of Morena and Gwalior (Fig. 21.3) clearly indicate that within only 30 km, there is a large variation in minimum and maximum temperatures and rainfall. Minimum temperature always remained lower in Morena than Gwalior, and even in years when Morena experienced −0.5 °C in winter, the temperature of Gwalior remained above 1.4 °C. Similarly, large variations were observed in maximum temperature, being 2–5 °C higher at Morena than at Gwalior. There have been large fluctuations in annual rainfall; at Morena it ranges 180–1200 mm, while at Gwalior it ranges between 500 and 1080 mm. It gives clear impact of ravines and physiographic variations on climatic variability.

## 21.3.2 Rehabilitation Work

### 21.3.2.1 Watershed-Based Management of Chambal Ravines

After establishment of Rajmata Vijayaraje Scindia Krishi Vishwa Vidyalaya (RVSKVV) in 2008, much needed research work on control and reclamation of ravines and their management for sustainable livelihood security was initiated under ICAR funded NICRA (National Initiative on Climate Resilient Agriculture) and Niche Area of Excellence projects at Aisha village, Morena district, MP. In both these projects, heavy emphasis was laid on plantation from the very beginning, and several species of native trees, fruit trees, medicinal plants, and different crops were evaluated on different slopping lands and leveled patches of land.

#### 21.3.2.2 Earth Work

Land under deep ravines is inaccessible, and afforestation/tree plantation alone does not check the advancement of deep ravines; therefore, slight land shaping is unavoidable. Further, after identification of main gully, its stabilization is to be achieved to control further deepening and formation of new ravines as well as stabilize the side ravines. Since a lot of natural vegetation did not help, construction of following types of structures was taken up to reclaim deep gullies:

(a) *Gabions*: Gully erosion is quite severe throughout Central India, particularly where concentrated runoff flows fast through cultivated land and along both the banks of river (Fig. 21.4). The construction of gabions worked successfully during the rainy season of first year, and a lot of silt got deposited against these, but the structure was not stable during next year, and the underground current caved through highly porous gabions, and a deep and wide trench appeared in the deposited silt. The gabions were too porous due to improper packing of poor-quality boulders. Composite gabions (lower part masonry while upper part of gabions using dressed boulders) worked better.

(b) *Masonry structures*: The site for raising the masonry structure should be at higher than the steep channel slope of larger area. Standard design is being followed with deep foundation of 45–60 cm below flow line (Fig. 21.5). In silt retention type check dams, provision of one or two holes (5–7 cm diameter) at 0.30–0.60 m below the crest level should be made for disposal of surpluss runoff.

**Fig. 21.4** Gabion structure for control of gully in ravine land

**Fig. 21.5** Construction of masonry structure

(c) *Bori bandhan*: The construction of stop dam is done with the help of soil-filled bags called *bori bandhan* in local language for control gully of ravine (Fig. 21.6). This technique is providing necessary irrigation facility after withdrawal of rains and checking the soil erosion. These were not successful like simple gabions due to caving under or runoff passing away through one or both ends.

(d) *Earthen check dams/gully plugs*: Earthen check dams were constructed in gullies to stabilize these by preventing their bed corrosion through deposition of silt on the bed. Depending upon the nature of the gully, water stood against the check dams (Fig. 21.7) for varying period. If the bed of the gully was pervious, the water percolated to charge the groundwater.

(e) *Anicuts against slope*: Construction of anicut, a small dam in the path of a stream for regulating the flow of rainwater, is a proper technique and is followed frequently.

#### 21.3.2.3 Land Shaping

Besides land shaping for accessibility, following approaches were tried to check the advance of deep gullies:

(a) Giving moderate slope (<6%) to vertical banks of gully head and plantation of strips with perennial tall grass like *Saccharum munja* and bushes such as Ziziphus nummularia, *Crotalaria burhia*, *Balanites roxburghii*, Carissa carandas, etc.

**Fig. 21.6** Bori bandhan in ravine land

**Fig. 21.7** Earthen check dam conserving rainwater in ravine land

(b) *Terracing*: Wherever feasible terracing is a good device of both soil and moisture conservation.

(c) *Multistep leveling system*: Various types of land-shaping systems were tried at different sites for a maximum period of 2 years (before termination of the project). Period of 2–3 years is not sufficient for such research work, and there is a need for testing and comparing these systems in different micro-watershed – a series of which are available adjacent to one another in the project area.

   (I) *Slopping land (3–6% slope) with shallow/medium soil:* Agronomical measures, i.e., strip/contour planting against slope (grass strips between crop strips for assuring low soil losses) with following crops:
   - Green gram
   - Pearl millet
   - Pigeon pea-cluster bean intercropping

   (II) *Mild slope land (>1–3% slope) with medium/deep soil:*
   - Green gram-taramira (*Eruca sativa*) cropping systems
   - Pearl millet-mustard cropping systems

   (III) *Flat land (1% or fewer slopes) with deep soil:* suitable for intensive cropping system, i.e.,
   - Sesame-taramira
   - Pearl millet-mustard
   - Green gram-wheat

## 21.4 Agroforestry for Carbon Sequestration and Climate Change

The climate change is recognized by the global scientific community as the greatest environmental threat to living organisms and ecosystem functioning. Carbon dioxide is the focus of global climate change studies, because it accounts for about 70–80% of the greenhouse gases. The practice of agroforestry on degraded lands has great potential for maintaining and increasing the land-based carbon sinks. Agroforestry can store carbon through enhancing buildup of soil organic matter and by biomass production that can pool more $CO_2$ from the air and can play an important role in reducing vulnerability, increasing resilience of farming systems, and buffering households against climate risks (Lorenz and Lal 2014). Sequestering carbon through tree-based system is now being considered as an attractive economic opportunity for carbon trading. In India, where majority of the population is dependent on agriculture, tree-based system provides a long-term strategy to increase the amount of carbon while still allowing for growing of food crops. Projections of ICRAF that the carbon market may exceed US$ 1 trillion by 2025 suggest that significant funds could potentially be available to finance sustainable rural development and prevention of climate change (ICRAF 2009). Ajit et al. (2013) studied that the agroforestry is not only improving livelihood of small and marginal farmers but also helping in mitigating global warming by enhancing carbon sequestration potential of Indian agriculture.

### 21.4.1 Details of Modules Tested for Carbon Sequestration

Five modules designed for ravinous land were tested on Chambal ravines in Morena under NICRA Project entitled Niche Area of Excellence on management of soil health and productivity in ravinous land. The experiments were started in 2011–2012 and continued till 2014–2015. Several species of forest and fruit trees, medicinal plants, and different crops were evaluated. The details of modules tested included following diversified cropping systems:

**$M_1$ (diversified cropping system module):** The cropping pattern included Pearl millet-Mustard, Pearl millet-Taramira (*Eruca sativa*), Green gram/Black gram-Mustard, Sesbania (*Sesbania acuminata*) - Mustard, Soybean-Mustard, and Pigeon pea + Cluster bean intercropping.

**$M_2$ (agri-horti module):** Intercrops grown with fruit trees included Aonla (*Emblica officinalis*) + Pearl millet – Sesamum; Custard apple (*Annona reticulata*) + Sesbania – Mustard – Maize fodder; Guava (*Psidium guajava*) + Green gram – Taramira; Ber (*Ziziphus mauritiana*) + Pigeon pea; Drumstick (*Moringa oleifera*) + Zinger (*Zingiber officinale*) – Taramira; and Pomegranate (*Punica granatum*) + Castor – Chickpea.

**$M_3$ (horti-medicinal/pastoral module)** (Fig. 21.8): Guava (*Psidium guajava*) + Anjan grass (*Cenchrus ciliaris*); Ber (*Ziziphus mauritiana*) + Stylo (*Stylosanthes scabra*); Aonla (*Emblica officinalis* ) + Guria grass (*Chrysopogon fulvus*); Ber (*Ziziphus mauritiana*) + Spanish grass (*Cymbopogon aciculatus*); Custard apple (*Annona reticulata*) + Satawar (*Asparagus racemosus*); and Pomegranate (*Punica granatum*) + Lampa grass (*Heteropogen contortus*).

**Fig. 21.8** Gooseberry (*Emblica officinalis*)-based pastoral system

**$M_4$ (silvi-medicinal module):** Multipurpose trees grown with medicinal plants included Khair (*Acacia catechu*) + Aloe vera (*Aloe barbadensis*); Karanj (*Pongamia pinnata*) + Satawar (*Asparagus racemosus*) + Lemongrass (*Cymbopogon citratus*); Guggal (*Commiphora wightii*) – Safed musli (*Chlorophytum arundinaceum*); Cotton tree (*Bombax ceiba*) – Ashwagandha (*Withania somnifera*); Bael (*Aegle marmelos*) – Castor (*Ricinus communis*); and Peelu (*Salvadora oleoides*) – Chandrasur (*Lepidium sativum*).

**$M_5$ (silvopastoral module):** Fodder grasses grown with multipurpose trees included Neem (*Azardirechta indica*) + Ginni grasses; Karanj (*Pongamia pinnata*) + Napier grass (*Pennisetum purpureum*); Khejri (*Prosopis cineraria*) + Marvel grass (*Dichanthium annulatum*); Arduso (*Ailanthus excelsa*) + Para grass (*Brachiaria mutica*); Siras (*Albizia lebbeck*) + Lampa grass (*Heteropogen contortus*); and Subabul (*Leucaena lencocephala*) + Anjan grass (*Cenchrus ciliaris*).

### 21.4.2 Carbon Production in Different Modules

The biomass carbon of different modules was estimated at the end of the years 2012–2013, 2013–2014, and 2014–2015 (Table 21.2). The biomass carbon was highest in agri-horticulture module ($M_2$) followed by horti-pastoral module ($M_3$), Silvi-medicinal ($M_4$), silvopastoral module ($M_5$), and diversified cropping system ($M_1$) during secon year (2013–2014) onward, whereas during first year of experimentation (2012–2013) maximum carbon production with diversified cropping system ($M_1$) followed by horti-pastoral ($M_3$), agri-horticulture ($M_2$), silvi-medicinal ($M_4$) and silvopastoral ($M_5$). The highest change in biomass carbon was 235% in module $M_2$ followed by 194% in $M_3$, 107% in $M_5$, and 78% in $M_4$ compared with diversified cropping system (Table 21.2) after 3 years of experimentation. In agroforestry/agri-horticulture systems, two major components, i.e., tree and crops, are mainly responsible for $CO_2$ sequestration. The total amount sequestered in each component differed greatly, and it was dependent largely on number of factors such as the type of system (the nature of components and age of plant), site quality, and previous land use and managements.

**Table 21.2** Average carbon production in different modules of ravine management

| | Carbon (Mg ha$^{-1}$) | | |
|---|---|---|---|
| Module | 2012–2013 | 2013–2014 | 2014–2015 |
| $M_1$ – Diversified cropping system | 2.12 | 2.59 | 2.91 |
| $M_2$ – Agri-horticulture | 0.99 | 3.96 | 9.74 |
| $M_3$ – Horti-pastoral | 1.64 | 3.79 | 8.56 |
| $M_4$ – Silvi-medicinal | 0.47 | 1.50 | 5.17 |
| $M_5$ – Silvopastoral | 0.24 | 1.24 | 5.03 |

**Table 21.3** Average cumulative carbon production by different types of system

| System | Carbon (Mg ha$^{-1}$) | | |
|---|---|---|---|
| | 2012–2013 | 2013–2014 | 2014–2015 |
| Fruit plants | 0.94 | 1.88 | 5.65 |
| Forest plants | 0.35 | 1.26 | 3.97 |
| Silvi-culture | 0.76 | 1.45 | 4.71 |
| Medicinal plants | 0.48 | 0.75 | 3.95 |
| Grasses | 0.38 | 1.08 | 1.88 |
| Crops | 1.82 | 1.97 | 2.15 |

### 21.4.3 Biomass Carbon in Different Types of Trees/Crop

Cumulative carbon production in different systems in experimental area was increased maximum during the third year of plantation compared with the first and second year (Table 21.3). The result showed that highest carbon accumulation was under fruit plants (5.65 Mg ha$^{-1}$) followed by silvi-culture (4.71 Mg ha$^{-1}$), forest plant (3.97 Mg ha$^{-1}$), medicinal plant (3.95 Mg ha$^{-1}$), crops (2.15 Mg ha$^{-1}$), and grasses (1.88 Mg ha$^{-1}$).

#### 21.4.3.1 Biomass Carbon in Different Trees

The scenario of carbon production in trees ranged from 0.02 to 1.60 Mg ha$^{-1}$ after 1 year of plantation, while after the second year, carbon production was 0.05–2.50 Mg carbon ha$^{-1}$, and after 3 years, cumulative values of carbon was 0.36–12.20 Mg carbon ha$^{-1}$ (Table 21.4). The results revealed that among the fruit trees, drumstick (*Moringa oleifera*) amassed the highest biomass carbon during 2014–2015, 2013–2014, and 2012–2013 followed by Neem, Shisham, Khamer, Arjun, Siras, Karanj, and Sagon during 2014–2015. The decrease in biomass carbon of pomegranate, guava, and custard apple in 2013–2014 was due to frost. All the forest trees (Khamer, Sagon, and Shisham) including forest and medicinal trees performed well during 2014–2015 as compared to 2013–2014 and 2012–2013. In fruit trees, the average biomass carbon was very low except drumstick and ber during study period.

#### 21.4.3.2 Biomass Carbon in Grasses

The biomass carbon was recorded highest in para grass (1.52 Mg ha$^{-1}$) followed by palmarosa (1.24 Mg ha$^{-1}$), lemongrass (10.8 Mg ha$^{-1}$), and Napier grass (0.93 Mg ha$^{-1}$) (Table 21.5). The higher carbon production was stored by para grass due to higher biomass cover, and the extensive root system not only increased biomass carbon but also improved soil aggregate stability and available soil moisture and in turn reduced soil erosion.

#### 21.4.3.3 Biomass Carbon in Crops

The average carbon production was highest in pearl millet (3.30 Mg ha$^{-1}$) in all the years followed by maize (2.82 Mg ha$^{-1}$), cluster bean (1.22 Mg ha$^{-1}$), sesame

**Table 21.4** Carbon sequestration of different types of trees

| Types of trees | Carbon ($Mg\ ha^{-1}$) | | |
|---|---|---|---|
| | 2012–2013 | 2013–2014 | 2014–2015 |
| Aonla (*Emblica officinalis*) | 0.20 | 0.65 | 1.24 |
| Guava (*Psidium guajava*) | 0.31 | 0.43 | 0.55 |
| Custard apple (*Annona reticulata*) | 0.15 | 0.42 | 0.51 |
| Ber (*Ziziphus mauritiana*) | 0.24 | 0.90 | 1.35 |
| Pomegranate (*Punica granatum*) | 0.22 | 0.41 | 0.51 |
| Drumstick (*Moringa oleifera*) | 1.60 | 2.50 | 12.20 |
| Arduso (*Ailanthus excelsa*) | 0.18 | 0.21 | 0.41 |
| Mahua (*Madhuca longifolia*) | 0.07 | 0.12 | 0.20 |
| Siras (*Albizia lebbeck*) | 0.18 | 0.68 | 3.20 |
| Babool (*Acacia nilotica*) | 0.19 | 0.29 | 0.49 |
| Khamer (*Acacia auriculiformis*) | 0.14 | 0.61 | 3.42 |
| Sagon (*Tactona grandis*) | 0.24 | 0.48 | 1.30 |
| Sheesham (*Dalbergia sissoo*) | 0.27 | 0.59 | 3.80 |
| Arjun (*Terminalia arjuna*) | 0.11 | 0.39 | 3.40 |
| Neem (*Azardirechta indica*) | 0.25 | 0.48 | 3.90 |
| Karanj (*Pongamia pinnata*) | 0.10 | 0.35 | 1.34 |
| Gugul (*Commiphora wightii*) | 0.02 | 0.05 | 0.36 |

**Table 21.5** Carbon production of different grasses during different years

| Types of grass | Carbon ($Mg\ ha^{-1}$) | | |
|---|---|---|---|
| | 2012–2013 | 2013–2014 | 2014–2015 |
| Pamarosa (*Cymbopogon martinii*) | 0.92 | 1.02 | 1.24 |
| Lemongrass (*Cymbopogon citratus*) | 0.73 | 0.91 | 1.08 |
| Napier grass (*Pennisetum purpureum*) | 0.61 | 0.78 | 0.93 |
| Para grass (*Brachiaria mutica*) | 1.18 | 1.47 | 1.52 |

(0.74 $Mg\ ha^{-1}$), black gram (0.66 $Mg\ ha^{-1}$), and soybean (0.65 $Mg\ ha^{-1}$) in *kharif* season, while during *rabi* season the average biomass carbon was highest in wheat (1.82 $Mg\ ha^{-1}$) crop followed by mustard (1.67 $Mg\ ha^{-1}$), chickpea (0.73 $Mg\ ha^{-1}$), and taramira (0.0.52 $Mg\ ha^{-1}$) during third year of experimentation (Table 21.6).

### 21.4.4 Total Carbon and Total Organic Carbon

To determine the C sequestration by soil, total carbon and total organic carbon in soil was estimated. The highest sequestration of total carbon and organic carbon took place in silvopastoral module ($M_5$) followed by agri-horticulture module ($M_2$), silvi-medicinal module ($M_4$), horti-pastoral module ($M_3$) and diversified agriculture module ($M_1$). Thus, highest C sequestration took place in module $M_5$ while the lowest in module $M_1$. Among all the natural techniques, agroforestry/agri-horticulture provides a win-win opportunity to achieve the objectives of carbon

**Table 21.6** Carbon production by different crops during different years

| Name of crop | Carbon (Mg ha$^{-1}$) | | |
|---|---|---|---|
| | 2012–2013 | 2013–2014 | 2014–2015 |
| *Kharif* (rainy season) crops | | | |
| Pearl millet (*Pennisetum glaucum*) | 3.29 | 3.41 | 3.50 |
| Maize (*Zea mays*) | 2.55 | 2.61 | 2.82 |
| Sesame (*Sesamum indicum*) | 0.61 | 0.69 | 0.74 |
| Black gram (*Vigna mungo*) | 0.46 | 0.52 | 0.66 |
| Cluster bean (*Cyamopsis tetragonoloba*) | 1.03 | 1.18 | 1.22 |
| Soybean (*Glycine max*) | 0.50 | 0.58 | 0.65 |
| *Rabi* (winter) crops | | | |
| Mustard (*Brassica juncea*) | 1.42 | 1.51 | 1.67 |
| Taramira (*Eruca sativa*) | 0.36 | 0.48 | 0.52 |
| Wheat (*Triticum aestivum*) | 1.58 | 1.62 | 1.82 |
| Chickpea (*Cicer arietinum*) | 0.51 | 0.62 | 0.73 |

sequestration and climate change mitigation and adaptation. Although agroforestry/agri-horticulture is not primarily designed for carbon sequestration, there are many recent studies to substantiate the evidence that these systems can play a major role in storing carbon in above-ground biomass (Nair et al. 2009).

### 21.4.5 Carbon Sequestration in Soil

Three years' results showed that there was significant increase in soil organic carbon content (Table 21.7) under different modules. The organic carbon which was just 0.11–0.12% at the start of experiment increased to a level of 0.22–0.36% under various modules in surface soil just within 3 years. During first year, it was maximum in $M_1$ and $M_5$, but there was more carbon sequestration in silvopastoral module ($M_5$) due to greater root mass with increasing age of trees/plants. The rate of soil carbon sequestration appears to be affected by tree density to some extent; however there are many other factors influencing it under systems, viz., precipitation, temperature, evapotranspiration, decomposition rate, litter fall, decay rates of tertiary tree roots, crop residue, species of tree/crop, chemical properties of the litter, etc.

## 21.5 Soil Organic Carbon Status in Chambal Ravines

The experiments carried out under Niche Area of Excellence (ICAR, New Delhi, India) program at RVSKVV, Gwalior (India), clearly indicated that the organic carbon status of ravine soils is very poor and mostly in active pool (Table 21.8), while passive pool is deprived of carbon indicating that this soil can work as a potential sink for organic carbon it can be increased to many folds. Experimental results showed that physical, chemical properties and organic carbon content of ravine land

**Table 21.7** Organic carbon at different depth of soil in different modules

| | Organic carbon in soil (%) | | | | | |
|---|---|---|---|---|---|---|
| | 2012–2013 | | 2013–2014 | | 2014–2015 | |
| Management module | 0–15 | 15–25 | 0–15 | 15–25 | 0–15 | 15–25 |
| $M_1$ – Diversified cropping system | 0.26 | 0.22 | 0.35 | 0.32 | 0.38 | 0.34 |
| $M_2$ – Agri-horticulture | 0.23 | 0.18 | 0.34 | 0.29 | 0.36 | 0.30 |
| $M_3$ – Horti-pastoral | 0.16 | 0.18 | 0.30 | 0.23 | 0.34 | 0.26 |
| $M_4$ – Silvi-medicinal | 0.22 | 0.23 | 0.33 | 0.31 | 0.36 | 0.33 |
| $M_5$ – Silvopastoral | 0.26 | 0.20 | 0.32 | 0.26 | 0.37 | 0.28 |

**Table 21.8** Different forms of carbon (%) available in surface soil (0–15 cm) under different modules

| | Active pool | | Passive pool | | | | |
|---|---|---|---|---|---|---|---|
| Module | VL | LL | VLL | NL | Organic | Inorganic | Total |
| Diversified cropping system | 0.12 | 0.02 | 0.01 | 0.01 | 0.16 | 0.874 | 1.034 |
| Agri-horticulture | 0.13 | 0.02 | 0.01 | 0.03 | 0.19 | 1.015 | 1.275 |
| Horti-pastoral | 0.14 | 0.02 | 0.02 | 0.03 | 0.21 | 0.902 | 1.112 |
| Silvi-medicinal | 0.13 | 0.02 | 0.02 | 0.03 | 0.20 | 0.971 | 1.171 |
| Silvopastoral | 0.15 | 0.01 | 0.01 | 0.00 | 0.17 | 1.255 | 1.425 |
| Mean | 0.132 | 0.0218 | 0.014 | 0.020 | 0.186 | 1.0036 | 1.2036 |
| Mg ha$^{-1}$ | 264.0 | 43.6 | 28.0 | 40.0 | 372.0 | 2007.2 | 2407.2 |

*VL* very labile, *LL* low labile, *VLL*,very low labile, *NL* non-labile

can easily be improved through integrated nutrient management, and there is tremendous scope for carbon sequestration in such soils consequently will promote crop production and resource conservations. Plantation through adopting various modules also helped to increase in organic carbon of soil pool.

## 21.6 Performance of Crops, Forest Trees, and Other Farming Systems in Managed Ravine Lands

### 21.6.1 Impact of Soil Fertility on Crops

The fertility treatments were imposed on different cropping systems on mild slopes as well as on flat lands after its reclamations. The following treatments were included: control or farmers' practice ($T_1$), farmyard manure (FYM) – 10 Mg ha$^{-1}$ ($T_2$), 75% recommended dose of fertilizer (RDF) + 2.5 Mg FYM ($T_3$), 50% RDF + 5 Mg FYM ($T_4$), 100% RDF ($T_5$), and 150% RDF ($T_6$). Data regarding the performance of various treatments on the yield of wheat grain (Table 21.9) clearly indicated that various levels of nutrients as well as farmyard manure produced significantly higher yield over control (1.91 Mg ha$^{-1}$). RDF gave significantly more yield over the rest of the treatments except superimposed (150% RDF) treatment

**Table 21.9** Effect of integrated nutrient management on yield of wheat grain of ravenous soil

| Treatment | Yield (Mg ha$^{-1}$) | % increase in yield over control |
|---|---|---|
| $T_1$ – control | 1.91 | – |
| $T_2$ – FYM (10 Mg ha$^{-1}$) | 2.46 | 29 |
| $T_3$ – 75% RDF+2.5 Mg ha$^{-1}$ FYM | 2.93 | 54 |
| $T_4$ – 50% RDF+5 Mg ha$^{-1}$ FYM | 2.97 | 57 |
| $T_5$ – 100% RDF | 3.51 | 84 |
| $T_6$ – 150% RDF | 4.27 | 124 |

FYM stands for farmyard manure and RDF for recommonded dose of fertilizers

which was most superior in its performance over different treatments including (100% RDF). Conjoint application of inorganic fertilizer with organic manure were at par but were superior in comparison to only organic manure. The results clearly suggest that inorganic source of nutrients is better than organic source. Increase in yield was enhanced to the tune of 124% over control with the application of 150% NPK; next in order was optimum nutrient level. Inorganic combined with organic improved the yield level by 57%. Marginal increase in yield was also recorded with FYM. Further these soils need regular monitoring and supplementation of organic manures to maintain fertility and yield potential.

### 21.6.2 Crop Production Systems

An attempt was made to cultivate popular crops (Table 21.10) of Chambal division in ravine lands with all the recommended practices followed in this region. Five crops, viz., pearl millet, sesame, black gram, cluster bean, and soybean in *kharif* (rainy season) and mustard, taramira (*Eruca sativa*), wheat, and chickpea during *rabi* (winter) season were grown. The recorded yield showed that yield of pearl millet and cluster bean during *kharif* were almost equally good as in normal soil of the area, while other crops were little poor. All the *rabi* crops have shown good performance and recorded satisfactory yield. Some of the crops grown in the ravine areas are shown in Fig. 21.9.

To work out most suitable and profitable cropping system, the system productivity of different systems was computed on the basis prevailing support/market price during the year (Table 21.11). Maximum system productivity was recorded for pearl millet-wheat and was closely followed by cluster bean-chickpea and next was pearl millet-mustard, while poor performance was given by sesame-mustard and soybean-mustard.

### 21.6.3 Performance of Forest Trees

The performance results showed that after 3 years of growth, maximum height was recorded in *Moringa oleifera* followed by *Terminalia arjuna*, *Tectona grandis*, *Dalbergia sissoo*, and *Ziziphus mauritiana*, while minimum height was recorded in

**Table 21.10** Grain and straw yield of *kharif* and *rabi* crops during 2013–2014

| *Kharif* crop | Grain (Mg ha$^{-1}$) | Straw (Mg ha$^{-1}$) | *Rabi* crop | Grain (Mg ha$^{-1}$) | Straw (Mg ha$^{-1}$) |
|---|---|---|---|---|---|
| Pearl millet | 3.33 | 8.32 | Mustard | 1.11 | 4.82 |
| Sesame | 0.21 | 0.90 | Taramira | 0.33 | 1.12 |
| Black gram | 0.43 | 1.67 | Wheat | 1.87 | 3.46 |
| Cluster bean | 1.25 | 1.13 | Chickpea | 0.94 | 1.25 |
| Soybean | 0.43 | 1.14 | | | |

**Fig. 21.9** Field crops cultivated after land shaping in the ravines. Top (left) pearl millet, (right) sesame; bottom (left) cluster bean, (right) taramira

*Commiphora wightii*. Similarly the stump diameter was maximum of *Moringa oleifera* followed by *Ziziphus mauritiana*, *Terminalia arjuna*, *Gmelina arborea*, and *Acacia nilotica*, while minimum value was obtained in *Annona reticulata*. The biomass of these trees is already mentioned while reporting carbon sequestration.

## 21.6.4 Goatary-based Livelihood System

It is easier to feed and maintain a herd of goats on poor ravine land, although many farmers object to their introduction on such land. Goats feed on grasses and shrubs which grow well even under arid conditions. Grasses can be grown on the bottom of ravine for controlled grazing, while trees and shrubs (which also give foliage used

**Table 21.11** Cropping system followed and computed systems productivity (equivalent to pearl millet-mustard) based on market prices (2013–2014)

| System | *Kharif* (INR $ha^{-1}$) | | *Rabi* (INR $ha^{-1}$) | | Total |
|---|---|---|---|---|---|
| | Grain | Straw | Grain | Straw | |
| Pearl millet-mustard | 41,612 | 16,638 | 33,916 | 7231 | 99,397 |
| Pearl millet-wheat | 41,612 | 16,638 | 28,110 | 20,784 | 107,144 |
| Pearl millet-taramira | 41,612 | 16,638 | 8476 | 1681 | 68,407 |
| Cluster bean-chickpea | 65,415 | 4827 | 29,223 | 2496 | 101,961 |
| Cluster bean-taramira | 65,415 | 4827 | 8476 | 1681 | 80,399 |
| Black gram-wheat | 18,490 | 1667 | 28,110 | 20,784 | 69,051 |
| Black gram-mustard | 18,490 | 1667 | 33,916 | 7231 | 61,304 |
| Soybean-wheat | 10,931 | 1139 | 28,110 | 20,784 | 60,964 |
| Soybean-mustard | 10,931 | 1139 | 33,916 | 7231 | 53,217 |
| Sesame-wheat | 9225 | 891 | 28,110 | 20,784 | 59,010 |
| Sesame-mustard | 9225 | 891 | 33,916 | 7231 | 51,263 |

The cost (in INR) of grain is for pearl millet, 1250; sesame, 1500; black gram, 4300; cluster bean, 5250; soybean, 2560; wheat, 1500; mustard, 3050; taramira, 2600; and chickpea, 3100. Cost (in INR) of straw is for pearl millet, 200; sesame, 100; black gram, 100; cluster bean, 150; soybean, 100; wheat, 600; mustard,150; taramira, 150; and gram, 200

as fodder) such as *Hardwickia binata*, *Parkinsonia aculeata*, *Prosopis cineraria*, *Sesbania sesban*, *Cassia auriculata*, *Ziziphus nummularia*, and *Z. mauritiana* can be planted all over the ravinous land, particularly where moisture regime is not favorable for fruit trees.

### 21.6.5 Runoff Collection and Recycling

Runoff collection and its recycling and its use for recharge of underground water is a very important component of ravine reclamation and control technology. Runoff collection can be made in several ways depending upon the site in watershed. It can be stored temporarily even against peripheral bund, contour/graded bunds, and submersible check dams, the construction of which will be necessary for the control of the ravine formation. At suitable sites, farm ponds or tanks can be constructed to store runoff. If at any site, construction of a peripheral bund is not feasible, construction of a bund along the boundary of two farmers both having ravenous land will be very appropriate to hold silt and water upstream. The bund must be of an appropriate cross section according to the catchment above bund.

It should also be compacted and protected by planting species such as *Saccharum munja*, *Carissa carandas*, *Agave sislana*, fodder grasses, etc. Further, it must be provided with a waste weir of appropriate type and size and at an appropriate point on the bund to dispose of excess of runoff. This bund will serve double purpose of holding water (for recycling/charging) and reclaim and control ravine formation. Formation of gullies which was taking place earlier will stop because as soon as runoff water from tableland meets the stored water, it loses its velocity/abrasive power.

At certain sites, just below the tableland, the rainwater causes small landslides forming deep pits. It will be quite appropriate to construct a bund in a way that many pits are converted into a tank. The tank will not only store runoff but will also stabilize the tableland. On land with deep ravines if it is not possible to afforest, suitable dam can be constructed against several deep and U-shaped ravines to create a good water reservoir fit for fish culture. The dam will have to be ably designed and constructed.

### 21.6.6 Establishment of Agro-based Cottage Industries

For generating income particularly for marginal farmers and the landless, it is necessary to have some source of additional income. Agro-based industries should be set up using locally available raw material, e.g., (a) preparation of milk products from surplus milk; (b) promotion of self-help groups for making tomato catchup, *tenti* (*Capparis decidua* raw fruit) pickle, aonla *murabba*, mango pickle, *bari* and *papad* making, etc.; (c) sausage making; (d) carpet making; and (e) other small articles such as winnowing trays, local chairs, etc. can also be prepared.

Other techniques such as (a) mushroom cultivation, (b) backyard poultry, (c) backyard nursery for raising seedlings of vegetables, flowers, etc. and (d) bee keeping can also be adopted as livelihood security systems.

Thus, a very low cost and very effective way of checking ravine development comprising of silvopastoral and crop production systems and partially changed topography was innovated for in situ water conservation and restoring of soil health. The introduction of various modules in ravine also improved physicochemical properties. There was tremendous improvement in organic carbon content, while introduction of agri-horticulture and horti-pastoral module has shown marked improvement in soil properties. These studies clearly indicate that reclamation of ravine lands is feasible by using the management techniques judiciously.

## References

Ajit DSK, Newaj R, Handa AK, Prasad R, Alam B, Rizvi RH, Gupta G, Pandey KK, Jain AK, Uma (2013) Modeling analysis of potential carbon sequestration under existing agroforestry systems in three districts of Indo-Gangetic plains in India. Agrofor Syst 87(5):1129–1146

Anonymous (2003) Guide lines on rehabilitation of gullied and ravine lands. Department of Land Resources, Ministry of Rural Development, Government of India

Chaturvedi OP, Handa AK, Kaushal R, Pandey VC (2016) Potential of agroforestry for natural resource conservation higher productivity from ravines and degraded lands. Global ravine conference on management ravines for food and livelihood security. 7–10th March, 2016. RVSKVV, Gwalior, pp 74–84

Coway TJ, Tans PP (2012) Trend of atmospheric carbon dioxide: http://www.esrl.noaa.gov/gmd/ccgg/trends. Last Access Apr 2012.

Dhruvanarain VV, Babu R (1983) Estimation of soil erosion in India. J Irri Drai Engg ASCE 109:419–434

Goswami BN, Venugopal V, Sengupta D, Madhusoodanan MS, Prince K, Xavier (2006) Increasing trend of extreme rain events over India in a warming environment. Science 314:1442–1445

ICAR report (2010) Degraded and wastelands of India status and spatial distribution. Indian Council of Agricultural Research, New Delhi

ICRAF (2009) Transforming lives and landscapes through agroforestry science world agroforestry centre (ICRAF). World Agroforestry Centre Medium-Term Plan 2010–2012. United Nations Avenue, Gigiri P. O. Box 30677-00100, Nairobi, Kenya. Website: www.worldagroforestry.org. World Agroforesty Centre

Intergovernmental Panel on Climate Change (IPCC) (2007) Climate change 2000: the scientific basis. Oxford University Press, Oxford

Lorenz K, Lal R (2014) Soil organic carbon sequestration in agroforestry systems. Agron Sustain Dev 34:443–454

Nair PKR, Nair Vimala D, Mohan KB, Haile Solomon G (2009) Soil carbon sequestration in tropical agroforestry systems: a feasibility appraisal. Environmental Science & Policy 12(8):1099–1111

# Management of Ravines Through Anicuts and Afforestation

22

Akhilesh Singh and S. K. Verma

**Abstract**

Ravines, the product of natural erosion, indicate the worst form of land degradation. The principal types of water erosion are sheet and gully erosion. Sheet erosion causes the removal of surface soil in thin uniform layers, year after year, and, therefore, usually goes unnoticed. Erosion damages rangelands, croplands and infrastructures. The locked-up production potential of ravines in only three states of Uttar Pradesh, Rajasthan and Gujarat alone is estimated to 3 million tons of food grain annually. Efficient reclamation schemes can be had through sustainable agricultural practices, controlled grazing, construction of peripheral bunds, contour bunding, contour trenches, gully plugging, spillways, diversion channel, breaking the length of slopes into smaller units and by plantation. This will prevent further damage to the valuable land as well as increase the production from reclaimed lands.

Different types of mechanical structures were constructed in the in 20 ha area at research site. These structures were *bori bandhan*, earthen bunds, masonry structure, gabion structure and modified gabion. The total sediment stored against the structures constructed till date is 7346 $m^3$ out of which 3100 $m^3$ against the earthen structures (16 earthen and 1 *bori bandhan*), 1705 $m^3$ against the gabion structures (three in numbers) and 2541 $m^3$ in permanent structure (four in numbers). The study revealed that *bori bandhan* conserved 43.5 Mg $ha^{-1}$soil, and these structures were partially fitted with moderate rainfalls; the earthen bunds were average 25.3 Mg $ha^{-1}$ soil stored in each structure, and not suitable as breaks frequently and suitable only for very low rains. Whereas the masonry

A. Singh (✉)
Department of Soil Science and Agricultural Chemistry, College of Agriculture, RVSKVV, Gwalior, India

S. K. Verma
Department of Soil Science & Agricultural Chemistry, RVSKVV, Vishwa Vidyalaya, Gwalior, Madhya Pradesh, India





J. C. Dagar, A. K. Singh (eds.), *Ravine Lands: Greening for Livelihood and Environmental Security*, https://doi.org/10.1007/978-981-10-8043-2_22

structures were found very good and most suitable to check the soil erosion, average 915.4 Mg $ha^{-1}$ soil stored against the masonry structures, gabion structures were tested, and results show that gabion stored 784.5 Mg $ha^{-1}$ soil with high spillage and found that there was problem of pilferage from below and the side. Modified gabions were constructed to resolve the problem of gabion and found that these structures are economical, effective with minimum soil loss and stored average 1204 Mg $ha^{-1}$ soil. The soil and water conservation through structures and afforestation is one of the most important components for management of ravine. This paper discusses the soil and water conservation from the anicuts and afforestation interventions that were found promising for improving productivity and reducing land degradation in different regions of India.

## 22.1 Introduction

Soil degradation is a major global threat for sustainable agriculture. At global scale (as per GLASOD survey), it has affected nearly 2 billion hectares of land area since the middle of the twentieth century (Scherr Sara 1999; Biggelaar et al. 2003). One of the main processes of soil degradation is accelerated erosion by water, which accounts for more than half of the area (1094 million hectares). Erosion by water removes topsoil and nutrients, exposes less fertile subsoil, reduces the available water holding capacity, reduces soil structural stability, causes surface sealing and reduces soil infiltrability.

Across the globe, 24 billion Mg of topsoil is annually from farmland, leading to the projection that 30% of the world's arable land could be depleted within 20 years. Soil in India is being damaged or irreversibly lost, directly through construction, mining or erosion and indirectly through pollution and mismanagement. These losses lead to a reduction in the productive capacity of land, loss of biodiversity and the historic archive as well as 'downstream' impacts of sedimentation. Although more than 99% of the world's food comes from the soil, experts estimate that each year more than 10 m ha (25 m acres) of cropland are degraded or lost as rain and wind sweep away topsoil (Mandal et al. 2009).

The productivity of some lands has declined by 50% due to soil erosion and desertification (Table 22.1). Yield reduction in Africa due to past soil erosion may range

**Table 22.1** Direct effects of accelerated erosion on crop yield in three states of India

| | MoA | NRSC | | | | IISWC |
|---|---|---|---|---|---|---|
| States | 1976 | 2000 | 2003 | 2005–2006 | 2008–2009 | 2014 |
| Gujarat | 0.40 | 0.10134 | 0.039 | 0.04 | 0.0339 | 0.110 |
| Madhya Pradesh (MP) | 0.683 | 0.75691 | 0.5274 | 0.1502 | 0.1453 | 0.312 |
| Rajasthan | 0.452 | 0.49528 | 0.66 | 0.1885 | 0.1525 | 0.274 |
| Uttar Pradesh (UP) | 1.230 | 0.28065 | 0.325 | 0.1481 | 0.1199 | 0.340 |
| Total | 2.765 | 1.63418 | 0.5514 | 0.5268 | 0.4516 | 1.036 |

Source: Pratap (2016). Depictions: *MoA* Ministry of Agriculture, *NRSC* National Remote Sensing Centre, *IISWC* Indian Institute of Soil and Water Conservation

from 2% to 40%, with a mean loss of 8.2% for the continent. In South Asia, annual loss in productivity is estimated at 36 million Mg of cereal, with equivalent values at US$5400 million by water erosion and US$1800 million due to wind erosion.

## 22.2 Soil Erosion Status of India

It is estimated that about 5334 million Mg of soil is lost annually which works out to 16.35 Mg $ha^{-1}$ (Dhruva Narayana and Ram Babu 1983) of which 29% is lost permanently into the sea, 10% gets deposited in the reservoirs decreasing their capacity by 1–2% every year and the remaining 61% is displaced from one place to another or redistributed. Among different land resource regions, highest erosion rate occurs in the black soil region (23.7–112.5 Mg $ha^{-1}$) followed by Shiwalik region (80 Mg $ha^{-1}$), Northeast region with shifting cultivation (27–40 Mg $ha^{-1}$) and the least in north Himalayan forest region (2.1 Mg $ha^{-1}$).

## 22.3 Impact of Soil Erosion

Soil erosion by water has both positive and negative impacts. Formation of alluvial plains and availability of construction materials like sand, gravels, debris, etc. are the positive outcomes of soil erosion by water loss of crop productivity and nutrients; reduced storage capacity of reservoirs, deterioration in water quality and loss of biodiversity are the negative impacts of erosion. In general, however, the accelerated erosion has some severe on-site negative effects. Principal processes that lead to decline in crop yield are attributed to numerous interacting processes and mechanisms (Table 22.2). On global scale, the on-site effects of soil erosion are undermining the productivity of about 33% of the world cropland (Brown and Young 1990).

**Table 22.2** Direct effects of accelerated erosion on crop yield

| Process | Mechanism of effect |
|---|---|
| Reduction in effective rooting depth | Water and nutrient imbalance |
| | Unfavourable soil temperature regime |
| | Edaphological unfavourable subsoil |
| Loss of plant nutrients | Nutrient deficiency |
| | Elemental toxicity |
| | Unfavourable soil pH |
| | Low effective rooting depth |
| Loss of plant available water | Reduced water infiltration |
| | High losses due to runoff |
| | Low effective rooting depth |
| Loss of land area | Land area lost due to permanent or ephemeral gullies |
| | Unfavourable conditions in depositional sites |
| Damage to seeding | Washing away of seeds and seedlings |
| | Seedling burial by deposition of water-borne sediments |

## 22.4 Generic Features and Problems of Ravines Areas

### 22.4.1 Soils and Geology

Soils of ravine are generally deep alluvial with extensive network of gullies running parallel to each other along riverbanks in all the ravine area of the state. The texture of soil varies from sandy loam to clay loam in texture and neutral to mildly alkaline in reaction. The peripheral lands are free from carbonates and pH is from 6.5 to 7.5. The soils are moderately well-drained, and '*kankar*' (calcareous layer) is observed throughout the profile depth. In some ravenous areas, dark clay texture soils are also found. The parent material is basaltic rocks which are broken into small fragments on exposure to rain. Hard pans are also found in certain districts at varying depths, which seriously inhibit a forestation work especially if the rotations are longer than 10 years. *Kachar* soils, i.e. soil on the Riverian terraces at the mouth of ravines, are medium-textured and moderately well-drained and have neutral to mildly alkaline reactions. The soils are severely eroded on steep slopes exposing *kankar* nodules. Due to excessive runoff, the soil is almost dry throughout the year. The calcareous nature of soil, moisture stress and uncontrolled grazing has ruined the lands with vital vegetal cover.

A ravine is generally a fluvial slope landform of relatively steep (cross-sectional) sides, in the order of 20–70% in gradient. Ravines manifest highly degraded landforms are discernible through satellite imageries. Utter neglect of fragile ravenous ecosystem both by the state and people, left to abuse and exploitation, results in sever degradation. These degraded ecosystems exacerbate extreme state of poverty intertwined with land degradation thus inducing social evils. Ravines are safe hideouts and shelter for bandits whose persistence antagonizes economic development and prosperity in the region. Eventually, ravines form natural drainage system in river basins, an inevitable necessity for a landscape. Ravine occur crisscross throughout the length and breadth of the country; however, these turn into a dilemmatic form particularly in northern flood plains due to weak geomorphology, erodible nature of alluvium, bare steep slopes open to intense rainfall, overgrazing and deforestation. Faulty agricultural practices on contiguous land and uncontrolled disposal of runoff further accentuate extension of gully heads and ravines.

The ravine lands are one of the most degraded and vulnerable ecosystems with severe on-site and off-site implications. A ravine system is a network of gullies developed by ephemeral streams with steep banks and a nearly vertical gully head and are deeper than 0.3 m. Apart from land and soil quality degradation in gullied and inter-gullied areas and declined productivity for food, fuel fodder and biomass, these sites are major sediment-producing hot spots, hence increasing the risk for flooding and sedimentation. The subsistence farming practised by resource constraint communities in ravine areas contributes to progressive degradation of these fragile ecosystems. Several natural and manmade factors have been responsible for ravine formation. Erratic, short-duration and high-intensity rainfall, erodible nature of soil, weak geology of alluvium, steep slopes and uneven terrain, faulty agricultural practices, illicit cutting of trees and bushes and overgrazing are some of the factors responsible for the formation of ravines.

Generally, ravine lands have developed in India along the major river systems on highly productive deep alluvial soils. Out of total ravine lands, 2.36 million hectares (64%) are spread over in the states of Uttar Pradesh, Madhya Pradesh, Bihar, Rajasthan and Gujarat (Dhruvanarayana 1993). A rough estimate suggests that about 8000 ha is added to these ravines annually (Sharma et al. 1980). The Yamuna-Chambal ravine zone is the largest and most severely degraded area. The ravines flank the Yamuna river for nearly 250 km and in Agra and Etawah attain a depth of more than 80 m. Nearly 389,000 ha are affected along the Yamuna in southern Uttar Pradesh. The Chambal ravines flank the river Chambal in a 10 km wide belt, which extends southwards from the Yamuna confluence to 480 km to the town of Kota in Rajasthan. Ravines also affect basins of several Chambal tributaries, viz. Mej, Morel, Kalisindh, etc. Altogether, about 5000 $km^2$ area is affected. In Gujarat, ravine belt covers 500,000 ha and extends from the southern bank of the Tapti, banks of the Narmada, Watrak, Sabarmati and Mahi basins. Besides these river basins, ravines are also found in Chota Nagpur, Mahanadi and upper Sone Valley, Indo-Gangetic Plain, Shiwalik and Bhabar tract and Western Himalayas even up to the Kashmir Valley (Dhruvanarayana 1993).

More careful planning is required for ravines deeper than 3 m. A successful ravine rehabilitation program for medium and deep ravine needs to focus on (a) gully head stabilization, (b) stabilization of side slopes, (c) gully bed stabilization (drainage line treatment for flattening the gully beds and (d) vegetative establishment. The three ravine centres of IISWC located at Agra, Kota and Vasad have evaluated suitable conservation measures for the stabilization of gully head, side slopes and gully beds, and recommendations on design specifications have been made. Once erosion process is arrested and gully system is stabilized, these lands can be used to contribute to livelihood and environmental security for local inhabitants. Some of the recommended land use systems for these lands are listed here.

## 22.5 Reclamation of Shallow Ravines for Agriculture

Generally, about 33% of the ravine systems are shallow ravines (<1 m deep) which can easily and safely be reclaimed for cultivation. If left untreated the ravines tend to deteriorate further and ingress into the adjacent marginal and fertile table lands. Steps involved in reclaiming shallow ravines are as follows.

### 22.5.1 Installation of Marginal Bunds

The first step for shallow ravine reclamation is planning and establishing marginal bunds at the downstream periphery of shallow ravines. These are also named as peripheral bunds. These bunds are intended to channelize runoff for providing safe disposal into the deeper ravines from the adjacent marginal lands through a suitably designed spillway. Depending on the expected runoff volume and vertical fall in downstream side, either straight drop, chute or pipe inlet type of spillway is installed at the draining points of peripheral bunds. Generally, for Chambal, Yamuna and Mahi, ravine marginal bunds of 1.0 $m^2$ cross-section (height 0.75 m; top width 0.50 m; bottom width 2.50 m; side slope 1:1.5) with 0.1–0.2% grade are recommended at a distance (equal to double the depth of gullies) from the gully heads to arrest the uninterrupted high velocity flow of surplus runoff from the arable lands to the adjoining gully heads to check the extension of gullies into the arable lands. These bunds are reinforced by planting grasses. Rooted slips of locally useful grasses, e.g. *Dichanthium annulatum*, *Cenchrus ciliaris*, *Cenchrus setigerus* and *Panicum antidotale*, are planted during the rainy season. Once the grass grows well, it provides protection to the bund against erosion and yields significant quantities of good fodder grass. The construction of these bunds also leads to moisture conservation to the tune of 15–20% in the adjoining lands.

### 22.5.2 Levelling or Terracing

Levelling and slope smoothening on the upstream side up to marginal bund are required for promoting in situ rain water conservation and improving water and nutrient use efficiency of cropping systems. After smoothening of sloping area and levelling of small humps and depressions, earthen bunds shall be installed at regular interval on contours (contour bunds) or with 0.1–0.2% grade (graded bunds). For most ravine systems of India, 0.60 $m^2$ cross-sectional bunds with 0.40 m top width, 1.6 m bottom width and 0.6 m height are recommended. Steeper slopes (>3%) or multidirectional slope may necessitate formation of terraces. It comprises of construction of steplike fields along contours usually by half-cutting and half-filling procedure. Original steep is converted into level fields, and thus the hazards of erosion are minimized. By adopting bench terracing, both degree and length of slope are reduced which help in soil moisture conservation for enhanced crop production. Cost of levelling in shallow ravine area is in the range of ₹35,000–45,000 per ha (at 2014 year prices).

### 22.5.3 Earthen Check Dams

Earthen check dams are constructed in the drainage channel at regular interval with the horizontal spacing ranging from 20 to 60 m depending on the bed slope of gully and height of dam between two earthen check dams. For construction of earthen check dams, 1.25–1.75 m height, 1.5–2.0 m top width, 5.25–7.25 m bottom width and 1:1.5 side slopes are adopted. Minor levelling of the irregular side slopes of the gully is carried out to convert the gully bed into terrace for agriculture. A pipe or small masonry spill is provided at one end of the dam to discharge surplus runoff. The up and down slope of the earthen dam are stabilized by developing vegetative cover. These check dams hold considerable amount of soil to facilitate the establishment of vegetation and are also known as soil savers. They facilitate the raising of economically useful crops in the gullied lands and lead to better survival and growth of planted vegetation.

## 22.6 Reclamation of Medium and Deep Ravines

More careful planning is required for ravines deeper than 3 m. The best use for these lands is to retire them for permanent vegetation. A successful ravine rehabilitation program needs to focus on following aspects:

- *Gully head stabilization*: Arresting the process of gully heads extension
- *Stabilization of side slopes*: Stabilization of gully sides
- *Gully beds stabilization*: Drainage line treatment for flattening the gully beds
- *Vegetative establishment*: Establishment of suitable tree and grass spp. for productive utilization of these lands

### 22.6.1 Slope Stabilization Measures

Diversion Drains

Diversion drains are made across the slope to divert excess runoff water away from an unstable area and discharge it safely into a natural waterway or grassed water course. The following points need to be kept in mind while designing a diversion drain:

- The bed slope (grade) of the drain should be such that it is non-erosion as well as non-silting one.
- The gradient of diversion drain should preferably be kept within 0.5%.
- A narrow and deep drain does not get silted up as rapidly as a broad and shallow drain of the same cross-sectional area and is, therefore, self-maintaining.
- Design of a diversion drain is like that of a grassed waterway

Contour Trenching

Contour trenching implies excavating a trench along the contour or along a uniform level. The excavated soil is heaped on downstream side of the trench in the form of a bund. Contour trench is used both on hill slope and barren waste lands for soil and moisture conservation as well as for revegetation purposes. These trenches break the slope length, reduce the runoff and consequently retard its scouring action and carrying capacity. The water retained in trenches helps in conserving the moisture and, therefore, provides advantageous site for sowing, planting and augmentation of groundwater recharge.

Types of Trenches

Contour trenches are broadly classified into (i) continuous and (ii) staggered, as discussed below:

i. *Continuous contour trenches*

Continuous contour trenches (CCTs) are the ones where there is no break in their length, and they can be 10–20 m long across the slope depending on the width of the field. The cross-section of the trench generally varies from 30 × 30 cm to 45 cm. They are constructed for moisture conservation in low rainfall areas receiving storms of mild intensities. These trenches demand high skill for construction and layout on contour. It has been observed that CCTs are prone to breaching if they are not constructed perfectly on contour or during high-intensity storms. Hence, these should be planned and constructed with care. Equalizers of 20–25 cm width and suitable height are placed at regular intervals to avoid concentration of flow and to prevent breaching of the trenches.

ii. *Staggered contour trenches*

In medium rainfall areas with highly dissected topography, staggered contour trenches (SCTs) are adopted. The length of the trenches is kept short, i.e. 2–3 m, and the spacing between the rows may vary from 3 to 5 m. The chances of breaches of SCTs are less as compared to CCTs. Staggered contour trenches have been found suitable along the banks of Chambal, Yamuna, Kalisindh and Mahi and their tributaries in the state of Rajasthan, UP, Madhya Pradesh and Gujarat (Pratap 2016).

Contour Wattling

Long and steep slopes of deep gullies, landslides, roadsides (on the mountains) and mine spoils are very unstable due to the presence of large quantities of loose friable material. Wattling is used for mechanical stabilization and revegetation of such steep slopes. It is a technique of breaking length of slope into shorter portions in which wattles are constructed at a vertical interval of 5–7 m up to 33% slope and 3 m up to 66% slope.

Crib Structures

Steep slopes (more than 40%) can be stabilized by constructing log wood crib structures filled with stone/brushwood. Crib can be made with eucalyptus poles of 2–3 m in length and 8–12 cm in diameter. These poles are joined together with the help of 20–25 cm long nails. Poles are buried to a depth of 50–75 cm and are erected in two lines, 1 m apart line to line and pole to pole. The height of the structure is kept 1.5–2 m above the ground depending upon the land slope. Horizontal braces of eucalyptus poles are fitted at 45 cm centre to centre.

Geotextiles

Natural geotextiles in the form of woven nets made from fibres of jute or coir have been used for stabilization of degraded slopes in landslides and mine spoil areas and cut slopes along roadsides. The geotextiles are basically used for initial establishment of vegetation on highly degraded sloping lands by holding the vegetation in place and conserving moisture and fine soil for its growth. The geotextile material biodegrades in about 2 years period by which time the vegetation should establish and take control of the erosion problem.

### 22.6.2 Drainage Line Treatment Structures

Drainage channels/gullies are the carries of runoff and sediment in a watershed. Steep bed gradient (slope) of a channel causes high runoff velocities with associated heavy sediment flow. Hence, the channel gradient needs to be reduced to bring the runoff velocities within permissible limits. Drainage line treatment (DLT) structures are classified as (i) temporary, (ii) semi-permanent and (iii) permanent structures from their functional point of view. Check dams are basically employed for DLT.

**Function of Check Dams**

- Check dams when constructed in series across gullies reduce the steep gradient of gully as they transform the longitudinal gradient of into series of step with low risers and long flat/low slope treads.
- Reduction in slope reduces the velocity of flow and silt-carrying capacity of stream causing the deposition of silt in gully bed.
- Reduction in velocity of flow increases the time of concentration thereby increasing the scope of percolation and groundwater discharge.
- Deposition of silt and conservation of moisture in the gully promotes vegetation growth.

Components of a Check Dam

A check dam should have the following essential elements:

(i) Spillway adequate to carry a selected designed flow
(ii) A key that anchors the structure into the bottom and sides of the gullies
(iii) An apron that absorbs the impact of water falling from spillway to prevent undercutting of the structure

Brushwood Check Dams

Brushwood check dams are constructed in small gullies (1.2–2.1 m deep) where wooden posts are abundantly available. Such check dams may be (i) single row-post check dam and (ii) double row-post brush dam, depending on the need as described below.

Earthen Gully Plugs

An earthen gully plug, constructed of local soil across the gully at suitable sites, can also be used for gully stabilization. Earthen gully plugs are suitable under following situation:

- In the upper catchment areas where there is scope for water storage.
- Suitable soil for the embankment is available.
- The depth of gully is less than 2 m.
- Gully bed slope is less than 10%.
- The site is having facility for side spillway.
- Design specifications of earthen gully plug are given below:

| | |
|---|---|
| Top width | : 0.6 m |
| Side slopes on u/s and d/s sides | : 2 H:1 V |
| Maximum height | : 3.0 m |
| Minimum height | : 1.0 m |
| Stone-pitching | : Up to FSL (full supply level of water) on u/s |

Loose Rock-Fill Dam

Construction of loose rock dams begins by converting the slope of side wall of gully into 1:1 at proposed location of the dam. A trench is then dug across the floor of the gully and into the banks into which the large stones are placed to form toe of the structure. The dam is built upward from the toe, using flatter rocks on downstream face. Rocks smaller than 100 mm diameter should not be used as they will be quickly washed out. On the other hand, a dam made of larger rocks will leave larger voids in structures through which water jets may flow thereby weakening the structure.

### 22.6.3 Semi-permanent Structures

These structures usually have a longer life (up to 10–15 years) and normally do not require any maintenance. They do not require the assistance of vegetative growth for controlling the gully. They can be constructed in places where materials are available in large quantity. These dams are very effective in steep gullies traversing hilly and mountainous regions. They are also useful in ravines where passage for livestock must be provided for.

Gabion Check Dams

Stone wire crate (Gabion) as semi-permanent structure are commonly used for drainage line treatment. The gabion check dams/cross-barriers may be constructed in second and third order (main) gullies/channels for retention of debris and soil accumulation without ponding. The debris-carrying capacity within the inter-structure reach reduces due to the reduction in channel gradient. The check dam encourages good plant cover not only along the bank but also in the bed of the stream due to increased moisture regime.

Cross-barriers have been found more suitable in main drainage channels of the mine spoil areas for debris collection and grade stabilization. Cross-barriers differ from check dams in only that no weir opening is provided and runoff can through almost the entire width of the structure. This reduces the risk of erosion on downstream side by runoff flowing through a narrow weir portion of the structure.

Katiyar and Dhruva Narayana 1987 have recommended specifications for gabion check dams for different channel gradients in mine spoil areas. Such structures have been found stable in drainage channels having slope up to 40%.

Gabion Structure with Ferro-cement Impervious Barrier

A ferro-cement gabion has an impervious wall of ferro-cement (about 2.5 cm thick) at the centre of the structure which goes below the ground level up to the hard strata. This ferro-cement partition supported by gabion should be able to retain the water as well as withstand the force of runoff water. G.I. wire of good quality should be used for wire mesh fabrication to ensure durability and strength to the structure.

Construction Procedure

The site first cleaned followed by the following steps:

- The foundation trench is excavated up to the hard strata.
- Lay down the concrete in the proportion of 1:3:6 and thickness of 15 cm for the ferro-cement of the foundation.
- Construct the ferro-cement wall at the centre of the foundation trench up to 7.5 cm below the ground level. Fill up the upstream and downstream sides of the ferro-cement wall with sand (about 15 cm thick) and the remaining trench with available soil. The sand layers protect the ferro-cement from swelling action of the surrounding soil and also allow the water which is sprinkled on the top for curing to go down to the bottom.
- Next, provide a 7.5 cm thick concrete bed in 1:2:4 proportion on the ferro-cement wall. Place on it the G.I. wire mesh of size 15 × 15 cm. Lay the remaining 7.5 cm thick concrete. A ferro-cement wall up to the full supply level of the structure should then be built above the concrete.
- Construct a stone bund keeping the ferro-cement wall at the centre.
- The whole structure including ferro-cement core wall should go up to 0.3–0.6 into the stable portion of the gully side prevent end-cuttings.
- Raise the end portion of the structure to the level equal to the flood depth plus free board. Aim is to prevent scouring of the *nalla* (channel) banks.
- On the downstream side, provide 2–2.5 m wide gabion/rubble soling along the full structure.

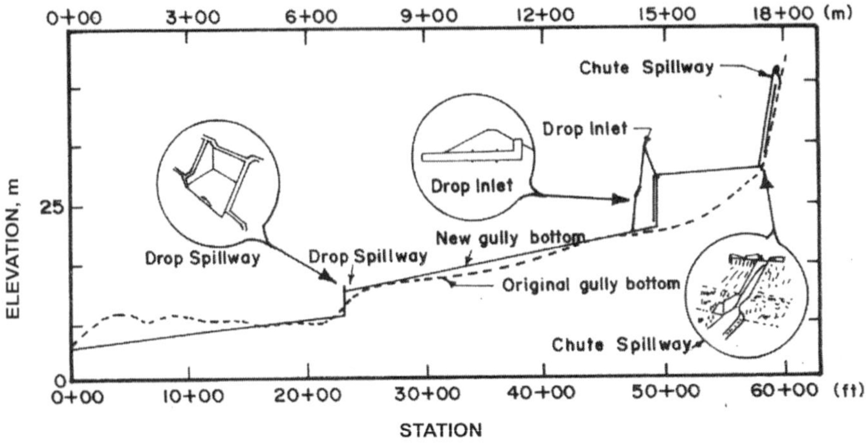


Source: Sharda et al. (2007)

**Fig. 22.1** Location of gully control structure (Source: Sharda et al. 2007)

### 22.6.4 Permanent Gully Control Structures

Permanent structures consist of reinforced concrete, concrete, concrete blocks, stone masonry, metal or earth and are used in gullies where temporary measures are either inadequate or impracticable. The main functions of a permanent structure are to halt the advancement of gully head at overfall, to control and stabilize the gully grade and to convey the water safely from grassed maintenance. Therefore, the construction of a permanent structure should be avoided except in special cases. They should be constructed under the following conditions:

- Where the volume and peak rate of runoff to be handled are very large and cannot be controlled by vegetative measures and simple field structures
- Where high degree of safety against the loss of life and property is warranted
- Site to be protected is inaccessible and regular maintenance of structure is not possible. Permanent structures are constructed when the benefits from such structures are justifiable compared to the cost of construction. General requirements of the permanent structure for gully control are:
  - They should be constructed with permanent material.
  - They should have adequate capacity to handle the peak rate of runoff.
  - They should help in stabilizing the gully and store water where necessary.

Location of Gully Control Measures

The suitable location of these spillways on typical gully reach is shown in Fig. 22.1. Chute spillways are located at the gully head to convey the water safely to the gully bed. The drop spillways are built along gully bed to act as control points so that gully bed is not eroded below the level of the structures. The drop inlet spillways are used in a gully for storage of water.

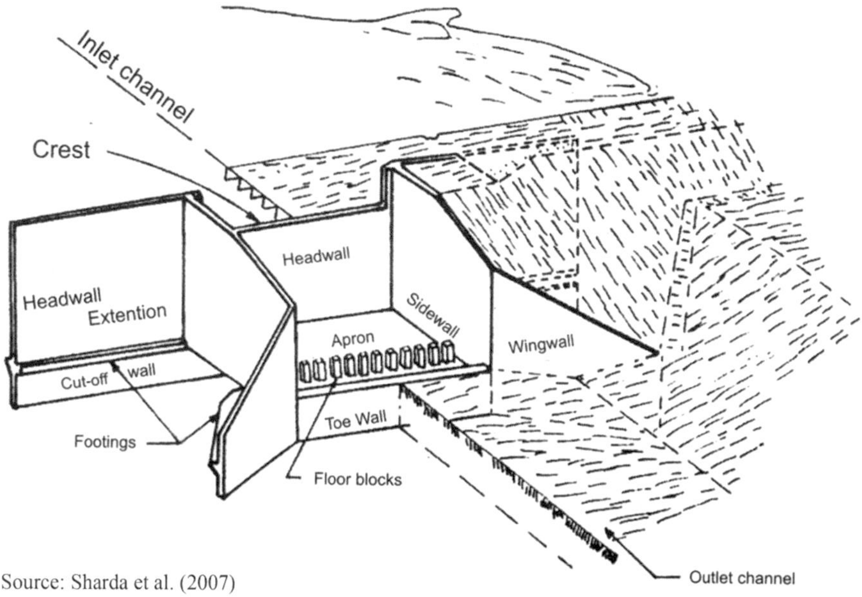


**Fig. 22.2** Drop spillway structure (Source: Sharda et al. 2007)

Drop Spillway

Drop spillway is a weir structure. Flow passes through the weir opening, drops to an approximately level apron or stilling basin and then passes into the downstream channel. The different components of the drop spillway are (1) head wall and head wall extension, (2) side walls, (3) wing walls, (4) apron, (5) longitudinal sills, (6) end sill and (7) cut-off walls (Fig. 22.2).

**Functional Uses**

- Gully and ravine stabilization
- Erosion control structures for stabilization of landslide and mined areas
- Protection of fields, roads, hutments, etc. from gullies
- Grade control for stabilizing channels and waterways
- Reservoir spillway where the total drop is relatively low
- Control of irrigation water

Adaptability

The drop spillway is an efficient structure for controlling low heads, normally up to 3 m.

**Advantages**

- The likelihood of serious structure damage is less than for other types of structures.
- The rectangular weir is less likely to be clogged by debris than the openings of other structures of comparable discharge capacities.
- They are relatively easy to construct.

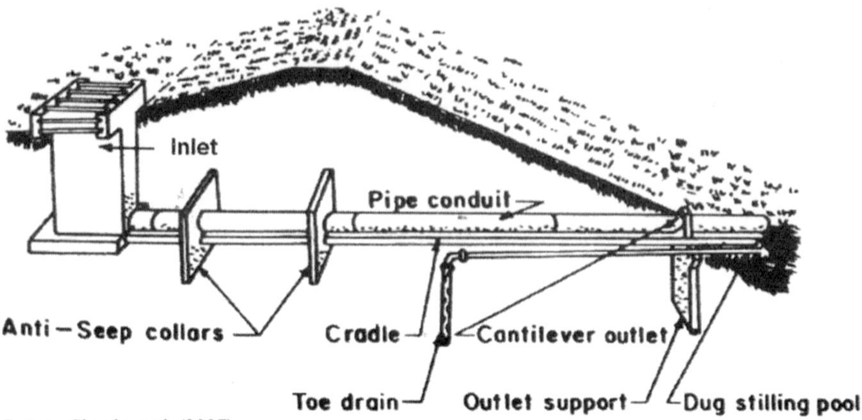


Source: Sharda et al. (2007)

**Fig. 22.3** Drop inlet spillway (Source: Sharda et al. 2007)

**Limitations**

- It is costlier than some other types of structures where the required discharge capacity is less than three comics and the total head or drop is greater than 3 m.
- It is not a favourable structure where temporary spillway storage is needed to obtain a large reduction in discharge.
- A stable grade below the structure is essential.

(A) Drop inlet spillway

A drop inlet spillway is a closed conduit generally designed to carry water under pressure from above an embankment to a lower elevation. An earthen embankment is required to direct the discharge through the spillway. Thus, the usual function of a drop inlet is to convey a portion of the runoff through or under an embankment without erosion. Vegetative or earthen spillways around one or both ends of the embankment should also be used in conjunction with drop inlet spillways. Nomenclature for various parts of a drop inlet spillway are given in Fig. 22.3. The riser of a drop inlet spillway may be of plain concrete, masonry or pipe. The barrel may be of reinforced Concretum concrete or clay tile or smooth metal pipe having water tight joints. In India RCC pipes are generally used.

**Functional Uses**

- Principal spillway for farm ponds and reservoirs
- Grade stabilization
- Principal spillways for debris basin
- Flood prevention structures
- Roadway structures
- Surface water inlet for drainage or irrigation

**Adaptability**

- It is a very efficient structure for controlling relative high heads usually above 3 m.
- It is well adapted to sites providing an appreciable amount of temporary storage above inlet.
- It may also be used in conjunction with relatively low heads as in the case of drop inlet on a road culvert in passing surface water through a spoil bank along a drainage ditch.

**Advantages**

- It requires less material than a drop spillway for the same head.
- Where an appreciable amount of storage is available above the inlet, the design discharge capacity of the structure can be materially reduced.
- It is the most efficient structure for flood prevention and channel grade stabilization.

**Limitations**

- Small drop inlets are susceptible to be choked by debris.
- It is limited to locations where satisfactory earthen embankments can be constructed.

(B) Chute spillway

It is an open channel like structure in which flow is carried down a steep slope at super critical velocities. It usually consists of an inlet, vertical curve section, steep slope channel and outlet. The major part of the drop-in water surface takes place in a channel. Flow passes through the inlet down to the taped channel to the floor of the outlet. Nomenclature for various parts of chute spillway are shown in Fig. 22.4.

**Functional Uses**

- Control the gradient in natural or artificial channels
- Convey runoff from upstream areas into the gully safely without erosion
- Serve as a spillway for flood protection, water conservation and sediment collection structure

**Adaptability**

- It is particularly adopted for gully head controls up to 5–6 m.
- It is suitable for sites where construction of check dams is not possible.
- It can also be constructed in combination with check dams and other retention-type structures.

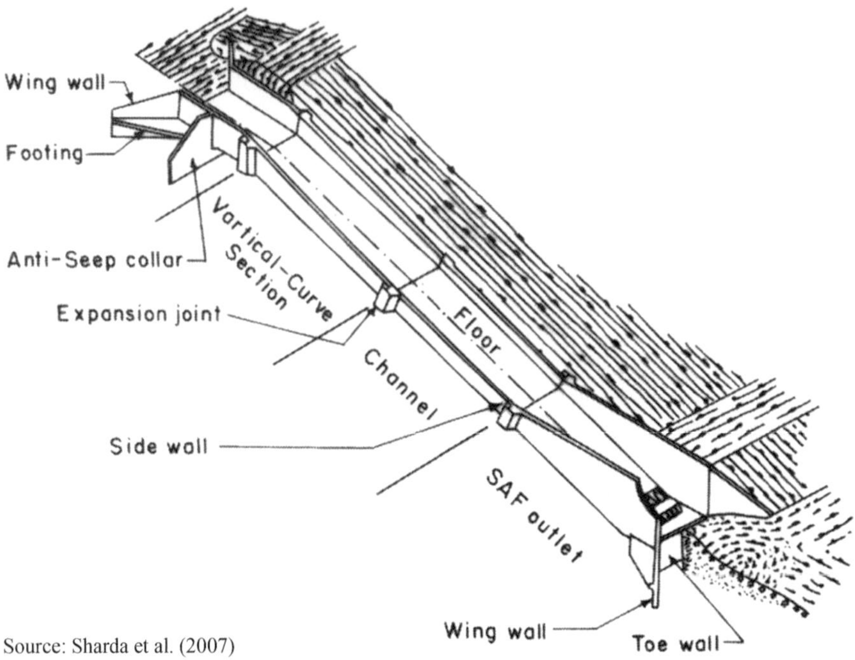


Source: Sharda et al. (2007)

**Fig. 22.4** Chute spillway and the component (Source: Sharda et al. 2007)

Advantages

Chute structure requires less construction material than drop structure of the same capacity and is hence more economical for high heads.

**Limitations**

- There is considerable danger of undermining of the structure by the rodents.
- In poorly drained locations, seepage tends to weaken the foundation. In such locations, if the construction of chute structure is very essential, provision to control the seepage is essential.
- It does not provide for any storage of water upstream of the structure.

#### 22.6.4.1 Guidelines for Selection of Permanent Gully Control Structures

Proper selection of permanent gully control structures for a given location is the key to successful and economical control of erosion or runoff. Each structure has its utility range for a given set of conditions. Sometimes site conditions will allow the use of more than one type of structure. However, in most of the cases, only a single structure will serve the desired purpose.

Usually, the degree of control of protection and the size of the watershed are the main considerations in selection of the structure. A spillway with adequate

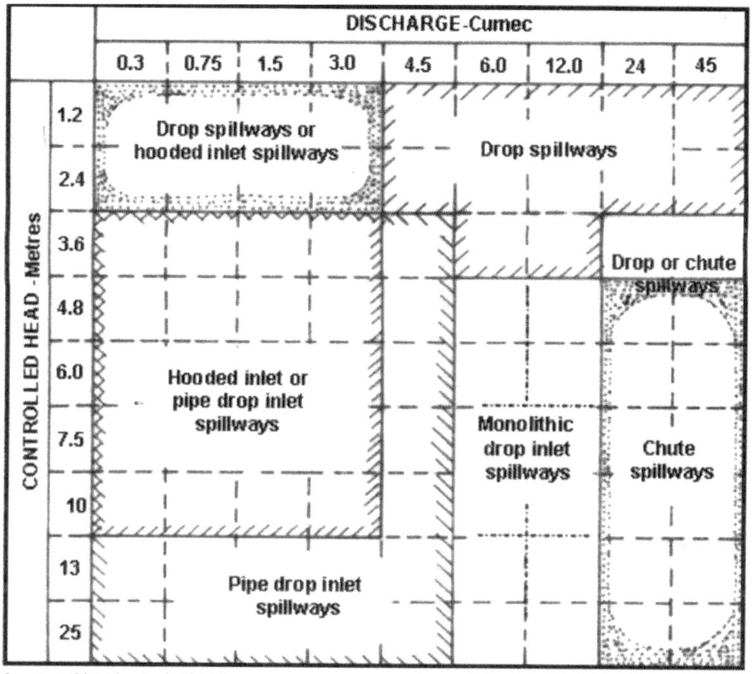


Source: Sharda et al. (2007)

**Fig. 22.5** Guidelines for selection of permanent gully control structures (Source: Sharda et al. 2007)

dimensions should be provided in the structure for safe disposal of design discharge; otherwise water will flow over the sides and damage the structure.

Selection of structure depends on the design discharge and the drop at the site. Based on this, a selection diagram (Fig. 22.5) has been developed which can be used for determining the type of structure required. This diagram is applicable for average field conditions and is based of the most economical structure for the given head and discharge, provided the site conditions allow installation of the structure. Site and foundation conditions are important factors for proper selection of structures.

The foundation material should have the desired supporting strength and resistance to sliding or piping. Further, the foundation material should be reasonably homogeneous to avoid differential or uneven settlement of the structure.

The outlet of the spillway should be designed in such a way that its function or stability will not be reduced by scour or deposition in the exit channel. The channel grade below the spillway should be strong enough to avoid undercutting of the outlet toe wall of cantilever support. The possibility of sediment deposition in the channel below the spillway should be found out. If sediment is a problem, the spillway outlet should be so designed that deposition will not interfere with the flow of water during the expected life of the structure.

## 22.7 Chambal Ravines (Case Studies)

The mechanism of Ravine Development on Chambal River is quite different as compared to any other river of the area. The formation of Chambal ravines does not follow the ways described in theories, and it is always deep (more than 10 m) in nature right from adjoining table land. The data shows that the Chambal ravines are still advancing at the rate of about 800 ha $year^{-1}$. The major role in its formation is played by soil water energy balance and accelerated by profile's soil texture, elevation difference between river bed and adjoining table land, low particle-binding energy and high-water transmission properties. Sometimes, elevation difference from river bed to table land is more than 30 m. The soil profile is alluvial in nature and sand content increases with depth (reaches 95% at a depth of 6 m) with very high-water intake rate (more than 1.25 m $day^{-1}$). The soil is very poor in organic matter content (less than 0.2%) and particle-binding energy (less than −40 mV), and this causes a very high dispersion in fine sand particles with the moving water in saturated and transient water flow. The infiltrating water achieves a very high hydrodynamic energy after reaching a certain depth and, if further accelerated with depth, causes displacement of soil along with horizontally passing waters. The water column near the vertically exposed profile has horizontally a very high hydraulic head drop and dispersed phase of sandy texture which causes pit-hole formation in adjoining area of recently formed ravine making a path for its further advancement. The formed pit-hole further becomes big to bigger in continuously dazzling water and makes a larger cavity at lower depth, the whole profile collapses and dispersed soil gets lost with flowing subsurface water. The continuous horizontal subsurface water flow with relatively higher rate and energy accelerates the formation of deep ravine.

### 22.7.1 Soil Conservation Strategies for Control Measures of Chambal Ravines

The technologies developed by various researchers are not suitable for Chambal ravines due to its different nature and posing a serious challenge to check its further advancement and gobbling some other new villages. Some new approaches have been started by the project for various stages.

#### 22.7.1.1 Management of Shallow Ravines

Shallow ravines (<1 m deep) constitute about one third of area in marginal lands. These can be reclaimed by levelling and utilized for cultivations. Since these shallow ravines are under potential threat of degradation, construction of graded peripheral bund, earthen check dams in the drainage line and masonry spillways to facilitate safe disposal of runoff into ravine is necessary. Soil and water conservation measures, i.e. in situ moisture conservation and strip and cover cropping, are recommended. Different types of temporary structures were constructed to check soil erosion from shallow ravines (Figs. 22.6 and 22.7). The impact of different conservation structures is given in Table 22.3.

**Fig. 22.6** Gabion structure for drainage line treatment in Chambal ravine (Photos by SK Verma)

**Fig. 22.7** Permanent soil and water conservation structure at Chambal ravines (see regeneration of vegetation along slopes) (Source: Verma et al. 2016)

**Table 22.3** Suitability in terms of conservation and physical term of different structures

| Type of structure | Soil conserved | Fitness | Remark |
|---|---|---|---|
| 1. Bori bandhan (gunny bag dam) | 43.5 Mg $ha^{-1}unit^{-1}$ | Partially | Suitable only for moderate rains |
| 2. Earthen bunds | 25.3 Mg $ha^{-1}unit^{-1}$ | For low rains | Not suitable as breaks frequently |
| 3. Masonry structures | 915.4 Mg $ha^{-1}unit^{-1}$ | Good | Most suitable structure |
| 4. Gabions | 784. 5 Mg $ha^{-1}unit^{-1}$ | High spillage | Pilferage from below and side arms |
| 5. Improvised gabion | 1204.2 Mg $ha^{-1}unit^{-1}$ | Economical and effective | Minimum loss |

#### 22.7.1.2 Management of Medium and Deep Ravines

Medium and deep gullies/ravines (>3 m deep) necessitate stabilization of gully head, side slopes and gully bed with drainage line treatment in conjunction with vegetative measures. Site-specific design specifications have been developed. The slope of medium gullies is eased by terracing, which can be utilized for horti-pastoral system, agroforestry with suitable soil working. The gully beds are stabilized by earthen gully plugs with or without pipe outlets or *bori* bunds composite check dams at suitable intervals.

Different soil conservation structures were constructed to manage Chambal ravines and conserve the natural resources. It includes masonry structures, gabions, submersible dams, earthen dams and plugging of gullies and temporary bunds *(bori bandhan)*. Initially all the measures normally failed due to leakage from the side or gushing of water under the structures; after some repairing, these structures worked for 2–3 years. The submersible dam has some edges over other structures provided it has been built at least at a depth of 1.00–1.25 m. A modified form of submersible dam plus gabion was introduced to check runoff and reduce the cost, and it is working satisfactorily.

#### 22.7.1.3 Performance of Different Structures to Control Erosion

The total sediment stored against the structures constructed up to December 2015 was 7346.2 $m^3$ out of which 3100 $m^3$ against the earthen structures (16 earthen and 1 *bori bandhan*), 1705 $m^3$ against the gabion structures (three in numbers) and 2541.1 $m^3$ in permanent structure (four in numbers).

#### 22.7.1.4 Terracing for Management of Deep Chambal Ravines

On steep sloping and very undulating lands of deep ravines, intensive farming can only be adapted after constructing the multilevel terraces (Fig. 22.8) which is a very innovative, simple and practical methodology for reclamation of the deep Chambal ravine. It comprises of construction of steplike fields usually by half-cutting and half-filling procedure. Original steep slope was converted into level fields, and thus the hazards of erosion are minimized. By adapting bench terracing, both degree and length of slope are reduced which help in soil moisture conservation for enhanced crop production. The partial levelling and construction of peripheral bund is required in each terrace to avoid runoff from the terrace. The reclaimed flat land is usable for crop cultivation (35–50%), modified slopes can be used for medicinal/fruit trees/grasses/silvopastoral system (30–45%) and plugged deep gullies are good for water storage.

### 22.7.2 Economics of Ravine Reclamation

Based on project experience and expenditure, the project has estimated reclamation cost for management of different types of ravine as detailed below:

**Fig. 22.8** Terracing for management of medium and deep Chambal ravine (Source: Verma et al. 2016)

1. *Shallow ravines*: The shallow ravines are up to 1.5 m deep, and their reclamation cost including earth work and bunding per hectare is mentioned in Table 22.4.
2. *Medium ravines*: Under this category ravines of 1.5–5 m deep come, and their per-hectare cost for reclamation is given in Table 22.5.
3. *Deep and very deep ravines*: Deep ravines are between 5 and 10 m deep, and more than 10 m deep ravines known as very deep ravines and their stage-wise reclamation cost are given in Tables 22.6. and 22.7.

The data regarding reclamation cost clearly showed that the deep ravines can be managed very comfortably through this technique, which also takes care of getting rid of any type of erosion losses, early earning, maintenance of agro-ecosystem, biodiversity and water recharging to underground aquifer of the area (Table 22.8).

**Table 22.4** Reclamation cost of shallow ravines

| S. No. | Work | Approximate h | Approximate cost (@ ) | Total cost (₹) |
|---|---|---|---|---|
| 1. | JCB machine | 50 h | ₹700 $h^{-1}$ | 35,000 |
| 2. | Tractor for levelling | 20 h | ₹400 $h^{-1}$ | 8000 |
| 3 | Bunding | 50 man-days | ₹200 man-day$^{-1}$ | 10,000 |
| 4. | *Total cost per ha* | | | 53,000 |

**Table 22.5** Reclamation cost of medium ravines

| S. No. | Work | Approximate h | Approximate cost (@ ₹) | Total cost (₹) |
|---|---|---|---|---|
| 1. | Hitachi machine | 50 | ₹ 1600 $h^{-1}$ | 80,000 |
| 2. | Dozer | 25 | ₹ 1000 $h^{-1}$ | 25,000 |
| 3. | Tractor for levelling | 25 | ₹ 400 $h^{-1}$ | 10,000 |
| 4. | Masonry structure | 5 nos. | ₹ 10,000 each | 50,000 |
| 5. | Bunding | 50 man-days | ₹ 200 man-days$^{-1}$ | 10,000 |
| 4. | *Total cost per ha* | | | 175,000 |

**Table 22.6.** Reclamation cost of terracing in deep ravines

| S. No. | Work | Approximate h | Approximate cost (₹)@ | Total cost (₹) |
|---|---|---|---|---|
| 1. | Hitachi machine 210 | 100 | 1700 $h^{-1}$ | 170,000 |
| 2. | JCB machine | 50 | 700 $h^{-1}$ | 35,000 |
| 3. | Tractor for levelling | 25 | 400 $h^{-1}$ | 10,000 |
| 4. | Masonry structure and gabion structures | 5 nos. | 20,000 each | 100,000 |
| 5. | Small structures | 5 nos. | 5000 each | 25,000 |
| 6. | Bunding | 100 man-days | 200 man-days$^{-1}$ | 20,000 |
| 4. | *Total cost per ha* | | | 370,000 |

**Table 22.7** Reclamation cost of half-mounting in deep ravines

| S. No. | Work | Approximate h | Approximate cost (₹) @ | Total cost (₹) |
|---|---|---|---|---|
| 1. | Hitachi machine 210 | 50 h | Rs 1700 $h^{-1}$ | 85,000 |
| 2. | JCB machine | 50 h | Rs 700 $h^{-1}$ | 35,000 |
| 3. | Masonry and gabion structures | 5 nos. | Rs 10,000 each | 50,000 |
| 5. | Bunding | 100 man-days | Rs 200 man-days$^{-1}$ | 20,000 |
| 4. | *Total cost per ha* | | | 200,000 |

**Table 22.8** Reclamation cost of deep ravines (multistep levelling/zero runoff model)

| S. No. | Work | Approximate h | Approximate cost (₹)@ | Total cost (₹) |
|---|---|---|---|---|
| 1. | Hitachi machine 210 | 20 | Rs 1700 $h^{-1}$ | 34,000 |
| 2. | Tractor for levelling | 10 | Rs. 400 $h^{-1}$ | 4000 |
| *Total cost per ha* | | | | 38,000 |

*Common expenses*: Besides expenses on reclamation, there are other expenses to make these ravines green. This includes the cost of planting material, irrigation facility, tube well and others, which is approximately ₹250,000 per ha.

## 22.8 Afforestation/Agroforestry for Management of Ravines

Afforestation has a potential for erosion control through the soil cover provided by tree canopy and litter, in addition to the role of trees in relation to the runoff barrier function (Nair 1993). The role of trees and shrubs in erosion control could be direct or supplementary. In direct use, the trees are themselves the means of checking runoff and soil loss. In supplementary use, control is achieved primarily by other means (grass strips, ditch-and-bank structures and terraces); the trees serve to stabilize the structures and to make productive use of the land, which they occupy. Nair (1993) and Young (1989) supported that leguminous trees have shown potential of reducing soil erosion through five principal ways: interception of rainfall impact by tree canopy, surface runoff impediment by tree stems, soil surface cover by litter mulch, promotion of water infiltration and formation of erosion resistant soil structure. Udawatta et al. (2002) reported that agroforestry and contour strip had a combined significant effect on runoff, sediment and nutrient loss reduction as compared with non-afforestation treatments. Okigbo and Lal (1985) reported that the cover measure involving the use of vegetation for soil protection maintains the hydrological balance in which the surface runoff component in the hydrological cycle would be minimized. Juo and Thurow (1998) also reported that vegetative barriers are generally used in combination with mechanical land treatments such as microcatchments. Thus, once the tree and grass species inside and around the microcatchments are established, a combined system of land treatments can increase infiltration and control of erosion. This could in turn improve physical, chemical and biological attributes of the soil for fertility maintenance.

Implication to sustainable reclamation agroforestry involves two stages. In the first stage, tree and/or shrub species are introduced onto degraded forestland together with any necessary mycorrhizal or rhizobial symbionts, with the objective of checking erosion and restoring soil organic matter and fertility status. In the second stage, the cover may be selectively removed and agricultural production introduced (Young 1989; Kessler 1992). However, time is needed to build up the enlarged plant-litter-soil nutrient cycle (Kessler 1992), a period during which exploitation of the vegetative biomass should be kept low with necessary protection from grazing, etc. The initial tree removal can be along contour aligned strips, with belts of trees remaining in between, led by stages towards hedgerow intercropping (Young 1989). Other options include fodder incorporation along strips or multi-storey systems (Young 1989). Rehabilitation of the world's degraded lands is important for several reasons. First, increasing crop yields is crucial to meeting the needs of the growing human population for food, feed, biomass energy, fibre and timber (in the absence of a massive increase in the equity of global resource distribution) (Perlin 1989). Second, anthropogenic changes in land productivity have deleterious impacts on major biogeochemical cycles that regulate greenhouse gas fluxes and determine Earth's total energy balance. Third, biodiversity preservation depends, in part, on increasing yields on human-dominated land to alleviate pressure to convert remaining natural habitat. Currently, land degradation is one of the paramount important requirements for the globe that is why the need is increasing towards the solution. Bureshi and Tian (1998) reported that loss of soil fertility, soil erosion and land degradation have forced to search for more sustainable systems. Agroforestry (AF) as a land use system is receiving greater attention in many countries to protect the land from various types of degradation. AF technologies/practices offer considerable benefits forth long-term agricultural sustainability. UNCCD (2003) stated that AF is a tool for achieving sustainable agricultural farming and improving the quality of life of the affected communities while simultaneously reversing the process of environmental and land degradation. Blay et al. (2004) also reported that AF can be practised in any of the ecological zone. It can be a way to reduce deforestation or land clearing and to increase crop yields (food, fodder, fibres, etc.) and the diversity of products grown, but an additional benefit is the creation of a carbon sink that removes $CO_2$ from the atmosphere or the maintenance of carbon in existing vegetation and, therefore, has implications for climatic change. Nair (1993) reported that AF systems like improved fallows, contour hedgerows and other systems involving permanent cover play an important role in arresting and reversing land degradation via their ability to provide permanent cover, improve organic carbon content, improve soil structure, increase infiltration and enhance fertility and biological activity through the provision of high biomass production, nitrogen fixation, a well-developed rooting system, high nutrient content in the biomass including roots, fast or moderate rate of litter decay and other benefits of AF to successfully improve soil properties. AF involves management systems that incorporate a tree or shrub component in the agricultural landscape, and it can increase both the carbon storage and biodiversity in areas where annual crops or degraded lands are predominant (Yousif and Raddad. 2006). Young (1989) also reported that AF has shown promising results in the rehabilitation of degraded lands. With its low level of inputs and multipurpose tree species focus, and as a land restoration strategy, AF shows a significant potential for small-scale subsistence farmers in dry land and developing regions. The presence of trees in an agricultural system can have a significant influence by increasing the soil fertility and ecosystem production capacity. Although AF systems have been considered a general solution for the reclamation of degraded lands, it is important to note that the ultimate success depends on the ability to increase the related knowledge among all partners involved and on the acceptance by farmers and local communities (Nair 1993). How could the rehabilitation of degraded land become successful? Apart from committed, sustained and proactive community participation in certain projects, activities focussing on rehabilitation of degraded lands with other drivers of successful dry land rehabilitation programs in denuded landscapes have been identified (Blay et al. 2004). The same author also mentioned some of solutions to make the situation successful:

- There should be some short-term benefits, whether monetary or material, in addition to anticipated future beneficial impacts.
- Local people's attitudes, behaviour and perceptions should be in line with the principle of the rehabilitation program – the people must accept and internalize that change will be in the long run be to their advantage.
- Comprehensive understanding of the interconnectedness, peculiarities and complexities of dry land ecosystems, especially dry land forests and woodlands.
- Rehabilitation should lead to improvements in soil fertility, hydrological processes, etc.
- Existing land uses should be analysed, and land attributes should be matched with land uses so as to determine drivers of degradation.

AF has a great role at sustainable land management. Sustainable land uses are those land uses that produce public goods and services for consumption by the people while at the same time ensuring the protection of the natural resource based upon which those modes of production or land uses are anchored. AF can contribute to the evolution of sustainable land use in dry lands. This is possible because in the first place, AF concerns about ecological and economic sustainability, resilience of environment and diversity of income. It is a system that merges production with ecosystem services. Dry land AF is aimed at increasing diversity of options available for mitigating the impacts of changing ecological circumstances and worsening economic environments and is more stable than monocropping or livestock rearing alone. The intensive production of agricultural and forestry monocultures is unique to advanced developed countries, while worldwide the separation of agriculture and forestry has proven to be difficult. Up to the present, AF has remained the primary land use approach in many parts of the developing world. These principles are true to the ravine lands also. Following AF systems may be adopted for management of ravine lands:

Agri-horti System

The field crops are grown in the interspaces of the fruit trees planted in the block planting system, or fruit trees are integrated with the crops in wider spaces. *Moringa oleifera*, *Citrus* spp. *Emblica officinalis*, *Psidium guajava*, *Punica granatum*, *Ziziphus mauritiana*, *Aegle marmelos*, *Cordia myxa*, *Carissa carandas*, etc. are frequently used for horti-agriculture system in reclaimed shallow ravine lands/marginal lands along the ravines, as they are compatible with most field crops in dry ecologies and can very well withstand the low resource conditions.

Horti-pastoral System

Suitable fruit trees and grasses are grown in combination. *Psidium guajava*-, *Ziziphus mauritiana*- and *Emblica officinalis*-based pastoral combinations for ravine lands include grasses such as *Cenchrus ciliaris*, *Chrysopogon fulvus*, *Cymbopogon* spp. (essential oil-yielding) and many other perennial grasses along with legumes like *Stylosanthes scabra*.

Silvopastoral System

Silvopastoral land use is the best land use for management of degraded ravine lands. Suitable trees such as *Azadirachta indica*, *Pongamia pinnata* (karanj), *Prosopis cineraria*, *Albizia lebbeck*, *Acacia nilotica*, *A. tortilis* and grass species (*Cenchrus* spp., *Panicum antidotale*, *Pennisetum pedicellatum*, *Dichanthium annulatum*, etc.) have been found useful in Chambal and Yamuna ravines.

Bamboo-based Production System

Bamboo is known to be one of the fastest-growing plants in the world, has proved to be the most effective soil conservation species for degraded forests and medium and deep gullies and has been demonstrated to be very useful species for economic utilization of ravines. A high carbon sequestering plant, it acts as a barrier and filter strip for overland flow, promotes silt deposition and slows down water flow along riverbanks. About one third area of ravines in India can be planted with bamboo. The most popular bamboo species in India is *Dendrocalamus strictus*, which covers about 45% of the area. The benefit-cost ratio of *Dendrocalamus strictus* plantation in ravines (1.98) is quite lucrative with an economic return of 19.3% over a 20-year period (Rao et al. 2011).

## 22.9 Performance of Different Forest and Fruit Trees in Chambal Ravines

Six fruit tree species [drumstick (*Moringa oleifera*), gooseberry/aonla (*Emblica officinalis*), *ber* (*Ziziphus mauritiana*), guava (*Psidium guajava*), custard apple (*Annona squamosa*) and pomegranate (*Punica granatum*)] and ten tree species of medicinal and timber values [neem (*Azadirachta indica*), arjuna (*Terminalia arjuna*), mahua (*Madhuca indica*), gugul (*Commiphora wightii*), karanj (*Pongamia pinnata*), adulsa (*Adhatoda vasica*), teak (sagon) (*Tectona grandis*), siras (*Albizia lebbeck*), Khamer (*Gmelina arborea*), Sheesham (*Dalbergia sissoo*)] were planted in pits of 30 × 30 × 45 cm in five different cropping systems, namely, diversified cropping system ($S_1$), agri-horti-system ($S_2$), horti-medicinal-pastoral system ($S_3$), silvi-medicinal system ($S_4$) and silvopastoral system ($S_5$). Out of these six species of fruit plants, drumstick, *ber,* gooseberry (*aonla*) and guava were found most suitable in Chambal ravines; while the growth of pomegranate and custard apple is very slow, trees such as *siras*, *arjuna*, *Sheesham*, *Khamer* and teak (*sagon*) are fast growing (Fig. 22.9). The soil loss from different management systems was recorded and computed with standard procedure, and the minimum silt loss was observed in silvopastoral system followed by silvi-medicinal system (Table 22.9).

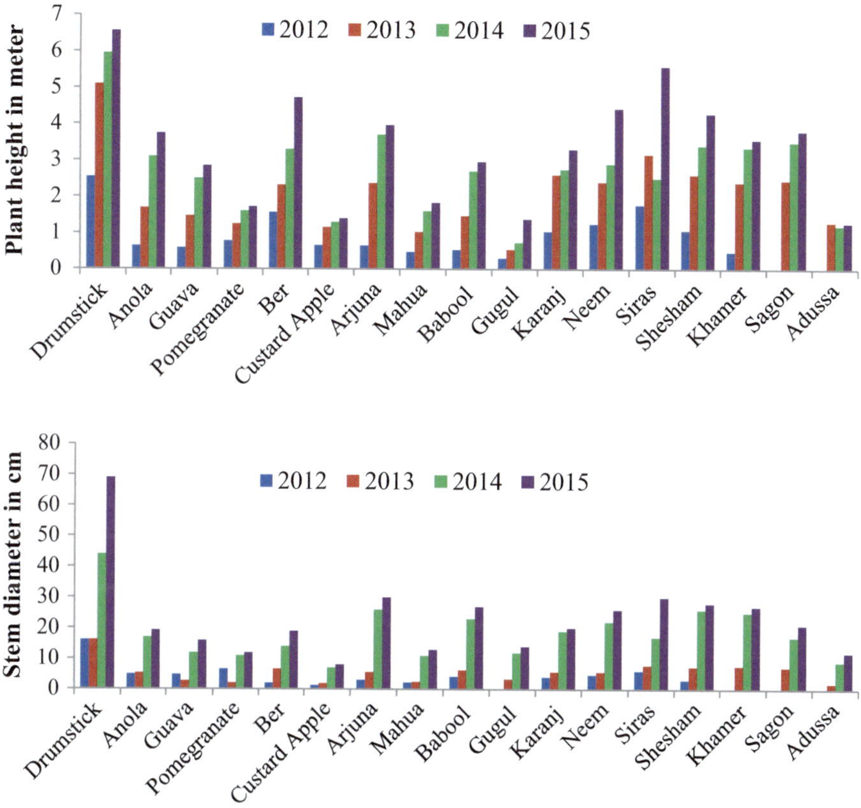


**Fig. 22.9** Growth of plants during different years in Chambal ravine (For scientific names of plants, see text above)

**Table 22.9** Erosion losses computed and recorded under different system

| | Computed | | | | Recorded | | | |
|---|---|---|---|---|---|---|---|---|
| | Silt loss (Mg ha$^{-1}$ year$^{-1}$) | | Runoff (cm) | | Silt loss (Mg ha$^{-1}$ year$^{-1}$) | | Runoff (cm) | |
| Systems | 2014–2015 | 2015–2016 | 2014–2015 | 2015–2016 | 2014–2015 | 2015–2016 | 2014–2015 | 2015–2016 |
| Control | 18.24 | 17.82 | 32.42 | 30.61 | 28.7 | 28.03 | 30.8 | 29.98 |
| $S_1$ | 07.54 | 07.19 | 26.94 | 25.10 | 16.2 | 15.93 | 25.6 | 25.41 |
| $S_2$ | 02.87 | 02.96 | 25.60 | 25.33 | 24.0 | 22.87 | 25.8 | 23.96 |
| $S_3$ | 14.37 | 12.73 | 25.21 | 23.45 | 18.6 | 18.65 | 25.9 | 24.21 |
| $S_4$ | 11.50 | 10.90 | 25.90 | 24.68 | 14.8 | 14.51 | 24.6 | 24.41 |
| $S_5$ | 11.50 | 10.56 | 26.37 | 24.86 | 08.4 | 08.33 | 24.2 | 23.83 |
| MSL | Nil | | Nil | | Nil | | | Nil |

$S_1$ diversified cropping system, $S_2$ agri-horti-system, $S_3$ horti-medicinal-pastoral system, $S_4$ silvi-medicinal system and $S_5$ silvopastoral system

## References

Biggelaar CD, Rattan L, Weibe K, Breneman V (2003) The global impact of soil erosion on productivity. I: absolute and relative erosion-induced yield losses. Adv Agron 81:1–48

Blay D, Bonkoungou ES, Chamshama AO, Chikamai B (2004) Report of Special Programme for developing countries rehabilitation of degraded lands. Forestry Research Network for Sub-Saharan Africa International Union of Forest Research Organizations

Brown LR, Young JE (1990) Feeding the world in the nineties. In: Brown LR, Alan D, Flavin C, Hilary F, Jodi J, Marcia L, Sandra P, Michael R, Linda S, John Y (eds) State of the world 1990: a world watch Institute Ed. L.R. Report on progress toward a sustainable society. World Watch Inst, Washington, DC, pp 59–78

Bureshi RJ, Tian G (1998) Soil improvement by trees in Sub-Saharan Africa. Agrofor Syst 38:51–76

Dhruva Narayana VV (1993) Soil and water conservation research in India. Indian Council of Agricultural Research, Krishi Anushandhan Bhavan, Pusa, New Delhi, p 454

Dhruva Narayana VV, Babu R (1983) Estimation of soil erosion in India. J Irrig Drain Eng, ASCE 109:419–434

Juo ASR, Thurow TL (1998) Sustainable technologies for use and conservation of steep lands. Texas University and Texas Agricultural Experiment Station College Station. Available online from www.infofftc.agnet.org

Katiyar VS, Dhruva Narayana VV (1987) Sahastradhara mined area rehabilitation research project. Technical Bulletin, CSWCRTI, Dehradun

Kessler JJ (1992) The influence of karate (*Viteilariaparadoxa*) and ne're' (*Parkiabiglobosa*) trees on sorghum production in Burkina Faso. Agrofor Syst 17:97–118

Mandal D, Sharda VN, Kumar A (2009) Soil loss tolerance limit for conservation planning in different states of India. Bulletin No. T-56/D-35. Central Soil and Water

Nair PKR (1993) An introduction to agroforestry. Kluwer Academic Publishers in Cooperation with International Centre for Research in Agroforestry (ICRAF)

Okigbo BN, Lal R (1985) Soil conservation and management in developing countries working paper no. 6; role of cover crops in soil and water conservation, FAO, United Nations, viadelle Termedi Caracalla, Rome, Italy

Perlin J (1989) A forest journey. Norton, New York

Pratap Narain (2016) Sustainable management of ravine ecosystem in context of livelihood opportunity. Soil and Water Conservation Bulletin No. 1, published by Indian Association of Soil and Water Conservationists, Dehradun. pp 35–44

Rao BK, Singh AK, Parandiyal AK (2011) Annual report of Project "Hydrologic and economic evaluation of bamboo plantation in major ravine systems of India"

Scherr Sara J (1999) Soil degradation a threat to developing country food security by 2020? 2020 vision. Food, Agriculture and the Environment Discussion paper 27. International Food Policy Research Institute, Washington, DC

Sharda VN, Juyal GP, Chandra Prakash, Joshi BP (2007) Training manual "Soil Conservation and Watershed Management". Published by the Director, CSWCRTI, Dehradun

Sharma AK, Pradhan IP, Nema JP, Tejwani KG (1980) 25 years research on soil & water conservation in ravine lands of Gujarat. CSWCR&TI, Research Centre, Vasad

Udawatta RP, Krstansky JJ, Henderson GS, Garrett HE (2002) Agroforestry practices, runoff and nutrient loss: a paired watershed comparison. J Environ Qual 31:1214–1225

UNCCD (2003) Community level workshop on best practices in agroforestry and soil conservation, Palapye, Botswana, 2–4 December 2003. Available from Secretariat website: www.unccdint. Accessed 15 Sept 09

Verma SK, Singh Akhilesh, Tomar PS (2016) Management of Chambal ravines for food and livelihood security. Technical Bulletin No. RVSKVV/61/2016. Published by Rajmata Vijayaraje Scindia Krishi Vishwa Vidyalata (RVSKVV), Gwalior

Young A (1989) Agroforestry for soil conservation. CAB International, Wallingford

Yousif EA, Raddad A (2006) Analysis of system based on *Acacia senegal* in the Blue Nile region: tropical dry land agroforestry on clay soils, Helsinki

# Rehabilitation of Degraded Lands in Semiarid and Subhumid Ecologies in India

# 23

V. P. Singh, L. Verchot, C. Martius, D. Nayak, and J. Rizvi

**Abstract**

Land degradation is a complex phenomenon caused by natural and anthropogenic factors. These factors create and accelerate soil erosion and contribute to the loss of biodiversity and degradation of soil quality and hydrological features. Adoption of a gamut of appropriate soil and water conservation measures, protection of vegetation from overgrazing, assisted regeneration of vegetation, etc. maybe one of the viable options to protect, rehabilitate, and rejuvenate the degraded lands. This paper presents the results achieved 10 years after the application of these measures in the semiarid (Rajasthan) and subhumid (Odisha) ecologies in India. There were considerable changes and improvements in ecological features in both the ecologies. Reduced level of soil erosion, some increase in soil depth, enhanced vegetation cover, improved vegetation composition and intraspecies density, and hydrological conditions were positively altered. The hydrological changes were visible not only at the study site but also in the villages inhabited in downstream of the study sites, where inhabiting communities claim increased and for longer duration water level in their wells. Due to increased water availability, these communities have successfully ventured in growing more crops, indicating increased cropping intensity on their land.

The changes have also been noticed on the improvement in the composition of floating avian and wild animal population at the rehabilitated sites, which have

V. P. Singh (✉) · D. Nayak · J. Rizvi
ICRAF, New Delhi, India
e-mail: V.P.Singh@cgiar.org

L. Verchot
Soils and Landscapes for Sustainability (SoiLS), International Center for Tropical Agriculture (CIAT), Cali, Colombia

C. Martius
Climate Change, Energy, and Low-Carbon Development, Center for International Forestry Research, Bogor, Indonesia



J. C. Dagar, A. K. Singh (eds.), *Ravine Lands: Greening for Livelihood and Environmental Security*, https://doi.org/10.1007/978-981-10-8043-2_23

attracted their in-migration, and appear capable of harboring them. The impact of rehabilitation efforts on soil chemical properties in both ecologies has been negligible during a decade or more of the rehabilitation efforts in this study. Therefore, any significant improvement in soil quality improvement in short term seems an unrealistic and fallacious expectation. The rehabilitation of degraded lands through social and biological measures, e.g., respectively, protection from grazing and assisted regeneration of vegetation, seems to have better impact than the physical/mechanical means. The details of the findings from the two cases – one from each ecology – are presented in this paper.

## 23.1 Introduction

Land degradation and desertification are complex phenomenon caused by natural and anthropogenic factors. Globally an estimated 3.5 billion ha of degraded lands lie unproductive/out of agricultural uses. In India alone there are estimates of over 120 million ha of such lands. Though they are called waste lands, in practical terms they are wasted lands because they can be brought into productive uses through appropriate rehabilitation measures.

Land degradation in various forms in India has been linked to a combination of factors like the area geomorphology; climatic conditions such as precipitation levels, temperature and wind velocity; removal of natural vegetation, uncontrolled grazing; and conversion of forests and grasslands into croplands. Land degradation leads to loss of soil, vegetation, soil nutrients, and affects hydrological features. Such a degradation impacts agricultural productivity, biodiversity, groundwater, and overall water availability. All these lead to decline in the quality of life, eventually affecting the socio-economic status of the region.

A gamut of appropriate soil and water conservation measures (Dube et al. 2011), protection from overgrazing, assisting regeneration of vegetation (Mebrat 2015; Ruecker et al. 1998; Shono et al. 2007), planting trees (Bremer and Farley 2010), and management of other resources may be one of the viable options to protect, rehabilitate, and rejuvenate the degraded lands (Aronson et al. 1993). Such practices may stimulate regeneration of natural vegetation (Paniagua et al. 1999), rejuvenate/reestablish soil biologically, and improve soil nutrient cycle (Lal 1996a, b; Nunes et al. 2012), rainwater intake, and subsurface water level. In addition, the habitations around rehabilitated areas may also benefit and improve their livelihoods. This study assessed the impact of the abovementioned rehabilitation efforts on degraded lands through the monitoring of selected vegetation, soil, and hydrological parameters in semiarid (Rajasthan) and subhumid (Odisha) ecologies. The assessment was done after a decade of applying the rehabilitation measures and using methods which could compare the control and intervention plots in a paired manner.

## 23.2 General Methodology

### 23.2.1 Sampling Plot Layout

The sampling design of the study used paired plots in control and rehabilitated/areas. Rehabilitated areas are those areas that have received any or a combination of interventions and managed as part of rehabilitation efforts. Some of the land rehabilitation activities involved protection of degraded lands from grazing through social as well as stone fencing and harvesting of naturally generated vegetation resources, erosion control, and water conservation measures through construction of anicuts, small check dams, and trenches and creation of rainwater reservoirs. Control areas were located adjacent to the rehabilitated areas but were not subjected to any biological, physical, or hydrological interventions. The control areas indicate the natural state of the land and vegetation without any rehabilitation activities. To avoid the effect of topographic variation between the sampling points in control and rehabilitated plots, they were chosen in a manner where distance between them was ensured to be less than 30 m. All measurements at each site were taken from six points in each plot which provided adequate number of replicates and sufficient inferences about the recovery trajectory of the land for comparison (Fig. 23.1). While the application of rehabilitation measures was carried out by the Foundation for Ecological Security (FES), the in situ measurements and sampling of various parameters were organized with the help of local institutions, including the farmers.

### 23.2.2 Soil Sampling and Other Measurements

A concentric plot of 12 m radius was laid out to sample soil and measure vegetation in each plot (Fig. 23.1). The general information on plot location in the form of geographical coordinates, above mean sea level elevation along with geographical errors, was duly recorded for each plot studied. In each plot, the soil samples were

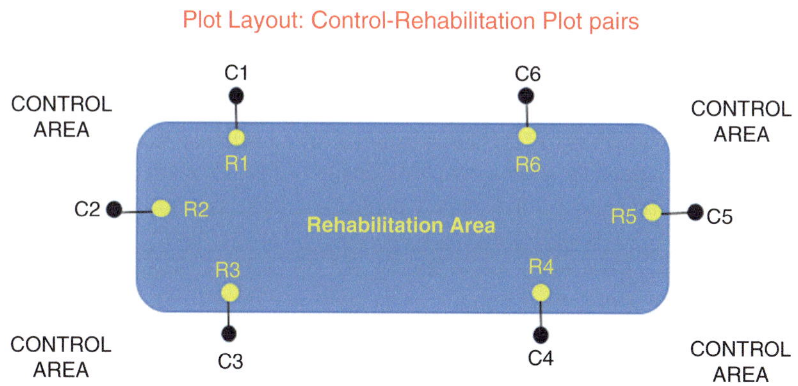


**Fig. 23.1** Schematic diagram of plot layout

**Table 23.1** Geographical location of sampling plots in semiarid ecology at Chitrawas, Udaipur, Rajasthan

| | GPS coordinates | |
|---|---|---|
| Plot pair | Control plot | Rehabilitated plot |
| CHW-C1R1 | N 24° 52′ 19.5″, E 73° 18′ 49.1″ | N 24° 52′ 16.5″ E 73° 18′ 48.6″ |
| CHW-C2R2 | N 24° 51′ 20.1″, E 73° 18′ 50.7″ | N 24° 51′ 21.6″ E 73° 18′ 51.5″ |
| CHW-C3R3 | N 24° 51′ 36.5″, E 73° 18′ 59.2″ | N 24° 51′38.4″ E 73° 18′ 55.9″ |
| CHW-C4R4 | N 24° 52′ 17.0″, E 73° 19′ 6.4″ | N 24° 52′ 13.2″ E 73° 19′ 8.7″ |
| CHW-C5R5 | N 24° 51′ 21.8″, E 73° 18′ 27.6″ | N 24° 51′ 22.9″ E 73° 18′ 28.0″ |
| CHW-C6R6 | N 24° 51′ 50.4″, E 73° 18′ 18.2″ | N 24° 51′ 51.8″ E 73° 18′ 27.6″ |

collected at four sampling points, first being at the center of the plot and second, third, and fourth, compulsorily at 0°, 120°, and 240°, respectively, with respect to plot center. Soil sampling and on-site observations/measurements on landforms, topography, soil physical properties, vegetation, and soil hydrology were recorded at all sites.

The soil samples from different soil depth ranges, e.g., 0–10 cm, 10–20 cm, and 20–50 cm from each sampling point, were collected. After compositing the soil samples of a specific depth range from each sampling point, a representative sample for each depth range was prepared. The collected soil samples with excess moisture from each soil depth were separately dried under shade and were sieved through 2 mm mesh. The soil samples thus prepared were analyzed for various parameters, such as the total carbon and total nitrogen; total phosphorus; exchangeable Al, Ca, Mg, Na, K; cation exchange capacity; pH; and EC.

Likewise, some of the land features and soil physical parameters, such as the physiographic position of the site in the general landscape (topo-sequence), slope, infiltration rate, and selected vegetation parameters, were measured in situ for all the study sites. Other soil measurements consisted were shear strength and texture, which were performed at every soil sampling point and for every soil layer. Some of the details of the methodology and schematic representation of the site and sample points can be seen in Tables 23.1 and 23.4 and Fig. 23.1.

### 23.2.3 Vegetation Measurements

All the trees falling within 12 m radius from the plot center were measured for the parameters, viz, height and diameter at breast height (dbh). Other details like species name, azimuth, and distance in respect to plot center were also recorded. The shrubs were measured by using plot center quadrant method or modified T-Square method within the radius of 6 m – for their density and dimensions. Sapling counts were also assimilated where sapling population existed considerably. Data on occurrence of some bamboo species and medicinal herbs in some of the sampled plots were also recorded.

### 23.2.4 Water Infiltration Measurements

Water infiltration measurements were performed nearby the center of the plot by using a standard sized infiltrometer, generally called water infiltration ring. The measurements were carried out in three plot pairs out of total six plot pairs in each site investigated. Standard procedures were followed while performing the measurements as described in CIFOR's sampling manual. Measurement data along with supplementary information were duly recorded and managed in a database.

### 23.2.5 Field Database Management

Soil samples collected from all the sites were analyzed for various properties as mentioned above at the Indian Arid Forestry Research Institute in Jodhpur. All the data including plot information in the form of geographical coordinates, elevation at above mean sea level, and year of rehabilitation, soil, land form, and vegetation were duly recorded for each plot studied in all the sampling sites in a well-designed format and were managed in the form of datasets.

Analytical procedures were employed to compare all measured parameters between the controls and rehabilitated sampling plots/points in each study site/ sampling plots. Some activities carried out during fieldwork were also photographed wherever necessary. Because of the massive size of the database, the presentations are made by site and by sampling points within a study/sampling site. Only the main findings and the observations are narrated in the text below.

## 23.3 Results and Discussion

The following results are for the sites studied in Rajasthan and in Odisha. At each site, there were measurements from six control plots and the adjoining six rehabilitated plots. To reduce the volume of the data and have a better understanding, the field data (land features, vegetation, and hydrology) results from each of the two categories (control and rehabilitated category) were summarized by pooling the results from all the six measurement points. Likewise, for the same reason while the soil chemical analysis was performed for each sample collected from all the sampling plots and the soil depths, the comparative results from the control and the rehabilitated plots for each site are presented in this paper.

### 23.3.1 Semiarid Ecology, Rajasthan

#### 23.3.1.1 Site Description of Chitrawas, Udaipur

This study site is situated at Chitrawas village under Sayara block of Udaipur district in Rajasthan, India. The site is some 75 km from Udaipur and is surrounded by a series of undulating hills representing southern part of Aravalli mountain range in

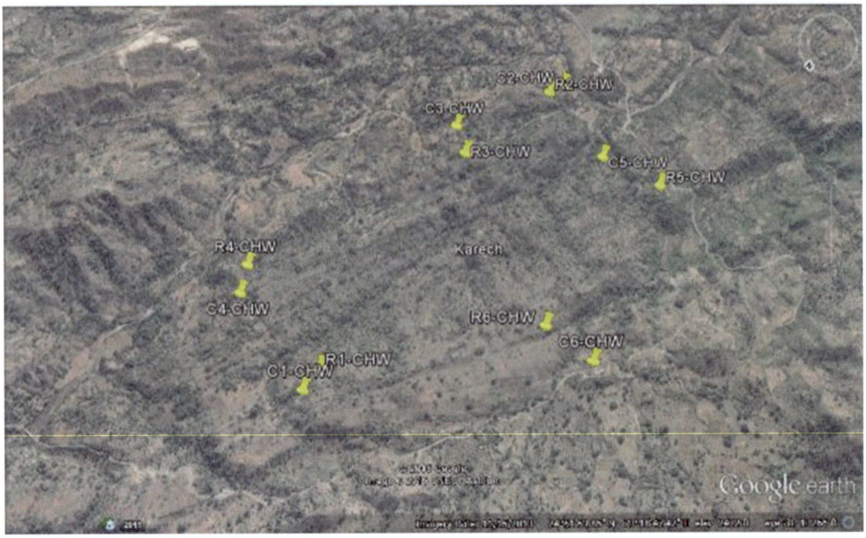


**Fig. 23.2** Sampling plots in semiarid ecology at Chitrawas, Rajasthan, as located on Google Earth Image

Rajasthan. The geographic coordinates of sampling points at this site are in Table 23.1. The rehabilitation area is spread over 250 ha (Fig. 23.2) and has undergone many interventions like trenches, gully stone plugging, check dams, and anicuts. Such interventions continued till year 2004.

The sampling plots mostly represent composite land form, and the sampling point pairs are located primarily on foot slopes and bottomlands in the toposequence. The sampling point pairs are located at elevations ranging from 716 to 840 m above mean sea level, with maximum upslope and downslope of 8° and 27°, respectively. Table 23.2 presents the results on land features, vegetation, and hydrology of this site.

#### 23.3.1.2 Land Features and Soil Physical Properties

At this site while the control plots were not managed, the rehabilitated plots were fenced and protected from free grazing for 6 months at least. Both types of plots at this site do not get flooded. They had trees, shrubs, and grasses. Forbs were generally not found in any of the plots.

In control area, the soil texture at 0–10 cm soil depth was medium clay, while it was heavy clay at depths 10–20 cm and 20–50 cm. In rehabilitated area, the soil texture was found to be medium clay to heavy clay in all the three soil depths.

The stone and gravel cover was one of the important features of both types of plots throughout this sampling site. The rocks and stones covered about 40% plot area in both cases. In control plots the stone/gravel cover ranged from 5% to 40%; in some cases it was more than 40% of the plot area. In rehabilitated area, the stone and gravel cover was relatively less (max. up to 40%) in most of the plots studied. There

**Table 23.2** Some important land features, vegetation, and hydrological parameters of control and rehabilitated plots in semiarid ecology at Chitrawas, Rajasthan

| Features | Control plots[a] | Rehabilitated plots[a] |
|---|---|---|
| Soil texture | Heavy clay, sandy clay | Medium clay – heavy clay |
| Slope, up and down, respectively | 8°, 27° | 7°, 16° |
| Major land form | Composite | Composite |
| Land form designation | Valley | Valley |
| Position in topo-sequence | Foot slope, bottomland | Foot slope, bottomland |
| Vegetation cover <5% for 10 months | No | No |
| Aquatic situation | No | No |
| Plots cultivated or not | No | Yes (plots fenced, no grazing allowed) |
| Vegetation types: | | |
| Trees | Yes | Yes |
| Shrubs | Yes | Yes |
| Grasses | Yes | Yes |
| Forbs | No | No |
| Woody type | Broad leaf, deciduous | Broad leaf, deciduous |
| Vegetation strata | Trees: scattered, dominated by *Wrightia tinctoria*<br>Shrubs: very less in population, dominated by *Annona reticulata* (custard apple) | Trees: scattered, dominated by *Wrightia tinctoria* |
| Visible erosion | Channel and mass | Rill |
| Rock/stone cover (% of plot area) | >40 | 5–40% |
| Conservation structures | None | 3–15 trenches |
| Auger depth restriction | 10 cm | 10 cm, in some none up to 50 cm |
| Shear strength | 3.5–4.0 | 4.5 |
| Major vegetation species | *Miliusa tomentosa* (1no.), *Wrightia tinctoria (1 no.)* , *Butea monosperma* (1 no.) | *Miliusa tomentosa* (2–6 no.), *Wrightia tinctoria* (3–7 no.), *Anogeissus latifolia* (4 no.), *Diospyros melanoxylon* (3–6), *Azadirachta indica* (1–2 no.) |
| Infiltration rate (cm $hr^{-1}$) | 4.2–22.1 | 2.5–16.4 |

[a]Results are summary from six sampling points in each category

was a visible erosion in both types of plots. The sheet, channel, and mass erosion was common in control plots, whereas the rill erosion was common in rehabilitated plots.

It is worth noting that the auger restriction in all the control plots was mostly within 10 cm depth. While in most of the rehabilitated plots, no auger restriction was observed up to 50 cm of soil depth, except in the sampling points R3 and R4 (Table 23.1). The mean soil shear strength recorded was 3.5–4.0 kg $cm^{-2}$ for the control and 4.5 kg $cm^{-2}$ for the rehabilitated area (Table 23.2). The land in the rehabilitated area seemed relatively more stable than the control as there was no visible mass or channel erosion in the sampling plots of this area.

#### 23.3.1.3 Vegetation

The vegetation cover in control area was generally composed of some trees and grasses. Shrubs and forbs were scanty in most of the control plots, whereas the rehabilitated plots were significantly populated with naturally regenerated perennial and annual grasses, shrubs, and forbs along with the trees. The tree species were scattered in both types of plots – the control and rehabilitated plots. In control plots, *Wrightia tinctoria*, *Butea monosperma*, and *Miliusa tomentosa* were the dominant species, whereas the rehabilitated plots, in addition to the species found in control plots, also had *Ailanthus excelsa*, *Anogeissus latifolia*, *Diospyros melanoxylon*, and *Azadirachta indica*. The relative abundance of these mentioned species was also higher in the rehabilitated plots than the control (Table 23.2). The control plots had fewer species and their lower density. Shrubs were less at this site, and scattered plants of *Annona reticulata* were noted in both types of plots. Herbaceous species were of mixed type at this site.

#### 23.3.1.4 Infiltration Rate

The water infiltration rate at this site varied from 4.2 to 22.1 cm $h^{-1}$ in the control plots and from 2.5 to 16.4 cm $h^{-1}$ in the rehabilitated plots. The mean water infiltration rate was 10.7 cm $h^{-1}$ for control area and 11.2 cm $h^{-1}$ for the rehabilitated area. The gravel and stone cover has possibly resulted in higher infiltration rate in both types of plots.

#### 23.3.1.5 Soil Chemical Properties

The soil properties analyzed included total carbon; total nitrogen; total phosphorus; exchangeable Al, Ca, Mg, Na, K; CEC; pH; and EC. Although some numerical differences could be seen between the control and rehabilitated plots at this site, the differences between the two were meager for most of the soil chemical properties at all the three depths. The soil pH at this site was between 7.2 and 7.6; the range covering both type of plots (Table 23.3). The total carbon content was in the range of 0.7–0.8%, whereas the total nitrogen content was in the range of 0.06–0.07%. These soils were very poor in carbon content, which is also reflected in very low nitrogen content.

There was a huge range in the total phosphorus content at this site; varying from 285 to 450 mg $kg^{-1}$ soil. There was also an irregular pattern with respect to the phosphorus content with soil depths. The cation exchange capacity varied between

**Table 23.3** Chemical properties of three soil layers (a, b, c) in control and rehabilitated sampling plots in semiarid ecology at Chitrawas, Rajasthan. Each value is an average of six sampling plots

| Sample code | Total C (%) | Total nitrogen (%) | Total P mg $kg^{-1}$ | EC ($\mu Scm^{-1}$) | pH | CEC cmol(+) $kg^{-1}$ | Exch. Al cmol(+) $kg^{-1}$ | Ca cmol(+) $kg^{-1}$ | Mg cmol(+) $kg^{-1}$ | Na cmol(+) $kg^{-1}$ | K cmol(+) $kg^{-1}$ | Total C (%) | Total nitrogen (%) | Total P mg $kg^{-1}$ | EC ($\mu Scm^{-1}$) | pH | CEC cmol(+) $kg^{-1}$ | Exch. Al cmol(+) $kg^{-1}$ | Ca cmol(+) $kg^{-1}$ | Mg cmol(+) $kg^{-1}$ | Na cmol(+) $kg^{-1}$ | K cmol(+) $kg^{-1}$ |
|---|---|---|---|---|---|---|---|---|---|---|---|---|---|---|---|---|---|---|---|---|---|---|
| Rajasthan sites | Control (average of 6) | | | | | | | | | | | Rehabilitated (average of 6) | | | | | | | | | | |
| Chitrawas (Udaipur) – a | 0.89 | 0.07 | 364.55 | 138.95 | 7.43 | 14.94 | 0.98 | 4.93 | 5.04 | 2.03 | 6.38 | 0.68 | 0.06 | 450.03 | 126.6 | 7.23 | 18.95 | 0.92 | 5.70 | 6.58 | 1.81 | 5.05 |
| Chitrawas (Udaipur) – b | 0.86 | 0.07 | 374.93 | 119.77 | 7.41 | 26.23 | 0.92 | 6.31 | 5.20 | 2.55 | 5.31 | 0.67 | 0.06 | 425.34 | 96.3 | 7.37 | 16.26 | 1.20 | 5.72 | 4.86 | 2.85 | 5.66 |
| Chitrawas (Udaipur) – c | 0.69 | 0.06 | 457.35 | 101.87 | 7.50 | 21.05 | 1.13 | 4.98 | 4.25 | 3.11 | 4.88 | 0.77 | 0.07 | 285.37 | 87.2 | 7.51 | 19.15 | 1.02 | 4.19 | 3.76 | 2.08 | 2.24 |
| Mean | 0.81 | 0.06 | 398.94 | 120.19 | 7.45 | 20.74 | 1.01 | 5.41 | 4.83 | 2.56 | 5.52 | 0.71 | 0.06 | 386.91 | 103.4 | 7.37 | 18.12 | 1.05 | 5.20 | 5.07 | 2.25 | 4.32 |

14 and 26 cmol kg$^{-1}$ soil, which was mostly contributed by the exchangeable cations, such as the Al, Ca, Mg, K, and Na. The concentration of these cations was 0.54–1.87, 4.9–6.0, 4.2–6.5, 1.8–3.0, and 2.2–5.8 cmol kg$^{-1}$, respectively. Magnesium content was relatively higher than other cations, and the sodium content was lowest of them all. The soils of this site also had low EC, generally in the range of 5–150 μSc m$^{-1}$.

The comparison of soil chemical properties between the two types of plots at this site shows that only magnesium content was slightly higher in the rehabilitated plots than the control plots, whereas the exchangeable Na and K and CEC were slightly lower in the latter plots. There was no substantial difference in other soil properties between the two types of plots (Table 23.3).

#### 23.3.1.6 Discussion

There are a couple of insights that can be glanced from the data collected and their analysis at this site. The foremost of them relate to the changes in some soil parameters and vegetation type and density.

The results indicate that this site has highly degraded soil and has very low vegetation cover. The land in the rehabilitated area seemed relatively more stable than in the control area as there was no visible mass or channel erosion observed in the sampling plots of the rehabilitated area. There was also some improvement in the soil depth and infiltration rates in the rehabilitated plots compared to the control plots. However, only meager differences could be noticed in the soil chemical properties.

It seems that in the semiarid climates, the improvement in chemical properties of the degraded soils is very slow process, and the nutrient addition is compromised by the losses through various means, such as the high level of oxidation, etc. That is why despite more than a decade of the rehabilitation efforts, the improvement in soil quality was decimal. On the other hand, however, there was a marked difference in the rejuvenated vegetation in the rehabilitated areas as compared to the control. The rehabilitated areas had various kinds of vegetation reestablished, including trees, shrubs, and grasses. And, owing to their higher number and better growth, their relative vegetative volume was also higher in the rehabilitated plots than the control plots. This trend indicates that for the soil to become productive, it does not necessarily have to have high fertility index. For the purpose of bringing such lands into productive uses, the recycling of the needed nutrients could be sufficient, and it is better to bring such lands in the productive uses through the rejuvenation of the native vegetation by protection from grazing and planting of trees, grasses, and legumes against going through the classical field crop agronomic practices.

### 23.3.2 Subhumid Ecology, Odisha

#### 23.3.2.1 Site Description of Kumuri, Angul, Odisha

This site is located at village Kumuri under Angul block, Odisha. The site falls in the periphery of Satkosia Wildlife Sanctuary and Forest Reserve and is topographically represented by medium and high gradient hills with considerable natural and

**Fig. 23.3** Sampling plots in subhumid ecology at Kumuri, Angul, Odisha, as located on Google Earth Image

**Table 23.4** Geographical location of sampling plots in subhumid ecology at Kumuri, Angul, Odisha

| | GPS coordinates | |
|---|---|---|
| Plot pair | Control (C) | Rehabilitation (R) |
| C1R1 | N 20° 43′ 29.0″, E 84° 47′ 34.9″ | N 20° 43′ 37.6″ E 84° 47′ 38.4″ |
| C2R2 | N 20° 43′ 26.0″ E 84° 47′ 41.0″ | N 20° 43′ 37.3″ E 84° 47′ 39.6″ |
| C3R3 | N 20° 43′ 47.2″ E 84° 47′ 17.4″ | N 20° 43′ 46.4″ E 84° 47′ 24.6″ |
| C4R4 | N 20° 43′ 49.9″ E 84° 47′ 26.3″ | N 20° 43′ 51.4″ E 84° 47′ 32.0″ |
| C5R5 | N 20° 43′ 53.3″ E 84° 47′ 23.7″ | N 20° 43′ 54.9″ E 84° 47′ 24.6″ |
| C6R6 | N 20° 43′ 54.5″ E 84° 47′ 18.6″ | N 20° 43′ 55.8″ E 84° 47′ 19.8″ |

planted vegetation cover (Fig. 23.3). The geographic coordinates of sampling points at this site are in Table 23.4.

#### 23.3.2.2 Land Features and Soil Physical Properties

The control and rehabilitated sampling plots at this site were in the upslope and downslope of <5°, <5° and $<5^{0}$, <26°, respectively. These plots are at the bottomland of low gradient slope and medium gradient hill on the slopping land form. Both types of plots have sandy clay loam and clay loam soil texture (Table 23.5). The control and rehabilitated plots were not aquatic in nature and were not cultivated. Both types of these plots have trees, shrubs, grasses, and forbs. The woody type vegetation was broad leaf and deciduous.

**Table 23.5** Some important land features, vegetation, and hydrological parameters of control and rehabilitated plots in subhumid ecology at Kumuri, Odisha

| Features | Control plots[a] | Rehabilitated plots[a] |
|---|---|---|
| Soil texture | Sandy clay loam, clay loam | Sandy clay loam, clay loam |
| Slope, up and down, respectively | $<5^{\circ}$, $<5^{\circ}$ | $<5^{\circ}$, $<26^{\circ}$ |
| Major land form | Slopping | Slopping |
| Land form designation | Low gradient slope | Medium gradient hill |
| Position in topo-sequence | Bottomland | Bottomland |
| Vegetation cover <5% for 10 months | No | No |
| Aquatic situation | No | No |
| Plots cultivated or not | No | No |
| Vegetation types: | | |
| Trees | Yes | Yes |
| Shrubs | Yes | Yes |
| Grasses | Yes | Yes |
| Forbs | Yes | Yes |
| Woody type | Broad leaf, deciduous | Broad leaf, deciduous |
| Vegetation strata | Trees, scattered population dominated by *D. melanoxylon*; shrubs, dominated by *L. camara*; herbs, different perennial and annual grasses accompanied by considerable forbs cover | Trees: *D. melanoxylon*, *Borassus flabellifer*, *Phoenix sylvestris*, and *Acacia leucophloea*<br>Shrubs: *Lantana camara*<br>Grasses: mixed grass cover |
| Visible erosion | None, sheet, rill | None, sheet, rill |
| Rock/stone cover (% of plot area) | <5 | <5 |
| Conservation structures | Nil | Stone gully pluggings (2), ponds, trenches |
| Auger depth restriction | None up to 30 cm | None up to 30 cm |
| Shear strength (kg $cm^{-2}$) | 5.0 | 5.0 |
| Major vegetation species | *Diospyros melanoxylon* (3–8 no.), *Holarrhena antidysenterica* (5 no.), *Terminalia alata* (2 no.), *Schleichera oleosa* (2)<br>Three plot had no trees | *Diospyros melanoxylon* (4–8), *Haldina cordifolia* (1–3), *Diospyros montana* (4), *Casearia elliptica* (2), *Syzygium cumini* (1–2), *Terminalia alata* (3), *Lagerstroemia parviflora* (3), *Chloroxylon swietenia* (2) |
| Infiltration rate (cm $hr.^{-1}$) | 8.9–16.4 | 2.82–8.8 |

[a]Results are summary from six sampling points in each category

The stone and gravel covered only <5% of the area, and there was no auger restriction up to 30 cm depth in both type of plots. The visible erosion in both types of plots was sheet and rill type, and some plots in both categories had no erosion (Table 23.5). The recorded mean soil shear strength was 5.0 kg/cm$^2$ for both, the control and the rehabilitated plots (Table 23.5).

#### 23.3.2.3 Vegetation

The vegetation stratum in control plots at this site included scattered population of trees dominated by *Diospyros melanoxylon*; shrubs, dominated by *Lantana camara*; and some herbs and different perennial and annual grasses accompanied by considerable forbs. In the rehabilitated plots, the vegetation stratum included trees, dominated by *Diospyros melanoxylon*, *Borassus flabellifer*, *Phoenix sylvestris*, and *Acacia leucophloea*; shrubs, mainly *Lantana camara*; and mixed grasses (Table 23.5).

The control plots had fewer number of plant species and their relative smaller abundance (density) as compared to the rehabilitated plots. The major dominant species in the control plots were *Diospyros melanoxylon (3–8 no.)*, *Holarrhena antidysenterica (5 no.)*, *Terminalia alata (2no.)*, and *Schleichera oleosa (2 no.)*. In rehabilitated plots they were *Diospyros melanoxylon (4-8no.)*, *Haldina cordifolia (1–3 no.)*, *Diospyros montana* (4 no.), *Casearia elliptica* (2 no.), *Syzygium cumini* (1–2 no.), *Terminalia alata* (3 no.), *Lagerstroemia parviflora (3 no.)*, and *Chloroxylon swietenia (2 no.)*. There were also liana, climbers, and some bamboo present in the rehabilitated plots (Table 23.5). Some plots in the control area had no trees, whereas in the rehabilitated area all the plots had trees, but no grasses.

#### 23.3.2.4 Infiltration Rate

The water infiltration rate at this site varied from 8.9 to 16.4 cm h$^{-1}$ in the control plots and from 2.82 to 8.8 cm h$^{-1}$ in the rehabilitated plots.

#### 23.3.2.5 Soil Chemical Properties

The soil pH at this site was between 5.3 and 6.6; the range covering both type of plots (Table 23.6). The total carbon content was in the range of 0.75 and 1.20%, whereas the total nitrogen content was in the range of 0.07–0.09%.

There was a huge range in the total phosphorus content at this site; varying from 360–520 mg/kg soil. The cation exchange capacity varied between15 and 25 cmol kg$^{-1}$ soil, ahigh latter value, which is corroborated by the number of exchangeable cations, such as the Al, Ca, Mg, K, and Na. The concentration of these cations was 0.7–1.20, 3.9–4.6, 5.3–7.0, 4.2–5.6, 1.7–3.7 cmol kg$^{-1}$, respectively. Magnesium content at this site was relatively higher than other cations; and the sodium content was lowest of them all. The soils of this site also had low EC, generally in the range of 31–118 μSc m$^{-1}$.

The comparison of soil chemical properties between the two types of plots at this site shows that only total carbon, phosphorus, pH, Mg, and K content was slightly higher in the rehabilitated plots than the control plots. There was no substantial difference in other soil properties between the two types of plots (Table 23.6).

**Table 23.6** Chemical properties of three soil layers (a, b, c) in control and rehabilitated sampling plots in subhumid ecology at Kumuri, Angul, Odisha. Each value is an average of six sampling points

| Soil layer | Total C (%) | Total nitrogen (%) | Total P mg $kg^{-1}$ | EC ($\mu Scm^{-1}$) | pH | CEC cmol(+) $kg^{-1}$ | Exch. Al cmol(+) $kg^{-1}$ | Ca cmol(+) $kg^{-1}$ | Mg cmol(+) $kg^{-1}$ | Na cmol(+) $kg^{-1}$ | K cmol(+) $kg^{-1}$ | Total C (%) | Total nitrogen (%) | Total P mg $kg^{-1}$ | EC ($\mu Scm^{-1}$) | pH | CEC cmol(+) $kg^{-1}$ | Exch. Al cmol(+) $kg^{-1}$ | Ca cmol(+) $kg^{-1}$ | Mg cmol(+) $kg^{-1}$ | Na cmol(+) $kg^{-1}$ | K cmol(+) $kg^{-1}$ |
|---|---|---|---|---|---|---|---|---|---|---|---|---|---|---|---|---|---|---|---|---|---|---|
| | Control (average of 6) | | | | | | | | | | | Rehabilitated (average of 6) | | | | | | | | | | |
| a | 1.06 | 0.09 | 379.87 | 50.85 | 5.85 | 24.46 | 0.94 | 4.60 | 6.11 | 1.70 | 4.73 | 1.19 | 0.09 | 428.92 | 51.50 | 5.84 | 15.46 | 0.84 | 3.92 | 5.67 | 1.90 | 5.23 |
| b | 0.91 | 0.08 | 406.78 | 47.93 | 5.97 | 17.14 | 0.85 | 4.62 | 6.24 | 2.80 | 5.84 | 1.01 | 0.08 | 519.68 | 52.48 | 6.05 | 22.84 | 0.75 | 4.67 | 6.95 | 2.34 | 5.14 |
| c | 0.75 | 0.07 | 458.54 | 53.10 | 6.08 | 17.84 | 1.11 | 3.90 | 5.35 | 3.74 | 4.28 | 0.84 | 0.07 | 364.03 | 54.52 | 6.16 | 22.29 | 1.15 | 4.37 | 7.06 | 2.05 | 5.62 |
| Mean | 0.90 | 0.08 | 415.07 | 50.63 | 5.97 | 19.81 | 0.96 | 4.37 | 5.90 | 2.75 | 4.95 | 1.01 | 0.08 | 437.54 | 52.83 | 6.02 | 20.20 | 0.91 | 4.32 | 6.56 | 2.09 | 5.33 |

#### 23.3.2.6 Discussion

The field measurements and the laboratory analytical results indicate that this site has poor soil texture and high infiltration rates. This site, however, had better soil fertility status and soil depth than the sites in the semiarid region. The Ca , Mg, and K levels at this site were in the ranges comparable to the sites in the semiarid climate. Although located in the humid climate with high rainfall, which leads to the leaching of the bases, the status of the bases at this site was not so poor.

Both types of plots at this site were visibly affected by the sheet and rill erosion. However, the rehabilitation measures have significantly contributed to lessening of the erosion. These trends indicate that the rehabilitation efforts, such as the gully plugging, digging ponds for holding the water, and constructing bunds along the slope, could be some of the effective measures to improve the degraded lands quickly, especially in terms of the vegetation reestablishment. The improvement in soil chemical properties at this site also, as in the sites in semiarid ecology, seems very slow and difficult to achieve in a decade's time span.

## 23.4 General Discussion

There are several inferences that can be drawn from the observations, field measurements, and laboratory analysis conducted in this study. There were two contrasting sites, one in the semiarid ecology and the other in subhumid ecology, that were studied for assessing the impact of rehabilitation efforts (varied means) in improving the degraded lands in terms of reestablishment of vegetation and improvement in soil quality, both in physical and chemical properties.

The comparison of rehabilitated plots with the control plots using the methodology of this study in both the ecologies could not establish (neither intended so) the true picture of what these sites initially were before the degradation. Also, the study could not separate the specific effect of each of the rehabilitation items. However, the methodology provided sound basis for assessing the overall effect of rehabilitation measures not only at a site but also the sites across ecologies.

Most of the degradation in the studied two ecologies was related to soil erosion through various processes, and these processes differed in type and magnitude across ecologies. Overexploitation of vegetation resources, such as through cut and carry and overgrazing, making the sites bare appeared to start the soil erosion and land degradation process.

The rehabilitation efforts through physical/mechanical means seem to have lesser impact in improving the degraded lands as compared to the social, e.g., protection from grazing, and biological means, such as assisted regeneration of vegetation, or planting trees. The achieved improvement was better in flatter lands than the slopping ones in the landscape. Therefore, slopping lands are the places where mechanical means, such as the terracing, bunding, and leveling, maybe effective, but at usurious cost implications, especially in areas having more than $45^{\circ}$ slopes.

The improvement in soil quality, especially in terms of chemical properties, was very slow and insignificant during the duration of the efforts, which ran almost over a decade in this case. Therefore, the expectation in soil quality improvement in such a duration seems unrealistic.

It is also noted that in spite of rehabilitation efforts, the rehabilitated area could not be brought into field crops cultivation at any of the sites in the two ecologies, though there were significantly higher biomass (through trees, shrubs, and grasses) produced in the rehabilitated plots. However, a substantial size of area close by the rehabilitated lands has been observed by the local communities to have improved availability of water, both in terms of volume and duration. This has improved the productivity and cropping intensity in some cases.

These findings indicate that the efforts to improve the degraded lands with the aim of bringing them into field crop production right away may have lesser chances of success. Such an aim may also have serious budgetary implications. Therefore, it may be better to start bringing the degraded lands into productive purposes through assisted regeneration of native vegetation and establishing/planting of trees, especially the fodder, fuel, and/or multipurpose species. These suggestions seem to involve lesser cost and covering of smaller land area requiring rehabilitation for planting trees (mostly the pit size area preparation). Such an endeavor is also expected to reduce the grazing and fuel wood collection pressure on the natural vegetation (forests), enhance biomass production and carbon sequestration, reduce further in situ degradation, improve soil quality in the long run, and ultimately culminate in stopping the degradation of additional areas.

## 23.5 Conclusion

This study has provided several insights and was able to dispel some misconceptions about the rehabilitation of the degraded lands, especially the expectation of quick restoration of soil fertility and bringing the degraded lands in to field crop production. The methodology adopted in this study seems to be effective in measuring the impact of the rehabilitation efforts and was suitable for different ecological conditions.

The information gathered in this study indicate that the rehabilitation of degraded lands is effectively possible through assisted regeneration of vegetation and other physical measures. However, the improvement may not reflect in the improvement of soil chemical quality in the expected span of time. One of the essential parameters, the community opinion, was found to be as important as the measurements in the field and laboratory and that the assessments should extend to also encompass some lands around the intervention areas.

The findings of this study also indicate that in practical terms, it may be better to start bringing the degraded lands into productive purposes through the succession of assisted regeneration of native vegetation; establishing/planting of trees, especially the fodder, fuel, and/or multipurpose species in a linear fashion; and intercropping tree rows with fodder grasses or hardy cereals, instead of embarking on field crop production.

## References

Aronson J, Floret C, Le Floc'h E, Ovalle C, Pontanier R (1993) Restoration and rehabilitation of degraded ecosystems in arid and semi-arid lands: a view from the south. Restor Ecol 1(1):8–17
Bremer LL, Farley KA (2010) Does plantation forestry restore biodiversity or create green deserts? A synthesis of the effects of land-use transitions on plant species richness. Biodivers Conserv 19(14):3893–3915
Dube R, Das S, Saxena V (2011) Chapter 6, India: rehabilitation of degraded lands. In: Youlin Y, Jin LS, Squires V, Kyung-soo K, Hye-min P (eds) Combating desertification & land degradation: proven practices from Asia and the Pacific. Published by the Korea Forest Service (KFS) and the Secretariat of United Nations Convention to Combat Desertification (UNCCD). Changwon, Republic of Korea. pp 67–82
Lal R (1996a) Deforestation and land-use effects on soil degradation and rehabilitation in western Nigeria. I. Soil physical and hydrological properties. Land Degrad Dev 7(1):19–45
Lal R (1996b) Deforestation and land-use effects on soil degradation and rehabilitation in western Nigeria. II. Soil chemical properties. Land Degrad Dev 7(2):87–98
Mebrat W (2015) Natural regeneration practice in degraded high lands of Ethiopia through area enclosure. Int J Environ Protect Policy 3(5):120–123. https://doi.org/10.11648/j.ijepp.20150305.11
Nunes A, Figueiredo A, Almeida A (2012) The effects of farmland abandonment and plant succession on soil properties and erosion processes: a study case in central of Portugal. Revista de Geografia e Ordenamento do Território, n.º 2 (Dezembro). Centro de Estudos de Geografia e Ordenamento do Território. pp 165–190
Paniagua A, Kammerbauer J, Avedillo M, Andrews MA (1999) Relationship of soil characteristics to vegetation successions on a sequence of degraded and rehabilitated soils in Honduras. Agric Ecosyst Environ 72(3):215–225
Ruecker G, Schad P, Alcubilla MM, Ferrer C (1998) Natural regeneration of degraded soils and site changes on abandoned agricultural terraces in Mediterranean Spain. Land Degrad Dev 9(2):179–188
Shono K, Cadaweng EA, Durst PB (2007) Application of assisted natural regeneration to restore degraded tropical forestlands. Restor Ecol 15(4):620–626

# Impact Assessment of Management Techniques Implemented in Chambal Ravines, Madhya Pradesh, India

24

V. P. Singh, R. B. Sinha, Y. P. Singh, D. Nayak, A. K. Singh, and J. Rizvi


**Abstract**

The impact of anicuts and afforestation and other techniques implemented at Bindwa, Morena district, and Himmatpur village, Bhind district, by the forest department of Madhya Pradesh was assessed using various parameters after 7 years of implementation. These efforts had remarkable positive effect on several parameters, most prominently on the seeded/planted as well as natural vegetation and in erosion control/reduction. As a result, there was an increase in soil depth along anicuts and improvement in some soil fertility parameters as well as in water volume in the wells. These improvements reflected in enhanced crop (field crops and vegetables) productivity, dairy and in the growth of trees. The outcome of these combined benefits shows four to five times improvement in farmers' income. Similar measure was also the calculated income from forest lands. This indicates a high potential of such efforts for improving farmers' livelihoods and income and the greening of ravines for achieving the environmental security. The benefit cost ratio for the anicuts and afforestation interventions for Bindwa village worked out to be 5.89:1 while that for Himmatpur village was 7.31: 1. Both these investments appear highly feasible and worth investing for improving the livelihoods of farmers in ravine lands. However, there are opportunities to enhance the outcome of such interventions by planning them at watershed level and greater participation of the community for validating and implementing them.



V. P. Singh (✉) · D. Nayak · J. Rizvi
ICRAF, New Delhi, India
e-mail: V.P.Singh@cgiar.org

R. B. Sinha
Department of Agriculture, Cooperation and Farmers Welfare, Ministry of Agriculture and Farmers Welfare, Government of India, New Delhi, India

Y. P. Singh · A. K. Singh
Rajmata Vijayraje Scindia Krishi Vishawa Vidhyalaya (RVSKVV), Gwalior, India







J. C. Dagar, A. K. Singh (eds.), *Ravine Lands: Greening for Livelihood and Environmental Security*, https://doi.org/10.1007/978-981-10-8043-2_24

## 24.1 Introduction

Chambal ravines represent the most degraded form of once cultivated land in India. The degradation resulted not only in loss of land and soil resources but also led to other processes, which are counterproductive to national economy. Some of these processes include floods in river basins, siltation in water reservoirs/tanks and the consequent loss of their storage capacity and damage to railway lines, roads and other public utility properties. The ravines have encroached upon the inhabited villages, which had to be shifted to new sites to avoid loss of lives. Besides these, an important socio-economic problem inflicted/associated with the ravines in Madhya Pradesh (MP), Uttar Pradesh (UP) and Rajasthan is the accelerated infestation of the area by dacoits through 'shelter in ravines', which are difficult to penetrate and inaccessible to public. The magnitude of this problem has been too serious to have missed the attention of not only the state and the national governments but also of the international agencies, such as the World Bank, European Economic Communities (now the European Union), German Technical Cooperation for International Development (GTZ, now GIZ) and others. It is worthwhile mentioning here that the analysis of the land features, soil quality and the problems faced by the farmers in Chambal ravines shows a strong similarity with the situation in ravines at Kisumu town around Lake Victoria in Kenya and some parts of Tanzania.

Under Gwalior State rule (before Independence) huge masonry structures to serve as submersible check dams were constructed at points where V-shaped wide and deep gullies met the main river Sank, a tributary of river Chambal. The check dams were too far apart and not designed properly to serve the purpose. Further, the built ones also gave up due to lack of maintenance. The government of M. P. has executed several ravine reclamation projects in Chambal Valley from 1955 to 1979 and with the World Bank assistance in 1980 and under 'Anti-Dacoity Program' during 1988–1992. While these latter projects were of pioneering nature, the benefits derived from them were not commensurate to the investments.

In 2007–2008, the forest department[1] of M. P. government undertook a program with two projects on the management of Chambal ravines through construction of anicuts and afforestation. It was an innovative approach undertaken by an enthusiastic team of field workers from forest department, local farming community and village leadership. These projects are known as the Tang Ghaati Pilot Project at Bindwa village, Joura Block, Morena District, and Pilot Project on Ravine Reclamation at Himmatpur village, Ater Block, Bhind District of the state. Scientists[1] of Rajmata Vijayaraje Scindia Krishi Vishwa Vidyalaya (RVSKVV), Gwalior and the International Centre for Research in Agroforestry (ICRAF) are also working on this problem and were keen to know about the impact of the program executed by the forest department. Accordingly, an impact study was initiated in July 2015. The results of the findings are presented in this paper.

[1] While the field interventions of anicuts and afforestation were implemented by the team led by Mr. R.B. Sinha, a team of scientists led by Dr. Y.P. Singh collected the field data for the impact assessment.

## 24.2 Materials and Methods

### 24.2.1 Study Location and Climate

The study was carried out in two villages: (1) Bindwa village, Joura Block, Morena District, and (2) Himmatpur village, Ater Block, Bhind District of Madhya Pradesh. The geographical locations of these villages, respectively, are 26° 45′ N latitude and 70° 30′ E longitude and 25° 54′ N latitude and 78° 12′ E longitude. The altitudes of these villages are 224 m and 190 m above mean sea level (amsl), respectively. Both the villages fall under a single agroclimatic gird, which is part of the vast Indo-Gangetic Plains.

The region in which these two villages are located has subtropical climate with three distinct seasons in the year, viz. hot and dry summer, cold and dry winters and monsoon season. The summer has maximum temperatures up to 49 °C and the winter as low as 0 °C in Bindwa village, Morena, while the maximum temperature of 48 °C and the minimum temperature of 2 °C in winters at Himmatpur village, Bhind. The high-temperature days in summer are longer than the low-temperature winter days at both the project sites. The average humidity in the morning and evening was 88–98% and 22–33%, respectively, at Bindwa village, and 84–97% and 21–34%, respectively, at Himmatpur village. The average annual rainfall at Bindwa, Morena, and at Himmatpur, Bhind, is 700 mm and 758 mm, respectively (Fig. 24.1). The range of annual rainfall and number of rainy days (from 2007–2008 to 2013–2014) at Bindwa, Morena, were 379–1030 mm and 32–68 days, respectively, whereas at Himmatpur, Bhind, these parameters were 471–1238 mm and 33–65 days, respectively.

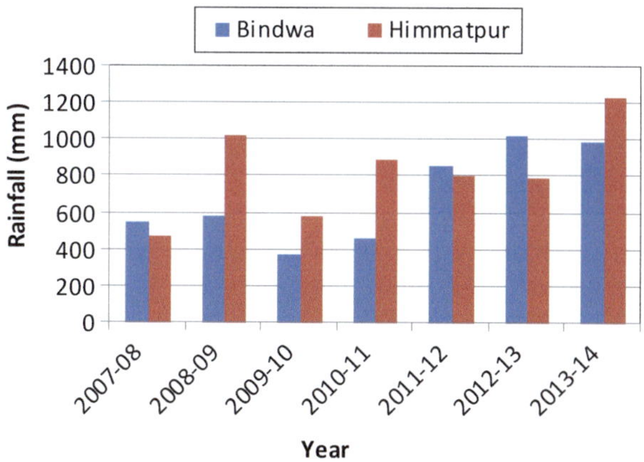


**Fig. 24.1** Rainfall pattern at Bindwa and Himmatpur villages in Madhya Pradesh

### 24.2.2 Soil

The study area has alluvial soils, with very high dispersion coefficient and fragile nature. The soil texture at Bindwa is loam while at Himmatpur is sandy loam. Bindwa soil contains 48.8% sand, 36.3% silt and 14.9% clay, while Himmatpur soil has 55.2% sand, 33.6% silt and 11.2% clay. These soils have bulk density between 1.52 and 1.58 Mg $m^{-3}$ and a high permeability/infiltration rate, varying from 28.1 to 35.9 mm $h^{-1}$. Similar soil analysis results from ravines have also been reported by Bhushan and Yadav (2000), Haigh (1984), Pani and Mohapatra (2001) and Ngoze et al. (2008).

Both the soils are neutral in reaction, have normal salt content and are slightly calcareous in nature (containing 1.6–2.2% $CaCO_3$). They also have highly eroded upper soil layers. With high dispersion coefficient, the soils are highly erodible. These soils are low in organic matter content (0.34%). Due to severe soil erosion and general land degradation, the soils have been depleted of organic matter and consequently of nitrogen as well as other macro- and micronutrients. Uncontrolled run-off continues to degrade the land into gullies, sub-gullies and sub-sub-gullies and ultimately into a network of gullied ravines, which are getting deeper with time and attaining depths up to 20–25 m.

### 24.2.3 Natural Vegetation

The ravenous lands at the study sites have several naturally growing grasses, herbs, shrubs, climbers and other plants. The natural vegetation, though in patches, consisted of grasses such as Doob (*Cynodon dactylon*), Anjana (*Cenchrus ciliaris*), Sama (*Panicum sumatrense*), Bhanjura (*Apluda mutica*) and Dab (*Chinensis cynosuroides*). The shrubs included Kareel (*Capparis decidua*), arni (*Clerodendrum phlomidis*), Phalsa/Chabeni (*Grewia asiatica*) and Heens (*Capparis sepiaria*). The herbs were Bhringraj (*Eclipta prostrata*) and Shankhpushpi (*Evolvulus alsinoides*), and the climbers were Satabar (*Asparagus racemosus*) and Kakora (*Luffa echinata*). Most of these species have disappeared due to overgrazing and exploitation by human beings who have no other means of living as they lost their once very productive land into ravines. The area gives a very deserted and fearful look. Sparsely some trees of Gugal (*Commiphora wightii*) (Fig. 24.2), Birbira (*Dichrostachys cinerea*), Chenkur/Khejari (*Prosopis cineraria*), Gurjan (*Dipterocarpus turbinatus*), Remaja (*Acacia leucophloea*), Semal (*Bombax ceiba*) and Vilayati Babul (*Prosopis juliflora*) can be seen on the horizon of the study sites.

### 24.2.4 Interventions Implemented/Works Executed

The construction of anicuts and afforestation have been widely promoted for the reclamation of ravine lands (Kelley 1983; Prasad 1988). These are also the two main activities undertaken for the management of ravines in this study. Afforestation of

**Fig. 24.2** Gugal (*Commiphora wightii*) tree in ravine

ravines in Bindwa village was taken up under 'Tang Ghaati (narrow valley) Pilot Project', while that in village Himmatpur was undertaken as a 'Pilot Project on Ravine Reclamation'.

#### 24.2.4.1 Anicuts

Anicut is a Tamil language word, meaning 'river bund', i.e. a bund across a stream. Anicut is an earthen bund constructed against the land slope to store run-off against it. It is so constructed as to hold all the run-off and silt against it without any other arrangements in general. However, some gully plugging was done at both the sites and six waste weirs at the Tang Ghaati Plantation Project area. The anicuts were spaced suitably, and their dimensions (length, width and height) are kept big enough to stand against the run-off. Obviously, the size of anicuts depended on site situation, size of ravine, its catchment area and the expected run-off in a season.

Twenty-three anicuts were constructed in Tang Ghaati Pilot Project area covering 20 ha land in revenue department lands and privately owned lands. The average size of anicuts in this project was 30.03 m in length, 8.54 m in width at bottom, 3.15 m in width at top and a height of 2.51 m. The lengths of various anicuts varied from 6.5 to 80.0 m with an average value of 30.03 m. The width of anicuts at bottom varied from 3.0 to 13.5 m with an average of 8.54 m. The height of anicuts varied from 1.03 to 7.02 m with an average of 2.51 m. The volume of earth work undertaken varied from 40.86 to 1048.06 $m^3$. In addition to the anicuts, 6 waste weirs and 20 gully plugs were also constructed above the larger anicuts. The cost of

**Table 24.1** Details of the Ravenous Land Management Project implemented in Bindwa village, Morena, and Himmatpur village, Bhind, Forest Department, Madhya Pradesh, India

| S. no. | Treatment/units | Work sanctioned<br>Location: Bindwa, Morena | Location: Himmatpur (Jamsara), Bhind |
|---|---|---|---|
| 1. | Name of project | Tang Ghaati Pilot Project (Pariyojana) | Pilot Project on Ravine Reclamation |
| 2. | Implementing agency | Forest Department, Morena (MP) | Forest Department, Bhind (MP) |
| 3. | Budget allotted (lakhs) | 2.69 | 8.89 |
| 4. | Budget spent (lakhs) | 2.69 | 8.89 |
| 5. | Period of activity implementation | January 2007 to October 2008 | January 2007 to November 2008 |
| 6. | Chambal ravine area taken up (ha) | 20.0 revenue land | 83.0 forest land |
| 7. | Slope (%) | 39% (2–95%) | 24% (2–75%) |
| 8. | Soil texture | Loam | Sandy loam |
| 9. | **Earth works** | | |
| 10.1 | Check dam (no.) | 23 (size, 30.03 m L × 8.54 m W at base; 2.51 m H from base to top ×3.15 m W at top) | 12 (size 14.5 m L × 30.0 m W at base; 5.5 m H from base to top × 6.0 m W at top) |
| 10.2 | Waste weir | 06 | – |
| 10.3 | Gully plugs (m) | 20 | 150 |
| 10.4 | Ring trench (m) | Nil | 3000 |
| 10.5 | Bunding of fields by farmers (m) | 12,000 | 21,000 |
| 11. | Afforestation through seeding (ha) in 2007–2008 | 20.0 | 72.0 |
| 12 | Impact analysis year | July 2014 | July 2014 |

construction of works/anicuts varied from ₹1115 to 23,581 per anicut. The total investment on all the 23 anicuts was ₹2, 59,800, i.e. about ₹2.6 lakhs approximately at this site.

In Himmatpur Pilot Project area, 12 big anicuts were constructed in the forest department lands and privately owned lands. The size of each anicut was 14.5 m in length, 30.0 m in width at bottom, 6.0 m in width at top and a height of 5.5 m. The anicuts were big enough to hold all the run-off against them, and, therefore, no waste weir was provided on any anicuts at this site. However, many gully plugs were constructed above each anicut to retain some run-off against them. In all, 150 gully plugs were provided on all anicuts in the whole project area of 83 ha. Since these anicuts were big, ring trenches along the anicut band on upstream side were also dug to trap silt in run-off before the run-off could reach the anicut. The total length of ring trenches was 3000 m. The total cost of anicut construction and gully plugging at this site was ₹8.89 lakhs. The details of works done at the two sites are presented in Table 24.1.

**Fig. 24.3** Filling of questioners for impact study from locations

#### 24.2.4.2 Afforestation

The Tang Ghaati Project area (20 ha) was a revenue land, and afforestation by seeding was done on this land. The Himmatpur Pilot Project area (83 ha) belonged to forest department and had scrubby forest due to the over-exploitation of vegetation by the local people. Plantation through seeding was done on 72 ha in this Pilot Project. The tree species planted by seeding in both the locations were Deshi babool (*Acacia nilotica*), Shisham (*Dalbergia sissoo*), Vilayati babul (*Prosopis juliflora*), Neem (*Azadirachta indica*), Hingote (*Balanites aegyptiaca*), Kher (*Acacia catechu*) and Subabool (*Leucaena leucocephala*) during the month of July 2007.

### 24.2.5 Impact Assessment

Socio-economic data in terms of changes in land use pattern, cropping patterns, cropping intensity, productivity, employment generation and income, including income from different sources, annual income of villagers per family, etc. were collected through pretested questionnaires and interviews in intervention villages (privately owned lands in the villages adjoining intervention sites; Fig. 24.3). For collecting biophysical data on the soil, samples from surface soil (0–15 cm) were taken from treated and untreated areas of the project and analysed for physico-chemical properties.

**Fig. 24.4** Taking observations of forest plants

To estimate the growth of trees after 7 years, observations were recorded on various growth parameters of 100 forest trees selected at random in each village location. Survival rate, diameter at stumped height (5 cm above ground level) and green and dry weight of the trees were recorded (Fig. 24.4). The total count and biomass of major shrubs, herbs, climbers and trees and only biomass of grasses were calculated on per ha basis from both the project sites. Team of scientists and experts interacted with farmers from time to time (Fig. 24.5).

The cost of rehabilitation in both the projects was calculated from input cost of the projects (expenditure per ha). Outputs in terms of monetary value at existing market price were obtained for shrubs, herbs, climbers, trees, etc. Similarly, the economics of cultivated crops, dairy and horticultural crops were also calculated for adjoining villages by considering the cost of labour, tillage, seeding, irrigation, fertilizer, pesticides, weeding, harvesting, etc. Gross returns (GRs) were calculated by multiplying the yields of crops, dairy and horticultural crops by minimum support price offered by the Government of India for crops (Economic Survey of India, 2013), milk price of '*Sanchi ugdha Sangh*' (*a local dairy cooperative*) and existing rates of vegetables in the local markets. The value of straw of wheat was calculated using current market rates. Net returns (NRs) were calculated as the difference between gross returns and total cost of cultivation (TCC) (NR = GR − TCC). The benefit/cost ratio was computed by dividing the GR by the total cost of cultivation (TCC).

**Fig. 24.5** Scientists and government officials interact with farmers at Himmatpur project area

## 24.3 Results and Discussion

### 24.3.1 Impact of Anicuts and Afforestation on Intervention Villages (Households in Villages Adjoining Intervention Sites)

This component of the study hereafter is also referred as impact on adjoining villages.

#### 24.3.1.1 Changes in Land Use Pattern

The construction of anicuts, stabilized by afforestation (tree plantation) and rejuvenation of natural vegetation including grasses and trees, held up almost all the run-off in Bindwa Tang Ghaati Project area, as breaches, if any, in the anicuts were taken care of immediately. The anicuts along with ring trenches also completely stopped the run-off in Himmatpur Pilot Project area. As a result, the free flow of run-off from intervention area to downstream farmers' fields in the adjoining areas, which was eroding their fields into ravines previously, was checked completely. This change encouraged the farmers to level their land and construct field bunds to hold the falling rainwater on the fields. Survey results show that the farmers constructed a total of 12,000-m-long field bunds in Bindwa village area and 21,000-m-long field bunds in Himmatpur village area. This work brought about many positive changes in cultivators' land as shown in Table 24.2.

**Table 24.2** Changes in land use pattern in Bindwa village, Morena, and Himmatpur village, Bhind, under Ravenous Land Management Project, Forest Department, Madhya Pradesh, India

| | Pre-project status | | After project status | | Absolute change | | Relative change (%) | |
|---|---|---|---|---|---|---|---|---|
| Change in land use pattern | A | B | A | B | A | B | A | B |
| Total geographical area (ha) | 176 | 305 | 176 | 305 | – | – | – | – |
| Ravenous land | 34.8 | 46.0 | 19.8 | 28.0 | −15.0 | −18.0 | −43.1 | −39.1 |
| Agricultural land | 120 | 251 | 135 | 269 | +15.0 | +18.0 | +12.5 | +7.1 |
| Net cultivated area (ha) | 108 | 246 | 122 | 262 | +14.0 | +16.0 | +13.0 | +6.5 |
| Pasture land (ha) | 10 | 04 | 10 | 04 | – | – | – | – |
| Irrigated land (ha) | 64.8 | 106 | 111.6 | 187 | +46.8 | +81 | +72.2 | +76.4 |
| Rain-fed area (ha) | 43.2 | 145 | 11.4 | 101 | −31.8 | −44 | −73.6 | −30.3 |
| Total farm families | 26 | 55 | 31 | 62 | +5 | +7 | +19.2 | +12.7 |
| Change in cropping intensity (%) | 110 | 115 | 190 | 167 | +80 | +52 | +72.7 | +45.2 |
| No. of tractor | 1 | 2 | 2 | 4 | +1 | +2 | +100 | +100 |
| No. of bullock | 6 | 14 | 0 | 2 | −6 | −12 | −100 | _143 |
| Large animals | 101 | 175 | 133 | 225 | +31 | +50 | +31.7 | +28.6 |
| Small animals | 11 | 132 | 30 | 224 | +19 | +92 | +172.7 | +70.0 |
| Water table (m) | 42.4 | 53.1 | 51.6 | 175 | +3 | +1.5 | +7.1 | +2.8 |

A: Bindwa village; B: Himmatpur village

The data presented in Table 24.2 show that 26 farm families in Bindwa village and 55 families in Himmatnagar village were benefited by the project work. The ravenous land reclaimed was 15 ha in Bindwa village and 18 ha in Himmatpur village, and consequently, the net cultivated areas increased in both the villages. There was also a rise in the water table, by 3 m in Bindwa village and by 1.5 m in Himmatpur village. The increase in water table increased the irrigated area by more than 70% in both the villages.

With increase in irrigated area, cropping intensity of Bindwa village rose by 72.7% while in Himmatpur village by 45.2%. Due to enhanced intensive farming, the animal draught (mainly bullocks) power was replaced by mechanical power (tractors), which doubled in just 2 years. The tractors were also available for custom hire service, which immensely benefited other farmers who did not own one. There was also an increase in the number of large frame as well as small frame livestock, mostly for dairy and meat purposes, indicating increased supply of fodder and prosperity of farmers of both the villages.

#### 24.3.1.2 Changes in Cropping Pattern

Due to the increase in irrigated area, not only the cropped area increased; there was also a change in the cropping pattern of this area (Table 24.3).

The area under long-duration pigeon pea was reduced in both the villages. In Bindwa village, it was replaced by short-duration varieties of pigeon pea, sesame, green gram and cluster bean whereas in Himmatpur village by sesame and cluster bean. In both the villages, the area under pearl millet increased; however, the increase was higher in Bindwa village. As regards the area under rabi crops, it increased considerably in case of wheat, berseem and chickpea and only moderately in case of mustard in Bindwa village. In Himmatpur village, the area under wheat, berseem and mustard increased, while that under chickpea was reduced.

#### 24.3.1.3 Change in Productivity of Major Crops

The degree and extent of soil erosion were reduced, and soil fertility status was improved in the project area. Consequently, the productivity of all kharif and rabi crops was higher as sown by the data in Table 24.4.

The increase in productivity of pigeon pea was the least, while that of sesame was the highest, followed by that of green gram. The increase in productivity of cluster bean and mustard was much higher in Himmatpur village than Bindwa village. However, the increase in the productivity of wheat was much higher in Bindwa village as compared to that in Himmatpur village.

#### 24.3.1.4 Generation of Employment

The data on average employment of farmers' families in both villages are presented in Table 24.5. The percent increase in employment in crop production activities was 9.0 in Himmatpur village and 10.2 in Bindwa village, whereas the increase in employment in dairy sector was 46.5% in Himmatpur and 50.0% in Bindwa village. The lower employment increase in crop production in comparison with dairy, vegetable, etc. was due to the increased level of mechanization in both the villages. The

**Table 24.3** Changes in area of major crops in Bindwa village, Morena, and Himmatpur village, Bhind, under Ravenous Land Management Project, Forest Department, Madhya Pradesh, India

| Major crops | Pre-project status (ha) | | Post-project status (ha) | | Absolute change | | Relative change (%) | |
|---|---|---|---|---|---|---|---|---|
| | A | B | A | B | A | B | A | B |
| Kharif crops | | | | | | | | |
| Pearl millet | 21 | 81 | 90 | 125 | +69 | +44 | +329 | +54 |
| Long- duration pigeon pea | 14 | 17 | 2 | 12 | −12 | −5 | −86 | −29 |
| Short- duration pigeon pea | – | – | 7 | – | +7 | – | – | – |
| Sesame | 7 | 11 | 11 | 38 | +4 | +27 | +57 | +246 |
| Green gram | 2 | 12 | 3 | 10 | +1 | −2 | +50 | −17 |
| Cluster bean | 8 | 14 | 9 | 36 | +1 | +22 | +12 | +157 |
| Rabi crops | | | | | | | | |
| Wheat | 20 | 42 | 51 | 87 | +31 | +45 | +155 | +107 |
| Mustard | 41 | 72 | 54 | 104 | +13 | +32 | +32 | +44 |
| Chickpea | 10 | 25 | 2 | 11 | −8 | −14 | +80 | −56 |
| Berseem | 2 | 19 | 4 | 15 | +2 | +6 | +100 | +67 |

A: Bindwa village; B: Himmatpur village

**Table 24.4** Change in productivity of major crops in Bindwa village, Morena, and Himmatpur village, Bhind, under Ravenous Land Management Project, Forest Department, Madhya Pradesh, India

| Major crops | Pre-project status (q/ha) | | Post-project status (q/ha) | | Absolute change | | Relative change (%) | |
|---|---|---|---|---|---|---|---|---|
| | A | B | A | B | A | B | A | B |
| Kharif crops | | | | | | | | |
| Pearl millet | 13.4 | 14.0 | 24.3 | 23.75 | +10.90 | +9.75 | +81.3 | +69.6 |
| Long- duration pigeon pea | 12.5 | 14.25 | 13.9 | 16.25 | +1.4 | +2.0 | +11.2 | +14.0 |
| Short- duration pigeon pea | – | – | 17.95 | – | – | – | – | – |
| Sesame | 2.46 | 2.65 | 3.50 | 3.90 | +1.04 | +1.25 | +42.3 | +47.2 |
| Green gram | 4.0 | 3.6 | 5.50 | 6.2 | +1.50 | +2.6 | +37.5 | +37.2 |
| Cluster bean | 15.5 | 12.8 | 17.0 | 16.7 | +1.50 | +3.9 | +9.7 | +30.5 |
| Rabi crops | | | | | | | | |
| Wheat | 36.6 | 38.4 | 46.38 | 40.13 | +9.78 | +1.75 | +26.7 | +4.51 |
| Mustard | 18.8 | 13.42 | 22.11 | 17.50 | +3.31 | +4.08 | +17.6 | +30.4 |
| Chickpea | 14.48 | 15.25 | 18.07 | 19.25 | +3.59 | +4.00 | +24.8 | +26.2 |
| Berseem | 349 | 408 | 455 | 512 | +106 | +104 | +30.4 | +25.5 |

A: Bindwa village; B: Himmatpur village; q = 100 kg

**Table 24.5** Average employment (days per farmer household per year) in Bindwa village, Morena, and Himmatpur village, Bhind, under Ravenous Land Management Project, Forest Department, Madhya Pradesh, India

| | Pre-project status | | Post-project status | | Absolute change | | % change | |
|---|---|---|---|---|---|---|---|---|
| Particulars | A | B | A | B | A | B | A | B |
| Agriculture | 440 | 610 | 485 | 665 | +45 | +55 | 10.2 | 9.0 |
| Dairy | 180 | 215 | 270 | 315 | +90 | +100 | 50.0 | 46.5 |
| Vegetables | – | – | 58[a] | 85[b] | +58 | – | – | – |
| Fuel wood[c] | 95 | – | 120 | – | +25 | – | 26.3 | – |
| Custom hire service[d] | 75 | 160 | 220 | 415 | +145 | +60 | 193.3 | 159.4 |
| Over all | 627 | 831 | 775 | 1014 | +148 | | 23.6 | 22.0 |

A: Bindwa village; B: Himmatpur village
[a]Only one family is engaged in vegetable cultivation
[b]Five families engaged in potato cultivation
[c]One family engaged in collection of fuel wood and marketing
[d]Two families in Bindwa and four families in Himmatpur engaged in custom hire service

**Table 24.6** Impact on annual income (₹ per house hold) in Bindwa village, Morena, and Himmatpur village, Bhind, under Ravenous Land Management Project, Forest Department, Madhya Pradesh, India

| | Pre-project status | | After project status | | % change | |
|---|---|---|---|---|---|---|
| Sources | Bindwa | Himmatpur | Bindwa | Himmatpur | Bindwa | Himmatpur |
| Crops | 18,550 | 24,000 | 65,500 | 85,400 | 253.1 | 255.8 |
| Vegetables | – | – | 9500 | 116,000 | – | – |
| Dairy | 11,250 | 16,500 | 56,000 | 75,000 | 397.8 | 354.55 |
| Sale of firewood | 3400 | – | 20,000 | – | 488.2 | |
| Custom hire service | 22,000 | 31,500 | 80,300 | 94,000 | 363.0 | 198.4 |
| Overall | 30,777 | 41,646 | 127,326 | 175,819 | 313.7 | 322.2 |

employment in vegetable production increased as the vegetable farming became more intensive, requiring more labour days in operations such as sowing, harvesting and packaging, etc. than in pre-project previous years. Similarly, due to the increase in fodder production, including berseem (*Trifolium alexandrinum*), dairying developed, and the dairy and other animals increased in number, which created more work and employment.

#### 24.3.1.5 Annual Income

Better crop yields and higher milk and vegetable production added significantly to the annual income of the households in the Bindwa and Himmatpur villages (Table 24.6). Overall, the annual income per household in Bindwa village increased to ₹1,27,326 from ₹30,777 and in Himmatpur village to ₹1,75,819 from ₹41,646, (an average increase in the order of more than 300%) in both the cases.

## 24.3.2 Effect on Vegetation

### 24.3.2.1 Growth of Seeded/Planted Trees

As already stated above, tree seeds were sown in over 20 ha ravine land in Bindwa village forest and on 72 ha in Himmatpur village forest during July 2007. The results of this intervention on forest vegetation are presented in Table 24.7.

#### Survival of Trees

Because of longer duration of water availability and better moisture regime throughout the year (Table 24.7) and less soil erosion, the germination and survival percentage of trees of different species was 80.9–96.1% in both project sites, a reasonable high percentage in such a situation. The highest survival rate (96.1%) was of *Prosopis juliflora*, followed closely by that of *Acacia nilotica* (92.3%), *Balanites aegyptiaca* (88.0%), *Azadirachta indica* (85.6%), *Acacia catechu* (84.7%), *Dalbergia sissoo* (81.6%) and *Leucaena leucocephala* (80.9%).

#### Tree Girth

After 7 years of plantation, trunk/stem girth of forest tree species was measured, and it was found that the diameter of stumped height (DSH) was like gains in tree height (Table 24.7). The range in DSH was between 61.7 and 128.7 mm at both the locations. The DSH was maximum in case of *Acacia nilotica* (128.7 mm plant$^{-1}$), followed by *Prosopis juliflora* (123.5 mm plant$^{-1}$), *Azadirachta indica* (111.4 mm plant$^{-1}$), *Leucaena leucocephala* (109.7 mm plant$^{-1}$), *Dalbergia sissoo* (105.8 mm plant$^{-1}$), *Balanites aegyptiaca* (88.4 mm plant$^{-1}$) and the lowest in case of *Acacia catechu* (61.7 mm plant$^{-1}$). Though the DSH of the tree species is smaller, it is not below the expected in such harsh environments. The DSH values also relate well with height of the plants as explained.

#### Tree Height

In regard to various growth parameters, the tree height of plants after 7 years of growth showed that *Acacia nilotica* and *Prosopis juliflora* were the fastest growing species and *Acacia catechu* the slowest growing one (Table 24.7) in these areas. In this period, *Acacia nilotica* attained a height of 7.52 m and *Acacia catechu* 3.56 m. The growth rate of *Dalbergia sissoo* and *Azadirachta indica* was between these two values. The height of trees observed was in the following order: *Acacia nilotica* (highest), followed by *Prosopis juliflora*, *Azadirachta indica*, *Dalbergia sissoo*, *Leucaena leucocephala*, *Balanites aegyptiaca* and *Acacia catechu*. The tree height values relate well with the DSH values in such environments.

#### Tree Biomass

The average green weight per plant was highest in case of *Acacia nilotica* (113.3 kg/plant), followed by *Azadirachta indica*, (95.0 kg plant$^{-1}$), *Prosopis juliflora* (58.9 kg/plant), *Leucaena leucocephala* (57.4 kg plant$^{-1}$), *Dalbergia sissoo* (50.9 kg plant$^{-1}$), *Balanites aegyptiaca* (43.3 kg plant$^{-1}$) and the lowest in case of *Acacia catechu* (30.6 kg per plant), as shown in Table 24.8. A similar trend was also observed with

**Table 24.7** Growth parameters of forest trees established in Bindwa village, Morena, and Himmatpur village, Bhind, under Ravenous Land Management Project, Forest Department, Madhya Pradesh, India

| Name of species | Survival (%) | | | Diameter (mm per plant) | | | Height (m) | | |
|---|---|---|---|---|---|---|---|---|---|
| | A | B | Average | A | B | Average | A | B | Average |
| Deshi babool (*Acacia nilotica*) | 90.7 | 93.9 | 92.3 | 132.3 | 125.0 | 128.7 | 6.77 | 8.27 | 7.52 |
| Shisham (*Dalbergia sissoo*) | 76.7 | 86.5 | 81.6 | 111.4 | 100.2 | 105.8 | 5.85 | 4.90 | 5.38 |
| Vilayati babul (*Prosopis juliflora*) | 93.8 | 98.4 | 96.1 | 113.6 | 133.2 | 123.5 | 6.38 | 7.44 | 6.91 |
| Neem (*Azadirachta indica*) | 80.1 | 91.1 | 85.6 | 112.8 | 110.0 | 111.4 | 6.10 | 6.60 | 6.35 |
| Hingote (*Balanites aegyptiaca*) | 85.3 | 90.7 | 88.0 | 86.7 | 90.8 | 88.8 | 4.84 | 4.20 | 4.52 |
| Kher (*Acacia catechu*) | 82.5 | 86.9 | 84.7 | 65.9 | 57.5 | 61.7 | 3.37 | 3.75 | 3.56 |
| Soobabool (*Leucaena leucocephala*) | 78.3 | 83.5 | 80.9 | 115.7 | 103.7 | 109.7 | 5.67 | 4.96 | 5.32 |

A: Bindwa village; B: Himmatpur village

**Table 24.8** Weight per plant of forest trees established in Bindwa village, Morena, and Himmatpur village, Bhind, under Ravenous Land Management Project, Forest Department, Madhya Pradesh, India

| Name of species | Green wt. plant$^{-1}$ (kg) | | | Dry wt. plant$^{-1}$ (kg) | | |
|---|---|---|---|---|---|---|
| | Bindwa | Himmatpur | Average | Bindwa | Himmatpur | Average |
| Deshi babool (*Acacia nilotica*) | 118.91 | 107.71 | 113.31 | 86.21 | 78.09 | 82.15 |
| Shisham (*Dalbergia sissoo*) | 49.58 | 52.17 | 50.88 | 35.95 | 37.82 | 36.89 |
| Vilayati babul (*Prosopis juliflora*) | 56.80 | 61.08 | 58.94 | 41.18 | 44.28 | 42.73 |
| Neem (*Azadirachta indica*) | 90.64 | 101.35 | 95.00 | 65.71 | 73.48 | 69.60 |
| Hingote (*Balanites aegyptiaca*) | 40.75 | 45.83 | 43.29 | 29.54 | 33.23 | 31.39 |
| Kher (*Acacia catechu*) | 25.50 | 35.71 | 30.61 | 18.49 | 25.89 | 22.30 |
| Subabool (*Leucaena leucocephala*) | 58.91 | 55.89 | 57.40 | 42.71 | 40.52 | 41.62 |

regard to the dry weight content per plant. The maximum dry weight was observed in case of *Acacia nilotica*, i.e. 82.2 kg per plant, and the lowest dry weight was of *Acacia catechu*, i.e. 22.3 kg plant$^{-1}$. The biomass results showed varied correlation with DSH and plant height values. In the overall assessment, the *Acacia nilotica* was the best suited in these conditions. This species is leguminous, fixing nitrogen and is an excellent fodder for small ruminants.

#### 24.3.2.2 Growth of Natural Vegetation

This section deals with the naturally occurring vegetation at the two study sites. The comparative results at treated and untreated site are presented below.

##### Plant Population and Green Biomass

After seeding of trees was done, the area was protected from stray animals and wood collectors. Consequently, the natural plants, shrubs and grasses regenerated and grew faster on treated ravenous land as compared to untreated (unprotected) area (Table 24.9). Among shrubs, a visible improvement in the plant population density (number of plants) and green weight per unit area (e.g. hectare) was recorded. The total number of shrubs, herbs and climbers is 59, 28 and 49 plants ha$^{-1}$, respectively, in untreated ravenous land, whereas the same are 89, 67 and 89 plants ha$^{-1}$ in treated area of Bindwa village. Similarly, the existing number of shrubs and climbers is 34 and 79 plants ha$^{-1}$ in untreated ravenous land, whereas 79 and 122 plants ha$^{-1}$ in the treated land at Himmatpur village, respectively. Himmatpur village did not have herbs in the treated as well as untreated locations.

The treated land also produced higher total green biomass in comparison with the untreated ravenous land at both the study sites. At Bindwa village, the available total green biomass weight of shrubs, herbs, climbers and grasses in the untreated land was 945.1, 1.26, 23.0 and 1,0630 kg ha$^{-1}$, respectively, whereas in the treated land at this site, their biomass weight was 1567.1, 3.04, 57.5 and 11,080 kg ha$^{-1}$, respectively. Similarly, the total green biomass weight of shrubs, climbers and grasses in the untreated land at Himmatpur village was 531, 48 and 5930 kg ha$^{-1}$, respectively, and in the treated area was 2112.2, 68.9 and 7475 kg ha$^{-1}$, respectively, at this village. The overall green biomass weight (shrubs, herbs, climbers and grasses) was 23.6% higher in the treated land in comparison with untreated land. Himmatpur village did not have herbs in the treated as well as untreated locations.

##### Number and Dry Weight of Forest Plants (Natural and Established)

Only a small number of trees, such as Chenkur/Khejri (*Prosopis cineraria*), Birbira (*Dichrostachys cinerea*), Gurjan (*Dipterocarpus turbinatus*), Semal (*Bombax ceiba*), Remaja (*Acacia leucophloea*), Gugal (*Commiphora wightii*) and Sahjan/drum stick (*Moringa oleifera*) could survive at both the study sites. They grew faster in the treated area in comparison with untreated ravenous land (Table 24.10).

The treated area produced higher number of trees and total green biomass weight in comparison with untreated ravenous land at both the locations. The total number of natural forest plants was 59 ha$^{-1}$ in untreated ravenous land, whereas it was 89 ha$^{-1}$ in the treated ravenous land at Bindwa location. Similarly, the existing

**Table 24.9** Green weight (kg ha$^{-1}$) of natural shrubs, herbs and grasses in treated and untreated land portions in Bindwa village, Morena, and Himmatpur village, Bhind, under Ravenous Land Management Project, Forest Department, Madhya Pradesh, India

| | | Untreated ravenous land | | | Treated ravenous land | | |
|---|---|---|---|---|---|---|---|
| Name of plants | Output | A | B | Average | A | B | Average |
| Shrubs | | | | | | | |
| Kareel (*Capparis decidua*) | Fruit | 1.1 (3)[a] | 5 (10)[a] | 3.2 (7)[a] | 2.1 (5)[a] | 5.2 (13)[a] | 3.6 (9)[a] |
| arni (*Clerodendrum phlomidis*) | Fodder | 401 (18)[a] | Nil | – | 516 (22)[a] | Nil | – |
| Chabeni | Fuel wood | 235 (24)[a] | 18 (2)[a] | 127 (13 kg)[a] | 250 (35)[a] | 48 (5)[a] | 149 (20)[a] |
| Heens (*Capparis sepiaria*) | Fuel wood | 308 (14)[a] | 508 (22)[a] | 1073 (37)[a] | 799 (27)[a] | 2059 (71)[a] | 1429 (49)[a] |
| Herbs | | | | | | | |
| Bhringraj | Bark | 1.15 (23)[a] | Nil | – | 2.8 (56)[a] | Nil | – |
| Shankhpushpi | Leaf | 0.11 (5)[a] | Nil | – | 0.24 (11)[a] | Nil | – |
| Climber | | | | | | | |
| Satabar (*Asparagus racemosus*) | Root | 2 (4)[a] | 7 (14)[a] | 4.5 (9)[a] | 2.5 (9)[a] | 8.9 (16)[a] | 7.5 (18)[a] |
| Kakora | Fruit | 21 (45)[a] | 41 (65)[a] | 31 (55)[a] | 55 (80)[a] | 60 (106)[a] | 73 (93)[a] |
| Grasses | | | | | | | |
| Bans (*Phyllostachys* spp.) | Stick | 70 (2)[a] | Nil | Nil | 85 (5)[a] | Nil | – |
| Doob (*Cynodon dactylon*) | Fodder | 1500 | 1180 | 1340 | 1900 | 1325 | 1612 |
| Anjana (*Cenchrus ciliaris*) | Fodder | 5600 | 1500 | 3550 | 6500 | 2125 | 4312 |
| Sama (*Panicum sumatrense*) | Fodder | 250 | Nil | – | 280 | Nil | – |
| Bhanjura (*Apluda mutica*) | Fodder | 1900 | 850 | 1375 | 2400 | 1050 | 1725 |
| Dab (*Desmostachya bipinnata*) | Fodder | 1380 | 2400 | 1890 | 2560 | 2975 | 2767 |

A: Bindwa village; B: Himmatpur village
[a]No. of plant ha$^{-1}$

natural forest plants were 34 ha$^{-1}$ in untreated land and 79 ha$^{-1}$ in treated ravenous land at Himmatpur location. The overall improvement was 78.7% in treated ravenous land as compared to the untreated ravenous land.

The total green biomass weight of natural forest plants in untreated ravenous land was 945.1 kg ha$^{-1}$ at Bindwa and 531 kg ha$^{-1}$ at Himmatpur, whereas in the treated ravenous land, the green biomass availability was 1567.1 kg ha$^{-1}$ at Bindwa and 2112.2 kg ha$^{-1}$ at Himmatpur site. The overall improvement in green biomass

**Table 24.10** Dry weight of wood (kg ha$^{-1}$) of natural and established forest trees in treated and untreated land portions in Bindwa village, Morena, and Himmatpur village, Bhind, under Ravenous Land Management Project, Forest Department, Madhya Pradesh, India

| Name of plants | Untreated ravenous land | | | Treated ravenous land | | |
|---|---|---|---|---|---|---|
| | Bindwa | Himmatpur | Average | Bindwa | Himmatpur | Average |
| Natural plant | | | | | | |
| Chenkur (*Prosopis cineraria*) | 308 (22)[a] | 133 (12)[a] | 220 (17)[a] | 488 (30)[a] | 255 (15)[a] | 352 (23)[a] |
| Birbira (*Dichrostachys cinerea*) | 180 (95)[a] | Nil | – | 365 (180)[a] | Nil | – |
| Gurjan | 160 (4)[a] | Nil | – | 360 (9)[a] | Nil | – |
| Semal (*Bombax ceiba*) | 19 (3) | Nil | – | 39 (4)[a] | Nil | – |
| Remja (*Acacia leucophloea*) | 24 (1)[a] | 44 (2)[a] | 44 (2)[a] | 34 (1)[a] | 75 (3)[a] | 32 (3)[a] |
| Gugal (*Commiphora wightii*) | 29 (8)[a] 2 kg gum) | 22 (6)[a] 1.5 kg gum | 25.5 (7)[a] 1.8 kg gum | 63 (13)[a] 4.1 kg gum | 35 (8)[a] 2.4 kg gum | 49 (11)[a] 4 kg gum |
| Sehajan | 110 (5)[a] | Nil | – | 220 (8)[a] | 55 (2)[a] | 138 (25)[a] |
| Established plant | | | | | | |
| Deshi babool (*Acacia nilotica*) | Nil | Nil | Nil | 4400 (51)[a] | 5625 (72)[a] | 5013 (62)[a] |
| Shisham (*Dalbergia sissoo*) | Nil | Nil | Nil | 288 (8)[a] | 455 (12)[a] | 372 (10)[a] |
| Vilayati babul (*Prosopis juliflora*) | 450 (28) | 611 (38)[a] | 530 (38)[a] | 2677 (65)[a] | 1860 (420)[a] | 2269 (54)[a] |
| Neem (*Azadirachta indica*) | Nil | Nil | Nil | 395 (6)[a] | 590 (8) | 493 (7)[a] |
| Hingote (*Balanites aegyptiaca*) | Nil | Nil | Nil | 770 (26) | 1070 (32)[a] | 920 (29)[a] |
| Kher (*Acacia catechu*) | Nil | 85 (7)[a] | – | 355 (19)[a] | 545 (21)[a] | 450 (20)[a] |
| Subabool (*Leucaena leucocephala*) | Nil | Nil | Nil | 130 (3)[a] | 325 (08)[a] | 228 (06)[a] |

[a]No. of plant per ha

weight of natural forest plants was 149.2% higher in treated ravenous land in comparison with the untreated ravenous land.

The total number of established forest plants was 28 ha$^{-1}$ in untreated ravenous land and 178 ha$^{-1}$ in treated ravenous land at Bindwa location. Similarly, the number of established forest plants was 45 ha$^{-1}$ in untreated ravenous land and 195 ha$^{-1}$ in treated ravenous land at Himmatpur location. The overall improvement in number of forest plants was 408% on treated ravenous land as compared with untreated ravenous land.

The total green biomass weight of forest plants in untreated ravenous land was 450 kg ha$^{-1}$ at Bindwa location and 696 kg ha$^{-1}$ at Himmatpur location. After treatment of ravenous land, it was 9015 kg $ha^{-1}$ at Bindwa location and 10,470 kg $ha^{-1}$ at Himmatpur location. The overall green biomass recorded from forest plants was 1085.5 kg $ha^{-1}$ from untreated ravenous land and 10,721 kg $ha^{-1}$ from treated ravenous land, showing an increase of 887%.

### 24.3.3 Changes in Physicochemical Properties of Soil

On execution of earth works, e.g. construction of check dams, anicuts, gully plugs, ring trenches and field bunding, run-off from treated areas was negligible, soil erosion was within permissible limits, and considerable deposits of fine particles, i.e. clay and silt, got deposited against anicuts/check dams and in the ring trenches (Table 24.11). Consequently, changes were inflicted in some soil physicochemical properties due to deposit of clay and silt particles and their downward movement. The texture of surface soil at downstream sites gradually changed from sandy loam to silt loam. The clay content also increased to 28.67% from initial 13.1%, at the pre-project period.

The soils were highly permeable and rich in $CaCO_3$ when untreated. After treatment, their permeability decreased to some extent due to the deposition of silt and clay, especially in the check dams. Also, the concentration of $CaCO_3$ in the deposited soil was observed to be lesser due to silt and clay deposition in treated area, while the organic matter content was noted to be slightly higher. Thus, soil physical properties were slightly improved. About soil reaction and conductivity, there was no significant change, and these were within normal limits.

#### 24.3.3.1 Soil Fertility Status

Because of reduced soil erosion and deposition of soil particles, the fertility status of soil was slightly improved (Table 24.11). Not only the content of available nitrogen, phosphorus, potassium and sulphur was higher in the treated area, the contents of available micronutrients (Cu, Fe, Zn and Mn) were also higher. The increased available macronutrients (kg $ha^{-1}$) were in the order of nitrogen 83, phosphorus 9.3, potassium 40 and sulphur 7.2 over the initial values of nitrogen 99, phosphorus 15.9, potassium 322 and sulphur 16.4. Similarly increase in available micronutrients (mg $kg^{-1}$) were copper 0.95, iron 6.7, zinc 0.36 and manganese 6.34 over the average initial values of copper 0.35, iron 17.2, zinc 0.34 and manganese 3.52.

## 24.4 Total Returns from Treated vs Untreated Ravenous Land

Pande et al. (2012) have done an economic analysis of bamboo plantations in three types of ravines in India. Using similar approach, the total returns from trees, shrubs and grasses, both natural and planted ones, were estimated from the yield of wood, medicinal products (bark, leaves, fruits, etc.) and fodder and their market prices. The data for different shrubs, grasses, herbs, climbers and trees are presented in Table 24.12, for both project villages, viz. Bindwa and Himmatpur.

**Table 24.11** Physicochemical properties of untreated and treated land portions in Bindwa village, Morena, and Himmatpur village, Bhind, under Ravenous Land Management Project, Forest Department, Madhya Pradesh, India

| | Untreated ravenous land | | | Treated ravenous land[a] | | | |
|---|---|---|---|---|---|---|---|
| Soil properties | A | B | Average | A | B | Average | ± change over untreated |
| Sand (%) | 48.8 | 55.2 | 52.0 | 34.2 | 30.3 | 32.3 | −19.7 |
| Silt (%) | 36.3 | 33.6 | 34.9 | 38.1 | 40.2 | 39.1 | +37.0 |
| Clay (%) | 14.9 | 11.2 | 13.1 | 27.7 | 29.5 | 28.6 | +15.0 |
| Soil texture | Loam | Sandy loam | – | Clay loam | Clay loam | | |
| $pH_2$ | 7.80 | 8.11 | 7.96 | 7.85 | 7.82 | 7.84 | −0.12 |
| $EC_2$ ($dS\ m^{-1}$) | 0.22 | 0.30 | 0.26 | 0.32 | 0.44 | 0.38 | +0.12 |
| $CaCO_3$ (%) | 1.56 | 2.18 | 1.87 | 0.54 | 0.71 | 0.63 | −1.24 |
| BD $Mg\ m^{-3}$ | 1.52 | 1.58 | 1.55 | 1.48 | 1.45 | 1.47 | −08 |
| IR ($mm\ h^{-1}$) | 28.1 | 35.9 | 32.0 | 3.7 | 1.4 | 2.6 | −29.4 |
| Organic carbon (%) | 0.12 | 0.25 | 0.19 | 0.44 | 0.42 | 0.43 | +0.24 |
| Available macronutrients ($kg\ ha^{-1}$) | | | | | | | |
| Nitrogen | 79 | 118 | 99 | 175 | 189 | 182 | +83 |
| Phosphorus | 11.2 | 20.6 | 15.9 | 19.9 | 30.5 | 25.2 | +9.3 |
| Potassium | 315 | 328 | 322 | 368 | 356 | 362 | +40 |
| Sulphur | 11.8 | 21.0 | 16.4 | 24.9 | 22.2 | 23.6 | +7.2 |
| Available micronutrients ($mg\ kg^{-1}$) | | | | | | | |
| Copper | 0.34 | 0.36 | 0.35 | 1.53 | 1.07 | 1.30 | +0.95 |
| Iron | 15.7 | 18.7 | 17.2 | 21.9 | 25.9 | 23.9 | +6.7 |
| Zinc | 0.32 | 0.35 | 0.34 | 0.83 | 0.56 | 0.70 | +0.36 |
| Manganese | 3.92 | 3.12 | 3.52 | 11.80 | 7.91 | 9.86 | +6.34 |

A: Bindwa village; B: Himmatpur village

[a]Soil characteristics of inner side of check dam are pH – 1:2 soil/water ratio, *EC* electrical conductivity 1:2 soil/water ratio, *BD* bulk density. Figure in parenthesis indicate mean value

**Table 24.12** Total economic outcome (₹ ha$^{-1}$) from natural and established (shrubs, herbs and forest plants) in treated and untreated land portions in Bindwa village, Morena, and Himmatpur village, Bhind, under Ravenous Land Management Project, Forest Department, Madhya Pradesh, India

| Name of plants | Untreated ravenous land | | Treated ravenous land | |
|---|---|---|---|---|
| | Bindwa | Himmatpur | Bindwa | Himmatpur |
| Shrubs | | | | |
| Kareel (*Capparis decidua*) | 22 | 100 | 42 | 110 |
| arni (*Clerodendrum phlomidis*) | 401 | Nil | 516 | Nil |
| Chabeni | 1410 | 108 | 1500 | 288 |
| Heens (*Capparis sepiaria*) | 1848 | 3048 | 4794 | 8236 |
| Total | 3681 | 3256 | 6852 | 8634 |
| Grasses | | | | |
| Bamboo | 700 | Nil | 850 | Nil |
| Total | 700 | – | 850 | |
| Herbs | | | | |
| Bhringraj | 58 | Nil | 140 | Nil |
| Shankhpushpi | 6 | Nil | 12 | Nil |
| Total | 64 | – | 152 | |
| Climber | | | | |
| Satabar (*Asparagus racemosus*) | 160 | 560 | 200 | 712 |
| Kakora | 420 | 820 | 1100 | 1200 |
| Total | 580 | 1380 | 1300 | 1912 |
| Tree | | | | |
| Gugal (*Commiphora wightii*) | 574 | 432 | 1198 | 690 |
| Chekur (*Prosopis cineraria*) | 3133 | 865 | 3172 | 1658 |
| Birbira (*Dichrostachys cinerea*) | 648 | Nil | 2190 | Nil |
| Gurjan (*Lannea coromandelica*) | 960 | Nil | 2160 | Nil |
| Semal (*Bombax ceiba*) | 114 | Nil | 234 | |
| Remja (*Acacia leucophloea*) | 144 | 264 | 204 | 450 |
| Sehajan (*Moringa oleifera*) | 660 | Nil | 1320 | 330 |
| Deshi babool (*Acacia nilotica*) | Nil | Nil | 28,600 | 46,698 |
| Shisham (*Dalbergia sissoo*) | Nil | Nil | 4608 | 7280 |
| Vilayati babul (*Prosopis juliflora*) | 2925 | 3972 | 17,401 | 12,090 |
| Neem (*Azadirachta indica*) | Nil | Nil | 2963 | 4425 |
| Hingote (*Balanites aegyptiaca*) | Nil | Nil | 5005 | 6955 |
| Kher (*Acacia catechu*) | Nil | 553 | 2308 | 3543 |
| Subabool (*Leucaena leucocephala*) | Nil | Nil | 780 | 1950 |
| Total | 9158 | 6086 | 69,983 | 86,069 |
| Grand total | 14,183 | 10,722 | 79,137 | 96,615 |

Rate of local market kareel fruit at ₹ 20 kg$^{-1}$; fodder at ₹1 kg$^{-1}$; fuel wood at ₹ 6 kg$^{-1}$; bhringraj bark and shankhpushpi leaf at ₹50 kg$^{-1}$; satabar root at ₹ 80 kg$^{-1}$; kakora fruit at ₹20 kg$^{-1}$; mature wood in Deshi babool, vilayati babul, chekur, hingote and kher at ₹6.5 kg$^{-1}$; shisham at ₹16 kg$^{-1}$; neem at ₹7.5 kg$^{-1}$; and other miscellaneous at ₹6 kg$^{-1}$

The calculated gross revenue from all trees, shrubs, etc. from untreated areas was ₹14,183 in case of Bindwa and ₹10,722 for Himmatpur village. On the other hand, the calculated gross revenue from treated areas was ₹79,137 and ₹96,615 from Bindwa village and Himmatpur, respectively. The calculated earnings of Bindwa village from tree, shrubs and grasses are to be 5.57 times higher from the treated area, while at that of Himmatpur village by more than 9.01 times in comparison with untreated (control) area. Evidently, such level of revenue from non-crop interventions will add to the overall income of the farmers from crop, dairy and other enterprises.

## 24.5 Economic Viability of the Project

The economic analysis of wasteland afforestation in India by Balooni (2003) provides a methodology for assessing the economic viability of such projects. The economic analysis of this project is presented in Table 24.13.

The calculated net returns from Tang Ghaati Project, Bindwa, and Pilot Project Himmatpur were ₹65,703 $ha^{-1}$ and ₹83,404 $ha^{-1}$, respectively, as compared with those from control (untreated natural ravine) ₹14,183 for Bindwa and ₹10,722 $ha^{-1}$ for Himmatpur. The additional returns by construction of anicuts and afforestation were ₹64,954 $ha^{-1}$ for Bindwa and Rupees 85,893 $ha^{-1}$ for Himmatpur. As shown in Table 24.13, the cost/benefit ratio for such an investment for Tang Ghaati Pariyojana works out to be 1:5.89 while for that of Pilot Project Himmatpur to be 1:7.31. Both these investments appear highly feasible and worth investing for improving the livelihoods of farmers in ravine lands. The findings of Pande et al. (2011) about incentives for soil and water conservation on farm in ravines of Gujarat also support this suggestion for future adoption.

**Table 24.13** Economic impact of Ravenous Land Management Project in Bindwa village, Morena, and Himmatpur village, Bhind, Forest Department, Madhya Pradesh, India

| Project name/villages | Cost of project (₹$ha^{-1}$) | Gross return (₹$ha^{-1}$) | Additional return (₹$ha^{-1}$) | Net return (₹$ha^{-1}$) | C:B ratio |
|---|---|---|---|---|---|
| Control (natural ravine of Bindwa, Morena) | Nil | 14,183 | – | 14,183 | – |
| Tang Ghaati Pariyojna, Bindwa, Morena | 13,434 | 79,137 | 64,954 | 65,703 | 1:5.89 |
| Control (natural ravine of Himmatpur, Bhind) | Nil | 10,722 | – | 10,722 | – |
| Pilot Project, Himmatpur, Bhind | 13,211 | 96,615 | 85,893 | 83,404 | 1:7.31 |

## 24.6 Conclusion

Afforestation and earth works executed in forest land improved the vegetation cover, soil properties and water resources in the study area. These efforts also resulted in stabilized ravines and improved productivity of forest land and also of the protected farmers' land in the adjoining area. The expected calculated revenue from the treated area appeared several folds higher than from the untreated area, which indicates a high potential of such efforts for improving farmers' livelihoods and income and the greening of ravines for achieving the environmental security. The earth works and afforestation model were highly successful for a specific situation in Madhya Pradesh. However, the results would have been better if the planning was done on watershed basis and the communities were more intensively involved in the implementation. For which, the macro-watershed could be divided into a number of micro-watershed for implementation of the interventions. In each micro-watershed, larger (huge) anicuts could be replaced by smaller masonry check dams. Also, construction of tanks at suitable sites (gully heads) should have been tried. Our experience shows that the afforestation/tree plantation does improve ravines and degraded lands, but it is not so effective in checking the advancement of deep gullies and the formation of ravines through them. The experiences of anicut construction, paying heavy dividends, need to be extended for evaluation on large-scale farmers' fields in a macro-watershed.

## References

Balooni K (2003) Economics of wastelands afforestation in India, a review. New For 26(2):101–136

Bhushan LS, Yadav RC (2000) Management of wastelands with special reference to ravines. Indian Farming 50(5). CAB Direct2: 22–24):27–31

Haigh MJ (1984) Ravine erosion and reclamation in India. Geoforum 15(4):543–561

Kelley HW (1983) Keeping the land alive: soil erosion – its causes and cures, Food and Agriculture Organization (FAO) of the United Nations Soils Bulletin, vol 50. FAO, Rome, pp 1–100

Ngoze S, Riha S, Lehmann J, Verchot L, Kinyangi J, Mbugua D, Pell A (2008) Nutrient constraints to tropical agroecosystem productivity in long-term degrading soils. Glob Chang Biol 14(12):2810–2822

Pande VC, Kurothe RS, Singh HB, Tiwari SP (2011) Incentives for soil and water conservation on farm in ravines of Gujarat: policy implications for future adoption. Agric Econ Res Rev 24:109–118

Pande VC, Kurothe RS, Rao BK, Gopal K, Parandiyal AK, Singh AK, Ashok K (2012) Economic analysis of bamboo plantation in three major ravine systems of India. Agric Econ Res Rev 25(1):49–59

Pani P, Mohapatra SN (2001) Delineation and monitoring of gullied and ravenous lands in a part of lower Chambal Valley, India, using remote sensing and GIS. Paper presented at the 22nd Asian Conference on Remote Sensing, 5–9 November, 2001, Singapore, organized by Centre for Remote Imaging, Sensing and Processing (CRISP), National University of Singapore; Singapore Institute of Surveyors and Valuers (SISV); Asian Association on Remote Sensing (AARS)

Prasad R (1988) Effectiveness of aerial seeding in reclamation of Chambal Ravines in Madhya Pradesh. Indian Forester 114(1):1–18

# Modeling Soil Erosion by Water

# 25

M. K. Hardaha

**Abstract**

Soil erosion is a natural process that occurs when the force of wind, raindrops or running water on the soil surface exceeds the cohesive forces that bind the soil together. In general, vegetation cover protects the soil from the effects of these erosive forces. However, land management activities such as ploughing, burning and heavy grazing may disturb this protective layer, exposing the underlying soil. About 1 billion hectares of land in the world is estimated to be eroded by water. Water erosion is resulted due to dispersive and transporting power of the water, as is evident from splash erosion where first of all the soil particles are detached from the soil surface by the action of raindrop and then transported with surface runoff. There is a direct relationship between the soil loss and runoff volume. There are many climatic and biotic factors responsible for soil erosion. Many erosion process models are available and used to estimate various soil erosion parameters. Their use depends upon the data required in the model and the data available. In the present chapter, some of these models such as the Universal Soil Loss Equation, Soil Loss Equation Model for South Africa, Soil Loss Equation Model for Mediterranean Region, Modified Universal Soil Loss Equation, Morgan-Morgan-Finney Model, Quasi-Three-Dimensional Runoff Model and Limburg Soil Erosion Model (LISEM) are discussed.

M. K. Hardaha (✉)
Directorate of Extension Services, Jawaharlal Nehru Krishi VishvaVidayalaya, Jabalpur, Madhya Pradesh, India



J. C. Dagar, A. K. Singh (eds.), *Ravine Lands: Greening for Livelihood and Environmental Security*, https://doi.org/10.1007/978-981-10-8043-2_25

## 25.1 Introduction

Although the term "erosion" was in use in the nineteenth century, the term "soil erosion" was introduced at the beginning of the twentieth century and did not come into general use until the 1930s. The word erosion is of Latin origin being derived from the word *erodere* – to eat away (*rodere* – to chew). The term erosion was first used in geology to describe the forming of hollows by water, the wearing away of solid material by action of river water, while surface wash and precipitation erosion were called "ablation" (*abatio* – to carry away). In addition to erosion and ablation, many other terms "corrasion" (*corradere* – to chew together), "corrosion" (*corodere* – to chew to pieces), "abrasion" (*abrodere* – to scrape off) and "denudation" (*denundere* – to strip) were also used. Soil erosion, however, is defined as the detachment, transportation and deposition of soil particles from one place to another by the action of wind, water or gravity forces. Actually, soil erosion is a natural process that occurs when the force of wind, raindrops or running water on the soil surface exceeds the cohesive forces that bind the soil together. In general, vegetation cover protects the soil from the effects of these erosive forces. However, land management activities such as ploughing, deforestation, burning and heavy grazing are responsible to disturb the protective vegetation cover exposing the underlying soil. Soil erosion can be classified based on rate, agent causing erosion and process of erosion. Table 25.1 presents the classification of soil erosion based on agents causing erosion.

The erodibility of the underlying soil depends on its type, location and degree of exposure to erosive processes. The process of soil erosion and transport involves not only the detachment of soil particles from a particular site but also their transfer and deposition elsewhere in the catchment in the form of sediment (Meyer and

**Table 25.1** Classification of soil erosion based on the active factors

| | | Term | |
|---|---|---|---|
| S.N | Factor | English | International |
| 1 | Water | Water erosion | Aquatic erosion |
| 1.1 | Precipitation, rain | Precipitation erosion, rain erosion | Pluvial erosion |
| 1.2 | River | River erosion | Fluvial erosion |
| 1.3 | Torrent | Torrent erosion | Torrential erosion |
| 1.4 | Lake | Lake erosion | Limnic erosion |
| 1.5 | Reservoir | Reservoir erosion | Lacustrine erosion |
| 1.6 | Sea | Sea erosion | Marine erosion |
| 2 | Glacier | Glacier erosion | Glacial erosion |
| 3 | Snow | Snow erosion | Nival erosion |
| 4 | Wind | Wind erosion | Aeolian erosion |
| 5 | Organisms | Biological erosion | Organogenic erosion |
| 5.1 | Plants | Erosion caused by plants | Phytogenic erosion |
| 5.2 | Animals | Erosion caused by animals | Zoogenic erosion |
| 5.3 | Man | Erosion caused by man | Anthropogenic erosion |

Source: Zachar (1982)

Wischmeier 1969). Our discussion in this section is confined to water erosion and here after erosion shall be understood as water erosion unless specified.

## 25.2 Factors Affecting Erosion

Erosion is resulted due to dispersive and transporting power of the water; as in case of splash erosion, first, the soil particles are detached from the soil surface by the action of raindrop and then transported with surface runoff. There is a direct relationship between the soil loss and runoff volume. The major factors, which affect the amount of soil erosion in large extent, can be summarized as:

### 25.2.1 Climatic Factors

The climatic factors which affect the soil erosion are rainfall characteristics, atmospheric temperature and wind velocity. Rainfall characteristic is one of the most effective factors among them which includes amount, intensity, frequency and duration of rainfall. High rainfall intensity has bigger raindrop size, which has higher kinetic energy and more erosive power. Rainfall characteristics, which produce more runoff amount and rate, cause more erosion. Frequent rain maintains the soil moisture in a desirable range, which reduces infiltration capacity of soil and thus produces more runoff. A uniform distribution of rainfall throughout the year always reduces the total erosion by maintaining the soil moisture within the optimum range for good vegetation over the land surface.

### 25.2.2 Soil Factors

Soil erosion has direct relation with runoff. All soil properties responsible for higher infiltration rate like low bulk density, high porosity, large particle size, low moisture content, granular structure, etc. cause low runoff and soil erosion. High cohesive force between soil particles results into lower detachability of particles as in case of clayey soil. On the other hand, large particle size reduces transportation of the detached particles.

### 25.2.3 Topographic Factors

The land slope and length of the slope are the two topographic factors, which strongly influence the soil erosion. As the length of the land in the direction of slope increases, detachment of soil particles goes on increasing. When the slope of the land is doubled, the particle size that can be transported increases 16 times.

### 25.2.4 Vegetation

Vegetation plays an important role in reducing soil erosion. In the presence of good vegetation cover, soil erosion can be reduced significantly. The vegetation helps in reducing soil erosion in the following ways:

- Kinetic energy of falling raindrops is absorbed by the canopy, i.e. leaves and stems of the vegetation, which reduces the detachment of soil particles.
- Some of the rainfall is intercepted by leaves and stems in case of trees and reduces runoff.
- Vegetation physically obstructs the velocity of flowing runoff.
- Fine roots of the vegetation bind soil particles.
- In the presence of good vegetation, evapotranspiration rate is faster; thus soil moisture reduces and infiltration increases.
- Decayed roots increase porosity of soil.
- Growth of certain soil fauna like earthworm is accelerated in presence of vegetation. These soil faunas pulverize soil and increase porosity.

## 25.3 Modelling Soil Erosion and Transport Within a Catchment

A variety of models have been developed related to soil erosion processes and sediment transport in a catchment area. Based on erosion processes and utility, these have been categorized as statistical, process-based, physically based and spatially distributed models. Linda May and Chris Place (2005) gave a comprehensive account on various soil erosion and transport models that included those published by Yamamoto and Anderson (1967), Beasley (1977), Foster et al. (1977), Foster (1982), Singer and Walker (1983), Beven et al. (1984), Moore and Burch (1986), O'Loughlin (1986), Sharma and Correia (1987), Oslin et al. (1988), Vertessy et al. (1990), Govindaraju and Kavvas (1991), Hairsine and Rose (1992a, b), Carling et al. (1993), Moore et al. (1993), Moussa and Bocquillon (1994), Nearing et al. (1994), Rose (1994) and Quinn et al. (1995). Existing models range from having mathematically and conceptually simple approaches to complex models predicting a wide range of erosion-related processes and complex applications. Some of these are reviewed below in relation to their suitability for use at the catchment scale.

### 25.3.1 Statistical Models

The Universal Soil Loss Equation (USLE) is probably the most widely used and simple statistical model in the world dealing with soil loss due to water erosion along a sloping land. It is based on a series of empirical relationships that describe the effect of a range of factors on soil erosion. These include rainfall intensity, soil erodibility, length and steepness of slope, land cover and land management practices. When combined with a sediment delivery ratio, the potential sediment loading to the catchment drainage system can be estimated (Dickinson et al. 1986; Hession and Shanholtz 1988).

The approach is simple and easy to apply but has a number of limitations. Some important limitations have been outlined by Zhang et al. (1996) which include the following: (1) gullying or mass movement (sediment load) is not considered; (2) deposition of sediment cannot occur within the area under consideration; and (3) the slope length and steepness factors must be determined only on the area that is contributing to runoff. Further, it cannot be extrapolated in applications outside the range of conditions for which it was developed (Foster 1982; Millington 1986; Nearing et al. 1994).

Despite its problems, Zhang et al. (1996), however, concluded that the USLE model could be used to evaluate erosion at any point within a catchment. Many workers have used USLE as base in developing several agricultural transport models. For example, models such as CREAMS [Chemicals, Runoff and Erosion from Agricultural Management Systems] (Knisel 1980), GLEAMS [Groundwater Loading Effects of Agricultural Management Systems] (Leonard et al. 1987) and EPIC [Erosion-Productivity Impact Calculator] by Williams et al. (1984) used the USLE, or modifications of it, to estimate soil erosion and to quantify the sediment yield from catchments (Ferro and Minacapilli 1995; Ferro 1997; Kothyari and Jain 1997; Ferro et al. 1998; Di Stefano et al. 1999). Further, Spanner (1983) and Jain and Kothyari (2000) stated that when the USLE (or one of its variants) is combined with gridded data from a geographical information system (GIS), the predictive power of this approach is increased.

### 25.3.2 Process-based Models

Process-based models consider all the three major processes or stages of soil erosion caused by runoff across a catchment. These are sheet erosion (removal of soil over the entire surface), rill (small channel) erosion and gully/ravine (large channel) erosion. Zhang et al. (1996), while reviewing a range of these models, concluded that these models could not be applied with any confidence to conditions not reflected in the calibration dataset of the original model because of the reason that the range of processes used in the model were oversimplified and many complex feedback processes were ignored. These models also ignore many important spatial interactions that occur within a catchment.

### 25.3.3 Physically-based Models

Physically-based models, such as the Water Erosion Prediction Project [WEPP] Model (Laflen et al. 1991; Flannagan and Nearing 1995; Sharma et al. 1995), are intended to represent the essential mechanisms that control erosion, such as interrill and rill erosion. These models also take into account other parameters such as plant growth, vegetation and climate. Zhang et al. (1996) explaining merits and demerits of these models concluded that these are very powerful models because they can represent a synthesis of the individual components that affect erosion, including the complex interactions between various factors and their temporal variability. However, it is generally very difficult to obtain the unique set of optimal parameters needed to drive physically based models using parameter estimation methods. It is also very difficult to apply such a model that has been developed for a small area to a much larger area, such as a catchment.

### 25.3.4 Spatially Distributed Models

This category of models has been explained through many names such as TOPMODEL, TOPOG, AGNPS and Erosion Hazard Index models. These require a spatial representation of the topography of an area using three-dimensional coordinates as input (Zhang et al. 1996). Long back such a dataset was expressed as a Digital Elevation Model (DEM) by Miller and Laflamme (1958). These data provide detailed information on attributes that affect erosion and sediment transport rates, such as elevation, slope, aspect, drainage, flow direction, flow accumulation and contributing catchment area. In addition to the above, spatial information on soil type, land cover, rainfall and other catchment characteristics may be added as separate data layers, where these are required to calculate erosion rates.

The data provided by the GIS-based DEM can be loosely linked to a soil erosion model, as in the case of the Areal Nonpoint Source Watershed Environment Response Simulation (ANSWERS) model described by Beasley (1977), or fully integrated with a soil erosion model, as in the Limburg Soil Erosion Model (LISEM) given by De Roo et al. (1996). However, it is important to take into account that the underlying erosion models may not have been developed for application at the wider scale of a catchment. The USLE, for example, was developed for use at the plot scale and is not applicable to large areas (such as entire catchments) unless it has been extended to include a sediment delivery term (Ferro and Minacapilli 1995). This is because the original equation does not simulate sediment deposition. CREAMS model applies only to field-sized areas (Donigian and Huber 1991).

The choice of soil erosion model to link to a GIS-based, spatially distributed model of soil erosion and sediment transport can have a big impact on the accuracy of the output. There is a general assumption that complex, physically based models should provide better results than the simpler soil erosion models; however, De Roo (1998) confirmed that this was not so and compared the output from a range of distributed models of varying complexity with detailed field measurements of soil erosion. The study showed that predictions from spatial models based on relatively simple erosion models, such as the USLE, compare favourably with those based on more complex, physically based, soil erosion models at the local scale. When scaled up to the catchment level, the simpler models tended to give better results than the more complex models. This is because the latter require a large number of input variables (e.g. infiltration capacity, soil texture, soil moisture content) to be estimated or measured at the catchment scale, introducing many sources of uncertainty and error into the simulations (Jetten et al. 1999). De Roo (1998) further recommended that simple models that encompass only the dominant processes operating in a catchment should be further developed in this context.

## 25.4 Commonly Used Models

Some of the commonly used soil loss models include the Universal Soil Loss Equation, Soil Loss Equation Model for South Africa, Soil Erosion Model for Mediterranean Regions, Modified Universal Soil Loss Equation, Morgan-Morgan-Finney Model, Water Erosion Prediction Project Model, Quasi-Three-Dimensional Runoff Model for soil erosion, and The LISEM Model. These are discussed here in brief.

### 25.4.1 Universal Soil Loss Equation

There have been sincere attempts to develop soil loss estimation models, beginning from the 1960s of the twentieth century. The most effective model on the soil loss was presented by Wischmeier and Smith (1965) and improved by Wischmeier and Smith (1978). The model is popularly known as Universal Soil Loss Equation (USLE). This also opened a new chapter for research in this field and formed the basic structure of most of the soil loss models after this period. Ghanshyam Das (2000) stated that USLE appears to have been based on the Musgrave (1947) equation with necessary modifications. The USLE can be written as

$$A = RKLSCP$$

where

$A$ = estimated gross soil erosion (Mg ha$^{-1}$ year$^{-1}$)
$R$ = rainfall erosivity factor (jouleha$^{-1}$ year$^{-1}$or Mg-m-cm/ha-s)
$K$ = soil erodibility factor (Mg ha$^{-1}$unit$^{-1}$ of R)
$S$ = slope factor
$L$ = slope length factor
$C$ = crop management or vegetative cover factor
$P$ = supporting conservation practice factor

#### 25.4.1.1 Rainfall Erosivity Factor (*R*)

It is the measure of erosive capability of rainfall. It is the function of rainfall intensity and amount of rainfall. Many indices have been developed for the R. $EI_{30}$ index is most commonly used to represent the rainfall erosivity:

$$\mathrm{EI}_{30} = \sum_{i=1}^{n} E_i.I30_i / 100$$

where

$E_i$ = total kinetic energy of ith rainstorm
$I_{30i}$ = maximum intensity of ith rainstorm for 30 min duration
$n$ = number of rainstorm having rainfall more than 2.5 mm

$$\mathrm{E} = \Sigma KE.\mathrm{P}$$

where

KE = kinetic energy of segment of rain having uniform intensity
$P$ = amount of segment of rain having uniform intensity

$$\mathrm{KE} = 210.3 + 89 \log I$$

where

$I$ = rainfall intensity, cm $h^{-1}$

Calculation of $EI_{30}$ index requires rainfall data recorded from automatic recording rain gauge. A simplified equation has been developed by Hardaha et al. (1996) for the Malwa region of Madhya Pradesh. The equation uses total erosive rainfall (ER), i.e. storms having rainfall more than 12.5 mm and expressed as

$$\mathrm{EI}_{30} = 9.524\,\mathrm{ER} + 5.60$$

Rambabu et al. (1978) prepared an iso-erodent map of India from which the approximate erosion value can be obtained directly.

#### 25.4.1.2 Soil Erodibility Factor

The soil erodibility factor ($K$) converts units of $R$ to amount of erosion. It is the average soil loss from a standard plot with 9% slope, 22.1 m long kept fallow by periodic tillage up and down the slope per unit value of $R$. Value of K can be determined experimentally for any soil on standard plot. The value of $K$ is estimated based on soil characteristics as given by Foster et al. (1981):

$$K = 2.8 \times 10^{-7} M^{1.14} (12 - a) + 4.3 \times 10^{-3} (b - 2) + 3.3 \times 10^{-3} (c - 3)$$

where

$K$ = soil erodibility, Mg/ha/(MJ-mm/ha-h)
$M$ = particle size parameter (%silt + %very fine sand) (100-%clay)
$a$ = organic matter content, %

**Table 25.2** Values for soil erodibility factor for different types of soil

| | *K* based on percent organic matter in soil | | |
|---|---|---|---|
| Soil type | 0.5% | 2.0% | 4.0% |
| Fine sand | 0.36 | 0.31 | 0.22 |
| Very fine sand | 0.94 | 0.81 | 0.63 |
| Loamy sand | 0.27 | 0.22 | 0.18 |
| Loamy very fine sand | 0.98 | 0.85 | 0.67 |
| Sandy loam | 0.60 | 0.54 | 0.42 |
| Very fine sandy loam | 1.05 | 0.92 | 0.74 |
| Silt loam | 1.07 | 0.94 | 0.74 |
| Clay loam | 0.63 | 0.56 | 0.47 |
| Silty clay loam | 0.83 | 0.72 | 0.58 |
| Silt clay | 0.56 | 0.51 | 0.43 |

$b$ = soil structure code (very fine granular, 1; medium or coarse granular, 3; blocky, platy or massive, 4)
$c$ = profile permeability class (rapid, 1; moderate rapid, 2; moderate, 3; slow to moderate, 4; slow, 5; very slow, 6)

Table 25.2 can be used when organic matter content and textural class of the soil are known.

#### 25.4.1.3 Slope Length Factor (*L*)

The length of the slope on which the overland flow occurs affects the rate of soil erosion. On large slope length, there is a higher concentration of overland flow and also a higher velocity of flow, which triggers a higher rate of soil erosion. Slope length factor is defined as the ratio of soil loss from a given slope length to that from a land having slope length equal to 22.1 m, if all other conditions remain unchanged. Mathematically it can be expressed as

$$L = \left(L_p / 22.1\right)^M$$

where

$L_p$ = actual unbroken length of slope ($m$)
$M$ = an exponent equal to:
 0.5 for slope >5%
 0.4 for slope 4–5%
 0.3 for slope <3%

#### 25.4.1.4 Slope Gradient Factor (S)

Soil erosion is greatly influenced by slope of the land. On steep slope, the flow velocity of runoff is high resulting in increased scouring, cutting and transportability of soil. Soil gradient factor is defined as the ratio of soil loss from a given degree

**Table 25.3** Crop management factor for some crop/grass covers at few ICAR centres in India

| Crop | Centres | | |
|---|---|---|---|
| | Kota | Agra | Lucknow |
| Green gram | 0.39 | – | 0.45 |
| Gram | 0.54 | – | – |
| Groundnut | 0.41 | – | 0.42 |
| Soybean | 0.42 | – | – |
| Cluster bean | 0.59 | 0.42 | – |
| Cluster bean + pigeon pea | – | – | 0.35 |
| Maize | 0.50 | – | – |
| Sorghum | 0.62 | 0.64 | – |
| Sorghum + pigeon pea | 0.33 | – | 0.28 |
| Sorghum + gram | – | 0.32 | – |
| Pearl millet | – | 0.61 | – |
| Sesamum | – | 0.51 | 0.39 |
| Natural vegetation | 0.14 | – | – |
| Grass (*Cynodon dactylon*) | 0.22 | – | – |
| Grass (*Dichanthium annulatum*) | 0.01 | 0.13 | – |

Source: Gurmel Singh et al. (1981)

of slope to that from land having 9.0% slope, if all other conditions remain unchanged. Empirically it can be expressed as

$$S = \left(0.43 + 0.30\,s + 0.043\,s^2\right) / 6.613$$

where

$s$ = slope of the filed in percentage

#### 25.4.1.5 Crop Management Factor (*C*)

Crop management factor or vegetative cover factor is defined as the ratio of soil loss from a land with given crop rotation and cover to that from a land clean tilled and kept permanent fallow. Vegetative cover acts many ways in reducing the soil erosion as discussed earlier. Gurmel Singh et al. (1981) compiled value of factor $C$ for some of the vegetative covers/crops grown at different centres (Table 25.3).

#### 25.4.1.6 Conservation Practice Factor (*P*)

A bare fallow land surface causes maximum soil erosion, especially when it is cultivated along the slope. Conservation practice factor is the ratio of soil loss from a land having specified conservation practice to the land ploughed up and down the slope, if all other conditions remain unchanged. Any practice adopted to reduce runoff amount and its velocity shall reduce the soil erosion. Table 25.4 presents value of $P$ for Indian conditions.

**Table 25.4** Values of conservation practice factor for different types of conservation practice

| Land slope (%) | Conservation practice factor | | |
|---|---|---|---|
| | Contour cultivation | Contour strip cropping | Bench terracing |
| <1 | 0.80 | – | – |
| 1–2 | 0.60 | 0.30 | – |
| 2–4 | 0.60 | 0.25 | – |
| 4–7 | 0.50 | 0.25 | – |
| 7–12 | 0.60 | 0.30 | – |
| 12–18 | 0.70 | 0.35–0.40 | – |
| >18 | 0.80–0.90 | 0.40–0.45 | 0.28 |

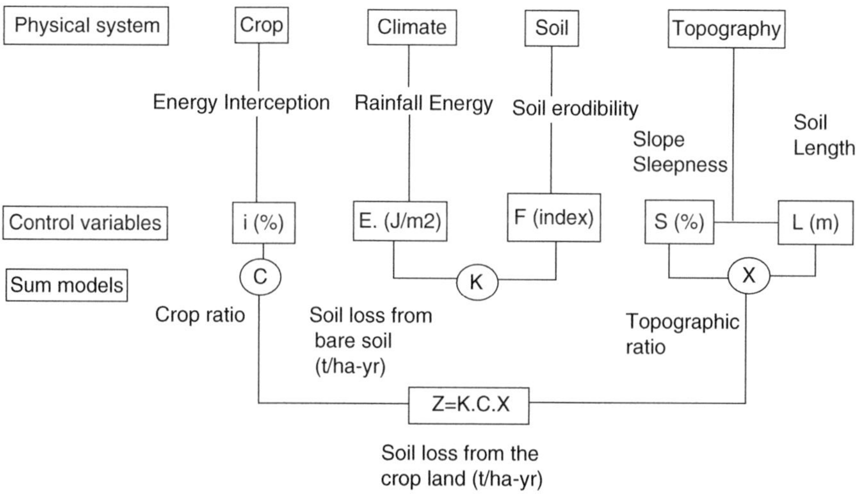


**Fig. 25.1** Framework of SLEMSA model (Elwell and Stocking 1982)

### 25.4.2 Soil Loss Equation Model for South Africa (SLEMSA)

SLEMSA was developed by Elwell (1978) for the southern region of Africa, and it was a modification over USLE. This model was designed to predict mean annual soil loss, rising from sheet erosion on area of arable land. Framework of SLEMSA is presented in Fig. 25.1. Bhargav (1999) has modified the SLEMSA model for Indian conditions for conservation practices in use by incorporating conservation practice factor (P). The modified model is $Z = \mathrm{KC} \times P$.

### 25.4.3 Soil Erosion Model for Mediterranean Regions (SEMMED)

A soil erosion model SEMMED (Soil Erosion Model for Mediterranean Regions) was developed for the test site Ardèche, France, by De Jong (1997). It comprises several modules, each of which describes a part of the erosion process such as soil

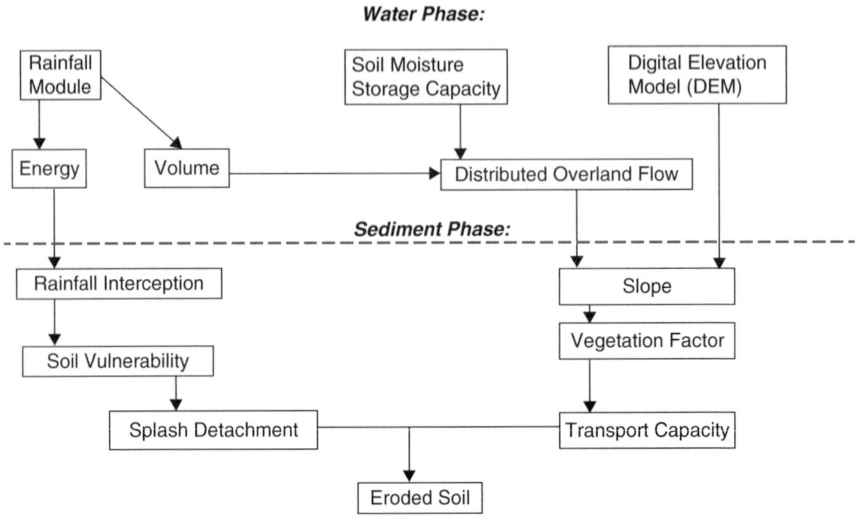


**Fig. 25.2** SEMMED Model (Source: De Jong 1997)

particle detachment, moisture storage in the top soil and transport of soil particles by overland flow. SEMMED uses (multi-temporal) Landsat TM images to account for vegetation properties, and it uses a digital terrain model in a GIS to account for topographical properties. Spectral vegetation indices allow a pixel-by-pixel assessment of vegetation properties, and the multi-temporal approach enables the assessment of the change of vegetative cover in one growing period. Figure 25.2 shows flow chart of the model.

### 25.4.4 Modified Universal Soil Loss Equation (MUSLE)

USLE was modified by Williams (1975) for predicting sediment yield by replacing its rainfall erosivity factor with runoff factor. The model can estimate sediment yield on a per-storm basis against the average soil loss on annual basis. The MUSLE is

$$Y = 11.8\left(Q q_p\right)^{0.56} \text{ KLSCP}$$

where

$Y$ = sediment yield from an individual storm
$Q$ = storm runoff volume
$q_p$ = peak runoff rate

Estimation of sediment yield from very large watershed is not very accurate due to variations in climatic factors, soil characteristics, land slope, crop management, erosion control practices and watershed hydraulics within the watershed area. Such watershed is divided into sub-watersheds of less than 25 $km^2$, and sediment yield can be computed using routing model as

$$\mathrm{RY} = \sum_{i=1}^{n} Y_i e -^{\mathrm{BTi}} \left(D50_i\right)^{0.5}$$

where

RY = sediment yield from the entire watershed, *t*
$Y_i$ = sediment yield from *i*th sub-watershed, t
$B$ = routing coefficient
$T_i$ = travel time from sub-watershed *i* to the watershed outlet, h
$D50_i$ = median particle diameter of the sediment for sub-watershed *i*, mm

Das and Chouhan (1990) observed that the value of *B* is equivalent to 1/*K* where *K* is the storage coefficient.

### 25.4.5 Morgan-Morgan-Finney Model

Morgan et al. (1984) developed a model for estimating annual soil loss from field-sized area on hill slopes. Inputs and flow chart of the model are illustrated in Fig. 25.3. For determination of annual rate of soil loss, the model compares the prediction of splash detachment and transport capacity of the overland flow. The lower of these two is considered as annual rate of soil loss. Some of the limitations of the model are:

- The model is more sensitive to change in the annual rainfall and soil parameters, when erosion is transport limited and also sensitive to changes in rainfall interception and annual rainfall, when erosion is detachment sensitive.
- It requires precise information on rainfall and other associated parameters, for having accurate prediction.
- This model cannot be employed for predicting the sediment yield from the drainage basin.
- Like USLE, it is also not suitable for predicting the soil loss, resulting from an individual storm.

### 25.4.6 WEPP Model

The Water Erosion Prediction Project (WEPP) Model (Nearing et al. 1989) has capability of predicting spatial and temporal distribution of net soil loss/gain for the entire hill slope for any period of time. It contains its own process-based hydrology, water balance, plant growth, residue decomposition and soil consolidation models

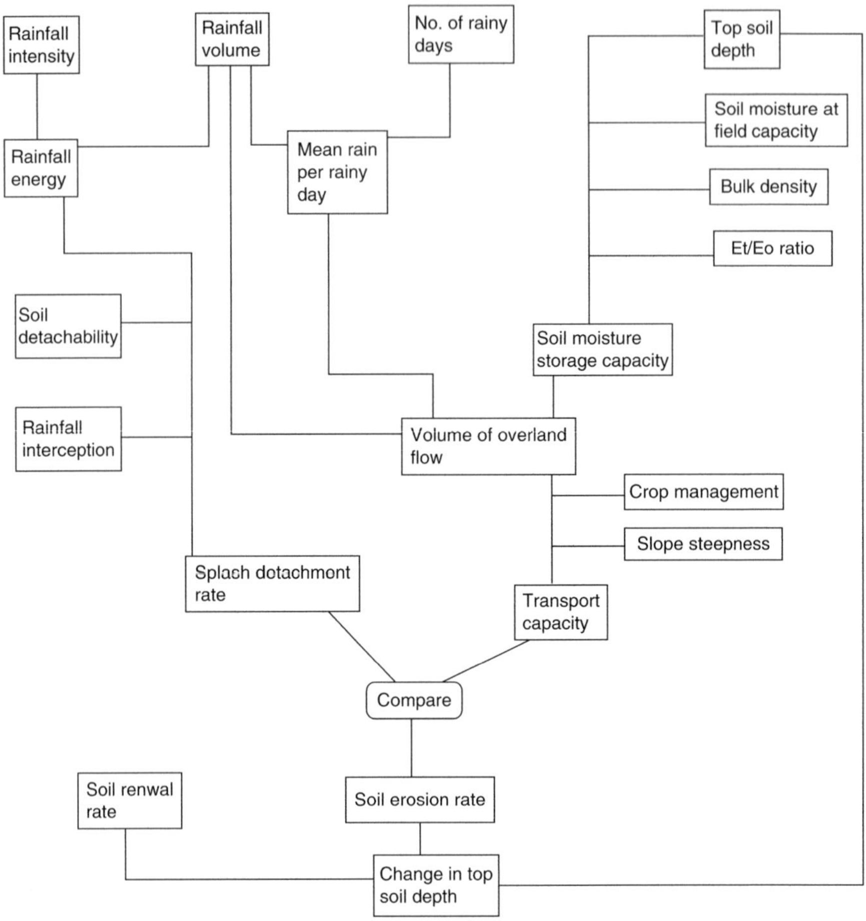


**Fig. 25.3** Morgan-Morgan-Finney Model for soil erosion (Source: Based on Morgan et al. 1984)

as well as a climatic generator and many other components that broaden its range of usefulness. The basic equation used for estimation of erosion from land is represented as

$$\frac{dG}{dx} = D_f + D_i$$

where

$G$ = sediment concentration
$x$ = distance down slope
$D_i$ = inter-rill erosion
$D_f$ = rill erosion

$$D_i = K_i.I_e.\sigma_{ir}.\mathrm{SDR}_{\mathrm{rr}}.F_{\mathrm{nozzle}}.(\mathrm{Rs}/W)$$

where

$K_i$ = inter-rill erodibility
$I_e$ = effective rainfall intensity
$\sigma_{ir}$ = inter-rill runoff rate
$\mathrm{SDR}_{\mathrm{rr}}$ = sediment delivery ratio
$F_{\mathrm{nozzle}}$ = adjustment factor to account for sprinkler irrigation nizzle impact variation
Rs = rill spacing
$W$ = width of the rill

$$D_f = Dc(1 - G/Tc)$$

where

$Dc$ = rill detachment capacity = $Kr\ (\tau_f - \tau_c)$
$Tc$ = transport capacity of flow in rill
$Kr$ = rill erodibility of soil
$\tau_f$ = flow shear stress
$\tau_c$ = critical shear stress

Tiwari et al. (2000) compared the WEPP predictions with the measured natural runoff plot data and found that the model efficiency is 0.71% in terms of annual soil loss with average magnitude of error 2.01 kg m$^{-2}$. It was concluded that WEPP is comparable with USLE and MUSLE.

### 25.4.7 Quasi-Three-Dimensional Runoff Model for Soil Erosion

Victor Demidov (2001) used the Quasi-Three-Dimensional Runoff Model for soil erosion modelling. The developed soil erosion model allows to simulate the temporal and spatial variations in erosion by raindrop impact and overland flow, sediment transport and deposition.

#### 25.4.7.1 Structure of the Model

The Quasi-Three-Dimensional Runoff Model of Rainfall Runoff Formation – a physically based model of rainfall runoff formation – is based on using differential equations which describe the processes of overland, groundwater, subsurface, channel flow as well as vertical moisture transfer in soil. The catchment is represented in the horizontal plane by rectangular grid squares. The main channel and the tributaries of different orders are represented by the boundaries of grid squares.

The model describes the following processes:

- Vertical moisture transport in the unsaturated zone (the one-dimensional Richard's equation is used; the calculations are carried out for each grid square of hill slope)
- Groundwater flow and the interaction of surface and groundwater on the hill slope and in the river channel (the two-dimensional Boussinesq equations are used)
- Overland flow (the two-dimensional kinematic wave equations are applied)
- Unsteady flow in the river network (the one-dimensional kinematic wave equations are used)

The organization of the interaction between components of the hydrological modelling system allows taking feedback into account. Coupling of the calculations of the vertical moisture transport with the overland and groundwater flow is accomplished by means of a special procedure.

#### 25.4.7.2 Modelling Soil Erosion and Sediment Transport in the River Basin

A soil erosion and sediment transport model was developed as a separate block of the hydrological modelling system. The soil erosion model describes the temporal and spatial variations of the soil erosion and the sediment transport in the river basins during flood events (erosion by raindrop impact and overland flow, sediment transportation and deposition).

The erosion rate by raindrop impact, $D_r$ (kg m$^{-2}$ s$^{-1}$), is expressed by the following equation:

$$D_r = K_r K_s i F_r R^{\beta}$$

where

$K_r$ = soil erodibility factor for erosion by raindrop impact
$K_s$ = fraction of bare soil
$i$ = ground surface slope
$R$ = rainfall intensity (çm/s)
$\beta$ = an exponent
$F_r$ = the factor reflecting the influence of the water depth on erosion by raindrop impact that is expressed as

$$F_r = \exp\left(1 - hD^{-1}\right) \text{ if } h > D$$
$$= 1 \qquad \text{if } h \le D$$

where

$h$ is the flow depth (m) and $D$ is the median diameter of raindrops that is determined from $D = 0.0193\ R^{0.182}$.

The erosion rate by overland flow impact, De (kg m$^{-2}$ s$^{-1}$), is calculated as

$$\begin{aligned} \mathrm{De} &= \mathrm{Ke}\left(\tau / \tau_c - 1\right) \quad \text{if}\ \tau > \tau_c \\ &= 0 \qquad\qquad\qquad\quad \text{if}\ \tau \le \tau_c \end{aligned}$$

where

Ke is the overland flow soil erodibility coefficient.
$\tau$ is the shear stress (kg m$^{-2}$ s$^{-1}$).
$\tau_c$ is the critical shear stress, which is taken to be

$$\tau_c = \rho\, g\, i\left(\mathrm{n}i^{-0.5} V_p\right)^{-1.5}$$

where

$\rho$ is the water density (kg/m$^3$).
$g$ is the acceleration of gravity (ms$^{-2}$).
$n$ is the Manning roughness coefficient.
$V_p$ is the pickup velocity (m/s) that is determined by the equation

$$V_p = 1.14\left(g\, a\, d\right)^{0.5}$$

where

$a$ (= $P_T\,\rho^{-1}-1$) and $P_T$ is the sediment density (kf/m$^3$).
$d$ is the grain diameter (m).

The sediment transport capacity, $G_T$ (kg m$^{-1}$ s$^{-1}$), is calculated by means of the Engelund-Hansen equation:

$$G_T = 0.04\left(V\,V*^2\,P_T\right) / \left(\psi\, a\, g\right)$$

where

$V$ is the flow velocity (m/s).
$V*$ is the shear velocity (m/s).
$\psi$ is the criterion which is $\psi = a\, d\, h^{-1}\, i^{-1}$.

The sediment transport by the overland flow is described by two-dimensional sediment continuity equation:

$$\frac{\partial(\mathrm{hC})}{\partial t} + \frac{\partial\left(G_x\right)}{\partial x} + \frac{\partial\left(G_y\right)}{\partial y} = E$$

$$E = -(1-\varepsilon)P_T \frac{\partial(z)}{\partial t}$$

where

$C$ is the sediment concentration (kg m$^{-3}$).
$G_x$ and $G_y$ are the sediment transport rates in the $x$ and $y$ direction, respectively.
$\varepsilon$ is the soil surface porosity.
$z$ is the soil surface elevation (m).
$E$ is the erosion or deposition rate on surface slope (kg m$^{-2}$ s$^{-1}$).

Sediment routing in channels is described by the one-dimensional sediment continuity equation. Numerical integration of these equations is carried out by an implicit finite-difference scheme.

### 25.4.8 LISEM

The LISEM (De Roo et al. 2001) is one of the first examples of a physically based model that is completely incorporated in a raster geographical information system. Incorporation means that there are no conversion routines necessary; the model is completely expressed in terms of the GIS command structure. Furthermore, the incorporation facilitates easy application in larger catchments, improves the user-friendliness and allows remotely sensed data from airplanes or satellites to be used. If required, the model can be linked easily with another GISs. Processes incorporated in the model are rainfall, interception, surface storage in micro-depressions, infiltration, vertical movement of water in the soil, overland flow, channel flow, detachment by rainfall, detachment by overland flow and transport capacity of the flow. Also, the influence of tractor wheelings, small paved roads (smaller than the pixel size) and surface sealing on the hydrological and soil erosion processes is taken into account.

After rainfall begins, some of it is intercepted by the vegetation canopy until such time as the maximum interception storage capacity is met. Besides interception, direct through fall and leaf drainage occur, which, together with overland flow from upslope areas, contribute to the amount of water available for infiltration. The amount of water remaining after infiltration begins to accumulate on the surface in micro-depressions. When a predefined amount of depressions is filled, overland flow begins. Overland flow rates are calculated using Manning's n and slope gradient, with a direction according to the aspect of the slope. When rainfall ceases, infiltration continues until depression storage water is no longer available. Soil detachment and transport can both be caused by either raindrop impact or overland flow. Whether or not a detached soil particle moves depends upon the sediment load in the flow and its capacity for sediment transport. When water and sediment reach an element with a channel, they are transported to the catchment outlet. Sedimentation within a channel appears when the transport capacity has been exceeded.

When there are no sufficient field measurements available, the distribution of a desired input variable can be derived from digitized soil or land use maps. A raster-based GIS is the ideal tool to serve needs and fulfil requirements associated with the DEM and the geo-statistical interpolation techniques. Further advantages of using a GIS are:

1. The possibilities of rapidly producing modified input maps with different land use patterns or conservation measures to simulate alternative scenarios
2. The ability to use very large catchments with many pixels, so the catchment can be simulated with more detail
3. The facility to display the results as maps

A series of maps can be produced showing the variation with time of spatial patterns of soil erosion, sedimentation and runoff over the catchment. These maps can be compared by subtraction to yield maps indicating how erosion or sedimentation might be affected by certain control measures within the catchment, or they can be viewed successively to create a video of the modelled process. Runoff can also be displayed as an overlay on the landform surface.

The main advantage of incorporating models in GIS is that the "source code" of the model then resides on the comprehensible abstraction level of one or two lines of source code, a GIS command, per process (e.g. interception, infiltration and sediment routing). Such a high level of abstraction simplifies model modification, maintenance and reusability of parts of the model in other models. The current implementation of LISEM is less than 200 lines (exclusive comments).

## 25.5 Input for LISEM

LISEM needs a number of input files and maps to run. These inputs are described below:

*Rainfall file*: Data from multiple rain gauges can be entered in an input data file. A map is used as input to define for each pixel which rain gauge must be used. For every time increment during the simulation of a storm, the model generates a map with the spatial distribution of the rainfall intensity. Thus, the model allows for spatial and temporal variability of rainfall. In the future, this approach allows for the input such as radar data indicating rainfall intensity patterns changing in space and time, e.g. to simulate a thunder storm which moves over a catchment.

*Tables for the soil water model*: Within the catchment, soil profiles are defined. The vertical soil water movement is simulated by subdividing a soil profile in a user-defined number of layers. For each characteristic soil horizon, the measured K-t3-h relations are read from the horizon-specific tables.

*Maps of relevant topographical, soil and land use variables:* To run LISEM, a number of maps are needed in the PCRaster format which include:

- A group of maps which describe the catchment morphology:
  - An "area.map", in which the main catchment is defined
  - An "id.map", which defines the spatial rainfall pattern
  - A map with the locations of the main outlet and sub-catchment outlets
  - A map with the "Local Drain Direction", which refers to aspect
  - A map with slope gradient
  - A map with the Manning's n for overland flow
  - A map with slope gradient of the main channels
  - A map with the Manning's n for channel flow
  - Two maps which describe the channel morphology
  - A map with the location and width of roads
- A map with the location and width of wheel tracks from tractors
- A group of maps needed for the soil water sub-model:
  - Map with the soil profile types, referring to the conductivity tables
  - A similar map but then for profiles under wheel tracks
  - Maps with the initial soil metric suction for each soil layer
- A group of maps with soil and land use variables:
  - A map of the Leaf Area Index
  - A map with the soil coverage by vegetation
  - A map with the crop height
  - A map with the Random Roughness of the soil surface
  - A map with the aggregate stability of the soil
  - A map with the soil cohesion
  - A map with the soil cohesion of channels

*Command file*: When the model is run, the user is prompted for the selection of the catchment, the rainfall event, a few tuning parameters and the desired output. Alternatively, the user can specify this information in a command file. This interface empowers the user to:

- Select the catchment by specifying the directory of the topographical, soil and land use map database.
- Select the soil water model parameters by specifying the directory of the soil water tables. Separating the map database and the soil water tables permits optional sharing of the soil water tables between different catchments.
- Select the rainfall event by specifying the rainfall file; select the starting and ending time of the simulation.
- Select the overall simulation time step and the minimum time step for the soil water sub-model.
- Select a precision factor of the soil water sub-model.
- Select a number of parameters and coefficients used in the detachment and transport formulas, such as settling velocity of the soil particles and a splash delivery

ratio. If necessary, a few of these parameters could be used for calibrating the sediment part of the model.

- Select names of the output files, e.g. hydrograph files (main outlet and outlets of predefined sub-catchments), runoff maps at several times, soil erosion map and the "results" file with totals.

### 25.5.1 Output for LISEM

The results of the LISEM consist of:

- A text file with totals (total rainfall, total discharge, peak discharge, total soil loss, etc.)
- An ASCII data file which can be used to plot hydrographs and sedigraphs
- PCRaster maps of soil erosion and deposition, as caused by the event
- PCRaster maps of overland flow at desired time intervals during the event

### 25.5.2 Validation of LISEM

The model results are compared with observed data (validation). Statistical criteria determine the "goodness of fit". The model user has to decide whether the results are satisfactory. If so, the simulations end and the 'final results" are produced. If the validation is not satisfactory, there are several options:

- Modify the model.
- Recalibrate the model.
- Change the resolution (pixel size or simulation time step).
- Collect more data.
- Collect better data (measurement errors).
- Collect different data (other variables).

This procedure is repeated until satisfactory results are obtained.

There are various erosion process models available, and the use depends upon the data required in the model and the data available.

## References

Babu R, Tejwani KG, Agrawal MC, Bhusan IS (1978) Determination of erosion index and isoerodent map of India. Indian J Soil Conserv 6(1):1–12

Beasley DB (1977) Answers: a mathematical model for simulating the effects of land use and management on water quality. Unpublished PhD. thesis. Purdue University, West Lafayette. 266 pp

Beven KJ, Kirby MJ, Schoffield N, Tagge A (1984) Testing a physically-based flood forecasting model TOPMODEL for three UK catchments. J Hydrol 69:119–143

Bhargav KS (1999) A modified SLEMSA model for Naurar sub-catchment of Ramganga river. Unpublished master's thesis. Submitted to GBPAUT, Pantnagar
Carling PA, Glaister MS, Flintham TP (1993) Soil erosion and conservation on land cultivated and drained for afforestation. Hydrol Process 7:317–333
Das G (2000) Hydrology and soil conservation engineering. Prentice Hall of India, New Delhi
Das G, Chouhan HS (1990) Sediment routing for mountainous Himalayan region. Trans ASAE
De Jong SM (1997) DeMon – satellite based desertification monitoring in the Mediterranean Basin. www.geog.uu.nl/fg/demon.html
De Roo APJ (1998) Modelling runoff and sediment transport in catchments using GIS. Hydrol Process 12:905–922
De Roo APJ, Offermans RJE, Cremers NHDT (1996) LISEM: a single event physically based hydrological and soil erosion model for drainage basins: I theory, input and output. Hydrol Process 10:1119–1126
De Roo APJ, Wesseling CG, Cremers NHDT, Offermans RJE, Ritsema CJ, van Oostindie K (2001) Lisem: a physically-based hydrological and soil erosion model incorporated in a gis. www.odyssey.maine.edu/gisweb/spatd6/egis/eg94023.html
Victor Demidov (2001) Modeling soil erosion and sediment transport on watersheds with the help of quasi three-dimensional runoff model. www.epa.gov/OWOW/watershed/proced/demidov.html
Di Stefano C, Ferro V, Porto P (1999) Modelling sediment delivery process by stream tube approach. Hydrol Sci J 44:725–742
Dickinson WT, Rudra RP, Clark DJ (1986) A delivery ratio approach for seasonal transport of sediment. In: Hadley RF (ed) Drainage basin sediment delivery. Proceedings of the symposium of the international commission on continental erosion. IAHS Publication No. 159, Alberqueque
Donigian AS, Huber WC (1991) Modeling of nonpoint source water quality in urban and non-urban areas, EPA/600/3-91/039. US Environment Protection Agency, Athens
Elwell HA (1978) Modelling soil loss in southern Africa. J Agric Eng Res 23:111–127
Elwell HA, Stocking MA (1982) Developing a simple yet practical method of soil loss estimation. Trop Agric 59(1):43–48
Ferro V (1997) Further remarks on a distributed approach to sediment delivery. Hydrol Sci J 42:633–647
Ferro V, Minacapilli M (1995) Sediment delivery processes at basin scale. Hydrol Sci J 40:703–717
Ferro V, Porto P, Tusa G (1998) Testing a distributed approach for modelling sediment delivery. Hydrol Sci J 43:425–442
Flanagan DC, Nearing MA (1995) USDA-water erosion prediction project: hillslope profile and watershed model documentation, NSERL report No. 10. USDA-ARS National Soil Erosion Research Laboratory, West Lafayette. 47097-1196
Foster GR (1982) Modelling erosion processes. In: Hann CT, Johnson HP, Brakensick DL (eds) Hydrological modeling of a small watershed. American Society of Agricultural Engineers, St Joseph, pp 297–380
Foster GR, Meyer LD, Onstad CA (1977) An erosion equation derived from basic erosion principles. Trans Am Soc Agric Eng 20:678–687
Foster GR, McCool DR, Renard KG, Mobdenhaur WC (1981) Conversion of universal soil loss equation to SI metric units. J Soil Water Conserv 36:355–359
Govindaraju RS, Kavvas ML (1991) Modeling the erosion process over steep slopes; approximate analytical solutions. J Hydrol 127:279–305
Hairsine PB, Rose CW (1992a) Modeling water erosion due to overland flow using physical principles: 1, sheet flow. Water Resour Res 28:245–250
Hairsine PB, Rose CW (1992b) Modeling water erosion due to overland flow using physical principles: 2, rill flow. Water Resour Res 28:237–243
Hardaha MK, Kale VS, Nema RK (1996) Erosive rains and erosion index for Indore. Indian J Soil Conserv 24(3):193–196
Hession WC, Shanholtz VO (1988) A geographical information system for targeting nonpoint-source agricultural pollution. J Soil Water Conserv 1988:264–266
Jain MK, Kothyari UC (2000) Estimation of sediment yield using a GIS. Hydrol Sci J 45:771–786

Jetten V, De Roo A, Favis-Mortlock D (1999) Evaluation of field scale and catchment-scale soil erosion models. Catena 37:521–541
Knisel WG (1980) CREAMS a field scale model for chemical, runoff and erosion from agricultural management system. USDA Conservation Report No. 26
Kothyari UC, Jain SK (1997) Sediment yield estimation using GIS. Hydrol Sci J 42:833–843
Laflen JM, Lane LJ, Foster GR (1991) WEPP – a new generation of erosion prediction technology. J Soil Water Conserv 46:34–38
Leonard RA, Knisel WG, Still DA (1987) GLEAMS: groundwater loading effects of agricultural management systems. Trans Am Soc Agric Eng 30:1403–1418
May L, Place C (2005) A GIS-based model of soil erosion and transport. Fresh Water Forum 23:48–61
Meyer LD, Wischmeier WH (1969) Mathematical simulation of the process of soil erosion by water. Trans Am Soc Agric Eng 12:754–758
Miller CL, Laflamme RA (1958) The digital terrain model – theory and application. Photogramm Eng 24:422–433
Millington AC (1986) Reconnaissance scale soil erosion mapping using a simple geographical information system in the humid tropics. In: Siderius W (ed) Land evaluation for land use planning and conservation in sloping areas, pp. 64–81. International Workshop, Enschede, The Netherlands, 17–21 December 1984
Moore ID, Burch GJ (1986) Modeling erosion and deposition: topographic effects. Trans Am Soc Agric Eng 29:1624–1630
Moore ID, Gessle PE, Nielson GA, Peterson GA (1993) Soil attribute prediction using terrain analysis. Soil Sci Soc Am J 57:443–452
Morgan RPC, Morgan DDV, Finney HJ (1984) A predictive model for the assessment of soil erosion. J Agric Eng Res 30:245–253
Moussa R, Bocquillon C (1994) TraPhyC-BV: a hydrologic information system. Environ Softw 9:217–226
Musgrave GW (1947) The quantitative evaluation of factors in water erosion, a first approximation. J Soil Water Conserv 2:133–138
Nearing MA, Foster GR, Lane LJ, Finkner SC (1989) A process based soil erosion model for USDA water erosion prediction project. Trans ASAE 32(5):1587–1593
Nearing MA, Lane LJ, Lopes VL (1994) Modeling soil erosion. In: Lal R (ed) Soil erosion research methods. Soil and Water Conservation Society and St. Lucie Press, Ankeny, pp 127–156
O'Loughlin EM (1986) Prediction of surface saturation zones in natural catchments by topographic analysis. Water Resour Res 22:794–804
Oslin AJ, Westsmith RA, Morgan DS (1988) STREAMS: a basin and soil erosion model using CADD remote sensing and GIS to facilitate watershed management. In: Modeling agricultural, forest and rangeland hydrology. American Society of Agricultural Engineers, St Joseph, pp 470–477
Quinn PF, Beven KJ, Lamb R (1995) The ln(α/tanβ) index: how to calculate it and how to use it within the TOPMODEL framework. Hydrol Process 9:161–182
Rose CW (1994) Research progress on soil erosion processes and a basis for soil conservation practices. In: Lal R (ed) Soil erosion research methods. Soil and Water Conservation Society and St. Lucie Press, Ankeny, pp 159–178
Sharma KD, Correia JF (1987) An upland soil erosion model derived from the basic principles. Earth Surf Process Landf 12:205–210
Sharma PP, Gupta SC, Foster GR (1995) Raindrop-induced soil detachment and sediment transport from interrill areas. Soil Sci Soc Am J 59:727–734
Singer MJ, Walker PH (1983) Rainfall-runoff in soil erosion with simulated rainfall, overland flow and cover. Aust J Soil Res 21:109–122
Singh G, Babu R, Chandra S (1981) Soil loss prediction research in India, Bull T 120/D-9. CSWCR & TI, Dehradun
Spanner MA (1983) The use of pre-existing digital data base for soil erosion prediction, Proceedings, CERMA Energy Resource Management Conference. CERMA, San Francisco
Tiwari AK, Risse LM, Nearing MA (2000) Evaluation of WEPP and its comparison with USLE and RUSLE. Trans ASAE 43(5):1129–1135

Vertessy RA, Wilson CJ, Silburn DM, Connolly RD, Ciesiiolka CA (1990) Predicting erosion hazard areas using digital terrain analysis. In: Proceedings IASH/AISH international symposium on research needs and applications to reduce erosion and sedimentation in tropical Stepp land, Suva, Fuji, pp 298–308

Williams JR (1975) Sediment yield prediction with universal equation using runoff energy factor. In: Sediment yield workshop Oxford, Nov.1972. Present and prospective of technology for predicting sediment yield and sources. Proceeding: Oxford, USDA Sedimentation Lab

Williams JR, Jones CA, Dyke PT (1984) A modelling approach to determining the relationship between erosion and soil productivity. Trans Am Soc Agric Eng 27:129–144

Wischmeir WH, Smith DD (1965) Predicting rainfall erosion losses from cropland east of the rocky mountains, USDA Handbook No. 282. USDA, Washington, DC

Wischmeir WH, Smith DD (1978) Predicting rainfall erosion losses – a guide to conservation planning, USDA Handbook No. 537. USDA, Washington, DC

Yamamoto T, Anderson HW (1967) Erodibility indices for wildland soils of Oahu, Hawaii, as related to soil forming factors. Water Resour Res 3:785–798

Zachar DC (1982) Soil erosion. Scientific Publishing Co, Amsterdam

Zhang L, O'Neill AL, Lacey S (1996) Modelling approaches to the prediction of soil erosion in catchments. Environ Softw 11:123–133

# Gender Challenges and Socio-economic Strategies for Livelihood Security in Ravine Areas

# 26

Reeta Mishra, Yagya Dev Mishra, P. P. Singh, and B. P. S. Raghubanshi

**Abstract**

Ravines result not only in the loss of nonrenewable land and soil resources but also lead to other socio-economic and gender challenges along with livelihood problems especially for farm women. Many factors such as rainfall, available cropland, implements, fertilizers, and off-farm activities affect household food security in the ravines. The gender bias is an important cause of poverty in ravine areas because it limits the women's capacity to contribute to food production and economic growth. Due to illiteracy of women and lack of knowledge regarding the new technologies, they are unable to understand the technologies properly. Programs seeking to increase food production, ensuring stability of food supplies, generating employment, and promoting accessible food for all are the main elements of food security in these regions. Integrated agriculture with crop, animal husbandry, agriproduct processing, and agribusiness is the base of the livelihood and ecological security systems of the ravines, which is the foundation for nutritional food security systems. Hence, there is a need to create awareness and provide training to farm women and rural youth in ravine areas on the latest tools and techniques. Skill development is a key for improving household productivity, employability, and income-earning opportunities for rural women and for enhancing food security and promoting environmentally sustainable rural development and livelihoods. Reduction in drudgery of women would facilitate women to get involved in development works and decision-making process. Hence in ravine areas, periodic training programs should be organized to emphasize on educating farm women regarding recognition of muscular-skeletal disorders and importance of rest pauses and maintaining proper posture while performing agricultural activities. Transfer of technology through extension

R. Mishra (✉) · P. P. Singh · B. P. S. Raghubanshi
RVSKVV- Krishi Vigyan Kendra, Morena, Madhya Pradesh, India

Y. D. Mishra
Directorate of Extension Services, RVSKVV, Gwalior, Madhya Pradesh, India





J. C. Dagar, A. K. Singh (eds.), *Ravine Lands: Greening for Livelihood and Environmental Security*, https://doi.org/10.1007/978-981-10-8043-2_26

system should be planned and executed on need-based priorities of farm women with the objective to reorient the system. Women of the ravine areas should be sensitized on the importance of involving in income-generating activities as this will help them to be self-reliant. Moreover, higher adoption of the income-generating activities will enhance the income of farm women, which in turn lead to the sustainable livelihood and improved quality of life of the farm women of ravine areas.

## 26.1 Introduction

In India, ravines are spread over an area of about 4 million hectares (Mha) largely along the rivers and their tributaries. Serious ravine intrusions occur along the banks of Beas in Punjab; Chambal in Madhya Pradesh, Rajasthan, and Uttar Pradesh; Kali Sindh, Banas, Morel, and Gambhir in Rajasthan; Yamuna in Uttar Pradesh; and Mahi, Sabarmati, Narmada, and Tapti in Gujarat. Chambal region is the most important division of Madhya Pradesh which comprises three districts, namely, Morena, Bhind, and Sheopur. Various socio-economic indicators have been proposed with special reference to Chambal region (Table 26.1).

These studies are site-specific for Morena, which is one of the districts that forms a part of the Chambal ravine system in Madhya Pradesh in India. About 50% of the total geographical area is under cultivation, while 59% of the cultivable area in the

**Table 26.1** Socio-economic indicators of Chambal region

| Indicator | Unit | State of Madhya Pradesh | Chambal region |
|---|---|---|---|
| Population | No. (000) | 60,385 | 3574 |
| Rural population | Percentage | 73.33 | 78.45 |
| Density of population | Persons per 100 $km^2$ | 196 | 282 |
| Female population | No. per 1000 male | 927 | 845 |
| Literacy | Percentage | 64.08 | 60 |
| Geographical area | $Km^2$ | 307,450 | 16,140 |
| Forest area | $Km^2$ | 88,090 | 3430 |
| Agricultural sown area | Percent | 49 | 46 |
| Ratio of irrigated area to sown area | Percent | 28 | 45 |
| Production of food grain | Thousand Mg | 152,471 | 943 |
| Total irrigated area | $Km^2$ | 56,690 | 4000 |
| Production of food grains | Thousand Mg | 152,471 | 943 |
| Gross value of agricultural products | US$ per ha | 600 | 2500 |
| Gross value of agriculture production | US$ | 113,400 | 7150 |

Source: Mudgal (2005)

district is irrigated by various sources. Canal is the main source of irrigation accounting for 43% of the total irrigated area (Thomas et al. 2011). Agriculture is practiced in both *kharif* (rainy) and *rabi* (winter) seasons during the year. Pearl millet is the major *kharif* crop, while wheat and mustard are grown during *rabi* season. Of all the issues that influence livelihood security in ravine areas, none is more profound than gender. The gender bias is an important cause of poverty, because it limits the women's capacity to contribute to food production and economic growth.

## 26.2 Socio-economic and Gender Challenges in Ravine Areas

Environmental considerations affect virtually all aspects of social, cultural, and economic development. Ravines result not only in the loss of nonrenewable land and soil resources but also leads to other socio-economic and gender challenges along with livelihood problems especially for farm women. The socio-economic conditions of agricultural women reflect their status in the society (Sobha 2001). Some of the bottlenecks are listed below:

### 26.2.1 Poor Access to Land

Poor households depend on agriculture for both food and cash to sustain their standards of living (Carr 2008). It is not very easy for the households to attain both food and cash. In most cases, they tend to attain food other than cash, mainly due to lack of access to land. This may lead to insufficient farming practices that may not allow a farmer to sell the products to attain cash because for a sufficient farming to occur it requires a large space of land so that there can be more products that are produced by the farm. Normally the poor households can only afford to produce for household consumptions. Because of poor access to land, the households are not economic units as much as it is a social unit (Carr 2005).

### 26.2.2 Low Literacy Rate of Women

The major challenges faced by the rural women entrepreneurs are illiteracy, lack of vital information, fear to take risks, lack of experience and training, feeling of insecurity, rampant corruption, lack of infrastructure lack of finance, etc. In addition, there are structural constrains in the form of inequality, limited purchasing power, condemnation by local elite, etc. (Mansuri 2011).

Due to illiteracy of women and lack of knowledge regarding the new technologies, they are unable to understand the technologies properly. Due to low level of illiteracy, they are lacking of proper training and information regarding new technologies which are being developed in the agricultural sector.

### 26.2.3 Male Dominant Society

Per World Bank (2012), "Gender refers to the social, behavioral, and cultural attributes, expectations and norms that distinguish men and women. Gender equality refers to the extent to which men's and women's opportunities and outcomes are constrained—or enhanced—solely based on their gender." Women's contribution to agricultural production varies from country to country, crop to crop, and task to task (Fabiyi et al. 2007). The assumption that the only place for women is the kitchen and the belief that women are subordinate and their husband's property have not only impeded their status but also handicapped their contribution to wealth creation and economic development. Underlying the assumption is the lack of recognition that in the household, women have distinct rights and obligation as well as distinct functions with regard to economic activities. Their situation is determined by their multiplicity for the care and well-being of their families and communities, household tasks and farm and nonfarm income-generating activities that cannot readily be monetized (Marilee 2009).

Most of the activities in which women engaged in their livelihood strategies are not defined as "economically active" employment in national account systems, yet are crucial to the well-being of household members (FAO 2010). Much of women's work is also undervalued because it is typically under-remunerated and often confined to the domestic or household realm (FAO 2011). Caring for children, the elderly, and the ill, collecting water and fuel for cooking and heat, and maintaining households and preparing food are the responsibilities which are mostly taken up by women and girls (FAO 2011). Yet women's substantial contribution continues to be undervalued in conventional agricultural and economic analyses and policies, while men's contribution remains the central, often sole focus of attention. Various studies have concluded that social constitutional constraints faced by women in terms of equal access to economic opportunities and productive resources such as land, credit (loan) technology, and market information tend to make them even more vulnerable (FAO 2010). Morena district is a kind of male-dominant society, wherein women are not allowed to work independently and take up any entrepreneurial ventures. Women are not allowed to go outside and work without family support.

Women are considered domestic workers by nature. If a woman earns money or takes part in social activities, she doesn't only suffer from the double burden of working at home and outside the home; her income is likely to be considered supplementary income, even though her earnings may be larger than the husband's. This double burden stands as one of the causes contributing to undernutrition among children, because of women lacking sufficient time to prepare food and feed their kids (WFP and Tropmed 2010). In Madhya Pradesh and Orissa, women strictly carry out many nonfarm activities, while other jobs are exclusively undertaken by men (Pandey et al. 2002; Rath et al. 2002; Som et al. 2002). Women are rarely involved in enterprise management and in higher-level positions in the public sector. Generally, both men and women work in agriculture, but average female participation rates in the nonfarm sector are low compared to those for men (Coppard 2001).

Furthermore, women have limited access to a wide range of services, and the movement of women is still restricted in many societies (World Bank 2012).

Most rural household livelihoods are practiced by women as a strategy to survive, but those livelihoods are not regarded as feasible (Ellis 2000). The primary reason might be that women don't have equal resources as men do, for example, they don't have access to productive resources and assets such as land, capital, credit, labor, and education (Abdulai and Crole-Rees 2001). Lack of these resources might be hypothesized to be culturally and traditionally motivated. Additionally, women can't have access to inheritance such as land, livestock, and money, or any other important capitals, but men do. Similarly, lack of access to resources might contribute to women's livelihoods to be economically unviable (Ellis 2000; Abdulai and Crole-Rees 2001).

### 26.2.4 Lack of Access to Resources

Women lack in access to need-based resources like raw material, capital, land, and other subsidiary resources. Rigid procedures of government and formalities in obtaining financial resources hinder the growth of women entrepreneurs in Morena. Gender inequalities in household resource management and the extra burdens that women face because of their gender roles can limit women's ability to manage household resources. In some countries, women obtain rights to use land for household and personal crops through men. This may result in women not being able to produce viable agricultural products and strengthening their household livelihoods. Consequently, women may often lack credit and other resources to improve their livelihoods (FAO 2003). Lack of access to resources coupled with gender suppressive tendencies because of tradition could lead to high levels of poverty among women (Wombeogo 2007).

### 26.2.5 Lack of Motivation and Support from Family Members

Lack of motivation and moral support from family members is the first and foremost factor responsible for hindering the growth of women entrepreneurs. When women want to start an enterprise, she will not get the required support from her family and other organization.

### 26.2.6 Lack of Suitable Infrastructural and Marketing Facilities

Lack of proper storage facilities in ravine areas at reasonable price and market facilities are the important factors which are hindering the growth of entrepreneurs.

These entrepreneurs lack proper marketing facilities and marketing channels. Hence, they are forced to sell their produce and value-added products at lower price at local level to local agents and middlemen.

Rural women often face great constraints in accessing goods and services and moving their own produce and products to market, and they face higher prices for consumer goods because of the distance and fuel costs involved in transporting goods to rural areas and low selling prices for their produce for the same reason. All these factors combine to hamper rural women's establishment and growth of on- and off-farm enterprises and productivity (FAO/IFAD/ILO 2010).

### 26.2.7 Drudgery Involved During Agriculture and Livestock-related Work

Rural women play a significant role in agricultural activities such as seeding, transplanting, weeding, fertilizer application, plant protection, harvesting, processing, storage, etc. Several of these operations were exclusively carried out by women only. Thus, they directly or indirectly influenced the course of agriculture by participating all above agricultural activities (Singh et al. 2004). Many of these operations are traditionally done in varying body postures, some of which if done for long duration are not only inconvenient but also cause serious health hazards.

Due to sociocultural traditions, the rural women have inaccessibility to modern technology, credit, training, and other facilities available to male workers and farmers. The women continue to perform farm operations which are full of drudgery such as transplanting, weeding, winnowing, shelling, decortications, milling, etc. (Singh and Nayak 2011). Women are overburdened while they still do their domestic work (Kunfaa et al. 2001). In ravine areas, drudgery of farmers and farm women occurs in the following field activities and home activities:

1. Farm operations like weeding, harvesting, cleaning, grading, threshing, and grain shelling
2. Post-harvest practices like grain processing, fruits and vegetable preservation and processing, storage of food grains
3. Dairy cattle rearing, cleaning, fodder collection, feeding, and milking

All these factors result in drudgery by causing physical and mental fatigue, monetary hardships, exploitation, pain, economic stress, etc. Majority of the respondents perceived farm activities as either moderately difficult or difficult. The farm women are employed in the operations which either not mechanized or least mechanized and involve a lot of drudgery (Singh et al. 2010). In ravine areas, during weeding farm women usually adapt squatting posture, and they continue to work in this posture for long duration without adapting any other posture due to which they reported severe pain in lower back and knees.

## 26.3 Strategies for Livelihood Security in Ravine Areas

As per the International Centre for development-oriented Research in Agriculture (ICRA 2012), a livelihood (making a living) is largely about generating income. This also includes aspects of food security (the ability to feed oneself and one's family), providing a home, health, security (reduced vulnerability to climatic, economic or political shocks, and so forth), sustainability (the ability to continue to make a satisfactory living), power (the ability to control one's own destiny), and others. It is important not to lose sight of long-term goals when analyzing the issue of livelihood. Gender equality, therefore, is the current subject of concern among the policy makers of the world because of enhancing economic growth and maintaining continuity of the development process as well (CPD-UNFPA 2002). Rural women are critical to the well-being of farm households, as they play a vital role both in the management of domestic work as well as farm-related work. They work for 14–16 h a day in trying to balance competitive demands in agriculture production, household-focused activities, and income generation.

The following strategies may help in empowering women in difficult areas such as ravine-dominating areas:

### 26.3.1 Holistic Approach of Gender Equality for Sustainable Livelihoods

Women's participation on equal terms with men in domestic decision-making makes them be recognized in the family. The educational attainment and participation in to gainful economic pursuit make women strong and determined. The financial freedom that comes with their employment opportunity can't be denied, and these two aspects generally make women capable of asserting their voice both in the family and in the society (Das 2006). It is important for rural women to be included in the management of resources and in decision-makings. They should be regarded as the users and managers of resources, especially natural resources (Adetunji and Adepoju 2009). Women are responsible for providing foods for the households, they secure overall family welfare, and they are somehow the backbone of smallholder of agricultural production (Ellis 1998; Adetunji and Adepoju 2009). Additionally, management of resources by women is crucial because it is their responsibility to sustain and maintain the standard of the household (Schmeer 2005; Adetunji and Adepoju 2009).

### 26.3.2 Awareness and Skills Development of Farm Families in Ravine Areas

There is a need to create awareness and provide training to farm women and rural youth in ravine areas on latest tools and techniques (Fig. 26.1). Keeping their physiology, women specific ergonomically comfortable equipment also needs to be

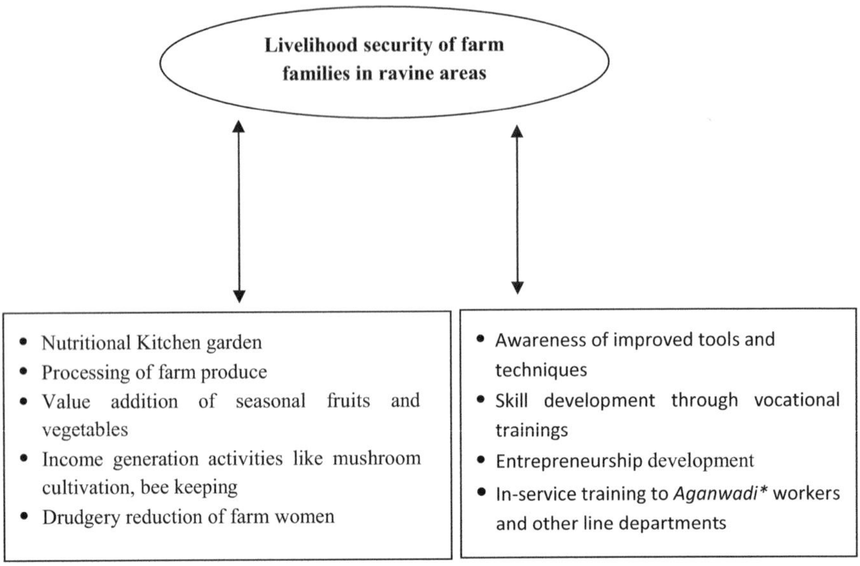


**Fig. 26.1** Integrated and holistic approach for livelihood security of farm families in ravine areas. *A program for women run by state governments in rural areas

developed (Singh and Nayak 2011). Skills development improves output, quality, diversity, and occupational safety and improves health, thereby increasing incomes and livelihoods of the poor.

As per human capital theory, the better educated the agricultural labor, the higher their productivity (Atchoarena and Gasperini 2003). Since then, skill is one of the main elements of entrepreneurship, which help community and organization move toward development. Skills development is a key for improving household productivity, employability, and income-earning opportunities for rural women and for enhancing food security and promoting environmentally sustainable rural development and livelihoods (Anuradha and Reddy 2013). The success of any enterprise serving as income-generating activities for women groups or organizations would entirely depend on training for sound economic growth and development (Hyytinen and Pajarinen 2005). Training is the engine of economic growth and development for women entrepreneurs engaged in business activities. Education and training enhances economic and employment opportunities in the nonfarm sector by providing the rural poor with marketing skills over and above their labor (Scott 2000).

### 26.3.3 Training Need to Farm Women in Ravine Areas

Training to farm women and rural youth particularly in value addition and other agricultural fields will ensure them:

| |
|---|
| Household food security by kitchen gardening and nutritional gardening |
| Processing of locally available fruits and vegetables |
| Design and development of low-/minimum-cost diet |
| Designing and development for high nutrient efficiency diet |
| Minimization of nutrient loss in processing |
| Storage loss minimization techniques |
| Location specific drudgery reduction technologies |
| Value addition of locally available fruits and vegetables |
| Mushroom cultivation |
| Tailoring program |
| Nursery raising |
| Handy craft from waste material |
| Dairy farming |
| Broiler poultry farming |
| *Agarbatti* (fragrant sticks used during worship) making |

Well-designed vocational training leads to better paid work and does not concentrate women in low-wage and low-skill work or reinforce occupational segregation between women and men (Tornqvist and Schmitz 2009). Economic empowerment is the capacity of women and men to participate in, contribute to, and benefit from growth processes in ways which recognize the value of their contributions, respect their dignity, and make it possible to negotiate a fairer distribution of the benefits of growth. Women's economic participation and empowerment are fundamental to strengthening women's rights and enabling women to have control over their lives and exert influence in society (Sweden, MFA 2010).

### 26.3.4 Processing and Value Addition of Ravine Produce Along with Underutilized Fruit Crops for Rural Livelihoods and Economy

Food security is much more than accessibility to food as it also involves availability and affordability of nutritious food. To this end, storage and preservation come into play because most of the food produced by the rural dwellers perishes due to poor storage and low adoption level of value addition technologies. This is because good processing, preservation, and value addition technologies are not yet within the reach of the rural dwellers (Okorji et al. 2003). The idea of integrated rural development includes growth in agricultural production, processing, marketing, and creation of employment opportunities for rural farmers (Ferris et al. 2006). Entrepreneurial activities in rural areas are based on agricultural and natural resource products; farm families broaden their income strategies by including value chain operations (Reardon et al. 2001).

An agro-enterprise not only enhances national productivity and generates employment but helps to develop economic independence and personal and social capabilities among rural women (Kuratko et al. 2011). In addition, agro-enterprises engaged in value-added products might also benefit from increased attractiveness of food products grown under eco-friendly management systems identified with a farm or rural community (Best 2003). Promoting agro-enterprise and small business is increasingly seen as a tool of generating meaningful and sustainable employment opportunities particularly for those at the margins of the economy such as women, the poor, and people with disabilities. Micro-, small-, and medium-sized enterprise recognized a way to promote women's economic empowerment while fighting poverty (Giovannucci 2005).

The underutilized fruits like gooseberry (*Emblica officinalis*), *bael* (*Aegle marmelos*), jamun (*Syzygium cumini*), karonda (*Carissa carandas*), passion fruit (*Passiflora edulis*), phalsa (*Grewia asiatica*), pomegranate (*Punica granatum*), pumpkin (*Cucurbita pepo*), tamarind (*Tamarindus indica*), wood apple (*Feronia limonia*), etc. are the main sources of livelihood for the poor and play an important role in overcoming the problem of malnutrition (Gajanana et al. 2010). Fruit processing is necessary where it ensures fair returns to the growers to improve their economic condition. It also helps to mitigate the problem of underemployment during off-season in the agricultural sectors. The perishable fruits are available as seasonal surpluses during certain parts of the year in different regions and are wasted in large quantities due to absence of facilities and know-how for proper handling, distribution, marketing, and storage. Furthermore, massive amounts of the perishable fruits produced during a specific season results in a glut in the market and become scarce during other seasons. Food preservation has an important role in the conservation and better utilization of fruits to avoid the glut and utilize the surplus during the off-season. It is necessary to employ modern methods to extend storage life for better distribution and processing techniques to preserve them for utilization in the off-season in both large and small scale (Bhattacharyya and Bhattacharjee 2007; Jena 2013).

### 26.3.5 Improving Livelihood and Economy of Farm Families Through Income-Generating Activities in Ravine Areas

Livelihoods are a dynamic realm that integrates both the opportunities and assets available to a group of people for achieving their goals and aspirations as well as interactions with and exposure to a range of beneficial or harmful ecological, social, economic, and political perturbations that may help or hinder a group's capacity to make a living (Mudege and Ezeh 2009). In fact, livelihoods diversification appears to be a strategy out of poverty. The livelihoods diversification can include agricultural work like crop and livestock farming, nonagricultural work like fishing, local trade, self-employment, and permanent employment. By diversifying their livelihoods, most households tend to achieve their goals (Ellis 2000).

Similarly, diversification maybe associated with success at achieving sustainable livelihoods under improving economic and social conditions. Generally, diversification is associated with off-farm activities (Niehof 2004). Diversification options are limited by economic constraints (e.g., household assets, such as land, and skills), influenced by social, cultural, and political factors, particularly caste and gender (Ellis 1998, 2000). There is an increasing awareness that livelihoods diversification plays a strategic role in rural livelihoods systems and rural households (Niehof 2004). Further, Ellis (1998) stated that livelihood diversification might be associated with a success at achieving livelihood security under improving social and economic conditions. Diversification can have an equalizing effect on rural incomes and wealth (Ellis 2000). The sustainable rural livelihood implies that any development intervention for the rural people should be congruent with their existing livelihood strategies and ability to adapt. The farmers and farm women in the ravine areas of Morena district are mainly dependent on agriculture as their source of livelihood. Therefore, to improve the livelihood of the community, certain additional source of income is most important. Some of these may include mushroom cultivation, beekeeping, and poultry production.

#### 26.3.5.1 Mushroom Cultivation

Mushroom cultivation is such a practice which can significantly serve the purpose of nutritional and livelihood security of the community. Through mushroom cultivation, people can easily generate some decent amount of extra income with least investment. It has been described as the most versatile and prolific agricultural venture all over the world (Ikeji 2010). Mushroom contains many essential amino acids and is thus a good source of protein. It also contains some unsaturated fatty acids and provides several of the B vitamins and vitamin D as well as the minerals potassium, phosphorus, calcium, and magnesium (FAO 2000; FIIRO 2009). Increased recognition of the nutritional value of many species, coupled with the income-generating potential of mushrooms, has generated desirable interest in the artificial cultivation of mushrooms. Commercial production of edible mushrooms is also growing because of its potential contribution to economic growth and as a source of cheap protein (Onuoha 2007). Mushroom cultivation is environmental friendly. Mushroom plays a pivotal role in sustaining eco-energy balance in nature (Jumpponen et al. 2004).

The capacity building program helps to improve awareness and cultivation of edible mushroom which currently is limited in ravine areas of Morena district. As mushroom is highly nutritious, therefore, it can be produced in commercial quantity in households of ravines through women's and rural youth's training. Mushroom cultivation can help to reduce poverty and strengthen livelihoods security through the low cost locally available substrate. Since it does not require much access to land, mushroom cultivation is a viable and attractive activity for both rural farmers and farm women.

#### 26.3.5.2 Beekeeping and Poultry Production

Beekeeping has been contributing to household incomes hence livelihoods in terms of food security and poverty alleviation. It has been advocated for by development agents, because of its nature: low input requirement, cheap technology based, and gender friendly and independent from environmental changes. It does not depend on soil, and it can be a single means of living for families with very little or no soil (Kizilaslan and Kizilaslan 2007). Poultry production enterprise is a potential area for women groups to harness income, create job opportunities, and improve quality life and standard of living for women residing in rural communities of the ravines. The participation of women groups can enhance poultry production for income generation in rural areas of ravines. Thus, rural poultry farming not only generates income and employment opportunities to small farmers including women but also brings about improvement necessary to uplift the life of rural communities (Gminder 2003).

### 26.3.6 Improved Household Health Through Nutritional Kitchen Gardens in Ravine Areas

Food insecurity is primarily a problem of low household incomes; hence farmers in rural areas aim at increasing their food production and their subsistence crops (Gladwin 2001). As per FAO (2003), food security refers to the availability of food and one's access to it. Seemingly, a household is considered food secured when its occupants do not live in hunger or fear of starvation. Food security for a household means access by all members at all the times to enough food for an active, healthy life. As per Thomson and Matz (1997) and FAO (2003), these include at a minimum the ready availability of nutritionally adequate and safe foods and an assured ability to acquire acceptable foods in socially acceptable ways (i.e., without resorting to emergency food supplies, scavenging, stealing, or other coping strategies). Additionally, food security exists when all people always have physical and economical access to sufficient, safe, and nutritious food to meet their dietary needs and food preferences for an active and healthy life (FAO 2003).

Agricultural production may not be their only source of income; to be food secured the rural households need to have multiple livelihood strategies (Pearce et al. 1996; Thomson and Matz 1997; Gladwin 2001). The reason for diversifying and relying on multiple livelihood is that these strategies are their poverty reduction mechanisms, and they are also capable of improving and sustaining their standards of living (Gladwin 2001). The use of home gardens has been suggested numerous times as a solution to some of the issues surrounding food security (Yu et al. 2009; FAO 2009; Collier et al. 2008; IICA 2009). A kitchen garden is one way that the individual or small community in ravine areas can provide themselves with a steady supply of fresh vegetables, local fruits, and fuel wood throughout the year.

### 26.3.7 Promoting Women Entrepreneurship Through Self-Help Groups (SHGs)

Sidhu and Kaur (2006) revealed that SHGs entrepreneurship is the only solution for the growing employment among the rural youth. It helps to generate employment for number of people within their own social system. To assist and develop rural women entrepreneurship, women at grassroot level should be involved through the women's organizations. Such organizations can be instrumental in identifying women for income-generating activities (Mansuri 2011). Organizing women into groups is an effective way of learning new ideas to become more competitive and gain economic of scales (Dixie 2005).

### 26.3.8 Drudgery Reduction of Farm Women for Improving Efficiency

Drudgery is a term used to represent the dissatisfactory experiences that constrain work performance in any activity (Technical module/AICRP- FRM/DRWA 2009). Majority of rural women in ravine areas take part in agricultural field operations. They, as agricultural laborer's, participate in several activities such as seeding, transplanting, weeding, fertilizer application, plant protection, thinning, harvesting, processing, selling, winnowing, storing, looking after animals, kitchen gardening, etc. (Sudharani and Raju 1991). Rural women perform wide range of activities like fetching water for household use which physiological stress is involved during pumping water and while carrying the vessel filled with water back to home (Oberoi et al. 2001). During these activities, stress is found beyond the acceptable limits of the women resulting in low work output along with deterioration of their health.

Jyotsana et al. (2005) stated that during wheat harvesting activity from morning till evening, women usually adapt squatting posture, and they continue to work in this posture for long duration without adapting any other posture due to which they reported severe pain in the lower back and knees. Musculoskeletal injuries and diseases likely affect the production agriculture workforce more frequently during their working years than any other safety and health problem. Disability due to musculoskeletal injuries and diseases incurred during their working years affect the production agriculture workforce more frequently and more severely than any other safety and health problem during the remainder of their working years and, for many, for the balance of their lives. Sadangi and Pattanaik (2006) also reported that farm operations like weeding, cutting, and carrying harvested crop were the moderately heavy activities.

For minimizing the drudgery involved in any of such activities, it would require introduction and adoption of labor-saving and drudgery-reducing technologies enabling rural women to participate more energetically with increased efficiency. Drudgery in ravine areas can be reduced by providing gender-friendly farm tools and equipment which increase the productivity of farm women with safety and comfort. To ensure better health and safety and to improve work efficiency and to reduce the drudgery of women, it is important to understand the occupational workload of farm women by studying the efficiency, heart rate, energy expenditure, and their fatigability, which could be assessed through physiological parameters by assessing heart rate and by estimating energy expenditure of some drudgery-prone activities the rural women perform in farms, in homes, and in allied fields. Reduction in drudgery of women would facilitate women to get involved in development works and decision-making process. Hence in ravine areas, periodic training programs should be organized to emphasize on educating workers regarding recognition of muscular-skeletal disorders and importance of rest pauses and maintaining proper posture while performing agricultural activities.

### 26.3.9 Access to Extension Services Through Information and Communication Technologies

Knowledge, information, and communication have been strategic focus of the developmental agencies and governments alike for over half a century (Chapman and Slaymaker 2002). The effective provision of information on food and agriculture is fundamentally important as it informs both the livelihood strategies of the rural poor themselves and the policies and strategies of agencies and institutions responsible for reducing rural poverty and food insecurity. In each case, it is only through improved information that individuals and institutions can make informed choices about the opportunities and constraints associated with agricultural development strategies (FAO 2000). An effective extension program might be a tool for carrying out income-generating activities to train and educate the farm women. Therefore, agricultural extension services need to develop suitable mechanism for imparting knowledge and skills on various aspects of income-generating activities for farm women of ravine areas.

The flow of communication and sharing of information among women groups engaged in agro-enterprise as income-generating activity is a salient indicator for sustainability. The issue of information and communication technology (ICT) can improve the process of collection, analysis, and dissemination among potential buyers and sellers of agricultural products. This may shorten the time of purchase, inventory, and pricing decision (Meera et al. 2004). Improved communication systems can enable individuals to organize themselves, use information to hold institutions accountable, and put pressure on relevant authorities to deal with their problems (CTA 2001). Agricultural researchers, extension providers, and educators form the knowledge triangle within which farmers are located and interact in a two-way information flow process with each of the three institutions (FAO/World Bank 2000). In many remote rural areas, the use of ICTs will depend as much on technical capacity building for local institutions to use them effectively, as on the infrastructure itself (Ballantyne et al. 2000).

Another means of communication or receiving information is the use of radios in rural communities. Barua and Yoshimura (2005) feel that the mobile phone is the most applicable information and communication instrument, and, in the future, as phones become more widely available, it will become an essential trading tool especially for women entrepreneurs. Davis and Rylance (2005) indicated that not only should agro-enterprise development could lead to skills in income generation through building local enterprises but would also seek to promote core values of gender equity of benefits. For the community participation, women's group engaged in agro-enterprise should be involved in planning and implementation phase, as this will facilitate the transfer of marketing skills (Rushtonand and Ngongi 2002).

## 26.4 Conclusion and Recommendations

The rural women of ravine areas are economically backward. Various factors such as age, level of education, access to credit, and membership of an association directly and significantly affect the involvement of women in income-generating activities, while family size inversely affects their involvement in income-generating activities. Transfer of technology through extension system should be planned and executed on need-based priorities of farm women with the objective to reorient the system. Skill development among women is the need of the hour to make them confident and to develop in them the ability to be a part of decision-making at home and outside.

Women of the ravine areas should be sensitized on the importance of involving in income-generating activities such as beekeeping and poultry as this will help them to be self-reliant. Training for value addition to farm produce or collection from wild such as making pickles from fruits of *Capparis decidua*, *Cordia rothii*, and *Carissa carandas* and collection of medicinal herbs can bring a sea of change in their livelihood. Moreover, higher adoption of the income-generating activities will enhance the income of farm women, which in turn leads to the sustainable livelihood and improved quality of life in ravine areas. The extent to which rural women participate in technology research and development, together with their access to new technologies, is an important factor that has impact on women's economic empowerment, both in relation to agricultural production and other enterprises like value addition, processing, and storage. So, it is essential to empower women through technology so that they can have higher efficiency with work output and reduced drudgery and health problems. Technologies, ergonomically designed, are labor saving, drudgery reducing, income generating, and productivity increasing and must be given wide publicity, and their use is encouraged through pragmatic extension.

## References


Abdulai A, Crole Rees A (2001) Determinants of income diversification amongst rural households in Southern Mali. J Food Policy 26:437–452

Adetunji MO, Adepoju A (2009) Household resource management by rural women in Ibarapa East local government area of Oyo State. J Appl Sci Res 5(6):636–644

Anuradha R, Vijaya Reddy G (2013) Lokanadha empowerment of rural women through vocational education and training. Conflux J Educ 1(2):8

Atchoarena D, Gasperini L (2003) Education for rural development: towards new policy responses. FAO, UNESCO, Rome

Ballantyne P, Labelle R, Rudgard S (2000) Information and knowledge management: challenges for capacity builders, Policy management brief 11. European Centre for Development Policy Management, Maastricht

Barua A, Yoshimura Y (2005) Rural poultry keeping in Bangladesh. Worlds Poult Sci J 53:387–394

Best R (2003) Farmer participation in market research to identify income-generating opportunities. Centro Internacional de Agricultura Tropical (International Center for Tropical Agriculture), Kampala, p 2. www.InternationalCenterforTropicalAgriculture.cgiar.org/

Bhattacharyya BK, Bhattacharjee D (2007) Bacteriocin: a biological food preservative. J Food Sci Technol 44(5):459–464

Carr ER (2005) Development and the household: missing the point? J Geojournal 62(1):71–83

Carr ER (2008) Men's crops and women's crops: the importance of gender to the understanding of agricultural and development outcomes in Ghana's Central Region. J World Dev 36(5):900–915

Chapman R, Slaymaker T (2002) ICTs and rural development: review of the literature. Current interventions and opportunities for action', ODI working paper 192. Overseas Development Institute, London

Collier AK, Hayden-Smith R, Torres C, Tagtow A (2008) High price of the rising cost of food in the United States: an overview. Food & Society Policy Fellows Program. http://groups.ucanr.org/victorygrower/files/59876.pdf

Coppard D (2001) The rural non-farm economy in India: a review of the literature, NRI report no. 2662. NRI, Chatham

CPD-UNFPA Paper 19 (2002) Gender-related development index. For 64 Districts of Bangladesh

CTA (2001) ICT update wireless: a help line for agricultural development? Issue 4 June 2001. A current awareness bulletin for ACP agriculture. Technical Centre for Agricultural and Rural Cooperation, Wageningen

Das S (2006) Increasing gender gap and socio – economic status of women in North – East India. IIPS International Conference, Mumbai

Davis J, Rylance C (2005) Addressing poverty through local economic and enterprise development. A review of conceptual approaches and practice. Working Paper Chatham Natural Resources Institute, Kent, 2005

Dixie G (2005) Horticultural marketing FAO marketing extension guide. FAO, Rome. 2005

Ellis F (1998) Survey article: household strategies and rural livelihood diversification. J Dev Stud 35(1):1–38

Ellis F (2000) Rural livelihood diversity in developing countries: analysis, policy and methods. Cambridge University, Cambridge

Fabiyi EF, Danladi BB, Akande KE, Mahmood Y (2007) Role of women in agricultural development and their constraints. Pak J Nutr 6(6):676–678

FAO (2000) FAOs corporate communication policy and strategy. Food and Agriculture Organization of the United Nations, Rome

FAO (2003) Gender and development plan of action. FAO, Rome

FAO (2009) Improving nutrition through home gardening. FAO: Nutrition and Consumer Protection Department, Rome. Available from http://www.fao.org/ag/agn/nutrition/household_gardens_en.stm

FAO (2010) Women in infrastructure works: boosting gender and rural employment policy brief.; 2010. Food and Agricultural Organisation, Rome

FAO (2010–2011) The state of food and agriculture; 2010–2011. Food and Agricultural Organisation, Rome

FAO/IFAD/ILO (2010) Making migration work for women and men in rural labour markets, Gender and rural employment policy brief #6. FAO, Rome

FAO/World Bank (2000) Agricultural knowledge and information systems for rural development: strategic vision and guiding principles. Food and Agriculture Organisation of the United Nations, Rome

Ferris RSB, Best R, Lundy MC, Ostertag MV, Wandsschaneider T (2006) Strategy paper: a participatory and area-based approach to rural agro enterprise development. Rural Agro Enterprise Development Project, Centro International de Agricultura Tropical (International Center for Tropical Agriculture), Cali., 2006, p 44

FIIRO (2009) Make money by growing mushrooms. Federal Institute of Industrial Research (FIIRO), Oshodi. ftp://ftp.fao.org/docrep/fao/011/i0522e/i0522e00.pdf

Gajanana TM, Gowda IND, Reddy BMC (2010) Exploring market potential and developing linkages–a case of underutilized fruit products in India. Agric Econ Res Rev 23:437–443

Giovannucci D (2005) Organic agriculture and poverty reduction in Asia: China and India focus: office evaluation. International Fund for Agricultural Development, Rome. 2005

Gladwin CH (2001) Addressing food security in Africa via multiple livelihood strategies of women. J Food Policy 26(2):177–207

Gminder CU (2003) Gender-oriented entrepreneurship promotion: strategies and tools along the project cycle: a manual for practitioners. Part II toolkit. The Swiss Agency for Development and Cooperation, Bern. 2003
Hyytinen A, Pajarinen M (2005) Financing of technology intensive small businesses: some evidence on the uniqueness of the ICT sector. Inf Econ Policy 17(1):115–132
ICRA (International Centre for development oriented Research in Agriculture) (2012) http://www.icra-edu.org/objects/anglolearn/ACFtTAGCm.pdf. Retrieved, Dec 2012. pp 1–8
Ikeji I (2010) Exploring potentials in mushroom farming. Business Day Monday 04 January 2010
Inter-American Institute for Cooperation in Agriculture (IICA), Barbados (2009) IICA's contribution to the development of agriculture and rural communities in Barbados: annual report 2008 IICA, Costa Rica. 35 pp
Jena S (2013) Development of a preserved product from underutilized passion fruit and evaluation of consumer acceptance. J Food Res Technol 1(1):11–19
Jumpponen A, Claridge WA, Trappe MJ, Label T, Claridge ID (2004) Ecological relationships among hypogenous fungi and trees: inferences from association analysis integrated with habitat modeling. Mycologia 96(3):510–525
Jyotsana KR, Singh K, Mehta M (2005) Ergonomic evaluation of rural women while performing wheat harvesting activity. J Hum Ecol 18(4):309–311
Kizilaslan H, Kizilaslan N (2007) Factors affecting honey production in apiculture in Turkey. J Appl Sci Res 3(10):983–987
Kunfaa EY, Dogbe HJ, Mackay C, Marshall (2001) Empty pockets. In: Voices of the poor. From many lands. Ghana Publishing Corporation, Accra
Kuratko D, Richardand F, Hodgetts M (2011) Entrepreneurship a contemporary approach. Harecourt College Publishers, New York
Mansuri BB (2011) Rural women entrepreneurs in India: opportunities and challenges. Kurukshetra 59(11):18
Marilee K (2009) Crucial role of women in food security in Oyun local government area of Kwara State, Nigeria. World J Agric Sci 9(3):23–29
Meera SN, Jhamtani A, Rao DUM (2004) Information and communication technology in agricultural development: a cooperative analysis of three projects from India. p 135
Moni T, Pankaj S, Atul S, Zakir H (2011) Biodiversity and livelihood options of people in Chambal ravine of Morena district, Madhya Pradesh, India. J Trop For 41. July–Sept 2011
Mudege N, Ezeh A (2009) Gender, aging, poverty and health: survival strategies of older men and women in Nairobi. J Aging Stud 23:245–257
Mudgal MK (2005) Socio-economic impact of ravine lands- a case study of River Chambal basin of state of Madhya Pradesh, INDIA. Geophys Res Abstr 7:00529
Niehof A (2004) The significance of diversification for rural livelihood systems. J Food Policy 29(4):321–338
Oberoi K, Gupta R, Verma P (2001) Physiological stress on the rural women while performing the activity of fetching water for household use. In: Ray GG, Ganguly AK, Nag PK (eds) Proceedings of Huminizing Work and Work Environment (HWWE). IIT, Bombay, pp 512–518
Okorji EC, Eze CC, Eze VC (2003) Efficiency of cassava processing techniques among rural women in Owerri, Imo State, Nigeria. JASR 3(2):84–96
Onuoha CI (2007) Cultivation of the mushroom (Pleurotus tuber regium) using some local substrates. http://www.sciencepub.net/life/life0404/13_life0404_58_61_Cultivation.pdf
Pandey R, Cannon T, Jena G, Wandschneider T (2002) Rural non-farm employment in Nayagarh District of Orissa: findings from a participatory rural appraisal. NRI, Chatham. (draft)
Pearce J, Ngwira A, Chimseu G (1996) Living on the edge. Save the children. UK, London
Rath A, Cannon T, Wandschneider T (2002) Rural non-farm employment in Bolangir District of Orissa: findings from a participatory rural appraisal. NRI, Chatham
Reardon T, Berdegue J, Escobar G (2001) Rural non- farm employment and incomes in Latin America: overview of issues, patterns and determinants. World Dev 29(3):395–409
Sadangi BN, Pattainaik S (2006) Drudgery reduction – a step towards women empowerment. Indian Farming: 7–9 and 14

Schmeer KK (2005) Married women's resource position and household food expenditures in Cebu, Philippines. J Marriage Fam 67(21):399–409
Scott JT (2000) An assessment of the small business innovation research program in New England: fast track compared with non-fast track projects. In: Wessner CW (ed) The small business innovation research program: an assessment of the department of defense fast track initiative. National Academy Press, Washington, DC, pp 104–140
Sidhu K, Kaur S (2006) Development of entrepreneurship among rural women. J Soc Sci 13(2):147–149
Singh YV, Nayak J (2011) Status of rural women in agriculture. Kurukshetra 59(11):6
Singh P, Jhamtani A, Bhadauria C, Srivastava R, Rahul, Singh J (2004) Participation of woman in agriculture. Indian J Ext Educ 40(3 & 4):24–26
Singh A, Gautam US, Pannace S, Singh A (2010) Ergonomic evaluation of farmwomen during maize shelling. Indian Res J Ext Educ 10(3):41–44
Sobha I (2001) Women in agriculture: a case study. Yojana 45:40–43
Som R, Kleih U, Kumar Y, Jena SK (2002) Rural non-farm employment in Madhya Pradesh: findings of a participatory rural appraisal in 8 villages, NRI report no. 2694. Natural Resources Institute, Chatham
Sudharani P, Raju VT (1991) Participation of women in agricultural operations. Indian J Ext Educ 28(1 & 2):54–59
Sweden, Ministry for Foreign Affairs (2010) On equal footing: policy for gender equality and the rights and role of women in Sweden's international development cooperation 2010–2015. MfA, Stockholm
Technical module/AICRP- FRM/DRWA/ (2009) Trainers' training module on drudgery reducing technology interventions for women in agriculture. DRWA, Bhubaneswar
Thomson AM, Matz M (1997) Implications of economic policy for food security: a training manual. FAO, Rome
Tornqvist A, Schmitz C (2009) Women's economic empowerment: Scope for Sida's Engagement. Sida (Swedish International Development Cooperation Agency) Working Paper, Sida, Stockholm
WFP & SEAMEO TROPMED (2010) Nutrition security and food security in seven districts in NTT Province, Indonesia: status, causes and recommendations for response. The report of joint assessment SEAMEO TROPMED (Southeast Asian Ministers of Education Organization – Tropical Medicine and Public Health) Regional Centre for Community Nutrition at the University of Indonesia. Jakarta, p 23
Wombeogo M (2007) Gendered poverty in Northern Ghana: multiple problems, few solutions. Africanus 37(11):36–53
World Bank (2012) Toward gender equality in East Asia and the Pacific: a companion to the World Development Report—conference edition
Yu B, You L, Fan S (2009) A typology of food security in developing countries under high food prices. International Association of Agricultural Economists Conference, Beijing, China, August 16–22. http://ageconsearch.umn.edu/bitstream/51043/2/IAAE%20food%20security%20typology.pdf

# Socio-economic and Conservation Measures in Ravine-affected Areas of Gujarat: Policy Interventions

27

V. C. Pande, R. S. Kurothe, H. B. Singh, B. Krishna Rao, Gopal Kumar, and P. R. Bhatnagar

**Abstract**

Most of the agricultural land resources are under the threat of degradation. The lack of scientific management practices, socio-economic issues, and non-implementation of policy initiatives are considered to be the major causes of degradation of resources. Further, the stakeholders are economically not in a position to invest in conservation technologies, which are too costly to afford for resource-poor stakeholders. In fact, poor economic condition of the stakeholders inhabiting ravine ecosystem prohibits them to make large-scale investments on the conservation of these resources. Since we have to meet the needs of ever-increasing population for food and other livelihood requirements, therefore, society can ill afford to leave any piece of land including degraded ravines unattended due to environmental benefits accrued from this ecosystem. A productive utilization of such degraded lands is called for with conservation of natural resources as the major objective. This warrants investment on rehabilitation of degraded eroded lands in general and ravines in particular. During successive plan periods beginning the Second Five-Year Plan, funds have been allocated by governments for land development including resource conservation under various government schemes. These have benefited some areas however; the wider impact of these investments is yet to be observed on the ground. One of the reasons, among many, has been reported to be the slow pace of adoption of conservation measures by marginal and small farmers who are in majority in ravines. While policy interventions to address the problem of degraded land have largely been a part of broader programs and schemes of the government, the present study builds a case for policy interventions specific to ravine ecosystem.

V. C. Pande (✉) · R. S. Kurothe · H. B. Singh · B. Krishna Rao · G. Kumar · P. R. Bhatnagar
ICAR-Indian Institute of Soil & Water Conservation, Research Centre, Vasad, Gujarat, India



J. C. Dagar, A. K. Singh (eds.), *Ravine Lands: Greening for Livelihood and Environmental Security*, https://doi.org/10.1007/978-981-10-8043-2_27

## 27.1 Introduction

Gullied or ravine lands, classified as wasteland based on ecological understanding, have drawn the attention of the government for a long time. Policy interventions to address the problem of ravine land have largely been a part of broader programs and schemes of the government. During the 1980s, several programs related to reclamation of degraded lands were initiated, which also included ravine-affected watersheds. The National Wasteland Development Board, established under the Ministry of Environment and Forests, was given responsibilities for the development of wastelands involving different stakeholders including local people at every stage of development. The major objectives of the programs included checking land degradation, putting wastelands to sustainable use, reclamation of degraded lands for agricultural use particularly increasing biomass and fuelwood production through raising trees and cultivation of grasses, and restoring ecological balance.

The land degradation issues and adoption of conservation agriculture have been widely documented. Many workers (Barbier 1988; Barbier and Burgess 1992; Coxhead Ian 1995; Coxhead and Shively 1995) explored socio-economic factors to understand the basic reasons of the land degradation problems and the statistics of input and output correlations during reclamation and found out the viable remedial measures including policies on subsidizing the initial investment costs and incentives to conserve the soil. Some workers dealt with the farm-level incentives and government intervention in agricultural markets (Repetto 1988; Barrett 1989; Shiferaw and Holden 2001). The ravine ecosystem is inhabited by farmers with small land holdings and landless farmers. Many of them possess scattered and undulating holdings giving very low yield and practice subsistence farming adjoining a system of gullied lands, and some of these factors do have direct relevance for policy interventions. Based on the classification of gullies in terms of land capability classification and feasible production system, the ravine land can be put to different production uses for different land holdings. Ravine ecosystem is very fragile and prone to excessive soil erosion. This is the reason that due to anthropogenic exploitations such as overgrazing and removal of vegetation has caused extensive damage to this ecosystem since long. The ravine development program needs government help, hence policy interventions.

The urgency of a nationwide policy for dealing with various soil degradation problems including soil conservation and ravine reclamation was understood from the early periods and also emphasized in the First Five-Year Plan of the country. The government efforts were visible from the fact that as early as in 1953, the Ministry of Food and Agriculture set up the Central Soil Conservation Board to initiate, organize, and coordinate research in soil and water conservation and to develop capacity by training personnel and assist states in carrying out soil conservation programs. A sum of about ₹16 million was spent in the First Five-Year Plan (FYP) on implementing soil conservation programs. These funds were mainly utilized for contour bunding and terracing of about 700,000 acres (283,300 ha) of agricultural lands.

In the Second Five-Year Plan also, government spent about ₹180 million on carrying out soil conservation works. Contour bunding and terracing made good progress, especially in the erstwhile Bombay State, and an area of about 2 million acres (810 thousand ha) was benefited with conservation work. The land- and water-related projects and schemes of the government were implemented under various programs, and funds were allocated under three heads, land development, minor irrigation, and soil conservation. This program also included ravine rehabilitation in different regions. Interestingly, in the early 1960s, when the third plan document was prepared, it was made sure that the document recorded the importance of soil and water conservation program. Reviewing the earlier experience, it noted that during the first and the second plans, 16 thousand hectares of land was covered under contour bunding. As per the estimate made then, a total of 40 thousand hectares of land needed soil and water conservation treatment in Gujarat. It was, therefore, thought that institutional finances should be mobilized to expand the work. Hence, in 1978, a separate corporation by the name of Gujarat State Land Development Corporation (GSLDC) was formed, which up to the year 1983 handled the ravine reclamation work only. Later it started covering all watershed areas needing soil conservation work in a phased manner. The absolute plan size in Gujarat grew very fast especially after the Fifth FYP. The amount to be spent under soil conservation and land reclamation program during the Third FYP was ₹2600 million which increased gradually in each plan and reached to ₹4502 million in the Fourth FYP and ₹7353 million in the Fifth FYP. However, it shot up to ₹37,600 million in the Sixth FYP, ₹60,000 million in the Seventh FYP, and ₹102,000 million in the Eighth FYP. Because of the sustained government efforts, a large part of the ravine-affected land could be reclaimed; however, a substantial part of the land remained under the threat of ravine extension due to faulty agricultural practices, and still a large area is under ravine-affected areas along the river course in Gujarat and the rest of the erosion-affected states.

Ravines, in Gujarat, are largely spread along Mahi, Sabarmati, Narmada, and Tapti rivers, among others. Ravine formation has significantly increased soil loss from agricultural lands and severely impacted the agricultural productivity. The loss of production due to nondevelopment of ravines has been estimated to be 3 million tons per annum. There is, therefore, an urgency to arrest the loss of this precious resource. Ravines have been classified into small, medium, and deep and narrow gullies, depending upon their depth, width, and side slope in India. Accordingly, their reclamation measures and the investment on these measures also differ. While small and medium gullies are reclaimed for cultivation safely with minor leveling and bunding, the deep and narrow gullies are recommended to be put under permanent vegetation of grasses and trees (Tejwani et al. 1975).

## 27.2 Technological Interventions and Socio-economic Analysis from Some Case Studies in Gujarat

A sizeable gullied land has been successfully treated with various engineering and bioengineering measures. In many watershed areas, the conservation measures with different production systems on these marginal lands have been economically evaluated. Earthen gully plugs of sizes varying from 1.1 to 2.2 $m^2$, depending upon runoff from gully top, have been recommended for gully bed. Easing the gully to a slope of 3:1 and sodding with *Dichanthium annulatum* and *Cenchrus ciliaris* grasses stabilize the gully slope. Gully top treated with peripheral bund of an average cross section of 0.9–1.3 $m^2$ with spillway appropriately located helps control runoff and soil erosion into gullies (Tejwani et al. 1975).

Technological interventions in ravine-affected lands include both arable and nonarable lands. Resource conservation interventions for arable land such as tillage and cropping systems under rain-fed conditions have proved beneficial. Mulch farming and ridge farming tillage systems have been tested and found quite effective in controlling runoff and soil erosion. As compared to conventional tillage system, mulch farming tillage system under sequential cropping of cowpea (*Vigna unguiculata*) followed by mustard (*Brassica campestris*) was observed to have lower runoff and soil loss by 38% and 51%, respectively. This system produced maximum gross financial return of ₹20,445 per ha and ₹25,064 per ha (1998–1999) under cowpea – mustard and cowpea + castor (*Ricinus communis*) cropping systems, respectively.

Improved conservation agronomic practices in some marginal watershed areas with vegetation management, comprising of contour tillage, preparing rough seedbed, cover crops, organic matter addition, and high plant density, were tested in farmers' field and reported to have almost double net returns as compared to the traditional cultivation. While the total expenditure incurred in improved and traditional practices was not markedly different, at the same time, the gross returns were higher by 26–49% from different crops following improved cultivation practices on the farm. The increase was attributed to the better crop management and use of high-yielding and certified seeds. Irrigation was one of the important factors, accounting for the largest share in the operational costs. Better water management in combination with conservation agronomic practices, therefore, resulted in higher crop productivity and net return. Economics in terms of net saving from various cropping systems were in the range of ₹6566–₹9765 per ha (1990–1991 prices). This showed the potential of generating farm income from marginal lands as well as additional employment (which increased by 16–100%) on the farms by adopting improved crop practices. Among different cropping systems studied, castor + sesame (1:3) followed by summer pearl millet gave the highest net return of ₹9765 $ha^{-1}$ $annum^{-1}$ with B:C ratio of 2.0. The favorable B:C ratio confirmed economic feasibility of conservation farming technology on reclaimed marginal lands of ravine.

Resource conservation measures for nonarable lands were also analyzed by Sharma et al. (1980) for Mahi ravine area. Silvopastoral agroforestry, consisting of tree (*Acacia nilotica*) and fodder species (*Cenchrus ciliaris* and *Dichanthium annulatum*), in deep and narrow ravines (in an area of 8 ha comprising hump, slope, and bottom), was established in 1960–1961 at the Research Center farm and was economically evaluated. Artificial regeneration of the fuelwood was done by direct sowing of *Acacia nilotica* seeds in staggered contour pits of 60 × 60 cm spaced at 3 × 3 m, 9 × 9 m, 13.5 × 13.5 m, and 18 × 18 m distance. Weeding and gap filling were continued for the first 2 years after sowing. Selective thinning of the tree was carried out on the 13th year after the first sowing. The natural regenerating species other than *Acacia nilotica* were also allowed to grow. Artificial regeneration of fodder grasses such as *Cenchrus ciliaris* was done by direct sowing (broadcasting of seeds at the rate of 4 kg $ha^{-1}$) and harrowing and that of *Dichanthium annulatum* by slip planting on staggered furrows at 45 × 45 cm spacing on slopes. Discontinuous furrows of 15 cm width and 1.82 m length at 1.21 m vertical interval between the furrows were made. All costs and benefits were considered, and a 15-year project life with 12% discount rate for computation of benefit-cost (B:C) ratio was considered in the analysis. The B:C ratio obtained worked out to be 1.38, and the internal rate of return obtained was 15%, whereas the payback period of the project was considered 15 years. The internal rate of return also fully supports the economic feasibility of such social projects. The payback period of 15 years shows that such projects are conducive to finance by the institutional financial organizations, which usually provide long-term credit for land development for a period of 20 years (Sharma et al. 1980).

Another socio-economic analysis was done in ravine-dominated watershed of an area of about 30 ha, which was reclaimed by constructing contour and peripheral bunds in the table lands, bench terraces in medium gullies and gully plugs, and composite check dams, grassed waterways, chute spillways, pipe outlets, etc. in deep gullies. These reclaimed and protected gullies were put under agriculture-, horticulture-, and forestry-based land uses. In these gullies, apart from arable crops and fruit trees, species of bamboo, fuelwood trees, and fodder grasses were raised. The economic analysis was carried out on watershed basis considering the period of analysis as 25 years and 12% annual interest rate. While calculating costs accrued due to crop production, the minor year-to-year changes in crop rotations on experimental basis were not considered. The benefits accrued due to the project efforts in terms of revenue obtained from sale of grasses and fuel crops and fruits were estimated based upon the average yield rates from various crops and horticultural plants under standard input conditions. The years of occurrence of the revenue from crops and horticulture were worked out at the actual time of accrual, whereas benefits from fuel and fodder plantation were estimated based upon the surveys conducted in 8 ha neighboring watershed. The results showed that the B:C ratio of the project on the given site and climatic conditions was 1.37:1.

Another example is from one 35 ha degraded gullied area, which was reclaimed by adopting integrated approach of soil and water conservation and afforestation measures along the right bank of the river Mahisagar near Sarnal in Kheda district of Gujarat by the National Tree Growers' Cooperative Federation (NTGCF). This was a revenue wasteland and had succumbed to the encroaching ravines over the years. Hence, to treat the area, each gully was considered as a unit from its head to the riverbed. Earthen gully plugs with loose boulder spillways were constructed along the gully beds at a vertical interval of 1 m. On the slopes, staggered contour trench pits, enriched with organic manure, were made, and various tree species were planted. Extensive data on capital investment and maintenance of forest plantation was collected and maintained by the National Tree Growers' Cooperative Society, Sarnal. This data was used for analysis. Local prices of inputs and outputs were used to compute cash inflow and outflow from the plantation. Field surveys were conducted on survival, height, and diameter at breast height (dbh) measurements of the tree species at the plantation site as well as similar site outside (opposite the plantation site), where afforestation conservation measures were not applied. The benefit-cost ratio worked out to be 1.71 at 10% discount rate considering 15-year project period. The net profit calculated at 1995–1996 prices was ₹8244 ha$^{-1}$, and internal rate of return was 18.1%, which showed the plantation to be viable. It was established that the plantation could repay the capital and other investments in 14 years.

Based on the analysis of these long-term studies, Kurothe et al. (2012) could conclude that deep and narrow gullies are recommended to be put under permanent vegetation of grasses and trees under viable silvopastoral system. Bamboo (*Dendrocalamus strictus*) plantation with *Cenchrus ciliaris* grass is recommended for productive and protective utilization of such degraded lands. This system was found to be effective in retaining 80–100% rainwater, thereby controlling runoff by 10–20 times in planted gullies as compared to untreated gullies. Besides, the system was also economically viable as it generated revenue from the sale of bamboo poles and grasses over a longer horizon of 40 years after establishment. The total cost over a period of 7 years was estimated to be ₹106,142 per ha (2010–2011 prices), out of which ₹44,422 per ha was spent in the first year. A net income of ₹30,550 per ha could be realized during the 7th year and ₹31,000 per ha from the 8th to 10th years after adjusting for costs, which increased to ₹43,000 per ha from the 11th year onward. The net present worth worked out to be ₹6010 ha$^{-1}$ year$^{-1}$, benefit-cost ratio of 1.85, and an internal rate of return 18.4% at 2010–2011 prices. Further, the value of nutrients saved through bamboo plantation was estimated to be ₹2126–5555 per ha and a soil carbon worth ₹41,000 per ha under a bamboo stand of 20 years (Pande et al. 2012). The study suggested that high cost of establishment might be met through financial incentives undertaken by village communities with a support of some financial organizations or through some government initiatives such as MGNREGA program or such similar schemes.

Bamboo-based bioengineering interventions were, similarly, tested for reclamation and productive utilization of degraded ravine lands by Rao et al. (2012). The interventions for gully bed stabilization included bamboo plantation with staggered contour trenches, bamboo plantation supported by *Bori* bunds, and bamboo plantation as live check dams. Similarly, bamboo-based interventions for preventing gully head extension and bank erosion were also tried by them. These interventions gave average annual net return varying from ₹64 thousand to ₹89 thousand, and benefit-cost ratio varying from1.96 to 2.09 showing the economic and ecological sustainability.

## 27.3 Socio-economic Analysis of Conservation Measures Taken in Mahi Ravines

Though the adoption of soil and water conservation measures is a necessity for increasing production of marginal watershed areas particularly affected by ravines and gullies, these measures demand capital inputs and are, therefore, a function of socio-economic analysis. Majority of the land holdings, in ravine-affected areas, are marginal and small having slope varying from 2% to 5%. Among these, majority of farms are adjacent to or close to gully mouth within 6–60 m range and, hence, prone to erosion. Less than half of these holdings did take some conservation measures (particularly bunding and some leveling) in the past. Further, among this group of holdings, a majority (more than 60%) got it done through state agency about 15–20 years back and did not invest subsequently to maintain it. Paucity of funds emerged as the major cause. Others, who spent for construction and/or maintenance, had made only small investment, except for a few large farmers. In fact, this group of farmers with conservation history also had parcels of land away from gully and put under cash crops like tobacco and wheat, which earned them some profits.

Pande et al. (2011) examined the farmers in the ravines of Gujarat, to understand the socio-economic factors affecting farming and conservation decisions on land tilled. It was observed that poor economic condition of farmers impeded their ability to make large-scale investment in conservation agriculture, in general, and improvement of land in particular. Hence, under the present price scenario, low farm profitability made the marginal farms vulnerable to the vicious poverty circle. Such farms, therefore, needed an initial help from the government or any funding agency. When comparing the farmers based on past history of reclamation, it was found that in fact, the farms benefiting from past conservation program of the state government had done better than those who were deprived of it. The former group of farms could lease-in better piece of land and took cash crop to raise their profit level. This fact was considered as advantageous in taking decision by them for investment in land reclamation programs. Thus, financial inclusion of these marginal farms could be an important incentive policy variable for adoption of conservation measures in this region. Further, farmers' total land holding size, land under crop production, and share of cropland as against ravine land in the total land holding positively influenced the farmers' decision to invest on land for reclamation. The choice of cash crops might also have positive relationship with investment decision. Farmers who were benefited from a previous conservation program of the state could lease-in good parcel of land and grow cash crop. This helped them reap farm profit, thereby affecting their decision to invest on land.

The average holding size was slightly bigger on farms with conservation history and so was the ravine land adjacent to these marginal table lands. As the holding size increased, the proportion of ravine land also increased in different classes of holdings. The land use pattern of the two groups of farms (those who followed conservation and those who did not) revealed that the distribution of arable and nonarable land was not markedly different within different landholding classes. Pearl millet-based cropping system was most prevalent across all the categories of farms. Those farmers who had irrigation facility also cultivated irrigated crops like wheat, tobacco, and summer pearl millet. The number of such farms was more under farms with conservation history as compared to the farms without conservation history. It was observed that some of the farms, particularly medium in the former category, had leased-in some better parcels of leveled land and took cash crops like tobacco and wheat.

Farmers reported the incidence of runoff and consequent soil loss during rains; however, as is expected, it was also reported during survey that the loss of topsoil on farms with conservation history was of relatively small magnitude having a little effect on yield loss in contrast to the farms with no conservation measures. Further, economic assessment of this damage due to soil erosion for a typical small-holder farm with land adjacent to ravine revealed that soil loss with farmers' present practice did not significantly affect crop productivity over a longer horizon of 30 years due to sufficient soil depth. Using on-farm trial data from research farm in Mahi ravines, yield-erosion relationship was developed and projected for the smallholder farms in Mahi ravines by Pande et al. (2013). Erosion damage function was also developed with local output prices and cost of production realized by the smallholder farms. Scenarios were developed about farmer's decision to switch over to conservation practice from the existing erosive practice at 10% discount rate considering 30 years of production horizon for different soil loss as case study. The most prominent pearl millet + pigeon pea cropping system on smallholder farms in Mahi ravines was considered for this study. Farmers' nonconservative erosive practice was compared with a conservation practice. Results revealed that conservative practices were more profitable and the decline in yield to the extent of 50% and 75% took much longer in case of conservative practice than the time span realized in case of erosive practice.

## 27.4 Participatory Resource Management

People's participation in ravine afforestation and conservation measures was examined at Sarnal village in Kheda district of Gujarat and reported by Pande et al. (2002). Factors affecting peoples' participation were explored to identify the factor(s) affecting participation in a ravine reclamation program. Peoples' Participation Index (PPI) in the afforestation program to reclaim community ravine land was estimated to be moderate (PPI, 51.3). Participation of landless (PPI, 53.6) was maximum followed by marginal farmers (PPI, 49.8) and small farmers (PPI, 44). This was natural considering the dependence of these groups of farmers on community land for livelihood sustenance. Among the factors affecting participation in community resource management, contact with external agency and frequency of benefits drawn by beneficiaries significantly affected peoples' participation. Frequent contacts led to higher motivation on the part of the stakeholders. Benefits drawn and its distribution across the cross section of stakeholders positively affected their motivation to participate in the community resource management.

## 27.5 Summary and Conclusion

Marginal lands in Mahi ravines largely practice subsistence farming, and income levels are not high enough to induce farmers to invest on land in general and soil and water conservation in particular. Majority of farmers who invested on conservation measures had received state help in the past. These farmers mainly invested on maintenance of earthen bunds (earthen field bund and leveling were reported to be the only conservation measures undertaken). Economic studies observed that real domestic price of pearl millet, the most prevalent cropping system on marginal lands of Mahi ravines, had not been favorable to farmers to give enough incentive. The ratio of average input to output price had also not been favorable in most of the years. The government's help in land leveling and bunding turned out to be a crucial factor governing investment on land, including soil and water conservation. Factors such as credit, income from cash crop, and share of cropland in total holding significantly affected investment decisions on marginal farms. Farms with conservation history were observed to have purchased/leased-in better land parcels and took cash crop such as tobacco, indicating positive effect of past government's assistance. Credit turned out to be a crucial factor suggesting availability of finance on easier terms. While marginal land could be put under profitable agroforestry production systems (such as drumstick- and gooseberry-based systems), higher initial investment could be supported with bank finance. Gullies can successfully be put under permanent vegetation such as bamboo which ensures profitability along with environmental security. High investment on irrigation and protection from wild life can be met from easier bank finance with long gestation period. This could also ensure raising profitable horticultural enterprises such as sapote (*Achrus zapota*) fruit trees on degraded ravine land with supportive conservation measures. Large-scale greening of ravines can be ensured by mobilizing *Kotar/Bihad Sudhar Samitis* at hamlet level under the overall management of *Panchayati Raj* Institutions. Ravine reclamation program can also be merged with various land-based government programs like MGNREGA for financial and other logistic support.

## References


Barbier EB (1988) The economics of farm-level adoption of soil conservation in the uplands of Java, Environment department working paper no. 11. World Bank, Washington, DC

Barbier EB, Burgess JC (1992) Agricultural pricing and environmental degradation, Background paper for world development report. World Bank, Washington, DC

Barrett S (1989) Optional soil conservation and the reform of agricultural pricing policies, LEEC paper 89–08. IIED/UCL London Environmental Economics Center, London

Coxhead Ian A (1995) Economic modelling of land degradation in developing countries, Staff paper 385. University of Wisconsin, Madison

Coxhead Ian A, Shively G (1995) Measuring the environmental impacts of economic change: the case of land degradation in Philippine agriculture, Staff paper 384. University of Wisconsin, Madison

Kurothe RS, Rao BK, Paradiyal AK, Singh AK, Gaur ML (2012) Conservation and production potential of bamboo in ravines lands. Anand Press, Anand, p 204

Pande VC, Khatik GL, Kurothe RS, Nambiar KTN (2002) People's participation in ravine reclamation through afforestation conservation measures – a case study. Indian J Soil Cons 30(1):77–82

Pande VC, Kurothe RS, Singh HB, Tiwari SP (2011) Incentives for soil and water conservation on farm in ravines of Gujarat: policy implication for future adoption. Agric Econ Res Rev 24(1):109–118

Pande VC, Kurothe RS, Rao BK, Kumar G, Parandiyal AK, Singh AK, Kumar A (2012) Economic analysis of bamboo plantation in three major ravine systems of India. Agric Econ Res Rev 25(1):63–73

Pande VC, Kurothe RS, Singh HB, Tiwari SP, Kumar G, Rao BK, Vishwakarma AK, Bagdi L (2013) Economic assessment of soil erosion damage on smallholder farms in marginal lands of Mahi ravines in Gujarat. Agric Econ Res Rev 26(1):63–71

Rao BK, Kurothe RS, Singh AK, Parandiyal AK, Pande VC, Kumar G (2012) Bamboo plantation based technological interventions for reclamation and productive utilization of ravine lands. CSWCRTI, Technical Bulletin No. T-62/V-4: 30pp

Repetto R (1988) Economic policy reform for natural resource conservation, Environment department working paper no. 4. World Bank, Washington, DC

Sharma AK, Pradhan IP, Nema JP, Tejwani KG (1980) 25 Years research on soil and water conservation in ravine lands of Gujarat, Monograph no.2. Central Soil and Water Conservation Research and Training Institute, Vasad

Shiferaw B, Holden ST (2001) Farm-level benefits to investments for mitigating land degradation: empirical evidence from Ethiopia. Environ Dev Econ 6(3):335–358

Tejwani KG, Gupta SK, Mathur HN (1975) Classification, control and reclamation of ravine lands. In: Soil and water conservation research (1956–1971). Indian Council of Agricultural Research Publication, New Delhi, pp 254–310

# Greening Ravine Lands: Policy Issues and the Way Forward

# 28

Jagdish Chander Dagar and Anil Kumar Singh

**Abstract**

Formation of ravines is considered the worst kind of soil erosion. The task of greening of ravines is both a challenge and an opportunity for scientific community. Research efforts across the globe have shown that the ravine lands can be managed as productive land use by adopting suitable engineering and biological techniques. In recent times, much attention is being paid towards inventions regarding gully erosion, and several research results have been published during the last two decades. Many techniques such as agroforestry-based farming systems involving highly remunerative components such as forest and fruit trees, arable crops, medicinal and aromatic crops, livestock, poultry, grasses and vegetables, along with minimum tillage, terracing and check dams, have proved to be effective for gully prevention and control. However, these techniques are rarely adopted by farmers and other stakeholders in the long run because of several bottlenecks in extension services and policy implementation. The rehabilitation of ravines is dependent and influenced by various economic policies like credit, trade, market, taxation, power, transport and above all requirement of costly engineering tools required for reclamation work. Moreover, it requires strong political will. Policy related to greening of ravines needs better state-centre coordination, participatory approach like watershed development programs and effective coordination mechanism between forest, agriculture and irrigation departments both at the national and state level for which establishment of an apex coordinating body or institutional support mechanism needs to be established. National Policy on Agroforestry is already at place, and protocols of Integrated Watershed Development Program are

J. C. Dagar (✉)
Indian Council of Agricultural Sciences, Krishi Anusandhan Bhavan-II, Pusa, New Delhi, India

A. K. Singh
RVS Krishi Vishwa Vidyalaya, R. V. S., Gwalior, India



J. C. Dagar, A. K. Singh (eds.), *Ravine Lands: Greening for Livelihood and Environmental Security*, https://doi.org/10.1007/978-981-10-8043-2_28

well documented, and ravine development mission may be a part of these programs. Soil scientists, environmentalists, agricultural scientists, policymakers and other experts including from other countries would have to work in cohesion with local people in designing effective land reclamation measures. Conservation of natural biodiversity of watersheds and ravines also needs priority.

## 28.1 Introduction

Soil erosion is considered to be one of the most serious problems in the world and has caused major economic, social and environmental disasters. It is estimated that out of about 2 billion hectares land affected by various forms of human-induced land degradation, the land eroded by water is the chief contributor worldwide covering about 1.1 billion hectares of degraded lands (Oldeman 1994). The most widespread form of water erosion is the loss of the topsoil, while terrain deformation by rills and gullies is estimated to occur on about 200 million hectares. Middleton and Thomas (1997) reported 467.4 Mha dryland susceptible to water erosion occurring in all the continents of which 141.0, 213.2 and 113.2 Mha falling in dry subhumid, semiarid and arid regions of the world, respectively. Almost 80% of the terrain affected by water erosion has a light to moderate degree of degradation which also includes gullies and ravines, while around 225 Mha is degraded to the extent that it is no longer suitable for agricultural land use.

Restoration of degraded ravine lands involves a wide range of approaches, from engineering to reforestation to agricultural interventions to reduce harmful practices such as excessive tillage and overgrazing. Agroforestry systems such as plantation-based system for specific outputs such as timber, fuel-wood, fruits, and items of daily needs; silvopastoral systems for livelihood security of different stakeholders; and alley cropping need to be developed to control runoff and soil erosion along slopes and produce fuelwood and crops or forages from the same piece of land, need to be developed. Multi-storeyed plantation-based agricultural systems also ensure richness of species and diversity of products. Lugo (1997) observed similar pattern in Puerto Rico monocultured tree plantations and found that there was high species richness of vegetation under the shade of monoculture stands. This led to the hypothesis that establishment of trees on degraded sites leads to richness of the species with arrested succession, and the same has also been proved at many other sites. But most accepted view is to grow native perennial species of trees, shrubs, grasses and other herbs adapted to the eroded habitats (Dagar 1999). The first step of greening the ravine lands is to protect from overgrazing and allow to establish natural vegetation. This is to be followed by agroforestry approach selecting suitable forest and fruit tree species, grasses and other herbaceous species. Silvopastoral systems play a major role in controlling soil erosion and consequently ravine formation.

Though viable soil conservation techniques have been developed, the adoption of tree-based practices is generally determined by the preference and involvement of farmers and implementation of economically rational policies. For example, when multi-storeyed mixed species systems are being replaced by monocultures, there is little or no extension support or financial incentives for practicing such systems. There are many different mechanisms that most of the governments can use to encourage farmers to take up different plantation or agroforestry systems. Pannell (2011) explained some of the main categories of policy approach relevant to tree-based agricultural systems indicating some key *pros* and *cons*. For instance, incentives in the form of subsidy, marketability and support from industries could help provide political support for development of these systems. At the same time, carbon mitigation and services provided by agroforests could help in promoting corporate deals. Nevertheless, this should be backed by legal instruments.

To help small farmers and other stakeholders in the catchment areas of ravine formation, policymakers (here government of the region) may come out with different holistic packages in the scenario of climate change, increasing globalization of trade and having different socio-economic adaptability of the stakeholders. In these areas, agroforestry-based packages need to be developed which include the payment of carbon sequestered in different tree-based components in ravines, environmental services rendered by these systems, and even to compensate the carbon loss through soil erosion along with runoff and sedimentation. This is a complex issue and needs different approach. Most public policies, however, do not include the environmental services rendered by these systems or even by the farmers into their frame-work. It is, therefore, envisaged that, in the global context of the challenges associated with food security, climate change mitigation, poverty alleviation and preservation of environment and biodiversity, a reorientation of the public policies in relation to reclamation programs including ravines through tree-based systems is warranted, as most of the small and marginal farmers, particularly in ravine areas, still rely on forest and fruit trees and grazing lands for their subsistence. In all, the policy options may include incorporating externalities, participation and inclusiveness, subsidiary and decentralization, transparency and accountability, partnerships with private business, integrated competitive market, environmental justice, economic sustainability, landless labour and crop insurance issues.

Moreover, to meet various diverse needs of ever-increasing human and animal population, particularly in developing and populated countries like India, we need to restore or rehabilitate all degraded lands including ravine lands. By applying appropriate planting and management techniques, various species of forest and fruit trees, forage grasses, medicinal and aromatic and other high value crops have been found equally remunerative and can be grown successfully in these habitats. It is evident from the previous discussion in various chapters of this book that tree-based technologies have additional environmental benefits including huge amount of carbon sequestration, biological reclamation, biodiversity restoration and mitigating climate change. To make the program greening of ravines successful, we need watershed-based participatory approach in mission mode, strong coordination of different departments, strong political will and involvement of scientists and policymakers both at national and state level, if necessary even players from outside the country. The mission 'greening the ravine lands' may find a place in the National Policy on Agroforestry and in Integrated Watershed Development Program, already in place. In the latter program, there is provision of gully plugging and land levelling, etc., but it needs specific attention and separate focussed program on ravine reclamation and restoration.

## 28.2 Constraints/Challenges

Despite obvious benefits of integrated reclamation practices followed worldwide particularly in utilizing the degraded lands for productive use and meeting the livelihood needs, tree-based approach could not get the place it deserves due to various constraints. Some of the challenges are discussed in the National AF Policy (2014), which are also true for ravine land development, and some of the concerns are discussed below:

### 28.2.1 Lack of Suitable Technologies for Identification of the Suitable Areas for Reclamation

There is a need to use remote sensing and GIS techniques to address the problems of land degradation and its mitigation. Rather than using it in a top-down framework, it would be better to make people central to a bottom-up approach to land-use monitoring and evaluation. In other words, there is a need to bring pixels closer to people (Pani 2016). Digitization and use of modern tools of remote sensing and GIS might help in solving some of the key problems in ravine areas, such as poor land records and land demarcation and boundary issues among local people. Further, it will help in monitoring of available natural resources in these habitats and taking proper decision in selection of land-use pattern. We need to identify the specific problematic areas based on land-use classification and apply suitable technologies which may be appropriate for that region. This may also help in identification of suitable species from already growing or identified and adapted vegetation. A lot of work has already been done on identification of suitable plant species for a particular situation in an ecological region.

### 28.2.2 Absence of Focused Land-use Policies and Institutional Mechanisms

Ravine reclamation policy has not gained the desired importance as a resource development tool due to various factors. Some of these factors include land ownership issues; delineation of boundaries in watersheds especially in ravine lands; agroforestry policy-related issues such as restrictive legal provisions for harvesting and transportation of trees planted on farmlands/ravine lands and use of non-timber produce, lack of quality planting materials, inadequate research on plantation models suitable across various ecological regions of the country, lack of post-harvest processing technologies, etc.; near non-existent extension mechanisms; lack of institutional and technical support mechanisms; inadequate marketing infrastructure and price discovery mechanisms; etc. This is also due to the reason that the ravine-effected areas fall under different departments such as agriculture, irrigation, forestry, rural development, etc. At times even ministries are different; hence lack of coordination is always a big bottleneck in implementing any program in such difficult areas.

### 28.2.3 Insufficient Research, Extension and Capacity Building

There are plenty research results on soil conservation and gully formation all over the world, but very limited site-specific technologies have been developed, which may be applicable directly on farmers' fields consisting of ravines and gullies. We need technologies on tree-based farming and rehabilitation of watersheds which are relevant for ravine lands. Even if we have technologies available in public and private domain, most of these do not regularly reach the farmers due to lack of a dedicated extension system. There is a serious lack of institutional mechanisms at all levels to promote land reclamation through these farming systems. Farming enterprise of small farmers needs to be understood, and it requires a convergent program which integrates trees, crops, water, livestock and other livelihood initiatives. This perspective of integration seems to be missing in the national initiatives in whatever form it may currently be. In most of the waterlogged areas, fish culture may be incorporated as state program to encourage farmers to increase their income from various enterprises such as poultry, livestock, fishery and forest and fruit trees along with high-income crops as has been proved in some isolated models discussed in this book. Most of the ravines in India fall in dry ecologies; therefore, we need strong silvopastoral systems based on control grazing or adopting cut-and-carry system, and introduction of palatable legumes along with perennial highly productive and nutritive grasses or forages. The details of impacts of overgrazing and management options are discussed in Chapter 3.

Further, there is not enough research on the reclamation models through vegetation, suitable for the ravines in diverse agroclimatic regions. We also lack information for the utilization of indigenous and multipurpose species particularly on the domestication of species in ravine lands. We must prepare special modules for ravine lands, and the local women must be trained for specific products, for example, pickle making from raw fruits of Tind (*Capparis decidua*), Karonda (*Carissa carandas*), Lasura (*Cordia rothii*), etc.; *murabba* from gooseberry (*Emblica officinalis*); insecticides from neem (*Azadirachta indica*); making plates from leaves of *Butea monosperma*; beekeeping; and processing of dairy-based products. Livestock-based farming systems can help in meeting socio-economic requirement of local farming community and sustaining livelihood. Farmers must be trained to preserve the fodder for lean period and must be educated on the importance of control grazing and cut-and-carry system so that they may increase productivity for better livelihood. They must be trained to identify and domesticate important medicinal plants found in ravines. Specific trainings such as honey production, gum products or silkworm, lac formation, etc. must be organized on regular basis in these areas.

### 28.2.4 Lack of Confidence Building in Farmers

Farmers growing forest or fruit trees are not sure that at harvest they will be able to get proper price of their produce. There should be back-buying guarantee and sufficient number of processing units at site of plantations. There should be sufficient industries at sites of plantation with appropriate regulations so that farmers can sell their product at reasonable price without being cheated. For example, bamboo is a very viable species for many ravine areas, and many of its products have space in rural-based economies; therefore, bamboo-based industry must find a frequent place in ravine areas. There are great scopes of growing medicinal plants in ravine areas provided farmers are assured the market and know-how of production, processing and preserving of these drugs.

### 28.2.5 Lack of Accredited Nurseries and Quality Planting Material

Planting materials such as seeds, seedlings, clones, hybrids, improved varieties, etc. are generally of mixed quality and not available commonly, particularly in the resource-poor regions. It is estimated that only about 10% of planting material is of high- quality in India, the rest without any guarantee for quality standards. This issue mainly relates to the production, handling, distribution and planting and supervision of high-quality planting material. There must be sufficient number of accredited nurseries nearby ravine areas to prepare required germ plasm. Local people must be involved in developing these nursaries. Many species found in ravines are used as medicinal value, these must be domesticated and appropriate techniques must be developed for their cultivation, particularly in stress environment as found in ravine areas. Different stakeholders must be trained for this.

### 28.2.6 Lack of Financial and Market Support

Farmers are not aware of the financial facilities they can avail from the government. For example, there is provision of insurance of the crop, but hardly any farmer comes forward to insure their crops in ravine areas. Moreover, this is also true that institutional finance in ravine areas has not been at par with its potential due to the lack of awareness of technical and economic data on different tree-based models and the techno-economic parameters required by financial institutions to evaluate finance needs and viability of the projects. There is urgent need at government level to understand the harshness being faced by the farming communities of these areas; they must be treated separately keeping all the risk involved in these regions. Similarly, little is done in developing and popularizing insurance products for vegetation-based ventures. Lack of awareness, unavailability of products suitable to growers, high cost of premium and unclear procedure of claim settlements are reported to be the factors responsible for this poor state of affairs. Most of the ravine areas are drought-prone or suffer due to floods; therefore, their crops must be insured on minimum charges from the farmers. All the products of farmers including tree components and animals must be insured.

The marketing infrastructure (market yard, etc.), including 'price discovery' mechanisms for tree-based produce in general, is unavailable in the country, particularly in ravine areas except in a few states which have either developed exclusive marketing infrastructure for agricultural produce or have dovetailed with the regulated agriculture commodity marketing systems. As a result, it is largely a buyer's market and the middlemen get the major share in profit. There is scope of gum production, feed production from pods of leguminous trees, rearing of lac insect, and production of products like honey, pickle or medicinal products. Moreover, the know-how and market facilities must be assured for different stake holders. If possible players from outside country may be involved for investment involving policy makers of the government.

### 28.2.7 Biodiversity Parks

Hardly there is any attention towards rich biodiversity available in various river basins and ravines (see Chap. 3), which can be explored for tourist spots. There is already Chambal River Sanctuary, consisting of a wealth of biodiversity, but this is a highly disturbed site due to sand mines and other exploitations. These areas must be highly protected from misuse and explored for tourism and educational picnic spots.

## 28.3 Socio-economic Considerations

As has been mentioned in initial chapters, soil and gully erosion cause major environmental and socio-economic disasters worldwide. Many urban and rural communities have been severely affected and landscapes and water bodies threatened. In many areas, the gullies have engulfed the roads, houses and establishments and affected many development projects. Human and animal populations, physical infrastructures, agricultural lands and socio-economic systems have adversely exposed to multidimensional hazards. In many developing countries, many villages and communities have been displaced and virtually disappeared because of the scourges of gully erosion. Mudgal (2005) reported that Chambal ravines have affected 948 villages in Bhind and Morena districts in Madhya Pradesh alone destroying the houses and wasting away their cultivable land. For example, Mrigpura village in Morena district is being gobbled up by the ravines so rapidly that all the land around it has turned into deep pits and the village is divided into 18 parts. One can imagine the life of these people when there is no land to till and no pasture to graze their animals and no work to do. Another village, Ratanbasai in Porsha block in Morena district, has split into eight new segments. The streets and roads have been destroyed, and it takes a tough walk across 3 km to cover all segments of the old village. Nayakpura, Rubara, Ajeetpura, Khadoli, Jaghana, Rithona, Mahuwa, Sursani and Gaushpur are among the innumerable ravine-affected villages in this region. Pani et al. (2011), Pani and Carling (2013) and Pani (2016) have expressed various issues related to land degradation and farmers' perception in Chambal region. There had been many social and unlawful activities in the region in the past, and the area has been dacoit-prone as the net of deep ravines provides good hideout to many outlaws. The socio-economic problems related to ravines are equally grave in many other regions also.

The extent of erosion and reclamation and management of ravines are social problems. The reclamation practices not only have to be scientifically robust but also be designed so as to be socially acceptable (Haigh 1984). The lack of environmental perception, poverty, landownership issues (land demarcation as big issue in ravenous area), fragmentation of land holding and lack of liaison between government officials and the farmers limit the extent to which any reclamation measures can be applicable and effective (Singh and Agnihotri 1987; Pani 2016). Moreover, the ravine reclamation issue falls under many ministries and departments; therefore, the problem of coordination is always there. The reclamation program must be brought under single department.

The major causes of land degradation in agricultural ecologies have been identified as socio-economic, policy factors and unsustainable management practices. Lack of economic power of marginal farmers to invest in conservation technologies has also been the cause of natural resource degradation. In fact, poor economic condition of the stakeholders inhabiting ravine ecologies prohibits their ability to make large-scale investments on the conservation of these resources. To meet the needs of ever-increasing population from limited land resources, one cannot afford to leave any piece of land unattended. Further, judicious management of these lands not only helps in enhancement of agricultural production, but environmental benefits are also accrued from ravine ecosystem; therefore, a productive utilization of such degraded lands is called for with conservation of natural resource as the major objective. This warrants investment on rehabilitation of ravine-affected lands. During successive plan periods, beginning second 5-year plan, funds have been allocated by governments for land development including resource conservation under various schemes. Though some areas have been benefitted, the wider impact of these investments is yet to be observed on the ground.

While policy interventions to address the problem of degraded land have largely been a part of broader programs and schemes of the government, the small farmers of ravine areas are usually deprived of these benefits. Agroforestry-based programs may change the socio-economic condition of both farmers and landless population living in ravine areas. For example, extending the production of bio-diesel through cultivation of tree-borne oilseed species (TBOS) such as *Azadirachta indica*, *Madhuca latifolia*, *Pongamia pinnata*, *Jatropha curcas*, *Simarouba glauca* and others and using the wood instead of energy-intensive materials may generate employment and livelihood opportunities in these areas. The promotion of oil industry unit in rural areas, when grown in plenty, could also provide a poverty alleviation option in the rural ravine areas. Farmers can use vacant, waste and marginally used land such as ravines or highly degraded land for growing such trees and benefit from the annual produce, which will add as their income. With the increased green cover, the environment will also benefit greatly.

Another interesting example is of *Prosopis juliflora*, which has been considered as an obnoxious weed in many areas of the country, and environmentalists advocate its complete eradication from all the sites. It is, however, now being explored in large areas in Andhra Pradesh, Rajasthan, Gujarat and Karnataka for its better utilization. The total area covered in three districts of Gujarat (Bhuj, Patan and Surendranagar) is estimated to be 17,000 ha, with a potential of 153,000 Mg of charcoal production and 186,000 man-days per rotation employment generation. Further, it has been reported that for fuel and charcoal making, an employment was generated for 6.34 million man-days and 7.03 million woman-days in Tamil Nadu alone (JC Tewari, personal communication). Further, dealing with socio-economic aspects of agroforestry systems, Dagar and Tewari (2016) mentioned that *Prosopis juliflora* is a major source of fuel for boilers of the power generation plants in Andhra Pradesh. A sum of about ₹1500 $Mg^{-1}$ is the price offered for its wood at factory gate depending on the season and moisture. Three plants in Andhra Pradesh have developed captive plantations of *Prosopis juliflora*. Commissioned biomass-based power plant capacity in Karnataka and Andhra Pradesh is assessed to be 266 MW with 46 power generation units in operation, with almost 78% in Andhra Pradesh alone. The possibilities of installation of *Prosopis*-based power houses in Rajasthan (India) are also being explored. This is most widely distributed in ravine lands and may be cultivated there as commercial tree. Further, its pods have recently been explored for animal feed, and many industries have been installed in Rajasthan for preparing feed for animals. Such enterprises need to be developed with other tree species in ravine areas. There can be many such entrepreneurs, which can be operative in these degraded lands.

Another interesting example of livelihood is from dry regions of Rajasthan, where local people are exploring the gum from *Acacia senegal*, which is also common tree in ravines of dry regions. The leaves of Palash tree (*Butea monosperma*), predominant in Kshipra ravines in Madhya Pradesh, are harvested by local people for making local plate cuplike structures, known as *dona pattal*. The locals also collect gum from this tree, and it is considered a good forage for lac insect. Thus, cottage industries like sericulture (on *Morus alba*), gums and resins and lac cultivation with improved technologies can be developed in these areas. Bamboo cultivation in ravines in many areas has been another success story. This tree-based approach also helps in mitigating climate change. As stated earlier, there is huge potential for cultivation of medicinal and aromatic plants such as *Aloe vera*, *Asparagus racemosus*, *Achyranthes aspera*, *Cassia angustifolia*, *Citrullus colocynthis*, *Lawsonia alba*, *Ocimum sanctum*, *Withania somnifera*, *Adhatoda vasica*, *Boerhavia repens*, *Solanum surattense*, *Tribulus terrestris*, species of *Cymbopogon*, *Vetiveria zizanioides* and so many others. These drugs have very big market, and adopting suitable program socio-economic conditions of the farmers can be improved to greater extent.

Further, as mentioned above in the previous section, many products can be obtained from these lands, and small cotton industries may be developed in rural areas for livelihood and environmental security. As discussed in Chap. 3, these areas are rich in biodiversity, which may prove as national wealth provided protected and utilized judiciously. More details on socio-economic aspects are given in Chap. 27 by Pandey and his associates in this book.

## 28.4 Policy Issues

At global level, watershed development research now has provided powerful technological and policy innovations that are rapidly spreading in Africa, Asia, Latin America and more recently in several developed countries. Now, integrated watershed developments are not only for sustainable production but are also for problem-solving mechanism. For example, for rehabilitation of degraded lands, mitigating climate change through carbon sequestration, employment generation and food and nutrient security through tree-based systems are playing a vital role in rehabilitation and biological reclamation of problem soils such as degraded dry lands prone to water and wind erosion including ravines and sand dunes. Profitable agriculture is possible if appropriate technologies and know-how are used judiciously. Biological barriers such as alley cropping and plugging in ravines, involving appropriate species, play important role in checking water erosion in sloping and ravine lands. Fruit-based agroforests involving trees and low water requiring crops such as cluster bean, pearl millet, mustard and barley as intercrops are most suitable for calcareous ravine lands. Non-conventional crops such as castor (*Ricinus communis*), *Aloe vera*, dill (*Anethum graveolens*), taramira (*Eruca sativa*), isabgol (*Plantago ovata*), senna (*Cassia senna*) and lemon grass (*Cymbopogon* spp.) can be cultivated successfully. Tree-based agriculture provides opportunities to meet the livelihood security of poor and landless masses and mitigate climate change and several other ecological services. For a successful action plan, we need agroforestry and farmer-friendly policies both at national and regional level. If implemented seriously there is no reason that problem of poverty and malnutrition is not solved at global level along with mitigating climate change and having environmental benefits. Moreover, we would have to rehabilitate all degraded lands including ravines to meet the demands of ever-increasing population.

In India, since the 1960s, the government has been coming out with various ravine reclamation programs but with very limited success. In 1967, a Central Ravine Reclamation Board was constituted to ensure ravine control and reclamation having two broad objectives (cited by Pani 2016): (a) to arrest further growth of ravine land and (b) to utilize the already reclaimed land for the productive purposes depending upon land capability and the practical consideration of social and economic conditions. In 1971, a mega ravine reclamation project was launched to develop 55,000 ha of ravine lands for agriculture and about 25,000 ha for forestry in three ravine-dominating states, viz. Rajasthan, Madhya Pradesh and Uttar Pradesh. The task was aimed to be completed in 25 years with a cost of ₹12,500 million, but only about 2000 ha of land was stated to be reclaimed (Trivedi 2015). The big setback was that only a few farmers were benefited by this activity in terms of land reclamation and ravine rehabilitation. Unfortunately, no attempt was made to go in detail of failure, and it appeared that no lession was learnt as is obvious from the fact that in the 1980s and 1990s some massive efforts were made through aerial seeding in ravines for species such as *Prosopis juliflora*, *Acacia nilotica* and *Pithecellobium dulce*. Obviously, this program also met setback and remained unsuccessful because the seeds dropped by the aircraft either washed away along runoff in the rains or get stuck at some place or germinate in the nearby fields making the situation more complicated.

The Department of Land Resources in the Ministry of Rural Development is administering area-based watershed programs in the country for development of wastelands/degraded lands to check the diminishing productivity of these lands which also include ravine lands. Under these programs an area of about 31.6 million hectares was to be covered, and since 1995 about ₹77,386 million have been released (NPCMT 2011). The success of the program is yet to be seen. Moreover, before investing huge money, one must evaluate the past performance, and the failures must be accounted for. The reclamation of ravines must be brought under a separate and focussed program as this is a sensitive area having a delicate ecological balance. In certain areas, it is possible to level the ravines, but for that a thorough Digital Elevation Model mapping should be done for identifying the proper slopes for levelling. We must not attempt those areas which have steep slopes. A thorough assessment of the impact of the change in land-use pattern on ecology of Chambal watershed and its hydrology and livelihood of the people, who live here, is critical and important. Further, the Chambal river system is home to several kinds of biodiversity such as fish, alligators and crocodiles, migratory birds and terrestrial animals (see Chap. 3) and a variety of plant species; so environmental considerations are also to be kept in mind.

It is pertinent to mention that most of the strategies followed by the government during the last two decades have been flexible and influenced by pseudo-economic indicators (Trivedi 2015). Now it is felt that there is a need for change in the basic thinking of such strategies. This much of large area with all its complex problems cannot be taken up by the government alone. Since long, it is felt that a bigger program on ravine reclamation must be started as a mass movement involving the local community as the key focal point and all other institutions (central, state, local governments, NGOs and the private sector) as support partners. Ravine ecosystem is a product of the erosive forces of water. In alluvial soil, it becomes necessary to follow the geographic unit of a catchment and to integrate its various aspects to produce economically viable returns to the community (RDR 2006). The guidelines have already been formulated for watershed development, which aim to create conditions conducive to higher agricultural productivity of the area while conserving natural resources which must be followed in reclaiming the ravines also with a few modifications if necessary. The entire work must be undertaken on mission mode involving skilled technical manpower and competent administration. The village *panchayats* and other stakeholders need to be taken into confidence. They must have understanding on how they will be benefited from this program.

The major activities to be undertaken by watershed development programs include soil and moisture conservation measures like construction of check dams, water harvesting structures, desilting of village ponds, treatment of drainage lines/gullies, land levelling, bunds on farms, treatment of problem soils, agroforestry including agri-horticulture, silvopastoral systems, organic farming, use of biofertilizers, value addition and marketing of produce through farmers' groups and training and capacity building of stakeholders. On paper, this is a complete package which can be advocated for any kind of land development and reclamation, but it needs complete planning to implement it at ground level. Ravine development program though fits in all these schemes, but this needs special attention as the problem is unique and complex to handle as compared to other reclamation programs. This requires involvement of most specialized persons, local communities and all other stakeholders. Moreover, we have now National Agroforestry Policy (National AF Policy 2014) at place, which must be taken into consideration while formulating tree-based programs in ravine catchments. If need be international collaboration may also be encouraged.

Ravine rehabilitation requires an integrated approach of treatment of table and marginal lands contributing runoff to the gullies on watershed basis. Using gullies according to land capability classes, following appropriate soil and water conservation measures and bringing land under permanent vegetation cover involving afforestation, fodder and fruit tree-based pasture systems and energy plantations are some of the simple and cost-effective ways of rehabilitation of the ravine areas. Despite technological demonstration of ravine land rehabilitation, so far their adoption by the farmers has not been very encouraging. Thus, there is immediate need for developing new affordable technologies, refinement of existing technologies, capacity building and technology transfer at field level. In addition, intangible benefits due to ravine rehabilitation also need to be demonstrated in clear terms to the stakeholders. Reclamation of every inch of land for productive utilization is emerging as a national priority. Therefore, ravine reclamation program must get priority. Farmer and other stakeholders must be supported both technically and financially through viable policy initiatives. Besides technological know-how, they must be given all initiatives such as cost of their innovations; availability of reliable germ plasm; cheap loans; special insurance schemes for drought, floods, failure of crops including trees and damage caused by natural calamities and wildlife; and assured market. Special rural industries, based on the ravine products such as bamboo, plywood, gum and resin, sericulture, lac production, food processing units, livestock-based products, etc., must be established on priority, as these will be based on livelihood security of people living in most disadvantage ecologies.

It is felt that most of the ravine lands can be brought under perennial vegetation adopting agroforestry practices which will help in controlling erosion, reducing runoff, improving *in situ* soil moisture conservation, increasing water table and improving productivity as well as profitability. Use of tree-based systems as a mitigation strategy for climate change is considered the major weapon in the hands of policymakers. This also creates a major opportunity for carbon sequestration. Concerted efforts are required to identify practical cost-effective means through changes in land-use practices using perennial vegetation as an option. The conversion of unproductive croplands and grasslands to agroforests has the best potential to soak up atmospheric carbon in ravine-affected areas.

Thus, for rehabilitation of ravines, we would have to adopt integrated approach of land levelling (where required), trenching, bunding in agricultural fields, gully plugging and using perennial vegetation involving scientists from different disciplines (soil, engineering, environment, social science, agroforestry, etc.), policymakers, administrators and local stakeholders. Capacity building at different levels must be the first step rather than the last step. Above all, the ravine reclamation program would have to be taken up on mission mode rather than with casual approach.

## 28.5 Conclusions and the Way Forward

Reclamation of ravine lands is a complex task which requires adequate technological, financial and policy support. There is a need to develop coordination among beneficiary groups, developmental agencies and policymakers for taking initiatives for undertaking developmental work for reclamation of ravine lands. However, the success of ravine reclamation program is linked with careful micro-level planning, adaption of economically viable production systems, effective community mobilization and efficient monitoring and policy support. Ravine area development program will not only contribute to land reclamation but also will help in bridging the growing demand and supply gap for food, fibre and woody produces to be obtained from these lands and in restoring livelihood security for resource constraint communities and mitigate climate change impacts. This may lead to development of cottage industries in rural areas to process the raw material such as wood, agroforestry products, gums and resins and livestock-based products. This will provide farmers more returns and widen the scope of employing more people. Hence, a pragmatic view of the ravine development is very important for ever-increasing population. To rear livestock, the demand for forage is increasing day by day. The degraded ravine lands may be put to productive use through silvopastoral systems to obtain economic returns while protecting the environment. Indeed, there is a need to plant fuel, fodder or multipurpose trees and grasses on ravine lands to provide fuel and forage for rural households and a supplementary source of regular income to the farmers besides protection to these lands against further degradation. Preliminary work under NICRA and Niche Area Project has made some progress on evolving technology for control and reclamation of deep/very deep ravines. Multistep levelling and superimposition of remunerative agroforestry/agri-horti/agri-medicinal plants have been found successful. Also, holding the sand/silt *in situ* through construction of modified gabions has met a grand success. All these techniques need to be tested on large scale in an Operational Research Project (ORP) for control and reclamation of deep/very deep ravines. ICAR may consider funding an ORP for at least a decade or so to show lasting results. Many cottage industries may be installed with proper technological inputs for both livelihood and environmental security.

The following recommendations may be useful for future action plan for developing programs for ravine reclamation:

- Modern remote sensing geospatial tool and GIS application along with adequate ground truthing should be used intensively to assess and monitor the ravine lands and developmental activities being undertaken in these regions.

- A State Ravine Research and Management Centre may be set up preferably near ravenous area of Morena (where Chambal ravines are predominant) to provide site-specific solutions.
- New affordable reclamation technologies must be developed, and extension of the same must be strengthened involving the participation of local communities.
- Site-specific agroforestry systems must be developed for ravines found in different ecologies integrating forest and fruit trees, grasses, medicinal and aromatic crops and livestock components.
- Cottage industries must be developed in rural areas for value addition and to process agroforestry products such as wood; pods to create feed; gums and resin; fruits for juice, jelly and pickle; other tree products; and livestock-based products.
- Ravines have a unique diversity representing a huge genetic pool. It is essential to conserve the existing flora and fauna of ravines to maintain the ecology and contribute towards livelihood of local people. Ravine biodiversity parks need to be set up in the ravine regions.
- Groundwater potential available with ravines is to be utilized through tube wells linked with micro-irrigation systems (sprinkler and drip systems fertigation) preferably using solar energy.
- Creation of market facilities ensuring the proper price of different commodities to farmers.
- Skill development of different stakeholders.
- Policy initiatives regarding the availability of cheap loans to the farmers for invention in ravine areas and insurance of crops (including trees) against drought, natural calamities including floods, hailstorm and wildlife damage.
- Initiatives to different stakeholders compensating them for their products contributing towards carbon sequestration and other environmental services rendered through tree plantations and biodiversity conservation.

## References

Dagar JC (1999) Soil erosion parameters with special reference to Kshipra ravines in Madhya Pradesh. Indian J For 22(3):210–221

Dagar JC, Tewari JC (2016) Agroforestry research developments: anecdotal to modern science. In: Dagar JC, Tewari JC (eds) Agroforestry research developments. Nova Science Publishers, New York, pp 1–45

Haigh MJ (1984) Ravine erosion and research in India. Geoform 15:543–561

Lugo AE (1997) The apparent paradox of re-establishing species richness on degraded lands with tree monocultures. For Ecol Manag 99:9–19

Middleton N, Thomas D (1997) World atlas of desertification, 2nd edn. Arnold Pub, London, p 182

Mudgal MK (2005) Socio-economic impact of ravine lands: a case study of river Chambal basin of state of Madhya Pradesh, India. Geogr Res Abstr 7:00529

National AF Policy (2014) National agroforestry policy 2014, vol 29. Government of India, Department of Agriculture & Cooperation, Ministry of Agriculture, New Delhi

NPCMT (2011) Irrigation: National Watershed Development Program-Agriculture Sector-National Portal of India; National Portal Content Management Team (NPCMT), Reviewed on 07-01-2011. http://www.archieve.india.gov.in/sectors/agriculture/index.php?id=7. Accessed 08/06/2017

Oldeman LR (1994) The global extent of soil degradation. Soil resilience and sustainable land use. CAB International, Wallingford, pp 99–118

Pani P (2016) Controlling gully erosion: an analysis of land reclamation processes in Chambal Valley, India. Dev Pract 26(8):1047–1059

Pani P, Carling P (2013) Land degradation and special vulnerability: a study of inter-village differences in Chambal Valley, India. Asian Geogr 30(1):65–79

Pani P, Mishra DK, Mohapatra SN (2011) Land degradation and livelihood in semiarid India: a study of farmers' perceptions in Chambal Valley. Asian Profile 39(5):505–519

Pannell DJ (2011) Policy perspectives on changing land management. In: Pannell DJ, Vanclay FM (eds) Changing land management: adoption of new practices by rural land holders. CSIRO Publishing, Canberra, pp 177–187

RDR (2006) Rajasthan development report, planning commission, government of India, New Delhi. Academic Foundation, New Delhi. ISBN-81-7188-464-4

Singh S, Agnihotri SP (1987) Rill and gully erosion in the subhumid tropical riverine environment of Teonthar Tahsil, Madhya Pradesh, India. Geografiska Annaler. Series A Phys Geogr 69(1):227–236. https://doi.org/10.2307/521380

Trivedi S (2015) Can Madhya Pradesh meet the Chambal challenge? Bus Stan 7:2015. www.business-standard.com/article/printer-friendly-version?article_id=115020700951_1

# Index

**A**
Aberrant weather, 34
Aboveground biomass, 101, 103, 105–107, 440
*Abrus precatorius*, 91
*Abutilon asiaticum*, 85
*Abutilon indicum*, 82, 84, 94, 95, 129
*Abutilon pannosum*, 89
*Acacia benthamii*, 79
*Acacia catechu*, 81, 84, 87, 123–126, 129, 136, 250, 291, 299, 319, 341, 344, 345, 389, 391, 466, 529, 537–539, 542, 545
*Acacia eburnea*, 80, 81, 93, 94, 98, 101, 127, 350, 355
*Acacia farnesiana*, 81
*Acacia jacuemontii*, 79
*Acacia leucophloea*, 79–81, 87, 101, 110, 121–127, 136, 139, 291, 295, 319, 349, 355, 366, 374, 376, 387, 516, 517, 526
*Acacia mearnsii*, 56, 337, 342
*Acacia nilotica*, 53, 79–81, 87, 93, 98, 101, 110, 121, 122, 124–127, 136, 139, 168, 230, 231, 246, 269, 290, 291, 293, 295, 296, 300, 301, 342, 347, 349, 350, 352, 355, 366, 374, 376, 378, 380, 387, 389, 390, 392, 438, 440, 468, 472, 502, 529, 537–540, 542, 545, 594, 610
*Acacia senegal*, 79, 93, 98, 101, 110, 111, 121, 124, 125, 171, 339, 349, 350, 366, 387, 389, 437, 609
*Acacia tortilis*, 53, 79, 98, 110, 136, 230, 291, 295, 301, 349, 350, 355, 378, 389, 390, 502
*Acalypha indica*, 82, 86, 128, 129
*Acalypha malabaricum*, 82
*Acanthospermum hispidum*, 82
*Achrus zapota*, 599
*Achyranthes aspera*, 81, 83, 86, 89, 94, 113, 609
Acid soils, 197, 301, 425
*Adhatoda vasica/zeylanica*, 88, 126, 129, 295, 502, 609
*Aegle marmelos*, 54, 81, 87, 99, 110, 128, 129, 228, 229, 250, 371, 377, 378, 438, 440, 466, 501
*Aeluropus lagopoides*, 94
Aeolian erosion, 550
*Aerva lanata*, 86, 129
*Aerva persica*, 79
*Aeschynomene indica*, 86
Afforestation, 57, 110, 112, 137, 138, 167, 233, 234, 258, 265, 267, 270, 271, 283, 311, 312, 315, 319, 321, 322, 324–326, 434, 456, 461, 524, 526, 528, 529, 531, 546, 547, 595, 596, 598, 612
*Agave americana*, 51, 269, 275
*Agave mexicana*, 88
*Ageratum conyzoides*, 82, 86, 89, 94
Aggregate analysis, 113
Agricultural landscape, 500
Agricultural mismanagement, 3, 32, 48, 200, 280, 362
Agricultural systems, 71, 165, 347, 386, 603
Agri-horticultural system, 52, 145, 228–229, 293, 294, 302, 369–373
Agri-horti-silvicultural systems, 293–295
Agriproduct-processing, 573
Agri-silvicultural system, 293
Agri-silvopastoral systems, 296
Agroforestry, 134–135, 282–283, 290–302, 327–328, 331, 332, 335–357, 361–381, 464–469, 499–502
Agroforestry interventions, 122, 134–135, 282, 290–297, 340, 344, 356
Agroforestry systems, 52, 163, 281, 283, 288, 290, 292–302, 331, 332, 336, 340, 341, 346–348, 361–381, 438, 441, 602, 614
*Ahaetulla nasuta* (Common Vine Snake), 75



J. C. Dagar, A. K. Singh (eds.), *Ravine Lands: Greening for Livelihood and Environmental Security*, https://doi.org/10.1007/978-981-10-8043-2

*Ailanthus excelsa*, 79, 87, 129, 136, 250, 291, 349, 355, 376, 466, 468
*Alangium salvifolium*, 87
*Albizia amara*, 53, 349, 352, 390
*Albizia chinensis*, 56
*Albizia lebbeck*, 56, 79, 81, 87, 129, 136, 250, 291, 293, 295, 301, 349, 352, 355, 374, 376, 389, 390, 466, 468, 502
*Albizia odoratissima*, 87
Alfisols, 200, 203, 328
*Alhagi pseudalhagi*, 86, 89, 94
Alley cropping, 293, 301, 332, 342, 348, 357, 374, 419, 602, 610
*Alloteropsis ciruicina*, 85
Alluvial fans, 32
Alluvial gullies, 9, 10, 26, 37, 39, 46
Alluvial plains, 11, 32, 71, 165, 218, 386, 479
Alluvial zones, 362, 434
*Alnus nepalensis*, 56, 300, 344, 352
*Aloe barbedensis*, 85
*Aloe vera*, 113, 129, 226, 250, 276, 295, 466, 609, 610
*Alternanthera pungens*, 85
*Alternanthera sessilis*, 86, 94
*Alysicarpus longifolius*, 81–83, 107, 122, 387
*Alysicarpus minima*, 79
*Alysicarpus monilifer*, 82, 83, 85, 97
*Alysicarpus rugosus*, 82, 83, 107, 122, 387
*Alysicarpus vaginalis*, 89
*Amaranthus gracilis*, 83, 85
*Amaranthus spinosus*, 85, 89
*Amaranthus viridis*, 89
Amazonas State, 204, 205
*Ammania baccifera*, 86
*Ammania multiflora*, 86
*Ampelocissus latifolia*, 91
*Amphiesma stolatum* (Buff Striped Keelback), 75
*Amphilophis glabra*, 103, 105
*Amphilophis odorata*, 105
Amphitheatre gullies, 39
*Anagallis arvensis*, 85, 95
*Andrographis paniculata*, 85, 297
*Andropogon ischaemum*, 103, 105, 169, 286
*Andropogon pumilus*, 56, 82, 98
*Anethum graveolens* (Dill), 610
Anicuts, 271, 274, 462, 478–504, 507, 510, 524, 526–529, 531, 543, 546, 547
Animal biodiversity, 70, 74–78
Animal husbandry, 281, 321, 323, 326
*Anisomeles indica*, 89
Anjan grass (*Cenchrus ciliaris*), 54, 296, 302, 355, 393, 394, 465, 466
*Annona reticulata*, 465, 468, 472, 511, 512
*Annona squamosa*, 54, 81, 110, 377, 502
*Anogeissus acuminata*, 81
*Anogeissus latifolia*, 84, 87, 93, 123
*Anogeissus pendula*, 87, 98, 122–126, 139, 349
Anthropogenic erosion, 550
Anthropogenic factors, 30–32, 35, 48, 96, 506
*Anthyllis cytisoides*, 102
*Antilope cervicapra* (Black Buck), 76
*Aonla* (*Emblica officinalis*), 87, 130, 269, 465, 468, 502
Aphrodisiac properties, 75
*Apluda aristata/mutica*, 80–83, 94, 95, 98, 103, 104
*Aponogeton natans*, 89
Aquatic erosion, 550
Araçagi gully, 210
*Arachis pintoi*, 187
Archaeological dating, 47
*Argemone mexicana*, 86, 89, 94, 95, 122, 128, 129
*Argyrogena fasciolata* (Banded Racer), 75
Aridisols, 200
Arid zone, 11, 71, 165, 386
*Aristida adscensionis*, 79, 85, 92, 98, 121, 365, 387
*Aristida cynantha*, 344
*Aristida depressa*, 92
*Aristida funiculata*, 80, 92, 98
*Aristida royaliana*, 365
*Arroyo*, 9, 25, 35, 386
*Arthraxon ciliaris*, 82
*Arundinella nepalensis*, 56
*Arundo donax*, 51, 57, 86, 114, 344
*Asparagus racemosus*, 79, 81, 85, 91, 113, 129, 465, 466, 526, 545, 609
*Asphodelus tenuifolius*, 89, 95
*Asteracantha longifolia*, 86
*Atylosia platycarpa*, 89
*Atylosia scarabaeoides*, 291
Available nutrients, 73, 111, 144, 171, 172
*Avenula bromoides*, 102
Avian fauna, 74
*Axis axis* (Chital), 74, 76
*Azadirachta indica*, 57, 81, 87, 101, 123–127, 129, 136, 168, 230, 291, 293, 295, 300, 301, 319, 349, 350, 355, 374, 376, 388–390, 437, 529, 537–539, 542, 545, 605, 608
*Azospirillum*, 175
Azotobacter, 145–158, 176, 177

**B**

Backyard nursery, 474
Backyard poultry, 474
*Bacopa monnieri*, 86, 89, 113

Badlands, 11, 23, 32, 34, 35, 41, 49, 446–449
Bael (*Aegle marmelos*), 228, 371, 466, 582
*Balanities aeegyptiaca/roxburghii*, 51, 54, 79–81, 93, 127, 295, 349, 350, 462
*Baliospermum montana*, 85
Bamboo-based agroforestry system, 301
Bamboo-based production system, 231–232, 502
Bamboo grooves, 399
Bamboo live check dam, 405, 406
*Bambusa arundinacea*, 81, 86, 115
*Bambusa bambos*, 137, 400
*Bambusa nutans*, 301
*Bandicota bengalensis* (Lesser Bandicoot Rat), 76
Bank gullies, 10, 37, 258
*Barleria acanthoides*, 79
*Barleria prionitis*, 81, 89, 295
*Batagur dhongoka* (Three-striped Roofed Turtle), 75
*Batagur kachuga* (Red-crowned Roofed Turtle), 75
Bater (*Coturnix coturnix*), 74
*Bauhinia purpurea*, 56, 345
*Bauhinia racemosa*, 81, 87, 126
*Bauhinia retusa*, 57
*Bauhinia variegata*, 81, 376
Bee-keeping, 583, 587, 605
Belowground biomass, 101, 105–107, 438
Bench terraces, 40, 268, 286, 328, 329, 595
Benefit-cost ratio (B:C ratio), 227, 228, 232, 502, 596
Berger-Parker dominance index, 123
Ber (*Ziziphus mauritiana*), 40, 52, 134, 226, 228, 293, 339, 373, 465, 468, 502
Bhabar grass (*Eulaliopsis binata*), 57, 105, 224, 289, 300, 341, 344, 345
B-horizon of soil, 29
Bio-climates, 96
Bio-diesel, 608
Biodiversity, viii, 4, 31, 38, 42, 70, 119–139, 162, 167, 282, 297, 300, 336, 347–348, 356, 357, 391, 394, 447, 451, 456, 478, 479, 497, 500, 506, 603, 609, 611
Biodiversity hotspots, 214
Biodiversity inventories, 74
Biodiversity park, 607, 614
Bio-edaphic serial stages, 100
Bio-energy, 350
Bioengineering, 211, 596
Bio-fertilizers, 173, 611
Bio-fuel, 350
Biogeography, 74
Biological activity, 500
Biological measures, 284, 289, 328
Biological nitrogen fixation, 174, 343
Biological spectrum, 96–97
Biomass, 51, 78, 106–109, 120, 158, 184, 190, 218, 265, 282, 289, 297, 301, 302, 343, 350, 367, 391, 417, 434, 437, 441, 442, 447, 450, 451, 464, 466–469, 472, 480, 499, 500, 520, 530, 540–542, 592
Biomass production, 52, 53, 107, 175, 265, 276, 289, 301–302, 343, 346, 350, 353, 440, 464, 500, 520
Biomass structure, 78
Biome, 184, 447
*Biophytum sensitivum*, 83, 85, 86, 89, 94, 95
Bio-reclamation, 80
Bioremediation, 345, 356
Biotic climax, 100
Biotic interference, 80, 97, 98, 122, 163, 222, 230, 342, 347, 381, 388, 390
Biotic pressure, 82, 96, 99, 120, 122, 127, 138, 365, 387, 435
Black gram (*Vigna mungo*), 40, 159, 225, 250, 288, 293, 320, 328, 349, 465, 468, 469, 471–473
*Blainvillea latifolia*, 85
*Blepharis asperrima*, 85, 89
*Blepharis maderaspatensis*, 82
Blue-green algae, 175
*Blumea eriantha*, 89
*Blumea lacera*, 82, 89
*Blumea membrana*, 82
*Bocorroca*, 9
*Boehmeria rugulosa*, 57
*Boerhaavia diffusa*, 79, 82, 83, 89, 126
*Boiga trigonata* (Common Cat Snake), 75
*Bonamia latifolia*, 79
Bori-bunds, 496
*Borreria stricta*, 82
*Boselaphus tragocamelus* (Neel Gai), 76
*Boswellia serrata*, 88, 93
*Bothriochloa intermedia*, 85, 349
*Bothriochloa pertusa*, 54, 80–83, 85, 93, 95, 98, 103, 104, 107, 109, 113, 127, 231, 296, 319
Boulder check dams, 288
*Brachiaria decumbens*, 56, 187
*Brachiaria humidicola*, 187
*Brachiaria mutica*, 54, 56, 230, 389
*Brachiaria ramosa*, 92
*Brachiaria reptans*, 85
*Brachypodium retusum*, 102
*Brassica juncea*, 40, 469
*Brassica nigra*, 464
Brazilian Central Plateau, 197
Brazilian savanna, 183
*Broussonetia papyrifera*, 344, 389

Brush wood, 134, 282, 342
Brush wood check-dams, 134, 167, 282
*Buchanania cochinchinensis*, 438
*Buddleja asiatica*, 89
Buffalo (*Bison bison*), 74
*Bulbostylis barbata*, 92
Bulk density, 49, 145, 146, 177, 300, 344, 526, 544, 551
Bundelkhand region, 123, 124, 338, 339, 352
*Bungarus caeruleus* (Common Krait), 75
*Butea monosperma*, 80, 81, 87, 93, 101, 110, 121, 127, 129, 139, 295, 355, 511, 605, 609

**C**

*Caesalpinia cristata*, 81
*Caesalpinia sepiaria*, 81
*Caesulia axillaris*, 86
*Cajanus cajan*, 40, 85, 341
Calcic soils, 203
*Calligonum polygonoides*, 98
*Calotes versicolor* (Oriental Garden Lizard), 75
*Calotropis gigantea*, 84, 88
*Calotropis procera*, 79, 85, 88, 93, 98, 121, 128, 129
*Canis aureus* (Common Jackal), 76
*Canis lupus pallipes* (Indian Wolf), 76
*Canna coccinea*, 89
*Cannabis sativa*, 88
Capacity building, 58, 323, 395, 409, 583, 586, 605, 611–613
*Capparis decidua*, 51, 53–54, 79, 81, 84, 88, 93, 94, 101, 110, 121–123, 128, 129, 139, 228, 319, 350, 366, 387, 474, 526, 587, 605
*Capparis grandis*, 81
*Capparis horrida*, 81
*Capparis sepiaria*, 80, 81, 88, 93, 94, 101, 121, 526, 541, 545
*Capparis zeylanica*, 121, 366, 387
*Capsella bursa-pastoris*, 86
*Caracal caracal* (Caracal), 76
Carbon sequestration, 24, 232, 271, 282, 296, 301–302, 346, 357, 393, 394, 410, 440–442, 451, 464–470, 472, 520, 603, 610, 612, 614
Carbon sink, 346, 434, 437–440, 464, 500
*Cardiospermum halicacabum*, 85, 91, 113
*Carex fedia*, 86
*Carissa carandas*, 53, 81, 84, 88, 98, 110, 122, 128, 130, 139, 228, 296, 297, 319, 377, 378, 462, 473, 501, 582, 587, 605
*Carthamus oxyacantha*, 95
*Cassia auriculata*, 98
*Cassia fistula*, 81, 87, 130
*Cassia javanica*, 81
*Cassia kleinii*, 90
*Cassia occidentalis*, 82, 90
*Cassia pumila*, 82, 95, 97
*Cassia senna*, 610
*Cassia siamea*, 87, 342, 349
*Cassia sophera*, 81, 85
*Cassia tora*, 90, 94, 121, 130, 366, 387
Castor (*Ricinus communis*), 374, 465, 466, 594, 610
Catchment area, 10, 22, 30, 31, 36, 42, 43, 52, 164, 261, 263, 280, 298, 345, 362, 486, 527, 552, 554, 603
*Catharanthus roseus*, 130
Cation exchange capacity, 329, 508, 512, 517
Causative agents, 2
*Cayratia trifolia*, 91
$^{14}C$ dating, 47
*Celosia argentea*, 85, 90
*Cenchrus biflorus*, 79, 98
*Cenchrus ciliaris*, 54–56, 79, 80, 98, 104, 110, 113, 134–136, 168, 169, 229, 286, 289–291, 295–297, 299, 349, 352, 355, 356, 376–378, 389, 390, 392, 393, 437, 438, 465, 466, 482, 501, 526, 541, 594–596
*Cenchrus setigerus*, 54, 79, 80, 85, 94, 98, 134, 230, 349, 376, 389, 390, 392, 482
Cenozoic sedimentary cover, 184
*Centella asiatica*, 85
*Ceropegia bulbosa*, 90, 122
Cerrado environment, 190
Cerrado vegetation, 187
*Chamaeleo zeylanicus* (Indian Chameleon), 75
Chamaephytes, 96
Chambal River, 34, 35, 42, 49, 53, 71, 72, 74–77, 80, 96, 164, 237, 239, 241–244, 257, 261–264, 311, 313, 322, 364, 370, 387, 437, 494, 607, 611
Chambal Valley, 35, 39, 71, 164, 255, 260–265, 268, 270, 280, 448, 524
Channel erosion, 9, 29, 30, 288, 512, 514, 553
Check dam, 51–53, 114, 115, 134, 137, 139, 222, 223, 233, 274, 275, 282, 283, 286–288, 299, 310, 324, 327, 339, 355, 404, 405, 408, 461, 462, 473, 482, 485, 486, 491, 507, 524, 543, 595, 596, 611
Check-dam system, 50, 52
*Chenopodium album*, 90, 126
*Chenopodium murale*, 85, 90
Childs equation, 240, 246

*Chinensis cynosuroides*, 526
*Chitra indica* (Narrow-headed Softshell Turtle), 75
*Chloris barbata*, 82, 92, 98, 104
*Chloris gayana*, 56, 85, 94, 105, 376
*Chlorophytum arundinaceum*, 84, 113, 466
*Chlorophytum tuberosum*, 90
C-horizon of soil, 29
*Chromolaena (Eupatorium) odoratum*, 342
*Chrozophora prostrata*, 86, 93, 94
*Chrozophora rottleri*, 86, 90, 94
*Chrysopogon fulvus*, 50, 56, 57, 82, 83, 85, 93, 95, 98, 103, 104, 107, 108, 113, 121, 122, 231, 250, 296, 341, 344, 345, 387, 389, 465, 501
*Chrysopogon montanus*, 289
Chute spillway, 50, 488, 491, 492, 595
*Cicer arietinum*, 40, 469
*Cissampelos pareira*, 92
*Citrullus colocynthis*, 92, 113, 128, 609
*Citrus aurantifolia*, 87, 371
*Citrus limon*, 40, 52, 81, 293, 339
*Citrus reticulata*, 354, 371
Civilization, 3
*Cleome brachycarpa*, 95
*Cleome viscosa*, 85, 95, 122
*Clerodendrum inerme*, 94
*Clerodendrum multiflorum*, 126
*Clerodendrum phlomidis*, 79, 81, 93, 94, 101
Climate and environmental changes, 11, 24
Climate change, vii, viii, 23, 26, 47, 49, 50, 58, 165, 256, 281, 282, 346, 362, 398, 442, 448, 456, 603, 609, 610, 612, 613
Climatic factors, 46, 551, 560
Climax, 80, 97, 98, 100
*Clitorea ternatea*, 92
*Coccinia grandis*, 92, 113
*Cocculus hirsutus*, 82, 92, 94
*Coelognathus helena* (Common Trinket Snake), 75
*Coix lacryma-jobi*, 86
Cold desert, 21, 35, 36
Colluvial gullies, 9, 10, 36, 50
*Commelina benghalensis*, 82, 86, 90
*Commelina forskalli*, 86
*Commelina haskarlii*, 86
*Commelina kurzii*, 82, 86
*Commiphora wightii*, 84, 123
Communities, vii, 7, 28, 32, 55, 58, 70, 78, 93, 96–98, 100, 103, 113, 122–124, 135, 167, 168, 218, 222, 229, 232, 233, 235, 265, 290, 296, 297, 314, 323, 327, 347, 351, 464, 480, 524, 547, 576, 580, 582–584, 586, 598, 605, 607, 613
Complex community, 97
Coniferous forest, 399, 406
Conservation agriculture, 436, 441, 592, 597
Conservation agronomic practices, 594
Conservation measure, 57, 138, 163, 167, 168, 224, 226, 248–250, 281–282, 284, 287, 290, 327, 336, 337, 339, 386, 391, 399, 442, 481, 494, 507, 567, 592–599
Conservation practice, 6, 186, 200, 222, 234, 269, 302, 353, 366, 555, 558, 559, 598
Construction of the palisades, 189
Continuous gullies, 288
Continuous scarp front gullies, 39
Contour cultivation, 338, 559
Contour hedgerows, 452, 500
Controlled grazing, 472
Convention on Biological Diversity (CBD), 132
*Convolvulus arvensis*, 85
*Convolvulus microphyllus*, 84, 90, 94
*Convolvulus pluricaulis*, 82
*Convolvulus prostratus*, 90
*Conyza leucantha*, 90
*Corbichonia decumbens*, 79
*Corchorus aestuans*, 86
*Corchorus bilocularis*, 86
*Corchorus depressus*, 90
*Corchorus olitorius*, 82
*Corchorus trilocularis*, 82
*Cordia dichotoma*, 87, 99, 371
*Cordia myxa*, 130, 228, 344, 377, 501
*Cordia rothii*, 81, 84, 98, 124, 127, 349, 587, 605
Cottage industries, 138, 474, 609, 613, 614
Cover crop, 52, 115, 162, 374, 440, 441, 594
Cow (*Bos taurus*), 74
*Crinum defixum*, 86
*Crinum latifolium*, 90
Critically endangered (CR), 74, 77
Crocodylus palustris (Mugger or Marsh Crocodile), 75
Crop geometry, 250
Crop management factor, 353, 558
Crop residues, 283, 351, 367–369, 440, 441
*Crotalaria burhia*, 79, 98, 121, 366, 387
*Crotalaria medicaginea*, 83, 85, 90, 98
*Crotalaria sericea*, 85
Crow (*Corvus splendens*), 74
*Cryptolepis buchanani*, 92
Cryptophytes, 96
C turnover, 440, 442
*Cucurbita pepo*, 582
Cultural attributes, 576
*Cuon alpinus* (Dhole), 76
*Curculigo orchioides*, 90

*Cuscuta reflexa*, 85, 91
*Cyanotis axilaris*, 82
*Cyathocline purpurea*, 85
*Cymbopogon citratus*, 466, 468
*Cymbopogon jwarancusa*, 98
*Cymbopogon martinii*, 81–83, 92, 93, 95, 98, 103, 104, 468
*Cymbopogon nardus*, 289
*Cynodon dactylon*, 50, 56, 80, 82, 84, 86, 94, 95, 103–105, 113, 114, 122, 126, 127, 130, 231, 289, 296, 387, 526, 541, 558
*Cynodon plectostachys*, 344
*Cynopterus sphinx* (Short-nosed Fruit Bat), 76
*Cyperus bulbosus*, 86
*Cyperus compressus*, 86
*Cyperus exallatus*, 86
*Cyperus iria*, 86
*Cyperus kyllingea*, 86
*Cyperus rotundus*, 82–85, 92, 94, 126
*Cyperus triceps*, 86

**D**
*Daboia russelii* (Russell's Viper), 75
*Dactylis glomerata*, 57
*Dactyloctenium aegyptium*, 79, 85, 86, 92
*Dactyloctenium sindicum*, 79, 85, 98
Dairy farming, 581
*Dalbergia sissoo*, 53, 79, 80, 87, 94, 98, 101, 110, 122–126, 136, 168, 291, 295, 337, 343–344, 349, 350, 355, 376, 388–391, 468, 471, 502, 529, 537–539, 542, 545
Danian stage, 72
Darcy's law, 240
Dating techniques, 26
*Datura fistuosa*, 88
*Datura innoxia*, 88
*Datura metal*, 85, 94, 130
*Datura strumarium*, 88
Deccan trap, 35, 71, 72, 164
Decision-making process, 585
Deep ravine, 49, 86, 93, 110, 145, 220, 222, 226–232, 243–244, 247, 250, 255, 259, 260, 266, 268, 270–271, 283, 286, 287, 310, 312, 316, 324, 325, 332, 366, 367, 369, 370, 375–377, 379, 380, 402, 434, 435, 450, 452, 461, 474, 481, 483–494, 496–498, 607, 613
Deer (*Axis axis*), 74
Deforestation, 3, 25, 32, 37, 47, 48, 56, 97, 139, 164, 184, 258, 280, 362, 390, 448, 452, 480, 500, 550
Degraded habitats, 31, 57
Dehydrogenase activity (DHA), 158–159, 178
*Delonix regia*, 81
*Dendrelaphis tristis* (Common Bronzeback Tree Snake), 75
Dendritic gullies, 39
*Dendrocalamus giganteus*, 301
*Dendrocalamus hookerii*, 301
*Dendrocalamus strictus*, 53, 79, 88, 130, 134, 136, 137, 169, 231, 232, 291, 295, 302, 342, 355, 356, 389, 391, 392, 400, 502, 596
*Dendrocalmus hamiltonii*, 400
Denudation of vegetation, 31
Denuded surfaces, 46
Desert climate, 99
Desert Development Program (DDP), 234
Desertification, 198, 478, 506
*Desmodium gangeticum*, 85
*Desmodium triflorum*, 82, 98
*Desmostachya bipinnata*, 79–80, 85, 86, 92, 98, 541
*Dichanthium annulatum*, 54, 230, 365, 376, 377, 387, 482
*Dichanthium caricosum*, 81, 83, 98, 103, 104, 107, 108, 113
*Dichrostachys cinerea*, 53, 81, 88, 93, 98, 101, 319, 389, 526, 540, 542, 545
*Digera alternifolia*, 85
*Digera muricata*, 90
Digital elevation model (DEM), 554, 611
Digital photogrammetric system, 46
*Digitaria adscendens*, 80, 86, 104
*Digitaria setigera*, 85, 86
*Dinebra arabica*, 85
*Diospyros melanaxylon*, 438, 511, 512, 516, 517
*Diospyros montana*, 81, 516, 517
*Dipteracanthus prostratus*, 93
Discontinuous furrows, 595
Discontinuous gullies, 31, 32, 38
Dispersible soils, 37
Distal gullies, 39
Distribution of species, 78
Diversity, 348
Diversity in richness, 78
Diversity indices, 122, 123, 348
Dominance, 78, 94, 110, 121, 123, 168, 365, 435, 448
Dominance index, 123
Drainage area, 23, 31, 37, 38, 48
*Dregea volubilis*, 85
Drop inlet spillway, 488, 490
Drop spillway, 50, 56, 286, 488, 489, 491
Drought, 34, 37, 120, 127, 132, 137, 139, 233, 234, 299, 327, 332, 340, 348, 395, 400, 416, 418–421, 426, 427, 456, 606, 612, 614

Drought-prone areas program (DPAP), 234
Droughts, 49, 207, 297, 378, 459
Drudgery in ravine areas, 585
Drudgery of women, 585
Dry deciduous forest, 100, 101
Dryland, 7, 8, 20, 40, 328, 419, 602
Dry-sub-humid areas, 602
DTPA-extractable micronutrients, 111, 170

**E**

Earthen gully plugs, 222, 270, 272, 403–405, 486, 496, 594, 595
Earthen marginal bunds, 52
Earthquakes, 35
*Echinochloa colonum*, 86, 95, 104
*Echinochloa crusgalli*, 82, 104
*Echinops echinatus*, 82, 84, 90, 93, 95, 98
*Echis carinatus* (Saw-scaled Viper), 75
*Eclipta prostrata*, 82, 84, 86, 90, 94, 526
Ecological adaptations, 78
Ecological balance, 218, 222, 366, 592, 611
Ecological restoration, 102, 104, 167
Ecological security, 393, 507
Ecological services, 282, 610
Ecological succession, 97, 190
Ecological sustainability, 596
Economic analysis, 232, 349, 441, 543, 546, 593, 595
Economic empowerment, 581, 582, 587
Economic growth, 575, 579, 580, 583
Economic returns, 229, 232, 233, 296–298, 326, 393, 394, 410, 502, 613
Economic sustainability, 501, 603
Ecosystem services, 104, 120, 128, 132, 135, 281, 347, 501
Edaphoecotropism, 102
*Ehretia acuminata*, 81
Electrical conductivity, 145, 148, 544
Electromagnetic induction sensors, 46
Elemental toxicity, 479
*Eleocharis atropurpurea*, 86
*Eleocharis dulcis*, 276
*Eleocharis plantaginea*, 86
*Elephantopus scaber*, 90, 126
*Eleusine compressa*, 79, 98
*Eleusine indica*, 86, 104
*Elytraria acaulis*, 90
*Emblica officinalis*, 40, 52, 53, 79, 81, 87, 110, 130, 134, 135, 228, 229, 250, 289, 293, 339, 355, 438, 465, 468, 501, 502, 582, 605
*Emilia sonchifolia*, 85, 90
Endangered species, 74, 332
Energy concept, 247
Energy matrix, 184
Energy plantation, 57, 134, 167, 282, 350, 364, 367, 434, 612
Engineering methods, 50, 354
*Enhydris sieboldii* (Siebold's Smooth-water Snake), 75
*Enicostema axillare*, 90
*Enicostema verticillatum*, 80, 82
Entisols, 200, 203, 260, 261
Environmental factors, 7, 42, 49, 336
Environmental security, vii, viii, 70, 218, 219, 221–229, 231–235, 357, 481, 547, 599, 609, 613
*Ephedra foliata*, 79
Ephemeral gully, 7, 9, 47
Ephemeral gully erosion model (EGEM), 47
*Ephidra foliata*, 84
Epiphytes, 96
*Eragrostiella bifaria*, 85
*Eragrostis ciliaris*, 79
*Eragrostis curvula*, 57, 289
*Eragrostis diarrhena*, 86
*Eragrostis pilosa*, 82
*Eragrostis poaeoides*, 85
*Eragrostis tremula*, 79
*Eremopogon foveolatus*, 121, 365, 387
*Erianthus (Saccharum) munja*, 51
*Erigeron canadensis*, 85
*Eriochloa procera*, 85
Erosion models, 6, 11, 24, 37, 47–48, 554, 555, 559, 563, 564
Erosion-prone gullies, 102, 239
Erosion rate, 7, 8, 40, 42, 43, 45–47, 101, 102, 144, 213, 345, 362, 479, 554, 564, 565
Erosive power, 27, 31, 165, 189, 262, 551
*Eruca sativa*, 251, 465, 469, 471, 610
*Erythrina indica*, 376
*Erythrina poeppigiana*, 340
*Erythrina suberosa*, 57, 344
*Eryx johnii* (Red Sand Boa), 75
*Eucalyptus camaldulensis*, 56, 79, 230, 342, 350, 355
*Eucalyptus citriodora*, 79
*Eucalyptus globulus*, 337, 342
*Eucalyptus grandis*, 56, 342
*Eucalyptus tereticornis*, 124, 125, 136, 230, 289, 291, 296, 336, 337, 350, 355
*Eucalyptus umbellata*, 87
*Eulaliopsis binata*, 57, 105, 289, 290, 299, 341, 344
*Eupatorium coelestinum*, 85
*Euphorbia antisyphilitica*, 295, 350
*Euphorbia dracunculoides*, 90

*Euphorbia geniculata*, 81–83, 86
*Euphorbia hirta*, 83, 85, 90, 94, 95
*Euphorbia hypericifolia*, 82, 83, 85
*Euphorbia thymifolia*, 83, 85, 90, 93
*Eutropis carinata* (Keeled Grass Skink), 75
*Eutropis innotata* (Blanford's Grass Skink), 75
*Eutropis macularia* (Bronze Skink), 75
Evolutionary pathways, 447
Evolution of gullies, 32–34
*Evolvulus alsinoides*, 82, 84, 90, 94, 526
Extension program, 410, 586
Extent of gully erosion, 19

**F**
Factors affecting gully formation, 30
*Fagonia cretica*, 79
Fallow, 113, 164, 282, 290, 299, 300, 341, 408, 457, 458, 500, 556, 558
Farm women, 575, 578–580, 583, 585–587
Farmers' perception, 607
*Farsetia hamiltonii*, 122
Father of soil conservation, 5, 21
Faunal biodiversity, 74, 348
*Felis chaus* (Jungle Cat), 76
*Felis silvestris* (Wild Cat / Desert Cat), 76
*Feronia limonia*, 79–81, 87, 99, 101, 110, 123, 228, 295, 349, 355, 582
Ferruginous crusting, 34
Fibrous root system, 110, 169, 299, 390, 399
*Ficus benghalensis*, 81, 87
*Ficus carica*, 81, 228
*Ficus glomerata*, 81, 115, 123
*Ficus hispida*, 87
*Ficus palmata*, 87
*Ficus racemosa*, 87
*Ficus religiosa*, 87
*Fimbristylis dichotoma*, 86
*Fimbristylis miliacea*, 86
*Flacourtia indica*, 87, 319
Flash floods, 446, 452
Floodplain catchment areas, 36
Floodplain water bodies, 36
Fluorescine diacetate (FDA), 158, 178
Fluvial erosion, 7, 198, 446, 550
Fodder grasses, 51, 52, 139, 230, 311, 337, 342, 376, 399, 466, 473, 482, 520, 595
Fodder species, 124–127, 594
*Foeniculum vulgare*, 225, 374
Food and environment security, 271
Food and nutritional security, 313
Forest cover, 25, 31, 46, 282, 398
Fossiliferous sedimentation, 35
Fox (*Valpes benghalensis*), 74
*Fumaria indica*, 86, 95
*Funambulus pennantii* (Five-striped Palm Squirrel), 76

**G**
Gabion, 56, 134, 167, 248, 269, 272–274, 276, 288, 339, 461, 462, 486, 487, 495, 496, 498, 613
Gabion check dams, 247, 282, 288, 486
*Gavialis gangeticus* (Ghariyal), 75
*Gazella gazella* (Chinkara), 76
Gender challenges, 574
Gender equality, 576, 579
*Geochelone elegans* (Indian Star Tortoise), 75
Geographic information systems (GIS), 7, 46, 233, 553, 554, 560, 566, 567, 604, 613
Geological advances, 46
Geologic time, 197
Geomorphic features, 24
Geomorphic threshold estimation, 46
Geomorphological evolution, 35
Geomorphological process, 196, 197, 213
Geophytes, 96
Geotextiles, 211, 484
Giant *Leucaena*, 414–417, 420, 424–427
Glacial erosion, 550
*Glinus lotoides*, 85, 94, 122
Global erosion hotspots, 196, 214
Global extent of water erosion, 20
Global rates of soil erosion, 8, 42
*Glomus aggregatum*, 421
*Gloriosa superba*, 92
*Glossocardia bosvallia*, 82
*Glycine max*, 85, 469
*Gmelina arborea*, 136, 291, 300, 344, 389, 472, 502
*Gnaphalium indicum*, 86, 94
*Gnaphalium luteo-album*, 86, 94
Goat (*Capra aegagrus*), 74
*Gomphrena celosioides*, 85
*Gongylophis conicus* (Common Sand Boa), 75
Gooseberry (*Emblica officinalis*), 40, 52, 228, 371, 376, 582
Gram (*Cicer arietinum*), 40
*Grangea maderaspatana*, 85
Grassed water-ways, 266, 269, 327, 352, 483, 595
Grassland communities, 78, 100
Gravitational erosion, 35
Grazing management, 37, 53
Grazing-resistant species, 52
Grazing susceptible number (GSN), 82, 85
Green gram (*Vigna radiata*), 40, 328

Greening of ravine lands, 169, 271, 456–474, 599, 603
Green manuring, 174, 355, 373
*Grewia asiatica*, 130, 228, 526, 582
*Grewia flavescens*, 88
*Grewia optiva*, 56, 349
*Grewia tenax*, 87, 121, 366, 387
Gross erosion, 42, 43, 346
Groundwater, 35, 37, 47, 169, 222, 233, 320, 338, 339, 419, 553, 563, 564
Groundwater Loading Effects of Agricultural Management Systems (GLEAMS), 47, 553
Groundwater tables, 35
Guggal (*Commifora wightii*), 84, 88, 123, 466
Guinea grass (*Panicum maximum*), 250, 290, 376, 391
Gullies, 24–57, 113–114, 183–192, 195–214, 247–248, 269, 272, 274, 286–287, 405, 488–493
Gully bed stabilization, 51, 222, 223, 226, 286, 291, 296, 481, 483, 486, 596
Gully channels, 24, 30, 31, 33, 38, 46, 187, 189, 288
Gully control, 31, 48, 51, 233, 268, 338, 488–493
Gully density, 7, 41
Gully dynamics, 23, 26
Gully erosion, 24–26, 30–31, 43–50, 195–214
Gully erosion models, 11, 24, 47–48
Gully evolution, 31–38, 46, 47, 203, 211, 212
Gully head, 5, 22, 23, 29, 33, 45, 46, 48, 50, 52–53, 113, 188–191, 203, 211, 218, 222, 226, 255, 257, 273, 274, 283–285, 287, 288, 324, 405, 462, 480–483, 488, 491, 496, 547, 596
Gully management, 38
Gully plugs, 48, 114, 115, 258, 270, 272, 286–287, 352, 355, 378, 403, 457, 462, 486, 496, 527, 528, 543, 594, 595
Gully prevention, vi, 2, 11, 24, 34
Gully stabilisation, 48, 191, 192, 226, 231, 269, 274, 290, 296, 378, 381, 481, 483, 486, 496
*Gymnosporia montana*, 79
*Gynandropsis pentaphylla*, 95

**H**

Habitat loss, 105
*Hachelochloa granularis*, 86
Halophytes, 94
*Hardella thurjii* (Crowned River Turtle), 75
*Hardwickia binata*, 87, 349, 352, 473
Hare (*Lepus ruficaudatus*), 74
Head cutting, 446
Hedge row, 289, 354, 419
*Helictotrichon filifolium*, 102
*Heliotropium curassavicum*, 94
*Heliotropium ellipticum*, 90
*Heliotropium indicum*, 95
*Heliotropium supinum*, 85
*Hemarthria compressa*, 86
*Hemiadalphis polyspermus*, 82, 94, 95
Hemi-cryptophytes, 96
*Hemidactylus brookii* (Brook's House Gecko), 75
*Hemidactylus flaviviridis* (Yellow-bellied House Gecko), 75
*Hemidactylus frenatus* (Common House Gecko), 75
*Hemidactylus leschenaultii* (Bark Gecko), 75
*Hemidactylus maculatus* (Northern Spotted Gecko), 75
*Hemidactylus triedrus* (Termite Hill Gecko), 75
*Hemidesmus indicus*, 92, 130
*Hemiechinus collaris* (Collared Hedgehog), 76
*Hemigraphis hirta*, 82, 85, 94, 95
Hen (*Gallus gallus domesticus*), 74
*Herpestes edwardsii* (Indian Grey Mongoose), 76
*Herpestes javanicus* (Small Indian Mongoose), 76
*Herpestes smithii* (Ruddy Mongoose), 76
Heterogeneity, 37, 78, 440
*Heteropogon contortus*, 50, 54, 56, 80–83, 92, 93, 95, 98, 100, 103, 104, 108, 110, 113, 121, 122, 289, 296, 365, 387
*Heteropsylla cubana* (*Leucaena* psyllid), 420
*Heylandia latebrosa*, 82, 83, 85, 93–95
*Hibiscus caesius*, 90
Highland, vii, 3, 8, 25, 41
Hillslope gullies, 9
*Hipposideros fulvus* (Fulvous Leaf-nosed Bat), 76
Historical studies on gully erosion, 11, 23–26
*Holoptelea integrifolia*, 79, 87, 94, 101, 319, 355, 376, 388
Home gardens, 348, 584
Horti-silvo-pastoral system, 297
House sparrow (*Passer domesticus*), 74
Humid regions, 6, 343
*Hyaena hyaena* (Striped Hyaena), 76
Hydraulic conductivity, 240
Hydraulic gradient, 32, 33
Hydrodynamic energy, 242, 494
Hydrodynamic pressure, 244
Hydrogeomorphic mechanisms, 37, 46

Hydrogeomorphic processes, 46
Hydrological balance, 32, 362, 419, 499
Hydrological processes, 48, 406, 501
Hydromorphic soils, 184
Hydrophytes, 85, 96
*Hygrorrhiza aristata*, 86
*Hystrix indica* (Indian Crested Porcupine), 76

**I**
Igneous rock, 35, 72
Impact assessment, 524–547
*Impatiens balsamina*, 86
Importance value index (IVI), 78, 81–84
Improved fallows, 348, 500
Improved gabions, 248
Inceptisol, 185, 200
Income-generating activities, 576, 580, 582–587
*Indigofera cordifolia*, 85, 98
*Indigofera enneaphylla*, 79
*Indigofera linifolia*, 79, 81, 83, 85, 94, 97
*Indigofera linnaei*, 82, 83, 90
*Indigofera trifoliata*, 82, 86, 98
*Indigofera trita*, 82
Indo-Gangetic plains, 32, 221, 446, 481, 525
Infilling of gullies, 37, 38, 48
Infiltration of nutrients, 70
Infiltration rate, 4, 28, 31, 107, 162, 163, 170, 203, 237, 239, 244, 268, 281, 300, 341, 347, 365, 387, 407, 408, 436, 508, 511, 512, 514, 516, 517, 519, 551
Information and communication technology (ICT), 586
Innovative techniques, 246–247
Institutional mechanisms, 604, 605
Integrated farming system, 323, 326–329, 332
Integrated nutrient management (INM), 150, 173–178, 470, 471
Integrated Wastelands Development Program (IWDP), 234, 603
Integrated watershed management (IWM), 285, 336–340, 353
Inter-rill erosion, 4, 29, 450
Inundation, 7, 37, 47
Inverted Berger-Parker dominance, 123
*Iphigenia indica*, 82, 84, 94
*Ipoemia aquatica*, 86
*Ipoemia aquatica fistulosa* (*carnea*), 86
*Ipomoea carnea*, 57, 88, 95, 114, 130, 344, 389
*Ipomoea fistulosa*, 81
*Ipomoea hederifolia*, 81, 92
*Ipomoea nil*, 81
*Ipomoea obscura*, 81
*Ipomoea pes-tigridis*, 92
*Ischaemum rugosum*, 82, 83, 86, 103, 104
*Iseilema anthephoroides*, 82, 98
*Iseilema laxum*, 54, 81–83, 85, 98, 103, 104, 107, 109, 113, 122, 296, 387
*Itapecuru* and *Barreiras* Formation, 203
IUCN Red List, 74

**J**
*Jasminum grandiflorum*, 81, 88
*Jasminum sambac*, 88
*Jatropha curcas*, 84, 88, 295, 437
*Jatropha gossypifolia*, 81
*Jatropha podagrica*, 88
*Juncus acutus*, 102
Jungli hen (*Gallus gallus*), 74
*Justicia diffusa*, 82, 83, 90, 97

**K**
*Kalmegh* (*Andrographis paniculata*), 297
*Kankar* (calcareous) pan, 355, 365
Karonda (*Carissa carandas*), 130, 228, 297, 371, 377, 582, 605
*Kickxia racemosa*, 82
*Kickxia ramossissima*, 93
Kinetic energy, 27, 28, 35, 340, 551, 552, 556
King fisher (*Coriacias benghalensis*), 74
Kinnow mandarin (*Citrus reticulata*), 371, 372
Kitchen gardens, 581, 584, 585
*Knoxia corymbosa*, 85
*Koa haole* (Common Leucaena), 414, 426
*Kochia indica*, 94
Koel (*Endynamys scolopacea*), 74
Kshipra ravines, 54, 81, 83–84, 122, 127, 366, 609
Kshipra watershed, 96, 98, 99
Kyoto protocol, 434

**L**
Lacustrine erosion, 550
*Lagascea mollis*, 82, 83, 86, 95
*Laggera aurita*, 86
Lake erosion, 550
Land capability class, 57, 166, 167, 283, 592, 612
Land degradation, 3, 4, 20, 24, 34, 71, 162, 163, 165, 168, 174, 195–198, 200, 201, 203, 211, 213, 218, 234, 235, 260, 265, 280, 297, 313, 336, 362–363, 386, 388, 398, 434, 441, 480, 500, 506, 519, 526, 592, 602, 604, 607, 608

Land levelling, 284, 338, 436, 440, 442, 603, 611, 612
Landsat data, 165, 280
Landsat imagery, 452
Land shaping, 272, 276, 461, 462, 464, 472
Land use conversion, 34
Langur (*Presbytis antellus*), 74
*Lannea grandis*, 57
*Lantana camara*, 81, 88, 121, 123, 366, 387, 516, 517
*Lantana indica*, 89
Laplace equation, 240
*Laptadenia pyrotechnica*, 79
*Lasora (Cordia dichotoma)*, 371
Late Pleistocene-Holocene, 34
*Lathyrus aphaca*, 85, 90
*Lathyrus sativus*, 85
*Launaea nudicaulis*, 82, 93–95
*Launea acaulis*, 90
*Launea aspleniifolia*, 82
*Lavaka*, 4, 9, 196
*Lawsonia alba/inermis*, 51, 84, 609
Leaf fodder, 293, 348, 369
Lemon (*Citrus limon*), 40, 52, 134, 229, 293, 339
*Lepidagathis cristata*, 82, 84, 90, 94
*Lepidagathis hamiltoniana*, 90
*Leptadenia pyrotechnica*, 79, 121
*Leptochloa fusca*, 94
*Lepus nigricollis* (Indian Hare), 77
*Leucaena collinsii*, 424
*Leucaena cruziana*, 423
*Leucaena diversifolia*, 424, 426
*Leucaena esculenta*, 420, 424, 426
*Leucaena greggii*, 424
Leucaena hybrids, 424, 426
*Leucaena lanceolata*, 423, 424
*Leucaena leucocephala*, 53, 56, 87, 110, 111, 124, 125, 127, 136, 168, 171, 269, 289, 291, 293, 296, 300, 341, 343, 349, 350, 374, 376, 390, 413–427, 437, 529, 537–539, 542, 545
*Leucaena leucocephala* subsp. *glabrata*, 414, 424–426
*Leucaena leucocephala* subsp. *Leucocephala*, 414
*Leucaena macrophylla*, 424
*Leucaena pallida*, 420, 424
*Leucaena psyllid*, 420
*Leucaena pulverulenta*, 423, 424, 426
*Leucaena retusa*, 424
Leucaena rhizosphere, 422
*Leucaena salvadorensis*, 424
*Leucaena shannonii*, 424
*Leucaena trichandra*
*Leucaena trichodes*, 424
*Leucas urticifolia*, 82
Lianas succulents, 96
LiDAR data, 46
Life form, 78, 96
Life-saving irrigations, 52, 225, 288, 292, 327, 350, 380
Limiting factor, 364, 368, 409
Limnic erosion, 550
*Lindenbergia indica*, 86
Linear gullies, 39
Lisem model, 7, 555, 566, 567, 569
*Lissemys punctata* (Indian Flapshell Turtle), 75
Litholic soils, 184
Lithology, 48, 447, 448
Livelihood, viii, 70, 120, 122, 127, 132–139, 218, 219, 221–229, 231, 232, 234, 235, 281, 296, 297, 327, 332, 337, 357, 389, 394, 464, 472–473, 506, 546, 547, 574–587, 598, 604, 605, 609, 611, 613, 614
Livelihood security, vii, 20, 58, 75, 120, 135–139, 235, 297, 351, 356, 362, 411, 460, 474, 575, 602, 610, 612, 613
Livestock herding, 74
Livestock population, 218, 368, 394
Livestock productivity, 368
Local communities, 124, 137, 500, 520, 611, 612, 614
Log check dams, 287
*Lolium perenne*, 57
Loose stone check dams, 287
Lopping, 126, 295, 369, 374
*Lusiurus sindicus*, 98
*Lutrogale perspicillata* (Smooth-coated Otter), 77
*Lycium barbarum*, 98
*Lycium europaeum*, 79
*Lycodon aulicus* (Common Wolf Snake), 75
*Lycodon striatus* (Barred Wolf Snake), 75
*Lygeum spartum*, 102
*Lygosoma punctata* (Spotted Supple Skink), 75

**M**

*Macaca mulatta* (Rhesus Macaque), 77
*Macfadyena unguis-cati*, 89
Macro-nutrients, 544
*Macropisthodon plumbicolor* (Green Keelback), 76
*Macroptilium atropurpureum*, 291
*Macrotyloma axillare*, 187
*Madhuca longifolia var latifolia*, 87

*Maerua oblongifolia*, 121
Magmatism, 35
Maina (*Acridatheres tristis*), 74
Maize (*Zea mays*), 40, 341, 469
*Malanocenchris jacquemontii*, 79, 82, 94, 97, 98, 104
Mammal, 74–77
Management options, 11, 24, 162, 282, 293, 605
Mango (*Mangifera indica*), 40, 52, 81, 87, 110, 134, 293, 371
*Manis crassicaudata* (Indian Pangolin), 77
Maranhão Island, 202, 207
Margalef index, 123
Marginal land, 57, 134, 139, 150, 162, 167, 170, 222, 224–226, 228, 230, 280, 284–286, 301, 316, 324, 327, 365, 370–374, 379, 434, 435, 437, 481, 494, 501, 593, 594, 598, 599, 612
Marília Formation, 184, 186
Marine erosion, 550
Martian gullies, 11
*Martynia annua*, 85, 90
Masonry check dams, 225, 288, 547
*Mavastrum coromandelianum*, 82
*Maytenus emarginata*, 79, 121
*Maytenus senegalensis*, 81, 89
Mean weight diameter (MWD), 146, 163, 177, 178, 229, 300
*Medicago polymorpha*, 85
*Medicago truncatula*, 423
Medicinal and aromatic crops, 332, 614
Medicinal herbs, 508, 587
Medicinal plants biodiversity, 127–128
Mediterranean environments, 38
*Megaderma lyra* (Greater False Vampire Bat), 77
Mega-phanerophytes, 96
*Melanocenchris jacquemontii*, 79, 82, 94, 97, 98
*Melhania hamiltonia*, 91
*Melia azedarach*, 81, 376
*Melilotus alba*, 85
*Melilotus indica*, 85
*Melilotus parviflora*, 95
*Melinus minutiflora*, 57
*Mellivora capensis* (Honey Badger), 77
*Melursus ursinus* (Sloth Bear), 77
*Mentha arvensis*, 297
*Merremia gangetica*, 86, 91
Meso-phanerophytes, 96
Mesophytic species, 381
Mesopotamia civilization, 3
*Meytenus emerginata*, 93
*Michelia oblonga*, 300, 344
Microbial biomass, 158
Micro-catchment, 228, 257, 354, 377, 499
Micronutrient content, 110, 155, 157, 170
Micro-nutrients, 179, 331, 526
Micro-phanerophytes, 96
*Miliusa tomentosa*, 88, 511, 512
*Millardia meltada* (Soft-furred Field Rat), 77
*Milletia hamata*, 89
*Millingtonia hortensis*, 81
*Mimosa hamata*, 80, 81, 98
*Mimosa himalayana*, 89
*Mimosa rubicaulis*, 81
Mimosine, 417–422, 424, 427
*Mimosops elengi*, 81
Minimum tillage, 436, 440
Mint (*Mentha arvensis*), 297
Modified Soil Loss Equation, 6
Mohenjodaro and Harappa civilizations, 3
Moisture conservation, 70, 233, 234, 292, 328, 336, 338, 340–345, 347, 353, 354, 356, 373, 377, 381, 398, 403, 440, 464, 482–484, 494, 496, 612
*Mollugo hirta*, 91
*Momordica charantia var. muricata*, 92
*Momordica dioica*, 79, 92, 113
Monoclimate theory, 100
Morgan, Morgan & Finney Model, 555, 561, 562
*Moringa oleifera*, 57, 81, 88, 110, 124, 130, 250, 293, 302, 374, 375, 465, 467, 468, 471, 501, 502, 540, 545
Morphological and hydraulic characteristics, 23
*Morus alba*, 54, 89, 131, 349, 354, 389, 609
*Morus indica*, 81
Mountain torrent, 144
*Mucuna pruriens*, 92
*Mukia maderaspatana*, 79
Mulching, 56, 187–189, 342, 391
Mulching contour wattling, 56, 484
Multi-purpose tree species (MPTS), 56, 124, 293, 369, 500
Multistep leveling system, 247
Multi-tier system, 297
*Mus musculus* (House Mouse), 77
Mushroom cultivation, 474, 581, 583
Mustard (*Brassica juncea*), 40, 175, 469

**N**

*Naja naja* (Spectacled Cobra), 76
Nano-phanerophytes, 96
National Bamboo Mission, 137, 139, 399, 410
National Chambal Sanctuary, 74, 166

National economy, 254, 524
National Policy on Agroforestry, 601, 603
National Remote Sensing Agency (NRSA), 165, 280, 387
National watershed development program, 234
Natural resource conservation, 231, 233, 297–301, 398
Natural succession, 52, 97, 167
Natural vegetation, 52, 79, 106, 121, 168, 271, 310, 316, 318, 352, 365–366, 387, 435–437, 461, 506, 520, 526, 531, 540–543, 558, 602
*Neel gai* (*Boselaphus tragocamelius*), 74
Neo-tectonic activities, 35
*Nepeta hindostana*, 91
*Nerium indicum*, 81, 89, 131
*Nerium oleander*, 102
*Nicotiana plumbaginifolia*, 86, 94
*Nicotiana tobacum*, 40
*Nilssonia gangetica* (Indian Softshell Turtle), 76
*Nilssonia hurum* (Peacock Softshell Turtle), 76
Nitrogen fixation, 295, 341, 416, 419, 500
Nitrogen mineralization, 173
Nival erosion, 550
Non-arable land, 53–57, 134, 144, 167, 232, 281, 337, 388, 594, 597
Nutrient cycing, 70, 120, 169, 174, 436, 499, 506
Nutrient deficiencies, 34, 163, 174, 178, 388, 479
Nutrient imbalance, 479
Nutrient management, 144–159, 224
Nutrient use efficiency (NUE), 150–152, 175, 178, 223, 482
Nutrition security, 313, 583
*Nyctanthes arbor-tristis*, 81

**O**

*Ocimum americanum*, 82, 84, 95, 295
*Ocimum canum*, 91
*Ocimum gratissimum*, 85, 91
*Ocimum sanctum*, 131, 295, 297, 609
*Ocimum tenuiflorum*, 91
*Oenanthe javanica*, 85
*Oldenlandia aspera*, 85
*Oldenlandia corymbosa*, 82, 84, 94, 95, 97
*Oligodon arnensis* (Common Kukri Snake), 76
*Oligodon taeniolatus* (Russell's Kukri Snake), 76
*Ophisops jerdonii* (Jerdon's Snake-eye), 76
*Ophisops microlepis* (Small-scaled Lacerta), 76
*Ophisops minor nictans* (Indian Dwarf Lacerta), 76
*Oplismenus burmannii*, 92
*Oplismenus compositus*, 92
*Opuntia dilleni*, 89
*Opuntia elatior*, 85
*Opuntia vulgaris*, 89
Organic fertilizers, 292
Organogenic erosion, 550
*Orobanche aegyptiaca*, 94
*Oropetium thomaeum*, 79, 85, 97
*Orthosiphone pallidus*, 91
Ottangan river, 122, 124
Over-grazing, 31, 48, 53, 97, 390, 602, 605
Owl (*Bubo zeylonensis*), 74
*Oxalis corniculata*, 86
*Oxalis martiana*, 86
Oxisols, 200

**P**

Palaeo-channel, 34, 35
Palatable species, 52, 169
*Panchayati Raj* Institutions, 599
*Pangshura tecta* (Indian Roofed Turtle), 76
*Pangshura tentoria* (Indian Tent Turtle), 76
*Panicum antidotale*, 50, 54, 85, 86, 104, 105, 134, 136, 169, 230, 231, 291, 352, 389, 482, 502
*Panicum coloratum*, 57
*Panicum maximum*, 54, 85, 134, 290, 341, 344, 352, 376, 389
*Panicum sumatrense*, 541
*Panicum trypheron*, 85
*Panicum turgidum*, 98
*Pannisetum purpureum*, 54
*Panthera pardus* (Leopard), 77
*Panthera tigris* (Tiger), 77
Papiya (*Clamator jacobinus*), 74
*Paradoxurus hermaphroditus* (Asian Palm Civet), 77
*Paraechinus micropus* (Indian Hedgehog), 77
Paraíba State, 204
Parasite, 85, 94, 96, 174
*Parkia roxburghii*, 300, 344
*Parkinsonia aculeata*, 473
Parrot (*Psittacula krameria*), 74
*Parthenium hysterophorus*, 86, 91, 94
Particle binding energy, 241, 494
Partridge (*Francolinus pondicerianus*), 74
*Paspalidium flavidum*, 82, 86
*Paspalum distichum*, 86
*Passiflora edulis*, 582
*Passiflora foetida*, 85
Passion fruit (*Passiflora edulis*), 582

Pasture, 4, 24, 37, 57, 164, 166, 167, 169, 198, 206, 230, 234, 282, 283, 296, 298, 300, 301, 311, 324, 336, 386, 389–391, 393, 419, 440, 450, 532, 607, 612
*Pavonia zeylanica*, 91
Peacock (*Pavo cristatus*), 74
Pearl millet (*Pennisetum typhoides*), 40, 328
*Pedalium murex*, 91, 113, 122
*Pedilanthus tithymaloides*, 91
Pedogenesis, 196
*Pennisetum clandestinum*, 56
*Pennisetum orientale*, 56
*Pennisetum purpureum*, 57, 86, 114, 134, 187, 341, 389, 466, 468
Perennial grasses, 37, 48, 50, 52, 70, 94, 95, 98, 107, 113, 250, 295, 296, 327, 332, 355, 501
Perennial vegetation, 10, 40, 51, 56, 170, 388, 434, 437, 450, 452, 612
*Pergularia daemia*, 92, 122
Peripheral bund, 223, 266, 269, 275, 324, 405, 473, 481, 482, 494, 496, 594, 595
Peripheral bunds, 50, 134, 167, 247, 282, 352
*Peristrophe bicalyculata*, 82, 83, 86
*Peristrophe paniculata*, 91
Permanent gullies, 9, 51, 488–493
Phalsa *(Grewia asiatica)*, 130, 526
Phanerophytes, 96, 97
*Pharagmetes australis*, 86, 102, 114
*Pharagmetes karka*, 86, 114
*Phaseolus trilobus*, 82, 83
Phenology, 78, 85, 106
*Phoenix sylvestris*, 80, 81, 516, 517
Phosphorus solubilizing bacteria (PSB), 145–158, 175, 178
Photosynthetic material, 70
*Phragmites australis*, 86
*Phyla nodiflora*, 84, 86, 94
*Phyllanthus fraternus*, 85
*Phyllanthus maderaspatensis*, 94
*Phyllanthus niruri*, 91, 131
*Phyllanthus viridis*, 91
*Physalis minima*, 85
Physico-chemical properties, 73, 111, 148, 150, 171, 300, 529, 543, 544
Physiognomy, 78, 184
Phytoclimate, 97
Phytogenic erosion, 550
Pigeon (*Columba livia*), 74
Pigeon pea (*Cajanus cajan*), 40, 85, 341, 351
Pilar Municipality, 204
*Pinus kesiya*, 300, 344
Pioneer species, 97, 100
Piping, 11, 37, 38, 49, 447, 448, 493
*Pipistrellus coromandra* (Indian Pipistrelle), 77
*Pipistrellus tenuis* (Indian Pygmy Bat), 77
*Piptatherum miliaceum*, 102
*Pithecellobium dulce*, 81, 342, 349, 610
Plant succession, 97
*Platanista gangetica* (Gangetic River Dolphin), 77
*Platyceps ventromaculatus* (Glossy-bellied Racer), 76
Pleistocene alluvium, 37
Pleistocene colluvium, 204
*Pluchea lanceolata*, 94
*Plumbago zeylanica*, 89
Plunge pool erosion, 34
Pluvial erosion, 550
Policy intervention, 409–410, 592, 608
Policy issues, viii, 11, 24, 602–614
Polyclimax, 100
*Polygala chinensis*, 82, 83, 94, 98
*Polygala erioptera*, 84, 94
*Polygala telephioides*, 84, 85, 94
*Polygonum glabrum*, 86, 94, 95, 114
*Polygonum plebeium*, 86, 94
Pomegranate (*Punica granatum*), 54, 88, 110, 131, 297, 465, 468, 501, 502
*Pongamia pinnata*, 79, 81, 88, 94, 95, 101, 110, 115, 127, 131, 168, 250, 295, 349, 355, 376, 380, 389, 391, 438, 466, 502, 608
*Populus ciliata*, 57
*Portulaca quadrifida*, 85, 91
*Portuleca oleracea*, 86, 94, 113
Post-harvest practices, 578
*Potamogeton crispus*, 91
Potential evapotranspiration, 20
Potential zone, 434, 437–440
*Potentilla supinea*, 86
Pot-hole, 113–114, 242
Poultry farming, 581
Poverty alleviation, 583, 603
Precipitation erosion, 2, 550
Precipitation index (PI), 28
Pre-climax, 99, 100
*Prionailurus bengalensis* (Leopard Cat), 77
*Prionailurus viverrinus* (Fishing Cat), 77
Problem of malnutrition, 582
Production constraints, 280, 281
*Prosopis cineraria*, 79, 81, 88, 93, 98, 101, 124, 126, 168, 250, 348–351, 380, 388, 466, 473, 502, 526, 540, 542, 545
*Prosopis juliflora*, 53, 79, 81, 88, 93, 100, 101, 110, 121, 122, 124, 125, 127, 168, 230, 266, 291, 295, 300, 349, 350, 366, 387, 389, 526, 529, 537–539, 545, 608

Protection from grazing, 51, 53, 54, 105–111, 381, 514, 519
Proximal gullies, 39
Pruning, 289, 293, 295
*Prunus domestica*, 269
*Psammophis leithii* (Leith's Sand Snake), 76
*Pseudosorghum halepanse*, 86
*Psidium guajava*, 88, 110, 228, 269, 371, 465, 468, 501, 502
*Psoralea corylifolia*, 85
*Psorellia coulifolia*, 82
*Pteropus giganteus* (Indian Flying Fox), 77
*Ptyas mucosa* (Indian Rat Snake), 76
*Pueraria hirsuta*, 344
*Punica granatum*, 54, 88, 110, 131, 228, 297, 465, 468, 501, 502, 582
*Pupalia lappacea*, 81, 85, 91
*Python molurus* (Indian Rock Python), 76

**Q**

Quasi-three-dimensional runoff model, 6, 555, 563
Quaternary loess, 35
Quaternary sediments, 199, 200

**R**

Radionuclides, 46
Rain erosion, 550
Rain intensity, 27, 28
Raindrop splash, 28
Rainfall energy, 4, 6, 21, 27
Rainfall erosion index (EI), 6, 27
Rainfall erosivity factor, 555, 560
Rainfall factor, 27
Rainfall intensity, 4, 27, 28, 267, 551, 552, 555, 556, 563, 564, 567
Rainfall simulation experiments, 5, 22
Rainfed cropping, 111, 171
Rain-induced erosion, 255, 256
*Ramphotyphlops braminus* (Brahminy Blind Snake), 76
*Rapa Nui* civilization, 3
*Rattus rattus* (House Rat), 77
Raunkiaer's normal spectrum, 96
Ravine classification, 3
Ravine ecosystem, 119, 166, 174, 178, 256, 442, 592, 608, 611
Ravine erosion, 9, 11, 23, 30, 32, 34, 39, 46, 260, 266, 281, 346, 386, 398, 446–449
Ravine extent, 134, 167, 219, 282, 354, 363, 593
Ravine formation, 11, 24, 26, 31, 32, 70–77, 164, 165, 168, 175, 219, 237–239, 243, 246, 255, 256, 265, 275, 315, 341, 356, 440, 446–449, 452, 473, 480, 593, 602, 603
Ravine land, vii, 10, 31, 70, 124, 144, 162, 218, 219, 221–229, 231, 232, 234, 235, 250, 254, 280, 310, 381, 395, 398, 434, 447, 457, 480, 526, 592, 602
Ravine management, 136–138, 250, 466, 478
Ravine reclamation, 58, 128, 222–232, 235, 265–271, 275, 284, 286, 290, 310–313, 321–323, 329–333, 352, 410, 441–442, 473, 481, 496–499, 524, 527, 528, 592, 593, 598, 599, 603, 604, 608, 610–613
Ravines, vii, 20–58, 70, 120, 144, 186, 219, 220, 247, 254–276, 280–283, 286, 289, 290, 292–299, 301, 302, 310, 336, 362, 386, 411, 433–442, 446, 456, 524, 574, 593, 602
Ravine thorn forest, 78, 121, 165, 168, 365, 387
Ravine top, 80, 134, 136, 167, 291, 301, 364, 378, 379, 387, 392
Ravine watersheds, 336–357
Reciprocal Simpsons Diversity, 123
Reclaimability classes, 9, 221
Reclamation, 222–232, 265–276, 279–302, 321, 323–324, 329–332, 441–442, 449–450, 481–493, 496–499, 604
Reclamation of ravines, 134, 220, 267, 268, 290, 298, 311, 319, 460
Recuperation of slopes, 190–191
Recycling of organic waste, 173, 174
Rehabilitation, vii, 9, 48, 50, 54, 57, 120, 133–139, 166, 168, 173, 191, 222, 226, 234, 283–289, 292, 314, 352, 370–378, 381, 385–395, 397–411, 441, 447, 449, 451, 460, 481, 483, 499–501, 505–520, 530, 593, 605, 608, 610, 612
Rejuvenation of Ravine Lands, 168
Reptiles, 32, 74–77
Reservoir erosion, 550
Restoration, 34, 97, 102, 104–115, 139, 162, 166, 168, 300, 302, 346, 434, 441, 447, 449–452, 500, 520, 602, 603
*Rhinolophus lepidus* (Blyth's Horseshoe Bat), 77
*Rhinopoma hardwickii* (Lesser Mouse-tailed Bat), 77
*Rhinopoma microphyllum* (Greater Mouse-tailed Bat), 77
Rhizobial-symbionts, 499
*Rhizobium*, 175, 416, 420–421

*Rhizobium tropici*, 420–421
*Rhynchosia capitata*, 81–83, 85, 122, 291, 387
*Rhynchosia minima*, 82, 83, 85, 93–95, 97, 122, 387
Ributaries, 162
*Ricinus communis*, 85, 89, 131, 349, 466, 610
Rills, vii, 4, 9, 21, 29, 33, 40, 49, 71, 204, 255, 258, 269, 275, 310, 313, 602
Riparian land, 398, 450–452
River backwater, 36, 37, 47
River erosion, 550
River siltation, 200
Riverine ecology, 313
Root density (RD), 102
Root development, 29, 51, 102, 103
Root ecology, 101–104
Root management, 293, 294, 373–374
Root ramification, 70
Root system, 31, 70, 94, 102–104, 110, 113, 128, 186, 299, 341, 343, 366, 370, 399, 401, 416, 417, 421, 467
Root-area-ratios (RAR), 102
Root-length density (RLD), 102
Rotational grazing, 295
*Rousettus leschenaulti* (Leschenault's Rousette), 77
*Ruellia tuberosa*, 84, 85
*Rumex dentatus*, 86, 94, 114
*Rungia pectinata*, 82, 86
*Rungia repens*, 82, 83, 86, 94, 95
Runoff, 275, 298–299, 331, 407–408, 473–474, 563–566
Runoff collection tank, 275
Runoff volume, 10, 37, 49, 482, 551, 560
Rural ecology, 281
Rural household livelihoods, 577
Rural women, 578–580, 582, 584, 585, 587
Rural women entrepreneurs, 575, 584
*Rusa unicolor* (Sambar), 77

**S**

*Saara hardwickii* (Hardwicke's Spiny Tailed Lizard), 75
*Saccharum benghalensis*, 79, 85
*Saccharum munja* (*benghalensis*), 50, 92–94, 113, 245, 269, 272, 275, 289, 296, 299, 300, 462, 473
*Saccharum spontaneum*, 86, 93, 94, 113, 114, 126, 269, 289
Safed-musali (*Chlorophytum arundinaceum*), 84, 90, 113, 466
*Sal* (*Shorea robusta*), 3
*Salix alba*, 56
*Salix tetrasperma*, 57, 355
*Salmalia malabarica*, 81, 344, 355
*Salvadora oleoides*, 79, 84, 89, 93, 98, 110, 121, 250, 349, 350, 374, 376, 466
*Salvadora persica*, 81, 93, 121, 123, 349
Sambar (*Carvus unicolor*), 74
Sand-mining, 74, 607
Sandstone, 72, 184, 186, 203, 283
São Luis City, Maranhão State (Brazil), 195–198, 200, 201, 203, 211, 213
*Sapindus mukorossi*, 81
Sapota (*Achrus zapota*), 297, 599
Saturated conductivity, 240
Savannah forest, 98
*Scirpus articulatus*, 86
*Scirpus littoralis*, 86
*Scirpus roylei*, 86
*Scleria stocksiona*, 82, 83
*Scotophilus heathii* (Greater Yellow House Bat), 77
*Scotophilus kuhlii* (Lesser Yellow House Bat), 77
*Scotozous dormeri* (Dormer's Bat), 77
Sea erosion, 550
Seasonal variations, 93–95
Secondary succession, 100, 122
*Securinega leucopyrus*, 89
Sediment, 42–43, 564–566
Sediment budgets, 37
Sediment deposition, 6, 42, 43, 341, 346, 354, 355, 493, 553, 554, 563, 564
Sediment load, 10, 35, 42, 43, 47, 164, 344, 553, 566
Sediment production, 42, 44–45, 52, 71
Sediment transport, vii, 6, 8, 9, 42, 448, 552, 554, 563–566
Sedimentary rocks, 56, 72, 197, 446
Sedimentation, 10, 20, 24, 35, 42–45, 174, 196, 224, 370, 450, 478, 480, 567, 603
Seepage forces, 7, 37, 47, 225, 492
*Sehima nervosum*, 28, 82, 83, 85, 93, 95, 98, 100, 103, 104, 107, 109, 113, 121, 122, 231, 290, 296, 376, 387
Seismic activity, 56
Semi-arid zone, 11, 71, 165, 386, 441
*Semnopithecus entellus* (Bengal Hanuman Langur), 77
*Senebiera didyma*, 95
Sensitive analysis, 175, 561, 611
Seral stage, 99
*Sericostoma pauciflorum*, 79
Serra Geral Formation, 184
*Sesamum mulayanum*, 122
*Sesbania bispinosa*, 86, 89

*Sesbania sesban*, 81, 88, 115, 473
*Setaria glauca*, 82, 83, 94, 95, 97, 104
*Setaria media*, 95
*Setaria pillosa*, 93
*Setaria sphacelata*, 56, 352, 419
*Setaria verticillata*, 82
Shallow ravine, 52, 222–224, 228, 232, 244, 255, 259, 266, 272, 316, 324, 366, 369, 372, 379, 435, 450, 481–483, 494–497, 501
Shannon diversity index, 123
Sharifa (*Annona squamosa*), 54, 81, 110, 377, 502
Sheep (*Ovis aries*), 74
Sheet erosion, 4, 6, 21, 27–29, 237, 255, 310, 313, 399, 553, 559
Shelterbelt, 245, 348
Shifting cultivation, 164, 329, 479
*Sibynophis subpunctatus* (Dumeril's Black-headed Snake), 76
*Sida cordifolia*, 85, 91, 131
*Sida glutinosa*, 85, 86
*Sida rhombifolia*, 85
Sierra da Capivara, 203
Silting of reservoirs, 4, 254
Silvopastoral system, 55, 57, 110, 127, 222, 229–231, 248, 250, 251, 295–296, 299, 329, 343, 345, 348, 349, 351, 355–357, 381, 474, 496, 502, 503, 596, 602, 605, 611, 613
*Simarouba glauca*, 349, 608
Simpsons diversity, 122, 123
*Sisymbrium irio*, 86, 95
*Sitana ponticeriana* (Fan-throated Lizard), 76
Site-specific nutrient management (SSNM), 172
Skill development, 233, 587, 614
Slope factor, 555
Slope gradient, 49, 446, 557, 566, 568
Slope length factor, 555, 557
Slope stabilization, 440, 442, 483–485
Snow erosion, 550
Societal value of soil carbon, 451
Socio-cultural traditions, 578
Socio-economic analysis, 593, 595, 596
Socio-economic and policy issues, 11, 24
Socio-economic indicators, 574
Socio-economic Strategies, 574
Sodicity, 174
Soil aggregate, 4, 28, 74, 107, 113, 162, 163, 175, 467
Soil amelioration, 354–355
Soil and water conservation, 6, 9, 21, 25, 57, 120, 134, 138, 144–159, 167, 219, 222, 234, 248–251, 282, 284, 310–333, 337, 339, 352, 370, 386, 393, 399, 450, 494, 495, 506, 546, 592, 593, 595, 596, 598, 599, 612
Soil and water conservation research, 5, 9, 27, 110, 169, 232, 399, 407
Soil binding capacity, 29, 103, 105
Soil carbon, 232, 302, 346, 408, 409, 434, 437, 441, 442, 469, 596
Soil conservation, viii, 5, 21, 29, 31, 43, 53, 127, 136, 170, 198, 213, 230, 232, 234, 244–248, 265, 267, 269, 281, 284, 286–289, 291, 297, 302, 310–313, 321, 325, 332, 336, 341, 346, 354, 392–394, 409, 411, 494–496, 502, 592, 593, 602, 605
Soil degradation, 8, 11, 23–24, 162, 163, 174, 184, 186, 200, 336, 340, 434, 440, 446, 478, 592
Soil erodibility factor, 5, 27, 353, 555–557, 564
Soil erosion, 26–30, 479, 549–569
Soil loss equation model, 6, 22, 559
Soil microbial biomass carbon (SMBC), 158, 178
Soil nitrogen, 374
Soil organic matter (SOM), 107, 163, 173, 175, 300, 328, 340, 343, 346, 434, 436, 437, 441, 442, 464, 499
Soil pollution, 174
Soil sealing, 203
Soil test crop response (STCR), 146–158, 172, 173, 177, 178
Soil texture, 29, 39, 205, 206, 210, 213, 214, 222, 241, 244, 494, 510, 511, 515, 516, 519, 526, 528, 544, 555
Soil water energy balance, 241, 494
Soil-binding factor, 103–105
*Solanum indicum*, 113
*Solanum nigrum*, 82, 83, 85, 91, 93, 128, 131
*Solanum surattense*, 85, 93, 94
*Solanum virginianum*, 91
*Sonchus arvansis*, 82
*Sonchus asper*, 95
Sorghum (*Sorghum durra*), 40
*Sorghum halepense*, 86, 104, 114
*Soymida febrifuga*, 136, 230, 291, 374, 376, 437
*Spalerosophis atriceps* (Black-headed Royal Snake), 76
Spatially distributed models, 552, 554
*Spergula arvensis*, 95
*Spergula pentandra*, 95
*Sphaeranthus indicus*, 91
Spillway, 50, 56, 134, 167, 223, 233, 282, 286, 355, 482, 485, 486, 488–494, 594, 595

Splash erosion, 4, 29, 255, 310, 313, 551
*Sporobolus coromandelianus*, 82, 98
*Sporobolus diander*, 79, 82, 93, 98, 104
*Sporobolus marginatus*, 82, 93, 94
Stability of a community, 78
Stabilization, 187
Staggered contour trenching (SCT), 137, 227, 288–289, 356, 484, 595, 596
STCR equation, 146–158
Steep slope, 10, 49, 82, 164, 198, 207, 219, 250, 280, 319, 435, 446, 448, 457, 480, 484, 491, 496, 611
*Stemodia viscosa*, 86
*Stizolobium deeringianum*, 291
Stratification, 35, 78, 103
*Striga euphrasioides*, 82, 83, 86
Strontium isotopic ratio, 46
*Stylosanthes guianensis*, 352
*Stylosanthes hamata*, 291, 390
*Stylosanthes scabra*, 250, 390, 465, 501
*Suaeda fruticosa*, 94
Subsistence farming, 218, 480, 592, 598
Sub-tropical regions, 41, 78, 414
Successional trend, 97–101, 122
Succulent, 96, 169, 350
*Suncus murinus* (Asian House Shrew), 77
Supply chain management, 410
Surface sealing, 478, 566
Surface water, 31, 48, 191, 238, 340, 436, 490, 491
*Sus scrofa cristatus* (Indian Wild Boar), 77
Suspended load, 43
Suspended sediment yield, 42, 43
*Sutera glandulosa*, 86, 94
Symbiosis, 416, 420–423
System productivity, 159, 300, 471
*Syzygium cumini*, 81, 110, 131, 228, 269, 516, 582
*Syzygium heyneanum*, 81

**T**
*Tabernaemontana divaricata*, 81
Tableland, 9, 29, 232, 267, 281, 362
*Tadarida aegyptiaca* (Egyptian Free-tailed Bat), 77
Tamarind (*Tamarindus indica*), 81, 88, 101, 110, 131, 228, 582
*Tamarix canariensis*, 102
*Tamarix dioca*, 296
*Taphozous longimanus* (Long-winged Tomb Bat), 77
*Taphozous melanopogon* (Black-bearded Tomb Bat), 77
*Taphozous nudiventris* (Naked-rumped Tomb Bat), 77
*Taphozous perforatus* (Egyptian Tomb Bat), 77
*Tatera indica* (Indian Gerbil), 77
Technological interventions, 218, 219, 221–229, 231, 232, 234, 235, 248, 352–357, 403–405, 593–596
*Tecomella undulata*, 79, 98, 350
*Tectona grandis*, 79, 88, 339, 341, 355, 389, 391, 471, 502
Tectonic plate, 35
Temperate climates, 30, 197
Tension crack development, 37
*Tephrosea farciformis*, 79
*Tephrosia candida*, 289
*Tephrosia purpurea*, 79, 82, 91, 94, 98, 121, 122, 131, 387
*Tephrosia tinctoria*, 91
*Terminalia arjuna*, 81, 88, 114, 131, 389, 468, 471, 472, 502
*Terminalia belerica*, 438
*Terminalia tomentosa*, 81
Terracing, 10, 35, 40, 56, 57, 70, 115, 187–188, 222, 227, 246, 266, 270, 272, 284–286, 288, 310, 312, 313, 316, 319, 322, 324, 328, 329, 375, 388, 435, 440, 450, 464, 480, 482, 496–498, 519, 559, 592, 595
Terrain, vii, 20, 40, 71, 162, 190, 219, 221, 229, 263, 310, 435, 446–448, 480, 560, 602
Terrestrial animal biodiversity, 70, 74–78
Terrestrial ecosystems, 105, 456
*Tetracerus quadricornis* (Four-horned Antelope), 77
*Thelepogon elegans*, 82, 83
*Themeda anathera*, 56
*Themeda triandra*, 54, 80–83, 95, 98, 103, 104, 107, 108, 121, 122, 296, 365, 387
Therophytes, 96, 97
*Thevetia peruviana*, 81
Thorn forest, 53, 78, 100, 101, 121
Threatened species, 86
*Thysanolaena maxima*, 93
Tillage system, 224, 594
*Tinospora cordifolia*, 84, 92, 113, 131
Tobacco (*Nicotiana tobacum*), 40, 225
Topographic factors, 31, 48, 95, 551
Torrent erosion, 344, 550
Torrential erosion, 550

*Tragia cannabina*, 82–84
*Tragus biflorus*, 81–83, 95, 104, 109, 113, 296
Transfer of technology, 323, 587, 612
Transient water flow, 239–242, 494
Trapezoidal gullies, 39
Tree-based pastoral systems, 134
Tree born oilseeds (TBOs), 283, 349
Trenching, 11, 25, 38, 134, 135, 165, 167, 227, 282, 288 289, 293, 297, 356, 376, 377, 386, 405, 406, 440, 483, 612
*Trianthema portulacastrum*, 85, 94
*Tribulus alatus*, 79
*Tribulus terrestris*, 79, 84, 91, 122
Tributaries, 22, 32–34, 42, 72, 80, 81, 122, 162, 223, 239, 258, 260–263, 280, 313, 315, 325, 355, 386, 434, 481, 484, 524, 563, 574
*Trichoderma amplexicaul*, 82
*Tricholepis radicans*, 91
*Trichosanthes anguina*, 92
*Trichosanthes cucumerina*, 85
*Tridax procumbens*, 82, 93, 131
*Trifolium alexandrinum*, 536
Trigger point, 448
*Triumfetta rhomboidea*, 84, 94, 98
Tropical climate, 184, 197, 343, 416
Tropical forests, 120, 184
Tropical region, 78
Tropical thorn forest, 53, 78, 100, 101, 121, 165, 168, 365, 387
Tunnel flow, 447
*Typha angustata*, 86

**U**
Ultisols, 200
Underground parts, 70
Universal soil loss equation (USLE), 6, 23, 353, 552–555, 560, 561, 563
Upper Cretaceous, 72
Urban soils, 201
*Urginea indica*, 91, 113
U-shaped gullies, 38, 39, 260, 261, 270

**V**
Valley-bottom gullies, 10, 37, 51
Value-added product, 578, 582
*Varanus bengalensis* (Common Indian Monitor), 76
*Varanus griseus*, 76
Vascular-arbuscular mycorrhizal fungi (VAMF), 421, 423
Vegetation ecology, 70–115, 122, 366
Vegetation structure, 78
Vegetative cover, 30, 31, 37, 105, 164, 246, 267, 268, 275, 276, 281, 297, 299, 319, 328, 347, 392, 482, 558, 560
Vegetative cover factor, 555, 558
Vegetative filte, 224, 327
*Verbascum chinensis*, 86, 94
*Verbena bipinnatifida*, 97
*Verbina bipinnata*, 82
*Vernonia cinerea*, 82, 84, 93, 94
*Veronica anagallis-aquatica*, 86
Vertebrate fauna, 74
Vertisols, 200, 260, 261, 328
*Vetiveria zizanioides*, 79, 85, 86, 93, 104, 113, 269, 290, 299, 300, 609
*Vicoa indica*, 80, 85, 95
*Vigna mungo*, 40, 328, 469
*Vigna radiata*, 40, 328
Village community, 320, 326, 596
Vindhyan hill range, 262
*Vitex negundo*, 57, 131, 344
*Viverricula indica* (Small Indian Civet), 77
*Voçoroca*, 4, 184, 196
Volcanism, 35
*Volutarella ramosa*, 82, 84
*Vossoroca*, 9
V-shaped gullies, 38, 261, 270, 365
Vulnerable lands, 411
Vulnerable species, 74
*Vulpes bengalensis* (Indian Fox), 77
*Vulpes vulpes pusilla* (Desert Fox), 77

**W**
Wasteland, 8, 123, 124, 164–165, 173, 229, 263, 300, 302, 312, 350, 352, 362, 394, 410, 436, 456, 457, 546, 592, 595, 611
Water conservation, 144–159, 248–250, 309–333, 336–340, 450
Water dispersible aggregates, 244
Water erosion, vii, 7–9, 20, 21, 23–25, 35, 40–42, 44–45, 71, 101, 138, 139, 144, 162, 164, 174, 200, 218, 254, 255, 280, 296, 302, 327, 345, 346, 356, 362, 364, 381, 386, 393, 434, 446, 452, 457, 479, 550–552, 602, 610
Water Erosion Prediction Project (WEPP) model, 6, 47, 553, 561, 563
Water holding capacity, 31, 34, 107, 169, 173, 365, 478
Water infiltration, 37, 185, 211, 288, 340, 343, 387, 419, 440, 479, 499, 509, 512, 517
Water stable soil aggregates, 107, 300, 340
Watershed, 268–269, 335–357, 460

Watershed approach, 267, 313, 333
Watershed-based management, 11, 24, 266, 268–269, 460–464
Watershed models, 326
*Wattakaka volubilis*, 92
*Wendlandia exserta*, 57
Wheat (*Triticum aestivum*), 40, 469
Wind erosion, 8, 144, 174, 327, 436, 448, 479, 550, 551, 610
*Withania somnifera*, 91, 113, 131, 295, 466, 609
Women empowerment, 233, 330, 581, 582, 587
Women entrepreneurs, 577, 580, 586
*Woodfordia fruticosa*, 115
Woody vegetation, 81, 96, 231

**X**

*Xanthium indicum*, 91, 121
*Xanthium strumarium*, 82, 86, 91, 94, 113, 121, 366, 387
*Xenochrophis piscator* (Chequered Keelback), 76
Xerophytic trees, 121, 366

**Y**

Yamuna ravine, 79, 80, 107, 127, 129–131, 134, 165, 169, 220, 226, 227, 229, 232, 257, 260, 262, 264, 266, 280, 290, 293–296, 298, 301, 302, 313, 315–317, 322, 331, 352, 355, 363–366, 370, 372, 373, 378, 386, 387, 391–393, 399, 403, 405, 437, 440, 446, 457, 481, 482, 502, 574

**Z**

*Zea mays*, 40, 469
Zero tillage, 436, 441
Zeta potential, 74, 244
*Ziziphus mauritiana*, 40, 52, 54, 79, 81, 84, 88, 101, 110, 122, 126, 131, 136, 226, 228, 229, 250, 293, 295, 297, 339, 350, 371, 390, 465, 468, 471, 473, 501, 502
*Ziziphus nummularia*, 79, 81, 82, 89, 93, 94, 98, 101, 121, 126, 319, 349, 351, 352, 365–366, 387, 462, 473
*Ziziphus oenoplia*, 89
Zoogenic erosion, 550

MIX
Papier aus verantwortungsvollen Quellen
Paper from responsible sources
FSC® C105338

If you have any concerns about our products,
you can contact us on
**ProductSafety@springernature.com**

In case Publisher is established outside the EU,
the EU authorized representative is:
**Springer Nature Customer Service Center GmbH**
**Europaplatz 3, 69115 Heidelberg, Germany**

Printed by Libri Plureos GmbH
in Hamburg, Germany